Broadband Switching

Architectures, Protocols, Design, and Analysis

1951-1991
40 YEARS OF SERVICE

IEEE COMPUTER SOCIETY
A member society of the
Institute of Electrical and Electronics Engineers, Inc.

BROADBAND SWITCHING
architectures, protocols, design, and analysis

CHRIS DHAS, VIJAYA K. KONANGI, and M. SREETHARAN

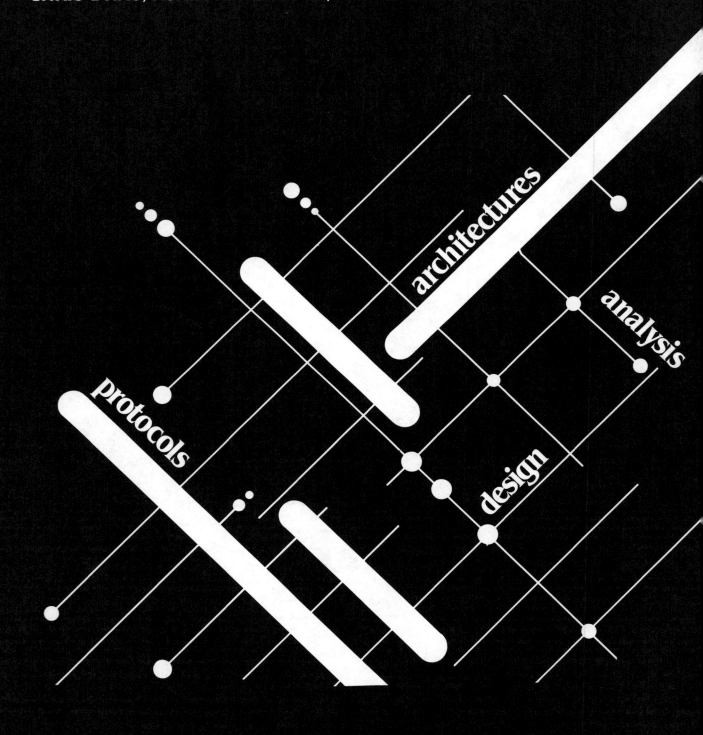

Broadband Switching

WITHDRAWN

Architectures, Protocols, Design, and Analysis

Edited by

Chris Dhas
Vijaya K. Konangi
M. Sreetharan

1951-1991

IEEE Computer Society Press
Los Alamitos, California

Washington ● Brussels ● Tokyo

IEEE COMPUTER SOCIETY PRESS TUTORIAL

Library of Congress Cataloging-in-Publication Data

Broadband switching: architectures, protocols, design, and analysis/
[compiled by] Chris Dhas, Vijaya K. Konangi, and M. Sreetharan.
p. cm. — (IEEE Computer Society Press tutorial)
Includes bibliographical references.
ISBN 0-8186-8926-9
1. Telecommunication — Switching systems. 2. Broadband communication systems.
I. Dhas, Chris, 1943- . II. Konangi, Vijaya K. III. Sreetharan, M. (Muthuthambi) IV. Series
TK5103.8B76 1991
621.382—dc20   91-16306
                                                                             CIP

Published by the
IEEE Computer Society Press
10662 Los Vaqueros Circle
PO Box 3014
Los Alamitos, CA 90720-1264

IEEE Computer Society Press Order Number 1926
Library of Congress Number 91-16306
IEEE Catalog Number EH0301-2
ISBN 0-8186-8926-9 (case)
ISBN 0-8186-6926-8 (microfiche)

Additional copies can be ordered from

IEEE Computer Society Press	IEEE Service Center	IEEE Computer Society	IEEE Computer Society
Customer Service Center	445 Hoes Lane	13, avenue de l'Aquilon	Ooshima Building
10662 Los Vaqueros Circle	PO Box 1331	B-1200 Brussels	2-19-1 Minami-Aoyama
PO Box 3014	Piscataway, NJ 08855-1331	BELGIUM	Minato-ku, Tokyo 107
Los Alamitos, CA 90720-1264			JAPAN

Third printing, 1992

Technical Editor: Jon T. Butler
Production Editor: Robert Werner
Copy Editor: Tom Culviner
Cover design: Jack Ballestero
Printed in the United States of America by Braun-Brumfield, Inc.

 THE INSTITUTE OF ELECTRICAL AND ELECTRONICS ENGINEERS, INC.

Preface

The early years of data communications were dominated by circuit switching technology. Then, as the cost of hardware dropped dramatically, bandwidth could be dynamically allocated to a call, rather than dedicated to it. This technique formed the basis for packet switching technology. Sophisticated protocols were used for error detection and correction, flow control, and packet sequencing.

In today's networking environment, most of the bursty traffic types use packet-switched networks, while voice traffic and other data traffic continue to use circuit-switched networks and channel-switched networks. This dedicated network approach, where each service has a separate network, not only increases the cost of communication and network operation, but also fails to take advantage of the inherent differences in the service characteristics.

Future applications, the bit rates of which range from Kbps to Mbps, need an integrated switching and networking methodology to exploit the high bandwidth made available by fiber-optic transmission technology. In this tutorial, the term *broadband switching* is used to describe such an integrated switching method.

The objective of this tutorial is to collect in one book not only papers on broadband switching but also supporting materials in the related areas of networking, performance analysis, and alternate technologies. This tutorial is intended for:

● *Voice and data communication managers.* The tutorial contains enough information to help managers understand broadband switching and networking issues.

● *Network planners, product managers, product planners, and technical specialists.* The tutorial covers the state of the art in broadband switching. In addition, it presents network architectures, broadband network requirements, protocols, and performance modeling.

● *Graduate students.* The tutorial covers a number of broadband switching architectures and performance modeling techniques. It contains sufficient material for use as a supplementary text in a course on ISDN systems or as the main text in an advanced course on broadband switching, in Computer Science, Computer Engineering, and Electrical Engineering departments.

The tutorial consists of nine chapters of original and reprinted material. Chapter 1 covers the various trends in network architectures

v

intended to meet the high-bandwidth requirements of users. Chapter 2 is concerned with the multistage interconnection networks. These interconnection networks provide parallel paths between the input and output ports. Chapter 3 describes various experimental architectures reflecting the widespread interest in broadband ISDN and the asynchronous transfer mode (ATM) technique.

Chapter 4 describes various forms of multistage interconnection networks for use in the switch fabrics for broadband networks. Chapter 5 highlights the architectural options available for switches for the broadband networks of the future. Chapter 6 covers the importance of packet replication and switching in broadcast switching systems.

Chapter 7 presents the important issues of bandwidth allocation and flow and congestion control. Performance modeling is the subject of Chapter 8. Chapter 9 provides material on photonic switching techniques and technology. Interested readers can use the extensive bibliography to delve further into the literature.

We extend our thanks to Mr. Ben Lisowski for his comments and critiques. We gratefully acknowledge the excellent secretarial help of Mrs. Janet Basch and Ms. Adrienne Fox.

Chris Dhas
Vijaya K. Konangi
M. Sreetharan

Table of Contents

Preface v

Chapter 1: Network Architecture 1

Network Protocols 5
   A.S. Tanenbaum (Computing Surveys, Vol. 13, No. 4,
   December 1981, pp. 453-489.)
A Layered Network Protocol for Packet Voice and Data Integration 42
   W.L. Hoberecht (IEEE Journal on Selected Areas in
   Communications, Vol. SAC-1, No. 6, December 1983, pp. 1006-1013.)
An Overview of FDDI: The Fiber Distributed Data Interface 50
   F. E. Ross (IEEE Journal on Selected Areas in
   Communications, Vol. 7, No. 7, September 1989, pp. 1043-1051.)
The QPSX MAN 59
   R.M. Newman, Z.L. Budrikis, and J.L. Hullett
   (IEEE Communications Magazine, Vol. 26, No. 4,
   April 1988, pp. 20-28.)
Switched Multimegabit Data Service and Early Availability Via MAN Technology 68
   C.F. Hemrick, R.W. Klessig, and J.M. McRoberts
   (IEEE Communications Magazine, Vol. 26, No. 4,
   April 1988, pp. 9-14.)
Packet Mode Services: From X.25 to Frame Relaying .74
   W.S. Lai (Computer Communications, Vol. 12,
   No. 1, February 1989, pp. 10-16.)
Broadband ISDN and Asynchronous Transfer Mode (ATM) 81
   S.E. Minzer (IEEE Communications Magazine,
   Vol. 27, No. 9, September 1989, pp. 17-24.)
The Evolution of Broadband Network Architectures 90
   D.V. Batorsky, D.R. Spears, and A.R. Tedesco
   (Proc. Globecom 1988, pp. 367-373.)
SONET: Now It's the Standard Optical Network 97
   R. Ballart and Y.-C. Ching (IEEE Communications
   Magazine, Vol. 29, No. 3, March 1989, pp. 8-15.)

Chapter 2: Interconnection Networks 105

Classification Categories and Historical Development of
Circuit Switching Topologies 107
   G. Broomell and J.R. Heath (Computing Surveys, Vol. 15,
   No. 2, June 1983, pp. 95-133.)
A Survey of Interconnection Networks 146
   R. J. McMillen (Proc. Globecom 1984, pp. 105-113.)
Packet Switching in N log N Multistage Networks 155
   D.M. Dias and M. Kumar (Proc. Globecom 1984, pp. 114-120.)

Chapter 3: Experimental Architectures 163

The "Prelude" ATD Experiment: Assessments and Future Prospects 167
    M. Devault, J.-Y. Cochennec, and M. Servel (IEEE Journal
    on Selected Areas in Communications, Vol. 6, No. 9, December
    1988, pp. 1528-1537.)
Definition of Network Options for the Belgian ATM Broadband Experiment 177
    M. De Prycker (IEEE Journal on Selected Areas in
    Communications, Vol. 6, No. 9, December 1988, pp. 1538-1544.)
Experimental Broadband ATM Switching System 184
    Y. Kato, T. Shimoe, K. Hajikano, and K. Murakami (Proc.
    Globecom 1988, pp. 1288-1292.)
Layered ATM Systems and Architectural Concepts for
Subscribers' Premises Networks 189
    J.P. Vorstermans and A.P. De Vleeschouwer (IEEE Journal
    on Selected Areas in Communications, Vol. 6, No. 9, December
    1988, pp. 1545-1555.)

Chapter 4: Switch Fabric Design and Analysis 201

A Fast Packet Switch for the Integrated Services Backbone Network 205
    P. Newman (IEEE Journal on Selected Areas in
    Communications, Vol. 6, No. 9, December 1988, pp. 1468-1479.)
Multipath Interconnection: A Technique for Reducing Congestion
Within Fast Packet Switching Fabrics 217
    G.J. Anido and A.W. Seeto (IEEE Journal on Selected Areas
    in Communications, Vol. 6, No. 9, December 1988, pp. 1480-1488.)
Architecture of a Packet Switch Based on Banyan Switching
Network with Feedback Loops 226
    H. Uematsu and R. Watanabe (IEEE Journal on Selected Areas
    in Communications, Vol. 6, No. 9, December 1988, pp. 1521-1527.)
A New Self-Routing Switch Driven with Input-to-Output Address Difference 233
    H. Imagawa, S. Urushidani, and K. Hagishima (Proc.
    Globecom 1988, pp. 1607-1611.)

Chapter 5: Switch Architectures 239

Burst Switching – An Update 243
    S.R. Amstutz (IEEE Communications Magazine, Vol. 27,
    No. 9, September 1989, pp. 50-57.)
Starlite: A Wideband Digital Switch 251
    A. Huang and S. Knauer (Proc. Globecom 1984, pp. 121-125.)
Dynamic TDM: A Packet Approach to Broadband Networking 256
    L.T. Wu, S.H. Lee, and T.L. Lee (Proc. ICC 1987, pp. 1585-1592.)
A Broadband Packet Switch for Integrated Transport 264
    J.Y. Hui and E. Arthurs (IEEE Journal on Selected Areas in
    Communications, Vol. SAC-5, No. 8, October 1987, pp. 1264-1273.)
The Knockout Switch: A Simple, Modular Architecture for
High-Performance Packet Switching 274
    Y.-S. Yeh, M.G. Hluchyj, and A.S. Acampora (IEEE Journal
    on Selected Areas in Communications, Vol. SAC-5, No. 8,
    October 1987, pp. 1274-1283.)

Synchronous Composite Packet Switching – A Switching Architecture for
Broadband ISDN 284
    T. Takeuchi, T. Yamaguchi, H. Niwa, H. Suzuki, and
    S.-I. Hayano (IEEE Journal on Selected Areas in
    Communications, Vol. SAC-5, No. 8, October 1987, pp. 1365-1376.)
Integrated Services Packet Network Using Bus Matrix Switch 296
    S. Nojima, E. Tsutsui, H. Fukuda, and M. Hashimoto (IEEE
    Journal on Selected Areas in Communications, Vol. SAC-5,
    No. 8, October 1987, pp. 1284-1292.)

Chapter 6: Broadcast Switching Networks 305

Design of a Broadcast Packet Switching Network 307
    J.S. Turner (IEEE Transactions on Communications, Vol. 36,
    No. 6, June 1988, pp. 734-743.)
A New Broadcast Switching Network 317
    C.-T. Lea (IEEE Transactions on Communications, Vol. 36,
    No. 10, October 1988, pp. 1128-1137.)
Nonblocking Copy Networks for Multicast Packet Switching 327
    T.T. Lee (IEEE Journal on Selected Areas in Communications,
    Vol. 6, No. 9, December 1988, pp. 1455-1467.)

Chapter 7: Bandwidth Allocation, Flow Control, and Congestion Control 341

Routing of Multipoint Connections 347
    B.M. Waxman (IEEE Journal on Selected Areas in
    Communications, Vol. 6, No. 9, December 1988, pp. 1617-1622.)
A Dynamically Controllable ATM Transport Network Based on the
Virtual Path Concept 353
    S. Ohta, K. Sato, and I. Tokizawa (Proc. Globecom 1988, pp. 1272-1276.)
Resource Allocation for Broadband Networks 358
    J.Y. Hui (IEEE Journal on Selected Areas in Communications,
    Vol. 6, No. 9, December 1988, pp. 1598-1608.)
Multichannel Bandwidth Allocation in a Broadband Packet Switch 369
    A. Pattavina (IEEE Journal on Selected Areas in
    Communications, Vol. 6, No. 9, December 1988, pp. 1489-1499.)
Flow Control Schemes and Delay/Loss Tradeoff in ATM Networks 380
    H. Ohnishi, T. Okada, and K. Noguchi (IEEE Journal on
    Selected Areas in Communications, Vol. 6, No. 9, December 1988, pp. 1609-1616.)
Access Interface Congestion Controls for Packetized
Voice Transport in Wideband Networks 388
    S.G. Eick (Proc. Globecom 1988, pp. 226-230.)
A Congestion Control Framework for High-Speed Integrated Packetized Transport 393
    G.M. Woodruff, R.G.H. Rogers, and P.S. Richards (Proc.
    Globecom 1988, pp. 203-207.)

Chapter 8: Performance Modeling 399

Performance Analysis of a Packet Switch Based on Single-Buffered Banyan Network 401
    Y.-C. Jenq (IEEE Journal on Selected Areas in
    Communications, Vol. SAC-1, No. 6, December 1983, pp. 1014-1021.)
A Simulation Study of Network Delay for Packetized Voice 409
    P.M. Gopal and B. Kadaba (Proc. Globecom 1986, pp. 932-938.)

Performance Comparison of Error Control Schemes in
High-Speed Computer Communication Networks 416
    A. Bhargava, J.F. Kurose, D. Towsley, and G. Vanleemput
    (IEEE Journal on Selected Areas in Communications, Vol. 6,
    No. 9, December 1988, pp. 1565-1575.)
On Cell Size and Header Error Control of Asynchronous Transfer Mode (ATM) 427
    D.P. Hsing, F. Vakil, and G.H. Estes (Proc. Globecom 1988, pp. 394-402.)

Chapter 9: Photonic Switching Systems 437

Communication Network Needs and Technologies – A Place for Photonic Switching? 441
    E. Nussbaum (IEEE Journal on Selected Areas in
    Communications, Vol. 6, No. 7, August 1988, pp. 1036-1043.)
Terminology for Photonic Matrix Switches 449
    R.I. MacDonald (IEEE Journal on Selected Areas in
    Communications, Vol. 6, No. 7, August 1988, pp. 1141-1151.)
Architectural Considerations for Photonic Switching Networks 460
    H.S. Hinton (IEEE Journal on Selected Areas in
    Communications, Vol. 6, No. 7, August 1988, pp. 1209-1226.)
Time Division Multiplexing Using Optical Switches 478
    A. Djupsjöbacka (IEEE Journal on Selected Areas in
    Communications, Vol. 6, No. 7, August 1988, pp. 1227-1231.)
Optical Self-Routing Switch Using Integrated Laser Diode Optical Switch 483
    R. Kishimoto and M. Ikeda (IEEE Journal on Selected Areas
    in Communications, Vol. 6, No. 7, August 1988, pp. 1248-1254.)
HYPASS: An Optoelectronic Hybrid Packet Switching System 490
    E. Arthurs, M.S. Goodman, H. Kobrinski, and M.P. Vecchi
    (IEEE Journal on Selected Areas in Communications, Vol. 6,
    No. 9, December 1988, pp. 1500-1510.)

Bibliography 501
Glossary 509
About the Authors 513

Chapter 1:

Network Architecture

1.1 Introduction

Current telecommunications services–which include circuit- or packet-switched data, private lines, or voice–are quite often provided by independent and parallel networks. Since each network is optimized for the application it supports, one is not efficient in supporting other applications.

In an era of worldwide competition, end users as well as network providers want the flexibility and versatility to respond quickly to the changing environment. The ability to define, manage, and monitor services, along with the ability to introduce new services tailored to specific applications, are characteristics expected of the networks of the future.

The idea of an integrated communications system that can support a diversity of services with different requirements has been discussed for some time. The Integrated Services Digital Network (ISDN) is an attempt to provide such an integrated access. ISDN specifies a standard interface based on digital transmission. It consists of a number of circuit-switched B channels for voice and bulk data, and a packet-switched D channel for signaling data. As such, ISDN is not a truly integrated network. Using a limited set of well-defined interfaces, ISDN allows user traffic to be integrated at the transmission level at the user network interfaces. Within the network, separate switching methods carry circuit-switched and packet-switched connections.

Providing a single network that concurrently serves many users with differing needs has a number of benefits because of the economy of scale in telecommunications. The network should be responsive to interactive traffic; provide high bandwidth for file transfer, high-speed data, and video applications; and have a low delay for packet voice. Each of these functions requires a different transport protocol and different data representation. The challenge in designing broadband networks and switches is to accommodate this diversity without increasing the complexity. What is needed is a structure with simple components, simple interfaces, and simple protocols.

The goals of an integrated network based on a single switching technique are good performance, effective resource utilization, and unified network management, operation, and maintenance to support various existing services uniformly.

Another important goal is to provide the flexibility to respond to and support future services.

The distributed processing capability requires fast access to on-line information and expensive computing resources. The use of graphics-based user interfaces with diskless processing and networked file systems places the network at the center of distributed processing. Within a limited geographical distance, local area networks, with a capacity of a few megabits per second, can support a few tens of computers and servers. The networks provide low delay, and their intrinsically low error rate allows them to use simple protocols. The persistent demand for higher bandwidth interconnections has led to the next generation of LANs, which can support higher bandwidths (100 Mbps). Although these LANs provide higher bandwidths, they have not solved the distance problem.

Therefore, a high-performance network covering a wider area–say with the capability to support thousands of end users scattered over an entire metropolitan area–is needed. Such metropolitan area networks (MANs) will likely be owned and operated by common carriers. Issues such as reliability and security are of greater importance than in a LAN environment, which is usually owned and operated by the organization that uses it.

Perhaps the most important factor spurring the change is the steady increase of transmission bandwidth available at an affordable cost. In the local area network environment, cheap bandwidth has already radically changed the structure of local area communication. In the future, both near-term and long-term, there will be a definite need for connecting high-bandwidth equipment with similar equipment at other locations, and with central information or processing facilities.

The frame-based transfer capabilities, such as frame relaying and switched multimegabit data service, are candidates for near-term high-speed data applications. Alternative solutions for multimedia applications range from the Fiber Distributed Data Interface for integrated services LANs and the Distributed Queue Dual Bus (DQDB) for MANs, to the asynchronous transfer mode (ATM) for wide area public networks. A communication structure based on ATM requires new design rules, new control strategies, and new architectural approaches.

At any given level of circuit technology, it is possible to transmit and receive data much faster than software, executed by the same circuit technology, can process all the communication protocols. With the exception of digital video, it is almost impossible for any one application to cope with all the data that can flow through optical fibers. Hence, the most economical approach would be to share the medium. With a shared-medium network, the switch would also be shared. Any station on the net would communicate directly with any other. It would be possible to have graceful growth, with the initial investment being mainly in the cabling.

A broadband network will enable a single network to support voice, video, image, and data services. There is, of course, uncertainty about the type of and the demand for new services, and the traffic mix of services at particular times. The effects of this uncertainty can be minimized by sharing the network resources across different services. As traffic peaks often occur at different times for different services, the network can be dimensioned for a desired service quality during busy hours. Thus, the flexibility of broadband integrated networks enables them to provide service for traffic with different characteristics and also to cope with the uncertainty of the traffic mix of the future.

Integrated access allows statistical multiplexing and dynamic reallocation of resources without reconfiguring the access channels, and also provides bandwidth on demand. For the network provider, an integrated network provides transmission efficiency and versatility. Network management and control are simple.

The layers, interfaces, and protocols in a network constitute the network architecture. The communications services and applications of the future will require higher bandwidths and flexible architectures to support multiple classes of traffic. In such a network architecture, a fiber-optic medium can provide cost-effective broadband transmission capabilities. The other components that play an important role are the switching and the protocols. This chapter presents the various trends in network architectures intended to meet the high bandwidth demanded by users.

1.2 Overview of Papers

Tanenbaum's paper, "Network Protocols," is a tutorial on protocols as well as networking. Tanenbaum presents an overview of the circuit and packet switching technology along with telephone, satellite, and local area networks. The network architecture model in this paper closely follows that of the International Organization for Standardization's reference model of Open Systems Interconnection (ISO OSI), and uses the ARPAnet terminology to make the abstract concepts of network architecture and protocols more concrete. Tanenbaum uses the X.21 standard to illustrate the physical layer protocol, and High-level Data Link Control (HDLC) to explain the data link layer protocol.

Tanenbaum provides a detailed explanation of the sliding-window protocol and its use in the X.25 packet-level protocol and discusses various routing schemes and congestion control techniques. The transport layer, which is responsible for providing an independent transport service to the session layer, is illustrated using the concept of a "transport station." Transport layer functions such as establishing and closing connections, flow control and buffering, and connection multiplexing are examined. The presentation layer, which performs a useful data transformation function, is explained using text compression and encryption.

Hoberecht's paper, "A Layered Network Protocol for Packet Voice and Data Integration," presents a layered packet protocol architecture that can be used for integrating voice and data in packet-switched networks. Hoberecht identifies network components and the factors that affect the protocol architecture, and defines a layered protocol for voice/data packet networks and the layer functions. After introducing the concept of an edge-to-edge error-recovery protocol, Hoberecht presents the delay analysis for data traffic using edge-to-edge and the traditional link-level error-correction protocols.

The Fiber Distributed Data Interface (FDDI) is the subject of Ross's paper. FDDI is a fiber-optic token-passing network that uses a dual-ring configuration for reliability. The rings are counterrotating, one clockwise and the other counterclockwise, and each station is connected to both the rings. Optical bypass switches can be used to remove any failed station from the network without disturbing the traffic flow to the other stations. FDDI is an upgraded version of the IEEE 802.5 token ring, with a transmission rate of 100 Mbps and a span of 100 kms. In addition to serving as a backbone LAN, it can be used to interconnect host computers in a computer-room environment. While FDDI operates only in the packet data mode, an upward-compatible enhancement, FDDI-II, will also provide circuit-switched capacity on demand.

Newman et al. describe the QPSX (queued packet and synchronous switch) MAN (metropolitan area network). It is a distributed switch/network based on a dual self-healing bus architecture. The head-ends generate time slots that propagate along the buses, and a user having data to send waits until an idle time slot appears at the sending location. Access to an idle time slot is determined by a distributed queueing algorithm, and the user writes data onto the appropriate bus to be transmitted toward the intended destination. This dual-bus architecture provides higher reliability by moving the logical gap in the system to the point of physical failure, so the system can operate using both the buses. QPSX provides high-speed packet and circuit switching capability at a transmission rate of around 150 Mbps. Although QPSX effectively provides both circuit and packet switching, the flexibility and versatility of the network are due to its packet switching protocol.

The switched multimegabit data service (SMDS) is discussed in the paper by Hemrick et al. The idea behind SMDS is to allow users to move high-speed data using a shared switched network. The motivation for this service is to interconnect LANs and handle the increase in applications based on distributed processing—for example, real-time access to files, desktop publishing, and shared high-resolution graphics in computer-aided design.

SMDS is a high-performance, connectionless, packet-switched data service designed to be easily integrated into the users' existing local environment. SMDS provides efficient and easy-to-implement protocols and procedures across the access interface. The connectionless packet transport is provided by the SMDS interface protocol, which provides error-detection, framing, and addressing functions within the network. The interfaces are designed to operate at 45 Mbps.

Lai's paper, "Packet Mode Services: From X.25 to Frame Relaying," traces the evolution of packet mode services from X.25 to frame relaying. While X.25 uses in-band signaling and control, frame relaying uses out-of-band signaling and control. The interworking between X.25 and frame relaying is also discussed.

The paper by Minzer, "Broadband ISDN and Asynchronous Transfer Mode," compares the limitations of the synchronous transfer mode (STM) with the flexibility of the asynchronous transfer mode (ATM), thus explaining the choice of ATM as the transfer mode for broadband ISDN.

The major technical considerations influencing the undefined ATM specifications are also discussed.

"The Evolution of Broadband Network Architectures," by Batorsky et al., explains the two major forces behind the emergence of broadband networks: the market for new services and technological advances. The authors describe the architecture of broadband networks for the long term and suggest approaches by which the initially available architectures might evolve toward the long-term architecture.

The last paper, "SONET: Now It's the Standard Optical Network," by Ballart and Ching, discusses the important technical attributes of the SONET (Synchronous Optical Network) standard. The advent of fiber-optic transmission systems with very large bandwidths has led to a new, simpler synchronous digital multiplexing hierarchy, which offers flexible multiplexing of high-speed channels. A major advantage of the system is the ease with which channels can be dropped and inserted at the network nodes. SONET defines standard optical signals, a synchronous frame structure for multiplexed digital traffic, and operations procedure. SONET forms part of the new synchronous signal hierarchy.

Network Protocols

ANDREW S. TANENBAUM

Wiskundig Seminarium, Vrije Universiteit, Amsterdam, The Netherlands

During the last ten years, many computer networks have been designed, implemented, and put into service in the United States, Canada, Europe, Japan, and elsewhere. From the experience obtained with these networks, certain key design principles have begun to emerge, principles that can be used to design new computer networks in a more structured way than has traditionally been the case. Chief among these principles is the notion of structuring a network as a hierarchy of layers, each one built upon the previous one. This paper is a tutorial about such network hierarchies, using the Reference Model of Open Systems Interconnection developed by the International Organization for Standardization as a guide. Numerous examples are given to illustrate the principles.

Key Words and Phrases: computer network, data communication, ISO OSI Reference Model, layered architecture, network, protocol

CR Categories: 1.3, 4.9, 6.9

INTRODUCTION

Ten years ago, only a handful of computer networks existed, mostly experimental networks built by research organizations. Today dozens of national and international networks and innumerable local networks operate on a commercial basis around the clock. From the beginning, many networks were designed hierarchically, as a series of layers, each one building on the one below. At first, each network design team started out by choosing its own set of layers. However, in the past few years, a consensus has begun to develop among network designers, a consensus embodied in the International Organization for Standardization's Reference Model of Open Systems Interconnection (ISO OSI). In this paper we present an informal introduction to computer networking using this model as a guide. A more thorough treatment of the ISO OSI model itself can be found in ZIMM80.

Before getting into the subject of network protocols, it is worth saying a few words about what we mean by a computer network. A computer network is a collection of computers, called *hosts*, that communicate with one another. The hosts may be large multiprogrammed mainframes or small personal computers. Networks can be classified as *local networks* or *long-haul networks*. The hosts on a local network are typically contained in a single building or campus and are connected by a high-bandwidth cable or other communication medium specifically designed for this purpose. Long-haul networks, in contrast, typically connect hosts in different cities using the public telephone network, an earth satellite, or both.

Local networks are nearly always completely owned by a single organization, whereas long-haul networks normally involve at least two organizations: the *carrier*, which operates the communication facility (telephone lines, microwave dishes, satellite, etc.), and the users, who own the hosts. This division of labor into (1) the provider of the communication facility and (2) the

"Network Protocols" by A.S. Tanenbaum from *Computing Surveys*, Vol. 13, No. 4, December 1981, pages 453-489. Copyright © 1981 by The Association for Computing Machinery, Inc., reprinted by permission.

CONTENTS

INTRODUCTION
   Protocols
   Overview of the ISO OSI Layers
1. THE PHYSICAL LAYER
   1.1 The Telephone System
   1.2 Communication Satellites
   1.3 Local Networks
   1.4 An Example Physical Layer Protocol: X.21
2. THE DATA LINK LAYER
   2.1 Stop-and-Wait Protocols
   2.2 Sliding-Window Protocols
   2.3 An Example Data Link Layer Protocol: HDLC
   2.4 Channel Allocation in Satellite Networks
   2.5 Channel Allocation in Local Networks
3. THE NETWORK LAYER
   3.1 Routing in Point-to-Point Networks
   3.2 Congestion Control in Point-to-Point Networks
   3.3 An Example Network Layer Protocol: X.25
4. THE TRANSPORT LAYER
   4.1 The Transport Station
   4.2 Establishing and Closing Connections
   4.3 Flow Control and Buffering
   4.4 Connection Multiplexing
5. THE SESSION LAYER
6. THE PRESENTATION LAYER
   6.1 Text Compression
   6.2 Encryption Protocols
   6.3 Virtual-Terminal Protocols
   6.4 File Transfer Protocols
7. SUMMARY
ACKNOWLEDGMENTS
REFERENCES

---

users of the communication facility has important ramifications for network architectures, as we shall see later.

The communication facility in a long-haul network is called the (*communication*) *subnet*, and often consists of a collection of minicomputers variously called *IMPs* (interface message processors), *nodes*, or *switches* connected by high-bandwidth leased telephone lines or a satellite. Figure 1 shows a network using telephone lines. Such a network is called a *point-to-point* or *store-and forward* network, as opposed to a *broadcast* network, such as a satellite network. The terms "host," "IMP," and "communication subnet" come from the U.S. Department of Defense's ARPANET, one of the first large-scale networks [McQu77]. We use this terminology gener-

ically because no consensus on nomenclature exists.

When the IMPs are connected by telephone lines, they are normally located on the carrier's premises, with each IMP servicing multiple hosts. To save on long-distance leased-line line charges, hosts and terminals are often funneled through remote concentrators. When the IMPs are connected by a satellite, the IMPs may be located on the customer's premises (e.g., on the roof). Local networks do not have IMPs; instead, each host has an interface card inserted into its backplane to control access to the network. This card is attached to the communication subnet, which is typically just a cable.

Although the ISO Reference Model can be used for both long-haul and local networks, it was designed primarily with the former in mind. Accordingly, in this paper we also treat both kinds of networks, but we emphasize slightly the long-haul variety, since issues such as routing and congestion control play a more prominent role in long-haul networks than in local networks.

In passing, we note that the subject of connecting distinct networks together is an increasingly important one, although it lies beyond the scope of this article. For an introduction to this subject see BOGG80 and POST80.

Protocols

As mentioned above, networks are almost always organized as a hierarchy of layers. Each layer performs a small set of closely related functions. The ISO Reference Model has seven layers:

(1) the physical layer,
(2) the data link layer,
(3) the network layer,
(4) the transport layer,
(5) the session layer,
(6) the presentation layer,
(7) the application layer,

as shown in Figure 2. All layers are present on the hosts, but only layers 1, 2, and 3 are present on the IMPs.

Each layer should be thought of as a program or process (possibly embedded in a hardware device) that communicates with

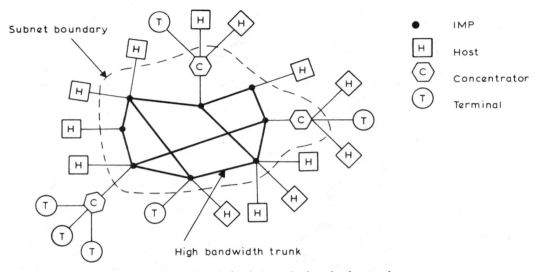

Figure 1. A typical point-to-point long-haul network.

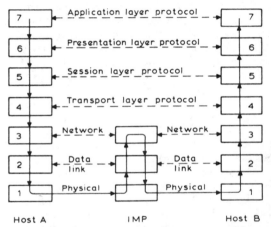

Figure 2. The seven-layer ISO Reference Model.

the corresponding process on another machine. In Figure 2, host layers 1, 2, and 3 think that they are communicating with their corresponding layers on the IMP, called *peers*. (In this example, hosts *A* and *B* are serviced by a common IMP; in general, multiple IMPs may intervene.) Layers 4–7, in contrast, communicate directly with their peer layers on the other host. The rules governing the layer *k* conversation are called the *layer k protocol*. The ISO model thus has seven protocols.

In reality, data are not transmitted horizontally, from machine to machine within a given layer, but are passed vertically down

the layers of the sending machine and up the layers of the receiving machine. Only in layer 1 does actual intermachine communication occur. When an application program, running in layer 7 on host *A*, wants to send a message to the application in layer 7 on host *B*, it passes the message down to the presentation layer on its own machine. The presentation layer transforms the data, adds a layer 6 *header* containing control information used by the layer 6 protocol, and passes the resulting message down to the session layer. The session layer then adds its own header and passes the new message down to the transport layer. The complete path from layer 7 on host *A* to layer 7 on host *B* is shown in Figure 2 by the solid line. The boundary between adjacent layers is called an *interface*. The layers, interfaces, and protocols in a network form the *network architecture*.

No layer is aware of the header formats or protocols used by other layers. Layer *k* on the sending machine regards its job as getting the bits that come in from layer *k* + 1 over to the receiving machine somehow (using the services of the lower layers). It neither knows nor cares what the bits mean.

A three-layer analogy may be helpful in understanding how multilayer communication works. Consider the problem of the

7

two talking philosophers. Philosopher 1 lives in an ivory tower in Kenya and speaks only Swahili. Philosopher 2 lives in a cave in India and speaks only Telugu. Nevertheless, Philosopher 1 wishes to convey his affection for *Oryctolagus cuniculus* to his Indian colleague (the philosophers are layer 3 peers). Since the philosophers speak different languages, each engages the services of a translator (layer 2 process) and an engineer (layer 1 process).

To convey his thoughts, Philosopher 1 passes his message, in Swahili, to his translator, across the 3/2 interface. The translator may convert it to English, French, Dutch, or some other language, depending only on the layer 2 protocol. The translator then hands his output to his engineer across the 2/1 interface for transmission. The physical mode of transmission may be telegram, telephone, computer network, or something else, depending only on the layer 1 protocol. When the Indian engineer receives the message, he passes it to his translator for rendition into Telugu. Finally, the Indian translator gives the message, in Telugu, to his philosopher.

This analogy illustrates three points. First, each person thinks of his communication as being primarily horizontal, with his peer (although in reality it is vertical, except in layer 1). For example, Philosopher 1 regards himself as conversing with Philosopher 2, even though his only physical communication is with translator 1. Second, actual communication is vertical, not horizontal, except in layer 1. Third, the three protocols are completely independent. The philosophers can switch the subject from rabbits to guinea pigs at will; the translators can switch from English to Dutch at will; the engineers can switch from telegram to telephone at will. The peers in any layer can change their protocol without affecting the other layers. It is for precisely this reason that networks are designed as a series of layers—to prevent changes in one part of the design (e.g., caused by technological advances) from requiring changes in other parts.

Overview of the ISO OSI Layers

The remainder of this article concerns the various layers in the ISO Reference Model, one section per layer. Before looking at the layers in detail, we first present a brief overview of each layer, to put the hierarchy in perspective.

The physical layer protocol is concerned with the transmission of a raw bit stream. Its protocol designers must decide how to represent 0's and 1's, how many microseconds a bit will last, whether transmission is full- or half-duplex, how the connection is set up and torn down, how many pins the network connector has, what each pin is used for, and other electrical, mechanical, and procedural details.

The data link layer converts an unreliable transmission channel into a reliable one for use by the network layer. The technique for doing so is to break up the raw bit stream into frames, each containing a checksum for detecting errors. (A *checksum* is a short integer that depends on all the bits in the frame so that a transmission error will probably change it and thus be detectable.) The data link protocol usually ensures that the sender of a data frame will repeatedly transmit the frame until it receives an acknowledgment frame from the receiver.

The network layer in a point-to-point network is primarily concerned with routing and the effects of poor routing, namely, congestion. In a broadcast network, routing is not an issue, since only one channel exists.

The task of the transport layer is to provide reliable host-to-host communication for use by the session layer. It must hide all the details of the communication subnet from the session layer, so that, for example, a point-to-point subnet can be replaced by a satellite link without affecting the session, presentation, or application layers. In effect, the transport layer shields the customer's portion of the network (layers 5–7) from the carrier's portion (layers 1–3).

The session layer is responsible for setting up, managing, and tearing down process-to-process connections, using the host-to-host service provided by the transport layer. It also handles certain aspects of synchronization and recovery.

The presentation layer performs generally useful transformations on the data to be sent, such as text compression. It also performs the conversions required to allow an interactive program to converse with any one of a set of incompatible intelligent terminals.

The content of the application layer is up to the users. Nevertheless, standard protocols for specific industries, such as airlines and banking, are likely to develop, although few exist now. For this reason we say no more about the application layer in this paper.

Although the ISO OSI Reference Model says nothing about how the layers are to be implemented, one possible configuration might have the physical layer in hardware, the data link layer in a special protocol chip, the network layer in a device driver, the transport and session layers in the operating system proper, the presentation layer in a set of library routines in the user's address space, and the application layer be the user's program.

At this point we have covered enough background material to say a little bit about the ISO OSI Reference Model itself. Basically, it is a framework for describing layered networks. It discusses the concept of layering in considerable detail, and introduces a uniform terminology for naming the various entities involved. Finally, it specifies the seven layers mentioned thus far, and for each layer gives its purpose, the services provided to the next higher layer, and a description of the functions that the layer must perform. The value of the model is that it provides a uniform nomenclature and a generally agreed upon way to split the various network activities into layers.

However, the ISO OSI Reference Model is *not* a protocol standard. By breaking a network's functions up into layers, it suggests places where protocol standards could be developed (physical layer protocols, data link layer protocols, and so on), but these standards themselves fall outside the domain of the model. With the model in hand, other organizations such as the Consultative Committee for International Telephony and Telegraphy (CCITT), the International Federation for Information Processing (IFIP), and the American National Standards Institute (ANSI) may develop specific protocol standards for the various layers. Although these standards may even-

tually be officially approved by ISO, such work is still in progress and, in any event, falls far outside the scope of the model.

As a final note, before plunging into the details of the various layers, we would like to point out that this article is about network protocols, with the ISO OSI Reference Model used as a guide; it is *not* an article about the model itself. We emphasize the communication algorithms and protocols themselves, a subject about which the Reference Model says nothing.

1. THE PHYSICAL LAYER

In this section we look at a variety of aspects related to the physical layer. Our emphasis is on the conceptual organization of the physical transmission facilities, not on the hardware details themselves. Point-to-point, satellite, and local networks are discussed. We conclude with a brief discussion of the X.21 physical layer protocol.

The function of the physical layer is to allow a host to send a raw bit stream into the network. The physical layer is in no way concerned with the way the bits are grouped into larger units, or what they mean. Nor does it rectify the problem of some bits being garbled by transmission errors. Recovery from such errors is up to the data link layer.

The communication subnet can be organized in one of two ways. In *circuit switching*, a fixed amount of transmission capacity (bandwidth) is reserved when the source initiates a conversation and released only when the conversation is over. The telephone system uses circuit switching. When someone calls a time-sharing service in a distant city, the connection is established after dialing and remains in force until one end hangs up. If the user goes out to lunch while still logged in, the connection remains intact and the charges continue to accumulate, even though the connection is actually idle.

With *packet switching*, in contrast, the user initially sets up a connection between his terminal or host and the nearest IMP, not the destination host. (We assume that the destination host also is connected to some IMP.) Whenever the user has data to send, he sends them to the IMP as a series of *packets*, typically 10–1000 bytes long.

Packets are routed from IMP to IMP within the subnet, until they get to the IMP which services the destination host. No circuits are reserved in advance within the subnet for the terminal-to-host connection (except the terminal-to-IMP and IMP-to-host circuits). Instead, the high-bandwidth IMP–IMP lines are dynamically shared among all the users on a demand basis; IMP–IMP bandwidth is only tied up when data are actually being transmitted.

Although the above discussion is cast in terms of a point-to-point network, the same considerations apply to broadcast channels. If a portion of the channel (e.g., one frequency band) is dedicated to a given conversation throughout its duration, without regard to actual usage, the network is circuit switched. If, however, the channel is dynamically requested, used, and released for every packet, the network is packet switched.

Circuit-switched networks are best suited to communication whose bandwidth requirements do not change much over time. Transmission of human speech is such an application, so it makes sense for the telephone network to be circuit switched. Terminal-to-computer and computer-to-computer traffic, however, is usually bursty. Most of the time there are no data to send, but once in a while a burst of data must be transmitted. For this reason, most computer networks use packet switching to avoid tying up expensive transmission facilities when they are not needed. However, in the future, all digital transmission systems will allow computers to dial a call, send the data, and hang up, all within a few milliseconds. If such systems become widespread, circuit switching may come back into favor.

1.1 The Telephone System

Since most existing long-haul networks use the telephone system for their transmission facilities, we shall briefly describe how the latter is organized. Most telephones are connected to a nearby telephone company switching office by a pair of copper wires known as a *local loop*. The switching offices themselves are connected by high-bandwidth *trunks* onto which thousands of unrelated calls are multiplexed. Although some trunks utilize copper wire, many utilize microwave relays, fiber optics, or wave guides as the transmission medium.

Because the bandwidth of the local loop is artificially limited to about 3000 Hz (hertz), it is difficult to transmit information over it by using, for example, +5 volts for a binary one and 0 volts for a binary zero. Such square wave signaling depends on high-frequency harmonics that are well above the 3000-Hz cutoff frequency. Only with very low date rates might enough information be below 3000 Hz to be intelligible. Instead, a device called a *modem* is inserted between the host and the telephone line. The input to the modem is pure digital data, but the output is a modulated sine wave with a base frequency of generally between 1000 and 2000 Hz. Since the modulated sine wave has fewer high-frequency components than the original square wave, it is affected less by the limited bandwidth.

A sine wave has three properties that can be modulated to transmit information: an amplitude, a frequency, and a phase. In amplitude modulation, two different amplitude values are used to represent 0 and 1. In frequency modulation, different frequencies are used for 0 and 1, but the amplitude is never varied. In phase modulation, neither the amplitude nor the frequency is varied, but the phase of the sine wave is abruptly switched to send data. In the most common encoding scheme, phase shifts of 45, 135, 225, and 315 degrees are used to send 00, 01, 10, and 11, respectively. In other words, each phase shift sends two bits. The three methods can be combined to increase the transmission capacity.

Many such transmission systems have been standardized and form an important class of physical layer protocols. Unfortunately, in many cases, the standards in the United States and Canada differ from those used by the rest of the world. For example, those ubiquitous 300-bit-per-second frequency modulation modems found near terminals around the world use different signaling frequencies in North America and Europe.

Probably the best known physical layer standard at present is RS-232-C, which specifies the meaning of each of the 25 pins on a terminal connector and the protocol governing their use. However, a new standard, RS-449, has been developed to replace this aging workhorse. RS-449 is upward compatible with RS-232-C but uses a 37-pin connector to accommodate the new signals. Unfortunately, 37 pins are insufficient, so users wishing to take advantage of all the features of RS-449 (notably the secondary channel) need a second 9-pin connector as well.

The transmission technology and protocols used on the interoffice trunks are different from those used on the local loop. In particular, digital rather than analog techniques are becoming increasingly widespread. The most common digital system is *pulse code modulation* (PCM), in which the analog signal coming in from the local loop is digitized by sampling it 8000 times per second. Eight bits (seven data and one control) are transmitted during each 125-μs (microsecond) sampling period. In North America, 24 such PCM channels are grouped together into 193-bit frames, with the last bit being used for synchronization. With 8000 193-bit frames per second, the gross data rate of this system, known as T1, is 1.544 Mbits/s (megabits per second). In Europe, the 1.544-Mbit/s PCM standard uses all 8 bits for data, with the 193rd bit (which is attached to the front rather than rear of the frame) used for signaling. Two different (and incompatible) 32-channel PCM standards running at 2.048 Mbits/s are also widely used outside North America. For more information about the telephone system see DAVI73 and DOLL78.

1.2 Communication Satellites

Although most existing long-haul networks use leased telephone circuits to connect the IMPs, satellite-based networks are becoming increasingly common. A communication satellite is a big repeater in the sky. Incoming signals are amplified and rebroadcast by a transponder on the satellite. The upward and downward signals use different frequencies to avoid interference. A typical communication satellite has 5–10 independent transponders, each with a capacity of about 50 Mbits/s.

Communication satellites are put into geosynchronous equatorial orbit at an altitude of 36,000 kilometers to make them appear stationary in the sky when viewed from the earth. Consequently, the ground station antenna can be pointed at the satellite when the antenna is installed and never moved. A moving satellite would require a much more expensive steerable antenna and would also have the disadvantage of being on the other side of the earth half the time. On the other hand, the great altitude required to achieve a 24-hour period implies an up-and-down propagation delay of 270 ms (milliseconds), which seriously affects the data link layer protocols and response time.

To avoid mutual interference, communication satellites using the 4/6-GHz (gigahertz) frequency band must be separated by an angle of 4 degrees as viewed from the earth. Since some orbit slots have been allocated by international agreement to television, military, and other use, the number of equatorial orbit slots available to data communication is limited. (As an aside, the allocation of orbit slots has been a political battleground, with every country, especially those in the Third World, asking for its fair share of slots for the purpose of renting them back to those countries able to launch satellites.) The 12/14-GHz band has also been allocated to data communication. At these frequencies, an orbit spacing of 1 degree is sufficient, providing four times as many slots. Unfortunately, because water is an excellent absorber of these short microwaves, multiple ground stations and elaborate switching are needed in order to avoid rain.

Three modes of operation have been proposed for satellite users. The most direct but most expensive mode is to put a complete ground station with antenna on the user's roof. This approach is already feasible for large multinational corporations and will become feasible for medium-sized ones as costs decline. The second approach is to put a small, cheap antenna on the user's roof to communicate with a shared satellite ground station on a nearby hill. The third approach is to access the ground station via a cable (e.g., a leased telephone circuit or even the same cable used for cable television).

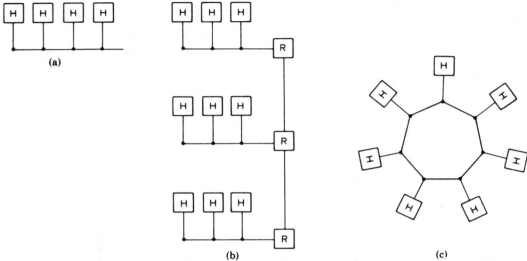

Figure 3. Local network topologies. (a) Linear cable with four hosts. (b) Segmented cable with repeaters and hosts. (c) Ring.

Physical layer satellite protocols typically have many PCM channels multiplexed on each transponder beam. Sometimes they are dedicated (circuit switched); at other times they are dynamically assigned as needed (packet switched). For more information about communication satellites see MART78.

1.3 Local Networks

In most local networks, the hosts are connected by a linear, tree-shaped, or ring-shaped cable, as shown in Figure 3. In Figure 3a, all hosts tap onto a common cable. In Figure 3b, multiple cables are used (e.g., one per floor of an office building), with repeaters connecting the segments. In Figure 3c, all hosts tap onto a unidirectional ring.

A widely imitated linear or tree-shaped local network is the Ethernet™ network [METC76]. The proper term for this kind of network is CSMA/CD (Carrier Sense Multiple Access/Collision Detect), although many people incorrectly use the term "Ethernet" (which is a trademark of the Xerox Corporation) in a generic sense. In these networks, only one packet may be on the cable at any instant. The cable is known as the *ether*, after the luminiferous ether through which electromagnetic radiation was once alleged to propagate. The principle behind CSMA/CD is simple: when a

host wishes to send a packet, it first listens to the ether to see if the ether is being used. If it is, the host waits until the current transmission finishes; if not, the host begins transmitting immediately.

The interface hardware must detect collisions caused by two hosts simultaneously starting a transmission. Collision detection is done using analog circuitry, in essence monitoring the ether to see if it agrees with the signal being transmitted. When a host interface (the analog of an IMP in this system, since the ether itself is totally passive) detects a collision, it informs the data link layer. The collision recovery action consists of aborting the current transmission, broadcasting a noise burst to make sure that everyone else detects the collision as well, waiting a random length of time, and then trying again. Collision detection is only feasible if the round-trip propagation delay is short compared to the packet transmission time, a condition that can be met with cable networks, but not, for example, with satellite networks.

Cable networks similar to the Xerox Ethernet network, but without the collision detect feature, also exist. Network designers can trade off the cost of collision detection circuitry against the time lost by not aborting colliding packets quickly.

Ring nets use a different principle: in effect, the whole ring is a giant circular shift

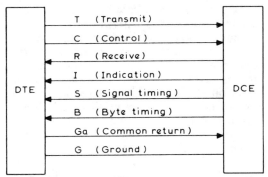

Figure 4. The DTE/DCE interface in X.21.

register. After each shift, the host interface can read or write the bit just shifted into it. Several different kinds of rings have been proposed [CLAR78, FARB72, FRAS75, LIU78, WILK79], differing primarily in their layer 2 organizations, which we describe later. Both CSMA/CD networks and rings typically operate at data rates of 1–10 Mbits/s. A substantial bibliography about local networks can be found in FREE80 and SHOC81.

1.4 An Example Physical Layer Protocol: X.21

At present, most physical layer standards, like RS-232-C and RS-449, utilize analog signaling. In the future, true digital interfaces will be needed. Recognizing this need, CCITT, the international standardization body for telephony, has developed a fully digital interface called *X.21*. X.21 is intended to be used to connect a host computer to a network. This connection remains established as long as the host wants to communicate with the network. Consequently, X.21 is a circuit-switched protocol, but host–host connections set up over the X.21 line may be either circuit switched or packet switched.

In X.21 terminology, the host is a *DTE* (Data Terminal Equipment) and the IMP is a *DCE* (Data Circuit-Terminating Equipment). The DTE–DCE interface consists of eight lines, as shown in Figure 4. The *S* line provides a clock signal to define bit boundaries. The (optional) *B* line provides a pulse every eighth bit, to allow byte alignment. The *C* and *I* lines are used for control signaling, analogous to the on-hook/off-hook signal on a telephone. The *T* and *R* lines are used for data and also for signaling.

To see how X.21 works, let us examine how a DTE calls another DTE, talks to it, and then hangs up. When the interface is idle, *T, R, C,* and *I* are all 1. The series of events is as follows (with a telephone analogy in parentheses):

(1) DTE drops *T* and *C* (DTE picks up phone).
(2) DCE sends "+ + + + + ··· + + +" on *R* (DCE sends dial tone).
(3) DTE sends callee's address on *T* (DTE dials number).
(4) DCE sends call progress signals on *R* (phone rings).
(5) DCE drops *I* to 0 (callee answers phone).
(6) Full duplex data exchange on *T* and *R* (talk).
(7) DTE raises *C* to 1 (DTE says goodbye).
(8) DCE raises *I* to 1 (DCE says goodbye).
(9) DCE raises *R* to 1 (DCE hangs up).
(10) DTE raises *T* to 1 (DTE hangs up).

The call progress signals in Step 4 tell whether the call has been put through, and if not, why not. The shutdown procedure in Steps 7–10 operates in two phases. After either party has said goodbye, that party may not send more data but it must continue listening for incoming data. When both sides have said goodbye, they then hang up, returning the interface to idle state, with 1's on all four lines. RS-449 and X.21 are described in more detail in BERT80 and FOLT80.

2. THE DATA LINK LAYER

As we have seen, neither X.21, RS-232-C, nor any other physical layer protocol makes any attempt to detect or correct transmission errors. Nor do these protocols recognize the possibility that the receiver cannot accept data as fast as the sender can transmit them. Both of these problems are handled in the data link layer. In the following sections we first discuss the relevant principles and then we give an example of a widely used data link protocol, HDLC (High-Level Data Link Control). Following the HDLC example, we look at some data link protocols for satellite and local networks.

As mentioned earlier, the approach used in the data layer is to partition the raw physical layer bit stream into frames so each transmitted frame can be acknowledged if need be. An obvious question is: "How are frames delimited?" In other words, how can the receiver tell where one frame ends and the next one begins?

Three methods are in common use on long-haul networks: *character count, character stuffing,* and *bit stuffing*. With the first method, each frame begins with a fixed-format frame header that tells how many characters are contained in the frame. Thus, by simply counting characters, the receiver can detect the end of the current frame and the start of the following one. The method has the disadvantage of being overly sensitive to undetected transmission errors which affect the count field; it also has the disadvantage of enforcing a specific character size. Furthermore, lost characters wreak havoc with frame synchronization. Digital Equipment Corporation's DDCMP (Digital Data Communication Message Protocol) uses the character count method, but few other protocols do. Use of character counts to delimit frames is likely to diminish in the future.

The second method for delimiting frames, character stuffing, is to terminate each frame with a special "end-of-frame" character. The problem here is what to do with "end-of-frame" characters that accidently appear in the data (e.g., in the middle of a floating point number). The solution is to insert an "escape" character before every accidental "end-of-frame" character. Now what about accidental "escape" characters? These are rendered as two consecutive escapes. Although these conventions eliminate all ambiguity, they do so at the price of building a specific character code into the protocol. IBM's BISYNC (BInary SYNchronous Communication) protocol uses character stuffing, but, like all other such protocols, it is gradually becoming obsolete.

Modern data link protocols for long-haul networks all use bit stuffing, a technique in which frames are delimited by the bit pattern 01111110. Whenever five consecutive one bits appear in the data stream, a zero bit is "stuffed" into the bit stream (normally by hardware). Doing so prevents user data from interfering with framing, but does not impose any character size on the data.

On local networks, one can use any of the above methods, or a fourth method: detecting frames by the presence or absence of a signal on the cable. This method is much more direct, but it is not applicable to long-haul networks.

Virtually all data link protocols include a checksum in the frame header or trailer to detect, but not correct, errors. This approach has traditionally been used because error detection and retransmission requires fewer bits on the average than forward error correction (e.g., with a Hamming code). However, with the growing use of satellites, the long propagation delay makes forward error correction increasingly attractive.

A simple checksum algorithm is: compute the Exclusive OR of all the bytes or words as they are transmitted. This algorithm will detect all frames containing an odd number of bits in error, or a single error burst of length less than the checksum, and many other combinations. In practice, a more complex algorithm based on modulo 2 polynomial arithmetic is used [PETE61, SLOA75].

2.1 Stop-and-Wait Protocols

As a first example of a data link layer protocol, consider a host A wishing to send data to another host B over a perfectly reliable channel. At first glance you might think that A could just send at will. However, this idea does not work, since B may not be able to process the data as fast as they come in. If B had an infinite amount of buffer space, it could store the input for subsequent processing. Unfortunately, no host has infinite storage. Consequently, a mechanism is needed to throttle A into sending no faster than B can process the data. Such mechanisms are called *flow control* algorithms. The simplest one calls for A to send a frame and then wait for B to send explicit permission to send the next frame. This algorithm, called *stop-and-wait,* is widely used.

More elaborate protocols are needed for actual channels that make errors. An obvious extension to our basic protocol is to

Sender

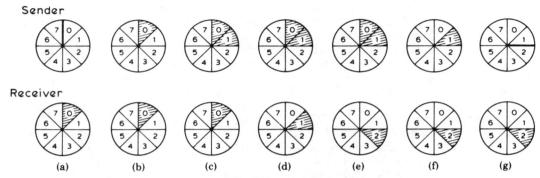

Receiver

   (a)   (b)   (c)   (d)   (e)   (f)   (g)

Figure 5. The sliding-window algorithm.

have A put a sequence number in each data link frame header and to have B put in each acknowledgment frame both a sequence number and a bit telling whether the checksum was correct or not. Whenever A received a negative acknowledgment frame (i.e., one announcing a checksum error), it could just repeat the frame.

Unfortunately, this protocol fails if either data or acknowledgment frames can be lost entirely in noise bursts. If a frame is lost, A will wait forever, creating a deadlock. Consequently, A must time out and repeat a frame if no acknowledgment is forthcoming within a reasonable period. Since each frame bears a sequence number, no harm is done if A has an itchy trigger finger and retransmits too quickly; however, some bandwidth is lost.

2.2 Sliding-Window Protocols

Stop-and-wait works well if the propagation time between the hosts is negligible. Consider, for a moment, how stop-and-wait works when 1000-bit frames are sent over a 1-Mbit/s satellite channel:

Time (ms)	Event
0	A starts sending the frame
1	Last bit sent; A starts to wait
270	First bit arrives at B
271	Last bit arrives at B
271	B sends a short acknowledgment
541	The acknowledgment arrives at A

For each millisecond of transmission, A has to wait 540 ms. The channel utilization is thus 1/541, or well below 1 percent. A better protocol is needed.

One such protocol is the *sliding-window* protocol, in which the sender is allowed to have multiple unacknowledged frames outstanding simultaneously. In this protocol, the sender has two variables, S_L and S_U, that tell which frames have been sent but not yet acknowledged. S_L is the lowest numbered frame sent but not yet acknowledged. The upper limit, S_U, is the first frame not yet sent. The current send window size is defined as $S_U - S_L$.

The receiver also has two variables, R_L and R_U, indicating that a frame with sequence number N may be accepted, provided that $R_L \leq N < R_U$. If $R_U - R_L = 1$, then the receiver has a window of size 1, that is, it only accepts frames in sequence. If the receiver's window is larger than 1, the receiver's data link layer may accept frames out of order, but normally it will just buffer such frames internally, so that it can pass frames to the network layer in order.

To keep sequence numbers from growing without bound, arithmetic is done modulo some power of 2. In the example of Figure 5, sequence numbers are recorded modulo 8. Initially (Figure 5a), $S_L = 0$, $S_U = 0$, $R_L = 0$, and $R_U = 1$ (receiver window size is 1 in this example). The current window is shown shaded in the figure. When the data link layer on the sending machine receives a frame to send (from the network layer), it sends the frame and advances the upper edge of its window by 1, as shown in Figure 5b. When it receives the next frame from the network layer, it sends the frame and advances the window again (Figure 5c). When the first frame arrives at the receiver, the receiver's window is rotated by advancing both edges (Figure 5d), and an acknowledgment is sent back. If frame 1 arrives at the receiver before the acknowledgment gets back to the sender, the state will be as shown in Figure 5e. When the first acknowledgment arrives, the lower edge of the sender's window is advanced (Figure 5f). Figure 5g shows the variables after both acknowledgments arrive.

As with stop-and-wait, the sliding-window protocol uses timeouts to recover from lost frames. The sender maintains a timer for each frame currently in its window. Whenever the lower edge of the window is advanced, the corresponding timer is stopped. Suppose, for example, that frames 0–4 are transmitted, but frame 1 is lost. The receiver will acknowledge frame 0, but discard frames 2–4 as they arrive, because they are outside the receive window (still size 1 in our example). Eventually, frames 1–4 will all time out and be retransmitted.

How many frames may our example sender have outstanding at any instant? The answer is seven, not eight, as might at first appear. To see why, consider the following scenario:

(1) The sender transmits frames 0–7.
(2) All eight frames arrive and are acknowledged.
(3) All eight acknowledgments are lost.
(4) The sender times out and retransmits the eight frames.
(5) The receiver unknowingly accepts the duplicates.

The problem occurs after Step 2, when the receiver's window has rotated all the way around and it is prepared to accept frame 0 again. Unfortunately, it cannot distinguish frame 9 from frame 0, so the stream of frames passed to the network layer will contain undetected duplicate frames.

The solution is to restrict the sender's window to seven outstanding frames. Then, after Step 2 above, the receiver will be expecting frame 7, and will reject all the duplicate frames, informing the sender after each rejection that it expects frame 7 next.

In the above example, whenever a frame is lost, the receiver is obligated to discard subsequent frames, even though they are correctly received. To avoid this inefficiency, we can allow the receiver's window

to be greater than 1. Now let us look at the lost frame problem again, with both the sender's and receiver's windows of size 7. When frames 2–4 come in, the receiver keeps them internally. Eventually frame 1 times out and is retransmitted. The receiver replies to the correct receipt of frame 1 by saying that it expects frame 5 next, thereby implicitly acknowledging frames 2–4 and preventing their retransmission. With frames 1–4 now safely in hand, the data link layer can pass them to the network layer in sequence, thus completely shielding the latter from the lost frame and its recovery. This strategy is often called *selective repeat,* as opposed to the *go back n* strategy implied by a receiver window size of 1.

Unfortunately, even with the window settings used above, the protocol can fail. Consider the following scenario:

(1) The sender transmits frames 0–6.
(2) All frames arrive; the receiver's window is now 7, 0, 1, 2, 3, 4, 5.
(3) All seven acknowledgments are lost.
(4) The sender times out and retransmits frames 0–6.
(5) The receiver buffers frames 0–5 and says it wants frame 7 next.
(6) The sender transmits frames 7–13 (sequence numbers 7, 0, 1, 2, 3, 4, 5).
(7) The receiver accepts frame 7 but rejects frames 0–5 as duplicates.

At this point the receiver has frames 7, 0, 1, 2, 3, 4, and 5 buffered. It passes them all to the unsuspecting network layer. Consequently, undetected duplicates sneak through again. To prevent this, the window size must be restricted to not more than half the size of the sequence number space. With such a restriction, the receiver's window after having received a maximum batch of frames will not overlap what it was before having received the frames. Hence no ambiguity arises about whether a frame is a retransmission or an original.

2.3 An Example Data Link Protocol: HDLC

As an example of a data link protocol that is widely used, we now briefly look at HDLC (High-Level Data Link Control). HDLC has many brothers and sisters (e.g., SDLC, ADCCP, LAP, LAPB), each having

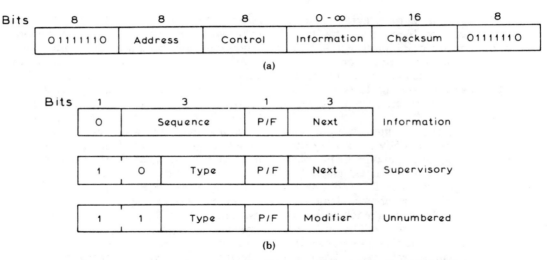

Bits   8   8   8   0 - ∞   16   8

| 01111110 | Address | Control | Information | Checksum | 01111110 |

(a)

Bits   1   3   1   3

0	Sequence	P/F	Next	Information
1	0	Type	P/F	Next	Supervisory
1	1	Type	P/F	Modifier	Unnumbered

(b)

Figure 6. (a) The HDLC frame format. (b) The control byte for the three kinds of frames.

minor, but irritating, differences in the control frames. How this situation came about has to do with how certain large bureaucracies view certain other large bureaucracies, a stone best left unturned here.

HDLC and its friends all use bit stuffing for delimiting frames. Their format is shown in Figure 6a. The *Address* field is used for addressing on multipoint lines (lines connecting more than two computers). The *Control* field is different for each of the three classes of frames (see Figure 6b). In *Information* frames (i.e., ordinary data), the *Sequence* and *Next* fields contain the sequence number of the current frame and of the next frame expected, respectively. When *A* sends a frame to *B*, the *Sequence* field is the number of the frame being sent and the *Next* frame is an acknowledgment to *B* saying that *A* has correctly received all frames sent by *B* up to but not including *Next*. Attaching an acknowledgment field to an outgoing data frame is widely known as *piggybacking*. The practice saves bandwidth by requiring fewer frames. Reducing the number of frames *sent* also reduces the number of frames *received*, and hence reduces the number of I/O interrupts on the receiving machine.

When no reverse traffic is present on which to piggyback acknowledgments, a *Type* = 0 supervisory frame is used. The other types of supervisory frames are for negative acknowledgment, selective repeat, and receiver temporarily not ready. The *P/F* bit stands for *Poll/Final* and has miscellaneous uses, such as indicating polling frames on multipoint lines and the final frame in a sequence.

Unnumbered frames consist of a hodgepodge of control information and comprise the area of greatest difference between the various HDLC-like protocols. Most of these frames are used to initialize the line and to report certain abnormal conditions.

Although Figure 6 depicts HDLC as having a 3-bit sequence number, an alternate format with 7-bit sequence numbers also exists, for use on satellite or other channels where large windows are needed to keep the channel busy. Gelenbe et al. [GELE78] have constructed a mathematical model of HDLC that can be used to calculate the throughput as a function of window size.

2.4 Channel Allocation in Satellite Networks

At this point we switch from the data link layer of point-to-point networks to that of broadcast networks, in particular, satellite and local networks. Broadcast networks are characterized by having a single channel that is dynamically requested and released by hosts for every packet sent. A protocol is needed for determining who may use the channel when, how to prevent channel overload, and so on. These problems do not occur in point-to-point networks. On the other hand, since every host receives every packet, broadcast networks usually do not have to make any routing decisions. Thus the main function of the network layer is not relevant.

As a consequence of these fundamental differences, it is not really clear where the channel-access protocol should be placed in the ISO OSI Reference Model, which does not mention the issue at all. It could be put in the data link layer, since it deals with getting packets from one machine to the next, but it could equally well be put in the network layer, since it also concerns getting packets from the source host to the destination host. Another argument for putting it in the network layer is that the main task of the access protocol is to avoid congestion on the channel, and congestion control is specifically a network layer function. Last, in most broadcast networks the transport layer is built directly on top of this protocol, or in some cases on top of an internetwork protocol, something lacking in the ISO OSI Reference Model. Nevertheless, we treat the subject as part of the data link layer because the IEEE local network standards committee (802) is probably going to put it there. By analogy, the contention resolution protocol for satellite channels also belongs in the data link layer.

A satellite link can be operated like a terrestrial point-to-point link, providing dedicated bandwidth for each user by time-division or frequency-division multiplexing. In this mode the data link protocols are the same as in point-to-point networks, albeit with longer timeouts to account for the longer propagation delay.

Another mode of operation, however, is to dynamically assign the channel among the numerous competing users. Since their only method of communication is via the channel itself, the protocol used for allocating the channel is nontrivial. Abramson [ABRA70] and Roberts [ROBE73] have devised a method, known as *slotted ALOHA*, that has some interesting properties. In their approach, time is slotted into units of a (fixed-length) packet. During each interval, a host having a packet to send can either send or refrain from sending. If no hosts use a given slot, the slot is just wasted. If one host uses a slot, a successful transmission occurs. If two or more hosts try to use the same slot, a collision occurs and the slot is also wasted. Note that with satellites the hosts do not discover the collision until 270 ms after they start sending the packets. Owing to this long delay, the collision detection principle from CSMA/CD is not applicable here. Instead, after detecting a collision, each host waits a random number of slots and tries again.

Clearly, if few hosts have packets to send, few collisions will occur and the success rate will be high. If, on the other hand, many hosts have packets to send, many collisions will occur and the success rate will be low. In both cases the throughput will be low: in the first case because of lack of offered traffic, in the second case because of collisions. Hence the throughput versus offered traffic curve starts out low, peaks, and then falls again. Abramson [ABRA73] showed that the peak occurs when the mean offered traffic is one packet per slot, which yields a throughput of $1/e$ or about 0.37 packets per slot. Hence the best one can hope for with slotted ALOHA is a 37 percent channel utilization.

Slotted ALOHA has another problem, in addition to the low throughput: stability. Suppose that an ALOHA system has many hosts. By accident, during one slot k hosts transmit and collide. After detecting the collision, each host decides to retransmit during the next slot with probability p (a parameter of the system). In other words, each host picks a random number between 0 and 1. If the number is less than p, it transmits; otherwise it waits until the next slot to pick another random number.

If $kp \gg 1$, many hosts will retransmit during each succeeding slot and practically nothing will get through. Worse yet, these retransmissions will compete with new packets from other hosts, increasing the number of hosts trying to use the channel, which just makes the problem worse. Pretty soon all hosts will be trying to send and the throughput will approach zero, collapsing the system permanently.

The trick to avoid collapse is to set the parameter p low enough that $kp < 1$ for the

k values expected. However, the lower p is, the longer it takes even to attempt retransmission, let alone succeed. Hence a low value of p leads to a stable system, but only at the price of long delay times.

One way to set p is to use a default value on the first retransmission, say 0.5, on the assumption that two hosts are involved in the collision. On each subsequent collision, halve p. Gerla and Kleinrock [GERL77] have another proposal; they suggest that each host monitor the channel all the time, just to measure the collision rate. When the collision rate is low, the hosts can set p high; when the collision rate is high, the hosts can set p low to minimize collisions and get rid of the backlog, albeit slowly.

A completely different way to avoid collisions is to attempt to schedule the slots in advance rather than have continuous competition for them. Crowther et al. [CROW73] proposed grouping slots into n-slot time slices, with the time slice longer than the propagation delay. In their system, contention is used initially, as described above. Once a host has captured (i.e., successfully used) a slot, it is entitled to use the same slot position in the next slice, forbidding all other hosts from trying to use it. This algorithm makes it possible for a host to transmit a long file without too much pain. If a host no longer needs a slot position, it sets a bit in the packet header that permits other users to contend for the slot the next time around.

Roberts [ROBE73] also proposed a method of reducing contention. His proposal also groups slots into time slices. One slot per slice is divided into minislots and used for reserving regular slots. To send a packet, a host must first compete for a minislot. Since all hosts see the results of the minislot contention, they can all keep track of how long the queue is and hence know who gets to send when. In effect, the use of minislots greatly reduces the amount of time wasted on a collision (like the CSMA/CD rule about aborting collisions as soon as they are detected).

2.5 Channel Allocation in Local Networks

As mentioned earlier, when a CSMA/CD host detects a collision, it jams the channel, aborts the current packet, waits a random time, and tries again. How long should it wait? Metcalfe and Boggs [METC76] decided to use a default maximum time interval on the first collision, with the actual waiting time being picked by multiplying a random number in the range 0.0–1.0 by the maximum time interval. On each successive collision the maximum time interval is doubled and a new random number is generated. They called their algorithm *binary exponential backoff*.

Various other algorithms have been proposed for CSMA/CD, including some that prevent all collisions. For example, Chlamtac [CHLA76], Chlamtac et al. [CHLA79], and Scholl [SCHO76] have suggested slotting time into intervals equal to the channel acquisition time (the round-trip propagation delay). After a successful transmission by host n, the next bit slot is then reserved for host $(n + 1)$ (modulo the number of hosts). If the indicated host does not claim its right to use the channel, the next host gets a chance during the succeeding bit slot, and so on. In effect, a virtual baton is passed from host to host, with hosts only allowed to transmit when holding the baton.

Rothauser and Wild [ROTH77] have also proposed a collision-free CSMA/CD protocol. To illustrate their suggestion, we shall assume that there are 1024 hosts, numbered from 0 to 1023 (in binary, although other radices can also be used). After a successful transmission, ten bit slots will be used to determine who goes next. Each host attempts to broadcast its 10-bit number in the ten slots, subject to the rule that as soon as a host realizes that a higher numbered host wants the channel, it must stop trying. If, for example, the first three bits are 011, then some host in the range 368–511 wants the channel, and so hosts below 368 must desist from competing on the current round. No host above 511 wants the channel, as evidenced by the leading 0 bit. In effect, the channel is allocated to the highest numbered contender. Since this system gives high-numbered stations an advantage, it is desirable to make the host numbers virtual, rotating them one position after each successful transmission.

Protocols that allow only a limited number of collisions have also been proposed [CAPE79a, CAPE79b, KLEI78]. Capetanakis'

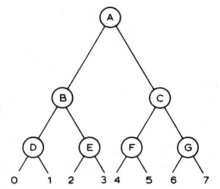

Figure 7. Eight machines organized in Capetanakis' (virtual) tree.

idea is illustrated in Figure 7 for a system with eight hosts. Initially, all hosts may compete. If a collision occurs, only those hosts under node *B* of the tree, namely, 0, 1, 2, and 3, may compete. If another collision occurs, the only descendants of node *D* may try, and so forth. As an example, suppose hosts 2, 3, and 4 all want the channel. After initial collisions for *A* (2, 3, and 4) and *B* (2 and 3), it will be node *D*'s turn and the channel will lie idle. Next comes node *E* and another collision. Then 2 and 3 each get a private slot, followed by *C*. At low load, the algorithm allows pure contention, but under high load it walks the tree looking for hosts that want to send.

Although more could be said about CSMA/CD protocols, we now turn our attention to ring networks. In one of the best-known rings [FARB72], an 8-bit token circulates around the ring when there is no traffic. When a host wants to transmit, it must first capture and destroy the token. Having done so, it may send its packet. When it is finished, it must put the token back, giving the next host downstream a chance to seize it.

If the token is ever lost (e.g., as a result of a ring interface malfunctioning), some mechanism is needed to regenerate it. One possibility is that each host wishing to send must monitor the ring. Having failed to see a token within the worst case interval—namely, all other hosts sending a maximum-length packet—the host generates a new token itself. However, with a little bad luck, two hosts might generate tokens si-

multaneously. Hence, it appears that token recovery in a ring net is similar to contention in CSMA/CD in systems. Clark et al. [CLAR78] have taken this observation to its logical conclusion and proposed a contention ring that is a hybrid of the token ring and CSMA/CD.

Yet another type of ring is exemplified by the Cambridge Ring [NEED79, WILK79]. This 10-Mbit/s ring contains several small slots around it, each slot consisting of 16 bits of data, an 8-bit source address, an 8-bit destination address, a bit telling whether the slot is full or empty, and a few other control bits. To transmit, a host interface just waits for a free slot and fills it up. When the slot arrives at the destination, the receiving interface sets the control bits telling whether it was accepted or not. About 10 μs after transmission, the slot comes back around again so that the sender can find out what happened to it. The sender is not permitted to reuse the slot immediately, as an antihogging measure. By having such small slots and preventing their immediate reuse, the ring guarantees an extremely short delay for small packets, but at the price of higher overhead than the token ring under heavy load.

Still another design is discussed in LIU78. In Liu's design, each ring interface has a shift register equal in length to the maximum packet size. When a host wants to send a packet, it loads up the shift register and inserts the shift register into the ring between two packets. This mechanism leads to low delay, since a host need only wait until the current packet has passed through. When the shift register becomes empty (through a period of low traffic), it can be removed from the ring.

From the above discussion, it should be obvious that many local network protocols have already been devised, with more being threatened all the time. Without standards, the most likely development would be a proliferation of local networks from various vendors, all incompatible; vendor A's terminal would not talk to vendor B's CPU because they would have different protocols embedded in their hardware. In an attempt to nip this incipient chaos in the bud, the Institute of Electrical and Electronics Engineers (IEEE) set up a committee in February 1980 to develop a standard for local network protocols. Although the standard, IEEE 802, was not completed at the time of this writing, the general picture looks as if it will probably be as follows:

The standard treats the physical and data link layers. The physical layer allows for base-band, broad-band, and fiber optics communication, and describes the interfacing of the host (DTE) to the cable. The data link layer handles channel access, as mentioned earlier, as well as addressing, frame format, and control.

The data link layer is split up into two sublayers, media access and data link control, with a third optional sublayer for internetworking (whose presence in the data link layer instead of in the network layer is certainly arguable). The media access sublayer handles channel allocation. It is here that a choice had to be made between CSMA/CD and some kind of ring. The arguments for CSMA/CD were that it was fair, easy to implement on a single chip, had six years of operational experience, and had three major companies (DEC, Intel, and Xerox) already publicly committed to it.

The token ring supporters' counterarguments were as follows: rings, unlike CSMA/CD, provide a guaranteed worst case access time (needed for real-time work, such as speech transmission); rings can be logical as well as physical, accommodating various topologies and allowing important hosts better access by inserting them into the logical ring in several places; and ring performance does not degrade at high load, as does CSMA/CD owing to the many collisions. Unfortunately, neither camp had the necessary two-thirds majority required by IEEE rules, and so a compromise was made in which both CSMA/CD and a token ring were included.

The data link control sublayer was designed to be as compatible with HDLC as possible, on the theory that the last thing the world needed was yet another brand-new data link protocol. Two types of service are provided for: connection oriented and pure datagram. In the former, the data link layer times out and retransmits lost frames, guarantees arrival in sequence, and regulates flow using the standard HDLC sliding-window protocol. In the latter, the data link layer guarantees nothing; once sent, the frame is forgotten (at least by the data link layer).

The major difference between the 802 frame and HDLC's is the presence in 802 of two addresses, source and destination, instead of the one address in HDLC, and the use of variable-length addresses (from 1 to 7 bytes), instead of fixed-length, 1-byte addresses. HDLC was designed for two-party, point-to-point lines, where no address is needed, and for multipoint master/slave lines, in which only the slave's address is needed. In contrast, 802 is aimed at multipoint symmetric lines, where any machine can send to any other machine, and so two addresses are required. The decision to have variable-length addresses up to 7 bytes is intended to allow processes to be designated by a worldwide unique address. Three of the 7 bytes are to be administered by an as-yet-unidentified international organization, and 4 are for local use. Most networks will only need 1- or 2-byte addresses for internal traffic.

3. THE NETWORK LAYER

When a frame arrives at an IMP in a point-to-point network, the data link layer strips off the data link header and trailer and passes what is left, called a *packet*, to the network layer. The network layer must then decide which outgoing line to forward the packet on. It would be nice if such decisions could be made so as to avoid having some lines congested and others idle. Hence congestion control is intimately related to routing. We first look at routing and then at congestion control, both for point-to-point networks. With the channel acquisition protocol for broadcast networks in the data link layer, the network layer for these networks is essentially empty.

Two opposing philosophies exist concerning the network layer. In most local networks and some long-haul networks, the network layer provides a service for delivering independent packets from source to destination with a high probability of success (although less than 1.0). Each packet carried is unrelated to any other packet, past or future, and must therefore carry a full destination address. Such packets are called *datagrams*.

The other approach, taken in many public data networks (especially in Europe), is to require a transmitter to first send a setup packet. The setup packet chooses a route for subsequent traffic and initializes the IMPs along the route accordingly. The user chooses, or is given, a *virtual circuit number* to use for subsequent packets going to the same destination. In this organization, data packets belonging to a single conversation are not independent, since they all follow the same route, determined by the virtual circuit number in them.

The advantage of using virtual circuits is that it guarantees that packets will be delivered in order and helps reduce congestion by making it possible to reserve resources (e.g., buffers) along the route in advance. The disadvantage is that a lot of IMP table space is taken up by idle connections and that there is a lot of overhead in setting up and closing down circuits, the latter a great concern in transaction-oriented database systems [MANN78]. With a datagram system, a query–response requires just two packets. With a virtual circuit system it requires six packets: setup, acknowledgment, query, response, close circuit, and acknowledgment.

3.1 Routing in Point-to-Point Networks

Many routing algorithms have been proposed, for example, BARA64, FRAT73, McQu74, RUDI76, SCHW80, and SEGA81. Below we sketch a few of the more interesting ones. The simplest algorithm is *static* or *directory* routing, in which each IMP has a table indexed by destination, telling which outgoing line to use. When a packet comes in, the destination address is extracted from the network layer header and used as an index into the routing table. The packet is then passed back down to the data link layer (see Figure 2) along with the chosen line number.

A variant algorithm provides two or more outgoing lines for each destination, each with a weight. When a packet arrives, a line is chosen with a probability proportional to its weight. Allowing alternatives eases congestion by spreading the traffic around. Note that when virtual circuits are used

within the subnet, the routing decision is only made for setup packets, not data packets.

Several proposals have been made for determining the routes to be put in the tables. Shortest path routing, which minimizes the number of hops (IMP–IMP lines), is an obvious candidate. In FRAT73 another method, based on flow deviation, is given.

The problem with static routing is just that—it is static—it does not adapt to changing traffic patterns and does not try to route packets around congested areas. One way to have the network adapt is to have one host function as a routing control center. All IMPs send it periodic reports on their queue lengths and line utilizations, from which it computes the best routes and distributes the new routing tables back to the IMPs.

Although seemingly attractive, centralized routing has more than its share of problems [McQu74]. To start with, if the routing control center malfunctions, the network will probably be in big trouble. Second, the complete optimal routing calculation for a large network may require a large dedicated host and even then may not be able to keep up with the traffic fluctuations. Third, since IMPs near the routing control center get their new tables before more distant IMPs do, the network will operate with mixed old–new tables occasionally, a situation that may cause traffic (including the new routing tables) to loop. Fourth, if the network is large, the traffic flow into and out of the routing control center may itself get to be a problem.

One of the earliest routing algorithms [BARA64] adapts to changing traffic, but does so without any central control. In *hot-potato routing*, when a packet arrives, it is assigned to the output line which has the shortest transmission queue. This strategy gets rid of the packet as fast as possible, without regard to where it is going. A much better idea is to combine static information about the suitability of a given output line with the queue lengths. This variant is known as *shortest queue plus bias*. It could be parameterized, for example, to use the shortest queue unless the line is going the wrong way, or to use the statically best line unless its queue exceeds some threshold.

Algorithms like this are known as *isolated adaptive* algorithms [McQu74].

Rudin's *delta routing* [Rudi76] combines some features from centralized and isolated adaptive algorithms. In this method, IMPs send periodic status reports to the routing control center, which then computes the *k* best paths from each source to each destination. It considers the top few paths equivalent if they differ (in length, estimated transit time, etc.) by an amount less than some parameter δ. Each IMP is given the list of equivalent paths for each destination, from which it may make a choice based on local factors such as queue lengths. If δ is small, only the best path is given to the IMPs, resulting in centralized routing. If δ is large, all paths are considered equivalent, producing isolated adaptive routing. Intermediate strategies are obviously also possible.

A completely different approach is distributed adaptive routing [McQu74], first used in the ARPANET, but replaced after ten years owing to the problems with looping discussed below. With this algorithm, each IMP maintains a table indexed by destination giving the estimated time to get to each destination and also which line to use. The IMP also maintains an estimate of how long a newly arrived packet would take to reach each neighbor, which depends on the queue length for the line to that neighbor.

Periodically, each IMP sends its routing table to each neighbor. When a routing table comes in, the IMP performs the following calculation for each destination. If the time to get the neighbor plus the neighbor's estimate of the time to get to the destination is less than the IMP's current estimate to that destination, packets to that destination should henceforth be routed to the neighbor.

As a simple example, consider a five-IMP network. At a certain instant, IMP 2 has estimates to all possible destinations, as shown in Figure 8a. Suddenly the routing table from IMP 3 (assumed to be adjacent to IMP 2) arrives, as shown in Figure 8b. Let us assume that IMP 2 estimates the delay to IMP 3 to be 10 ms, on the basis of the size of its transmission queue for IMP 3. IMP 2 now calculates that the transit

Destination

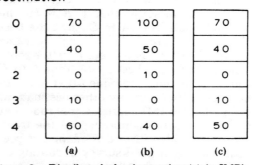

	(a)	(b)	(c)
0	70	100	70
1	40	50	40
2	0	10	0
3	10	0	10
4	60	40	50

Figure 8. Distributed adaptive routing. (a) An IMP's original routing table. (b) Routing table arriving from a neighboring IMP. (c) The new routing table, assuming a 10-ms delay to the neighbor.

time to IMP 0 via IMP 3 is 10 + 100 ms. Since this time is worse than the 70 ms for its current route, no change is made to entry 0 of the table. Similarly 10 + 50 > 40, and so no change is made for destination 1 either. However, for destination 4, IMP 3 offers a 40-ms delay, which, when combined with the 10-ms delay to get to IMP 3, is still better than the current route (10 + 40 < 60). Therefore, IMP 2 changes its estimate of the time required to get the IMP 4 to 50 ms, and records the line to IMP 3 as the way to get there. The new routing table is given in Figure 8c.

Although this method seems simple and elegant, it has a problem. Suppose that *A*, *B*, and *C* are connected by lines *AB* and *BC*. If number of hops is used as a metric, *B* thinks it is one hop from *A*, and *C* thinks it is two hops from *A*. Now imagine that line *AB* goes down. *B* detects the dead line directly and realizes that its delay to *A* is now infinite via *AB*. Sooner or later, however, *C* offers *B* a route to *A* of length two hops. *B*, knowing that line *AB* is useless, accepts the offer, and modifies its tables to show that *A* is three hops away via *C*. At this point *B* is routing packets destined for *A* to *C*, and *C* is sending them right back again. Having packets loop forever is not considered a good property to have in one's routing algorithm. This particular problem causes great anguish for the transport layer, as we shall see shortly.

To get around the problem of looping packets, several researchers (e.g., Chu78, Sega81) have proposed using the *optimal-*

ity principle to guarantee loop-free routing. This principle states that if B is on the optimal route from A to C, then the best route from B to C falls along the same route. Clearly if there were a better route from B to C, the best route from A to C could use it too. Consequently, the set of best routes to C (or any other destination) from all other IMPs forms a tree rooted at C. By explicitly maintaining all the trees, the routing algorithm can adapt but prevent looping. A good survey of routing algorithms can be found in SCHW80.

3.2 Congestion Control in Point-to-Point Networks

Now we turn to the problem of congestion in point-to-point networks. Actually, little is known about how to deal with it, and all the proposed solutions are rather ad hoc. Davies [DAVI72] suggested starting each network with a collection of special packets called *permits* that would roam about randomly. Whenever a host wanted to send a packet, its IMP would have to capture and destroy a permit before the new packet could be injected into the network.

This mechanism guarantees that the maximum number of packets in the network can never exceed the initial number of permits, which helps somewhat, but still does not guarantee that all the legal packets will not someday end up in one IMP, overloading it. Furthermore, no one has been able to devise a way to regenerate permits lost in IMP crashes (short of deadstarting the whole network, which is unacceptable). If these permits are not generated, carrying capacity will be permanently lost.

Another congestion control scheme is due to Irland [IRLA78]. This scheme calls for IMPs to monitor the utilization of each outgoing line. When a line utilization moves above a trigger value, the IMP sends a *choke packet* back to the source of each new packet needing that line, telling the source to slow down.

Kamoun [KAMO81] has proposed a congestion control scheme based on the observation that when packets must be discarded in an overloaded IMP, some packets are better candidates than others. In particular, if a packet has already made k hops, throwing it away amounts to discarding the investment in resources required to make those k hops. This observation suggests discarding packets with the smallest k values first. A variation of this idea that does not require a hop counter in each packet is to have IMPs discard newly injected packets from local hosts in order to salvage transit traffic with $k \geq 1$.

The limiting case of a congested network is a deadlocked network. If hosts A, B, and C are all full (no free buffers), and A is trying to send to B and B is trying to send to C and C is trying to send to A, a deadlock can occur, as shown in Figure 9.

Merlin and Schweitzer [MERL80a, MERL80b] describe several ways to prevent this kind of *store-and-forward deadlock* from occurring. One way is to provide each IMP with $m + 1$ packet buffers, where m is the longest path in the network. A packet newly arriving in an IMP from a local host goes into buffer 0. At the next IMP along the path it goes in buffer 1. At the following IMP it uses buffer 2. After having made k hops, it goes in buffer k. To see that the algorithm is deadlock free, consider the set of all buffers labeled m. Each buffer is in one of three states:

(1) empty,
(2) holding a packet destined for a local host,
(3) holding a packet destined for a distant host.

In Case 2 the packet can be delivered and the buffer freed. In Case 3 the packet is looping and must be discarded. In all cases the complete set of buffers labeled m can be made empty. Consequently, all packets in buffers labeled $m - 1$ can be either delivered or discarded, one at a time. The process can then be repeated, freeing the buffers labeled $m - 2$, and so on.

Other kinds of deadlocks in computer networks are discussed in GUNT81.

3.3 An Example Network Layer Protocol: X.25

To help standardize public long-haul networks, CCITT has devised a three-layer protocol of its own. The physical layer is X.21 (or X.21 bis, a stopgap analog interface to be used until the digital network arrives).

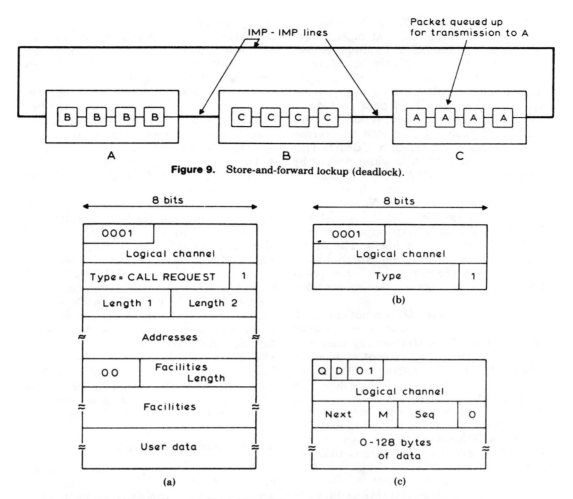

Figure 9. Store-and-forward lockup (deadlock).

Figure 10. X.25 headers. (a) CALL REQUEST packet. (b) Control packet. (c) Data packet.

The data link layer consists of two variants of HDLC (LAP and LAPB). Whether the next layer is network layer protocol or a transport layer protocol is a matter of some debate in the network community. Let us call it a network layer protocol and discuss it now.

X.25 (which is the collective name for all three layers) is virtual circuit oriented [RYBC80]. To set up a virtual circuit, a host (DTE) sends a CALL REQUEST packet into the network. The remote host can either accept or reject the incoming call. If it accepts it, the virtual circuit is set up; otherwise the circuit is cleared.

Figure 10a shows the format of the CALL REQUEST packet. The first 4 bits are 0001.

The next 12 bits are the virtual circuit number chosen by the originating host. The third byte is the type code of CALL REQUEST. The next byte gives the number of decimal digits in the caller's and callee's addresses, followed by up to 30 bytes containing the addresses themselves in binary coded decimal. (The telephone community has been using decimal numbers for 100 years, and old habits die hard.) The *Facilities* field is used to request services such as calling collect. Since the facilities field is variable length, a length field is needed. Finally, the user data field can be used in any way the user chooses, for example, to indicate which process within the called host expects the call.

25

When the CALL REQUEST packet arrives at the destination, that machine accepts or rejects the call by sending a packet of the form shown in Figure 10b. Acceptance or rejection is indicated in the *Type* field. Once the virtual circuit has been set up, both sides may send data packets at will, which makes the connection, by definition, full duplex. Either side may terminate the call by sending a CLEAR REQUEST packet, which is acknowledged by a CLEAR CONFIRMATION packet.

An ordinary data packet is shown in Figure 10c. The *Sequence* and *Next* fields are analogous to those in HDLC. Like HDLC, X.25 layer 3 also has an optional format with 7-bit sequence numbers. The M bit can be used by a host to indicate that more data follow in the next packet, thus partitioning the packet stream into multipacket units.

The meaning of the Q bit is not specified, but it is provided to allow the transport layer a means for distinguishing transport layer data packets from control packets. The D bit stands for Delivery confirmation. If a host sets it on all the packets sent on a certain virtual circuit, the *Next* field will contain a true acknowledgment from the remote host, producing an end-to-end confirmation. If, however, it is always set to 0, then the *Next* field just means that the local IMP (DCE) received the packet specified, not that the remote host did. Conceivably, when $D = 0$, the local IMP could write all the packets on magnetic tape to be mailed to the remote IMP for delivery in a couple of days (bargain basement service).

In the original version of X.25, only $D = 0$ was provided. That point generated so much controversy that delivery confirmation was added later, as was a pure datagram facility and something called *Fast Select*. With the Fast Select facility, the user data field in the CALL REQUEST packet is extended to 128 bytes and a similar field is added to the CLEAR REQUEST packet (used to reject incoming calls). Thus a host can send a short query in the CALL REQUEST packet and get the reply in the CLEAR REQUEST packet, without having to open a virtual circuit.

Because layers 2 and 3 in X.25 have so much overlap, it is perhaps useful to point out that the layer 2 sequence numbers and acknowledgments refer to the traffic between host and IMP for all virtual circuits combined. If a host sends the IMP seven packets (frames), each one for a different virtual circuit, the host must stop sending until an acknowledgment comes back. The layer 2 protocol is required to keep the host from flooding the IMP. In contrast, in layer 3, the sequence numbers are per virtual circuit and therefore flow control each connection separately.

X.25 layer 3 also has a few control packets. These include RESET and RESET CONFIRMATION, used to reset a virtual circuit; RESTART and RESTART CONFIRMATION, used to reset all virtual circuits after a host or IMP crash; RECEIVER READY, used for acknowledgments; RECEIVER NOT READY, used to indicate temporary problems and stop the other side even though the window is not full; and INTERRUPT and INTERRUPT CONFIRMATION, used to send out-of-band signals, such as breaks. All these control packets use the format of Figure 10b, in some cases augmented with an additional byte or two for additional information.

4. THE TRANSPORT LAYER

The network layer does not necessarily ensure that the bit stream sent by the source arrives intact at the destination. Packets may be lost or reordered, for example, owing to malfunctioning IMP hardware or software. The X.25 standard provides a mechanism (RESET and RESTART packets) for the network to announce to a host that it has crashed and lost track of both the current sequence numbers and any packets that may have been in transit. To provide truly reliable end-to-end (i.e., host-to-host) communication, another layer of protocol is needed: the transport layer. (Note that X.25 with $D = 1$ comes close to being end to end, but is not quite enough since it provides no way to transparently recover from network RESETs and RESTARTs.)

Another way of looking at the transport layer is to say that its task is to provide a network independent *transport service* to the session layer. The session layer should not have to worry about any of the implementation details of the actual network. They must all be hidden by the transport layer, analogous to the way a compiler must hide the actual machine instructions from the user of a problem-oriented programming language.

4.1 The Transport Station

The program within the host that implements the transport service is called the *transport station*. Its chief functions are to manage connection establishment and teardown, flow control, buffering, and multiplexing. Although a transport station might conceivably offer only datagram primitives to its users, most offer (and emphasize) virtual-circuit primitives. As a bare minimum, the following primitives or their equivalents are normally available:

connum = CONNECT(local, remote),
connum = LISTEN(local),
   status = CLOSE(connum),
   status = SEND(connum, buffer, bytes),
   status = RECEIVE(connum, buffer,
          bytes)

The primitives for establishing a transport connection, CONNECT and LISTEN, take *transport addresses* as parameters. Each transport address uniquely identifies a specific transport station and a specific *port* (connection endpoint) within that transport station. For example, CCITT has decreed that X.25 will use 14-digit numbers for addressing. The first three identify the country, and the fourth identifies the network within the country. (Multiple country codes have been assigned to countries that expect to have more than ten public networks.) The last ten digits of the X.25 address are assigned by each network operator, for example, five digits to indicate hosts and five digits for the hosts to allocate themselves.

In our example, the LISTEN command tells the transport station that the process executing it is prepared to accept connections addressed to the indicated local address. The process executing the LISTEN is blocked until the connection is established, at which time it is released, with the variable *connum* being set to indicate the connection number. The connection number is needed because multiple connections may be open at the same time and a subsequent SEND, RECEIVE, or CLOSE must be able to tell which connection is meant. If something goes wrong, an error number can be returned in *connum* (e.g., positive for connection established, negative for error).

The CONNECT command tells the transport station to send a message (e.g., X.25 CALL REQUEST) to another host to establish a connection. When the connection has been established (or rejected, for example, due to illegal addresses), the connection number or error code is returned in *connum*.

An important design issue is what should the transport station do if a CALL REQUEST packet comes in specifying a transport address for which no LISTEN is pending? Should it reject the request immediately, or should it queue the request in the hope that a LISTEN will be done shortly? If the request is queued, should it time out and be purged if no LISTEN is forthcoming within a reasonable time? If so, what happens if the LISTEN finally occurs after the timeout?

In the above example, both LISTEN and CONNECT are blocking primitives, that is, the caller is halted until the command completes. Some transport stations use nonblocking primitives. In other words, both calls complete immediately, perhaps only checking the syntactic validity of the addresses provided. When the connection is finally established, or definitively rejected, the respective processes are interrupted. Some transport stations that use nonblocking primitives also provide a way for a process to cancel an outstanding LISTEN or CONNECT, as well as a method for a listening process to inspect an incoming connection request before deciding to accept or reject it.

The primitive CLOSE speaks for itself. The status returned would normally be "OK" if the connection actually existed and "error" if it did not.

The SEND and RECEIVE primitives do the real work of message passing. For the sake of clarity, we refer to the entities exchanged here as "messages," to distinguish them from the "packets" of the network layer and "frames" of the data link layer. A message will be encased in a packet, which will be inserted into a frame before transmission, of course. SEND specifies the connection on which to send, the buffer address, and the number of bytes. RECEIVE has the same parameters, although here *bytes* might initially contain the buffer size and later be filled in with the size of the received message. Again, both of these could be provided in nonblocking as well as blocking versions.

A more elaborate transport station could offer commands to send and receive datagrams, to send and receive interrupt signals, to reset the connection in the event of error, and to interrogate the status of the other side, a facility particularly useful for recovering from network layer failures.

4.2 Establishing and Closing Connections

As we pointed out earlier, one consequence of adaptive routing is that packets can loop for an indefinite period of time. If a packet gets trapped, the sending transport station will eventually time out and send a duplicate. If the duplicate gets through properly, but the original packet remains trapped for a while, problems can arise when it finally escapes and is delivered. Imagine, for example, what would happen if a message instructing a bank to transfer a large sum of money were stored and later repeated, long after the transaction had already been completed.

A useful first step is to limit the amount of time that a packet can exist in the network. For example, a counter could be put in the packet header. Each time the packet was forwarded, the counter could be decremented. When the counter reached zero, the packet would be discarded. Alternatively, a timestamp in the packet could be used to render it obsolete after a certain interval.

The next step is to have the transport stations use a sequence space so large (e.g., 32 bits) that no packet can live for a complete cycle. As a result, delayed duplicates can always be detected by their sequence numbers. However, if all new connections always start with sequence number 0, pack-

ets from previous connections may come back to plague later ones. Therefore, it is necessary to have each new connection initialize its sequence numbers to a value known to be higher than that of any existing packet.

Unfortunately, not even these measures are enough. Since each host has a different range of sequence numbers outstanding, each one must specify the initial sequence number for packets it will send. Assume that sequence numbers are chosen during the call establishment phase. With some bad luck, the following scenario could occur at an instant when A wanted to set up a connection with sequence number 100 to B:

(1) A sends a CALL REQUEST packet with sequence number 100.
(2) The packet is lost.
(3) An old CALL REQUEST with sequence number 50 suddenly arrives at B.
(4) B's CALL ACCEPT packet, with sequence number 700, is lost.
(5) An old CALL ACCEPT from B with sequence number 650 suddenly arrives at A.

At this point the connection is fully established, with A about to send packet 100, but B expecting packet 50. Similarly, B intends to send packet 700, but A expects 650. The result is a deadlock.

Tomlinson [TOML75] proposed a connection establishment protocol that works even in the face of delayed control packets. It is called the *three-way handshake*. An example follows (S means sequence, A means acknowledgment):

(1) A sends a CALL REQUEST packet with S = 100.
(2) B sends a CALL ACCEPTED packet with S = 700, A = 100.
(3) A sends a packet with S = 101, A = 700.

Now consider what happens in the face of the same lost and duplicate packets that lead to deadlock above. When B receives the CALL REQUEST with S = 50, it replies with S = 700, A = 50. If this packet gets through, A sees the bad acknowledgment and rejects the connection. The only way A can be spoofed is for an old CALL AC-

CEPTED packet with A = 100 to appear out of the blue. Such a packet could only be generated in response to an *old* CALL REQUEST packet with S = 100, something A has not sent for a long time. Sunshine and Dalal (SUNS78] discuss this problem in more detail.

By now you should be convinced that the protocol required to establish a transport layer connection in the face of an unreliable network layer is nontrivial. What about closing a connection? Surely that, at least, is easy: A sends B a request to close, and B sends back a close acknowledgment. Unfortunately, things are not that simple. As an example, let us briefly consider the two-army problem.

Two divisions of the white army are encamped on the opposite walls of a valley occupied by the blue army. If both divisions attack simultaneously, the white army will win; if either division attacks alone, it will be massacred. The white army divisions must synchronize their attack using an unreliable channel (e.g., a messenger subject to capture). Suppose that white's A division sends the message: "Let's attack at teatime," and gets the reply "OK." Division B has no way of being sure that the reply got back. If it just goes ahead and attacks, it might get slaughtered. Furthermore, A is well aware of this line of reasoning and hence may be afraid to attack, even after having received an acknowledgment.

At this point you may be thinking: "Why not use a three-way handshake here?" Unfortunately, it does not work. A could confirm receipt of B's acknowledgment, but because this confirmation could get lost, A does not know which situation holds:

(1) B got the confirmation and the war is on.
(2) The confirmation got lost and the war is off.

How about a four-way handshake? This is no better. An *n*-way handshake? Still no good. In all cases, the sender of the last message cannot tell whether or not it arrived. If its arrival is essential to starting the war, the sender has no way of telling whether the receiver is going to attack or not. If its arrival is not essential to starting the war, one can devise an equivalent pro-

tocol not containing it and apply the above reasoning to the new protocol.

The implication of all this is that a closing protocol in which neither side can hang up until it is convinced that the other side also intends to hand up is, at the very least, more complicated than it at first appears. The issue is discussed further in SUNS78 and YEMI79.

4.3 Flow Control and Buffering

An important design issue in the transport layer is flow control. Since no transport station has an infinite amount of buffer space, some way must be provided to prevent a fast sender from inundating a slow receiver. Flow control is well known in other contexts, such as operating systems design, where it is known as the producer–consumer problem. Although it also occurs in the data link and network layers, some new complications are present in the transport layer.

To start with, the data link layer usually has one connection for each adjacent IMP—a handful at most—whereas the transport layer in a large multiprogrammed computer may have many connections open simultaneously. A stop-and-wait protocol in the transport layer is usually undesirable, since both sender and receiver would have to be scheduled and run for every message sent. If each machine had a response time of 500 ms between the moment a process became ready to run and the time it ran, the transport connection could support two messages per second, at best, and probably fewer. Consequently, large windows are needed to achieve high throughput, but the combination of large windows and many open connections necessitates many buffers, most of which are idle most of the time.

An alternative design is not to dedicate buffers to specific connections, but to maintain a buffer pool and pull buffers out of the pool and assign them to connections dynamically, as needed. This strategy entails some risk, since no buffer may be available when one is needed. To avoid this risk, a buffer reservation protocol is needed, increasing traffic and overhead. Furthermore, if the buffers are of fixed size and messages

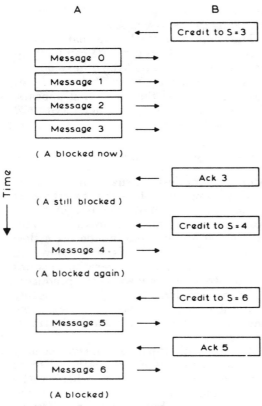

A   B

← Credit to S = 3

Message 0 →

Message 1 →

Message 2 →

Message 3 →

(A blocked now)

← Ack 3

(A still blocked)

← Credit to S = 4

Message 4 →

(A blocked again)

← Credit to S = 6

Message 5 →

← Ack 5

Message 6 →

(A blocked)

Figure 11. Flow control using credits.

vary from a few characters to thousands of characters, a pool of fixed-sized buffers is not attractive either. No one solution is best. Each transport station must make compromises appropriate to its expected work load.

Another important issue is the relation of flow control to error control. With the sliding-window protocol, an acknowledgment message has two distinct functions: to announce that a message has arrived and to grant the sender permission to send another message. In the transport layer, this coupling is not always desirable.

To see why, consider the dilemma of a transport station that is chronically short of buffer space. What should it do if a message arrives, but the process using the connection has no RECEIVEs outstanding? If it does nothing, the sending transport station will eventually time out and send it again. If it sends an acknowledgment, the other transport station may send

yet another message. The problem comes from the fact that the transport station has no control over the rate at which the user does RECEIVEs. Earlier, we more or less assumed that the network layer was always hungry for new packets—a reasonable assumption, since the network code has little else to do.

One way out of this dilemma is to decouple acknowledgments and flow control. To do so, we introduce two kinds of control messages: acknowledgments and credits. An acknowledgment simply says that a certain message (and by implication, all lower numbered messages) has arrived safely. Upon receiving an acknowledgment, the sender may release the buffers containing all the acknowledged messages, since none of them will ever be retransmitted. However, an acknowledgment does *not* imply permission to send any more messages.

Such permission is granted by a credit message. When a connection is established, the receiver grants some credits to the sender. These credits may be for so many messages or so many bits or both. Every time a message is sent, the credits for message count and/or bit count are decremented. When the credits are all used up, the sender must stop sending until more credits arrive. Such credits may be sent as distinct messages or they may be piggybacked onto data or acknowledgment messages. This scheme provides a simple and flexible mechanism for preventing unnecessary retransmissions in the presence of heavy and variable demands on limited buffer space. An example is given in Figure 11.

4.4 Connection Multiplexing

Multiplexing of connections plays an important role in several layers. In lower layers, for example, packets and frames ultimately destined for different hosts are multiplexed onto the same output lines. In the transport layer, two different forms of multiplexing occur. In *upward multiplexing* (shown in Figure 12a) several transport connections are multiplexed onto the same network connection (e.g., the X.25 virtual circuit). Upward multiplexing is often financially better, since some carriers charge by the packet and also by the second for each virtual circuit that is open.

Now consider the plight of an organization (e.g., an airline) that has 100 telephone operators to handle customer inquiries. If each operator is assigned to a separate virtual circuit, 100 virtual circuits to the central computer will be open all day. The other option would be to use a single virtual circuit to the computer, with the first byte of data being used to distinguish among the operators. The latter has the disadvantage that if the traffic is heavy, the flow control window may always be full, thus slowing down operation. With a dedicated virtual circuit per operator, full windows are much less likely to occur.

The other form of multiplexing, *down-ward multiplexing* (Figure 12b), becomes interesting when the network layer window is too small. Suppose, for example, a certain network offers X.25, but does not support the 7-bit sequence number option. A user with a large number of data to send might find himself constantly running up against full windows. One way to make an end run around the problem is to open multiple virtual circuits for a single-transport connection. Packets could be distributed among the virtual circuits in a round-robin fashion, first a packet on circuit 0, then a packet on circuit 1, then one on circuit 2, and so on.

Conceivably the two forms could even be combined. For n connections, k virtual circuits could be set up with traffic being dynamically assigned.

5. THE SESSION LAYER

In many networks, the transport layer establishes and maintains connections between hosts. The session layer establishes and maintains connections, called *sessions*, between specific pairs of processes. On the other hand, some networks ignore the session layer altogether and maintain transport connections between specific processes. The ISO OSI Reference Model is exasperatingly vague on this point, stating only that the session layer connects "presentation-entities" and that the transport layer connects "session-entities."

To keep the following discussion from

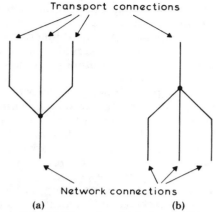

Figure 12. (a) Upward multiplexing. (b) Downward multiplexing.

vanishing in a linguistic fog, we assume that transport connections are between hosts and session connections are between processes. Thus, when a process wants to talk to another process, it makes its desires known to the session layer, which then engages the services of the transport layer to set up a transport connection to the remote host for use by the session.

A principal task of the session layer is to connect two processes together into a session. Since it is inconvenient for users to be aware of hard transport addresses, the session layer could allow them to refer to destinations by symbolic name, with the session layer doing the mapping onto transport addresses. For example, a user could say, in effect, "Give me a phototypesetter process," with the session layer worrying about where such beasts were to be found.

When a session is set up, an activity often call *session binding*, certain conventions about the coming session can be established. Typical conventions are half-duplex versus full-duplex data transfer, character codes, flow control window sizes, the presence or absence of encryption or text compression, and how to recover from transport layer failures.

Another task that the session layer can perform is particularly useful in networks where the user primitives for sending and receiving messages are nonblocking, and where the user may have multiple requests outstanding on the same session at any instant. Under these circumstances, replies may come back in an order different from that in which the requests were sent. The session layer's *dialog control* function can keep track of requests and replies and reorder them if need be to simplify the design of the user programs.

Another aspect of dialog control is bracketing groups of messages into atomic units. In many database applications it is highly undesirable that a transaction be broken off part way, as a result of a network failure, for example. If the transactions consists of a group of messages, the session layer could make sure that the entire group had been successfully received at the destination before even attempting to start the transaction.

Our discussion of the session layer is now complete. The brevity of this section is directly related to the fact that few networks make much of a distinction between the transport and session layers. In fact, many networks have neither a session layer nor any of the dialog control functions belonging to the session layer. While there are no internationally accepted standards for the transport layer yet, there are at least a few serious proposals that have been under discussion for several years [Depa76, INWG78]. Session layer protocols have not come as far yet. This situation has occurred because the protocol community has been tackling the layers more or less bottom up and is currently in the vicinity of layer 4. Higher layer standards will no doubt be forthcoming in the future.

6. THE PRESENTATION LAYER

The function of the presentation layer is to perform certain generally useful transformations on the data before they are sent to the session layer. Typical transformations are text compression, encryption, and conversion to and from network standards for terminals and files. We examine each of these subjects in turn.

6.1 Text Compression

Bandwidth is money. Sending thousands of trailing blanks across a network to be "printed" is a good use of neither. Although the network designers could leave the matter of text compression to each user program, it is more efficient and convenient to put it into the network architecture as one of the standard presentation services.

Obvious candidates for text compression are runs of repeated bits (e.g., leading zeros) and repeated characters (e.g., trailing blanks). Huffman coding is also a possibility. Since text compression is such a well-known subject outside the network context (see, e.g., Davs76), we do not consider it further here.

6.2 Encryption Protocols

Information often has great economic value. As an example, just think about the data transmitted back by oil companies from exploratory sites. With more and more data being transmitted by satellite, the problem of data security looms ever larger. The financial incentive to erect an antenna to spy on competitors is great and the cost is low. Furthermore, privacy legislation in many countries puts a legal requirement on the owners of personal data to make sure such data are kept secret. All these factors combine to make data encryption an essential part of most networks. The December 1979 issue of *Computing Surveys* [Comp79] is devoted to cryptography and contains several introductory articles on it.

An interesting question is: "In which layer does the encryption belong?" In our view, encryption is analogous to text compression: ordinary data go in and compressed or indecipherable data come out. Since everyone agrees that text compression is a presentation service, logically encryption should be too. For historical reasons and implementation convenience, however, it is often put elsewhere, typically the transport layer or the data link layer.

The purpose of encryption is to transform the input, or *plaintext*, into an output, or *ciphertext*, that is incomprehensible to anyone not privy to the secret *key* used to parameterize the transformation. Thus plaintext is converted to ciphertext in the presentation layer of the source machine and reconverted to plaintext in the presentation layer of the destination machine. In neither machine should the user programs be aware of the encryption, other than having specified encryption as an option when the session was bound.

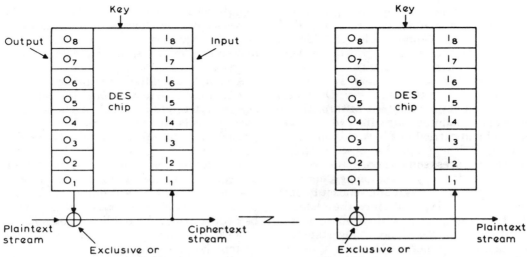

Figure 13. A stream cipher using DES. Data arrives from the left and is encrypted for transmission. The destination machine decrypts it and outputs the plaintext.

One of the best-known encryption methods is the *substitution cipher*, in which a unit of plaintext is converted into a unit of ciphertext. In a *monoalphabetic cipher*, each letter is converted into another letter according to a fixed rule. For example, "a" becomes "M," "b" becomes "R," "c" becomes "G," etc. In this example, the encryption key is MRG ..., that is, the ciphertext corresponding to the plaintext abc.... Although 26! different monoalphabetic substitutions exist, these ciphers can be broken by a clever ten-year-old using the frequency statistics of natural language.

Most computer ciphers use the same principle, but on a larger scale. The U.S. federal government has adopted a substitution cipher that is fast becoming a de facto standard for nongovernmental organizations as well. The *DES* (Data Encryption Standard) cipher takes a 64-bit plaintext input block and produces a 64-bit ciphertext output block. The transformation is driven by a 56-bit key. Conceptually, at least, one could prepare a big table, with 2^{64} columns, one for each possible input, and 2^{56} rows, one for each possible key. Each table entry is the ciphertext for the specified input and key.

DES can also be operated as a *stream cipher*, as shown in Figure 13. The input shift registers on both source and destination machine are initialized to the same 8-byte (random) number, $I_1, \ldots, I_8$. Data are presented for encryption 1 byte at a time, not 8 bytes at a time. When a byte arrives, the DES chip converts the 8 bytes $I_1, \ldots, I_8$ into the output $O_1, \ldots, O_8$. Then O_1 is Exclusive Or'ed with the input to form the ciphertext byte. The ciphertext byte is both transmitted and fed back into I_1, shifting I_2 to I_3, and so on. I_8 is shifted out and lost. Decryption at the other end is similar. Note that feeding back the ciphertext into the DES input register makes subsequent encryption dependent on the entire previous plaintext, and so a given sequence of 8 plaintext bytes will have a different ciphertext on each appearance in the plaintext.

DES has been the subject of great controversy since its inception [DAVA79, DIFF76a, HELL79, HELL80]. Some computer scientists feel that a wealthy and determined intruder who knew, for example, that a certain message was in ASCII, could determine the key by trying all keys until he found one that yielded ASCII plaintext (i.e., only codes 0–127 and not 128–255). If the ciphertext is k bytes long, the probability of an incorrect key yielding ASCII input is 2^{-k}. For even a single line of text, it is unlikely that any key but the correct one could pass the test.

The dispute centers about how much a DES-breaking machine would cost. In 1977, Diffie and Hellman [DIFF77] designed one and computed its cost at 20 million dollars.

The DES supporters say this figure is too low by a factor of 10, although even they concede DES cannot hold out forever against the exponential growth of very large-scale integrated circuits.

To use DES, both the source and destination must use the same key. Obviously the session key cannot be sent through the network in plaintext form. Instead, a master key is hand carried in a locked briefcase to each host. When a session is set up, a key manager process somewhere in the network picks a random key as session key, encrypts it using the master key, and sends it to both parties for decryption. Since the plaintext of this message is a random number, it is hard to break the cipher using statistical techniques. Numerous variations of the idea exist, typically with master keys, regional keys, local keys, and the like.

Shamir [SHAM79] has devised a clever way to share (master) keys in a flexible way among a large group of people, so that n arbitrary people can get together and assemble the master key, but $n - 1$ people can gain no information at all. Basically, each person is given a data point that lies on a degree $n - 1$ polynomial whose y intercept is the key. With n data points, the polynomial, hence the key, is uniquely determined, but with $n - 1$ data points it is not. Modulo arithmetic is used for obfuscatory purposes.

All the master key methods have a significant drawback, though: it is impossible for computers that have not previously had any contact with each other to agree on a session key in a secure way. Considerable academic research has been done on this topic in recent years (not without its own controversy—see SHAP77 and SUGA79), and some interesting results have been achieved. Merkle [MERK78a], for example, has suggested that two strangers, A and B, could establish a key as follows. A sends k ciphertext messages to B with the instruction to pick one of them at random and break it by brute force (i.e., try all possible keys until a plaintext starting with 64 0's appeared). The rest of the message consists of two random numbers, the key number and the key itself. Having broken the cipher, B then sends the key number back to A to indicate which message was broken.

Clearly an intruder will have to break $k/2$ messages on the average to find the right one. By adjusting k and the difficulty of breaking a message, A can achieve any degree of security desired.

A completely different approach to key distribution is that of *public key cryptography* [DIFF76b] in which each network user deposits an encryption key E in a publicly readable file. The user keeps the decryption key D secret. The keys must satisfy the property that $D(E(P)) = P$ for an arbitrary plaintext P. (This is essentially the definition of a decryption key.) The cipher system must be such that D cannot be deduced from the publicly known E.

With this background, the encryption system is obvious and trivial: to send a message to a stranger, you just encrypt it with his publicly known key. Only he knows the decryption key and no one can deduce it from the encryption key, so the cipher cannot be broken. The utility of the whole system depends on the availability of key pairs with the requisite properties. Much effort has gone into searching for ways to produce such key pairs. Some algorithms have already been published [MERK78b, RIVE78, SHAM80]. The scheme of Rivest et al. effectively depends on the fact that given two huge prime numbers, generating their product (the public key) is computationally easy, but given the product, finding the prime factors (the secret key) is very hard. In effect, their system takes advantage of the fact that the computational complexity of factorization is high.

Another area where cryptography plays a major role is in authentication. Suppose that a customer's computer instructs a bank's computer to buy a ton of gold and debit a certain account. The bank complies, but the next day the price of gold drops sharply and the customer denies ever having issued any purchase order. How can the bank protect itself against such unscrupulous customers? Traditionally, court battles over such matters have focused on the presence or absence of an authorized handwritten signature on a piece of paper. With electronic funds transfers and similar applications the need for "digital" (i.e., electronic) signatures is obvious.

With a slight additional restriction, public key cryptography can be used to provide these badly needed digital signatures. The restriction is that the encryption and decryption algorithms be chosen so that $D(E(P)) = E(D(P))$. In other words, the order of applying encryption and decryption must be interchangeable. The M.I.T. algorithm [RIVE78] has this property.

Now let us reconsider the ton-of-gold problem posed earlier. To protect itself, the bank can insist that a customer C use the following protocol for sending signed messages. First, the customer encrypts the plaintext message P with his secret key; that is, the customer computes $D_C(P)$. Then the customer encrypts this result with the bank's public key E_B, yielding $E_B(D_C(P))$. When this message arrives, the bank applies its decryption key D_B to get

$$D_B(E_B(D_C(P))) = D_C(P).$$

Now the bank applies the customer's public key, E_C, to recover P. The bank also saves P and $D_C(P)$ in case trouble arises.

When the angry letter from the customer arrives, the bank takes both P and $D_C(P)$ to court and asks the judge to decrypt the latter using the customer's public key. When the judge sees that the decryption works, he will realize that the bank must be telling the truth. How else could it have come into possession of a message encrypted by D_C, the customer's secret key? Since the bank does not know any of its customer's secret keys, it cannot forge messages (to generate commissions); hence customers are also protected against unscrupulous banks. While in jail, the customer will have ample time to devise interesting new public key algorithms.

Unfortunately, as Saltzer [SALT78] has pointed out, the public key digital signature protocol suffers from some nontechnical problems. For example, immediately after the price of gold drops, the management of the gold-buying company could run to the police claiming it had just become aware of yesterday's key burglary. Depending on local laws, the company might or might not be able to weasel out of obligations undertaken with the "stolen" key. (As an aside, note that the owner of a stolen credit card usually has only a small liability for its subsequent misuse.)

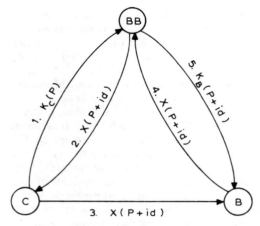

Figure 14. A digital signature protocol with conventional cryptography.

Saltzer also points out that a company is free to change its public key at will. Stronger yet, it may be company policy to do so regularly. If the company changes its key before accusing the bank of fabricating the purchase order, it will be impossible for the bank to convince the judge. This observation suggests that some central key registration authority may be needed. However, if such a central authority, call it Big Brother (BB), exists, conventional cryptography can also be used to achieve digital signatures [NEED78, POPE79].

The signature protocol using DES is illustrated in Figure 14. When a new customer C joins the system, the customer hand carries a secret (DES) key, K_C, to BB. Thus, BB has each user's secret key and can therefore send and receive secure messages from each user. In addition, BB has a secret key of its own, X, that it never discloses to anyone. The protocol for buying gold is as follows (P is the plaintext purchase order):

(1) The customer sends $K_C(P)$ to BB.
(2) BB decrypts the message and returns $X(P + \text{identification})$.
(3) The customer sends $X(P + \text{identification})$ to the bank.
(4) The bank sends $X(P + \text{identification})$ to BB.
(5) BB sends $K_B(P + \text{identification})$ back to the bank.

The "identification" appended to the message by BB consists of the customer's identity, something that BB can guarantee since the incoming message is encrypted by a key only known to one user, plus the date, time, and perhaps a sequence number. Messages encrypted by X can be freely sent through the network, since only BB can decrypt them, and BB is assumed to be trusted. If a dispute arises, the bank can go to the judge with $X(P + \text{identification})$, which the judge can then order BB to decrypt. The judge will then see the identification and know who sent the original message. While in jail, the customer will have ample time to devise interesting new signature protocols using conventional cryptography.

6.3 Virtual-Terminal Protocols

Dozens of brands of terminals are in widespread use, no two of which are identical. Needless to say, a network user who has just been told that the program or host he wishes to use does not converse with his brand of terminal is not likely to be a happy user. For example, if the program treats carriage returns and line feeds as equivalent and the user's terminal only has a "newline" key, which generates one of each, the program will perceive alternate lines as being empty.

To prevent such difficulties, protocols have been invented to try to hide terminal idiosyncracies from application (i.e., user) programs. Such protocols are known as *virtual-terminal protocols*, since they attempt to map real terminals onto a hypothetical network virtual terminal. Virtual-terminal protocols are part of the presentation layer.

Broadly speaking, terminals can be divided up into three classes: scroll mode, page mode, and form mode. Scroll-mode terminals do not have any intelligence. When a key is struck, the character is sent over the line and perhaps printed as well. When a character comes in over the line, it is just displayed. Most hard-copy terminals, and some of the less expensive CRT terminals, are scroll-mode terminals.

Even though scroll-mode terminals are simple, they still can differ in many ways: character set, line length, half duplex/full duplex, overprinting and the way line feed,

carriage return, tab, vertical tab, backspace, form feed, and break are handled.

Page-mode terminals are typically CRT terminals with 24 or 25 lines of 80 characters. Most of these have cursor addressing, so that the operator or the program can randomly access the screen. Some of them have a little local editing capability. They have the same potential differences as scroll-mode terminals, and, additionally, problems with screen length, cursor addressing, blinking, reverse video, color, multiple intensities, and the details of the local editing.

Form-mode terminals are sophisticated microprocessor-based devices intended for data entry. They are widely used in airline reservations, banking, and many other applications. In a typical situation the computer displays a form for the operator to fill out using cursor control and local editing facilities. The completed form is then sent back to the computer for processing. Sometimes the microprocessor can perform simple syntax checking, to make sure, for example, that a bank account field contains only numbers.

Two kinds of virtual-terminal protocols are commonly used. The first one is intended for scroll-mode terminals and is based on the ARPANET Telnet protocol [Davd77]. When this type of protocol is used, the designers invent a fictitious virtual terminal onto which all real terminals are mapped. Application programs output virtual-terminal characters, which are mapped onto the real terminal's character set by the presentation layer at the destination. Supporting a new kind of real terminal thus requires modifying the presentation layer software to effect the new mapping, but does not require changing any of the application programs.

Since most of the current research in virtual-terminal protocols focuses on page- or form-mode terminals, let us move on to them. A general model that has been widely accepted is the data structure model of Schicker and Duenki [Schi78]. Roughly speaking, protocols based on this approach use the model of Figure 15. Each end of a session has a data structure that represents the state of the virtual terminal.

The data structure consists of a collection

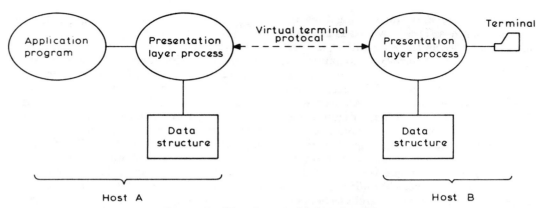

Figure 15. Virtual-terminal protocol model.

of fields, each of which contains certain attributes. Typical attributes are the size of the field, whether it accepts numbers, letters, or both, its rendition (an abstract concept used to model color, reverse video, blinking, and intensity), whether it is protected against operator modification or not, and so forth. The program is written using abstract operations on the data structure. Every time the program changes the data structure on its machine, the presentation layer sends a message to the other machine telling it how to change its data structure. The remote presentation layer is responsible for updating the display on the real terminal to make it correspond to the newly changed data structure. Similarly, changes made to the display by the human operator are reflected in the data structure on the operator's side of the session. Messages are then sent to bring the other side up to date. The protocol used for these messages is the virtual-terminal protocol.

Although a clever presentation layer can come a long way toward hiding the properties of the real terminal from the user program, it cannot work miracles. If the program needs a 24 × 80 screen with cursor addressing and four renditions, the presentation layer will be hard pressed to map everything onto a simple hard-copy terminal. Consequently, virtual-terminal protocols always have an option negotiation facility that is used to establish what each end of the connection is able to provide and what it wants from the other end.

This negotiation can be symmetric or asymmetric. In symmetric negotiation, each end announces its capabilities, inspects its partner's, and sets the parameters to the lowest common denominator. For example, if one end has a 24 × 80 screen, and the other has a 25 × 72 screen, the screen used will be 24 × 72. In asymmetric negotiation, one side makes a proposal and the other side accepts or rejects the proposal. If the proposal is rejected, the proposer may try again. Symmetric negotiation solves the problem of who should go first, but requires more complicated rules to determine what the result of an exchange is. It can also fail, for example, if both sides want to work in alternating (half-duplex) mode, and each wants to go first.

Another important design issue in virtual-terminal protocols is how to handle interrupts (attentions). When a user hits the "break" or "quit" key to terminate an infinite loop with a print statement in it, the presentation layer must purge the pipe of input already queued up; otherwise break will have no apparent effect. It is easy for the presentation layer on the terminal side to begin discarding input upon seeing a break, but it is much harder to determine when to stop discarding. Waiting for the prompt character does not work, since it might occur in the data to be discarded. A special out-of-band signaling protocol is needed. A survey of virtual-terminal protocols is given in DAY80.

6.4 File Transfer Protocols

The most common uses of computer networks at present are for logging onto remote machines and transferring files between machines. These two areas are similar in that just as there is a need for programs to talk to a variety of incompatible terminals, there is a need for programs to read a variety of incompatible files. In principle, the same approach can be used for file transfer as for terminals: define a network standard format and provide a mapping from and to each existing file format.

In practice, this approach seems to work fairly well for terminals, but less well for files, primarily because the differences between terminals are not as great as between file types. Mapping reverse video onto blinking is straightforward compared to mapping 60-bit CDC floating point numbers onto 32-bit IBM floating point numbers, especially when the numbers are strewn randomly throughout the file.

Files are transferred for four primary reasons:

(1) to store a file for subsequent retrieval;
(2) to print a remote file on the local printer;
(3) to submit a file as a remote job;
(4) to use a remote file as data input or output.

Each category of use has its own peculiarities.

When a file is stored for subsequent retrieval, it must be possible to produce an exact, bit-for-bit copy of it upon request. Clearly transmission must be fully transparent, without escape codes that do funny things. The number of bits in the file must be recorded in the stored file, to allow transport between machines with differing word lengths. The last word on the storage machine may be partially full, and so some record of how many bits are in use is required.

When a file is transferred to be printed, problems can arise as a result of different print conventions. Some machines store print files in FORTRAN format, with fixed-length records (with or without some fudge for trailing blanks), and carriage control characters in column 1. Other machines use ASCII style variable-length records, with line feeds and form feeds for indicating vertical motion. When the file is being

moved to be used for remote job entry, the same problems are present.

Moving data files containing mixtures of integers, floating point numbers, characters, etc., between machines is nearly impossible. In theory, each data item (e.g., integer, floating point number, character) could occupy one record in a canonical format, with the data type and value both explicitly stored. In practice the idea does not seem to work well, not only because of problems of interfacing existing software to it, but also because of the high overhead and the problems involved in converting floating point numbers from one format to another.

Another aspect of file transfer is file manipulation. Users often need to create, delete, copy, rename, and otherwise manage remote files. Most file transfer protocols tend to concentrate on this aspect of the problem because it is not as hopeless as the conversion aspect. Gien [GIEN78] has described a file transfer protocol in some detail.

7. SUMMARY

Computer networks are designed hierarchically, as a series of independent layers. Processes in a layer correspond with their peers in remote machines using the appropriate protocol, and with their superiors and subordinates in the same machine using the appropriate interface. The ISO OSI Reference Model has been designed to provide a universal framework in which networking can be discussed. Few existing networks follow it closely, but there is a general movement in that direction.

The seven-layer ISO model can be briefly summarized as follows. The physical layer creates a raw bit stream between two machines. The data link layer adds a frame structure to the raw bit stream, and attempts to recover from transmission errors transparently. The network layer handles routing and congestion control. The transport layer provides a network-independent transport service to the session layer. The session layer sets up and manages process-to-process connections. The presentation layer performs a variety of useful conversions. Finally the application layer is up to the user, although some industry-wide protocols may be developed in the future.

The literature on computer networks is huge. Readers unfamiliar with it, but wishing to continue their study of the subject, may be interested in the textbooks by Davies et al. [DAVI79] and Tanenbaum [TANE81], or the book edited by Kuo [KUO81].

ACKNOWLEDGMENTS

I would like to thank Yogen Dalal, Adele Goldberg, and an anonymous, but artistic, technical editor for their numerous and helpful comments.

REFERENCES

ABRA70   ABRAMSON, N. "The ALOHA system—another alternative for computer communications," in *Proc. 1970 Fall Jt. Computer Conf.*, AFIPS Press, Arlington, Va., pp. 281–285.

ABRA73   ABRAMSON, N. "The ALOHA system," in *Computer-communication networks*, N. Abramson and F. Kuo (Eds.), Prentice-Hall, Englewood Cliffs, N.J., 1973.

BARA64   BARAN, P. "On distributed communication networks," *IEEE Trans. Commun. Syst.* **CS-12** (March 1964), 1–9.

BERT80   BERTINE, H. V. "Physical level protocols," *IEEE Trans. Commun.* **COM-28** (April 1980), 433–444.

BOGG80   BOGGS, D. R., SHOCH, J. F., TAFT, E. A., AND METCALF, R. M. "Pup: An Internet architecture," *IEEE Trans. Commun.* **COM-28** (April 1980), 612–624.

CAPE79a   CAPETANAKIS, J. I. "Generalized TDMA: The multi-accessing tree protocol," *IEEE Trans. Commun.* **COM-27** (Oct. 1979), 1476–1484.

CAPE79b   CAPETANAKIS, J. I. "Tree algorithms for packet broadcast channels," *IEEE Trans. Inf. Theory* **IT-25** (Sept. 1979), 505–515.

CHLA76   CHLAMTAC, I. "Radio packet broadcasted computer network—the broadcast recognition access method," M.S. thesis, Dep. Mathematical Sciences, Tel Aviv Univ., Tel Aviv, Israel, 1976.

CHLA79   CHLAMTAC, I., FRANTA, W. R., AND LEVIN, D. "BRAM: The broadcast recognizing access method," *IEEE Trans. Commun.* **COM-27** (Aug. 1979), 1183–1190.

CHU78   CHU, K. "A distributed protocol for updating network topology information," Rep. RC 7235, IBM Thomas J. Watson Res. Cent., Yorktown Heights, N.Y., 1978.

CLAR78   CLARK, D. D., POGRAN, K. T., AND REED, D. P. "An introduction to local area networks," *Proc. IEEE* **66** (Nov. 1978), 1497–1517.

COMP79   *Computing Surveys* **11**, 4 (Dec. 1979).

CROW73   CROWTHER, W., RETTBERG, R., WALDEN, D., ORNSTEIN, S., AND HEART, F. "A system for broadcast communication: Reservation-Aloha," in *Proc. 6th Hawaii Int. Conf. Systems Science*, 1973, pp. 371–374.

DAVA79   DAVIDA, G. I. "Hellman's scheme breaks DES in its basic form," *IEEE Spectrum* **16** (July 1979), 39.

DAVD77   DAVIDSON, J., HATHAWAY, W., POSTEL, J., MIMNO, N., THOMAS, R., AND WALDEN, D. "The ARPANET Telnet protocol: Its purpose, principles, implementation, and impact on host operating system design," in *Proc. 5th Data Communication Symp.* (ACM/IEEE) (1977), pp. 4.10–4.18.

DAVI72   DAVIES, D. W. "The control of congestion in packet-switching networks," *IEEE Trans. Commun.* **COM-20** (June 1972), 546–550.

DAVI73   DAVIES, D. W., AND BARBER, D. L. A. *Communication networks for computers*, Wiley, New York, 1973.

DAVI79   DAVIES, D. W., BARBER, D. L. A., PRICE, W. L., AND SOLOMONIDES, C. M. *Computer networks and their protocols*, Wiley, New York, 1979.

DAVS76   DAVISSON, L., AND GRAY, R. (Eds.). *Data compression*, Dowden, Hutchinson & Ross, Stroudsburg, Pa., 1976.

DAY80   DAY, J. "Terminal protocols," *IEEE Trans. Commun.* **COM-28** (April 1980), 585–593.

DEPA76   DEPARIS, M., DUENKI, A., GLEN, M., LAWS, J., LEMOLI, G., AND WEAVING, K. "The implementation of an end-to-end protocol by EIN centers: A survey and comparison," in *Proc. 3rd Int. Conf. Computer Communication* (ICCC) (Aug. 1976), pp. 351–360.

DIFF76a   DIFFIE, W., AND HELLMAN, M. E. "A critique of the proposed data encryption standard," *Commun. ACM* **19** (March 1976), 164–165.

DIFF76b   DIFFIE, W., AND HELLMAN, M. E. "New directions in cryptography," *IEEE Trans. Inf. Theory* **IT-22** (Nov. 1976), 644–654.

DIFF77   DIFFIE, W., AND HELLMAN, M. E. "Exhaustive cryptanalysis of the NBS data encryption standard," *Computer* **10** (June 1977), 74–84.

DOLL78   DOLL, D. R. *Data communications.* Wiley, New York, 1978.

FARB72   FARBER, D. J., AND LARSON, K. C. "The system architecture of the distributed computer system—the communications system," in *Symp. Computer Networks*, Polytechnic Institute of Brooklyn, Brooklyn, N.Y., April 1972.

FOLT80   FOLTS, H. C. "Procedures for circuit-switched service in synchronous public data networks," *IEEE Trans. Commun.* **COM-28** (April 1980), 489–496.

FRAS75   FRASER, A. G. "A virtual channel network," *Datamation* **21** (Feb. 1975), 51–56.

FRAT73   FRATTA, L., GERLA, M., AND KLEINROCK, L. "The flow deviation method: An approach to store-and-forward communication networks," *Networks* **3** (1973), 97–133.

FREE80   FREEMAN, H. A., AND THURBER, K. J. "Updated bibliography on local computer networks," *Comput. Arch. News* **8** (April 1980), 20–28.

GELE78   GELENBE, E., LABETOULLE, J., AND PUJOLLE, G. "Performance evaluation of the HDLC protocol," *Comput. Networks* **2** (Sept.–Oct. 1978), 409–415.

GERL77   GERLA, M., AND KLEINROCK, L. "Closed loop stability controls for S-ALOHA satellite communications," in *Proc. 5th Data Communication Symp.* (ACM/IEEE), (1977), pp. 2-10–2-19.

GIEN78   GIEN, M. "A file transfer protocol (FTP)," *Computer Networks* **2** (Sept.–Oct. 1978), 312–319.

GUNT81   GUNTHER, K. D. "Prevention of deadlocks in packet-switched data transport systems," *IEEE Trans. Commun.* **COM-29** (April 1981), 512–524.

HELL79   HELLMAN, M. E. "DES will be totally insecure within ten years," *IEEE Spectrum* **16** (July 1979), 32–39.

HELL80   HELLMAN, M. E. "A cryptanalytic time-memory tradeoff," *IEEE Trans. Inf. Theory* **IT-26** (July 1980), 401–406.

INWG78   "A proposal of an Internetwork end to end protocol," in *Proc. Symp. Computer Network Protocols*, University of Liege, Belgium, Feb. 1978, pp. H:5–25.

IRLA78   IRLAND, M. I. "Buffer management in a packet switch," *IEEE Trans. Commun.* **COM-26** (March 1978), 328–337.

KAMO81   KAMOUN, F. "A drop and throttle flow control policy for computer networks," *IEEE Trans. Commun.* **COM-29** (April 1981), 444–452.

KLEI78   KLEINROCK, L., AND YEMINI, Y. "An optimal adaptive scheme for multiple access broadcast communication," in *Proc. ICC* (IEEE), 1978, pp. 7.2.1–7.2.5.

KUO81   KUO, F. F. (Ed.) *Protocols and techniques for data communication networks*, Prentice-Hall, Englewood Cliffs, N.J., 1981.

LIU78   LIU, M. T. "Distributed loop computer networks," in *Advances in Computers*, M. C. Yovits (Ed.), Academic Press, New York, 1978, pp. 163–221.

MANN78   MANNING, E. G. "On datagram service in public packet-switched networks," *Comput. Networks* **2** (May 1978), 79–83.

MART78   MARTIN, J. *Communications satellite systems*, Prentice-Hall, Englewood Cliffs, N.J., 1978.

McQU74   McQUILLAN, J. M. "Adaptive routing algorithms for distributed computer networks," Ph.D. dissertation, Div. Engineering and Applied Sciences, Harvard Univ., 1974.

McQU77   McQUILLAN, J. M., AND WALDEN, D. C. "The ARPA network design deci-

sions," *Comput. Networks* **1** (Aug. 1977), 243–289.

MERK78a   MERKLE, R. C. "Secure communications over insecure channels," *Commun. ACM* **21** (April 1978), 294–299.

MERK78b   MERKLE, R. C., AND HELLMAN, M. E. "Hiding information and receipts in trapdoor knapsacks," *IEEE Trans. Inf. Theory* **IT-24** (Sept. 1978), 525–530.

MERL80a   MERLIN, P. M., AND SCHWEITZER, P. J. "Deadlock avoidance in store-and-forward networks—I: Store-and-forward deadlock," *IEEE Trans. Commun.* **COM-28** (March 1980), 345–354.

MERL80b   MERLIN, P. M., AND SCHWEITZER, P. J. "Deadlock avoidance in store-and-forward networks—II: Other deadlock types," *IEEE Trans. Commun.* **COM-28** (March 1980), 355–360.

METC76   METCALFE, R. M., AND BOGGS, D. R. "Ethernet: Distributed packet switching for local computer networks," *Commun. ACM* **19** (July 1976), 395–404.

NEED78   NEEDHAM, R. M., AND SCHROEDER, M. D. "Using encryption for authentication in large networks of computers," *Commun. ACM* **21** (Dec. 1978), 993–999.

NEED79   NEEDHAM, R.M. "System aspects of the Cambridge ring," in *Proc. 7th Symp. Operating Systems, Principles* (ACM), 1979, pp. 82–85.

PETE61   PETERSON, W. W., AND BROWN, D. T. "Cyclic codes for error detection," *Proc. IRE* **49** (Jan. 1961), 228–235.

POPE79   POPEK, G. J., AND KLINE, C. S. "Encryption and secure computer networks," *Comput. Surveys* **11** (Dec. 1979), 331–356.

POST80   POSTEL, J. B. "Internetwork protocol approaches," *IEEE Trans. Commun.* **COM-28** (April 1980), 604–611.

RIVE78   RIVEST, R. L., SHAMIR, A., AND ADLEMAN, L. "A method for obtaining digital signatures and public key cryptosystems," *Commun. ACM* **21** (Feb. 1978), 120–126.

ROBE73   ROBERTS, L. G. "Dynamic allocation of satellite capacity through packet reservation," in *1973 Nat. Computer Conf.*, AFIPS Press, Arlington, Va., pp. 711–716.

ROTH77   ROTHAUSER, E. H., AND WILD, D. "MLMA—a collision-free multi-access method," *Proc. IFIP Congr. 77*, (IFIP) (1977), 431–436.

RUDI76   RUDIN, H. "On routing and delta routing: A taxonomy and performance comparison of techniques for packet-switched networks" *IEEE Trans. Commun.* **COM-24** (Jan. 1976), 43–59.

RYBC80   RYBCZYNSKI, A. "X.25 interface and end-to-end virtual circuit service characteristics," *IEEE Trans. Commun.* **COM-28** (April 1980), 500–510.

SALT78   SALTZER, J. H. "On digital signatures," *Oper. Syst. Rev.* **12** (April 1978), 12–14.

SCHI78   SCHICKER, P., AND DUENKI, A. "The virtual terminal definition," *Comput. Networks* **2** (Dec. 1978), 429–441.

SCHO76 SCHOLL, M. "Multiplexing techniques for data transmission over packet switched radio systems," Ph.D. dissertation, Computer Science Dep., UCLA, 1976.

SCHW80 SCHWARTZ, M., AND STERN, T. E. "Routing techniques used in computer communication networks," *IEEE Trans. Commun.* **COM-28** (April 1980), 539–552.

SEGA81 SEGALL, A. "Advances in verifiable fail-safe routing procedures," *IEEE Trans. Commun.* **COM-29** (April 1981), 491–497.

SHAM79 SHAMIR, A. "How to share a secret," *Commun. ACM* **22** (Nov. 1979), 612–613.

SHAM80 SHAMIR, A., AND ZIPPEL, R. "On the security of the Merkle-Hellman cryptographic scheme," *IEEE Trans. Inf. Theory* **IT-26** (May 1980), 339–340.

SHAP77 SHAPLEY, D., AND KOLATA, G. B. "Cryptology: Scientists puzzle over threat to open research, publication," *Science* **197** (Sept. 30, 1977), 1345–1349.

SHOC81 SHOCH, J. "An annotated bibliography on local compuer networks," Xerox Tech. Rep., Xerox PARC, April 1980.

SLOA75 SLOANE, N. J. A. *A short course on error correcting codes,* Springer-Verlag, Berlin and New York, 1975.

SUGA79 SUGARMAN, R. M. "On foiling computer crime," *IEEE Spectrum* **16** (July 1979), 31–32.

SUNS78 SUNSHINE, C. A., AND DALAL, Y. K. "Connection management in transport protocols," *Comput. Networks* **2** (Dec. 1978), 454–473.

TANE81 TANENBAUM, A. S. *Computer networks,* Prentice-Hall, Englewood Cliffs, N.J., 1981.

TOML75 TOMLINSON, R. S. "Selecting sequence numbers," in *Proc. ACM SIGCOMM/ SIGOPS Interprocess Communication Workshop* (ACM) (1975), pp. 11–23.

WILK79 WILKES, M. V., AND WHEELER, D. J. "The Cambridge digital communication ring," in *Proc. Local Area Communication Network Symp.*, Mitre Corp and NBS (1979), pp. 47–61.

YEMI79 YEMINI, Y., AND COHEN, D. "Some issues in distributed process communication," in *Proc. 1st Int. Conf. Distributed Computer Systems* (IEEE), 1979, pp. 199–203.

ZIMM80 ZIMMERMANN, H. "OSI reference model—the ISO model of architecture for open systems interconnection," *IEEE Trans. Commun.* **COM-28** (April 1980), 425–432.

Received January 1981; final revision accepted September 1981.

A Layered Network Protocol for Packet Voice and Data Integration

WILLIAM L. HOBERECHT

Reprinted from *IEEE Journal on Selected Areas in Communications*, Vol. SAC-1, No. 6, December 1983, pages 1006-1013. Copyright © 1983 by The Institute of Electrical and Electronics Engineers, Inc. All rights reserved.

Abstract —This paper discusses considerations in the design of packet protocols suitable for interactive voice and interactive data communication, and then outlines a potential layered protocol architecture for the internal communication of a long haul network that might support packet voice and packet data transport.

Following the protocol description, the paper compares potential delays for two voice/data packet network architectures: one using only link retransmission, the other using only edge retransmission for data (as included by the voice/data protocol). The underlying data traffic loads offered to the network are the same for the two methods, although they give rise to different traffic patterns. This preliminary analysis shows that the average delay using an edge-to-edge recovery discipline can be made comparable to the delays introduced with a link-by-link recovery discipline, if the network uses high speed transmission facilities (e.g., over 1 Mbits/s) having good error characteristics (e.g., one or less packets corrupted in each 1000), and sends up to 128 bytes of customer data as a single packet.

I. INTRODUCTION

THE similarity of packet voice and packet data transport, recognized by many [1]–[4], suggests further study in the area of designing an integrated voice/data network. Unfortunately, along with these similarities are several conflicting communication requirements. Data applications demand essentially error-free transmission and existing protocol standards (e.g., CCITT Recommendation X.25) reflect this need. On the other hand, voice is tolerant of occasional errors. The inherent characteristics of voice allow small amounts of lost or corrupted information to be reconstructed, or even omitted, without a severe degradation in voice quality. Voice, however does have more stringent requirements on the amount of delay permissible between the time an utterance is spoken and the time it is subsequently heard by the receiver. While the exact limits on the amount of acceptable delay are unknown, subjective tests have shown that excessive delays may not be tolerated by telephone users when nondelayed phone service is available [5], [6] (cost of service was not a parameter in these tests).

One recent study [7] addressed the issue of voice/data packet protocols and described a point-to-point protocol meeting the needs of both voice and data. Some earlier studies in the area of protocols for packetized voice [8], [9] have concentrated on end-to-end aspects of the protocol

Manuscript received April 1, 1983; revised July 1, 1983. A portion of the paper was presented at GLOBECOM '83, San Diego, CA, November 28–December 1, 1983.

The author is with Bell Laboratories, Naperville, IL 60566.

(i.e., peer communication between packet voice transmitters and receivers) and on using current packet data networks for packet voice transport. These efforts included a demonstration of packet voice over Arpanet, reported in 1978 [10], suggesting that it may also be possible to use existing packet data networks (or at least their protocol architecture) for packet voice. While this earlier work considered use of current network designs for packet voice transmission, this paper presents an alternative view of a layered protocol structure for the *internal* operation of a long haul packet network that can support packet voice and packet data transport. These discussions and the description of the protocol architecture represent preliminary thinking on concepts that might be incorporated into an integrated packet voice/data system. The intent is to stimulate further thought on possible protocol architectures for voice/data packet transport.

The remainder of this paper describes packet voice communication and a possible internal network packet voice/data protocol structure. The paper first presents some of the factors that have a pronounced effect on the protocol design and the placement of protocol functions, and then describes a layered protocol for the internal communication of a voice/data packet network. The final section presents an analysis of delay for data using the voice/data packet protocol and compares it to delay encountered using a traditional error recovery protocol.

Throughout the paper, a distinction is made between edge-to-edge protocols (completely contained within the network, exercised between packet network interfaces) and end-to-end protocols (exercised between network customers).

II. PACKET VOICE/PACKET DATA NETWORK COMPONENTS

Fig. 1 shows the components of a generic packet network:

1) Packet network interfaces (PNI) through which users (i.e., subscribers and other networks) connect to the network.

2) Interswitch links that carry voice, data, and signaling information.

3) Packet switches (PS) that connect to both packet network interfaces and other packet switches with high-speed lines.

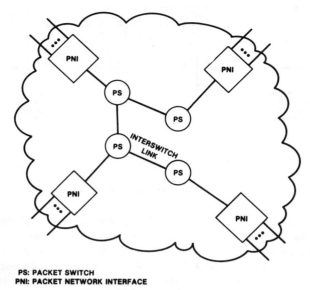

PS: PACKET SWITCH
PNI: PACKET NETWORK INTERFACE

Fig. 1. Packet network components.

III. Factors Influencing the Protocol Design

Several factors, all interrelated, have a pronounced effect on the protocol design. The primary underlying philosophy behind the design of a voice/data packet network and protocol is to consider these factors and design the network to take advantage of them. Three factors have been identified that significantly affect the packet voice/data protocol: characteristics of the applications using the network, performance requirements, and transmission facilities.

A. Characteristics of the Application

The transport of packetized voice is a major application of an integrated packet network. Because of the redundancy in this voice information, it may not require the robust error correction methods commonly found in packet protocols. Exercising conventional flow control on packets containing voice information seems inappropriate. Due to delays that can be introduced through the application of flow control procedures, voice information sent through the network may be rendered unusable. Degrading the level of service for established calls, a characteristic of such flow control procedures, is an undesirable practice in a voice network. Additionally, flow control practices have been developed to allow communication between devices operating at different rates. Because voice coding and decoding equipment would be operating at the same rate, such flow control practices are not necessary. This does not preclude the need for regulating the flow of information into the network to protect network resources.

B. Performance Considerations

The design objectives for a packet voice network place several constraints on the operation of the network. Two categories are particularly influential in the protocol design: loss performance (lost data/voice packets, lost signaling packets) and delay objectives [11].

Delay objectives for the network can be met, in part, through the elimination of *link* error recovery techniques that require retransmission of corrupted information. This has a twofold effect: first, one source of variable delays introduced by packet networks is eliminated, and second, network transmission procedures are simplified.

C. Quality and Speed of Transmission Facilities

The exclusive use of high-quality, high-speed (e.g., over 1 Mbit/s) digital transmission facilities in a packet voice network [1] encourages examination of nontraditional error correction techniques. Current methods, aimed at ensuring error-free information delivery, typically correct errors on each link. While this complexity and expense (in elapsed time, processor resources required, memory, and buffer utilization) can be justified for slow transmission facilities plagued with "high" bit error rates, the supporting reasons become less clear for the high-speed transmission facilities that would be used in a packet voice network.

The performance of the interswitch digital transmission facilities can be expected to provide sufficiently reliable performance to allow the responsibility of error correction to be moved to the packet network interfaces. As a higher level function, the error correction required for data communication would not be performed on each network link, but once across the network. This simplifies processing that must be done by network switches without degrading the performance seen by network users.

IV. Network Access Arrangements

The placement of functions (e.g., voice encoding) in a packet voice/data network is specified by the protocol architecture. For the generic packet voice/data network considered in this paper, two alternatives are possible for voice coder placement. In the first, the interface to the network is a circuit mode interface. The voice stream into the network is a deterministic stream of information. At the PNI, this stream is encoded, placed in packets, and transported across the network. At the distant edge of the network, the packet stream that was carried through the network is converted back into a circuit voice stream. The data interface to the network is packet mode, possibly using the X.25 protocol.

The alternative method uses a packet mode interface to the network for both voice and data traffic. Voice encoding and packetizing functions are performed by customer equipment, before information enters the network. The network does not process (e.g., code and packetize) the voice information, as it did in the first case, but only transports packets between network customers. As with the first type of interface, two grades of network transport are required: voice transport (fast, but errors are permitted) and data transport (error-free service is required). Other grades of network transport may be desirable, but this paper limits discussion to only these two possibilities.

The remainder of this paper focuses on a circuit-mode

Physical Layer
    Medium, Line Code, etc.
    Transmission Characteristics

Link Layer
    Delimitations
    Transparency
    Error Detection

Packet Network Sublayer
    Logical Channel Multiplexing
    Call Signaling
    Flow Control Indicator

Packet Transport Sublayer
    (Performed Edge to Edge)
    For Voice
        TIME STAMP
    For Data
        SEQUENCING
        WINDOW FLOW CONTROL
        SESSION SYNCHRONIZATION
        ERROR RECOVERY

Fig. 2. Protocol functions.

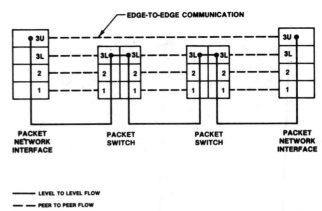

Fig. 3. Internal network protocol layers.

voice interface to the network and a packet-mode data interface. Under this architecture, the packet network interface performs voice encoding and packetizing at the edges of the network. Although this discussion concentrates on internal network operation and functions, these concepts can also be extended, migrating functions from the network to customer premises equipment.

V. A VOICE/DATA PACKET PROTOCOL

Interswitch communication in a voice/data packet network is according to the network transport protocol. Access to the network is not under the protocol specified for interoffice transmission.

Following the spirit of the draft open systems interconnection (OSI) model that the International Organization for Standardization is formulating [12], this partitioning of functions and rules is specified as a hierarchy of layers, each performing specific services, such as error detection. The packet transport protocol outlined here has been designed using the general principles that have been guiding the development of the OSI model. This voice/data packet protocol supports only transport functions (i.e., the transport of packets through the network); the higher level functions (end-to-end) discussed in the OSI model appear as higher layers above this basic transport capability and are not discussed here.

The protocol structure provides the functions necessary for flow control of data within the network and ensures reliable data transmission through the network. Under this structure, the network can be a reliable data-transfer medium. The basic transport mechanism is provided by the lower three levels of the protocol (1-, 2-, and 3-lower) applied to transmissions across each network link. The packet transport sublayer (level 3-upper), employed edge-to-edge within the network between packet network interfaces, provides functions specific to the application (Figs. 2 and 3). In this scheme, the lowest three levels provide packet transmission and routing through the network, but allow the possibility of packet corruption and packet loss. At the edges of the network, any detected errors are corrected with a peer communication between the entry

and exit packet network interfaces, thus providing reliable data communication through the network. This protocol structure allows the same link communication protocols to be used for voice and data, and employs an edge-to-edge protocol that is tailored to the specific service used by the network customer. A later section discusses delays experienced by data using this edge-to-edge recovery strategy.

Controlling the flow of information into the network can be accomplished with several mechanisms: call blocking if insufficient network resources are available at the time the call is requested; packet discarding during short periods of congestion; call clearing by the network to face occasional overloads caused by failures of network components (links, switches, etc.); window flow control for data; and other methods.

The following sections present a closer view of the each protocol level.

A. Level 1 — Physical Layer

The physical level of the protocol specifies the electrical characteristics and representation of transmitted bits. Additionally, bit time synchronization and performance characteristics are included in this level. Level 1 is, in practice, similar to the level 1 of conventional packet data protocols and is not discussed further here.

B. Level 2 — Link Layer

The link layer performs several functions necessary for successful transmission between network switches. These functions include frame delimiting, error detection, and bit pattern transparency. As a major departure from HDLC, error recovery procedures are not included in this level. Functions such as link set up, link disconnection, and link resynchronization procedures are not included in the packet voice/data level 2 protocol because no link state information is maintained at level 2.

The link layer protocol procedures for the packet voice network do not include provisions for error recovery or for flow control. When an error is detected in a frame, the corrupted frame is immediately discarded by the switch. No further actions are taken at level 2. These measures are aimed at reducing sources of network introduced packet delay.

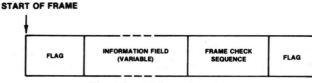

Fig. 4. Level 2 frame format.

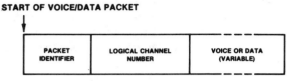

Fig. 5. Level 3-lower packet format.

Each level 2 frame carries a data, voice, or signaling message. The format for information frames (Fig. 4) includes a leading flag that delimits the start of an information frame, a variable-length information field, a frame check sequence for detecting frame errors, and a trailing flag that delimits the end of the information frame. A two byte frame check sequence could probably provide adequate error detection capability.

C. Level 3L — Packet Network Sublayer

Level three of the voice/data protocol has two sublayers: level 3-lower (3L) and level 3-upper (3U). The packet provides the unit of information transfer at the packet network sublayer of this protocol. The primary function provided here is the routing of packets along a path that is established at call-setup time. This path is fixed for the duration of the call.

Two types of packets are included in the level 3L portion of the network protocol: signaling packets and data/voice packets. Signaling packets are used in call establishment, call disconnection, call administration, and link maintenance. These signaling packets are exchanged between adjacent switches (link signaling packets) and between packet network interfaces in the call path (edge signaling packets). Data/voice packets carry information being sent between level 3U entities. Fig. 5 shows a possible format for a voice/data packet. Information contained in the level 3L header includes three fields: a packet identifier field indicating the type of packet and processing required by the switch, a logical channel number field that uniquely identifies a call on a link, and a variable-length voice/data field containing higher level information. Signaling packets would have similar format, with a signaling message indicator and additional parameters carried in the variable-length voice/data field.

D. Level 3U — Packet Transport Sublayers

Multiple types of sublevel 3U allow the network to support different classes of packet transport. The specific messages exchanged and actions taken at level 3U depend on the application being supported by the network (e.g., packet voice transmission and interactive data exchange).

For packet voice communication, level 3U functions are performed at the voice coder location (in the packet network interface). Similarly for data exchange, packet transport sublayer functions are also carried out between packet network interfaces (as an edge-to-edge protocol). However, the concepts presented here do not prelude alternative architectures that place these functions external to the network.

There are two distinct sets of level 3U functions: those for reliable data communication and those for voice communication.

1) Packet Transport Sublayer for Packetized Voice: For packetized voice only, level 3U messages provide synchronization of voice information. Voice that has been encoded and packetized must be time identified, in some manner, relative to either an absolute time reference or to other voice packets (both preceding and subsequent) to allow a smooth play back of voice at the receiver. This level 3U function provides a means of playing out, at the correct time, received voice information. Such a service is necessary because of the variable delays that can be encountered in a long haul packet network. Voice messages are never retransmitted with this protocol architecture.

A time stamp placed in the level 3U header could identify, relative to other voice messages for the call, the time at which the voice message was generated. The information contained in this field would be generated and used only by the two packet network interfaces in the call path. Tandem switching equipment does not include level 3U processing, so this field would not be modified by intermediate switches. Other time stamping schemes are possible [13].

2) Packet Transport Sublayer for Reliable Data Communication: For data communication only, level 3U includes the functions necessary to provide reliable, error-free data transmission. These functions include synchronization of level 3U entities, flow control, and detection and recovery of lost messages. Protocol standards, such as HDLC and CCITT Recommendation X.25, have evolved with the intention of providing reliable communication between devices that do not necessarily process data at the same rate. The packet voice/data protocol supports these functions at level 3U; the packet transport sublayer for data assures error free, sequential message delivery and includes provisions for flow control.

A possible message format for the packet transport sublayer, shown in Fig. 6, includes a message format (MF) field consisting of subfields that can indicate the logical relationship of sequential messages, use of normal or extended-sequence numbering, and other functions necessary to support the access protocol for data; a control field containing flow control and check pointing information; and a variable-length data field.

The set of supervisory messages includes messages necessary to establish and synchronize a level 3U data connection, reliably transfer data across the network, and clear the connection. Level 3U procedures might be thought of as a modified (and enhanced) set of HDLC control procedures exercised across the network (rather than across a single link).

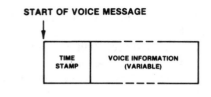

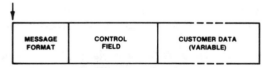

Fig. 6. Level 3-upper message formats.

VI. DELAY WITH RETRANSMISSIONS FOR CORRUPTED DATA

This section examines cross-network delay using two different error recovery disciplines. The first uses retransmission between adjacent network switches to recover corrupted data; the interswitch protocol guarantees reliable transmission across each network link. Average cross-network data transmission delays for this protocol are compared to the delays encountered under a second protocol. In this second protocol, lost or corrupted data are recovered by edge-to-edge retransmission procedures between the sending and receiving packet network interfaces (this is the method used in the voice/data packet protocol previously described); the interswitch protocol makes no attempt to recover corrupted data.

A. Link Recovery Model

This section describes a model for analyzing performance using a reliable link protocol to accomplish error-free transport across a network, with a concatenation of reliable links forming the error-free path.

A frame sent across a link may be lost with probability p, which is determined by the error characteristics of the transmission facilities and the size of the frame sent across the link. If this happens, the link is effectively out of service until the sender "pulls back" and retransmits the corrupted frame. The *cycle time*, T_C, is the elapsed time before the sender realizes that it must retransmit the corrupted frame. The cycle time has three components: 1) the time for the frame to cross the link, 2) the interframe arrival time (i.e., the time between reception of the corrupted frame and reception of the next error-free frame), and 3) the time for the reject to be returned to the sender. During this period, the receiver discards all frames it receives. To frames awaiting transmission, the link appears to be transmitting a single, long frame (called an *Xframe* in this paper) during the cycle time. Note that this Xframe is not an actual frame transmitted on the link, but serves to describe the amount of time the link is unavailable because of error recovery procedures.

As a preliminary effort to gauge the effect of the link unavailability during the cycle time, the link is modeled as an $M/G/1$ queue. For this model, link traffic has two components: frames containing useful data (which also contain embedded data acknowledgments for traffic in the reverse direction), and Xframes. Using an exponential service time distribution for user data and a deterministic service time for Xframes (this service time is exactly equal to the cycle time) completely describes all traffic on the link for the model.

Using an $M/G/1$ queue to approximate the average transmission time per link for data (T_L), a weighted first and second moment ($\bar{x}$ and μ_2) describes characteristics of the link traffic. D_{prop} is the propagation delay per link and ρ_L is the total utilization of the link; the utilization of the link attributed to data traffic (ρ_{data}) is slightly lower than ρ_L.

$$T_L = (D_{\text{prop}}) + \bar{x} + \frac{1}{2} \frac{\rho_L}{2(1 - \rho_L)} \frac{\mu_2}{\bar{x}}.$$

The average cross-network delay for a path of N_{links} links, for frames containing data, is

$$\text{Average Network Delay} = (N_{\text{Links}})(T_L). \qquad (1)$$

This delay equation will be used later in comparing edge-to-edge and link recovery strategies.

While the effect of Xframes can be negligible at low link speeds, it can become significant at higher link speeds (e.g., over 1 Mbits/s) because of the increasing size of the Xframe and the number of frames that must be retransmitted. For example, at 75 percent, total link utilization on a 1.5 Mbits/s link, each frame error causes the link to be unavailable for useful data transmission for over 30 frame transmission times, creating a delay for subsequent frames. The probability of the link corrupting a frame closely describes the amount of link traffic attributed Xframes. For the model, every time a frame is corrupted this Xframe appears on the link. Because a frame may be retransmitted several times (additional retransmission is necessary if the frame becomes corrupted when retransmitted), an attempt to transmit a useful data carrying frame across a link causes $p/1 - p$ Xframes to appear on a link. This information is used in determining the probabilities of normal frames and Xframes being present on a link, which in turn is used to find the weighted $\bar{x}$ and μ_2.

B. Edge Recovery Model

This section describes a model for analyzing performance in a network that uses a nonguaranteed-delivery link protocol, supplemented with a reliable edge-to-edge protocol across the network for error-free packet transport. With an edge-to-edge strategy, error recovery is between the corresponding edges of a call, and is done on a per call basis. In this section, *frames* are sent between adjacent switches on a link. Corrupted frames are simply discarded when a switch detects an error. *Messages* are sent between corresponding edges of a call. If a message is lost in traversing the network, it is retransmitted again from the sending edge.

The probability of having to invoke error recovery procedures is based entirely on the probability of successfully crossing the network. We find this using the probability p of a frame being corrupted on a link. From this the probability P of, anywhere in the network path, corrupting a frame containing this message is

$$P = 1 - (1 - p)^{N_{\text{links}}} \qquad (2)$$

Each link carries a mix of three types of traffic (as contrasted to the link model, in which each link carried only two types of traffic: information frames, which carry customer data, and X frames). The first type is data traffic, which has the same exponential distribution that is used for the link scenario. To allow a fair comparison with the link recovery model, the amount of traffic attributed to user data, ρ_{data}, is identical for the two models. For the edge recovery model, every data message has an explicit acknowledgment, which has a deterministic service time distribution. This is the second type of traffic. Discarded message traffic enters in as the third factor contributing to link utilization. Discarded messages that the sender must retransmit have exponentially distributed service times. The average service time for all traffic on a link is a weighted average of these three service times. The total link utilization for this model, ρ_E, reflects contributions from each of these three generators of link traffic.

In a call between transmitting edge T and receiving edge R, the delay between the occurrence of an error and discovery of this error by T has three components. These are analogous to the three outlined for link recovery: intermessage arrival time (the time between the moment the corrupted message would have arrived and the time the next uncorrupted message arrives), successful cross-network transit of a data message, and successful cross-network transit of the reject message in the reverse direction.

For a single call, receiver R discards $N_{\text{discarded}} = (T_C)(H) + 1$ messages if the sender continues transmitting messages after the error occurred. H is the user throughput class expressed in the message per second and T_C is the edge-to-edge cycle time (as defined in the link model discussion). Because of retransmissions, the event of corrupting a message can happen more than once in trying to get a message through the network. Each message sent through the network results in $(P/(1 - P))N_{\text{discarded}}$ extra messages. These extra messages directly increase the queueing delays experienced packets for other calls in the network.

Recall the definitions for ρ_L and ρ_E. ρ_L is the utilization of each network link in a network using link recovery. ρ_E is the utilization of these links if the network uses edge-to-edge recovery. In general, for a fixed data traffic load (i.e., fixed ρ_D), ρ_E is greater than ρ_L. This is for two reasons: the explicit data acknowledgments under the edge-to-edge protocol (that are not present in the link scenario) and the comparative error rates (2).

1) Cross-Network Delay (Including Effects of Retransmission Traffic): Computing the average transmission time across a link for data (T_{data}), including propagation delay and system time,

$$T_{\text{data}} = (D_{\text{prop}}) + \frac{1}{\mu_{\text{data}}} + \frac{1}{2} \frac{\rho_E}{(1 - \rho_E)} \frac{\mu_2}{\bar{x}}.$$

In this equation, $1/\mu_{\text{data}}$ is the service time for frames carrying user data; and the first and second moments (μ_2

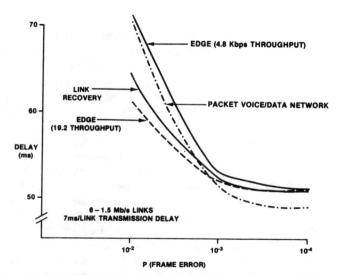

Fig. 7. Average cross-network data delay.

and $\bar{x}$) are found using the service time distributions of the three types of link traffic along with their probabilities of occurrence.

The total average delay incurred in successfully traversing the network one time (without error recovery procedures) for data is $D_{\text{data}} = (N_{\text{links}})(T_{\text{data}})$.

The average intermessage arrival time (I_M) is based entirely on the user transmission speed and amount of data that can be transported in a frame, causing an increase in error detection time at the receiving edge as user throughput class decreases.

For edge-to-edge error recovery, the equation for cross-network delay, including the effects of retransmission procedures, is

$$\text{average cross network delay} = \frac{(1 + P)D_{\text{data}} + PI_M}{(1 - P)}. \qquad (3)$$

This delay equation will be used later in comparing edge-to-edge and link recovery strategies.

C. Delay Comparison

Fig. 7 shows comparative delays for the two recovery disciplines in two types of networks, found using (1) and (3). These delays are for a cross network path of 6 1.5 Mbits/s links, a 7 ms/link propagation and processing delay, uncorrelated frame errors, and frames that contain 128 bytes of customer data not including the frame header. This analysis is not all encompassing; the intent here is to show the effect of X frames when high-speed facilities are used and show the effect of user throughput on average data delay when using the edge-to-edge recovery mechanism.

In Fig. 7 the amount of data traffic carried is the same for the link and edge recovery curves, ensuring equivalent traffic conditions for the comparison. However, this does not imply that the total link utilization is the same for the two models. (Less data traffic is offered to the voice/data packet network because that network must also carry voice packets.) The method used to generate the link utilizations forces the total link utilization for the link recovery method to be constant, independent of the probability of corrupting a frame. This means that as the probability of corrupting a frame decreases (moving from left to right in the figure), the ratio of useful data traffic to *X* frames increases, and the amount of data traffic increases. For the two edge recovery curves, the amount of useful data traffic increases to stay the same as for the link curve.

Fig. 7 shows that the delays using link recovery are close to the delays encountered using edge-to-edge recovery under the same data load. Not unexpectedly, as the probability of corrupting a frame decreases, the average delay decreases, because fewer retransmissions are required for either method. In the link case, the *X* frames caused by the retransmission procedures are large and could add significant delays when they appear on a link. With fewer retransmissions (i.e., fewer *X* frames), the total link utilization decreases, diminishing the magnitude of the queueing delays. For the edge-to-edge curves, a major component of cross-network delay is the time to notify the sender that a message must be retransmitted. With this occurring less frequently, the average cross network delay decreases. Additionally, less "extra" traffic (that must be retransmitted) is now appearing on the link, decreasing the queueing delays. However, this is a minor effect for low user throughput classes.

As user throughput increases, two factors affect the amount of "extra" messages present on the link for the edge-to-edge case. First, the number of these extra messages sent during the cycle time (i.e., time to notify the sender that an error has occurred) increases simply because more messages can be sent per second. Along with this, the intermessage arrival time decreases, causing the cycle time to decrease. The net result is that as throughput increases, the number of superfluous messages increases, and for a fixed data load this slightly increases the queueing delays. However, the decrease in time to detect the error significantly decreases the delay introduced by the retransmission protocol. Fig. 7 shows this in the two edge curves, demonstrating that the faster user throughput (19.2 kbits/s) has a lower average delay than the 4.8 kbits/s user throughput. (Note that for user throughputs well over 100 kbits/s, these superfluous messages can cause the link utilization to exceed available bandwidth if the offered data load is held constant. Factors such as link speed and error rate determine the actual customer bandwidth at which this occurs.)

The curve for the voice/data packet network shows average data cross-network delays for an integrated packet network that uses edge-to-edge recovery for data. Although the total link utilizations for this curve are comparable to the other curves, the amount of data traffic is less for this curve because voice traffic is also carried. Corrupted data introduces some extra traffic that must be retransmitted, but corrupted voice messages do not cause any additional network traffic. Because of this, the amount "extra" link traffic caused by retransmission procedures is not as great as it is for the two data-only networks. For this example, frames containing voice messages have a length of 250 bits with a deterministic distribution and outnumber useful data messages by 4 to 1. This ratio does not include discarded messages or acknowledgment messages.

D. *Extending the Models*

While this paper shows several concerns in finding delay and throughput for the two recovery methods, more can be done in this area. Possible efforts aimed at continuing this study include modifying the model to more accurately describe the probability of corrupting a frame. Additionally, including the effects of corrupted acknowledgment and reject packets will give a more accurate description of network traffic and cross network delay. An enhanced model can include provisions for various message length distributions, increasing the usefulness of the results.

E. *Conclusions for the Delay Models*

For link correction, two factors influence the effectiveness of the network: 1) unless the network uses selective reject, a frame error *prevents* the sending party on a link from transmitting useful frames for a short period, increasing the cross network delay for subsequent frames; 2) the amount of time required to detect and correct a frame transmission error depends on the link speed and utilization.

When using edge-edge recovery, network protocols are characterized as follows. First, when a message is lost, this prevents useful messages from being sent *for a single call* until the sender retransmits the corrupted message. This causes extra link traffic, increasing queueing delays for all calls on that link. Second, the amount of time it takes to correct for a message error is dependent on the user throughput class. (This does not include errors that are detected by the expiration of a timer.)

In a packet network that sends 128 byte messages as a single packet through the network, link recovery and edge-edge recovery methods give comparable performance as seen by a network customer. This conclusion is valid for networks using high speed transmission facilities (e.g., over 1 Mbits/s) having good error characteristics (e.g., corrupted packet rate of 10^{-3}) with reasonable link utilizations for data traffic (e.g., less than 60 percent).

In a combined voice/data packet network, an edge recovery mechanism can provide acceptable delay characteristics for data traffic.

VII. Summary

Several concepts outlining the philosophy of a possible protocol structure for integrated voice, data, and signaling have been presented. The protocol overview covers the rules for communication between adjacent network switches, and across the network between the corresponding packet network interfaces of a call. In this protocol, error correction procedures are not included at the link layer or packet network sublayer for data or voice information. Error recovery for data is performed edge-to-edge, across the network. The average cross-network delays for data packets under this edge-to-edge error recovery discipline are comparable to the expected delays under a more traditional link-by-link recovery protocol.

While this paper has centered discussion on a possible design for a voice/data *internal* network protocol, the area of accessing such a network with an integrated packet protocol is ripe for study. The concepts outlined in this paper of providing basic transport functions at levels 2 and 3L in the protocol, and tailoring the packet transport sublayer (3U) for the specific application (e.g., packet voice) can be extended to apply to network access protocols.

ACKNOWLEDGMENT

The material presented in this paper has emerged from discussions with members of the Advanced Switching Technology Laboratory at Bell Telephone Laboratories. M. Decina (while at Bell Laboratories on leave from the University of Rome), J. Turner, W. Montgomery, and L. Wyatt have provided their time in discussing packet voice communication, and their help is appreciated.

REFERENCES

[1] J. W. Forgie and A. G. Nemeth, "An efficient packetized voice/data network using statistical flow control," in *Proc. IEEE Int. Conf. Commun.*, 1977.
[2] G. J. Coviello, O. L. Lake, and G. R. Redinbo, "System design implications of packetized voice," in *Proc. IEEE Int. Conf. Commun.*, 1977.
[3] M. E. Ulug and J. G. Gruber, "Statistical multiplexing of data and encoded voice in a transparent intelligent network," in *Proc. 5th Data Commun. Symp.*, Sept. 1977.
[4] D. Minoli, "Issues in packet voice communication," *Proc. IEEE*, vol. 67, Aug. 1979.
[5] E. T. Klemmer, "Subjective evaluation of transmission delay in telephone conversations," *Bell Syst. Tech. J.*, vol. XLVI, July–Aug. 1967.
[6] —, "Subjective evaluation of delay and echo suppressors in telephone communications," *Bell Syst. Tech. J.*, vol. XLII, Nov. 1963.
[7] M. Listanti and F. Villani, "An X.25-compatible protocol for packet voice communications," *Comput. Commun.*, pp. 23–31, Feb. 1983.
[8] D. Cohen, "A protocol for packet switching voice communication," in *Proc. Comput. Network Protocols Symp.*, Liege, Belgium, Feb. 1978.
[9] —, "Issues in transet packetized voice communication," in *Proc. 5th Data Commun. Symp.*, Snowbird, UT, 1977.
[10] S. L. Casner, E. R. Mader, and E. R. Cole, "Some initial measurements of ARPANET packet voice transmission," NTC 1978.
[11] J. G. Gruber, "Delay related issues in integrated voice and data networks," *IEEE Trans. Commun.*, vol. COM-29, June 1981.
[12] H. Zimmermann, "OSI reference model—The ISO model of architecture for open systems interconnection," *IEEE Trans. Commun.* (*Special Issue on Computer Network Architectures and Protocols*), vol. COM-28, Apr. 1980.
[13] W. A. Montgomery, "Techniques for packet voice synchronization," *IEEE J. Select. Areas Commun.*, vol. SAC-1, pp. 1022–1028, Dec. 1983.

William L. Hoberecht received the B.S. degree in mathematics from Clarkson College of Technology, Potsdam, NY, in 1979, and the M.S. degree in computer, information, and control engineering from the University of Michigan, Ann Arbor, MI, in 1980.

He is with Bell Telephone Laboratories, Naperville, IL, where his initial assignments were in the areas of packet switching and the design of packet protocols for data and voice communication. He is currently investigating fault-tolerant architectures for advanced packet switching system designs. His technical interests include data communications, packet switching, and operating system design.

An Overview of FDDI: The Fiber Distributed Data Interface

FLOYD E. ROSS

Reprinted from *IEEE Journal on Selected Areas in Communications*, Vol. 7, No. 7, September 1989, pages 1043-1051. Copyright © 1989 by The Institute of Electrical and Electronics Engineers, Inc. All rights reserved.

Abstract—Fiber Distributed Data Interface (FDDI), which employs an optical fiber medium, is a 100 Mbit/s local area network (LAN) based on a token ring protocol. Updating earlier efforts on this subject by the author and others, this paper presents an overview of this technology and explores the basis for its success. Particular emphasis is placed on the technical specifications for an upwards-compatible version of FDDI, FDDI-II, which adds the capability for circuit-switched services to the packet services of the basic FDDI, thus creating an integrated services LAN.

I. INTRODUCTION

THE Fiber Distributed Data Interface (FDDI) is a 100 Mbit/s local area network (LAN). Using optical fiber as the medium, the FDDI protocol is based on a token ring access method. FDDI is being developed in Accredited Standards Committee (ASC) X3T9—chartered to develop computer input/output (I/O) interface standards. In the late 1970's, ASC X3T9 recognized the need for a new I/O channel standard as an alternative to Federal Information Processing Standards (FIPS) 60-63. By late 1982, work had been started on FDDI in the ASC X3T9.5 technical committee. In mid-1983, two proposals encompassing the Physical layer and the Media Acess Control for FDDI were submitted by Sperry.

Through this same time period, the Institute of Electrical and Electronics Engineers (IEEE) P802 standards project was developing local area network (LAN) standards with data rates up to 20 Mbits/s. FDDI followed the packet data architectural concepts of IEEE P802 and chose the emerging 4 Mbits/s token ring protocol of IEEE P802.5 as the starting point for the FDDI protocol. These choices placed FDDI in an ideal position to be both the backbone network and the follow-on network to the IEEE P802 LAN's.

Meanwhile, in another standards arena, the Open Systems Interconnection (OSI) model had been put in place. This layered the design of computer interconnections, allowing the development of separate standards for the different layers, and provided the proper framework for the development of a set of FDDI standards.

Another factor of significance in the development of FDDI was an increased use of high-performance video workstations for a variety of applications. These brought an emphasis on facilitating low-cost implementations.

Manuscript received May 15, 1988; revised March 1, 1989.
The author is with UNISYS, Malvern, PA 19355.
IEEE Log Number 8929561.

Driven by these forces, FDDI grew from the original Sperry proposals to satisfy the needs of many applications, including the back-end (I/O channel) interface, LAN backbone, and front-end high-performance LAN applications. In the course of its development, each new wave of supporters brought new demands which, in turn, were accommodated by the emerging FDDI standards. As a result, the set of services offered by FDDI is broad enough to allow individual optimization of FDDI networks to satisfy the needs of diverse environments.

One enhancement, FDDI-II, will offer signficantly increased services by integrating circuit-switched data traffic capabilities into what had originally been strictly a packet LAN. The impetus for FDDI-II came from the new generation of digital PBX's of the early 1980's. Their needs were similar to those of many real-time applications including digital voice and video networks as well as sensor and control data streams. All of these disciplines contributed to the emerging FDDI-II design definition which began in late 1984.

A. Outline of Paper

Section II of this paper addresses several issues of overall signficance to FDDI. Section III describes general FDDI characteristics, with the Physical layer, MAC layer, and SMT operation more fully described in Sections IV, V, and VI, respectively. Section VII supplies an overview of FDDI-II concepts, with Section VIII providing details on FDDI-II operation. The status of the FDDI standardiztion effect is covered in Section IX.

II. FDDI AS THE SOLUTION

The widespread acceptance of FDDI has been due to a number of factors. The IEEE P802 effort provided several medium-speed (1–20 Mbit/s) LAN's. In effect, IEEE P802 popularized the LAN, and by doing so, created a market for a higher speed LAN to perform the backbone function for lower speed IEEE 802 LAN's and to satisfy applications that require a higher performance LAN.

Contributing also to the acceptance of FDDI is the dramatic improvement in the price and performance characteristics of optical fiber and related components such as optical transmitters and receivers. With the many other advantages that the use of optical fiber offers—high data bandwidth, security, safety, immunity to electromagnetic interference, and reduced weight and size—the concept of an all-fiber LAN was most attractive. Because the FDDI design has been optimized to the use of optical fiber since its inception, FDDI is now in a leadership role in the development of optical fiber LAN technology.

FDDI has maximized the value of standardization. The true value of standards is being recognized in all marketplaces—from the individual customer to multinational corporations and the federal government. With emphasis firmly on standards, FDDI has conformed to the ISO Model. The rich functionality that has been integrated into FDDI to meet the needs of a number of market segments means that it can be the one standard satisfying the requirements of the broad, high-speed LAN marketplace. The value of one high-speed optical fiber LAN standard is immense.

A. Advantages of a Ring Design

No paper on FDDI would seem complete without at least a cursory summary of the advantages that a ring design offers. A ring can be shown to offer superior reliability, availability, and serviceability, even in the face of physical damage to the network. A ring topology can be designed to be capable of continued operation despite any projected failure.

Other advantages include the interconnect simplicity of the physical hardware at the interface level. The point-to-point connections around the ring not only provide an easy focus of standardization, but also allow different ring links to have different characteristics and optimization points. Optical fiber, which does not adapt well to bus configurations, can be easily accommodated. Optical fiber has sufficient bandwidth that bit-serial transmission may be used, thus significantly reducing the size, cost, and complexity of the hardware required by a network.

Ring topologies offer advantages in the ease of initial configuration and reconfiguration as the network requirements change. Failing stations or fiber links can be isolated through the use of appropriate protocols. These protocols also provide for the logical addition and deletion of stations without detrimental effects on existing ring traffic. Actual physical addition or removal of stations from the network is also facilitated because ring initialization, failure isolation, recovery, and reconfiguration mechanisms can provide for continued operation even while the cables are being rearranged.

Ring topologies inherently impose no restrictive logical limit on the length of ring links, the number of stations, or the total extent of the network that can be accommodated.

Ring topologies, and the protocols supported by them, offer significant performance advantages. These include insensitivity to load distribution, the ease of fairly allocating the available bandwidth, low arbitration times, bounded access delay, and no requirement for long preambles.

In today's technology, a ring topology appears to best satisfy the requirements of high-performance networks operating from 20 to 500 Mbits/s where high connectivity and large extents are required. Some of the references, and in particular [10], are recommended for an in-depth review of the advantages that a ring design offers.

III. FDDI CHARACTERISTICS

The initial version of FDDI uses optical fiber with light-emitting diodes (LED's) transmitting at a nominal wavelength of 1325 nm over multimode fiber. Connections between stations are made with a dual fiber cable employing a polarized duplex connector. A single-mode fiber (SMF) version of Physical Layer Medium Dependent (PMD) uses laser diode transmitters, with two power levels categories specified, the lower of which retains the same receivers as the basic PMD. SMF–PMD will allow individual links to be extended up to 60 or possibly even 100 km.

The data transmission rate is 100 Mbits/s. The effective sustained data rate at the data link layer can be well over 95 percent of this peak rate. The four out of five code used on the optical fiber medium requries a 125 Mbaud transmission rate. The nature of the clocking, which adjusts for accumulated jitter between frames, limits frames to 4500 octets maximum. Multiple frames may, however, be transmitted on the same access opportunity.

A total of 1000 physical connections and a fiber path of 200 km have been used as the basis for calculation of the default values of the recovery timers. Considering reconfiguration requirements, these choices allow a maximum configuration of 500 stations (each station represents two physical connections) linked by 100 km duplex cable. The choice of longer times than the default values for the recovery timers will allow larger networks to be configured. For smaller networks, performance can be optimized by choosing shorter times for the recovery timers. There is no minimum configuration requirement.

A. Station Organization

Fig. 1 shows the component entities necessary for an FDDI station. Identified components, conforming to both the IEEE 802 structure and the OSI concept of layering, are: Station Management (SMT), which specifies the local portion of the network management application process, including the control requried for the proper internal configuration and operation of a station in an FDDI ring; Media Access Control (MAC), which specifies the lower sublayer of the data link layer, including the access to the medium, addressing, data checking, and data framing; Physical Layer Protocol (PHY), which specifies the upper sublayer of the physical layer, including the encode/decode, clocking and framing for transmission; and Physical Layer Medium Dependent (PMD), which specifies the lower sublayer of the physical layer, including power levels and characteristics of the optical transmitter and receiver, interface optical signal requirements, the connector receptacle footprint, the requirements of confirming optical fiber cable plants, and the permissible bit error rates. Two alternative versions of PMD are shown; the basic PMD and SMF–PMD which allows the use of single-mode optical fiber.

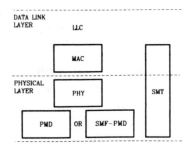

Fig. 1. FDDI relationship to OSI model.

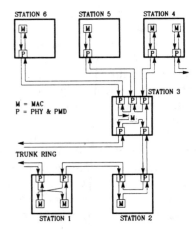

Fig. 2. FDDI topology example.

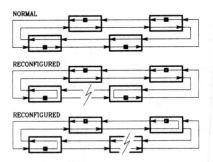

Fig. 3. Reconfiguration of counterrotating rings.

FDDI MAC provides a superset of the services required by the Logical Link Control (LLC) protocol developed by IEEE P802.2. Fig. 1 assumes the use of the IEEE 802.2 LLC as the upper sublayer of the data link layer. Any other appropriate LLC may be used.

These basic component entities allow a variety of station types as shown in the FDDI topology example of Fig. 2. As shown, the FDDI trunk ring consists of a pair of counterrotating rings. Two main classes of stations, depending on whether or not they are allowed to attach directly to the trunk ring, are specified. A dual attachment station has two PHY entities. It may attach directly to the trunk ring or indirectly via a concentrator. It may have one or more MAC entities. In the case of two MAC entities, one may be in each of the counterrotating rings or both may be in the same ring. A dual attachment station may have optical bypass switches to remove it from both rings, at the same time healing them, if the station is powered down or disabled by SMT.

A single attachment station has one PHY and one MAC, and therefore cannot be attached directly into the main FDDI ring. Instead, it must be attached to the ring via a concentrator. A concentrator is a special station that has extra PHY's used to attach other stations that are to be inserted into the main FDDI ring. Varying levels of functionality, including multiple MAC's, are permitted in concentrators.

In Fig. 2 stations 1, 2, and 4 are dual attachment stations. Station 3 is a concentrator connecting stations 4, 5, and 6 into the FDDI ring. Stations 5 and 6 are single attachment stations. Station 4, even though it is a dual attachment station, must perform as a single attachment station insofar as its attachment to the ring via the concentrator is concerned. Operation of the second PHY of station 4 on some other ring is allowed, but there must not be any interconnection of the two rings within this station at a level visible to MAC or PHY.

B. Reliability Provisions

Ring topologies allow for the isolation of failing attachments through several mechanisms. Counterrotating rings, as shown in Fig. 3, are basic to the FDDI structure. The counterrotating ring concept uses two rings connected to each station or concentrator—one clockwise and the other counterclockwise. When a failure in a link occurs, the stations on either side reconfigure internally as shown in the middle diagram. The functional stations adjacent to the break make use of the connection in the reverse direction to close the ring, thus eliminating the bad link. In this figure, the dark squares represent the logical (MAC) attachment within the stations. Should a station itself fail, as shown in the bottom diagram of this figure, the stations on either side reconfigure to eliminate the failing station and both of the links to it.

Stations may offer a bypass capability, whereby an optical switch is used to bypass a station's receiver and transmitter connections so that the signal from the previous station is passed directly to the next station. Bypassing may be activated by a station itself, at the instigation of its neighbor, by a human operator, automatically at the removal of power, or by some overall network-controlling function.

Yet another approach is the use of concentrators, as shown in Fig. 2, that are attached directly to the trunk ring and, in turn, provide drop connections for a number of stations—in this case, stations 4, 5, and 6. A concentrator may then monitor all of its slave stations and isolate any faulty station. It may also provide for the graceful logical insertion and removal of stations from the ring.

The use of all three techniques allows FDDI networks to be configured to tolerate a variety of station or link failures and physical network configurations without catastrophic consequences. When failures occur, the network automatically reconfigures, eliminating any failing element and maintaining ring operation. Continuous monitoring of the failed link or station allows the network to automatically reconfigure and restore normal operation when repair is effected. Any of these reconfigurations may result in the loss of individual frames, which then need to be retransmitted.

C. Data Encoding

Information on the medium uses a four out of five group code, with each code group called a symbol. As shown in Table I, in the 32-member symbol set, 16 symbols are data symbols, each representing four bits of ordered binary data. Three symbols are used for line-state signaling which is recognized by the physical layer hardware, two are used as control indicators, and four are used for starting and ending delimiters. One of the starting delimiters (L) is not used in the basic FDDI, but is reserved for FDDI-II. The remaining seven symbols of the symbol set are not to be transmitted since they violate code run length and dc balance requirements. The QUIET symbol is a necessary member of the line-state symbol set since it is used to indicate the absence of any functional signal.

D. Frame and Token Formats

Information is transmitted on the FDDI ring in frames which are variable in length. Tokens are special short fixed-length ''frames'' that are used to signify the right to transmit data. Fig. 4 shows the frame and token formats.

The Preamble (PA) field, consisting nominally of 16 IDLE symbols (a maximum frequency signal that is used for establishing and maintaining clock synchronization), precedes every transmission. The Starting Delimiter (SD) field consists of a two-symbol sequence (JK) that is uniquely recognizable independent of previously established symbol boundaries. The SD establishes the symbol boundaries for the content that follows.

The Frame Control (FC) is a two-symbol field that defines the type of frame and its characteristics. It distinguishes between synchronous and asynchronous frames, the length of the address field (16 or 48 bits), and the kinds of frame (e.g., LLC or SMT). One set of FC values is reserved for implementer frames that have no defined format and are to be repeated unchanged by all conforming FDDI stations. The FC field also provides for two kinds of tokens, restricted and nonrestricted. The latter is used in a special class of service which provides for extended dialogs among a limited set of cooperating stations. Two Ending Delimiter (ED) symbols (TT) complete a token.

The Destination Address (DA) and Source Address (SA) fields may be either 16 or 48 bits long, depending on the FC value. DA may be either an individual or a group address, the latter of which has the potential to be recognized by more than one station.

The 32-bit Frame Check Sequence (FCS) field is a cyclic redundancy check using the standard polynomial used in the IEEE 802 protocols. The information field of a frame, like the other fields covered by the FCS check,

TABLE I
SYMBOL CODING

DECIMAL	CODE GROUP	SYMBOL	NAME	ASSIGNMENT	
00	00000	Q	QUIET	LINE STATE SYMBOL	
31	11111	I	IDLE	"	
04	00100	H	HALT	"	
24	11000	J		STARTING DELIMITER	
17	10001	K		"	
05	00101	L		"	
13	01101	T		ENDING DELIMITER	
07	00111	R	RESET	CONTROL INDICATOR	
25	11001	S	SET	"	
30	11110	0		DATA SYMBOL	0000
09	01001	1		"	0001
20	10100	2		"	0010
21	10101	3		"	0011
10	01010	4		"	0100
11	01011	5		"	0101
14	01110	6		"	0110
15	01111	7		"	0111
18	10010	8		"	1000
19	10011	9		"	1001
22	10110	A		"	1010
23	10111	B		"	1011
26	11010	C		"	1100
27	11011	D		"	1101
28	11100	E		"	1110
29	11101	F		"	1111
01	00001	V	VIOLATION	NOT TRANSMITTED	
02	00010	V	VIOLATION	"	
03	00011	V	VIOLATION	"	
06	00110	V	VIOLATION	"	
08	01000	V	VIOLATION	"	
12	01100	V	VIOLATION	"	
16	10000	V	VIOLATION	"	

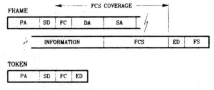

Fig. 4. Frame and token formats.

consists only of data symbols. Data symbols are not used in fields not covered by the FCS check.

The ED field of a frame is one delimiter symbol (T). It is followed by the Frame Status (FS) field that has a minimum of three control indicator symbols that are modified by the station as it repeats the frame. These indicate, when set, that an error has been detected in the frame by the station, that the addressed station has recognized its address, and that the frame has been copied by the station.

IV. PHYSICAL LAYER (PHY AND PMD) OPERATION

PHY provides the protocols and PMD the optical fiber hardware components that support a link from one FDDI station to another. PHY simultaneously receives and transmits. The transmitter accepts symbols from MAC, converts these to five-bit code groups, and transmits the encoded serial data stream on the medium.

The receiver recovers the encoded serial data stream from the medium, establishes symbol boundaries based on the recognition of a start delimiter, and forwards decoded symbols to MAC. Additional symbols (QUIET,

IDLE, and HALT) are interpreted by PHY and used to support SMT functions.

PHY also provides the bit clocks for each station. The total ring, including all stations and links, must remain the same apparent bit length (i.e., no bits may be created or deleted) during the transmission of a frame around the ring. Otherwise, an error would be generated in the frame as it is repeated around the ring. In the face of jitter, voltage, temperature, and component aging effects, such stability can only be realized through special provision. PHY provides an elasticity buffer which is always inserted between the receiver and the transmitter. The receiver employs a variable frequency clock, using standard techniques such as a phase-locked loop oscillator, to recover the clock of the previous transmitting station from the received data. The transmitter, in contrast, uses a local fixed-frequency clock. The elasticity buffer in each station compensates for the difference in frequency between the local clock and that of the upstream station by adjusting the bit delay through the station. The elasticity buffer in each station is reinitialized to its center position during the preamble (PA) that precedes each frame or token. This has the effect of increasing or decreasing the length of the PA, which is initially transmitted as 16 or more symbols, as it proceeds around the ring.

The transmitter clock has been chosen with 0.005 percent stability. With a minimum elasticity buffer of 10 bits, frames of up to 4500 octets in length can be transmitted without exceeding the limits of the elasticity buffer.

Simulations of early FDDI designs, presented to the technical committee, showed that rather than maintaining a nominal 16 symbol PA length, there was a tendency for the PA lengths in long rings to move toward a flat unbounded distribution. Longer PA's created no problem by themselves, but came at the expense of shorter, or even negative, length PA's. A negative PA indicates that the PA has been so severely shortened that it has completely disappeared and that symbols have consequently been lost off the end of a frame, resulting in the loss of the entire frame. This problem was solved by means of a smoothing buffer function incorporated into PHY. This examines the PA length between frames and either inserts or deletes, as the case may be, preamble symbols (or bytes) in order to maintain the PA near the nominal 16 symbol length, thus ensuring that the preamble never decreases below 4 or 5 bytes. Simulation work presented to the technical committee showed that the algorithm chosen, even under worst case conditions, reduced the probability of frame loss to less than 10^{-12}. Later work presented to the technical committee has reaffirmed this result with extended testing of a large physical ring configuration.

V. Token MAC Functional Operation

A major function of any station is deciding which station has control of the medium. MAC schedules and performs all data transfers on the ring.

The basic concept of a ring is that each station repeats the frame that it has received from its upstream neighbor to its downstream neighbor. If the destination address (DA) of the frame matches that MAC's address and there is no error indicated, then the frame is copied into a local buffer with MAC notifying LLC (or SMT) of the frame's arrival. MAC modifies the indicator symbols in the FS field of the frame as it repeats it to indicate the detection of an error in the frame, the recognition of its own address, and the copying of the frame. The frame propagates around the ring to the station that originally placed it on the ring. The transmitting station may examine the indicator symbols in the FS field to determine the success of the transmission. The MAC of this transmitting station is responsible for removing from the ring all of the frames that it has placed on the ring (a process termed stripping). MAC recognizes these frames for stripping by the fact that the SA contained in them is its own address. IDLE symbols are placed on the medium during stripping.

If MAC has a frame from LLC (or SMT) to transmit, it may do so only after a token has been captured. A token is a special frame that indicates that the medium is available for use. Priority requirements, necessary to assure the proper handling of frames, are implemented in the rules of token capture. Under these rules, if a given station is not allowed to capture the token, then it must repeat it (or in certain cases reissue a token) to the next station in the ring. Only after having captured a token and stripping it from the ring is MAC allowed to transmit a frame or frames. When finished, MAC issues a new token to signify that the medium is available for use by another station.

The FDDI MAC uses a Timed Token Rotation (TTR) protocol to control access to the medium. Under this protocol, each station measures the time that has elapsed since a token was last received. The initialization procedures establish the Target Token Rotation Time (TTRT) equal to the lowest value that is bid by any of the stations. Two classes of service are defined. Synchronous service allows use of a token whenever MAC has synchronous frames queued for transmission. Asynchronous service allows use of a token only when the time since a token last was received has not exceeded the established TTRT. Multiple levels of priority for asynchronous frames may be provided within a station by specifying additional (more restrictive) time thresholds for token rotation.

The use of the TTR protocol imparts some useful operational characteristics. It allows stations to request and establish (via SMT procedures) guaranteed bandwidth and response time for synchronous frames. It establishes a guaranteed minimum response time for the ring because, in the worst case, the time between the arrival of two successive tokens will never exceed twice the value of TTRT. It also provides a guaranteed level of ring utilizations equal to (TTRT–RL)/TTRT where RL is the physical ring latency—essentially the time for a token to propagate around the ring under no load conditions. Reference [15] and others by the same author have dealt extensively with these aspects of FDDI operation.

Low values of TTRT (e.g., 4 ms) may be used to establish an average token rotation time of 4 ms and a guaranteed response time not exceeding 8 ms. This would be useful in a time-critical application (e.g., packetized voice). Larger value of TTRT may be used to establish very high ring utilizations under heavy loads. For instance, using TTRT of 50 ms and a ring latency (RL) of 0.25 ms (reasonable for a ring consisting of 75 stations and 30 km of fiber), the above formula shows that a utilization of 99.5 percent can be achieved.

VI. Station Management (SMT) Operation

SMT is the local portion of the network management application process, including the control required for proper operation of an FDDI station in an FDDI ring. SMT monitors activity and exercises overall control of station activity. These functions include control and management within a station for such purposes as initialization, activation, performance monitoring, maintenance, and error control. Additionally, SMT communicates with other SMT entities on the network for the purpose of controlling network operation. Examples of these SMT functions include the administration of addressing, allocation of network bandwidth, and network control and configuration.

The SMT Connection Management (CMT) function establishes the physical connections between adjacent stations. For this function, CMT uses streams composed of QUIET, HALT, and IDLE symbols in low-level signaling protocols. Once a physical connection is established, CMT creates a logical configuration within the station by activating the appropriate paths between the PHY and MAC entities within that station. A large degree of flexibility is provided in the logical configuraiton, which may be established consistent with station functionality and desired station personality. This flexibility allows FDDI to support a wide range of topologies and applications.

VII. FDDI-II Concepts

FDDI-II is an upward-compatible enhancement of the basic FDDI that adds a circuit-switched service to the existing packet capability.

A packet service is a service where the elements of data to be transferred are placed in frames. Packets may vary in length and are self-defining in that each contains delimiters that mark its beginning and end and an address that specifies the target station. FDDI packets are called frames.

In contrast, a circuit-switched service provides a continuous connection between two or more stations. Instead of using addresses, the connection is established based upon some prior agreement, which may have been negotiated using packet messages or established by some other suitable convention known to the stations involved. This prior agreement typically takes the form of knowing the location of a time slot, or slots, that occur regularly relative to a readily recognizable timing marker.

A common timing marker used in North America is the Basic System Reference Frequency (BSRF), a 125 μs

clock used by the public networks. Use of this clock is assumed for FDDI-II. In local FDDI usage, this is referred to as the cycle clock and is signaled by the JK starting delimiter of the FDDI-II cycle format.

In FDDI-II, a circuit-switched connection is described as N bits beginning at byte M after the cycle clock marker in Wideband Channel (WBC) number X. The last descriptor is necessary because FDDI-II has 16 WBC's that may be independently assigned to either packet-switched or circuit-switched data. This definition allows connections at data rates of all multiples of 8 kbits/s (i.e., $N = 1$) up to the 6.144 Mbit/s data rate of a WBC. If need be, multiple WBC's may be used to accommodate higher data rates.

The data transferred in a circuit-switched mode is best described as a stream of data. The data rate is appropriate to the service being provided—with, for example, 64 kbits/s being used for a digital voice data stream. Other data stream rates, even up to many Mbits/s in the case of video, are used for other applications. Once a connection is established, the data rate remains constant.

The contrasting nature of packet-switched and circuit-switched data is of interest. Most packet data traffic occurs in random quantities and at random times. This is referred to as asynchronous traffic. Other packet traffic, more regular in nature and occurring in relatively predictable quantities on a regular time basis, is referred to as synchronous (packet) traffic. In contrast, isochronous data occur in precise amounts on a precise time basis. They typically represent a sequence of digital samples from a sensor (e.g., voice or video). More importantly, isochronous data must by synchronized with clock information to ensure the accurate regeneration of the sampling clock (as distinct from the bit clock) to minimize distortion in data reconstruction. Isochronous data are more easily transferred in a circuit-switched network.

Networks that carry isochronous data must maintain precise synchronism with the cycle clock. For the FDDI ring, this means that one station (called the cycle master) must insert a delay for all isochronous data so that the ring appears to be an exact multiple of 125 μs in length. FDDI incorporates this delay in the cycle master in such a way that it does not cause any delay in the packet traffic. This is essential in providing an integrated services network with acceptable packet service.

VIII. FDDI-II Operation

Fig. 5, much like Fig. 1, shows how FDDI-II is implemented using one additional standard—Hybrid Ring Control (HRC). HRC becomes the new lowest sublayer of the data link layer, taking its place between MAC and PHY. HRC multiplexes data between the (packet) MAC and the isochronous MAC (I-MAC). This requires that the (packet) MAC be able to transmit and accept data on a noncontinuous basis because packet data are interleaved with isochronous data.

FDDI-II is a network with 100 Mbits/s of bandwidth available. This bandwidth may be devoted totally to op-

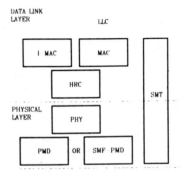

Fig. 5. FDDI-II relationship to OSI model.

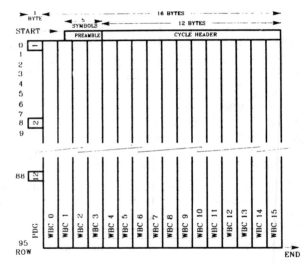

Fig. 6. FDDI-II cycle format.

eration as a packet network. Alternatively, portions of this bandwidth, in units of WBC's, may be dynamically separated for use with circuit-switched data. Up to 16 WBC's may be assigned. Each WBC is 6.144 Mbits/s, which is four times the North American and three times the European basic access rate to the telephone network. WBC's are full duplex and are independently allocatable and deallocatable. In effect, a broadband circuit capability has been provided with 16 available channels.

The WBC's provide a bandwidth division mechanism between the packet and isochronous traffic with a granularity of 6.144 Mbits/s. The allocation of virtual services within the isochronous traffic is allowed with an 8 kbit/s granularity. Once a station has been assigned a WBC or a number of WBC's, that station may suballocate the combined bandwidth of these WBC's as required. This suballocation may be in terms of any multiples of 8 kbit/s subchannels, including the commonly used 16, 32, 64 (B channel), 384, 1536, 1920, and 2048 kbit/s subchannels. Mixtures of these data rates in the same WBC are allowed. If preferred, the aggregate of any or all of the allocated WBC's may be used as one virtual service, satisfying the needs of such applications as high-resolution video. Thus, a multiplicity of virtual circuits may be provided within the same FDDI-II ring.

Assignment of all 16 WBC's, each at 6.144 Mbits/s, yields a total bandwidth of 98.304 Mbits/s. After allowance for the preamble and the cycle header, if all WBC's were allocated, a residual 768 kbit/s channel would then be left for packet traffic. This bandwidth, consisting of 12 bytes every cycle (125 ms), known as the Packet Data Group (PDG), is interleaved with the 16 WBC's as shown in Fig. 6. The order of transmission is left to right by row starting with the top row. Data steering logic in HRC augments the PDG with the bandwidth of any WBC's that are not assigned. Each WBC is one of the columns in Fig. 6 and represents a bandwidth of 6.144 Mbits/s or 96 bytes per cycle. This is an efficient system that allows the bandwidth of all unallocated WBC's to be used by the packet channel. Thus, for example, with eight (i.e., one half) of the WBC's assigned to isochronous service, 49.92 Mbits/s of bandwidth would be available for packet traffic.

A ring is initialized in basic (token) mode and switched to a hybrid mode of operation, combining both packet-switched and circuit-switched data capabilities, only after a station has negotiated for and won the right to be cycle master and has the synchronous bandwidth allocation required to support it. This cycle master then generates cycles at an 8 kHz rate (every 125 µs) and inserts the latency required to maintain an integral number of cycles synchronously on the ring. Alternative designs may allow the ring to be initialized directly in hybrid mode.

The cycle header format is shown in Fig. 7. It follows the Preamble (PA) which is nominally five symbols long. The Starting Delimiter (SD) is the same symbol pair (JK) used as the SD for frames when FDDI is operating in basic mode. In hybrid mode, frames use an IL symbol pair, readily recognized in the rigidly formatted context of hybrid mode, as their starting delimiter. The Cycle Sequence (CS) byte provides a modulo 192 cycle sequence count. The Maintenance Voice Channel (MVC) byte provides a 64 kbit/s voice channel for maintenance purposes.

The 16 symbols of programming information (P0–P15) determine whether the corresponding WBC (0–15) is allocated to packet or isochronous traffic. Thus, each of these (Pn) controls the multiplexing of one of the columns (WBCn) shown in Fig. 6. The C1 and C2 symbols are used for synchronization control in the transfer of this programming information to the other stations by the cycle master.

A. FDDI-II Priority Levels

Four kinds of traffic may coexist in an FDDI-II ring. Once WBC's have been allocated, the isochronous traffic within them has the highest priority.

Second highest priority is given to synchronous packet traffic where predictable units of data are to be delivered at regular intervals. Delivery is guaranteed with a delay not exceeding twice TTRT. These data may be transmitted following the capture of either a restricted or nonrestricted token.

CYCLE HEADER

| PA | SD | C1 | C2 | CS | P0 | P1 | ↗ | P15 | MVC |

Fig. 7. Cycle header format.

The bandwidth required for both isochronous and synchronous traffic is allocated from the available FDDI bandwidth. The allocation algorithm must ensure that the total allocation does not exceed 100 percent. Unallocated bandwidth is used on an as-available basis for asynchronous packet traffic.

Third highest priority is given to asynchronous traffic operating in restricted token mode. Such traffic may be transmitted upon the capture of either a restricted or a nonrestricted token. Cooperating stations may enter a restricted token mode of operation, which allows them to issue and use restricted tokens, only after negotiating an agreement using nonrestricted tokens. Restricted token mode operation allows stations to vie for available asynchronous bandwidth on a dialog basis.

Lowest priority is given to asynchronous traffic that may be transmitted only by capturing a nonrestricted token. This mode of operation allows stations to vie for the available asynchronous bandwidth on a single frame basis.

B. FDDI-II Applications

FDDI-II considerably expands the range of applications that may be addressed by FDDI rings. An FDDI-II ring may connect high-performance processors, mass storage systems, high-performance workstations, and perform the backbone function to a number of lower performance LAN's. The same ring may have some of its bandwidth allocated to isochronous services provided by the WBC's. This isochronous bandwidth may in turn be suballocated into a variety of virtual circuit services such as video, voice, and quite possibly, control or sensor data streams. The division of bandwidth between these two kinds of services may be adjusted based on the time of day or other requirements. In practice, no single instance of FDDI connects to all these types of equipment, and certainly not to all the equipment of a large site. Instead, multiple instances of FDDI-II rings will coexist with high traffic units, such as processors, attaching to multiple FDDI networks.

In a packet-switched mode, FDDI can be a backbone for bridges to a variety of other lower-speed LAN's, for example, the various IEEE 802 media access methods. It may also provide a backbone for gateways to the public data networks. In both cases, connections to processors and mass storage subsystems can be provided. Connections to high-performance workstations are likewise provided.

It is anticipated that FDDI-II implementations will follow those of the basic FDDI by one to two years. Committee X3T9.5 has specified the rules of coexistence of these two FDDI implementations to ensure interconnectability and interoperability of FDDI-II implementations operating in basic mode with the basic FDDI implementations. This also allows the use of FDDI-II chips for all applications once they become available.

IX. FDDI STANDARDS EFFORT AND STATUS

The FDDI standardization effort is taking place in the ASC X3T9.5 committee which meets bimonthly with an attendance of well over 100. The official voting membership, limited to one per corporation, is approximately 80. Ad hoc working meetings, scheduled as required, are presently focusing on SMT, FDDI-II, and SMF–PMD detail design definitions as well as issues concerning MAC-level bridging.

After completion by ASC X3T9.5, a draft proposed American National Standard (dpANS) is forwarded to ASC X3T9 for a technical letter ballot which approves forwarding to X3. X3 conducts a four-month public review, followed by an X3 letter ballot. Finally, the dpANS is forwarded to the Board of Standards Review (BSR) for formal approval. Upon approval, the standard is given a final editing to conform to ANSI requirements and published.

MAC (Rev. 10) is an American National Standard (ANSI X3.138-1987) and was published in July 1987. Current activities are focused on potential enhancements to MAC, including formalization of the 48-bit addressing structure used in IEEE 802 LAN's, reliability enhancement in the criteria for frame recognition, and incorporation of the FDDI-II requirements.

PHY (Rev. 15) is also an American National Standard (ANSI X3.148-1988) and was published in December 1988.

PMD (Rev. 8), X3.166-198x, completed the X3 four-month public review in late 1987. This initial forwarding had been intended to allow a full set of comments on the optical parameters documented in it. Five sets of comments received resulted in a number of refinements to the optical specifications of PMD. There were also a number of comments on the connector choice and its documentation. In April 1988, X3T9.5 finalized the optical specifications and reaffirmed its connector choice. PMD, having completed a second (two-month) public review in December 1988, is now being processed for final approval as a standard.

Work has progressed smoothly on the SMF–PMD project which was started in mid-1987. An X3T9 technical letter ballot approved the SMF–PMD document (Revision 4) and it has been forwarded to X3 for the four-month public review and final approval as a standard.

The technical definition of FDDI-II, long maintained in a working paper, has now been completed and is contained in Revision 5 of the Hybrid Ring Control (HRC) document. This document also passed an X3T9 technical letter ballot and has also been forwarded to X3 for public review and approval.

Participation in the working meetings on SMT has been high through much of 1987 and into 1989, with well over 100 attendees at the most recent meetings. Representing the meeting ground between the widely differing FDDI implementation and usage philosophies, SMT has proved a difficult task. Contentious items have included the use of concentrators without a MAC, link confidence test duration, physical layer link error monitoring, and physical layer signaling. But the committee's determination to produce an SMT stnadard assuring interoperability has prevailed and SMT is progressing toward a technical letter ballot in the latter part of 1989.

In a similar process designed to produce technically equivalent ISO standards for FDDI, the FDDI documents are being processed by ISO/IEC JTC1/SC 13. This process has resulted in the approval of MAC (IS 9314-2) and PHY (IS 9314-1) as ISO standards. PMD (DIS 9314-3) has passed the letter ballot for approval as an International Standard with publishing expected in late 1989. SMF–PMD and HRC are currently being processed for submission to ISO.

Led by the ASC X3T9.5, this FDDI standards effort has continued to reflect the FDDI product implementations of its many supporters, including a number of semiconductor manufacturers working towards FDDI chip sets. Several FDDI chip sets available either now or in the near future promise cost-effective implementations of these FDDI standards in the near future.

X. Conclusion

FDDI has proved a very successful standardization effort, far more successful than those who conceived it could ever have imagined. Users, system houses, chip designers, and component vendors have come together to produce a set of standards of lasting significance. With the widespread use of the basic FDDI will come a broad based market for FDDI-II which satisfies the much wider range of requirements typical of voice, video, and sensor/control based data streams.

References

[1] ANSI/IEEE Standard, 802.5-1985, "Token Ring Acess Method and Physical Layer Specifications."
[2] American National Standard, "FDDI Token ring Media Access Control (MAC)," ANSI X3.139-1987.
[3] American National Standard, "FDDI Token Ring Physical Layer Protocol (PHY)," ANSI X3.148-1988.
[4] Draft Proposed American National Standard, "FDDI Token Ring Physical Layer Medium Dependent (PMD)," ACS X3T9.5 Rev. 9, Mar. 1989 (ANSI X3.166-198x).
[5] Draft Proposed American National Standard, "FDDI Hybrid Ring Control (HRC)," ASC X3T9.5 Rev. 5, Apr. 1989.
[6] Draft Proposed American National Standard, "FDDI Token Ring Station Management (SMT)," ASC X3T9.5, Rev. 5, May 1989.
[7] Draft Proposed American National Standard, "FDDI Token Ring Single-Mode Fiber Physical Layer Medium Dependent (SMF-PMD)," ASC X3T9.5 Rev. 4, Apr. 1989.
[8] F. Ross, "FDDI—A tutorial," *IEEE Commun. Mag.*, May 1986.
[9] V. Iyer and S. Joshi, "FDDI's 100-Mbps protocol improves on 802.5 spec's 4-Mbps limit," *EDN*, May 2, 1985.
[10] F. E. Ross, "Rings are 'round for good," *IEEE Network Mag.*, Jan. 1987.
[11] M. J. Johnson, "Reliability mechanisms of the FDDI high bandwidth token ring protocol," *Comput. Networks ISDN Syst.*, vol. II, no. 2, pp. 121–131, 1986.
[12] H. Salwen, "In praise of ring architecture for local area networks," *Comput. Design*, pp. 183–192, Mar. 1983.
[13] R. M. Grow, "A timed token protocol for local area networks," presented at Electro'82, Token Access Protocol (17/3), May 1982.
[14] M. J. Johnson, "Fairness of channel access for non-time-critical traffic using the FDDI token ring protocol," in *Proc. Seminar Real-Time Local Area Networks*, Bandol, France, Apr. 1986, pp. 145–157.
[15] ——, "Proof that timing requirements of the FDDI token ring protocol are satisfied," *IEEE Trans. Commun.*, vol. COM-35, pp. 620–625, 1987.
[16] F. E. Ross, "FDDI—A perspective," in *1988 Fiber Optics Sourcebook*, 3rd annu. ed., pp. 193–210.
[17] M. J. Johnson, "Analysis of FDDI synchronous traffic delays," in *Proc. Syst. Design Networks*, Apr. 1988, pp. 65–72.
[18] ——, "Performance analysis of FDDI," presented at EFOC/LAN'88, June 1988.
[19] J. Hamstra, "FDDI design tradeoffs," presented at the 13th Conf. Local Comput. Networks, Oct. 1988.
[20] W. E. Burr, "The FDDI optical data link," *IEEE Commun. Mag.*, May 1986.

Floyd E. Ross received the B.A. degree in mathematics and physics from Mount Allison University in 1955 and did postgraduate work at McMaster University, Hamilton, Ont., Canada.

He is a Staff Engineer with UNISYS, Devon, PA, with responsibilities in communications networks and in LAN definition and implementation. He is an active contributor to accredited standards committee X3T9 on I/O Interfaces and to IEEE 802. He is Vice Chairman of X3T9.5 and has been the leader of the FDDI standards development effort since its inception in that committee. He has chaired many of the FDDI working meetings including those on SMT and FDDI-II and is technical editor of the four basic FDDI standards. He has also presented and authored a number of papers on both local area networks and FDDI.

The QPSX Man

R. M. Newman
Z. L. Budrikis
J. L. Hullett

Introduction

Public communications are catching up with demands set by information technology developments and will soon lead them. A number of standards forums, the CCITT, T1, the IEEE, the ISO and others are working on interrelated projects to define interfaces, networks and protocols that promise to make data communications over a wide spectrum as ubiquitous and as universally switchable as the telephone. More than that, from the user's prospective all communications will be fully integrated flowing over the one interface and access loop, with telephone just one of the many services.

The CCITT has already defined the integrated services digital network (ISDN) and is defining the broadband ISDN, or B_ISDN. T1 is defining a synchronous optical network (SONET), capable of carrying the large traffic within the network. ISO is standardising protocols and procedures so that different equipments can communicate with each other. The IEEE in P802 is bridging the gap between private and public networks. It is defining local area networks (LANs) suitable for customer premise or campuses, and metropolitan area network (MAN) that will interconnect LANs and provide other integrated services as part of the public network. In fact, the IEEE defined MAN can be a significant stepping stone on the path to a full B_ISDN.

The IEEE MAN is being defined in 802.6, a Working Group of P802. The project is actively supported by Telcos which gives assurance that it will measure up to public network requirements. A technology proposal that has particularly wide Telco support is one put forward from Australia. It is the Dual Bus QPSX MAN [1],[2].

QPSX (the queued packet and synchronous switch) is a distributed switch/network that will fulfill the requirements of a public MAN. The switch architecture of QPSX is based on two contra-directional buses as shown in Fig. 1. This allows full duplex communications between each pair of nodes of the switch. The communications on QPSX are integrated, providing isochronous (circuit switched) and non-isochronous (packet switched) services. The two switch modes function independently and the total capacity of the switch is shared flexibly and completely between them.

The bus architecture of QPSX is inherently highly reliable. The operability of the network is to all intents independent of the operation of the individual nodes. Furthermore, fault handling mechanisms have been developed that permit the isolation of potentially catastrophic faults with no service degradation in the case of a single failure, and minimal disruption even with multiple failures.

The access of packets to the buses of QPSX is controlled by a unique MAC (media access control) protocol referred to as distributed queueing [3]. The distributed queueing protocol provides near perfect access characteristics independent of network size and speed. In particular, the protocol supports 100 percent throughput and can provide average packet access delay equal to that of a perfect scheduler, the ultimate queueing system.

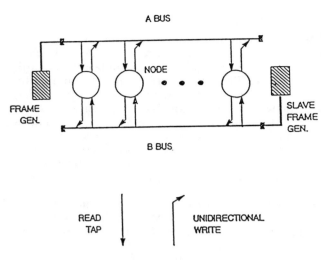

Fig. 1. QPSX dual bus architecture.

The operation of the protocol is based on two control bits, the busy bit which indicates whether a slot on the network is used and the REQ (request) bit which is sent whenever a node has a packet waiting for access. Each node by counting the number of REQs it receives and non-busied packets that pass it, can determine the number of packets queued (i.e., in line) ahead of it. This counting establishes a single ordered queue across the network for access to each bus. With such queued access, levels of priority can be established by operating a number of queues, one for each level. Within each level the perfect performance characteristics are maintained: the packets will gain access as soon as capacity becomes available, but priority is always given to packets in higher level queues.

QPSX as a Public Network MAN

The P802 forum started with development of LAN standards and much of the MAN work has evolved from this base. However, the MAN has very much greater functionality and a different role. The increased functionality includes the ability to cover much larger geographical areas, to operate at much higher speeds, and to provide switching for a mix of traffic services. These capabilities take the MAN well beyond the realm of in-house or building networks and into the domain of the public network.

QPSX MAN is particularly well suited to application in the public network. It has the necessary features of outstanding reliability and independence of network size and speed, to meet the requirements of all public network services. As well, it has the features, such as security, network administration, bandwidth control and charging facilities, which are also essential in the public network. A further advantage in relation to its application in the public network is that it has a frame structure that will be compatible with the evolving broadband ISDN. Hence, QPSX can be seen as an evolution path to this future public network technology.

The Dual Bus Architecture of QPSX

The principal components of the QPSX architecture are a head station, two unidirectional buses, and a multiplicity of access units (AUs). These are configured in the unidirectional bus structure shown in Fig. 1. The head station generates the frame synchronization on the forward bus and the end station generates the frame pattern at the same rate on the reverse bus.

Access units attach to both buses via read and write connections. The writing on to the bus is by logical OR of the data from upstream with the data from the unit. The read connection is placed ahead of the write connection and allows all data to be copied from the bus unaffected by the unit's own writing. The digital bus implementation of the AU connection is shown in Fig. 2.

The bus architecture of QPSX has been chosen for its reliability. As can be seen from the Fig. 2, the data bus passes directly by each node in the network. Data does not pass through the node as is the case in ring networks. Hence, in QPSX, nodes can fail or even be removed from the network with no consequence on the operation of the rest of the network. The network does not even need to reconfigure in this case.

The only fault conditions in QPSX that possibly may affect the operation of the network are a "jabbering" node or failure of the bus. The first fault is handled by a fault detection and isolation scheme which may be instituted, again without causing reconfiguration of the network. Failure within the bus is the only condition handled by network reconfiguration. The QPSX reconfiguration scheme is simple and very fast acting.

The Looped Bus Architecture

An extension of the dual bus architecture is the looped bus shown in Fig. 3. The only change in this from the original architecture is that the end points of the buses are co-located. Data do not flow through the head point of the loop. Thus, while the looped bus architecture may appear similar to a ring, it has no log-

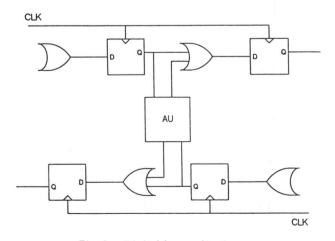

Fig. 2. Digital bus realization.

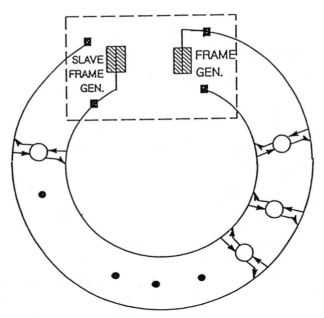

Fig. 3. QPSX looped bus architecture.

ical similarity to it. Hence, it provides all the reliability advantages associated with the QPSX bus architecture.

There are two purposes in looping the QPSX bus architecture. One is that the framing generator for the two buses is common and this point provides the entry point for the PSTN clock into the network. The second and more important reason for looping the buses is that it provides an automatic route diversity within the switch. In the case of a bus fault, the network can isolate the fault and close the data buses through the head point of the loop. Hence, the network is reconfigured and fully operational in the reconfigured state. This is achieved without the need for any redundant transmission hardware, which is in direct contrast with ring networks that require a second redundant ring to institute network reconfiguration.

The Architecture of the Public MAN

The role of the QPSX MAN in the public network is to provide interconnectability for a range of communication services not yet provided. The geographical scope of the MAN is such that it is planned to cover complete cities. In practice it is most sensible if the MAN covers an area that corresponds to the telephone number plan area, NPA. While the dual bus architecture of QPSX is such that a single network would operate over very large geographical areas, it will not be long before the capacity requirements of the attached customers will exceed 2×150Mbits/s capacity of the network. In fact the number of networks required in large cities could well approach the hundreds and even thousands. As a result, there is a need for bridged interconnection of networks to form a single large MAN.

In order to permit a manageable and efficient interconnection of the networks within the MAN, a simple

interconnection architecture and bridging algorithm is required. The architecture of the public network based on QPSX is very similar to that of the public telephone network and is illustrated in Fig. 4. It is based on a hierarchy with possible cross links between switches at a common level of the hierarchy. The cross links provide route diversity in the case of failure and also provide a short path between switches which have a high degree of cross traffic.

At the lowest level of the MAN hierarchy is the access network, shown as level one in the Fig. 4. This level provides for the customer's connection into the public network. The protocol operated over the access network is the same QPSX protocol. The access network will go from the customer to the nearest site owned by the public network provider. In terms of utilizing the cabling efficiencies associated with distributed switching, if the network provider has space leased at sites away from the central office, then connection to the public network switch will be at those points. Otherwise, all connections to the public switch will be in the central office. In this latter case, the length of the access network loops generally will be longer.

The point where the access network reaches the office level network is the edge of the public network. This point, called the network connection point (NCP), is responsible for maintaining and administering the customer's connection. The functions performed at NCP are illustrated in the Protocol architecture diagram, Fig. 5. There are two principal sets of functions performed at the NCP. These are routing and edge-to-edge services. The routing layer performs the function of converting the destination E.164 address indicated in packets sent by the customer to determine where the packet is to be routed. The routing layer will translate that address into an internal address used within the QPSX MAN. The edge-to-edge

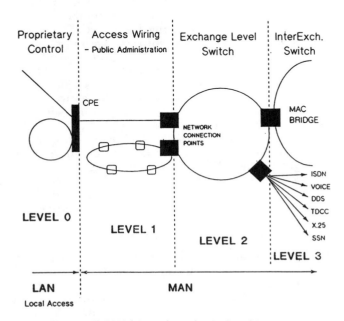

Fig. 4. QPSX hierarchy (physical architecture).

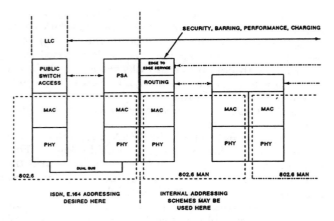

Fig. 5. Protocol architecture.

services layer will perform functions related to the network provider's management of the customer's connection, such as security provision, charging, performance monitoring barring, etc.

The NCP represents the point where the customer is physically and logically connected to the switch hierarchy of the QPSX MAN. The switch hierarchy consists of office level switches at the lowest level, interoffice switches at the next higher level and so on upwards to whatever depth is required. All switches are interconnected by bridges which can provide up to 150Mbits/s throughput. The same system hardware can be used at all levels of the hierarchy, though ultimately higher rate switches will be required at the higher levels. Nodes at all switch levels are connected to the QPSX bus at cluster points, and clusters may be either widely distributed, or adjacent thus forming a centralized switch realization.

Communications across MANs will be routed through the source MAN to a gateway to the wide area network (WAN). This gateway will again perform the necessary routing layer functions using the E.164 address contained in the packet to determine the path through the WAN to the destination MAN. It is not intended, at least in the first realizations of QPSX, for synchronous communications to be carried in the WAN or even in the interoffice switches of the QPSX hierarchy. Rather this traffic will be given over to the existing switch services provided for it.

The routing within the MAN is based on a simple protocol which permits very high throughput in the bridges of the MAN. In fact, a bridge can switch between two networks up to the full capacity of the QPSX dual bus. Route diversity within the bridging algorithm is also provided to circumvent bridge faults.

Frame Format and Synchronization

Communications on the QPSX buses are within the frame format shown in Fig. 6. The frame interval is 125 microseconds, matching that of digital telephony. The frame is subdivided into a fixed number of equal size units called slots. Slots provide the subdivision between the isochronous and non-isochronous traffic.

A slot may be allocated to contain either type of communication. Slots are allocated to isochronous use commensurate with the demand for isochronous capacity. The slots not allocated to isochronous use are available for packet communications.

For isochronous communications, each octet within a slot provides a 64kbit/s channel. By using multiple octets per frame higher rate circuits can be constructed. By the use of a multi-frame count in the frame header, circuits at sub-64kbps can also be constructed.

For non-isochronous communications, a slot once claimed for access by a node, is used to transfer a single segment of a packet. Access to slots is handled by the Distributed Queueing protocol. The segmentation of packets into slots and the reassembly of the packet at the destination is by a source-identifier-linked segmentation and reassembly protocol. Both protocols are discussed in later sections.

The frame structure for QPSX is very similar to that which is currently under consideration in CCITT as the structure for B-ISDN. It has already been decided that the B-ISDN frame will consist of fixed sized cells, these cells being a similar concept to QPSX slots. In the standardization effort of QPSX, it is expected that slot size and slot header size will be aligned with that adopted by CCITT for B-ISDN cell size and cell header. Further levels of compatibility such as the format of the header will be a future issue. However, it is expected that IEEE 802.6 will develop a MAN standard well in advance of the B-ISDN standard by the CCITT.

The Distributed Queueing Protocol

Distributed Queueing is a media access control (MAC) packet switching scheme that controls the access of fixed length data segments to the slots on the QPSX bus. The protocol overcomes many of the limitations of other media access protocols. The maximum throughput of the network is not constrained by network speed or size and the protocol operates on the highly reliable dual bus architecture.

The operation of Distributed Queueing is fundamentally different from all existing MAC protocols. In all other access schemes, no continuous record is kept in the nodes that explicitly indicates the state of the network. Hence, when in these protocols a node has a packet to transmit, it must first derive the information from the network as to when to access. This leads directly to a sensitivity in almost all protocols to the size of the network. The larger the network, the longer the delay in obtaining the access information. This limits the size over which the protocols may operate efficiently.

By contrast, with Distributed Queueing a current state record is kept in every node which holds the number of segments awaiting access to the bus. When a node has a segment for transmission, it uses this

Fig. 6. Frame format for a QPSX man.

count to determine its position in the distributed queue. If no segments are waiting, access is immediate, otherwise deference is given only to those segments that queued first.

With the queueing of segments for the bus, slots are never wasted. This guarantees minimum access delay at all levels of loading right up to the maximum network utilization of 100 percent. This performance is achieved with negligible control overhead, in fact only two bits per slot, and is effectively independent of network size and speed.

The Basic Distributed Queueing Algorithm

The operation of the basic Distributed Queueing algorithm is particularly simple, as illustrated in Fig. 7. The protocol uses just two control bits, busy and REQ, to control access to the slots on the forward bus. An identical, but independent arrangement applies for access to the opposite bus.

The Busy control bit is a marker at the head of each slot which indicates whether the slot is full and hence not available for access. The REQ control bit is the crux of the queueing mechanism and is used to signal when a segment has queued.

When a node has a segment for transmission on the forward bus it will issue a single REQ bit on the reverse bus. This bit will pass to all upstream nodes, where upstream is defined in relation to flow on the forward bus. This REQ bit serves as an indicator to the upstream nodes that an additional segment is now queued for access.

Each node keeps track of the number of segments queued downstream from itself by counting the REQ bits as they pass on the reverse bus, as shown in Fig. 8a). For each REQ passing on the reverse bus, the request (RQ) counter is incremented. One REQ in the RQ counter is canceled each time an empty slot passes on the forward bus. This is done since the empty segment that passes the node will be used by one of the downstream queued segments. Hence, we see that with these two actions the RQ counter keeps a precise record of the number of segments queued downstream.

When a node has a segment for access to the bus, it transfers the current value of the RQ count to its second counter, the countdown (CD) counter. This action loads the CD count with the number of downstream segments queued ahead of it. This, along with

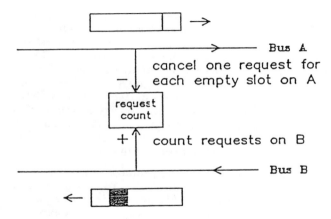

Fig. 8a). *Distributed queueing (Bus A). Node not queued to send.*

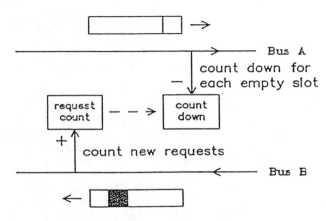

Fig. 8b). *Distributed queueing (Bus A). Node queued to send.*

the sending of the REQ for the node's segment, effectively places the segment in the distributed queue.

To ensure that the segments registered in the CD counter gain access before the newly queued segment in the given node, the CD counter is decremented for every empty slot that passes on the forward bus. This operation is shown in Fig. 8b). The given node can then transmit its segment in the first empty slot after the CD count reaches zero. The claiming of the first free slot ensures that no downstream segment that queued after the given segment, can push in and access out of order.

During the time that the node is waiting for access for its segment, any new REQs received from the reverse bus are added to the RQ counter, as shown in Fig. 8b). Hence, the RQ count still tracks the number of segments queued downstream and the count will be correct for the next segment access.

Thus, with the use of two counters in each node, one counting outstanding access requests and the other

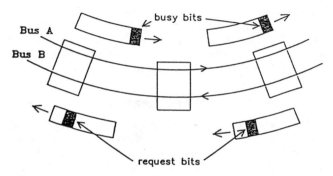

Fig. 7. *Distributed queueing. Queue formation on Bus A.*

counting down before access, a first-in-first-out queue is established for access to the forward bus. The queue formation is also such that a slot is never wasted on the network if there is a segment queued for it. This is guaranteed since the CD count in the queued nodes represents the number of segments queued ahead. Since at any point in time one segment must have queued first, then at least one node is guaranteed to have a CD count of zero. It is that node that will access.

Priority Distributed Queueing

The Distributed Queueing protocol can be extended to permit the assignment of priority to segment access. The priority access control is absolute in that segments with a higher priority will always gain access ahead of segments at all lower levels. This is achieved by operating separate distributed queues for each level of priority.

The operation of separate distributed queues for each level of priority consists of the use of a separate REQ bit on the reverse bus for each level of priority and separate RQ counters for each priority level. The counters operate much as before, except that account must be taken of REQs at higher levels. That is, a RQ counter operating at a particular level will count REQs at the same and higher priority levels. Thus the RQ count records all queued segments at equal and higher priorities.

The operation of the CD counter is also slightly altered. A CD counter operating at a particular priority level will, in addition to counting down the passing empty segments, be incremented for REQs received at higher priority levels. This allows the higher priority segments to claim access ahead of already queued segments.

With such a priority scheme, the access performance for high priority traffic is ideal. Immediate access can be gained without any delay due to already queued traffic. Such a feature is very important for network signaling. It would also be important for packet voice, should that be implemented.

Performance

The operation of the Distributed Queueing protocol provides two key advantages over the operation of other existing protocols. One is that, through its use of REQ counters, an ordered queue is established for access where capacity on the network is never wasted. The second advantage is through the use of segmented access, which allows finer granularity and control in the allocation of the bandwidth resource.

The access performance advantage for the Distributed Queueing protocol is best demonstrated by the graph in Fig. 9. This figure shows the access delay for packets on a typical MAN. To represent the delay best it has been normalized to the average length of the packets. Also, shown on the figure is the token ring protocol. Distributed Queueing has a very clear advantage in terms of access delay.

The advantage demonstrated in Fig. 9 is even further exaggerated when either the speed or the size of the network increases. The normalized delay performance for the Distributed Queue is unaffected by these network parameters, whereas the token ring normalized delay will continue to increase with the network size and speed. The increase in delay is clearly undesirable.

The limitation of the token ring is the delay in the transfer of the token. Even at very low utilization, a node wishing to access must wait for the token to pass to it. On average this delay is at least half of the ring latency at low loading and increases beyond this as load increases. Such a delay eventually limits the usable size of a token ring network.

The second performance advantage associated with distributed queue access is that of segmentation. Packets sent over the QPSX network are not sent as a whole but rather in segments to suit the slot size on the bus. Segmented transfer versus whole packet transfer is much like the comparison of time shared processing versus batch processing.

With whole packet transfer and likewise batch processing, the single processor is allocated to serve each packet unit at a time. Consequently, very large packets cause long delays for all other packet transfers. Most affected by this are short packets and high priority packets. Both must wait while the large packet completes access.

By contrast to this operation, with segmented access such as in the distributed queue the access is shared on a segment by segment basis. This means that even during the transfer of a very large packet, higher priority packets need only wait at most one segment before gaining access. Likewise, other packets at the same priority will be able to obtain immediately their share of the capacity. Demonstration of the access advantage for the short packets is illustrated also in Fig. 9.

Thus, segmentation provides a much finer granularity for bandwidth sharing and aligns communications switching with the performance expectations now commonly accepted in the computer industry.

The Segmentation and Reassembly Scheme

We have noted the performance advantages associated with segmented access of packets. Now we describe the scheme used to control the segmentation of packets at the source and the reassembly at the ultimate destination.

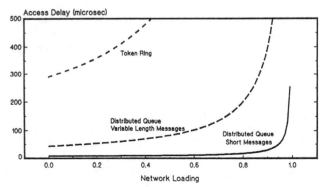

Fig. 9. Delay for distributed queueing and token ring (100 Mbits/s, 100km).

The segmentation process is straightforward and involves the fragmentation of the original packet to match the slot size on the bus. This process is shown in Fig. 10, where the segments of the packet are formed. This includes also the forming of a packet header segment which is necessary in routing the packet and instituting the reassembly process.

All segments consist of a header field along with the actual data as shown in Fig. 11. The header field consists of two sub-fields. The first is a single bit, the C/D subfield, which is used to indicate whether the segment is a control or a data segment. The second field is the source identifier (SI). This is used to provide the logical linking between slots of the same packet.

The SI is a fifteen bit field. Each node in a QPSX dual bus will have one or more unique SI number(s) related to its physical location along the bus. The SI is in addition to, and independent of, any network layer addressing. The allocation of the SI numbers is locally administered and only has significance within a single network. The purpose of the SI is to identify segments of a single packet. This information is used to reassemble the segmented message at the destination. To describe the operation of the packet transfer scheme, the segmentation of the packet at the source is described first and the action of the receiver is considered later.

The train of segments sent by the source is shown in Fig. 10. The first segment of a multi-segment message is a control segment. Within this segment a control field immediately after the segment header will indicate the beginning of message (BOM) code. This code signifies the start of a new packet transfer. The SI field of this segment is set to the source's value and the information field of the segment contains additional control information amounting to a packet header. Part of this information is routing information required to transfer the packet to the destination. All routing within the MAN is based on physical access unit addresses, not on the network layer addresses. Also, contained within the information field of the first segment is the length of the packet.

All segments of the packet until the last are each placed in the information fields of slots following the

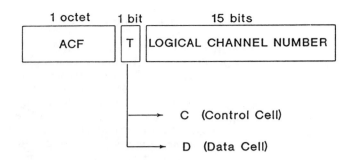

Fig. 11. QPSX segment header.

first control slot. In the header of each of these slots the C/D bit is reset to indicate data and the SI field is the same value as was sent in the first or control segment. The transfer of the multi-segment packet is completed by sending a further control slot which contains the end of message (EOM) code in the control field after the segment header.

For the transfer of a message that only requires a single segment, the SSM (single segment message) code is used in the control field of the first segment. The SI is not required in this case, however it is still used for consistency in operation.

To receive the packets segmented as described above, each AU will implement the state machine in Fig. 12. This state machine will handle the receipt of a packet from a single source at a time. To handle concurrent receipt of more than one message, multiple state machines and receiver buffers are used in the AU. In the diagram, the condition for a state transition is shown at the start of the transition line and the action taken is at the arrowhead of the transition line.

The receive machine consists of two states: IDLE and REASSEMBLE. In the IDLE state the machine is not currently receiving any message. Thus, in this state the machine will check for control slots with the control field equal to BOM or SSM. In the case a BOM code is received the machine will check the destination identifier in the first part of the information field. If the destination identifier is its own, then this indicates the packet is intended for that station and the state machine enters the REASSEMBLE state. In the case that SSM is detected in the IDLE state and the destination identifier field matches, the length and information fields are copied and the AU will indicate to the higher layers that a packet is received. The receive

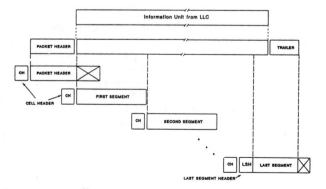

Fig. 10. Packet segmentation scheme.

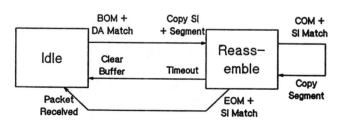

Fig. 12. Reassembly machine.

machine remains in the IDLE state after copying a SSM segment.

The REASSEMBLE state is used to receive the segments following the BOM control segment of a multi-segment message. In this state, the segments with the SI equal to that copied from the BOM segment will be received. The information fields of these following segments are appended together to form the complete packet. New packets addressed to the given station in this state are ignored by the state machine. Other idle receive machines are used to receive such messages. When the EOM segment of the packet is detected by the receive state machine, again with the same SI, the machine will pass the reassembled packet to the station and return to the IDLE state. This completes the receipt of the packet.

To guard against the loss of the EOM segment, which would cause the receive state machine to be locked in the REASSEMBLE state, a timer is used. This timer is started after each segment is received. If the timer expires before the next segment is received, a failure in the transfer is assumed. The machine will then clear all copied slots and return to the IDLE state. Loss of segments of the packet can be detected by comparing the length field given in the BOM slot with the number of slots received.

The SI can be reused by the source for its next packet transmission immediately after the EOM segment is sent. Even if the destination of the first packet did not receive its EOM slot, the immediate reuse of the SI does not present a problem. The BOM issued for the new packet can serve to terminate the receipt of slots by the first destination. As a further safeguard, knowledge of packet length can prevent the receipt of further segments not intended for the first destination.

Stations that need to send more than one message concurrently, such as bridges, may be allocated more than one SI. The number of packets that may be sent concurrently equals the number of SIs allocated.

Reliability Features of the QPSX Network

The reliability of QPSX stems both from its architecture and its Distributed Queueing medium access control protocol or MAC. There are two reasons for the inherent reliability of the bus architecture. First, the network nodes are logically adjacent to the bus not serially and logically connected as in a ring. As a consequence data do not pass through the nodes and passive MAC failures have no effect. Secondly, because the node writing function is by logical "OR," i.e., there is no

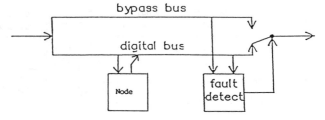

Fig. 13. Node fault detection.

NORMAL

PHYSICAL FAILURE

HEALED

MINIMAL SERVICE DISRUPTION

Fig. 14. Healing mechanism. Physical breaks are healed by repositioning the natural break in the loop.

overwriting, active MAC failures and bus failures are easily detected by input to output coincidence checking. An arrangement to implement this is shown in Fig. 13. Such failure points when they occur are bypassed at the node. The operation can be totally hardware based, with immediate response and causes no loss of synchronism, and minimal network disruption. With QPSX no node monitoring functions are required to check the MAC.

In this context a comparison with ring based networks is pertinent. With rings [4] both active and passive node MAC failures are catastrophic in that both invoke healing mechanisms associated with redundant rings. Such action involves considerable network disruption because of the need to reconfigure with attendant loss of synchronization. Any additional node failure can then produce isolated network islands.

With the QPSX architecture those faults involving node failure do not need to invoke any network healing mechanism which reconfigures the network. Faulty nodes are easily and non-disruptively bypassed.

Failures in the physical transport however do involve reconfiguration. But on healing the network captures the same synchronization and therefore there is minimal disruption to all synchronous communications. The healing mechanism is depicted in Fig. 14 and shows how physical faults or line breaks are healed by repositioning the natural break in the loop to the position of the break due to failure. Significantly there is no loss of network capacity involved in network reconfiguration. Here again the contrast with ring type networks is sharp. If a ring network carries traffic on a back-up ring then network reconfiguration, due to either node or transport failure, introduces a 50 percent reduction in packet data capacity. Alternatively if the back-up ring carries no data traffic, there is always a 50 percent inefficiency factor relative to the QPSX bus network. This is a critical factor in relation to cost efficiency, more particularly so in the public environment.

With QPSX it takes a real disaster, such as a backhoe through a physical transport, to invoke network healing. As explained, node failures do not invoke reconfiguration since faulty nodes are excised using very simple low complexity and low cost hardware. With rings however node failure invokes global healing.

Summary

QPSX is being developed as a public metropolitan area network and is in the IEEE Standards arena for 802.6 MAN Standardization. It has the inherent robustness and reliability necessary in a public network and it is being developed with management and healing features that materially enhance that reliability.

Next after reliability, an important issue is performance and in this respect QPSX is again quite outstanding. It uses a unique distributed queueing access control that permits fair access with the smallest possible delay and up to 100 percent loading. Moreover, the performance is independent of network rate or size.

The basic QPSX subnetwork is a pair of oppositely directed buses with station, bridge and gateway access units attached to both. Bus rates are up to 150 Mbits/s, with the streams divided into fixed length slots recurrently at the frame rate of eight kilohertz. Slots can be reserved for synchronous communications and those not reserved are available for packet-switched transfers.

A metropolitan area network will consist of many, possibly hundreds or even thousands, of QPSX subnetworks. These will be interconnected in an approximately hierarchical manner using bridges. MANs will be interconnectable to each other via gateways over a Wide Area Network. Customer premises networks or LANs will connect to a MAN also via gateways. Routing over the MAN and WAN will be by physical addresses of gateways, bridges and stations, related to E.164 addresses. Data packets up to eight kbytes in length will be transferred in segments over virtual circuits, a circuit normally existing only for the duration of a packet. Thus, with respect to packets the network will give a datagram or connectionless service. A connection-orientated packet service will also be possible.

Data transfer in QPSX resembles closely the transfer mode that has been decided by CCITT SGXVIII for B_ISDN making alignment of the two standards feasible. Complete alignment requires that the slot or cell as well as the segment or cell header sizes be equal in the two and that the headers encompass the product set of functions, expressed by identical codes and formats. Thus aligned, QPSX MAN would greatly speed the advent of B_ISDN.

References

[1] Draft proposed, IEEE Standard 802.6 metropolitan area network (MAN), QPSX dual bus media access control and physical layer protocol documents. Submitted July 1987.

[2] Z. L. Budrikis, J. L. Hullett, R. M. Newman, D. Econo-
mou, F. M. Fozdar, R. D. Jeffery, "QPSX: A Queued Packet and Synchronous Circuit Exchange," *Proc. 8th Int. Conf. Comp. Comm.*, Munich, 1986, pp. 288-293.

[3] R. M. Newman, J. L. Hullett, "Distributed Queueing: A Fast and Efficient Packet Access Protocol for QPSX," *Proc. 8th Int. Conf. Comp. Comm.*, Munich, 1986, pp. 294-299.

[4] F. E. Ross, "FDDI—a tutorial", *IEEE Comm. Mag.*, vol. 24, no. 5, pp. 10-17, May 1986.

Robert M. Newman was born on September 15, 1963. He received his Bachelor of Engineering with first class honors from the University of Western Australia in 1985, and has completed his Ph.D. research studies at the University of Western Australia. In both or these degrees Mr. Newman was involved in research work that led to the initial concept of QPSX switching and has also performed detailed modelling of the switch's operational characteristics.

Since April 1987, Mr. Newman has been employed as Projects Director, Network Systems for QPSX Communications Pty Ltd. In this role he has been responsible for the development of the initial QPSX switch products and has been actively involved in the standardization of dual bus technology in the IEEE 802.6 Committee. Mr. Newman is the co-inventor of the core QPSX technology.

Zigmantas L. Budrikis is a Director of QPSX Communications Pty Ltd and is on the Electrical Engineering Faculty in the University of Western Australia. He was born in Lithuania on September 29, 1932. He graduated B.Sc. majoring in physics in 1955 and B.E. with first class honors in electrical engineering in 1957, both from the University of Sydney, and Ph.D. in 1970 from the University of Western Australia.

Dr. Budrikis worked in the research laboratories of the PMG's Department (now Telecom Australia) from 1958 to 1960 and in Australia's Aeronautical Research Laboratories in 1961. Since 1962 he has been on the academic staff of the University of Western Australia. He was a visiting lecturer at U.C. Berkley in 1968, a visitor at Vilnius University at Technical University Munich in 1976-1977, an MTS at Bell Laboratories Holmdel in 1972 and 1974, and a consultant on four separate occasions at AT&T Bell Laboratories at Holmdel and Murray Hill over the period 1981 to 1985. He has worked extensively on problems of TV signal coding, human vision modeling, electromagnetic theory foundations, data communications and computer graphics. He is author of some forty papers and holder of several patents.

Dr. Budrikis is a Fellow of the Institution of Engineers, Australia and a member of IEEE, the New York Academy of Science, and the Optical Society of America.

John L. Hullett was born in Perth, Western Australia on December 20, 1942. He received his Bachelor of Engineering with first class honors in electrical engineering in 1965 and his Ph.D. in 1970 from the University of Western Australia.

Dr. Hullett worked in the research laboratories of Telecom Australia from 1969 to 1972 where from 1971 he was head of the Data Systems Division. Since 1972 he has been a member of the academic staff of the University of Western Australia and a member of the faculty of engineering. He has worked extensively in digital, data and optical communications, and networking and has published some forty five papers in these fields. He is a member of the IEEE.

Dr. Hullett is the co-inventor of the core QPSX technology and acts now as the Director of Strategic Planning and the Technical Division of QPSX Communications Pty Ltd. ∎

Switched Multi-megabit Data Service and Early Availability Via MAN Technology

Christine F. Hemrick
Robert W. Klessig
Josephine M. McRoberts

Reprinted from *IEEE Communications*, Vol. 26, No. 4, April 1988, pages 9-14. Copyright © 1988 by The Institute of Electrical and Electronics Engineers, Inc. All rights reserved.

The widespread penetration of local area networks (LANs) has firmly established very high-speed data networking (1–10Mbits/s) in the local premises environment. The success of LANs has been synergistic with the rapidly growing use of more powerful PCs and workstations for high performance applications such as remote file serving and distributed processing. These market and technology trends portend an emerging market for multi-megabit communications services that extend beyond the local premises, across metropolitan and wide area environments.

This paper discusses work currently proceeding at Bell Communications Research in the field of novel high-speed data services and supporting network requirements [1]. The service described herein, switched multi-megabit data service (SMDS), is one which could be offered by carrier networks to meet emerging customer high-speed data needs. The use of Metropolitan Area Network (MAN) technology to provide high-speed packet switching in carrier networks appears to provide a promising opportunity for early availability of SMDS.

In this article there follows a section providing an overview of the motivations behind the service development, followed by a discussion of the importance of the embedded customer base and the factors that significantly influence the nature of the service. Next there is a description of specific SMDS features and characteristics. Finally, there is a discussion of the use of MAN technology for the support of SMDS.

Motivations for SMDS

Multi-megabit data communications have rapidly penetrated the workplace because of two complementary technology trends: 1) the availability of local area network (LAN) technology that provides high capacity, low latency performance and a low per-port interconnection cost; and 2) the dramatic, continuing decline in the cost of powerful computing devices.

These trends have led to the development of distributed processing techniques, providing real-time, shared access to files and executable software. The growth of distributed processing applications, such as desktop publllishing and CAD/CAM (which, prior to these technology advancements, was restricted to "stand-alone" type configurations) is expected to continue, due to the real benefits they bring to users in the areas of productivity, management and control, and "user-friendliness." However, the key to user acceptance and satisfaction with such systems is the very high-speed communication between devices that allows functions to be performed either locally or remotely without the user perceiving a difference in performance.

The need to respond to these market and technology trends has provided the principal motivation behind our work to define SMDS and the means for its early availability. The initial market for wide-area multi-megabit data communications is emerging now among larger and more sophisticated business and public sector customers. Thus, there is a need to make SMDS available early to meet key customer demands. For early deployment, the use of MAN technologies appears to offer a promising approach. The MAN trials recently conducted by NYNEX/New England

Telephone [2] and U.S. West/Northwestern Bell [3] have demonstrated the market and technical feasibility of delivering high-speed data service by employing MAN technology in the carrier network environment. While early availability is a key objective, attention must also be paid to the continuing growth in the wide-area, high-speed data market that is expected to occur in concert with progress toward the addition of Broadband capabilities for multiple services (for example, high quality video, high-speed data, etc.) into carrier networks. Thus, SMDS is also planned to be an initial offering of "multi-service" Broadband carrier networks. In fact, a key aspect of the service is that it has been defined independent of the underlying architecture and technology used to support it in the network. This is essential so that customers do not incur service discontinuity as the supporting network evolves from the use of early availability technology for SMDS to the introduction of technology supporting multiple Broadband services.

In addition to being a timely service concept from a market perspective, SMDS is intended to take advantage of two of the carrier networks' unique strengths: the wide-spread, rapidly growing deployment of single mode optical fiber and an extensive operations infrastructure. Use of this fiber to support SMDS alongside existing services has the potential for significant economies, over the situation where facilities must be dedicated to a single customer or service. The existence of the operations infrastructure is the basis for the economical provision of quality service, consistent with carrier network standards and customer expectations.

The Importance of the Target Customer Environment and Factors Influencing Service Characteristics

The potential SMDS customer is likely already to be a user of high-speed data communications on the local premises and will already have a substantial investment in hardware (hosts, terminals, workstations, LANs), software, applications and operations support. This substantial investment by the customer dominates the cost of related communications facilities and services. To introduce successfully a new data communication service to such customers, it is absolutely essential that the service not require substantial changes in the customer's existing communications environment. Rather, the service must be compatible with, and enhance the value of, the customer's embedded base. These are key factors that have influenced our service decisions.

Figure 1 represents a typical scenario in which SMDS might be used. The typical SMDS customer environment is likely to contain multiple networks, which may be either closely located or geographically separated. Each network constitutes a "subnetwork" in the customer's overall multiple network configuration. Typically today, such subnetworks will include local area networks, sometimes interconnected using wide-area subnetworks such as leased lines or X.25 packet-switched networks. Multiple LANs are usually inter-

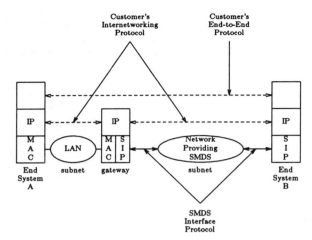

Fig. 1. *Example scenario of the use of SMDS.*

connected by means of "bridges" or "gateways." To achieve integration of the subnetworks into a larger internetwork, the customer equipment implements internetworking and transport protocols above the subnetworks in "gateways" and end systems. Commonly, the internetworking protocol will operate in a "connectionless" fashion, while the transport protocol will provide end-to-end reliability and control functions. Taken together, these provide a robust, error-controlled, flow-controlled service to the customer's distributed systems applications, which run over these protocols.

The most popular such protocol coupling, used for network interconnection, is one based on the DARPA-developed[1] internetworking protocols, popularly known as TCP/IP[2] [4], [5]. The TCP/IP protocol suite has long been the de facto standard for use in university and scientific and engineering environments, and in the Department of Defense networks for which it was originally developed. More recently, it has become an important factor in the commercial marketplace, emerging as the basis for building "multi-vendor" networks [6]–[8].

The TCP/IP architecture is very similar to the architecture for internetworking that comprises a part of the open systems interconnection (OSI) standards [9], [10]. In the OSI architecture an internetworking protocol, ISO 8473 [11], [12] positioned in the layer 3, network layer [13], [14] corresponds to the IP in TCP/IP. The OSI Transport protocol, ISO 8073 [15] includes one protocol class, "Class 4,"[3] that generally corresponds in functionality to TCP.

---

[1] DARPA is the Department of Defense Advanced Research Projects Agency

[2] "TCP" is an acronym for "transmission control protocol." "IP" is "internetwork protocol" or "internet protocol."

[3] The OSI transport protocol includes five classes of protocol, each of which provides a different degree of functionality and robustness. "Class 4" is the most robust, and includes the mechanisms for error detection and recovery, flow control, etc. that are required to provide an end-to-end, reliable transfer capability over a connectionless network environment.

A number of vendor-proprietary network architectures also closely correspond to the TCP/IP architecture and the OSI standards for internetworking. These include Digital Equipment Corp.'s DECNET/DNA, Xerox's XNS and Unisys' BNA. The type of architecture that these products reflect represents a now well-established and growing trend for internetworking, particularly where high-speed LANs are involved.

Role of SMDS in the Customer Environment

For SMDS to function in a subnetwork role as depicted, it must have a high degree of compatibility with the communications protocol architecture and the gateway philosophies presently used by its target customers. For example, in such environments it is important that gateways be relatively simple, and minimize the amount of resources (for example, buffers, processor cycles) used. Thus, we conclude that SMDS should offer a simple connectionless transfer mode, like a LAN, and that the interface to the network providing SMDS have simple procedures that are efficient for a gateway to utilize. Like the media access control (MAC) protocol of most LANs, the SMDS Interface Protocol should be easy to implement, by installing a new board interface and a simple software driver, for example.

Extension through SMDS

Not only should SMDS fit into the target customer's present environment, but its introduction into this environment has the primary goal of extending over a wide area the set of capabilities and performance the customer experiences in the local area.

The "end-user" applications found in the typical SMDS customer environment are very diverse. The availability of high-bandwidth communications on the local premises provides opportunities for functionality to be shared, with significant, positive consequences on the cost and method of doing business. For example, as mentioned earlier, applications that previously were accomplished by direct access to powerful processors (such as CAD/CAM, sophisticated graphics for publication layouts), can now be done using desktop workstations that communicate, via high-speed local networks, to supporting processors and peripherals. Files and executable programs can be stored on a powerful central "server" machine that can be accessed by multiple workstations. Additionally, if the user needs to access other software or data files, these can be quickly downloaded for use in the workstation, so that from a user perspective, they are as convenient and efficient as if they were resident in the workstation itself.

In today's environment, customers have limited choices, to extend the use of these applications beyond the local premises. One alternative is to resort to wide-area service offerings that support throughput at lower-than-LAN speeds, for example, leased line facilities at 56Kbits/s or 1.544 Mbits/s (T1). However, for the most part, these cannot meet the low delay requirements of high performance applications such as distributed processing. Another alternative is to acquire dedicated facilities that offer higher throughput and better delay characteristics but have equally higher cost, for example, private fiber optic networks or 44 Mbit/s (DS3) facilities. The objective of SMDS is to provide a more economical, shared network solution that would allow the use of high performance applications to be extended over a wide-area, and thus, no longer be limited by performance or by the feature/functionality discontinuity between widely separated subnetworks.

SMDS Requirements Summary

In summary, based on our view of the target customer, SMDS must exhibit the following general characteristics:

- features and characteristics similar to LANs;
- a simple, efficient interface; and
- high throughput and low delay performance.

Specific Features and Characteristics of SMDS

From the assumptions and objectives described above, we have derived a view of the specific features and characteristics that SMDS should exhibit in order to meet the target market.

Fundamentally, SMDS is a very high-speed connectionless or "datagram" packet-switched service. Typical of datagram services, it provides for the transfer of individual "units" of data, termed Service Data Units (SDUs), each of which has associated source and destination addresses. To efficiently serve high-speed data applications, SDUs have a variable length, up to a maximum of 8191 octets. Thus, the service can be used for applications which perform bridging between IEEE 802 standard LANs without the need to segment and reassemble MAC frames.

Source and destination addresses are delivered with each SDU, as specified by the sender. The network supporting SMDS validates all source addresses to ensure that the specified address is legitimately assigned to the interface from which the SDU was sent. Any SDU with a fraudulent source address is discarded.

Consistent with the support of SMDS by B-ISDN, addressing based on CCITT Recommendation E.164, the ISDN numbering scheme, is used to identify customer interfaces, with an additional provision for "group addressing."

Group addressing is analogous to the multi-casting feature of many LAN technologies, which is exploited by many equipment vendors in their communications software and applications. Its usage allows a single address to identify collectively a set of addresses. Incorporation of a group addressing feature in SMDS allows such vendor systems to adapt easily to using SMDS.

SMDS address screening can be used to allow a customer to restrict the reception of SDUs or the delivery of SDUs. Specifically, in *destination address screening*, all SDUs sent by the customer will have their destination addresses checked against a set of candidate destination addresses. In *source address screening*, all SDUs to be delivered to the customer will have their source addresses checked against a set of candidate sources. Any SDU that doesn't match one of the candidate source or destination addresses, respectively, is discarded. Alternately, the screening features can each be provisioned to discard SDUs when screened addresses *do* match the candidate addresses in the screening table.

When the destination address screening and source address screening features are used in conjunction, a "logical private network service" may be offered. All customers in the "logical private network service" will subscribe to both screening features and specify in the address screening tables the addresses with which they are allowed to communicate. By tailoring the screening tables and permissions, a customer can also create variants of this "logical private network service."

Other addressing features included in SMDS are the ability to assign multiple addresses to a single customer interface, and some degree of mobility for addresses.

Another key feature of SMDS is the provision and enforcement of "Access Classes." Since most transport layer implementations provide for end-to-end flow control, it is unnecessary to include an explicit flow control mechanism across the interface to the network supporting SMDS.* Moreover, such explicit flow control techniques impose penalties on the high performance of the service. However, since SMDS would be offered over very high-speed access paths, it is necessary to enforce limits on the rate of information transfer on the access, both in the subscriber-to-network direction, and in the network-to-subscriber direction. To provide for this enforcement, SMDS has various access classes and a "credit manager" mechanism. The credit manager is an algorithm implemented by the network which observes an implicit credit window based on predefined parameter values. The parameter values can be chosen to permit varying rates of sustained information transfer and different degrees of "burstiness." SDUs transferred from the subscriber to the network that violate the access class are discarded. In the network-to-subscriber direction, the network credit manager ensures that transfers always obey the access class. To avoid unnecessary data unit loss, the subscriber equipment can also implement and observe the access class parameters for transfers into the network.

The Use of MAN Technology for Early Availability of SMDS

To provide SMDS in a time-frame that will meet the emerging market, we have identified an approach which uses the BOC/IDC embedded base of multi-megabit fiber-based transmission systems, complemented by the use of MAN technology to provide very high-speed packet-switching. Figure 2 illustrates how MAN switching systems (MSSs) such as, switching systems based on MAN technology would be used as part of a network configuration to offer SMDS. The SMDS interface protocol (SIP) is defined across the subscriber-network interface (SNI). Within the net-

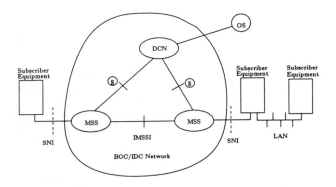

KEY

SNI: Subscriber-network interface
IMSSI: Inter-MAN switching system interface
(S): Generic interface for operations
MSS: MAN switching system
OS: Operations system
DCN: Data communications network
LAN: Local area network

Fig. 2. Network in support of SMDS.

work, MSSs are interconnected across an Inter-MSS Interface (IMSSI). Each MSS communicates with BOC/IDC Operations Systems (OSs) via a generic interface ($), across an operations Data Communications Network (DCN).

Subscriber systems' access to the network supporting SMDS is by means of a dedicated DS3 path from the subscriber system to an MSS. Because a standard DS3 rate channel is used, the location of the MSS that terminates the access may be in the subscriber's serving central office (CO) or may be located in another CO (perhaps at a fiber hub point) in the network. This flexibility in locating MSSs is a key factor in efficient early deployment of SMDS since the customer base is likely to initially be small, but potentially geographically dispersed. The transmission facilities used for the access will be optical fiber based, and although the initial interface is being defined as an electrical DS3 interface (DSX-3), it is planned to define an optical SONET (Synchronous Optical Network) interface as soon as equipment conforming to the SONET standard is available. The use of digital transmission hierarchy standards such as DS3 allows existing operations capabilities for BOC/IDC transmission systems to be exploited, resulting in significant advantages for overall SMDS operations support.

The DS3 electrical interface represents the first level in the three level interface protocol, referred to as the SIP, SMDS interface protocol, as depicted in Fig. 3. Level 3 of the SIP associates appropriate addressing information with each data unit of subscriber information that traverses the interface. Level 3 also contains the algorithm for enforcing the subscribed-to "access class" thereby enforcing the agreed to limits on sustained information transfer and burstiness. Since

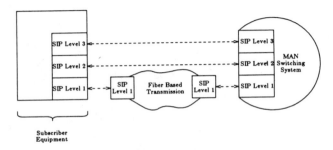

Fig. 3. SIP protocol stack.

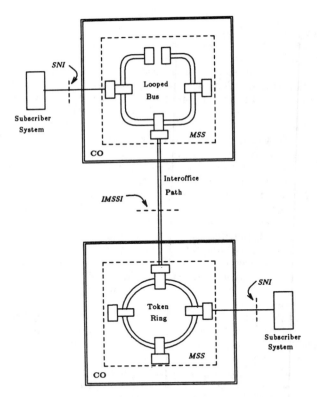

Fig. 4. Examples of MAN technology in MSSs.

this algorithm does not require explicit state synchronization between level 3 entities, the level 3 protocol is very simple. Framing of the level 3 protocol unit is performed by the level 2 of SIP which also includes a 32 bit CRC for error detection purposes.

Since the service (SMDS) expressed by SIP is "connectionless" or "datagram" in nature, no explicit flow control, sequencing, or retransmission procedures for errored data units are defined in the protocol. In fact, the interface that SIP provides to higher layer communications protocols in the subscriber systems is very similar to that available from most LANs. As a result, it is expected that existing and emerging CPE developed for use in high-speed LAN environments will be able to easily integrate and support the SIP for access to SMDS.

Inside the MSS, any one of several technologies could be used to meet the functional and performance criteria needed to offer SMDS. The network configuration illustrated in Fig. 2 is used to define the generic functions that an MSS must perform and the necessary interfaces it must meet, in particular the SNI and its protocol SIP, the IMSSI, and the generic interface for operations. No particular technology or architecture is prescribed for use inside an MSS and, in fact, the existence of a well-defined IMSSI provides for the deployment of MSSs that use different internal technologies. As an example, Fig. 4 illustrates two possible MSSs, one of which utilizes a shared medium ring topology and the other based on a looped bus arrangement. In each case, functionality of the MSS is distributed among components acting as stations on the shared media. (It is important to note that the shared medium which is used to transport the information of multiple subscribers is never directly accessible to any subscriber system. Components of the MSS terminate the shared medium MAC and deliver across the SNI only the information intended for the particular subscriber. It should also be noted that the focus of the MSS is solely on providing "data only" service and thus any capability of the MAN technology for supporting voice traffic is not exploited in providing SMDS.

As described here, our detailed work to date prescribing early availability of SMDS has focused on a precise technical service description, the framework architecture for describing generic functions and interfaces as shown in Fig. 2, details of the SIP, and use of the

existing BOC/IDC embedded transmission plant. Work is proceeding at Bellcore in additional areas such as the definition of the protocol between MSSs across the IMSSI and the operational requirements associated with the introduction of very high-speed packet-switching capabilities and the MSSs.

Summary

High-speed data networking across metropolitan and wide area environments represents an emerging market and a potential service opportunity for carrier networks. SMDS is a service concept that is intended to address this market by respecting the embedded and growing customer base of hardware, software and applications and extending the use of high performance applications beyond the limits of the local premises. The SMDS service concept has been defined so as to be suitable for deployment with different supporting network architectures and technology bases. Thus, the use of MAN technology to bring very high-speed packet switching to the carrier networks appears to provide a promising approach for early availability of SMDS.

Acknowledgments

The authors wish to acknowledge the other members of the technical service description team for SMDS, Jon Shantz, Kiho Yum, Patrick Sher, and Joe Lawrence.

References

[1] "Metropolitan area network generic framework systems requirements in support of switched multi-megabit data service," Technical Advisory TA-TSY-000772, Bell Communications Research, Inc., issue 1, Feb. 1988.

[2] NYNEX tests metro area network," *Telephony*, June 30, 1986.

[3] "Metro-area net to be tested," *Network World*, Sept. 22, 1986.

[4] "Internet Protocol," MIL-STD 1777, *DDN Protocol Handbook*, vol. 1, DCA, Dec. 1985.

[5] "Transmission Control Protocol," MIL-STD 1778, *DDN Protocol Handbook*, vol. 1, DCA, Dec. 1985.

[6] "COMNET '87: The 'Network Solution' show," *LAN: The Local Area Network Magazine*, Apr. 1987.

[7] "Big turnout expected at TCP/IP conference," *Communications Week*, March 9, 1987.

[8] "IBM embraces TCP/IP," *Network World*, May 25, 1987.

[9] ISO 7498-1984(E), Information processing systems—Open Systems Interconnection—Basic reference model, 1984.

[10] "Recommendation X.200, reference model of Open Systems Interconnection for CCITT applications," *CCITT Red Book*, vol. VIII.5, 1985.

[11] ISO 8473-1987(E), Information processing systems—Telecommunications and information exchange between systems—Protocol for providing the connectionless-mode network service, 1987.

[12] A. L. Chapin and D. M. Piscitello, "An International Internetwork Protocol Standard," *J. Telecommunications Networks*, vol. 3, no. 3, Fall 1984.

[13] ISO 8648-1987(E), Information processing systems—Telecommunications and information exchange between systems—Internal organization of the network Layer, 1987.

[14] C. Hemrick, "The internal organizzation of the Network Layer: Concepts, Applications and Issues," *J Telecommunications Networks*, vol. 3, no. 3, Fall 1984.

[15] ISO 8073-1984(E), Information processing systems—Open Systems Interconnection—Transport protocol, 1984.

Christine Hemrick is District Manager of Data Services Concepts in the Network Planning Area at Bell Communications Research. This district has responsibility for new service concept development, with a particular focus on high-speed data and services for broadband networks. She received a B.S. degree in mathematics from North Carolina State University in 1974.

Prior to joining Bellcore in 1985, Ms. Hemrick was a senior staff member with the National Telecommunications and Information Administration's (NTIA) Institute for Telecommunication Science (ITS) in Boulder, Colorado. While with NTIA, she served on behalf of the U.S. State Department as coordinator of U.S. participation in CCITT Study Group VII, and as head of the U.S. delegation to that study group. Previous to her NTIA experience, she held positions at Digital Equipment Corporation and GTE Telenet.

Robert Klessig is District Manager of Protocol Architecture Planning at Bell Communications Research. The district is involved in formulating internal network protocol architectures for both data transport and network control. He is also an active participant in IEEE Project 802.6, Metropolitan Area Network Standards. He received his B.S. in Engineering from Case Western Reserve University, and his M.S. in electrical engineering and Ph.D. in electrical engineering and computer sciences, both from the University of California, Berkeley.

Previous to the formation of Bellcore, he was a member of the technical staff and then supervisor at Bell Telephone Laboratories working in the areas of network facility planning, special services planning and data communications network planning. He has also been an acting assistant professor and visiting lecturer in the Electrical Engineering and Computer Sciences Department at the University of California, Berkeley.

He has published papers in the areas of nonlinear programming, optimal control, facility planning, and data communications protocols. He is a member of IEEE, SIAM, ORSA, and the Mathematical Programming Society.

Josephine McRoberts is a member of technical staff at Bell Communications Research, in the Data Services Concepts District. She has been involved in broadband services research, including video and high speed data services, with an emphasis on target applications. She attends the Broadband subworking groups of T1S1 and CCITT SGXVIII. She received her BSc in applied physics from the University of Strathclyde, Scotland in 1984 and a MSE in systems engineering from the University of Pennsylvania in 1985.

She is a member of IEEE and the U.K. Institute of Physics. ∎

Packet mode services: from X.25 to frame relaying

Wai Sum Lai discusses the relationship and interworking between frame relaying and X.25

Since its inception, the X.25 standard has become a widely accepted and deployed data service in both public and private networks. X.25 has also been defined as the packet mode service in Integrated Services Digital Networks, because of its current utility and the need to ensure compatibility with existing public data networks. The move to out-of-band call control with ISDN creates possibilities for the provision of additional packet mode services such as frame relaying. The relationship and interworking between X.25 and frame relaying services are discussed here.

Keywords: computer networks, packet switching, Integrated Services Digital Networks, X.25

The only standardized packet mode service currently being offered in an Integrated Service Digital Network (ISDN) is the X.25-based virtual circuit. Also under development at the CCITT, and various other standards bodies, is the provision of additional ISDN packet mode services.

The historical development of packet mode services at the CCITT is briefly traced, then the architectural and functional characteristics of the newly proposed additional ISDN packet mode services, with particular emphasis on frame relaying service, is presented. Finally, how frame relaying can interwork with the existing large installed base of X.25 is discussed.

HISTORICAL PERSPECTIVE

The utility and usage of computer equipment in the workplace has changed dramatically over the past two or three decades — from a centralized model of a large, single central processing unit with small numbers of

AT & T Bell Laboratories, Crawfords Corner Road, Holmdel, NJ 07733, USA

appending input/output devices, to a distributed model in a multivendor environment. As these changes occurred, the need for a cost-effective method of communication between dispersed machines arose. Packet switching addressed this need very well, with the inherent benefits of:

- interconnection flexibility between dissimilar types of equipment;
- data transmission integrity through error detection/correction;
- speed matching; and
- providing a shared communications facility over wide geographic areas.

Ever since the potential for packet switching was demonstrated by the Advanced Research Projects Agency (ARPA) network in the late 1960s and early 1970s, much international activity for the standardization of packet switching services was followed. As a result, in 1976, the CCITT published the then new X.25 Recommendation (*Interface between Data Terminal Equipment and Data Circuit-Terminating Equipment for Terminals Operating in the Packet Mode on Public Data Networks and Connected to Public Data Networks by Dedicated Circuits*) for packet mode communications based on virtual circuit service.

X.25 has gained wide acceptance. Since the mid 1970s, numerous X.25-based Packet Switched Public Data Networks (PSPDN) have been implemented all over the world. To ensure the global ubiquity of packet mode services, it is critical that these networks be interconnected in a standard way. To address this need, the X.75 Recommendation (*Terminal and Transit Call Control Procedures and Data Transfer System on International Circuits between Packet-Switched Data Networks*) was produced in 1980, where the internetwork signalling procedures between PSPDNs are specified.

With the advent of ISDN[1] in the 1980s, attention has focused on the provision of useful packet mode services in that environment. Because of the necessity to provide interworking with current packet data services, and the fact that X.25-based PSPDNs are well established worldwide, X.25 has been standardized as the packet mode service that can be offered in an ISDN (Recommendation X.31/I.462: *Support of Packet Mode Terminal Equipment by an ISDN*).

To ease the provision and use of different telecommunications services, out-of-band call control is employed in ISDNs. This control method is based on the principle of separation of the control and user planes in the ISDN protocol reference model (Recommendation I.320: *ISDN Protocol Reference Model*). The concept of separate control and user planes is used to distinguish interactions associated with control and signalling functions from those needed to transfer user data.

However, X.25 in the ISDN environment as specified in Recommendation X.31 uses in-band, rather than out-of-band, procedures for the control of virtual circuits. Such an approach is being taken for pragmatic reasons — to offer access to packet data services through an integrated physical interface while minimizing deployment and interworking difficulties.

The evolution of packet mode services in ISDN is currently under investigation in various national and international standards forums. In CCITT Recommendation I.122 (*Framework for Providing Additional Packet Mode Bearer Services*), the provision of additional packet mode services utilizing out-of-band call control has been proposed. Before discussing these new services, let us first review the current ISDN packet mode services of X.31, and the applicability of out-of-band call control.

X.31 PACKET MODE SERVICES

Recommendation X.31 defines the procedures for the support of existing packet mode terminals accessing an ISDN. In this support of X.25 Data Terminal Equipment (DTE), two scenarios are specified:

1 In the so-called *minimum integration scenario* (see Figure 1a), X.25 packet calls are handled transparently through an ISDN. In effect, the ISDN provides only a physical digital pipe between the X.25 user terminal adapter at the customer's premises, and an appropriate interworking port in an existing PSPDN. With the ISDN being operated as a Layer 1 (physical layer) relay, X.25 Layer 2 (link layer*) and Layer 3 (packet layer) functions are performed outside the ISDN.
2 In the *maximum integration scenario* (see Figure 1b), a packet handling function is provided within an ISDN. In conjunction with such a Packet Handler (PH), an ISDN appears to a PSPDN either as another PSPDN, or as an extension of the same PSPDN. As a result, once a physical connection is established between an X.25 terminal and a PH in an ISDN, the complete processing of the X.25 packet call can be carried out within the ISDN.

Thus, in the minimum integration case, the X.25 DTEs can be considered to be only physically connected to an ISDN, but logically connected to a PSPDN. In the maximum integration case, the DTEs are both physically and logically connected to the ISDN where the X.25 call is

---

*In Open Systems Interconnection (OSI) terms, Layer 2 is called the data link layer. Here it is referred to simply as the link layer.

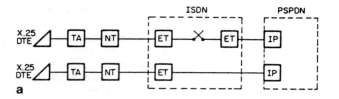

a

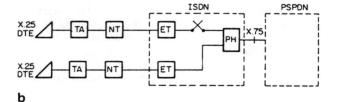

b

Figure 1. *X.31 packet mode services — access arrangements: (a) access to PSPDN services, (b) access to ISDN virtual circuit bearer service. TA: terminal adapter; NT: network termination; ET: exchange termination; IP: ISDN interworking port; PH: packet handler*

processed. In view of their different capabilities, these two scenarios have now been renamed as *access to PSPDN services* and *ISDN virtual circuit bearer service*, respectively.

In both types of service, existing in-band call control procedures in X.25 continue to be employed for the establishment and release of virtual circuits[†], i.e. the complete packet layer (Layer 3) of X.25 is used for both call control and data transfer phases of a packet call. What X.31 does is really just the encapsulation of existing X.25 packet mode services in an ISDN envelope at the physical layer.

With the X.25 packet layer procedures remaining intact, the number of modifications required to accommodate existing X.25 DTEs in an ISDN is minimized. In this way, a large population of PSPDN users can easily migrate to ISDNs as these networks are deployed. A smooth transition for both carriers and users is thus ensured as a result of the consistency and uniformity between PSPDN and ISDN packet mode services.

OUT-OF-BAND CONTROL

While in-band call control is being specified in Recommendation X.31 for the support of X.25 to minimize deployment and interworking difficulties, out-of-band call control has been adopted for the additional ISDN packet mode services currently under investigation by the CCITT, and described in Recommendation I.122.

As detailed in I.122, this use of out-of-band control is made in recognition of the fundamental principle of separation of the control and user planes for all telecommunications services in ISDNs. Adherence to this principle enables ISDNs to achieve the goal of service independence: the same signalling mechanisms as other ISDN bearer services will be employed by the additional

---

[†] In the case of circuit-switched B-channel access to packet mode service, out-of-band control based on Q.931 message exchange (*ISDN User-network Interface Layer 3 Specification*) over the signalling D-channel is first used for the establishment of a circuit-switched physical connection. After this, the in-band X.25 procedure is used for the establishment of a virtual circuit over the physical connection.

ISDN packet mode services in order to completely set up packet calls.

In most existing transport networks there is no clear distinction between the paths for the control and user information streams. The significant advantage of this approach is that, since network control tends to be intimately tied in with the associated information carriage, the resulting network design can have a high degree of optimization. The consequence of this is that different information carriages will require different forms of network control, because of different control requirements. A close coupling between information and control also leads to a lack of the flexibility needed to cater for new signalling and control needs.

With the ever-increasing demand for data communications within the business, and to some extent the residential, environment the role of the network is changing from that of a predominantly voice network to one which is more responsive to other new services. The acceptance and deployment of digital switching and digital transmission throughout the public telephone networks has primed the network with the basic capabilities to handle traffic other than voice. To ensure maximum efficiency of the network, separation of signalling control from the information carriage is desirable.

Major benefits derived from this separation include:

- there is the potential for the integration of control for voice, data and other new services. As a result, there can be sharing across all bearer services of functional units, providing, e.g. call control, supplementary services and higher layer capabilities. So, as new bearer services are added, the proliferation of signalling types and procedures can be avoided. Also, as new supplementary services are added, they could possibly be applied to all bearer services;
- since the information path no longer needs to worry about control, its logic can be substantially simplified — this implies that hardware implementation can now be exploited more fully;
- with the separation of signalling, the provision of multimedia calls can be accommodated — this type of call can have very different attributes associated with different portions of a single call, or for the two directions of a single call;
- different qualities of service are usually required by information transfer and signalling control. Their separation therefore allows for separate network provisioning to cater for their different needs, e.g. the design of functional units used for signalling functions are no longer constrained by the bearer channels with which they may be associated for call supervision. In the same vein, the attributes of requested bearer services can be based only on information transfer requirements, and not on signalling;
- in addition to independent optimization, independent evolution of control and transport networks can also be accommodated.

I.122 ADDITIONAL PACKET MODE SERVICES

In Recommendation I.122 four types of additional ISDN packet mode bearer services are defined: frame relaying 1; frame relaying 2; frame switching; and X.25-based additional packet mode.

From a high-level architectural viewpoint, this family of new services is built upon a common framework that aims at service independence by using the concepts of out-of-band signalling and link layer multiplexing. What differs among them is the extent of functionality provided by a network during the data transfer phase.

Control and user planes

As discussed above, all I.122 services adopt out-of-band signalling. Specifically, for the establishment and release of packet-mode calls, Q.931 messages over the signalling D-channel are used. These procedures are in the Control plane (C-plane). After call establishment, all procedures for the transfer of user data are in the User plane (U-plane).

According to Recommendation I.122, such a separation of C-plane procedures from those in the U-plane can occur in one of several ways: on a physically separate interface; on another channel (time slot) within the same interface; or on a separate logical link within the same channel (e.g. the D-channel).

Link layer multiplexing

In the U-plane, in order to facilitate the sharing of bandwidth resources among different user data streams, the concept of link layer multiplexing is used.

In a conventional packet network, a link entity typically does not provide any multiplexing function. Multiplexing in X.25 is achieved through the use of logical channels in the packet layer. So, effectively it is Layer 3 that provides the switching function.

In I.122 services, switching is done in the link layer through the statistical multiplexing of different *data link connections* on the same physical channel. This support of multiple data link connections in the link layer is based on the multiplexing operation specified in Recommendation Q.921/I.441 (*ISDN User-network Interface Data Link Layer Specification*) (Link Access Protocol on D-Channel (LAPD)). With LAPD as a basis, the creation of a new protocol is therefore not required.

Network functionality

With link layer multiplexing in the U-plane during the data transfer phase as a foundation, the family of I.122 services is offered by a network through different degrees of protocol support, i.e. different levels of protocol termination at the network edges after call establishment. Figure 2 shows the different protocol break points, or points at which a network can terminate the protocol(s) in order to provide the requested bearer services. It can be seen from this figure that, in terms of the network functionality provided, frame relaying requires the least, while X.25-based additional packet mode the most.

In *frame relaying service* (see Figures 3a and b), the network supports only the minimal, or the so-called *core*, set of functions at Layer 2 that are needed to transfer frames between communicating terminals during the data transfer phase. It suffices for now to note that these functions essentially ensure that only frames with a valid format and address (i.e. data link connection identifier) are delivered. All other Layer 2 functions, as well as those

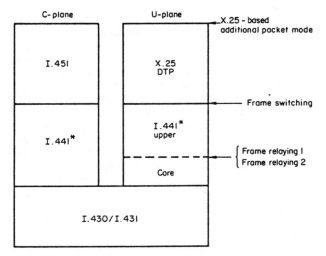

Figure 2. Functionality of an ISDN node at the user-network interface for the provision of I.122 services. Core: I.441 core functions; I.441*: I.441 extended to cover I.122 requirements; I.441* upper: elements of procedures specified in I.441* (i.e. upper part of I.441*); DTP: data transfer part of X.25 packet layer procedures

at Layer 3 and above, are transparent to the network. They are therefore operated on an end-to-end basis by the users themselves. (In addition to link layer multiplexing which is provided by the network, a user may also choose to perform Layer 3 multiplexing within a particular data link connection. But such an operation will be transparent to the network.)

There are two types of frame relaying service. In *frame relaying 1 service* (see Figure 3a), any user-specified link layer end-to-end protocol above the core functions can be used. For *frame relaying 2 service* (see Figure 3b), the elements of procedures as specified in Recommendation Q.921/I.441 are to be used end-to-end.

As a result of the fact that the network provides only the core functions of the link layer, a user's link layer can be viewed as split into two parts: the lower core functions and the upper end-to-end procedures. This decomposition is conceptually very similar to the Medium Access Control (MAC) and the Logical Link Control (LLC) sublayers of the link layer in the IEEE 802 Local Area Networks (LAN).

In *frame switching service* (see Figure 3c), a user's link layer protocol in the U-plane is specified to be Q.921, and this is fully terminated by the network. So, only Layer 3 and above are considered to be end-to-end.

In *X.25-based additional packet mode service* (see Figure 3d), Q.921 and the data transfer part of the X.25 packet layer procedures are specified to be a user's Layer 2 and 3 protocols, respectively, in the U-plane. Both these protocols are fully terminated by the network. Observe that only the data transfer part of the X.25 packet layer procedures is required here. This is because the Q.931-based out-of-band signalling method eliminates the need for the call control part of the X.25 packet layer procedures.

Relationships between different services

A closer examination of Figures 3a, b and c reveals some interesting relationships between the frame relaying 1,

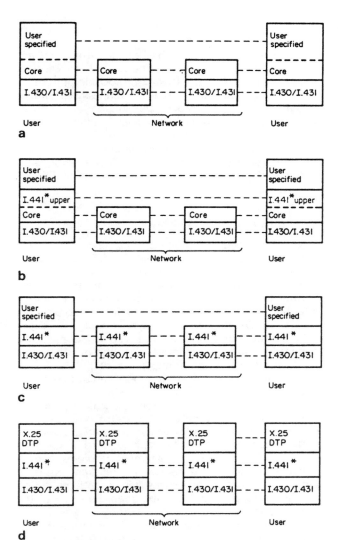

a

b

c

d

Figure 3. I.122 additional packet mode services: U-plane procedures. (a) Frame relaying 1 service, (b) Frame relaying 2 service, (c) Frame switching service, (d) X.25-based additional packet mode service. Core: I.441 core functions; I.441*: I.441 extended to cover I.122 requirements; I.441* upper: elements of procedures specified in I.441* (i.e. upper part of I.441*); DTP: data transfer part of X.25 packet layer procedures

frame relaying 2 and frame switching services. To see these we take two orthogonal viewpoints in the U-plane: a vertical cut across the protocol stack at the user-network interface; and a horizontal split along the boundary between Layers 2 and 3 of a user.

At the user-network interface, from the perspective of a network there is really no significant difference between frame relaying 1 and frame relaying 2, since only core functions are provided by the network in both cases. The difference between them mainly stems from the standpoint of OSI layer service. This is because, depending on the specific Layer 2 protocol being used, a different Layer 2 service will be provided to Layer 3. So, with the specification of Q.921 as the Layer 2 protocol, the Layer 2 service of frame relaying 2 is completely specified by I.122. For frame relaying 1 with the user-specified Layer 2 protocol, the Layer 2 service is defined, not by I.122, but by the users themselves. So, it can be different from one user connection to another.

In this sense, one can say that both frame relaying 2 and frame switching provide the same Layer 2 service, since Q.921 is specified as the Layer 2 protocol in both cases. However, as noted earlier, their difference arises from the different levels of protocol termination at the user-network interface. As a result of such a difference, even though the Layer 2 services provided by frame relaying 2 and frame switching respectively are the same in terms of functionality, there can be differences in the performance between the two Layer 2 services in terms of delay or throughput.

Why different services?

Having described the different I.122 services, it is now appropriate to discuss the rationale for defining the range of services as above. One reason is that such a family can probably be tailored appropriately to the needs of the growing diversity of user application. Indeed, the original intention behind developing this family was to offer not only an 'X.25-like' service to complement the existing X.31 packet mode services, but also to support new applications such as the interconnection of LANs, large file transfers, high resolution graphics, etc.

By defining a spectrum of services it can be envisioned that the users are given the 'flexibility' to choose, according to the needs of different applications, different terminations of protocols by the network in order to obtain the different grades of service desired. However, such a freedom can potentially lead to a complex interface with significant management problems for both network providers and users. The different types of protocol termination can also result in potential confusion on the user's part, and possibly inefficient use of the services provided by an ISDN.

It should be noted that one of the principles of ISDN is that a broad spectrum of applications should be supported by the same network using a limited number of connection types and multipurpose user-network interface arrangements.

With these considerations in mind, it is desirable to limit the number of bearer services provided by an ISDN. The current view is that not all of these services may need to be provided by a network. A new question for the CCITT in the next Study Period (1989–1992) is to determine which service(s) should be retained.

FRAME RELAYING SERVICE

The relative merits of the above services notwithstanding, one type of service which stands out most is frame relaying. When compared to the others, the protocol functionality required at the user-network interface for frame relaying is much reduced. This service would therefore permit less per-frame processing by a network. As a result, less delay and higher throughput could be expected. Because of this promise, frame relaying is currently attracting much attention.

Following this emphasis, the focus in the rest of the paper is on the frame relaying service. Furthermore, the discussion is based on the frame relaying 1 service in view of its generality in terms of the user link layer end-to-end protocol.

To simplify subsequent description, the term 'frame relaying 1 service' is abbreviated to *FR service*. By the term *Frame Relaying network (FR network)* we mean an ISDN that provides, on a per-call basis, FR service.

For the support of FR service, the link layer within a FR network provides only the so-called *core functions of I.441*. For discussion purposes, these core functions can be logically separated into two groups: frame integrity control and frame multiplexing.

Frame integrity control includes the following basic functions associated with an individual frame: frame delimitation, e.g. by High-level Data Link Control (HDLC) flags; transparency of user information, e.g. through the use of HDLC zero insertion/deletion; and frame integrity protection, e.g. by HDLC Frame Check Sequence (FCS).

These types of processing are done on a frame-by-frame basis. Building on this, the frame multiplexing function is responsible for the interleaving of, and discrimination between, the frames from different data link connections on the same physical channel.

To achieve multiplexing, each data link connection within a given physical channel is being assigned a unique identifier, called the *Data Link Connection Identifier* (DLCI). Such a DLCI only has a local significance, i.e. its value is only unique within a physical channel.

Switching in the link layer is achieved by binding the DLCIs to routing information at intermediate nodes to form a set of network edge-to-edge logical paths[2]. These bindings at intermediate nodes are established at call setup before the start of data frame transfer, and released at call clearing.

In this way, a connection-oriented link layer service is offered to the users of a FR network. The characteristics of this service are: preservation of the order of frame transfer from one network edge to another; non-duplication of frames; and a (very) small probability of frame loss.

INTERWORKING WITH X.25

To prevent isolated islands of subscribers from being created, the provision of connectivity to the existing user base must be a key aspect of any new service. So, a complete definition of a new packet mode service such as FR should include a specification not only of the services to be provided at the network access interface, but also of that provided when interworking with existing networks.

FR and X.25 are two different network access protocols: they provide different types of service to a user. The numbering plans for the two networks are also potentially different, since FR users can be identified according to the ISDN numbering plan E.164 (*Numbering Plan for the ISDN Era*) rather than the public data network numbering plan X.121 (*International Numbering Plan for Public Data Networks*).

An *Interworking Function (IWF)* is therefore required to resolve the differences so that a user on a FR network can communicate with a DTE on an X.25 network (by the term X.25 network either a PSPDN or an ISDN offering X.25 service based on X.31 is meant). Such an IWF can be provided either by a FR network or an X.25 network.

Conceptually, an IWF can also be configured as an intermediary between a FR network and an X.25 network. In this case, the IWF is separate from either network. However, such a non-integrated approach may cause complication in the management of internetwork connections, especially in the procedures for maintenance,

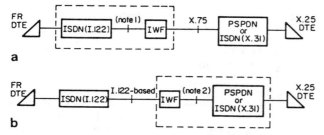

Figure 4. *FR-X.25 interworking arrangements. (a) IWF logically belongs to ISDN (I.122), (b) IWF logically belongs to PSPDN or ISDN (X.31). (Note 1: In the control plane, signalling system No 7, I.451 with appropriate extensions, or proprietary protocols with equivalent functions may be used. Note 2: X.25 or equivalent proprietary protocols may be used.) -----: logical grouping*

traffic planning and accounting. This strategy is therefore not considered any further.

FR-X.25 INTERWORKING ARCHITECTURES

We first discuss the case when the IWF logically belongs to a FR network (see Figure 4a). As described above, such an IWF must bridge the difference in layer services between a FR network and an X.25 network.

Using the terminology of Recommendation X.300 (*General Principles and Arrangements for Interworking between Public Data Networks and between Public Data Networks and other Public Networks*), the association of a FR network with an IWF could be considered globally as a *Type I subnetwork**. Such an association should offer the full capability of OSI (Open Systems Interconnection) connection-oriented network layer service. (An X.25 network is by definition a Type I subnetwork.)

As specified in X.300, the interworking arrangement between two Type I subnetworks should be based on Recommendation X.75, i.e. if a FR network is to provide the IWF, then the FR-X.25 interworking procedure should be X.75-based.

Alternatively, an X.25 network can offer an adapter which provides the IWF (see Figure 4b). Such an adapter would resemble a user requesting FR service from a FR network. Hence, I.122-based procedures, appropriately extended to cover network-to-network interface requirements, can form the basis for FR-X.25 interworking.

In this interworking arrangement, a FR terminal attached to a FR network first accesses an X.25 network by using such I.122-based procedures to set up a FR call via the FR network to the adapter in the X.25 network[†].

After the establishment of such a FR connection, the data path between the FR terminal and the adapter will be totally transparent to the FR network. As described above, any user-specified Layer 2 and Layer 3 end-to-end protocols can run atop the core functions provided by the FR network. In particular, the FR terminal can now establish an X.25 virtual circuit to a remote DTE on the

---

*Based on the network support for the OSI connection-oriented network layer service, four basic categories of subnetworks are considered in X.300. These are referred to as Types I, II, III and IV subnetworks. A Type I subnetwork is one that provides functionality equivalent to that of a PSPDN.

[†]In X.300, this method of interworking is called *port access*.

X.25 network via the existing in-band X.25 packet layer procedures for call control.

Similar two-stage calling procedures apply for calls originating from an X.25 DTE to a FR terminal. A FR connection from the adapter to the FR terminal must first be established over the FR network before any X.25 packet layer call requests can be passed over to the called FR terminal.

Note that under this scheme, X.25 virtual circuits are being multiplexed on a single FR connection. However, it is possible for both Layer 2 and Layer 3 multiplexing to co-exist, e.g. a FR terminal may simultaneously have multiple FR connections to the same or different X.25 networks on a single bearer channel, with each FR connection carrying one or many X.25 virtual circuits. Similarly, an X.25 network may have multiple FR connections to multiple FR terminals, over one or many bearer channels.

COMPARING X.75/I.122 INTERWORKING ARRANGEMENTS

To recapitulate, the strategies that can be used for FR-X.25 interworking are: OSI network layer interworking through X.75; and I.122-based interworking. Under X.75 interworking, a FR-X.25 internetwork connection consists of three segments: a FR segment inside the FR network; a virtual circuit segment through the X.75 internetwork link; and another virtual circuit segment inside the X.25 network. In contrast, there are only two segments in I.122-based interworking: a FR segment from the FR network user to the adapter, and a virtual circuit segment from the adapter to the X.25 DTE. Obviously, this makes the data transfer phase more efficient.

Another aspect is that X.75 is based on in-band signalling. I.122-based procedures provide out-of-band call control, at least for the establishment of FR connections. It is therefore in line with ISDN philosophy.

While there are advantages in using I.122 in the long-term, X.75 provides a number of immediate benefits. In particular, the X.75 strategy provides:

- backward compatibility: in principle, no changes to the installed X.25 networks will be required;
- an interworking protocol that is proven and widely supported (furthermore, E.164/X.121 interworking has essentially been resolved for X.75); and
- a basis for ensuring consistency of the OSI network layer services.

Therefore, X.75 should also be the basis of interconnection. This will ensure the interworking of any new FR networks with the large established base of X.25 networks.

In the I.122-based interworking scheme, X.25 networks are being accessed by multiplexing X.25 virtual circuits over a FR connection. Conceptually, such an access mechanism is very similar to the use of X.31 to access PSPDN services (i.e. the minimum integration scenario, see above). The essential differences between these two methods, when the X.25 calls are being carried on bearer channels, are the use of Q.931 to establish a FR connection rather than just a bearer channel, and the use of LAPD on the bearer channel instead of LAP on B-channel (LAPB). Therefore, with minor extensions to X.31, existing packet-mode terminals can use FR as an additional mechanism for accessing X.25 networks.

CONCLUSION

The evolution of packet mode services from X.25 with in-band control and packet layer multiplexing to FR with out-of-band control and link layer multiplexing has been presented.

FR is still currently under development at the CCITT and various other standards bodies. To ensure the success of FR, its interworking with the existing X.25 services must be developed, and for this purpose both X.75- and I.122-based interworking alternatives can be used.

ACKNOWLEDGEMENTS

The author would like to thank Wing-Man Chan, Ian Cunningham, David Delaney, Ken Haywand, Bam Liem, Stephen Ng, Andy Nichol and John O'Connel for their help in this work.

REFERENCES

1   **Rutkowski, A** *Integrated Services Digital Networks* Artech House, USA (1985)

2   **Lai, W S** 'Packet forwarding' *IEEE Commun. Magazine* Vol 26 No 7 (July 1988) pp 8–17

Wai Sum Lai *received his BSc degree in electrical engineering from the University of Hong Kong in 1970, his MSc degree in information and computer sciences from the University of Hawaii in 1974, and his PhD degree in systems and computer engineering from Carleton University, Ottawa, Canada in 1987. From 1970 to 1971 he was a consumer radio engineer for TransWorld Electronics Ltd., Hong Kong. Between 1971 and 1975 he was a research assistant for the ALOHA System at the University of Hawaii. From 1975 to 1988 he worked with Bell-Northern Research, Ottawa, Canada. He has recently moved to AT&T Bell Laboratories, Holmdel, NJ, USA.*

Broadband ISDN and Asynchronous Transfer Mode (ATM)

Steven E. Minzer

THE EMERGENCE OF HIGH-SPEED TELECOMMU-nications technologies has provided an impetus for the definition of Broadband ISDN (BISDN) [1–5]. BISDN is conceived as an all-purpose digital network. It will provide an integrated access that will support a wide variety of applications for its customers in a flexible and cost-effective manner [6]. The network capabilities will include support for:

* *Interactive and distributive services*—The network serves as a common carrier of both interactive and distributive communications, which may include audio, video, and data.
* *Broadband and narrowband rates*—Switched services at rates of up to about 140 Mb/s, with the possibility of raising this limit in the future.
* *Bursty and continuous traffic*—Some services (e.g., voice and video) may need guaranteed bandwidth to meet performance requirements (low information loss and delay); on the other hand, bursty data services may be provided more cost-effectively with a lower grade of service, using resources that can be shared statistically.

BISDN is conceived as an all-purpose digital network.

* *Connection-oriented and connectionless services*—Some applications, including most continuous-bit-rate communications, are best served by connection-oriented services. Connection-oriented services are characterized by separation of the procedures for connection establishment and the end-to-end transfer of user information. Connection establishment procedures, which must precede information transfer, determine a route and set up a path between users. Other services, including mail and other data-oriented communications, may not warrant separate connection establishment and information transfer phases, with route-related and user information carried in the same message.
* *Digital signal processing*—Some services require transparent transport of digital bit streams through the network; others may take advantage of code compression capabilities used to conserve internal network bandwidth or code conversion capabilities provided to overcome terminal incompatibilities.
* *Point-to-point and complex communications*—Some services require a single point-to-point connection, either unidirectional or bidirectional, between two end points; other "multimedia" applications benefit from parallel connec-

tions between end points, connections among multiple users, and even combinations of multiple-connection, multi-user calls. Some multi-user services may involve digital signal processing.

The goal of BISDN is to define a user interface and network that will meet these varied requirements. An objective of the network is to be able to accommodate volatile changes in service mixes, both at the level of the individual interface and the system as a whole. High-capacity and high-performance fiber-based transmission facilities are generally assumed to be required to support this environment. In addition, the network must have flexible switching capabilities and special-purpose service modules. This paper focuses primarily on the switching and multiplexing of BISDN, though the transmission aspects of the BISDN interface cannot be ignored. CCITT Study Group XVIII, the current focus for international standards on BISDN, generically calls the switching and multiplexing aspects the "transfer modes." The transfer mode chosen by the CCITT Study Group XVIII Task Group on ISDN Broadband Aspects (BBTG) as the basis of BISDN is called the Asynchronous Transfer Mode (ATM). ATM is a high-bandwidth, low-delay, packet-like switching and multiplexing technique. It is essentially a connection-oriented technique, though it is envisioned as a basis for supporting all services, connectionless as well as connection-oriented.

In retrospect, the choice of the term "Asynchronous" may have been unfortunate, for it has often been confused with asynchronous transmission. ATM is not an asynchronous transmission technique. Figure 1 depicts the hierarchical relationships of some of the functional layers required for information transfer across BISDN.[1]

The transmission layer, also referred to as the physical-medium-dependent layer, underlies the transfer modes. A wide variety of service modules can be supported as higher-layer functions above the transfer mode layer. Connectionless services, multimedia, and digital signal processing capabilities can all be included in the BISDN as service modules. User-network signaling for control of connection-oriented and complex services also belongs at this level. A point-to-point connection-oriented service may not require a service module above the transfer mode layer, since ATM can be used to perform the relaying functions required to transport the user information through the network.

---

[1]This illustration is not a complete picture of the BISDN architecture. It excludes the "adaptation" layer, which would appear directly above the transfer mode layer, its primary function being conversion between service-specific information units and the ATM format.

EH0301-2/91/0000/0081/$01.00 © 1989 IEEE

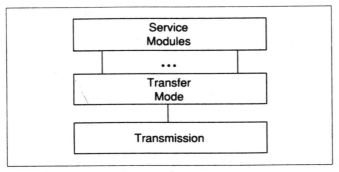

Fig. 1. BISDN functional layers.

The choice of ATM as the transfer mode for BISDN signifies a fundamental shift in principles. Two years ago, there were major disagreements on the most appropriate transfer mode for BISDN. The Synchronous Transfer Mode (STM), based on time division switching and multiplexing, was initially assumed to be the appropriate transfer mode for BISDN by many, though there was a relatively small but vocal group of advocates for "new" ATM techniques. Hybrid structures were, for a time, also considered as an option. Studies on these approaches were compartmentalized, with groups of CCITT Study Group XVIII experts working almost independently on each. There was no coherent overall plan to coordinate their efforts [7].

Fortunately, since then, the situation regarding the choice of transfer mode has improved. Over the next several meetings, the focus gradually shifted in the direction of ATM. CCITT Recommendation I.121 [8], a guideline for future BISDN standardization, designates ATM as the "target transfer mode solution for implementing a BISDN." It recognizes that ATM "will influence the standardization of digital hierarchies and multiplexing structures, switching and interfaces for broadband signals." There are only a few scattered references to other transfer modes, primarily with respect to network evolution. For example, I.121 states:

> In the early stages of the evolution of BISDN, some *interim* user-network arrangements (e.g., combinations of Synchronous Transfer Mode [STM] and ATM Techniques) may need to be adopted *in certain countries* to facilitate early penetration of digital service capabilities. (Italics are mine.)

I.121 contains no sentiment to sanction international interfaces other than those based exclusively on ATM.

Although there is a virtual consensus on ATM, a number of very significant issues remain unresolved. Failure to agree on these could derail the effort to define a unique international standard. These issues are related to ATM parameters and the underlying transmission system for the User-Network Interface (UNI), the interface connecting a customer to a BISDN public network.

One major issue is the degree to which voice service considerations should be a factor in determining ATM parameters. While it is recognized that voice will have to be interworked with existing network facilities for a considerable period of time, views differ on the implications of this requirement on the specification of ATM parameters. There are also other factors that must be taken into account in setting these parameters; these may be more significant for the long-term utility of the network (e.g., suitability for high-quality video and high-speed data).

Another key issue pertains to the use of a new optical transmission system, popularly referred to as SONET [9]. I.121 contains a general statement that "ATM can be supported by any digital transmission hierarchy or system" and specifically refers to the SONET recommendations (G.707 [10], G.708 [11], and G.709 [12]) as an example. Flexibility in choice of internal transmission facilities is an advantage (e.g., existing DS3 facilities might be used to carry some ATM-based services); but a single standard is desirable at the UNI for user equipment portability. The choice between an interface with all of its physical bandwidth organized as an ATM structure and one in which an ATM structure is embedded within another framework such as a SONET frame has been left unresolved in I.121.

Complicating matters even further is the fact that these issues cannot be settled independently of each other. Both transmission and ATM parameters affect the usable capacity on an interface, limiting service capabilities that can be provided. A BISDN interface should be thought of as a system constrained by a set of higher-level service capabilities at one end, lower-level transmission capabilities at another, and ATM capabilities sandwiched in between.

This paper is divided into two parts. In the first, the basic transfer mode concepts are explained and reasons for choosing ATM over STM are given. In the second part, the main technical issues having an impact on the ATM specifications are discussed.

Paradigms and Evolution

A paradigm [13] consists of the fundamental concepts, rules, and procedures used in circumscribing a system. A change of paradigm, which should occur relatively infrequently, involves the redefinition of familiar concepts. It happens when the existing paradigm has difficulty dealing with important problems. Frequently, a definition of a new paradigm is based on assumptions about the future, which are taken largely on faith, for most current needs are supported within the existing paradigm.

The choice of ATM over STM can be viewed as the selection of a new paradigm. Both STM and ATM deal with the rules for dividing up the usable capacity (bandwidth) of an interface and allocating it to user services. The fundamental concept in the division and allocation is the channel. A channel associates a portion of the usable capacity with a service. ATM redefines this fundamental concept.

Interworking allows for an extended migration, a mandatory requirement since the entire network cannot change at once.

The elaboration of the ATM concept is in part motivated by the perceived need to accommodate the wide variety of new services mentioned earlier. On the other hand, the substantial investment in the existing plant of a complex and expensive network must be a major consideration in any scenario, since a paradigm shift cannot succeed without a smooth migration strategy to lead from the network of today to the network of the future. I.121 includes a section entitled "Evolution Steps," recognizing the need for such a strategy; however, the recommendation does not indicate any specific steps. There is surprisingly little attention devoted to the subject of evolution in the text of the recommendation. The prevailing sentiment is that network administrators will follow their own course, as dictated by national experience and needs. The only identifiable "evolutionary requirement" is that BISDN services be interworked with services in other networks, which may be a tautology since BISDN, by definition, should be able to support any service.

Interworking allows for an extended migration, a mandatory requirement since the entire network cannot change at once. But network evolution, consisting of gradual migration of geographical sections and functional elements of the network (i.e., transmission, switching, services, signaling, and control modules), should be distinguished from planning to introduce a succession of devices on a step-by-step basis to perform a particular function (e.g., an STM switch to be replaced by a hybrid switch to be replaced by an ATM switch). A series of small steps is not necessarily the best way to reach a known target, though undoubtedly some catering will be needed to meet short-term demand. Since major network overhauls take a long time and are expensive, it is best to take as few major steps as possible.

The Limits of STM

STM, the old paradigm, was once a new paradigm. It superseded Frequency Division Multiplexing (FDM). The rule for subdivision and allocation of bandwidth using STM is: allocate time slots within a recurring structure (frame) to a service for the duration of a call. An STM channel is identified by the position of its time slots within a synchronous structure, replacing identification by frequency band of the older FDM paradigm.

FDM primarily supports analog communication systems. STM thrives in an environment dominated by digital telephony. It is the foundation of the digital transmission hierarchies and was introduced into ISDN with the $2B + D$ channel structure of the basic ISDN interface and the $nB + D$ ($n = 23$ in North America, $n = 30$ in Europe) of the primary rate interfaces.[2]

Initially, it was generally assumed by most experts in the standards community that STM would serve as the paradigm for BISDN as well. The interface structure for a BISDN could be expressed as:

$$iH_4 + jH_3 + kH_2 + lH_1 + mH_0 + nB + D$$

The coefficients $i,...n$ indicate the number of occurrences of a particular type of bearer channel on an interface. A bearer channel can carry user information end to end. There is only one signaling channel, the D-channel, which is used primarily for carrying user-network control messages, though it may also be used for user information but not at the full channel rate. Proposed and standardized rates for the bearer channels appear in Table I. Values either standardized or firmly agreed on appear in bold numerals.[3]

The STM interface structure expressed in Table I is an extension of the narrowband ISDN structure, obtained by adding higher-bit-rate channels to the previously standardized $nB + D$ structures. It was thought that STM structures would facilitate early deployment of BISDN systems, since they could be based on familiar concepts associated with existing technology. Compatibility with existing systems was a primary objective and it was envisioned that newer services would be carried via overlay structures and networks. The decision to specify channel rates for BISDN, though channel rates cease to have the same significance in the ATM paradigm, was in part based on a feeling that we should not burn all of our gateways behind us, facilitating possible interworking with STM-based fabrics.

The suitability of the STM paradigm for BISDN was first questioned because of a concern about its flexibility in meeting the needs of the future. Existing STM-based interfaces tend to be rigidly structured. In the digital hierarchies, primarily used for interoffice transport, interfaces aggregate a fixed number of identical channels of the next level of the hierarchy. The basic $2B + D$ ISDN interface has little flexibility in its channel structure, largely because of its limited capacity. The primary rate interface can potentially carry a mix with other channel types (H_0 and possibly H_1 channels). But demand circuit-switched services at these higher rates are not essential services of an ISDN. Many digital switches, originally designed before ISDN was conceived, support only 64-kb/s circuits, making it

highly unlikely that higher-rate circuit switching fabrics will become ubiquitous.

Flexible assignment of sets of time slots to channels for switched services, with each channel consisting of one or more slots per frame, is possible. But it requires coordination of relatively complex mapping functions by both the user and network sides of an interface. There are approximately 2,000 usable 8-bit time slots per frame in a 150-Mb/s interface, and even more slots in higher-rate interfaces. Given a large number of slots and even the modest variety of channel types that were being discussed, it would require relatively sophisticated functionality to manage the mapping of all possible combinations of slot assignments for a dynamically changing mix of services (all the sets of values for $i,...n$).

In order to somewhat simplify the mapping function, STM advocates divided the usable capacity into a limited number of fixed STM-based partitions, called containers. Each container would be permanently assigned a set of time slots. For example, a proposed interface at about 150 Mb/s may be based on an interface structure of:

$$H_4 + 4H_1 + nB + D$$

In order to increase flexibility, the containers for the H_4 and H_1 channels may be further subdivided (e.g., the H_4 container could be divided into H_2 containers, which in turn could be subdivided to carry more H_1 channels). Channels could not span containers. Strict specification of the rules for subdivision and assignment of slots to channels might provide some flexibility without all of the complexity associated with unconstrained assignment of slots to channels.

Table I. Bearer Channel Rates (kb/s)

Channel	2.048 kb/s Hierarchy	1.544 kb/s Hierarchy
B	64	
H_0	384	
H_1	1,920	1,536
H_2	32,768	about 43,000-45,000
H_4	132,032-138,240	

However, the subdivision of containers can have some undesirable consequences. For example, given the above partitioned structure, a user might request a call needing a fifth H_1 channel, which can only be completed if the capacity for the available H_4 container is carved up. After allocating the fifth H_1 channel from within the H_4 container, some of the original H_1 calls might be disconnected. Even though the total idle usable capacity on the interface is sufficient, an H_4-rate call cannot be completed because the capacity for the H_4 channel does not exist within the H_4 container. If lower rate connections using the H_4 partition have long holding times, they can effectively lock out H_4 service. To remedy this situation, it is conceivable for these slots to be reassigned dynamically to an H_1 container during a call, though this has not normally been part of the functionality of an STM-based system. Historically, STM systems never needed to reassign slots because all channels at a given level of the digital hierarchy were the same size.

Multiple-rate STM also complicates the switching system. In terms of utilization of switching bandwidth on a per-connection basis, it would be more efficient to deploy a separate switching fabric for each channel rate. However, deployment of multiple fabrics is undesirable from the perspective of the network as a whole, complicating management, provisioning, and maintenance. A multi-rate switching fabric, though more complex than a single-rate fabric, is conceivable, but it is difficult to build an efficient broadband switching capability out of 64-kb/s modularity. Thus, even with multi-rate switching capability, deployment of several fabrics may be the ultimate fate of an STM-based network.

---

[2] A B-channel is a 64-kb/s bearer channel, with the capacity required to carry Pulse Code Modulated (PCM) voice. The D-channel, which is 16 kb/s in the basic and 64 kb/s in the primary rate interface, carries signaling and possibly other packetized services.

[3] H_3 has been omitted because there is currently little interest in defining a channel in the 60-70-Mb/s range.

So far, we have focused on problems that may arise if STM-based solutions are used to carry a dynamically changing mix of services at a variety of fixed channel rates. STM is at its best when it comes to fixed-rate services. BISDN, however, needs to support an entirely different class of services—bursty services—as well. STM techniques cannot transport these services efficiently. STM channels must accommodate peak transfer rates. Higher-layer multiplexing and switching (yet another type of switching fabric) would be required to efficiently use an STM channel.

The Flexibility of ATM

ATM techniques attempt to eliminate these limitations. Usable capacity can be dynamically assigned on demand. A single fabric could conceivably be used to switch all services. And ATM-based networks may be engineered to take advantage of statistical gain among bursty services while guaranteeing acceptable performance for continuous-bit-rate services.

The ATM paradigm did not, strictly speaking, originate with BISDN. Some of its fundamental concepts can be traced back to packet switching systems. But ATM differs significantly from older store-and-forward networks. It supports a wider variety of services by adopting procedures that ensure higher grades of performance.

STM and ATM have different rules for allocating usable capacity to channels. In ATM, specific periodic time slots are not assigned to a channel. Usable capacity (bandwidth) is segmented into fixed-size[4] information-bearing units called cells. Each cell consists of a header and an information field (see Figure 2). These cells can be allocated to services on demand.

The header contains a Virtual Channel Identifier (VCI) field. The VCI is a label that is used like the STM time slot position for channel identification. Like a time slot number, a VCI value is locally significant to an interface and a cell may undergo VCI translation, analogous to time slot interchange, before it is transported over another interface.

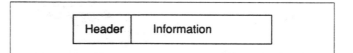

| Header | Information |

Fig. 2. ATM cell.

An ATM interface structure consists of a set of labeled, not positioned, channels. Labeled channels do not have to be restricted to a small set of fixed-rate values. The service mix and information transfer rates are decoupled from characteristics of the switching fabric. Mapping of time slots to channels at call establishment is not performed. Partitioning of the interface into containers is unnecessary, low- and high-bit-rate services sharing the same homogeneous "container."[5] Thus, ATM can be responsive to short-term fluctuations and long-term changes in service demand.

---

[4]Variable-size units at the ATM level were considered, but relatively small, fixed-size units were chosen. Small fixed-size cells may reduce queuing delay for a high-priority cell, since it waits less if it arrives slightly behind a lower-priority cell that has gained access to a resource. It was also thought, based on the state of existing experimental fast packet switching technology, that fixed-size cells could be switched more efficiently.

[5]While not necessary, an interface could still be partitioned using synchronous multiplexing for transmission purposes. It may be desirable to multiplex several ATM streams using STM techniques to achieve transmission interfaces exceeding the rate of operation of ATM switches and multiplexers. SONET is designed to provide this STM multiplexing capability.

In comparison to STM equipment, ATM multiplexers and switches are less dependent on considerations of bit rates for particular services. ATM equipment can flexibly support a wide variety of services with different information transfer rates without the complications associated with terminating a multiplicity of positioned channels. Consequently, multiple overlay rate-dependent circuit fabrics are unnecessary, enhancing network integration.

ATM may be suitable for both bursty and continuous-bit-rate services. ATM, operating in a deterministic mode, can support real-time continuous-bit-rate services by circuit-mode emulation, provided that sufficient switching and transport bandwidth is guaranteed. Bandwidth is allocated as part of a connection establishment signaling procedure that precedes end-to-end information exchange. The information transfer mode of a deterministic service must meet minimal end-to-end performance requirements for delay and loss so that a receiver can reconstitute a periodic signal in which the time interval separating instances of the signal are equal.

ATM may also be able to operate in a statistical mode to concentrate bursty data. Bursty traffic may be assigned a lower priority so that it does not adversely affect deterministic services. While ATM may not be as "efficient" as STM in terms of bandwidth utilization and total delay for a specific continuous-bit-rate service (because of additional header overhead and the time to collect a full cell of information), these factors may not significantly impair resource utilization or performance, and may be offset by ATM's overall advantages. ATM may be more efficient in carrying bursty services. Each service may not require continuous allocation of its peak bandwidth, allowing more services to statistically share the resource. There are still significant technical issues [14] and concerns over the tradeoff between statistical gain and quality of service. Some feel that the basic ATM capabilities should be deterministic, requiring peak bandwidth allocation for bursty services, with a possibility for statistical multiplexing above ATM [15]. ATM, in their view, is justified solely in terms of its support of deterministic services at any bit rate. Others are trying to deal with the statistical gain problem in order to optimize resource utilization and avoid introducing an additional level of multiplexing and switching.

In comparison to STM equipment, ATM multiplexers and switches are less dependent on considerations of bit rates for particular services.

Hybrid Solutions

Even as the merits of the new approach became increasingly apparent, it was still common to view STM as a necessary transitional link. It was assumed that a hybrid structure might facilitate gradual change from an STM-based to an ATM-based network, though the technical basis for this belief was rarely closely scrutinized.

The term hybrid has been used in several different contexts in standards discussions. There is the "hybrid network," which means that the network might contain internally separate STM-based and ATM-based fabrics. Individual interfaces could still be either ATM or STM. The UNI might be based on ATM, but some services may be segregated within the network and transported over STM overlay fabrics. Internal segregation of different types of traffic may be advantageous.

The term hybrid has mostly been used in discussions of interfaces, not networks, but even in this context, its usage has not always been clear. The distinction between hybrids on the horizontal and vertical plane has been frequently glossed over. Employing STM and ATM techniques at different levels of an interface is not the same thing as having both techniques coexist at the same level. An example of a vertical hybrid would be an interface that carries ATM cells within the payload capacity of a SONET transmission frame. It is hybrid in the sense that SONET uses STM techniques to allocate time slots between transmission overhead and usable payload. At the transfer mode level, ATM techniques are used for multiplexing user channels. A vertical hybrid permits use of the STM multiplexing capabilities of SONET to build higher-rate interfaces consisting of several ATM streams.

An example of a horizontal hybrid is an interface supporting user services directly on both STM and ATM channels. Some channels are identified by time slot, others by labels. In the past, a variety of arguments had been made in favor of this approach, some based on alleged limitations of ATM. For example, it was argued that it might not be possible to switch video services at the H_4 rate. Even if this were true, and recent evidence indicates that it is not, video could be circuit switched in cell format and the UNI payload could still be based entirely on an ATM multiplexing structure. It was also argued that time slots for B-channels were needed for voice in order to avoid the delay incurred in collecting a full cell of voice samples; however, other means for dealing with voice delay, which are to be discussed in the next section of this paper, are now being favored.

The reluctance, internationally, to embrace horizontal hybrid interfaces indicates a commitment to looking forward to the full ATM access and integrated ATM fabric, avoiding the institution of aspects of the STM paradigm into the user payload structure of BISDN. It eliminates the prospect of two modes of voice service, which conceivably might coexist on a single interface, one for STM voice and another for ATM voice. Perpetuation of a duality complicates termination of information-bearing and signaling channels.

Some Unresolved Issues

The selection of ATM has resulted in a common paradigm for the transfer mode to be used at the UNI. But it has not resolved all issues concerning the elaboration of the ATM concept. ATM specifications are influenced by considerations stemming from other functional layers of the BISDN architecture. At this time, the most salient differences confronting standardization of the UNI for BISDN are concerned with ATM parameters and the transmission structure underlying the ATM layer.

I.121 indicates that two BISDN interface rates will be standardized, one at about 150 Mb/s, the other at about 600 Mb/s. This section focuses on the 150-Mb/s interface because the constraints on interface parameters are more severe at this rate. Use of an ATM-like cell structure is not necessarily restricted to BISDN interfaces. IEEE 802.6 is defining a protocol for the Media Access Control (MAC) layer for a Metropolitan Area Network (MAN) called Distributed Queue Dual Bus (DQDB) (originally QPSX) [16], with a cell structure that is intended to conform with that of ATM. While DQDB will initially be used over a DS3 rate (44.736 Mb/s) physical interface, conforming to the ATM structure is intended to facilitate migration to BISDN.

The discussion of the technical issues is organized for presentation purposes into two "positions." One position is being formulated within T1S1, a body commissioned by the American National Standards Institute (ANSI) to develop positions on ISDN standards for North America. The other characterizes a position defined by members of the European Telecommunications Standards Institute (ETSI),[6] a regional organization that coordinates telecommunications policies in Europe.

These two positions were chosen to provide a vehicle for understanding technical issues. The standards community is not divided into well-defined, monolithic camps. There are other significant actors, including participants from Japan and Australia, that are also involved.

Table II presents system parameters pertaining to the 150-Mb/s interface for each of these two positions. The row headings in the table and the next three subsections of this paper each correspond to a layer in the architecture depicted in Figure 1. The service mix and usable transmission parameters apply only to the 150-Mb/s interface, but the ATM parameters carry over to other BISDN interfaces, i.e., the 600-Mb/s interface.

Parameter definitions for Table II are:
- C_{info}: Total information payload provided to the user by ATM in Mb/s
- H_4: Capacity in Mb/s for an H_4-rate service
- D: Capacity in Mb/s for user-network signaling
- x: Capacity in Mb/s for some narrowband services
- S_h: Length of the header portion of a cell in octets
- S_i: Length of the information portion of a cell in octets
- I: Information payload capacity in 125 μs in octets
- n: A positive integer

Table II. Interface Parameters

	T1S1	ETSI
Service Mix	$C_{info} = H_4 + D + x$	
ATM Parameters	$S_h = 5$	$S_h = 4$
	$S_i = 64$	$S_i = 32$
Usable Transmission Capacity	$I = 2,346$	$I = n \cdot [S_h + S_i]$
	$C_{info} = 10^{-6} \cdot 8,000 \cdot [8 \cdot S_i] \left[\dfrac{I}{S_h + S_i} \right]$	

Service Demand

The service mix requirements for the 150-Mb/s interface are expressed in terms of a set of fixed-rate channels (see Table II). As previously noted, the specification of BISDN channel rates is in a sense a throwback to the STM paradigm. But channel rates still have value as a guideline for service rates. These rates may serve as a basis for internally deployed overlay networks. The H_4-rate may also serve as a peak value for services, ensuring that some capacity remains available for multiplexing other services and control messages.

The service mix equation in the table gives the total ATM information payload capacity available to the user. The right side of the equation includes some narrowband services, x, which take up the slack left over by an H_4-rate service and signaling. Precise values have not been selected for any of the terms of the right side. The range of values being considered for H_4 appears in Table I. Some groups primarily interested in high-quality video services, such as Europe's Joint Committee on Television Transmission (CMTT), want H_4 as high as possible; however, selection of a smaller H_4 increases x and/or makes a lower value of C_{info} feasible. The capacity requirement for the D-channel is generally thought to be relatively low, and other lower-priority traffic may statistically share it.

There have been few recent discussions on the value of x. Arguments for letting $x = 4H_1 + 30B$, popular among STM

---

[6]Until recently, the European Conference of Postal and Telecommunication Administrations (CEPT) performed this function.

advocates, are not as compelling for an ATM-based interface. In ATM, the fifth H_1 anomaly discussed previously disappears. Thus, it may be possible to allocate all broadband and narrowband services from a slightly smaller amount of undifferentiated usable payload. The value of x is not likely to be a major stumbling block.

ATM Parameters

Table II shows differences in proposed values for ATM parameters. The discrepancies in the header size (S_h) can be attributed to different concepts of ATM functionality and differences in some field sizes. The difference in the information field size (S_i) stems from two views on the implications of interworking voice service between ATM and existing telephony fabrics.

The ATM Cell Header

In theory, the label is the only field that is essential in the ATM cell header in order to perform switching and multiplexing functions. However, other considerations, some related to enhanced performance, warrant the inclusion of other fields.

The five-octet cell header format shown on the left side of Figure 3 (each line is one octet) is the product of a coordinated effort by T1S1 and IEEE 802.6. The header contains an 8-bit Access Control Field (ACF) for multi-access interfaces, a 20-bit label, all of which is VCI, a 2-bit payload type field (TYPE) for maintenance, a 2-bit priority field (PRIORITY) and an 8-bit Header Check Sequence (HCS) field. The right side of Figure 3 shows a four-octet header format conforming to the ETSI position. It calls for a 24-bit label field containing Virtual Path Identifier (VPI) and VCI subfields, a 5-bit HCS, a 1-bit TYPE, and 2 bits of "spare capacity" (SPARE).

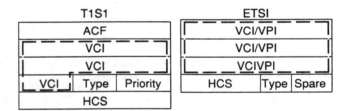

Fig. 3. ATM header formats.

One significant difference is the appearance of the ACF field in only the T1S1 format. The purpose of the ACF is to provide an efficient and reliable mechanism for multiple BISDN terminals to be connected to a shared medium. This option makes it possible to define BISDN interface configurations similar to those employed for Local Area Networks (LANs) and MANs. While T1S1 favors having the shared medium option at the UNI, many CCITT delegations want to restrict BISDN interfaces to point-to-point interfaces only. This would require multiplexing and switching of BISDN terminals on a single UNI to be performed as a centralized function. According to this view, a shared medium interface is not subject to standardization. ACF functionality for a multi-access interface is dependent on the terminal configuration, they contend, and it is undesirable to define an ACF in the ATM layer that is geared to a particular topology.

Another significant difference is the inclusion of the VPI in the ETSI label. The virtual path concept, originally conceived for cell-based cross-connect systems internal to the network, is now being considered for the UNI as well. The VPI defines an additional layer of multiplexing on a per-cell basis underneath the VCI, making it possible to group VCIs

into a virtual path that can be switched as a unit. The label field, which combines VCI and VPI, is 24 bits. The size of the subfields may vary at different interfaces but will be fixed at a particular interface, such as the UNI. Sizes for the VCI and VPI subfields, the role for a VPI at the UNI, and the relationship between virtual paths and BISDN signaling and control functions need more study.

In theory, the label is the only field that is essential in the ATM cell header in order to perform switching and multiplexing functions.

The selection by T1S1 of a 20-bit label over a 16-bit label, the first value under consideration, was based less on a firm belief that all the extra bits are needed at the UNI, but on a feeling that they provide a margin of safety. 16 bits may be sufficient for communications passing between a single subscriber and the network at the UNI. However, the additional space may be needed for other purposes. For example, some of the label space may be used exclusively for local communications, or the network may require a larger number space between remote electronics and the central office. All 20 bits are designated as VCI, but it is conceivable that some bits may be assigned for VPI functions if it is agreed that the virtual path concept is the appropriate way to cross-connect virtual channels.

The role of priority in the ATM layer is another point that remains unsettled and has caused some confusion. The confusion is partly semantic, for priority can pertain to both delay and loss. Delay priority is related to the order in which queued cells are served. Higher-priority cells would, on the average, incur less delay between two given end points, since they are served first. Changing the delay priority within a given VCI can result in resequencing of cells, which contradicts the common notion of circuit.

A second type of priority, referred to as loss priority, may not be related to the order in which cells are delivered. Loss priority determines the order in which the network discards cells in overload situations when buffer overflow is impending. Some application may be designed to take advantage of such a capability, changing the loss priority for cells belonging to a single VCI on a per-cell basis. For example, a video application may use subband coding and may be willing to accept a higher cell loss rate for some information. Some loss algorithms may be more complicated than those for delay priority. The implications of instituting loss priority on services not requiring this feature must also be studied.

The ETSI header does not include an explicit field for priority. Priority can be selected when a connection is established, with its value stored where needed (e.g., in routing tables). Loss priority cannot be changed on a per-cell basis. When the value is not carried explicitly, tables containing priority for each circuit might need to be maintained at several points throughout the network. An explicit field can reduce the number of locations; however, the network may not be able to trust a user-supplied value, especially in the delay priority case, and may need to verify it at some point against the value specified at call establishment, if applicable.

An HCS field that provides one-bit error correction capability on the label significantly improves the quality of services for virtual connections emulating circuit-mode connections. Correction of random bit errors in the VCI can prevent the loss of an entire cell. Loss of a cell, in turn, could result in loss of frame in an emulated circuit service. The HCS can also be used to detect burst errors. A 5-bit HCS can provide single-bit correction over a 24-bit field; a 6-bit HCS would be required to protect 32 bits. A longer HCS buys additional error detection capability [17]. ETSI proposed a 5-bit HCS, France recommended raising it to 6 bits, and most other delegations advocate an 8-bit field.

The Payload type field can be used for maintenance purposes. Choice of size is not likely to become a significant issue.

The Cell Information Field

The effect of the delay incurred in collecting a full cell of voice samples has been a major consideration in discussions on cell size. The current performance standards of the existing network do not tolerate much additional delay in the local access portion of a voice connection without possible degradation in the quality of service in the guise of signal loss or echo. If additional delay is incurred, measures such as deploying echo cancellation devices must be taken.

ETSI favors a 32-octet information field, as shown in S_i of Table II. This cell size would incur an additional 4 ms to collect 32 1-octet samples using 125-μs-based voice protocols. Their view is that this additional delay can be compensated for by reducing the maximum distance of connections that do not require echo cancelers. Larger information-field sizes would require echo cancelers for shorter distances.

In the telephone network in the United States and elsewhere, echo cancelers would be required in interworking units even if more than 16 octets of information are collected. The United States has stricter standards than many European countries, limiting signal loss in calls over short distances, so there would be a problem even in local calls. Consequently, an alternative method for dealing with delay, independent of the information field size, is needed. Either cells are not completely filled with voice samples of a single connection or echo cancelers must be used. Partial filling of cells may inflate the required value of x in the service mix equation of Table II. Various techniques, sometimes called composite techniques, may allow sharing of a cell among several connections. But these add an additional level of multiplexing and run counter to the objective of basing all network services on a single, homogeneous transfer mode.

One advantage of a longer cell is that it gives ATM processing units more time to process a header, though it is technically feasible to build equipment based on either size. Another is that, given the header sizes under consideration, studies [17] have shown some improvement in capacity utilization for 64 octets over the smaller-size information fields. Studies confined to the ATM layer may in fact be underestimating the cost of small cell sizes since there may be additional per-cell overhead for segmentation/reassembly and other functions in an "adaptation layer" above ATM.

Transmission and Its Impact on Aggregate Usable Capacity

The usable transmission capacity rows of Table II define the aggregate usable capacity available in a 150-Mb/s interface. The different forms of the expressions on the first line of these rows reflects an underlying disagreement on the fundamental transmission structures. Figure 4 depicts the structure of the two approaches being considered, the SONET approach favored by T1S1 and the Asynchronous Time Division (ATD) approach favored by some in ETSI.

Each rectangle in Figure 4 represents a 125-μs snapshot of a transmission signal. Transmission occurs from left to right, row by row. Cells are bounded by solid lines, and in the case of the structure in Figure 4a, a single cell can be di-

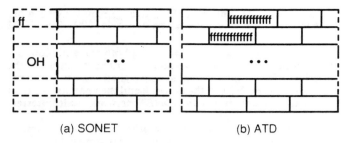

| (a) SONET | (b) ATD |

Fig. 4. Framing structures.

vided by SONET transmission Overhead (OH). If SONET is used, a framing pattern would not be required in ATM cells since cell boundaries can be derived from the 125-μs framing (ff) in the SONET overhead. All cells can be used to carry user information. In contrast, ATD, a physical interface consisting solely of cells, requires a framing pattern within the ATM stream. Since framing is essential for delimiting cell boundaries, 100% occupancy of cells is not possible. Any additional transmission and operations OH required must also appear in cells, making it difficult to determine a precise value for the usable capacity. The C_{info} value computed for an ATD interface is slightly inflated, because it does not account for OH occupying information portions of cells. One method of framing in ATD fills empty cells with a special pattern (depicted as ffffff...). When the interface is overloaded, that is, there is a 100% offered load, a framing cell would have to be periodically "forced."

T1S1 favors SONET over ATD because:

- It can be used to carry either ATM-based or STM-based payloads, making it possible to initially deploy a high-capacity fiber-based transmission infrastructure for aggregated narrowband and data services. This early deployment, which can be justified purely in pre-BISDN terms, preadapts the network for BISDN and ATM. Assuming widespread penetration of SONET for these purposes, it is sensible to continue to use SONET for BISDN rather than a new structure to perform the same functions.

- SONET offers cross-connect capabilities that can be used to transport aggregations of ATM connections within the network. This is an alternative to the VPI-based cross-connect function mentioned previously. A SONET-based cross-connect, in contrast to a virtual path cross-connect, may not be quite as flexible in meeting instantaneous changes or in statistically allocating bandwidth. These features, however, may not be desirable in a cross-connect system, since the objective in its design is to avoid complexity and potential complications if congestion is allowed to occur.

- The cells of a specific connection may be "segregated" and circuit switched using a "SONET channel." For example, an H_4-rate video connection may be demultiplexed and then mapped into its own exclusive payload envelope of the SONET STS-3c (155 Mb/s) signal, which can be circuit switched in the network. It is possible to define payload envelopes at a variety of rates and the channels can be subdivided using SONET's virtual tributary capabilities [9]. The payload can remain in cell format, making STM/ATM format conversion unnecessary, or it may be converted to a more STM-oriented format. STM-like switches of this type can be used internally as an interim measure, before ATM switches are available. It may even prove to be more cost effective than ATM switching for handling some services, such as commercial video distribution.

- SONET synchronous multiplexing capabilities can be used to combine several ATM streams to build interfaces at rates greater than those supported by the ATM layer. This technique may be employed inside the network and at the UNI. For example, a 622-Mb/s (STS-12) interface can be built from four 155-Mb/s (STS-3c) modules, each one carrying a separate ATM multiplex. This arrangement may be more cost effective than one using a single 600-Mb/s ATM stream.

ATD proponents argue that basing both transmission and transfer mode functions on a common structure simplifies the interface. This contention must be weighed against the versatility of using SONET in the network and the benefits of a UNI that has commonality with SONET network node interfaces.

The differences in the first line of the usable transmission capacity rows of Table II can be explained in terms of the structures in Figure 4. SONET provides a fixed-rate payload envelope, I, which is computed by subtracting SONET overhead from the SONET interface rate. In ATD, there is greater freedom in determining I. In the past, it has been suggested that an integer number of cells occur in a 125-μs period, making I a multiple of the cell size and an integer value n. This condition is not essential for ATM, since there is no requirement that synchronization cells be transmitted periodically, let alone fall on a 125-μs boundary; however, periodic framing makes it possible to align the framing patterns for synchronous multiplexing without introducing an external framing pattern and 125-μs periodicity makes it simpler to derive 125-μs clock from the signal, which may simplify provision of some services.

The second line of the usable transmission capability row computes C_{info}, the total usable capacity provided by ATM to its users. Adopting a SONET-based pointer scheme proposed by AT&T for delimiting ATM cells [18], T1S1 uses all 2,346 octets of payload in a SONET STS-3c frame for carrying ATM cells, yielding a C_{info} of 139.264 Mb/s. The ETSI ATD approach, in contrast, makes it possible to derive I from the other parameters. Thus, issues dealing with service mix and ATM parameters may be settled first. The ATM parameters favored by ETSI, $S_h = 4$ and $S_i = 32$, incur a relatively high percentage of ATM header overhead, yielding a C_{info} of only 133.120 Mb/s in a SONET STS-3c.

The ETSI system is molded by their perspective on interworking voice communications with shorter cells. By and large, they have ignored the implications of choosing SONET for a UNI on their selection of ATM parameters. T1S1, on the other hand, takes SONET at the UNI as a given and looks beyond the voice delay problem, seeing no significant difference in the implications of 32- versus 64-cell information fields for voice and emphasizing other applications for its selection of ATM parameters.

Conclusions

Some substantial issues stand in the way of achieving the CCITT objective of a unique worldwide interface for BISDN. However, standards organizations in the past have pulled things off when least expected. The agreement on the SONET network node interface is a good example. CCITT Study Group XVIII Working Party VII overcame many obstacles to make it acceptable to the international community [9].

The CCITT Study Group XVIII Broadband Task Group has laid the foundation for an agreement on a unique international standard for the BISDN UNI by reaching consensus on the selection of the ATM paradigm, but significant differences remain. The standards organizations are working against the clock. Business pressure to provide high-bandwidth communications over wider areas is growing. Inability to agree in a timely fashion increases the likelihood of regional solutions. CCITT called an interregnum meeting for January 1989 to try to expedite international agreement, hoping to achieve a standard in 1990. Compromise solutions were explored at the January meeting, but not much progress was made. This too is reminiscent of the SONET experience.

The focus of discussions in the Broadband Task Group have tended to emphasize transmission, switching, and

multiplexing. Discussions of other aspects of BISDN are in the embryonic stage. In the future, the emphasis is likely to shift more in the direction of higher-layer capabilities, including adaptation of higher-layer functions to ATM, new service-specific functions, and signaling. There has been little discussion of multimedia calls and services, which have a tremendous potential for thriving in a BISDN environment. Multimedia services may require rethinking some of our concepts [19].

Reaching agreement on the fundamental transmission, multiplexing, and switching aspects of BISDN is an important step in building the intelligent broadband network of the future.

Reaching agreement on the fundamental transmission, multiplexing, and switching aspects of BISDN is an important step in building the intelligent broadband network of the future. Defining the transport capabilities for BISDN is a complex problem, requiring a solution taking transmission, switching, and services factors into account. Involvement of many different organizations in this process has complicated the situation further. The task of achieving a solution that is both forward-looking and backward-compatible is not trivial. The payoff for coming up with a workable solution could be immense.

Addendum

The results of the latest CCITT Study Group XVIII meeting, held in June 1989, include agreements on some ATM and transmission parameters. ATM cells will consist of a 5-octet header (S_h) and a 48-octet information field (S_i). For the UNI, the header includes:

- A 4-bit "generic flow control" field, which may be used to perform some of the functionality of an ACF field, though it can also be used for point-to-point
- A 24-bit label, with up to 12 bits for VPI and the remaining 12 to 16 bits for VCI
- A 2-bit payload type field
- A 2-bit reserved field
- An 8-bit header error check field

There is no explicit priority indication.

CCITT will standardize two physical interfaces to BISDN, one based on SONET, the other a variation of ATD. The UNI interface rate for both of these was set at 155.520 Mb/s, with an interface transfer capacity equivalent to 149.760 for carrying cells containing user information. Thus, the value of I is 2,346 octets on both interfaces, and C_{atm} is about 135.979 Mb/s. The H_4 rate may be further constrained by additional per-cell overhead at the adaptation layer. If a 1-octet-per-cell adaption layer header for continuous-bit-rate services is added, C_{atm} is effectively reduced to about 133.146 Mb/s.

References

[1] J. L. Gimlett, M. Stern, R. S. Vodhanel, and N. K. Cheung, "Transmission Experiments at 560 Mb/s and 140 Mb/s Using Single-Mode Fiber and 1300-nm LEDs," *Electron. Lett.*, vol. 21, pp. 1,198–1,200, Dec. 1985.

[2] G. Hayward, L. Linnell, D. Mahoney, and L. Smoot, "A Broadband ISDN Local Access System Using Emerging-Technology Components," *Proc. of Int'l. Switching Symp. '87*, vol. 3, pp. 597--601, 1987.

[3] J. S. Turner, "New Directions in Communications," *IEEE Commun. Mag.*, vol. 24, no. 10, pp. 8–15, Oct. 1986.

[4] J. P. Coudreuse and M. Servel, "Prelude: An Asynchronous Time-Division Switched Network," *Proc. of the IEEE Int'l. Conf. on Commun. '87*, vol. 2, pp. 769–773, 1987.

[5] C. Day, J. Giacopelli, and J. Hickey, "Applications of Self-Routing Switches to LATA Fiber Optic Networks," *Proc. of Int'l. Switching Symp. '87*, vol. 3, pp. 519–523, 1987.

[6] D. Spears, "Broadband ISDN—Service Visions and Technological Realities," *Int'l. J. for Analog and Digital Cabled Syst.*, vol. 1, no. 1, pp. 3–18, Jan. 1988.

[7] S. Minzer, "Broadband User-Network Interfaces to ISDN," *Proc. of the IEEE Int'l. Conf. on Commun. '87*, vol. 1, pp. 364–369, June 1987.

[8] CCITT Recommendation I.121, "Broadband Aspects of ISDN," Blue Book, Geneva, Switzerland, 1989.

[9] R. Ballart, Y. C. Ching, "SONET: Now It's the Standard Optical Network," *IEEE Commun. Mag.*, vol. 29, no. 3, pp. 8–15, Mar. 1989.

[10] CCITT Recommendation G.707, "Synchronous Digital Hierarchy Bit Rates," Blue Book, Geneva, Switzerland, 1989.

[11] CCITT Recommendation G.708, "Network Node Interface for the Synchronous Digital Hierarchy," Blue Book, Geneva, Switzerland, 1989.

[12] CCITT Recommendation G.709, "Synchronous Multiplexing Structure," Blue Book, Geneva, Switzerland, 1989.

[13] T. Kuhn, *The Structure of Scientific Revolutions*, 2nd ed., Chicago, IL: University of Chicago Press, 1970.

[14] J. Gechter and P. O'Reilly, "Conceptual Issues for ATM," *IEEE Network*, vol. 3, no. 1, pp. 14–16, Jan. 1989.

[15] B. Eklundh, I. Gard, and G. Leijonhufvud, "A Layered Architecture for ATM Networks," *Proc. of IEEE Globecom '88*, pp. 409–413, Nov./Dec. 1988.

[16] R. M. Newman, Z. L. Budrikis, and J. L. Hullett, "The QPSX MAN," *IEEE Commun. Mag.*, vol. 26, no. 4, pp. 20–28, Apr. 1988.

[17] D. P. Hsing, F. Vakil, and G. H. Estes, "On Cell Size and Header Error Control of Asynchronous Transfer Mode (ATM)," *Proc. of IEEE Globecom '88*, pp. 394–399, Nov./Dec. 1988.

[18] T1S1.1/88-390, "Mapping BISDN ATM Cells into SONET STS-3c Payload," AT&T Technologies, Oct. 1988.

[19] S. E. Minzer and D. R. Spears, "New Directions in Signaling for Broadband ISDN," *IEEE Commun. Mag.*, vol. 27, no. 2, pp. 6–14 and 52, Feb. 1989.

Biography

Steven E. Minzer is a Member of Technical Staff in the Multimedia Communications Research Division of Bellcore, Morristown, New Jersey. He has been involved in Broadband ISDN standards activities since 1985, participating in Committee T1 and CCITT Study Group XVIII meetings. He has also been involved in protocol design for an experimental broadband network at Bellcore. Before joining Bellcore, he was with AT&T Bell Laboratories, designing and implementing software for the DMERT operating system. He also worked in database administration and software development for GTE Automatic Electric. He obtained degrees in political science from Cornell University (B.A.) and Ohio University (M.A.) and in computer science from Northern Illinois University (M.S.). His publications include several papers and a Bellcore Special Report on BISDN and articles in *ISS '87* and *ICC '87* on protocols for the experimental broadband network.

The Evolution of Broadband Network Architectures

D. V. Batorsky, D. R. Spears and A. R. Tedesco

Bellcore
Red Bank, NJ 07701

ABSTRACT

Market pull and technology push are the major forces driving future broadband networks. Firstly, this paper examines these two driving forces. Secondly, it describes a long-term broadband network architecture. Finally, it describes possible evolutionary paths from early-availability architectures through the longer-term.

1. BROADBAND SERVICES MARKET PULL

New service opportunities are a major driving force for future broadband networks [1] [2] [3]. Business customers have a growing need for information technology and increasingly sophisticated telecommunications. Local Area Networks (LAN) have increased the intrapremises data rates from kbit/s to Mbit/s over the past decade, but interconnection of these LANs is still a problem.

Residential customers are demanding more selectivity in video entertainment, evidenced by the increasing number of channels demanded on cable television systems and the increasing rentals of video cassettes. Future entertainment services may range from viewing normal broadcast video sources, to viewing special events such as first-run movies on a pay-per-view basis, to video-on-demand in which a user is connected directly to a remotely-located video source with complete control over that source.

This section discusses some of the existing and future services that a broadband network should support.

1.1 Existing Services

A broadband network should support existing voice telecommunications services. Other voice services, such as 7 kHz and 15 kHz speech communications, may also become important, especially for business customers.

A broadband network should support existing data communications services, such as those that provide features for X.25 and SNA applications. In addition, higher-speed data services, discussed in the next section, will become increasingly important.

A broadband network should support existing private line services, including the DS-0, DS-1 and DS-3 rates. New innovations in the signaling protocol for broadband networks may enhance these capabilities by allowing private virtual networks to be defined dynamically.

1.2 High-Speed Data Services

The widespread penetration of LANs has firmly established very high-speed networking (1-10 Mbit/s) in the local premises environment. The success of LANs can be attributed to the growing use of more powerful PCs and workstations for high performance applications such as remote file serving and distributed processing. The growing need for LAN interconnection and such applications as desktop publishing and CAD/CAM suggest an emerging market for multi-megabit communications services that extend beyond the local premises, across metropolitan and wide area environments. Switched Multi-megabit Data Service (SMDS) is a high-speed connectionless data service proposed for broadband networks to meet these future data communications needs [4].

1.3 Video and Image Services

The availability of high-speed communications can completely change the character of business interactions. In desktop teleconferencing applications, windows on the user's terminal can display full-motion images of the other parties in a conference. Viewgraphs, sketches, photographs, and text files can be exchanged and either displayed or printed. Other parties can be added to a conference as needed for consultation, and subconferences created for private conversations.

Electronic mail, commonly used today for transmitting text messages, can be stimulated by the availability of broadband networks. Integrating multiple media, including text, voice, image and full-motion video, can increase the effectiveness of these systems.

Switched access television service, the delivery of entertainment video to residential customers over a switched broadband network, can tap a large existing market and offer several advantages over current delivery alternatives. These advantages include access to a wider variety of programming, elimination of the need for scrambling equipment, opportunities for new services such as pay-per-view, and higher quality video such as high definition television (HDTV).

Recent increases in the sale and rental of video recorders, players and cassettes show that users want to control their own programming. Video-on-demand service provides the user with ultimate control over video programming without having to purchase and maintain a video cassette player and without having to pick up and return video cassettes. Instead, a user calls a video vendor and selects a program from a list of available video sources. Video information from the vendor is switched to the user via the broadband

EH0301-2/91/0000/0090/$01.00 © 1988 IEEE

network, and control signals from the user (play, stop, pause, rewind, etc.) are transmitted to the vendor over a data connection.

1.4 Other Future Services

Other future broadband services include access to reference information. Text, graphics and images can be viewed on a student's terminal, printed on a local printer, or stored in a disk file on the student's PC. Some people predict that broadband networks will completely change the way data is viewed. Just as the telephone system displaced the telegraph in the nineteenth century, *telesophy*[1] may displace today's simplistic data communications [5]. Telesophy advocates predict a new mass communications system that will explicitly support the sharing of knowledge within a global information community.

Video browsing may become popular for shopping. For example, by calling a vendor via a broadband network, a shopper can view a full-motion demonstration of a riding lawn mower illustrating its maneuverability around trees and its ease of operation. With interactive capability, the seller can tailor the presentation to the level of detail desired by the buyer. A complete home merchandising system can provide cost and availability information to the buyer, allow the buyer to order the product and specify shipping information, and allow the seller to update an inventory system automatically. Real estate sales is another possible application. A person transferring to a different city can access video information remotely about homes for sale, schools, churches, shopping centers or maps of the new community.

2. BROADBAND TECHNOLOGY PUSH

Over the past few years, broadband technology has emerged as an attractive alternative for future telecommunications networks. It promises an integrated access structure with flexible allocation of bandwidth, capable of providing existing services and positioned to serve a broad range of future services. It promises a more integrated switching fabric, eliminating the need for a switching fabric for each service type and the associated complexity and high operations costs. Finally, it promises to simplify interoffice transmission and switching facilities.

Several parallel advances are responsible for the shift in economics of integrated broadband networks.

2.1 Optical Fiber

Optical fiber is used extensively in the interoffice plant. It is now starting to be used in the loop between the Central Office (CO) and subscriber carrier locations. The cost of single-mode fiber, which offers virtually unlimited bandwidth, dropped from 5 dollars/m in 1982 to about 23 cents/m in 1987. With continued progress some experts predict that the cost will drop to 4 cents/m by 1992.

---

1. Telesophy is a word coined by Bruce Schatz of Bellcore. It comes from the words telephony and philosophy and means "wisdom at a distance". The "tele" part indicates that the access can be made independent of physical location in the communications network. The "sophy" part indicates that filtering can be done independent of physical data type.

2.2 Low-Cost Optical Devices

Until recently, single-mode fiber implied the need for laser transmitters. However, single-mode laser sources had inadequate reliability for use in the loop and were not compatible with near-term cost objectives. On the other hand, lower cost light-emitting diodes (LED) as used in low-bit-rate multi-mode systems were known to meet loop reliability requirements. It has now been demonstrated that light emitting diodes can be used with significant lengths of single-mode fiber at bit rates in the 600 Mbit/s range [6].

More recent work has been aimed at reducing the cost of laser sources through simple packaging and relaxed yield criteria [7]. As laser technology for loop applications matures and as production quantities reach levels characteristic of the loop market, laser costs and reliability should improve. Thus, both LED and laser sources now show the potential for use as low-cost optical sources for integrated broadband networks.

2.3 Low-Cost Broadband Circuit Switches

Much progress has been made in broadband circuit switching technology. Just a few years ago, video switches used discrete relays or *pin* diodes to implement an analog switching fabric and were both large and costly.

Several experimental devices, using both CMOS and ECL technologies, have been fabricated for space-division circuit switching of broadband channels. Using such VLSI technologies and new packaging methods, it is now possible to create a compact 64x64 switching subsystem with a 4x4 matrix of 16x16 CMOS crosspoint chips on a single card measuring four inches square [8]. In addition to the advantages of digital switching and smaller size, this board represents about a 10:1 reduction in the size and the cost per crosspoint over that of the analog switching fabric.

2.4 Synchronous Optical NETwork (SONET)

The advances in synchronous multiplexing technology, the widespread deployment of incompatible fiber systems, and the need for flexible management of the bandwidth of fiber transmission systems led to the formulation of the Synchronous Optical NETwork (SONET) concept [9]. SONET is a network hierarchy of octet-interleaved synchronous signals. This hierarchy forms a family of standard electrical synchronous transport signals (STS) that can be simply converted to a family of standard optical carriers (OC). Through these STS and OC interfaces, SONET provides a set of standard transmission interfaces for customer premise equipment, switches, distribution and interoffice transport systems.

The STS rates are integer multiples of 51.730 Mb/s. The interface structure contains separate overhead and information payload. As a result operations, administration and maintenance capabilities are an integral part of the transport systems. Currently, there is a significant interest in introducing SONET based transport in exchange carrier networks when the equipment becomes available over the next several years. Moreover, SONET is proposed as a basis for defining broadband user-network interfaces for ISDN.

2.5 Label Multiplexing

Position multiplexing, also called Synchronous Transfer Mode (STM), is used for multiplexing channels into a common transmission stream, such as DS-1 or DS-3. An STM stream has a fixed channel structure; each bit within the payload is associated with a specific channel, which has a specific information transfer rate.

With label multiplexing, also called Asynchronous Transfer Mode (ATM), each frame is divided into cells that are similar to small packets [10] [11]. Each cell contains a fixed-length header and a fixed-length payload. A label in each header explicitly identifies the channel. Therefore, multiplexing and demultiplexing hardware is simplified and the structure of the transmission stream can be changed dynamically. Furthermore, due to the fixed size of the cells, STM can be emulated by ATM. ATM has been proposed for use in the SONET payload of integrated broadband networks.

2.6 ATM Cell Mergers and Sorters

Cell mergers and sorters provide flexible handling of ATM cells and have a variety of uses [12]. For example, cell mergers in the Interface Module (IM) can multiplex information from many different users onto a single STS-3c signal for transmission to the CO; a cell sorter in the CO can provide the demultiplexing function. A cell sorter in the IM can separate video information from lower speed signals and route it to a circuit switch. A cell merger in the IM can combine a circuit-switched video stream with lower-speed voice and data streams for delivery over a single fiber to the customer.

2.7 ATM Cell Switching

The technology required for broadband cell switching is much different from that used in lower speed packet switching networks. One objective of conventional store-and-forward packet switching networks is improving the error performance of noisy facilities. Copies of packets are held at the transmitting side of a link until acknowledged by the receiving side of the link. Using optical fiber transmission, the error performance of a broadband network is expected to be much better than that of current networks, so retransmission is much less likely. The primary objectives for broadband cell switching are low delay for traffic such as speech, high throughput for traffic such as still frame video, low cell loss rate, and low internal buffer requirements for minimizing equipment costs.

One promising technology for broadband cell switching is the banyan routing network. A cell processor inserts a routing field, with the most significant bit first, into each cell delivered to the input of the banyan switch. The first stage of the banyan switch examines the first bit of the routing field and routes to the top or the bottom half of the second stage based on the value of this bit. The second stage examines the second bit of the routing field and routes to the corresponding fourth of the network. This process continues through each of the N stages. A switch with N stages has 2^N inputs and 2^N outputs and is composed of $N * 2^{N-1}$ switching nodes, with two inputs and two outputs each.

If a banyan routing network is preceded by a Batcher sorting network, the combination is non-blocking internally. A 32x32 Batcher sorting network and a 32x32 banyan routing network have been implementing in VLSI; the current chips operate at over 100 Mbit/s per port with the next version expected to operate at greater than 155 Mbit/s [13].

2.8 Broadband System Prototypes

Several broadband system experiments are proceeding, including The Prelude switch in France[14], the Elastic Basket switch in Japan [15], the Knockout switch at Bell Laboratories [16], the Experimental Research Prototype at Bellcore [17] [18] [12], and the PARIS switch at IBM [19]. Each of these experiments focuses on different aspects of broadband packet switching - the interface structure, the services, the switching fabric, the protocols and the addressing - and each experiment represents a different solution to broadband packet switching.

3. A LONG-TERM BROADBAND ARCHITECTURE

Figure 1 shows a view of a long-term target architecture for a broadband network. An interworking unit (IWU) at the customer's premises provides the conversion between various customer interfaces and an integrated user-network interface. This user-network interface uses SONET transmission at the STS-3c basic rate of 155.52 Mbit/s or at a higher rate such as 622.08 Mbit/s, depending on the service needs of the customer. ATM cells within the SONET payload provide flexible multiplexing in order to support a wide range of existing and future services.

In the distribution and drop plant between the Remote Electronics (RE) and the customer, single-mode fiber is dedicated to the customer, forming a star topology[2]. Two fibers can be provided to each customer, one for upstream and one for downstream transmission, or both directions of transmission can share the same fiber using WDM techniques. The cost difference between these two alternatives is slight, with short loops favoring two fibers and longer loops favoring one fiber [20]. Operations costs, while not quantified, may favor the use of a single fiber.

Single-mode fiber is also used in the feeder plant between the RE and the CO, but operates at a higher rate near 2.4 Gbit/s. Although the RE-to-CO distances are longer than the customer-to-RE distances, the feeder costs are offset by the concentration provided by the RE and by the higher transmission rate in the feeder and represent less than 5% of the total cost [20].

Customers near the CO will be served directly from an IM in the CO. Customers further from the CO may be more economically served from an IM located remotely [20], trading off the additional cost of placing the IM in a remote location against the cost of longer lengths of dedicated fiber.

Several topologies are possible for interconnecting the RE locations to the CO. In a star topology, dedicated fiber connects each RE to the CO. In a ring topology, a fiber ring

---

2. A recent T1S1.1 contribution proposed another level, called a Remote Multiplexer (RM), between the RE and the customer in order to increase sharing of the distribution plant. The fiber between the RE and the RM is called the sub-feeder plant and can use either a star, bus, or ring topology.

starts at the CO and passes through each RE using add/drop multiplexers to drop and insert channels. A combination star/ring is also possible for the feeder topology in which switched channels are interconnected in a star topology and broadcast channels in a bus topology. The latter alternative may offer some savings, but the effect on the overall cost is slight since the feeder cable represents only a small portion of the total cost.

The physical implementation of a Next Generation Switch (NGS) [21] may incorporate different switch fabrics for different traffic types. For example, ATM cells carrying video information may be separated from cells carrying lower-speed information; video cells may routed to a circuit switching fabric while cells of other traffic are concentrated by ATM mergers in the IM and routed to a self-routing Batcher-banyan switching fabric. Although the physical implementation may incorporate multiple switching fabrics, the fabric appears to be integrated from the point of view of switch control and from the customer's perspective.

User-network and network-network signaling are integrated, with the same signaling protocol being used for both circuit and packet communications. Multi-media and multi-party communications are an integral part of these signaling procedures, facilitating the development of future interactive applications.

Hub offices provide the interconnections between CO locations. Like the feeder plant the interoffice plant connecting the CO locations to the Hub office may be connected in various topologies, such as star, ring or bus. If the community of interest is high enough, two COs may also have direct trunk groups in addition to connections to the Hub office.

Narrowband networks will continue to exist for many years. Interworking units (IWU) in central offices and hub offices provide the necessary conversions for interconnecting to other networks. In addition, fiber loops may be installed in exchanges with existing narrowband offices. An IWU in these offices provides conversion between the SONET interface on the fiber and an Integrated Digital Loop Carrier (IDLC) interface to the existing switch.

Integrated network operations and maintenance systems provide continuous surveillance of the network, reconfiguring the network during failure conditions to provide continuous service, isolating the cause of the failure, and dispatching personnel to repair the failed component. Integrated administration systems provide rapid service provisioning and accurate billing.

If coherent optics or dense WDM systems become practical and low-cost, Fig. 1 may change as follows. The active electronics in the RE may be moved back to the CO and multiple wavelengths used on the RE-to-CO fibers. This provides dedicated channels from the CO to the RE without increasing the number of fibers in the feeder plant. From a first cost standpoint, this alternative trades off the added cost of WDM equipment on a per-line basis against the cost of 2.4 Gbit/s multiplexers on a per-feeder basis. Also, if only passive components remain in the RE, the operations costs may be lower. In either case, the physical architecture of the installed fiber is identical independently of whether the RE contains active electronics.

4. EVOLUTIONARY PATHS TOWARD A BROADBAND NETWORK ARCHITECTURE

The evolution toward the broadband network architecture described in the previous section will be constrained by several factors. First, standards for broadband are still being discussed, although much progress has been made over the past year, particularly in standardizing SONET. Second, even after broadband standards are complete, the development of a next generation switch (NGS) is likely to require several years. During this time, some components of the NGS will become available earlier than others, resulting in network architectures using combinations of old and new technology. Third, initial broadband systems will be more expensive since they are at the beginning of the cost learning curve. Finally, even after the cost of broadband systems reach the lower part of the learning curve, capital constraints will limit the rate at which broadband systems can be installed.

In light of these factors, an evolutionary strategy for introducing broadband network elements is needed that satisfies several conditions. First, it should allow some broadband services that are already in demand, such as high-speed data services, to be offered using early availability technology. Second, it should allow fiber to be installed for growth in the loop plant instead of copper, at a cost competitive with copper, in order to be prepared for the provision of future broadband services. Finally, the evolutionary path should allow cost effect upgrades as NGS systems become available.

The next two sections outline a possible evolutionary strategy for residential and business network architectures.

4.1 Evolution of Network Architectures for Business Services

The desire to provide early availability of high speed data services coupled with current development of Metropolitan Area Networks (MAN) equipment, leads to the possible architecture implementation shown in Fig. 2. A gateway, G, provides the conversion from a LAN protocol to the Switched Multimegabit Data Service (SMDS) protocol carried in a DS-3 rate transmission stream. The DS-3 electrical signal is carried to the CO by an optical transmission system with a proprietary interface. In the CO, the DS-3 signal is cross connected to a multiplexer and carried to a hub office, again using a transmission system with a proprietary interface. At the hub office the DS-3 signals are connected to a MAN switching system (MSS), where each SMDS packet is segmented into cells and routed to the destination port.

As SONET transmission systems become available, the architecture of Fig. 2 allows for evolution of the network to include SONET transmission systems with standard interfaces. Furthermore, when ATM mergers and sorters become available, the architecture shown in Fig. 2 evolves to that shown in Fig. 3. Standard SONET transmission is used on the optical fibers with ATM cells carried in the SONET payload. Segmentation and reassembly of SMDS packets into ATM cells is done at the customer's premises. ATM mergers and sorters in the COs reduce the interoffice transmission requirements. Finally, the MSS can be based on an ATM merger/sorter pair operating in a MAN emulation mode with static translation tables.

As the demand for SMDS service increases, the throughput capacity of the ATM merger in the hub office is reached. At that point, a higher capacity ATM cell switch can be placed between the ATM sorter and merger, as shown in Fig. 4. A cell switch with N ports can route cells from any of the N mergers at the input ports to any of the N sorters at the output ports, increasing the capacity of the system by a factor of N. As the data traffic further increases, cell switches are placed in the COs to handle intra-CO traffic, reducing the load on the interoffice links. Cell switches in the COs also allow for direct trunks between COs with a high community of interest, further reducing the load on the interoffice links.

Voice services can also be provided over the same fibers in each of the three architectures shown above. In Fig. 2 the proprietary transmission can provide multiple DS-3 channels over the fiber. For example, one DS-3 channel may carry voice traffic to the existing voice switch via an Integrated Digital Loop Carrier (IDLC) interface. In Fig. 3 the voice information is carried in ATM cells on the user-network interface, split off from the data traffic in the CO using ATM sorters, and routed to the existing voice switch. When ATM cell switches are installed in a CO, the voice traffic may be carried by the cell switch, with the interworking unit moved to the trunk side of the office. This evolution is discussed further in the following section.

4.2 Evolution of Network Architectures for Residential Services

For growth of voice services in the loop, optical fiber is becoming a cost-effective alternative to copper. In addition, it positions the loop for providing future broadband services.

Several suppliers are now developing equipment for providing POTS service over fiber using proprietary interfaces. After completion of the broadband standards for SONET and ATM, standard equipment will become available for providing POTS service using the architecture shown in Fig. 5. The IWU provides the BORSCHT functions and places speech samples into ATM cells. At the RE, the cells from several users are merged together into a common stream, reducing the transmission requirements between the RE and the CO. At the CO, the aggregated SONET/ATM interface is converted to an IDLC interface compatible with the existing voice switch.

As the amount of fiber in an exchange increases, the existing voice switch may be replaced with an ATM cell switch capable of switching both voice and data services. The architecture shown in Fig. 5 is easily evolvable to that shown in Fig. 6, since the electronics in the RE are unchanged. When the cell switch is installed in the CO, an IWU on the trunk side of the office replaces the IWU on the line side.

A loop based on optical fiber has the capability to support video distribution services in addition to voice and data services. The architecture shown in Fig. 7 is one alternative for early availability video distribution. This architecture is similar to that shown in Figs. 5 and 6, except a second wavelength is multiplexed onto the fiber using WDM techniques. This second wavelength may carry several analog video signals using techniques such as frequency-modulated (FM) microwave subcarriers [22]. The incremental cost of adding video service in this architecture is the cost of the optical source and the WDM equipment in the RE, and the cost of the tuner at the customer's premises.

A second alternative for providing video distribution service is shown in Fig. 8. A channel-change message from the user results in a crosspoint being closed in the circuit switch. Thus, the selected channel is switched to the input of the ATM merger. This alternative uses ATM for multiplexing the voice, video, and data signals into a single transmission stream carried by a single wavelength. The incremental cost for providing video service with this architecture is the cost of the merger and the allocated cost of the circuit switch in the RE, and the cost of an ATM cell sorter and a codec at the customer's premises.

The architectures shown in Figs. 7 and 8 provide different service capabilities. The first architecture provides broadcast access to a limited number of video signals, but does not constrain the number of channels simultaneously accessible from that limited set. The second architecture provides switched access to an unlimited number of video sources, but limits the number of sources available simultaneously. The incremental cost of accessing an additional channel with the first architecture is the cost of an additional tuner. The incremental cost of accessing an additional channel with the second architecture is the allocated cost of an additional port on the circuit switch and the cost of an additional codec. Also, depending on the video coding algorithm, the line rate requirements may change. For example, a 155.52 Mbit/s signal is expected to carry one HDTV signal or three NTSC signals, while a 622.08 Mbit/s signal carries up to four HDTV signals.

Other early-availability architectures are also possible for carrying video signals to residential customers, but are not discussed here.

5. CONCLUSIONS

This paper described possible services providing a market pull for broadband networks and recent technological advances providing a technology push. A long term broadband network architecture was described based on a double star topology, optical transmission, and high-speed circuit and packet switching. Finally, evolutionary paths for broadband network architectures were described, which gradually introduce broadband network components and capabilities as they become available in order to provide early-availability of some broadband services.

6. ACKNOWLEDGEMENTS

We would like to acknowledge the work of Tom Helstern, Craig Valenti and Mih Yin and the helpful comments of Howard Bussey, Harry Cykiert, Lloyd Linnell, Stagg Newman, Howard Sherry, and Steve Walters.

REFERENCES

1. P. E. White, "The Broadband ISDN - the Next Generation Telecommunications Network," *Proc. ICC*, June 1986.

2. D. R. Spears, "Broadband ISDN Switching Capabilities from a Services Perspective," *J. Select. Areas in Commun.*, Oct. 1987.

3. D. R. Spears, "Broadband ISDN - Service Visions and Technological Realities," *Int. J. Digital and Analog Cabled Systems*, Vol. 1, April 1988.

4. C. F. Hemrick, R. W. Klessig and J. M. McRoberts, "Switched Multi-megabit Data Service and Early Availability Via MAN Technology," submitted to *IEEE Communications Magazine*.

5. B. R. Schatz, "Telesophy - System for Manipulating Knowledge," *Proc. Globecom*, Nov. 1987.

6. J. L. Gimlett, M. Stern, R. S. Vodhanel and N. K. Cheung, G. K. Chang, H. P. Leblanc, P. W. Shumate and A. Suzuki, "Transmission Experiments at 560 Mbit/s and 140 Mbit/s Using Single-Mode Fiber and 1300nm LEDs," *Electron. Lett.*, Vol. 21, Nov. 1985.

7. L. A. Reith, P. W. Shumate and Y. Koga, "Laser Coupling To Single-Mode Fibre Using Graded-Index Lenses And Compact Disc 1-3 μm Laser Package," *Electron. Lett.*, Vol. 22, July 1986.

8. G. Hayward, A. Gottlieb, S. Jain and D. Mahoney, "CMOS VLSI Applications in Broadband Circuit Switching," *IEEE J. Select. Areas Commun.*, Oct. 1987.

9. "American National Standard for Optical Interface Rates and Formats," ESCA Committee, T1 Draft, Sept. 1987.

10. M. W. Beckner, T. T. Lee and S. E. Minzer, "A Protocol and Prototype for Broadband Subscriber Access to ISDN's," *Proc. ISS*, March 1987.

11. L. T. Wu, S. H. Lee and T. T. Lee, "Dynamic TDM: A Packet Approach to Broadband Networking," *Proc. ICC*, June 1987.

12. H. E. Bussey and F. D. Porter, "A Second Generation Prototype for Broadband Integrated Access and Packet Switching," *Proc. Globecom*, Nov. 1988.

13. C. Day, J. Giacopelli and J. Hickey, "Applications of Self-Routing Switches to LATA Fiber Optic Networks," *Proc. ISS*, March 1987.

14. J. Coudreuse, M. Servel, "Prelude: An Asynchronous Time-Division Switched Network", *Proc. ICC*, June 1987.

15. S. Morita, T. Katsuyama, I. Kazuhiko, and H. Hayami, "Elastic Basket Switching - Application to Distributed PBX", *Proc. ISS*, March 1987.

16. K. Eng, M. Hluchyj and Y. Yeh, "A Knockout Switch for Variable-Length Packets", *Proc. ICC*, June 1987.

17. G. Hayward, L. Linnell, D. Mahoney and L. Smoot, "A Broadband ISDN Local Access System using Emerging-Technology Components," *Proc. ISS*, March 1987.

18. L. R. Linnell and D. R. Spears, "A Broadband System Prototype," *Proc. NCF*, Sept. 1987.

19. I. Cidon, I. Gopal and H Heleis, "PARIS - An approach to integrated private networks", *Proc. ICC*, June 1987.

20. K. W. Lu and R. S. Wolff, "Cost Analysis for Switched Star Broadband Access", *Proc. Globecom*, Nov. 1988.

21. "Next Generation Switch Symposium," Bellcore Special Report, SR-TSY-000755, Issue 1, June 1987.

22. R. Olshansky and V. A. Lanzisera, "60-Channel FM Video Subcarrier Multiplexed Optical Communications System," *Electron. Lett.*, Vol. 23, No. 22, Oct. 1987.

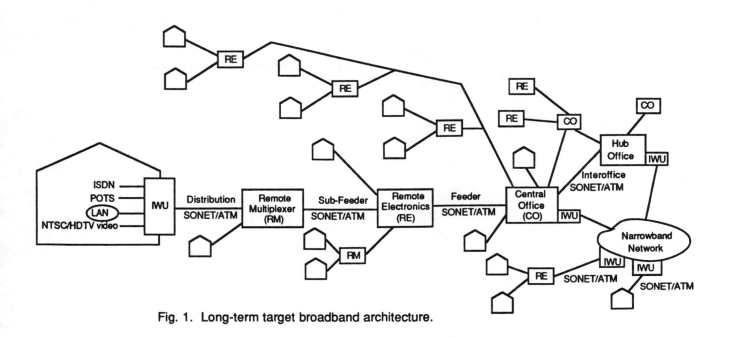

Fig. 1. Long-term target broadband architecture.

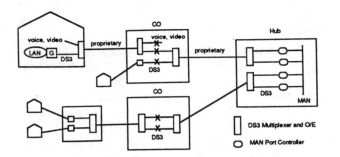

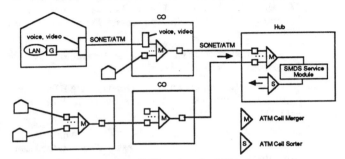

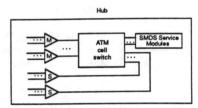

Fig. 2. Early availability architecture for providing SMDS service using MAN technology.

Fig. 3. Architecture for providing SMDS service using ATM technology.

Fig. 4. Expanding the capacity of a switch for providing SMDS service.

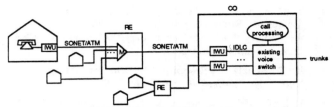

Fig. 5. Architecture for POTS-on-fiber using ATM and existing voice switches.

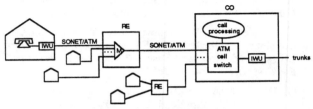

Fig. 6. Architecture for POTS-on-fiber using ATM cell switch.

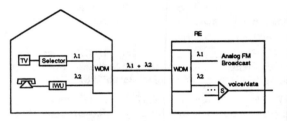

Fig. 7. Architecture for adding broadcast analog video.

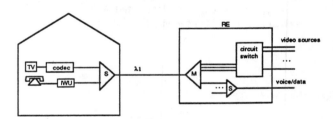

Fig. 8. Architecture for adding switched digital video.

Reprinted from *IEEE Communications*, Vol. 29, No. 3, March 1989, pages 8-15. Copyright © 1989 by The Institute of Electrical and Electronics Engineers, Inc. All rights reserved.

SONET: Now It's the Standard Optical Network

Ralph Ballart
Yau-Chau Ching

SONET (SYNCHRONOUS OPTICAL NETWORK) IS the name of a newly adopted standard, originally proposed by Bellcore (Bell Communications Research) for a family of interfaces for use in Operating Telephone Company (OTC) optical networks. With single-mode fiber becoming the medium of choice for high-speed digital transport, the lack of signal standards for optical networks inevitably led to a proliferation of proprietary interfaces. Thus, the fiber optics transmission systems of one manufacturer cannot optically interconnect with those of any other manufacturer, and the ability to mix and match different equipment is restricted. SONET defines standard optical signals, a synchronous frame structure for the multiplexed digital traffic, and operations procedures.

SONET defines standard optical signals, a synchronous frame structure for the multiplexed digital traffic, and operations procedures.

SONET standardization began during 1985 in the T1X1 subcommittee of the ANSI-accredited Committee T1 to standardize carrier-to-carrier (e.g., NYNEX-to-MCI) optical interfaces. Clearly, such a standard would also have an impact on intracarrier networks, and for that reason has been a subject of great interest for many carriers, manufacturers, and others. Initial T1 standards for SONET rates and formats and optical parameters have now been completed. The history and technical highlights of the SONET standard and its applications are the subject of this paper.

Since it began in the post-divestiture environment, SONET standardization can be thought of as a paradigm for the development of new transmission signal standards. Bellcore's original SONET proposal was not fully detailed because all the technical questions were not yet answered. However, some aspects of the proposal have been carried through the entire process and are now part of the final standards. These include:

- The need for a family of digital signal interfaces, since the march of technology is going to continually increase optical interface bit rates
- The use of a base rate SONET signal near 50 Mb/s to accommodate the DS3 electrical signal at 44.736 Mb/s
- The use of synchronous multiplexing to simplify multiplexing and demultiplexing of SONET component signals, to obtain easy access to SONET payloads and to exploit the increasing synchronization of the network
- Support for the transport of broadband (> 50 Mb/s) payloads
- Specification of enough overhead channels and functions to fully support facility maintenance

As standardization progressed, two key challenges emerged, the solution of which gave SONET universal application. The first was to make SONET work in a plesiochronous[1] environment and still retain its synchronous nature; the solution was the development of payload pointers to indicate the phase of SONET payloads with respect to the overall frame structure (see "SONET Signal Standard—Technical Highlights"). The second was to extend SONET to become an international transmission standard, and thereby begin to resolve the incompatibilities between the European signal hierarchy (based on 2.048 Mb/s) and the North American hierarchy (based on 1.544 Mb/s). Toward the latter goal, the International Telegraph and Telephone Consultative Committee (CCITT) standardization of SONET concepts began in 1986 and the first Recommendations (standards) were completed in June 1988.

This paper will not present a full technical picture of the national and international SONET standards. Instead, we will concentrate on those aspects of the standards and standardization process that are of particular interest. In the next section, a brief and instructional history of the SONET standard is presented. As philosopher George Santayana said, "Those who cannot remember the past are condemned to repeat it." We will then discuss key technical aspects of the SONET standard. Finally, an outline of future work is given in the final section.

---

[1] As defined in CCITT, corresponding signals are plesiochronous if their significant instants occur at nominally the same rate, any variation in rate being constrained within specified limits.

A History of SONET in T1 and CCITT

The standardization of SONET in T1 started in two different directions and in three areas. First, the Interexchange Carrier Compatibility Forum (ICCF), at the urging of MCI in 1984, requested T1 to work on standards that would allow the interconnection of multi-owner, multi-manufacturer fiber optic transmission terminals (also known as the mid-fiber meet capability). Of several ambitious tasks that ICCF wanted addressed to ensure a full mid-fiber meet capability, two were submitted to T1. A proposal on optical interface parameters (e.g., wavelength, optical power levels, etc.) was submitted to T1X1 in August 1984 and, after three and a half years of intensive work, resulted in a draft standard on single-mode optical interface specifications [1]. The ICCF proposal on long-term operations was submitted to T1M1 and resulted in a draft standard on fiber optic systems maintenance [2].

In February 1985, Bellcore proposed to T1X1 a network approach to fiber system standardization that would allow not only the interconnection of multi-owner, multi-manufacturer fiber optic transmission terminals, but also the interconnection of fiber optic network elements of varying functionalities. For example, the standard would allow the direct interconnection between several optical line terminating multiplexers, manufactured and owned by different entities, and a digital cross-connect system. In addition, the proposal suggested a hierarchical family of digital signals whose rates are integer multiples of a basic module signal, and suggested a simple synchronous bit-interleaving multiplexing technique that would allow economical implementations. Thus, the term Synchronous Optical NETwork (SONET) was coined. This proposal eventually led to a draft standard on optical rates and formats [3]. For the remainder of this paper, the focal points are the history and highlights of the rates and formats document. However, one should always be reminded that this document is only one part of the inseparable triplet: optical interface specifications, rates and formats specifications, and operations specifications.

As it turned out, the notions of a network approach and simple synchronous multiplexing had been independently investigated by many manufacturers. Some of them were already developing product plans, thus complicating the standards process. With the desire of the network providers (i.e., the OTCs) for expedited standards, SONET quickly gained support and momentum. By August 1985, T1X1 approved a project proposal based on the SONET principle. Because the issues on rates and formats were complex and required diligent but timely technical analyses, a steady stream of contributions poured into T1X1. Several ad hoc groups were formed and interim meetings were called to address them. The contributions came from over thirty entities, representing the manufacturers and the network providers alike.

In the early stage, the main topic of contention was the rate of the basic module. From two original proposals of 50.688 Mb/s (from Bellcore) and 146.432 Mb/s (from AT&T), a new rate of 49.920 Mb/s was derived and agreed on. In addition, the notion of a Virtual Tributary (VT) was introduced and accepted as the cornerstone for transporting DS1 services. By the beginning of 1987, substantial details had been agreed upon and a draft document was almost ready for voting. Then came CCITT.

The SONET standards were first developed in T1X1 to serve the U.S. telecommunications networks. When CCITT first expressed its interest in SONET in the summer of 1986, major procedural difficulties appeared. According to the established protocol, only contributions that had consensus in T1X1 were forwarded, through U.S. Study Group C, to CCITT. As a result, some aspects of U.S. positions in CCITT appeared to lack flexibility without input from T1X1. Addi-

TABLE I. CCITT Rec. G.702 Asynchronous Digital Hierarchies (in Mb/s)

Level	North America	Europe	Japan
1	1.544 (DS1)	2.048	1.544
2	6.312 (DS2)	8.448	6.312
3	44.736 (DS3)	34.368	32.064
4	—	139.264	97.728

tionally, the views of other administrations in CCITT were not thoroughly understood in T1X1. There were also differences in schedule and perceived urgency. CCITT runs by a four-year plenary period and their meetings are six to nine months apart, while T1 approves standards whenever they are ready and its technical subcommittees meet at least four times a year. While T1X1 saw the SONET standard as a way to stop the proliferation of incompatible fiber optic transmission terminals, no such need was perceived by many other nations whose networks were still fully regulated and non-competitive.

The procedural difficulties were partially resolved when representatives from the Japanese and British delegations started to participate in T1X1 meetings in April 1987. These representatives not only gave to T1X1 the perspectives of two important supporters of an international SONET standard, they also served as a conduit between T1X1 and CEPT, the European telecommunications organization.

Separately, interests in an international SONET standard also gained support in the US. Spearheaded by Bellcore, informal discussions in search of an acceptable solution took place in a variety of forums, and contributions in support of this standard were submitted to both T1 and CCITT. Many of these informal discussions had the highest level of corporate support from several U.S. companies, including manufacturers and network providers.

In July 1986, CCITT Study Group XVIII began the study of a new synchronous signal hierarchy and its associated Network Node Interface (NNI). The NNI is a non-media-specific network interface and is distinct from the-user-network interface associated with Broadband ISDN. The interaction between T1X1 and CCITT on SONET and the new synchronous hierarchy was fascinating to the participants and will probably alter the way international standards are made in. the future. The U.S. wanted an international standard, but not at the price of scrapping SONET or seriously delaying an American national standard upon which OTC networks were planned. The CCITT was not used to working so quickly on so complicated an issue, but was concerned about being supplanted by the T1 committee in the development of new standards.

The U.S. first formally proposed SONET to CCITT for use in the NNI at the February 1987 Brasilia meeting; this proposal had a base signal level (rate) near 50 Mb/s. Table I shows that the European signal hierarchy has no level near 50 Mb/s, and therefore CEPT wanted the new synchronous hierarchy to have a base signal near 150 Mb/s to transport their 139.264 Mb/s signal.

Thus, the informal European response was that the U.S. must change from bit interleave to byte interleave multiplexing to provide a byte organized frame structure at 150 Mb/s. However, there was still no indication from many administrations that an international standard was either desirable or achievable. It took T1X1 three months and three meetings to agree to byte-interleaving and the results were submitted to CCITT as a new T1X1 draft standard document. Thus, T1X1 never gave up the responsibility of developing a SONET standard for the U.S. and, while conceding changes to CCITT wishes, progress was made in other areas of the U.S. standard.

After CCITT met again in Hamburg in July 1987, a formal request was made to all administrations to consider two alternative proposals for an NNI specification near 150 Mb/s. The U.S. proposal was based on the SONET STS-3 frame structure; the STS-3 frame could be drawn as a rectangle with 13 rows and 180 columns of bytes. CEPT proposed, instead, a new STS-3 frame with 9 rows and 270 columns. (Commonly referred to as the 9-row/13-row debate, this prompted one amateur poet to chide that neither conforms to the correct SON(N)ET format of 14 lines.) An NNI near 150 Mb/s received unanimous support because it was assumed that future broadband payloads would be about that size. A North American basic module near 50 Mb/s could be easily derived in both proposals, with a frame structure of either 13 rows and 60 columns or 9 rows and 90 columns.

The Europeans wanted a 9-row frame structure to accommodate their 2.048 Mb/s primary rate signal. This signal has 32 bytes per 125 μs, but in the 13-row proposal could only be accommodated in the most straightforward way using three 13-byte columns, or 39 bytes. The Europeans decried this waste of bandwidth, and refused to consider any alternative (and more efficient) mapping of the 2.048 Mb/s signal into the 13-row structure. Their 9-row frame structure could carry the U.S. 1,544 Mb/s primary rate signal (requiring about 24 bytes/l25 μs) in 3 columns of 9 bytes and the 2.048 Mb/s signal in 4 columns of 9 bytes.

The CEPT 9-row proposal called for changes in both the rate and format in the U.S., just as T1X1 was about to complete the SONET standard. However, the request also carried an attractive incentive from a CEPT subcommittee, who stated in a letter that these were the only changes necessary for an international agreement. In addition, the text of the CEPT proposal was based largely on the T1X1 draft document, so that it was complete. Therefore, after the Hamburg meeting, there was tremendous international pressure on the U.S. to accept the 9-row proposal. After some intense debates in T1X1, the U.S. agreed to change.

Unfortunately, the CEPT proposal did not have unanimous support from all CEPT administrations. While some administrations were anxious to get an international standard, a few became concerned that the 9-row proposal favored the U.S. DS3 signal over the CEPT 34.368 Mb/s signal. A CEPT contribution to the November 1987 CCITT meeting stated that it was too early to draft Recommendations on a new synchronous hierarchy. Little progress was made at that meeting and the international SONET standard was in serious jeopardy.

Many T1X1 participants were upset at the apparent change in CEPT's position. Since there were no alternative proposals from CCITT at its November meeting, T1X1 decided to approve the two SONET documents for T1 letter balloting. However, the balloting schedule was deliberately set such that it fell between the CCITT meeting at the beginning of February 1988 and the T1X1 meeting at the end of February 1988. This scheduling allowed a last ditch attempt for an international agreement. In CCITT, a mad rush to rescue the international standard also took place. In addition to a series of informal discussions, a pre-CCITT meeting was held in Tokyo to search for a compromise. Under the skillful helmsmanship of Mr. K. Okimi of Japan, the CCITT meeting in Seoul proposed one additional change to the U.S. draft standards. The new proposal called for a change in the order that 50 Mb/s tributaries are byte-multiplexed to higher SONET signal levels. It also put more emphasis on the NNI as a 150 Mb/s signal by including optional payload structures to better accommodate the European 34.368 Mb/s signal. The U.S. CCITT delegates eventually viewed this proposal as a minor change to the U.S. standard (minor to the extent that equipment under development would probably not require modification) and agreed to accept it. An

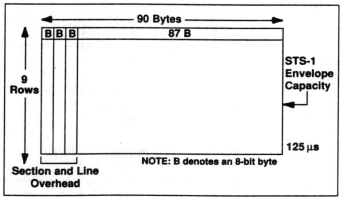

Fig. 1. STS-1 frame.

extensive set of three CCITT Recommendations was drafted and approved by the working party plenary. The U.S. acceptance of these changes was predicated on the understanding that no additional changes of substance would be considered in approving the final versions of the Recommendations.

In February 1988, T1X1 accepted the new changes at its meeting in Phoenix. T1 default balloting based on the change was completed in May and the final passage of the American national standard is expected this summer. Editorial corrections to the CCITT Recommendations [4–6] were completed in June during the Study Group XVIII meeting and with their final approval later this year, the international SONET standard will be born!

SONET Signal Standard—Technical Highlights

In this section, we describe the technical highlights of the American national standards related to SONET. We use U.S. rather than CCITT terminology, although everything described is consistent with both the American national standards and the CCITT Recommendations.

SONET Signal Hierarchy

The basic building block and first level of the SONET signal hierarchy is called the Synchronous Transport Signal—Level 1 (STS-1). The STS-1 has a bit rate of 51.84 Mb/s and is assumed to be synchronous with an appropriate network synchronization source. The STS-1 frame structure can be drawn as 90 columns and 9-rows of 8 bit bytes (Figure 1). The order of transmission of the bytes is row by row, from left to right, with one entire frame being transmitted every 125 μs. (125 μs frame period supports digital voice signal transport, since these signals are encoded using 1 byte/125 μs = 64 kb/s.) The first three columns of the STS-1 contain section and line overhead bytes (see the following subsection). The remaining 87 columns and 9-rows are used to carry the STS-1 Synchronous Payload Envelope (SPE); the SPE is used to carry SONET payloads including 9 bytes of path overhead (see next section). The STS-1 can carry a clear channel DS3 signal (44.736 Mb/s) or, alternatively, a variety of lower-rate signals such as DS1, DS1C, and DS2.

No physical interface for the STS-1 signal has been defined as yet; the Optical Carrier—Level 1 (OC-1) is obtained from the STS-1 after scrambling (to avoid long strings of ones and zeros and allow clock recovery at receivers) and electrical-to-optical conversion. The OC-1 is the lowest-level optical signal to be used at SONET equipment and network interfaces.

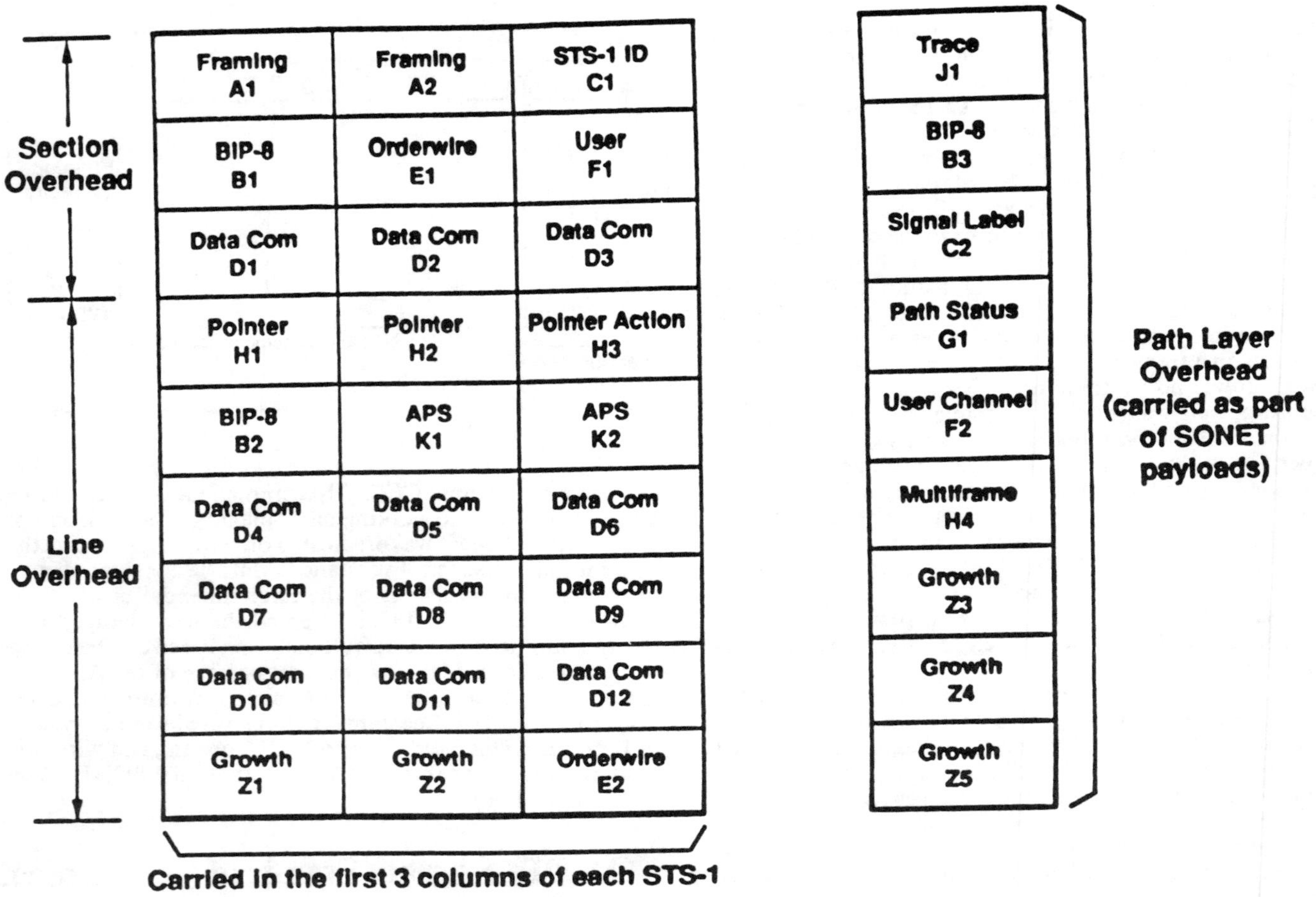

Fig. 2. SONET overhead bytes.

SONET Overhead Channels

The SONET overhead is divided into section, line, and path layers; Figure 2 shows the overhead bytes and their relative positions in the SONET frame structure. This division clearly reflects the segregation of processing functions in network elements (equipment) and promotes understanding of the overhead functions. The section layer contains those overhead channels that are processed by all SONET equipment including regenerators. The section overhead channels for an STS-1 include two framing bytes that show the start of each STS-1 frame, an STS-1 identification byte, an 8-bit Bit-Interleaved Parity (BIP-8) check for section error monitoring, an orderwire channel for craft (network maintenance personnel) communications, a channel for unspecified network user (operator) applications, and three bytes for a section level data communications channel to carry maintenance and provisioning information. When a SONET signal is scrambled, the only bytes left unscrambled are the section layer framing bytes and the STS-1 identification bytes. The second (link) layer of the section data communications channel protocol is LAPD while ISO 8473 is under study for the third (network) layer; higher layers of the protocol will be defined in future updates of the standard.

The line overhead is processed at all SONET equipment except regenerators. It includes the STS-1 pointer bytes (discussed below), an additional BIP-8 for line error monitoring, a two-byte Automatic Protection Switching (APS) message channel (both $1 + 1$ and 1 by N protection are supported), a nine-byte line data communications channel, bytes reserved for future growth, and a line orderwire channel. The higher lay-

ers of the line data communications channel are not specified in the current version of the SONET standard.

The path overhead bytes are processed at SONET STS-1 payload terminating equipment; that is, the path overhead is part of the SONET STS-1 payload and travels with it. The path overhead includes a path BIP-8 for end-to-end payload error monitoring, a signal label byte to identify the type of payload being carried, a path status byte to carry maintenance signals, a multiframe alignment byte to show DS0 signaling bit phase, and others.

Multiplexing

Higher rate SONET signals are obtained by first byte-interleaving N frame-aligned STS-1s to form an STS-N (Figure 3). Byte-interleaving and frame alignment are used primarily to obtain a byte-organized frame format at the 150 Mb/s level that is acceptable to the CCITT; as discussed below, frame alignment and byte-interleaving also help an STS-N to carry broadband payloads of about 150 or 600 Mb/s. All the section and line overhead channels in STS-1 #1 of an STS-N are used; however, many of the overhead channels in the remaining STS-1s are unused. (Only the section overhead framing, STS-1 ID, and BIP-8 channels and the line overhead pointer and BIP-8 channels are used in all STS-1s in an STS-N.) The STS-N is then scrambled and converted to an Optical Carrier—Level N (OC-N) signal. The OC-N will have a line rate exactly N times that of an OC-1. Table II shows the OC-N levels allowed by the American national standard.

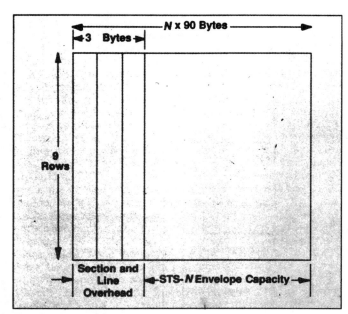

Fig. 3. STS-N frame

SONET STS-1 Payload Pointer

Each SONET STS-1 signal carries a payload pointer in its line overhead. The STS-1 payload pointer is a key innovation of SONET, and it is used for multiplexing synchronization in a plesiochronous environment and also to frame align STS-N signals.

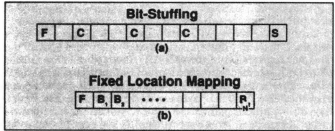

Fig. 4. Payload multiplexing methods.

Pointers and Multiplexing Synchronization

There are two conventional ways to multiplex payloads into higher-rate signals. The first is to use positive bit-stuffing to increase the bit rate of a tributary signal to match the available payload capacity in a higher-rate signal. As shown in Figure 4a, bit-stuffing indicators (labeled C) are located in fixed positions with respect to signal frame F and indicate whether the stuffing bit S carries real or dummy data in each higher-level signal frame. Examples of bit stuffing are the multiplexing of four DS1 signals into the DS2 signal and the multiplexing of seven DS2 signals into the asynchronous DS3 signal. Bit-stuffing can accommodate large (asynchronous) frequency variations of the multiplexed payloads. However, access to those payloads from the higher-level multiplexed signal is conceptually difficult, since the tributary signal must first be destuffed (real bits separated from the dummy bits) and then the framing pattern of the payload must be identified if complete payload access is required.

TABLE II. Levels of the SONET Signal Hierarchy

Level	Line Rate (Mb/s)
OC-1	51.84
OC-3	155.52
OC-9	466.56
OC-12	622.08
OC-18	933.12
OC-24	1244.16
OC-36	1866.24
OC-48	2488.32

The second conventional method is the use of fixed location mapping of tributaries into higher-rate signals. As network synchronization increases with the deployment of digital switches, it becomes possible to synchronize transmission signals to the overall network clock. Fixed location mapping is the use of specific bit positions in a higher-rate synchronous signal to carry lower-rate synchronous signals; for example, in Figure 4b, frame position B2 would always carry information from one specific tributary payload. This method allows easy access to the transported tributary payloads, since no destuffing is required. The SYNTRAN DS3 signal is an example of a synchronous signal that uses fixed location mapping of its tributary DS1 signals. However, there is no guarantee that the high-speed signal and its tributary will be phase-aligned with each other. Also, small frequency differences between the transport signal and its tributary signal may occur, due to synchronization network failures or at plesiochronous boundaries. Therefore, multiplexing equipment interfaces require 125-μs buffers to phase-align and slip (repeat or delete a frame of information to correct frequency differences) the tributary signal. These buffers are undesirable because of the signal delay that they impose and the signal impairment that slipping causes.

In SONET, payload pointers represent a novel technique that allows easy access to synchronous payloads while avoiding the need for 125-μs buffers and associated slips at multiplexing equipment interfaces. The payload pointer is a number carried in each STS-1 line overhead (bytes H1, H2 in Figure 2) that indicates the starting byte location of the STS-1 SPE payload within the STS-1 frame (Figure 5). Thus, the payload is not locked to the STS-1 frame structure as it would be if fixed location mapping was used but instead floats with respect to the STS-1 frame. (The STS-1 section and line overhead byte positions determine the STS-1 frame structure; note in Figure 5 that the 9-row-by-87-column SPE payload maps into an irregular shape across two 125-μs STS-1 frames.)

Any small frequency variations of the STS-1 payload can be accommodated by either increasing or decreasing the pointer value; however, the pointers cannot adjust to asynchronous frequency differences. For example, if the STS-1 payload data rate is high with respect to the STS-1 frame rate, the payload pointer is decremented by one and the H3 overhead byte is used to carry data for one frame (Figure 6). If the payload data rate is slow with respect to the STS-1 frame rate, the data byte immediately following the H3 byte is nulled for one frame and the pointer is incremented by one (Figure 7). Thus, slips and their associated data loss are avoided while the phase of the STS-1 synchronous payload is immediately known by simply reading the pointer value. Thus, SONET pointers combine the best features of the positive bit-stuffing and fixed location mapping methods. Of course, these advantages come at the cost of having to process the pointers; however, pointer processing appears readily implementable in today's Very Large Scale Integration (VLSI) technologies.

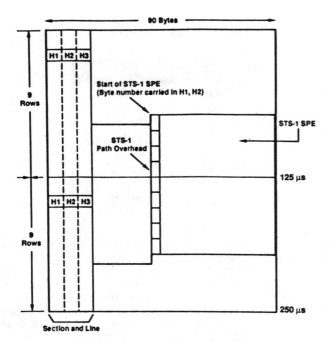

Fig. 5. STS-1 SPE in interior of STS-1 frame.

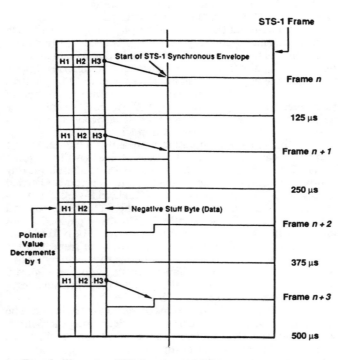

Fig. 6. Negative STS-1 pointer adjustment operation.

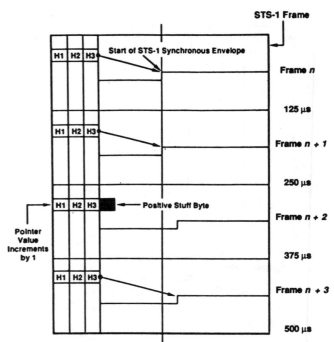

Fig. 7. Positive STS-1 pointer adjustment operation.

and line overhead bytes, two phase-aligned signals (A and B) are formed. A and B can then be byte-interleaved to form a higher level STS-*N* signal. As shown, this can be done with minimum payload buffering and signal delay.

With frame alignment, the STS-1 pointers in an STS-*N* are grouped together for easy access at an OC-*N* receiver using a single STS-*N* framing circuit. If it is desired to carry a broadband payload requiring, for example, three STS-1 payloads, the phase and frequency of the three STS-1 payloads must be locked together as the broadband payload is transported through the network. This is easily done by using a "concatenation indication" in the second and third STS-1 pointers. The concatenation indication is a pointer value that indicates to an STS-1 pointer processor that this pointer should have the same value as the previous STS-1 pointer. Thus, by frame aligning STS-*N* signals and using pointer concatenation, multiple STS-1 payloads can be created. The STS-*N* signal that is locked together in this way is called an STS-*N*c, where the "c" stands for concatenated. Allowed values of STS-*N*c are STS-2c, STS-3c, STS-6c, STS-9c, etc. For broadband User Network Interfaces (UNI), STS-3c and STS-12c are of particular interest.

Broadband Payload Transport with Payload Pointers

As discussed above, STS-1 payload pointers can be used to adjust the frequencies of several STS-1 payloads in multiplexing to the STS-*N* signal level. As this is done, the various STS-1 section and line overhead bytes are frame-aligned. In Figure 8, two hypothetical and simplified SONET frames (A and B) are out of phase with respect to the arbitrary, outgoing (multiplexed) SONET signal phase. By recalculating the SONET pointer values and regenerating the SONET section

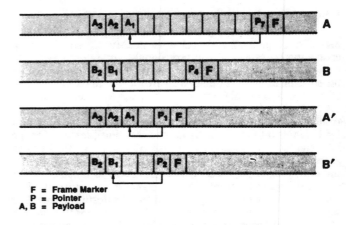

F = Frame Marker
P = Pointer
A, B = Payload

Fig. 8. Frame alignment using pointers.

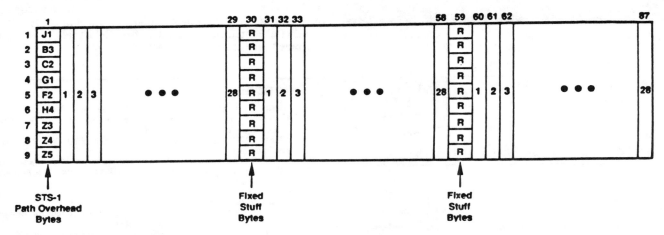

Fig. 9. VT-structured STS-1 SPE: all VT1.5.

As discussed in the section on the history of SONET standards, the Europeans had no interest in using the SONET STS-1 signal. Instead, they were interested in using a base signal of about 150 Mb/s to allow transport of their 139.264 Mb/s electrical signal and for possible Broadband ISDN applications. As the above discussion shows, the technical solution to this problem is the use of the STS-3c signal. In the U.S., we can continue to think of this signal as three concatenated STS-1 signals. In Europe and the CCITT, the STS-3c is considered as the basic building block of the new synchronous hierarchy and is referred to as the Synchronous Transport Module—Level 1 (STM-1).

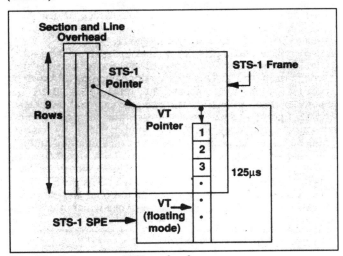

Fig. 10. Pointers and VT payload access.

Sub-STS-1 Payloads

To transport payloads requiring less than an STS-1 payload capacity, the STS-1 SPE is divided into payload structures called virtual tributaries (VTs). There are four sizes of VTs: VT1.5, VT2, VT3, and VT6, where each VT has enough bandwidth to carry a DS1, CEPT-1 (2.048 Mb/s), DS1C, and DS2 signal, respectively. Each VT occupies several 9-row columns within the SPE. The VT1.5 is carried in three columns (27 bytes), the VT2 in 4 columns (36 bytes), the VT3 in six columns (54 bytes), and the VT6 in twelve columns (108 bytes).

A VT group is defined to be a 9-row-by-12-column payload structure that can carry four VT1.5s, three VT2s, two VT3s, or one VT6. Seven VT groups (84 columns), one path overhead column, and two unused columns are byte-interleaved to fully occupy the STS-1 SPE. Figure 9 shows the STS-1 SPE configured to carry 28 VT1.5s. VT groups carrying different VT types can be mixed within one STS-1. As discussed in the section on history, the ability of the 9-row format structure to flexibly carry both the 1.544 and 2.048 Mb/s signals was a necessary step in reaching an international agreement on SONET.

Two different modes have been adopted for transporting payloads within a VT. The VT operating in the "floating" mode improves the transport and cross-connection of VT payloads. A floating VT is so called because a VT pointer is used to show the starting byte position of the VT SPE within the VT payload structure. In this sense, the operation of the VT pointer is directly analogous to that of the STS-1 pointer, and has the same advantages of minimizing payload buffers and associated delay when mapping signals into the VT. Figure 10 shows conceptually how the STS-1 and VT pointers are used to locate a particular VT payload in an STS-1. The other VT mode is the "locked" mode. The locked VT does not use the VT pointer, but instead locks the VT payload structure directly to the STS-1 SPE. (Of course, the STS-1 SPE still floats with respect to the STS-1 frame.) The locked mode improves the transport and cross-connection of DS0 signals by maintaining the relative phase and frequency of DS0 signals carried in multiple locked VT's. When VT-organized, each STS-1 SPE carries either all floating or all locked VTs.

More than one specific payload mapping is possible with each of the VT modes described above. Asynchronous mappings are used for clear channel transport of nominally asynchronous signals using the floating mode of operation; conventional positive bit-stuffing is used to multiplex these signals into the VT SPE. "Byte synchronous" mappings have been defined in both the locked and floating modes for the efficient, synchronous transport of DS0 signals and their associated signaling; conventional fixed position mappings are used to carry the DS0's in the VT SPE (floating mode) or VT (locked mode). "Bit synchronous" mappings are used in both the locked and floating modes for the clear channel transport of unframed, synchronous signals. The VT mappings that have been defined in the current version of the American national standard are given in Table III.

TABLE III. Sub-STS-1 Mappings

Mappings	VT (Virtual Tributary) Modes	
	Floating	Locked
Asynchronous	DS1 CEPT-1 DS1C, DS2	— .
Byte Synchronous	DS1 CEP-1	DS1 CEP-1 SYNTRAN
Bit Synchronous	DS1 CEP-1	DS1 CEP-1

Optical Parameters

The SONET optical interface parameters were developed in parallel with the SONET rates and formats. The optical parameters specified in the American national standard include spectral characteristics, line rate, power levels, and pulse shapes; jitter specifications will be developed in the next phase of the standard. The current optical specifications extend up to OC-48 (see Table II). It is expected that as more experience is gained with high data rate systems, the optical parameters associated with OC-18, OC-24, OC-36, and OC-48 will be updated.

The intent of this first optical interface standard is to provide specifications for "long reach" fiber transmission systems, i.e., systems using lasers. The second phase of SONET standardization in T1 will address "short reach" specifications for fiber transmission systems based on LEDs and low-power, loop lasers.

Conclusion and Future Work

The Synchronous Optical Network concept was developed to promulgate standard optical transmission signal interfaces to allow mid-section meets of fiber systems, easy access to tributary signals, and direct optical interfaces on terminals, and to provide new network features. The basic SONET signal format can transport all signals of the North American hierarchy up to and including DS3, and also future broadband signals. SONET will soon be an American national standard and a CCITT transmission signal hierarchy standard. The second phase of SONET T1 standardization will fully specify the data communications channel protocols, specify short-reach SONET optical interfaces for use in intra-office applications, and update SONET optical parameters for selected levels above OC-12.

SONET represents a successful test case for standards-making in the post-divestiture environment. Of course, the ultimate test for any standard is the development of products and services that are compliant with the new standard. For specific implementations, requirements beyond those contained in the standard are often needed. For example, Bellcore has issued a series of Technical Advisories giving additional requirements for SONET multiplexes, digital cross-connect systems, and digital switch interfaces. The first field trials of SONET equipment are expected in 1989.

References

[1] ANSI T1.106-1988, "American National Standard for Telecommunications—Digital Hierarchy Optical Interface Specifications, Single Mode," to be issued.
[2] T1M1.2/87-37R2 "Functional Requirements for Fiber Optic Terminating Equipment."
[3] ANSI T1.105-1988 "American National Standard for Telecommunication—Digital Hierarchy Optical Rates and Formats Specification," to be issued.
[4] CCITT Recommendation G.707, "Synchronous Digital Hierarchy Bit Rates," to be issued.
[5] CCITT Recommendation G.708, "Network Node Interface for the Synchronous Digital Hierarchy," to be issued.
[6] CCITT Recommendation G.709, "Synchronous Multiplexing Structure," to be issued.

Biography

Ralph Ballart is District Manager of Network Node Equipment Requirements at Bell Communications Research, Inc. (Bellcore), responsible for the development of requirements for SONET digital cross-connect systems and trunk side digital switch interfaces. Ralph is also Chairman of the T1X1.5 U.S. CCITT Drafting Sub-working Group that has responsibility for drafting contributions related to SONET to CCITT Study Group XVIII.

He received his B.S. degree in physics from the then Polytechnic Institute of Brooklyn in 1973 and his Ph.D. in physics from the University of Arizona in 1980. He joined Bell Telephone Laboratories in 1980 and worked on fundamental network planning tools and network applications for digital cross-connect systems. In 1984, he joined Bellcore and worked on requirements for SYNTRAN equipment. He was promoted to District Manager in 1985; his district participated in the technical development of SONET and its international standardization in CCITT. Dr. Ballart is a member of IEEE Communications Society.

Yau-Chau Ching is District Manager of SONET Interface Standards, responsible for strategic planning to support the SONET project in its initial applications as well as its later extension to Broadband ISDN. He is the Co-Chairman of T1X1.5 Working Group (Optical Hierarchy), where much of the SONET standardization work takes place. Before the inception of T1X1.5, he was the Co-Chairman of T1X1.4 Subworking Group on Optical Rates and Formats and an active member of T1X1.4 Subworking Group on Optical Interface. These latter groups were the predecessors of T1X1.5.

He received his B.E.E.E. degree from City College of New York in 1966 and his Ph.D. from New York University in 1969 under a National Science Foundation fellowship. From 1969 to 1984, he was with Bell Laboratories, where he did a series of exploratory development work in data compression, such as interframe video coding, digital speech interpolation, embedded ADPCM coding, and Fast Packet Network. From 1980 to 1983, he supervised a group responsible for the implementation of the Access Interface to the Fast Packet Network.

Since the divestiture of the Bell System, he has been with Bellcore, where his district was responsible for the generic requirements of fiber optic transmission systems. His district was also the originator of the SONET concept in Bellcore and its strong advocate in T1X1. Dr. Ching is a member of IEEE, Tau Beta Pi, and Eta Kappa Nu, and he holds 11 patents.

Chapter 2:

Interconnection Networks

2.1 Introduction

An interconnection subsystem is an essential component in multiprocessor computer systems and in switching systems. Examples of interconnection subsystems range from buses and rings to the crossbar. In multiprocessor systems with multiple processors and memories, efficient communication between processors and memory is critical for high performance. Interconnection subsystems such as the centralized bus and crossbar schemes are impractical for high-performance systems. This has motivated the search for interconnection schemes that offer the connection characteristics of a crossbar but without the associated disadvantages. A class of interconnection schemes called multistage interconnection networks has been used in high-end telephone switching systems.

An interconnection network can be classified on the basis of its mode of operation, control strategy, switching methodology, and topology. Typical operational modes are synchronous and asynchronous. Although a hybrid mode of operation is possible, it is uncommon.

An interconnection network consists of a number of switching nodes and links connecting these nodes. The control of the switch nodes may be either centralized or distributed.

Circuit switching and packet switching are the two most common methods for transferring information between the source and the destination. In a circuit-switched interconnection network, a physical path is established between the source and destination and the connection is maintained for an arbitrary length of time. On the other hand, in an interconnection network using packet switching, fixed-length packets of information, each carrying the destination address, are routed through the interconnection network without setting up a physical path from the source to the destination. The switching decisions are distributed, and each switch node selects the path to the next stage on the basis of the destination address.

In general, the interconnection network topologies tend to be regular and can be either *static* or *dynamic*. Dynamic interconnection networks may use a single stage or multiple stages to provide a connection between the source and destination. A single-stage interconnection has a single switching stage, whereas a multistage interconnection network has more than one switching stage. The multistage interconnection network can connect an arbitrary input to an arbitrary output.

The capability for providing multiple concurrent connections between disjoint input-output pairs varies among the different multistage interconnection networks. An interconnection network is said to be *nonblocking* if any desired connection between unused ports can be established immediately without interference from any arbitrary existing connections. An interconnection network is classified as *rearrangeable* when a desired connection between unused ports may be temporarily blocked, but can be established if one or more of the existing connections are rerouted or rearranged. An interconnection network is said to be *blocking* if there exist connection sets that will prevent additional desired connections from being established between unused ports, even with the rearrangement of existing connections.

A multistage interconnection network consists of a number of switching stages and links that connect the outputs of one stage to the inputs of the next stage. The input links to the switch nodes in the first stage are called the *network input ports*, and the output links of the switch nodes in the last stage are called the *network output ports*. A switch node can consist of n inputs and k outputs. The most commonly used switch node consists of two inputs and two outputs (2x2). At any instant, none, one, or two connections can be established, and these connections can in turn be bidirectional.

2.2 Overview of Papers

The paper "Classification Categories and Historical Development of Circuit Switching Topologies," by Broomell and Heath, is a tutorial on interconnection networks that use circuit switching, where a connection is established before information is exchanged between the source and the destination. After identifying the areas where circuit-switched interconnection networks can be used in parallel/distributed processing environments, it goes on to discuss topology, control, and other related issues. A classification scheme based on connection capability, topology, and basis of development is presented. The authors also describe a number of example interconnection networks.

In "A Survey of Interconnection Networks," McMillen presents a comprehensive survey of multistage interconnection networks and compares the topologies of the principal interconnection networks. Techniques for the incorporation of fault tolerance in interconnection networks are also set forth.

Dias and Kumar, in "Packet Switching in N log N Multistage Networks," survey packet-switched interconnection networks used to connect multiple processors and memories. The throughput and the delay are used as the metrics to evaluate and compare various interconnection networks. Finally, the authors identify a number of communication-related applications that can take advantage of this technology.

Classification Categories and Historical Development of Circuit Switching Topologies

GEORGE BROOMELL AND J. ROBERT HEATH

Department of Electrical Engineering, University of Kentucky, Lexington, Kentucky 40506

A broad tutorial survey is given of the various topologies available for use in circuit switching systems for tightly coupled parallel/distributed computer systems. Terminology and issues of circuit switching as related to parallel/distributed processing are first discussed. Circuit switching networks are then classified according to connection capability, topological geometry, and basis of development. Topological relationships of specific networks are addressed.

Categories and Subject Descriptors: A.1 [**General Literature**]: Introductory and Survey; B.3.4 [**Hardware**]: Input/Output and Data Communications—*interconnections (subsystems)*; C.1 [**Computer Systems Organization**]: Processor Architectures; C.2.1 [**Computer Systems Organization**]: Network Architecture and Design—*network topology*; C.2.4 [**Computer Systems Organization**]: Network Architectures and Design—*distributed systems*

General Terms: Theory

Additional Key Words and Phrases: Circuit switching, computers, distributed/parallel processing, interconnection networks, topology

INTRODUCTION

Anyone designing a parallel or distributed architecture computer system immediately encounters a central, major obstacle: the "interconnection network" problem. Most parallel/distributed architectures require some type of interconnection network to link the various elements of the computer system.

In the broadest sense, the term "interconnection network" for computers can encompass everything from nationwide networks, linked by telephone or satellite, to manual switches, which allow a few microcomputers to interchange files. For parallel/distributed computer architectures, neither of these extremes is applicable. Instead, a parallel or distributed architecture computer requires a network capable of providing very rapid data transfer among many processing elements, memories, I/O ports, etc., in close proximity under automatic control. Although long-distance data networks can often provide delays of as little as 250 milliseconds, which could be considered "fast" in some applications, closely linked computer systems may require delays of microseconds or less for efficient operation. Distributed computers, as compared to large computing systems, are physically compact, generally fitting in one room or one cabinet, so that the physical distance to be spanned is relatively short.

The difficulties in implementing such a network increase rapidly with the number of processors to be interconnected and the "tightness" or data transfer speed requirements of the system. The problems of

"Classification Categories and Historical Development of Circuit Switching Topologies" by G. Broomell and J.R. Heath from *Computing Surveys*, Vol. 15, No. 2, June 1983, pages 95–133. Copyright © 1983 by The Association for Computing Machinery, Inc., reprinted by permission.

CONTENTS

INTRODUCTION
1. CIRCUIT SWITCHING SYSTEMS IN PARALLEL/DISTRIBUTED PROCESSING
   1.1 Circuit and Packet Switching
   1.2 Applications of Circuit Switching Systems
   1.3 Topology
   1.4 Control
   1.5 Other Related Topics
2. CLASSIFICATION OF CIRCUIT SWITCHING NETWORKS
   2.1 Connection Capability
   2.2 Geometric Relationships
   2.3 Single-Stage and Multistage Networks
   2.4 Basis of Development
3. TOPOLOGICAL RELATIONSHIPS OF SPECIFIC NETWORKS
   3.1 Crossbar-Based Networks
   3.2 Cell-Based Networks
4. ANALYSIS OF SWITCHING SYSTEMS
5. CONCLUSIONS
ACKNOWLEDGMENTS
REFERENCES

---

tightly interconnecting tens (or thousands) of processing elements can be complex. Design of a switching system to accomplish this interconnection requires many major decisions, including what method of data transfer to use, what topology to use for the system, and what method of control to use. For a tightly coupled system, a circuit switching system that provides direct linkage between the resources is often required for faster data transfer rates than that obtainable with other types of switching systems, such as packet switching.

When the problem of designing a circuit switching system is first encountered, one's immediate response is to survey existing literature. Recently there has been a great deal of interest in all types of interconnection networks, and a large number of papers have appeared in the literature. A brief review of the papers available may leave a newcomer confused by the many types of networks available and the specialized terms used in discussing them. To help alleviate some of these problems, we attempt to provide a tutorial survey of a limited area of the field: the topology of circuit switching systems. In this paper, we first cover terminology and issues related to the use of circuit switches in parallel/distributed processing. Second, we classify circuit switching networks according to connection capability, topological geometry, and basis of development. Third, we cover topological relationships of specific networks. We conclude with a brief overview of the approaches for analyzing switching systems. One major intent is to provide the newcomer with background and knowledge to better understand and utilize the many specialized papers now appearing on circuit switching.

Many excellent survey papers of switching systems for parallel/distributed processing already exist [Feng 1981; Joel 1979; Marcus 1977; Masson et al. 1979; Siegel 1979a; Siegel et al. 1979; Thurber 1974, 1978]. Most cover a large variety of switching systems in a general overview. We, on the other hand, focus only on topics related to the topology of networks that can be used for direct circuit switching applications.

1. CIRCUIT SWITCHING SYSTEMS IN PARALLEL/DISTRIBUTED PROCESSING

1.1 Circuit and Packet Switching

Parallel/distributed processing exists when the actual processing of a problem is shared by several (or many) processors working simultaneously on parts of the problem. The application of parallel/distributed processing may take different forms. In one form, several computers are used to process a problem, and the problem is sufficiently large that each computer processes a relatively large, complex part. Linkage between these computers often uses I/O ports. Computers connected in this loosely coupled manner form a "computer network," and the interconnection used is often a packet network.

A parallel/distributed processing system also exists when an array or system of closely linked microprocessors (or other small computing elements) each performs small, simple parts of a problem. In this form of parallel/distributed processing, circuit switching systems are often required in order to provide higher data transfer rates.

From a comparison standpoint, a packet or message switching network exists when fixed-length data packets or variable-length messages are transmitted through a routing network from point to point. Such networks often operate in "store-and-forward" modes, in which the entire message/packet is stored at the various switching points until an appropriate path opens; thus there is never any clear continuous path from input to output. On the other hand, a circuit switching system actually completes direct circuit paths between inputs and outputs. In circuit switching systems, data can be transferred directly from the input to the output at high data rates, which depend solely on the electrical characteristics of the path. This differs from the packet/message switching system, in which the data transfer rate often depends on the system's traffic load, since packets may be held at intermediate switching elements waiting for a clear path when traffic is heavy. In circuit switching systems, the amount of traffic and the demand for system resources affect the capability to establish the desired connection at all. This will be discussed in greater detail in Section 2.1.2.

Circuit switching is defined as the ability to establish a direct connection between points or between two or more network ports. This can consist of either a direct electrical connection or a direct logic path through gates. Within a full circuit switching system, an established connection may consist of either a single path for serial data transfer or several paths for parallel data transfer.

In referring to such systems, the terms "circuit switching system" and "circuit switching network" are used almost synonymously, with minor connotative differences possibly existing in the minds of users. We effectively use the terms as equivalents. The term "circuit switch" is often used as the equivalent of the entire circuit switching network, although it is also commonly used to mean a switching element inside the network, especially when the element is a relatively large one, such as a crossbar switch.

In the early days of circuit switching, direct electrical paths in switching networks were normally established using electromechanical relays that operated crosspoints or contact pairs. Today, even though switching is normally done by all solid-state electronics, the early terms remain in common use. Although "contact pair" may be a better term, "crosspoint" is the term more commonly used to denote an electronic circuit that can electrically connect or disconnect two conductors in response to some external control signal. Many different electronic circuits can fulfill this purpose.

Logic gates are also commonly used for switching in digital systems. Although there is no direct conductor path between points in a gate system, there is a direct, immediate logic path from point to point, which is generally considered analogous to the circuit path in a digital system.[1]

We shall discuss only topologies for a single plane or one-bit-wide path between the connected port sets. The measures that we give of the systems, especially the measures of complexity and cost, are based on this single-bit path assumption. To completely switch an 8-bit data bus would require 8 bidirectional or 16 unidirectional switching paths, which requires many copies of the single-conductor topology. Designers may reduce hardware costs (and trade off data speed) by using either multiplexing or serial transfer through switching systems in order to reduce the number of bit-wide paths which must be switched.

In general, the systems that we discuss are intended for use in closely coupled computer systems with only short physical distances (not more than a few meters) between the elements. Thus we assume that conductor paths linking the switching elements are short and add essentially no cost or signal delay to the system.

Even though the topologies of packet systems and circuit switches are often similar, this paper is limited specifically to networks that can be used with a circuit

---

[1] Switching systems built from gates have the additional complication of being either unidirectional or bidirectional, depending on the actual gate arrangements used and the needs of the system. An address-bus switching system, for example, could consist of unidirectional logic, whereas a data bus switching system would necessarily consist of either true conductor paths or a bidirectional logic system. Since network topology is the main issue here, the choice of implementation is not especially important.

switching protocol in support of closely linked parallel/distributed processing.

1.2 Applications of Circuit Switching Systems

There are many ways in which a circuit switching system can be applied in a parallel/distributed processing system. Among them are

(a) *Data transfer.* Data must be transferred between the processing elements (PEs), during computations, in the form of communications. ("Data" here means both computational data and data necessary for overall control of the distributed system.) The PEs within the system may be interconnected by a circuit switching system.

(b) *Memory access.* A parallel/distributed processing system can have many separate memory banks/modules that may need to be accessed by many or all PEs. The PEs and memories may be interconnected with a circuit switching system, and a second system may possibly be used to interconnect the PEs and I/O peripherals.

(c) *Data alignment.* Many parallel/distributed processors are designed as arrays that operate efficiently on vector or matrix data. In such cases a circuit switching system may be used as a data alignment network to ensure that the right data get to the right processor.

(d) *Partitioning.* Some parallel/distributed processors are designed to be "partitioned" into independent subsystems of different sizes [Siegel et al. 1979]. This function may be accomplished by using a circuit switching system.

(e) *Miscellaneous needs.* Circuit switching systems also find application in other related uses, such as in variable-word-length computer architectures, macrolevel dynamic pipeline computer architectures, and a variety of other computing applications in which electrical connections must be switched. The potential uses of circuit switching systems are too varied for a full list to be given in detail.

1.3 Topology

In most cases, "efficient" switching networks are composed of many switching subsystems, interconnected in a regular pattern to minimize the number of required switching elements. *Topology* is the geometric pattern, usually defined by an interconnection algorithm, that is used to interconnect the various switching elements in the system. Because a system's topology directly affects which circuit connections can be established within it, the study of topology necessarily includes the study of limitations, if any, on the switching capabilities of the various systems. (Terms relating to the connecting ability of the various systems are discussed in Section 2.1.)

The topology of a system is also closely related to the capabilities of the different subswitches, or switching nodes, within the system. Node capability may have a major effect on system switching capabilities, and will be mentioned as the various networks are discussed.

Not all of the topologies discussed were originally envisioned as circuit switching systems, although many have become closely associated with that use. Examples are the Waksman permutation pattern [Waksman 1968] and the fast Fourier transform butterfly pattern [Pease 1968], which are basically algorithms for mapping input sets onto output sets, but also have excellent application as circuit switching topologies.

1.4 Control

The control of a switching system becomes very important during actual implementation. The control of switching systems is a complex problem, and is the subject of much research at present. A discussion of the topic of control, even if limited to circuit switching systems, could require a separate paper of equal or longer length. The main topic of our paper is system topology; to hold our presentation to manageable size, we give the subject of control less attention than it deserves. Many of the papers referenced in the text contain some discussion of control, and a few with more thorough discussions are indicated at appropriate places.

1.5 Other Related Topics

To give perspective in the complex field of computer switching systems we briefly mention some other major topics, and show their close relationship to circuit switching.

1.5.1 Time Division Multiplexing

Time division multiplexing (TDM) is the use of a single electrical path to carry different signals at different times within a short time interval. Circuit switching system connections can be used to carry TDM signals. Simple examples in computer applications include serial data transfer over single paths, and address/data time sharing of conductor paths, as is done in some of the new-generation microprocessor families. Use of TDM has little relation to the topology of the switching system beyond the requirement of adequate frequency response.

1.5.2 Frequency Division Multiplexing

Frequency division multiplexing (FDM) is the use of an electrical path to carry two or more signals at different frequencies. One computer-related example is the common full-duplex modem, which carries signals to and from terminals, etc., encoded on audio-frequencies. Paths in circuit switching systems could be used to carry FDM signals, but it is unlikely that FDM will find much use in short-distance circuit switching applications since the multiplexing and demultiplexing is relatively expensive.

1.5.3 Time–Space Trade-Offs

Time–space trade-offs exist in which some of the spatial topologies discussed in this paper may be replaced by time equivalents. As discussed in Section 1.5.1 above, TDM is sometimes considered a trade-off between "space," in the sense of the physical space used by switching components, and time. (Two signals time-multiplexed on a single path replace two paths that would otherwise be required.) More complex time–space trade-offs in switching systems are considered by Marcus [1970a]. One specific case relates to systems called time-slot interchangers [Marcus 1970b], which are shown to be equivalent in some respects to some of the topologies discussed in this paper.

1.5.4 Computer Networks

Computer networks, in the sense of linking large computers together, are generally relatively long-distance and low-speed networks when compared to the circuit switching networks discussed here. Communication in such cases is often by message or packet. Computer network topology is generally quite different from circuit switching network topology since in long-distance applications the conductor links are often more costly than the switching modes—essentially the reverse of short-distance networks.

1.5.5 Sorting Networks

Sorting networks, such as Batcher's [Batcher 1968], are used for sorting data. In such networks, data are input to the network in an "unordered" spatial relationship and output from the network sorted into a specified spatial order. Although the topologies of such networks may be similar to those of circuit switches, their operation may be different.

1.5.6 Single-Stage Recirculating Data Manipulation Networks

Single-stage recirculating data manipulation networks pass data repeatedly through the same network in order to route them from an input to an output. In this case, the data must enter the network, be passed through the network for the number of times necessary to route them to the correct output, and then exit the network. One well-known example is the "shuffle–exchange" system [Lang 1976; Lang and Stone 1976; Lawrie 1975; Stone 1971]. Although such networks obviously cannot be used directly as circuit switches, the shuffle–exchange concept, in particular, has had a great impact on circuit switching topology and is discussed again in Section 3.2.4.

2. CLASSIFICATION OF CIRCUIT SWITCHING NETWORKS

Circuit switching networks may be classified by several qualities. Figure 1 shows the several classifications that are discussed below.

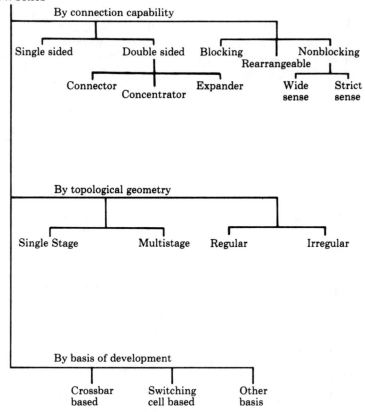

CIRCUIT
SWITCHING
NETWORKS

By connection capability

Single sided   Double sided   Blocking   Nonblocking

Rearrangeable

Connector   Concentrator   Expander   Wide   Strict
                                          sense   sense

By topological geometry

Single Stage   Multistage   Regular   Irregular

By basis of development

Crossbar   Switching   Other
based   cell based   basis

Figure 1. Some possible network classifications.

2.1 Connection Capability

Circuit switching networks can be classified by the connection capabilities of the network and the relation of network inputs and outputs. The most basic distinction concerns whether inputs and outputs are differentiated or whether all ports are treated equally.

2.1.1 Single-Sided and Double-Sided Networks

Single-sided or *one-sided* networks (Figure 2) have only one set of connections or ports, all of which are treated equally. In these networks, any port can be connected to any other port. In a single-sided network, conductor paths must be bidirectional since any port can be either a source or a destination. A telephone network can be viewed as a single-sided network since any phone can call or be called by any other. Single-sided networks have not been heavily covered in the literature and seem to have fewer applications than double-sided networks. A few specific single-sided networks will be discussed where appropriate.

Double-sided or *two-sided* networks (Figure 3) connect two distinct sets of ports, usually referred to as "inputs" and "outputs." Ports in one set generally may be connected only to ports of the other set. A good example of a double-sided network application in computers might be to interconnect several processors at the inputs to memory blocks at the outputs, allowing any processor to use any memory block. This example illustrates the fact that the terms "input" and "output" are seldom clear-cut,

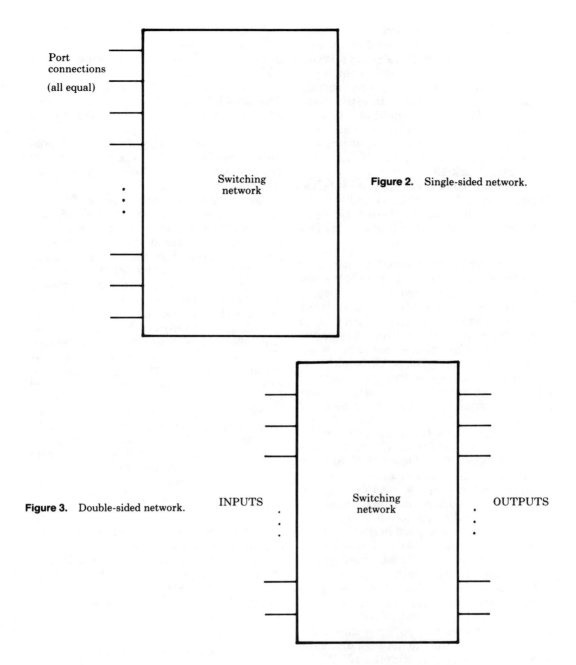

Port
connections

(all equal)

Switching
network

Figure 2. Single-sided network.

Figure 3. Double-sided network.

INPUTS

Switching
network

OUTPUTS

and often just differentiate between two classes of entities to be interconnected. In connecting processors to memories, we presume that data flow into and out of the memories at different times, so that neither the memories nor the processors are clearly inputs or outputs. For similar reasons, the conductor paths in two-sided networks could be either unidirectional or bidirectional, depending on the actual application. Thus, the fact that a circuit switching network is double sided states little except that the network is used to interconnect two distinct sets of devices or port types.

The distinction between single- and double-sided networks is not always clear.

Telephone networks, which in the large view are clearly one sided, are usually analyzed as two-sided networks [Beneš 1965], with the calling phone viewed as the input and the called phone viewed as the output. In computer applications, interprocessor communication systems could also be considered one sided. Actually, most are two-sided networks connecting output ports of all processors to input ports of the same processors [Kuck 1978].

Double-sided networks may be subclassified by connection capability and are usually divided into three classes [Masson 1976; Masson et al. 1979; Thurber and Masson 1979]:

(a) *Connectors* or *connection networks* are networks in which any specific unused input may be connected to any specific unused output. A conceptual example is the common telephone system (viewed as a double-sided network) in which any phone may call any other phone that is not busy. A processor–memory network would normally be a connector, so that any processor may be connected to any single memory bank not already being used by some other processor.

One term which has been adopted for connection networks is the $(N, M, C)_x$ *connector* [Marcus 1977; Masson 1976; Masson et al. 1979]. This refers to a connector with N inputs and M outputs, capable of establishing C input–output connections at one time. The χ indicates nonblocking (n) or rearrangeable (r) discussed in Section 2.1.2. The definition places no real restrictions on the size of N or M, but C must be less than or equal to the smaller of the two.

Among networks that may already be familiar, the "permutation network" for permuting N inputs onto N outputs is an (N, N, N) connector. A "data alignment network" such as the flip network discussed in Section 3.2.6 is also often an (N, N, N) connector, at least topologically, since it allows (certain) limited permutations of data over various outputs. Note that the C term simply states how many connections may exist simultaneously and does not measure the ability of a network to actually perform all possible permutations of inputs onto outputs.

(b) *Concentrators* or *concentration networks* are networks able to connect any specific unused input to any unspecified or arbitrary unused output.

A conceptual example of such a connection would be a call to an operator in a telephone system where the caller has no reason to prefer one operator over any other. In computer applications, one possible example would be the connection of processors to arbitrary "scratch pad" memory blocks as needed.

Using the same terminology as for connectors, these networks are $(N, M, C)_x$ concentrators. For these networks, $N \geq M$ since there is no reason to have more nonspecific outputs than can be used at any one time by the available inputs.

(c) *Expanders* or *expansion networks* are networks in which a specific, not necessarily unused, input can be connected to any specific unused output. Put simply, a specific input can be connected to several specific outputs. A computer-oriented example might be a network that connects processors to a set of several memory banks and I/O ports, selected from a large group of memories and ports, in order to establish a computing system from a subset of available resources. A second example could be a network used to "broadcast" data from one input to several outputs.

Once again, these networks can be described using the standard terminology as $(N, M, C)_x$ expanders.

Even though these connection classes are dependent to some extent on network topology, they are much more dependent on the control of the network. The network topology has much more impact on the ability of the network to establish desired input–output connections.

2.1.2 Ability to Establish Connections

Networks are also classified by their ability to establish desired connections between an input and an output. To a large extent, this is the ability to add a new arbitrary connection to a network with arbitrary existing connections. Basically, either a network may allow any arbitrary connection to be established at any time or the establishment of some connection may be blocked by an existing connection which uses a path or contact pair that is needed for the new connection. There are variations in the ability of networks to establish new connections, and several terms have evolved to describe them [Joel 1979; Marcus 1977; Thurber and Masson 1979]:

(a) A network is said to be *nonblocking in the strict sense* or *strictly nonblocking* if any desired connection between unused ports can be established immediately without interference from any arbitrary existing connections.

(b) A network is said to be *nonblocking in the wide sense* or *wide-sense nonblocking* if any desired connection between two unused ports can be established immediately without interference from arbitrary existing connections, *provided* that the existing connections have been inserted using some routing algorithm (peculiar to the network). If the algorithm has not been followed, some attempted connections may be blocked.

(c) A network is said to be *rearrangeable* when a desired connection between unused ports may be temporarily blocked, but can be established if one or more existing connections are rerouted or rearranged.

(d) A network is said to be *blocking* if there exist connection sets that will prevent some additional desired connections from being established between unused ports, even with rearrangement of the existing connections.

In the $(N, M, C)_\chi$ terminology system, the optional placeholder χ becomes "n" for nonblocking networks and "r" for rearrangeable networks. (Although little used, "b" indicates blocking.) Networks of any type (connector, concentrator, expander) may be nonblocking, rearrangeable, etc. Thus when this system was used, a full network description would be something like $(N, N, N)_n$ connector. In many discussions, a two-term description, such as "nonblocking connector" is more commonly used unless a more mathematical description is needed.

The classes of networks relevant to blocking are a direct result of the various network topologies, where the major-trade-off is between blocking and crosspoints. In general, the more crosspoints, the more alternate routes possible and the less chance of blocking. Much of the effort in studying and developing these network topologies goes toward obtaining some required connecting capability with a minimum number of crosspoints.

The optimum network for a particular application depends on the connection capability requirements. In applications in which not all inputs and outputs are in use at once, and delays in connection can be tolerated (as in the telephone system), a blocking network may suit the purpose and require relatively few crosspoints. Some of the computer-oriented networks discussed below are blocking connectors, but have been carefully laid out so that they are capable of establishing the most needed connections. Blocking can be tolerated in connections that are seldom used, if at all. In other cases, such as a very high throughput random connection network, it is possible that no blocking at all could be tolerated and a strictly nonblocking connector would be required regardless of higher hardware cost.

A great deal of study has been given to rearrangeable and nonblocking networks since they are considered to be more powerful and useful than blocking networks. We shall summarize some of the results of this study in relation to specific networks.

2.2 Geometric Relationships

Drawings of networks often bring out qualities that could be called geometric and which may be used to classify networks in ways different from those discussed in the last two sections.

A circuit switching network is generally a set of switching elements interconnected in some pattern. These elements are arranged into one or more *stages*, as illustrated in Figure 4. From the geometric viewpoint, stages are generally represented as columns of switching elements that must be operated to establish an input–output connection. In networks in which the number of stages is different for different paths, the number of stages is the maximum number required for any connection.

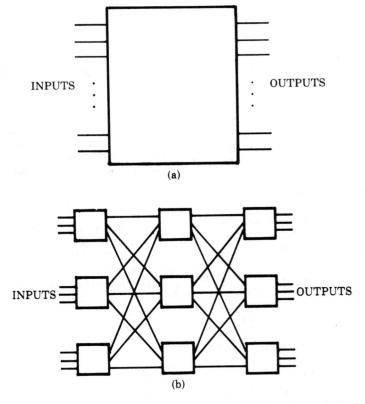

Figure 4. Single-stage and multistage networks: (a) single-stage network; (b) three-stage network.

The number of stages is closely related to what is usually called the *delay* of a switching network. This is the measure of the number of switching elements through which the signal must pass and in which it is thus subject to gate or other time delays. This assumes no time delay in the paths interconnecting the switching elements.

2.3 Single-Stage and Multistage Networks

The only true single-stage circuit switching network is the "crossbar switch" (also called "rectangular switching array"), which will be discussed more thoroughly in Section 3.1.1. A crossbar switch has a crosspoint for each input–output pair, and only one contact pair needs to be closed to establish any input-to-output connection.[2]

Most switching networks consist of several stages, and are said to be *multistaged*. These range from networks with as few as two stages, through some very powerful three-stage networks, up to networks with a very large number of stages. Several repercussions occur as a result of "staging" a network. The most immediate is that staging generally decreases the number of crosspoints required for large N's. Staging also increases the delay, which becomes significant in the many-staged networks that some topologies require to interconnect large numbers of inputs and outputs.

Another major consideration concerns the control of the network. There are sim-

---

[2] This single-stage network is not to be confused with the recirculating single-stage networks discussed briefly in Section 1.5.6. The single-stage crossbar can perform any input-to-output permutation with one set of connections, making it suitable for circuit switching, whereas the recirculating networks may require several passes through the single stage to perform a given input-to-output permutation.

ple control procedures for some simple staged networks, but in general staged networks are difficult to control efficiently. The overall field of efficient control of networks is complex and remains a good area for further work.

2.3.1 Regular and Irregular Networks

Most networks consist of elements connected together in a pattern. This pattern can be either *regular* or *irregular*. There is no set of tight definitions, but in general, a regular network (which is sometimes called a *uniform* network [Masson 1976]) looks symmetric about either the vertical or the horizontal axis, or both.

Even though irregular networks often tend to look somewhat unordered, almost all useful irregular networks do follow some pattern. The pattern, however, may vary from stage to stage, unlike a regular network, in which the pattern is uniform from stage to stage. Often irregular networks are derived from, or are very similar to, regular networks. The primary benefit of irregular circuit switching networks is that they require fewer crosspoints than a similar regular network with identical connection capabilities.[3]

One simple measure of a network's regularity is whether all possible input–output connections have the same delay. The importance of nonuniform delay varies with individual applications. In a network used to switch a data bus among various memories for high-speed access, regularity and uniform delay could be very important. In a network used for switching asynchronous serial communications links among processor ports, it would probably be of little or no consequence and an efficient irregular network would be acceptable.

2.4 Basis of Development

Networks are generally developed from some basic starting point, which is often one of the network's most obvious qualities.

Most starting points seem to fall into one of only a few families. The families are not well defined and they overlap, especially with networks based on algorithms that can be applied in different ways. However, we consider the analysis of networks, according to their basis of development, useful, and have chosen that structure for this discussion of networks.

Although it is very difficult to draw clean lines between the groups, and others might reasonably disagree with our definition of the basis of some networks, we argue that most networks fall into two major groups: *crossbar based*, or *mode* or *cell based*. There are a few other named networks, based primarily on mathematical properties, which are discussed briefly.

2.4.1 Crossbar-Based Networks

Crossbar-based networks have evolved from telephone switching technology in which many inputs were interconnected by large banks of crossbar switches.

In 1953, Clos [1953] showed that crossbar switches could be interconnected in stages so that far fewer crosspoints could achieve the same nonblocking switching capability as a full crossbar switch. The Clos networks have been followed by modified and related networks based on the premise of interconnecting "small" crossbars as parts of a large switching network. A typical crossbar array is shown in Figure 5, which is based on the general model from Masson and Jordan [1972]. The term "crossbar" seems firmly entrenched in connection with such networks, but is no longer descriptive. No one today envisions such a network as actually consisting of electromechanical crossbar arrays. Instead, what is termed a "crossbar switch" consists of a rectangular (or square) array of electronic switches capable of connecting any input to any output. Having so noted, we shall continue to use the term "crossbar," since the term "rectangular switch array"—which may be more correct—is cumbersome.

2.4.2 Node- or Cell-Based Networks

Most staged networks, including the crossbar-based networks just discussed, can be considered as networks of similar or iden-

---

[3] In long-distance communication networks, on the other hand, very irregular topologies result from the need to minimize cable distances and conform to geographical, political, and other constraints that have no effect on most short-distance switching systems.

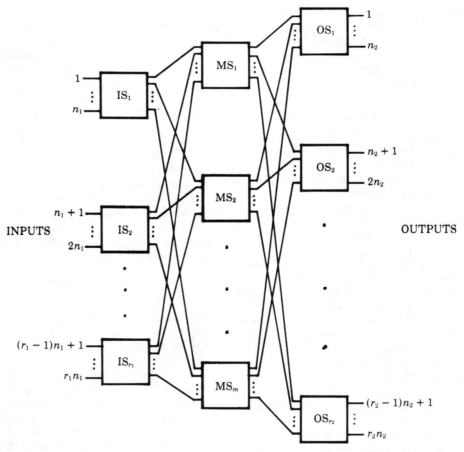

Figure 5. Typical crossbar switching array (IS, input switch; MS, middle switch; OS, output switch).

tical switching cells or nodes. In this sense, almost all networks are node or cell based.

There is, however, a large group of networks that have been developed specifically to use two-input, two-output switching cells as the basic network element. (The actual definition of the cell sometimes differs for different networks.) There are also networks that have been developed assuming a three-input, three-output cell or some other simple switching cell rather than the general $N \times M$ crossbar. All of these networks are included in what we call node- or cell-based networks.

A typical 2×2 node-based network is shown in Figure 6. This type of network is discussed in more detail in Section 3.2, along with specific examples.

2.4.3 Other Types of Networks

Several networks have been discussed in the literature using graph theory for their analysis. Some of these networks have been only loosely translated into hardware, if at all. Most of the networks developed from graph theory can be translated into a network quite similar to cell-based or even crossbar-based networks.

A typical network is shown in both graph form and circuit (schematic) form in Figure 7. Interpretation of a graph network compared to a more schematic representation can often be confusing. Their appearance is similar, but the interpretation of elements in each is different. In the schematic form, the lines represent conductors which interconnect the switching cells. In graph

118

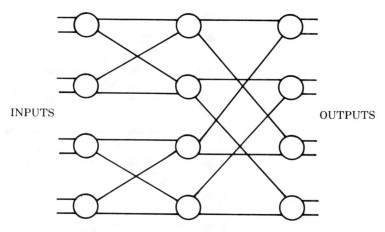

Figure 6. Typical 2 × 2 cell-based network.

form, the lines, or edges, represent a *possible* connection between the conductors represented by dots, or nodes. Thus the edge or line patterns of the graph represent the switching elements, but appear to be quite similar to the conductor pattern in the schematic form. In the graph, only one edge connecting to each node is normally allowed to carry a signal (if the element cannot "broadcast" to multiple outputs). In the circuit diagram all conductors can carry signals simultaneously in most networks. These relationships must be kept in mind when comparing apparently similar graph representations and schematic representations of networks.

As already mentioned, there are many networks which, although they are not circuit switches, are closely related topologically. The Batcher bitonic sorting network [Batcher 1968], mentioned in Section 1.5.5, is shown in Figure 8, primarily to illustrate this strong geometrical similarity.

3. TOPOLOGICAL RELATIONSHIPS OF SPECIFIC NETWORKS

In this section, we discuss many specific networks in a broad overview. The networks are presented in the two major families discussed above: crossbar based and cell based. Within each group, we discuss the networks primarily from a historical viewpoint. The classifications and ordering of networks have been selected to give an overview of the relationships among different topologies.

3.1 Crossbar-Based Networks

The electromechanical crossbar switch has been one of the major elements in telephone switching. The concept of one contact pair for each input–output combination that it represents is sufficiently powerful that many networks have been developed using the crossbar as a basis. Most of the important ones can be placed in one large group related to a basic concept developed by Clos [1953].

3.1.1 Single-Stage Crossbar

The single-stage crossbar is an array of individually operated contact pairs, in which there is one pair for each input–output combination, as shown in Figure 9. Figure 9 also shows typical block figures for the crossbar. A crossbar that has the same number of inputs as outputs is said to be "square" or "$N \times N$," whereas one with differing numbers is "rectangular" or "$N \times M$." "N" is the number of inputs to a switching system and "M" is the number of outputs, if different from N.

The single-stage crossbar, or general rectangular switch array, is the most straightforward strictly nonblocking circuit switching network. Since each input–output combination has an individual contact pair, any input can be connected to any output, at

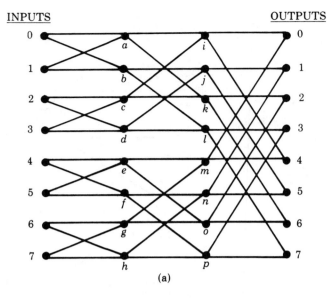

(a)

Figure 7. Typical simple network: (a) graph form; (b) circuit (schematic) form.

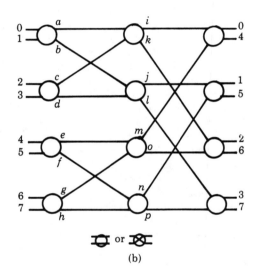

(b)

any time, with no possibility of blocking. The crossbar, although generally considered to be a connector, with suitable control could function as a concentrator or expander as well. The single-stage crossbar is the only network with a true uniform delay of 1, since there is only one switching element in any path from input to output. A single-stage crossbar network is a very capable switching network.

This capability comes at high cost. "Cost" in switching networks is generally related to what is sometimes referred to as "complexity," the number of switching elements required. The cost of a crossbar, then, is NM (or N^2 for a square switch). As a general magnitude, this is "on the order of N^2," which is commonly abbreviated as simply "$O(N^2)$." This "cost" is not necessarily directly related to dollar cost, which is a function of many variables over the entire system. The concept of cost here is, instead, only a general measure of the total number of switching elements required for

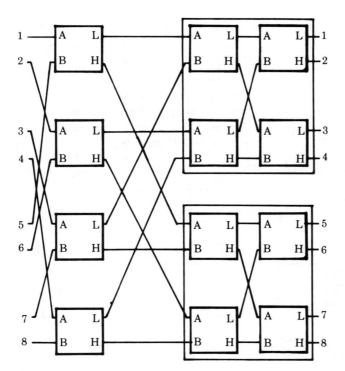

Figure 8. Batcher bitonic sorting network (A and B are data inputs; L indicates output of lower valued data; H indicates output of higher valued data). [After Batcher 1968.]

a network topology, usually expressed as a function of inputs (and sometimes outputs). This cost measure is one method of comparing the relative complexity of various networks.

The crossbar is considered a useful connector when the network has few inputs, in which case the total number of crosspoints is reasonable. Although the crossbar is the most costly network for large numbers of inputs, it is actually less costly than the Clos-type networks (discussed in Section 3.1.4) for less than about 25 inputs [Clos 1953; Marcus 1977]. For numbers much higher than 25, the cost of the crossbar becomes higher than that of well-designed staged networks. (Note again that cost here is not dollars but a measure of the number of switching elements.) The crossbar is a familiar, known entity, readily controlled, possessing full nonblocking switching capability. The lack of good measurement criteria for many aspects of constructing real networks, such as control requirements, makes it very difficult to determine which type of network is most cost effective in actual situations.

3.1.2 The Triangular Crosspoint Array

As a single-sided network, the crossbar may be reduced to a triangular array of crosspoints, as shown in Figure 10, in both detailed and block form. Like the two-sided crossbar, this network has one contact set available for each possible pair of ports and is thus the most powerful (most costly, least blocking) single-sided connector. Since there are few single-sided networks to compare it with, we do not discuss it in detail.

3.1.3 #5 Crossbar

The #5 Crossbar, occasionally mentioned in the literature, was one system used for reducing the number of crosspoints required in telephone networks many years ago. It can be considered a staged network, but does not give the benefits of the Clos-type staged networks discussed in the next section. As Beneš [1964a, 1965] shows, the #5 Crossbar gives essentially no benefit except for a reduction of the number of crosspoints required. The number of crosspoints is still high, and the network is highly blocking. The #5 Crossbar is not a useful network by current standards, nor

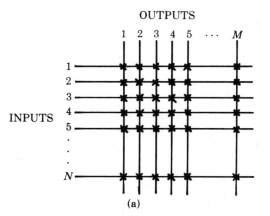

OUTPUTS

1 2 3 4 5 ··· M

INPUTS

1
2
3
4
5
·
·
·
N

(a)

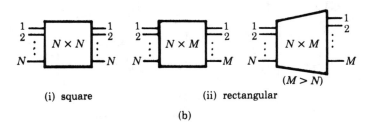

$N \times N$

$N \times M$

$N \times M$

$(M > N)$

(i) square

(ii) rectangular

(b)

Figure 9. Crossbar networks: (a) crossbar or rectangular switch array (×, switching element, which may be open or closed); (b) block diagrams for crossbars.

when compared to the other networks discussed.

3.1.4 Clos Networks

Clos [1953] proposed the use of a network with three or more stages of relatively small crossbar switches, which sharply reduces the number of crosspoints required to interconnect a large number of inputs and outputs. His network arrangement was proposed for telephone system use and, like the single-stage crossbar it replaces, is strictly nonblocking. Although the literature often refers to *the* Clos network, Clos actually proposed a system of constructing networks that includes many different specific networks. The type of network most commonly referred to as "the Clos network" is shown in Figures 11 and 12. In Figure 11, the network is shown in detailed form, and in Figure 12, it is shown in the more common block form, labeled to suit the work of Masson and others discussed in the next section.

Clos's general three-stage nonblocking arrangement uses rectangular crossbars in all stages. In a network with an equal number of inputs and outputs, the middle switches are square. Each crossbar has one output connected to an input of each crossbar of the following stage; thus in an unused network there exists a possible path through each of the middle-stage switches between any output and any input. Clos shows that if at least $(n_1 + n_2 - 1)$ middle-stage crossbars are used, the network will be totally nonblocking. This becomes $m \geq (2n - 1)$ for a square network, a basic criterion for nonblocking three-stage crossbar networks.

Construction procedure for this type of network is quite flexible. To use identical crossbar in each of the input and output stages, n_1 must be a factor of N and n_2 must be a factor of M. However, Clos shows examples of a slightly modified network using one nonidentical input or output crossbar switch to accommodate prime numbers or the use (for whatever reason) of arbitrarily sized crossbars, which are not factors of the numbers of inputs or outputs.

Since the network has the switching capability of a general crossbar, any of the crossbars in the network can be decomposed into a Clos-type network itself, retaining the network's nonblocking capabilities and sometimes reducing the number of crosspoints in large networks. Decom-

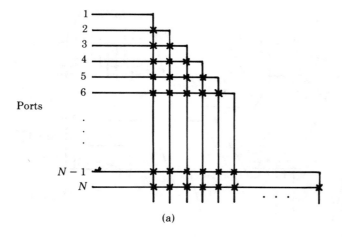

Ports

(a)

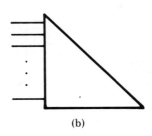

(b)

Figure 10. Single-sided crossbar switching network: (a) triangular crosspoint array ($\times$, 1×1 switching element); (b) block diagram.

position of all middle-stage switches leads to a five-stage network, and if repeated, to seven-, nine-, etc. stage networks. A sample five-stage network is shown in Figure 13. Decomposition of the input and output stages can lead directly to a seven-stage network.

For large N, the number of crosspoints required in a Clos network is much smaller than the $O(N^2)$ for a single-stage crossbar. The exact number is a function of the specific network used. Clos shows that for a square three-stage network, for example, the number of crosspoints is

$$(2n - 1)\left(2N + \frac{N^2}{n^2}\right),$$

where n is a factor of N and $n = n_1 = n_2$ in Figure 12. He tabulates the number of crosspoints for a number of specific networks, and shows, for instance, that a three-stage square network results in a savings for N greater than about 25. (This is also done by Marcus [1977].)

The same principle may be applied to single-sided networks and some analysis of them is also done by Clos. A sample of such a network is shown in Figure 14.

All of Clos' original work was done with respect to the networks used as nonblocking connectors.

3.1.5 Generalized Multistage Crossbar Networks

Clos' concept of the three-stage network is quite powerful. Since his original proposal, much work has been done in refining the concept. Masson, in particular, together with Jordan, has done a great deal to give a general structure to the concept [Masson and Jordan 1971; Masson 1973, 1976; Thurber and Masson 1979]. They have defined a general three-stage network, as shown in Figure 12, and have adopted the terminology "$v(m, n_1, r_1, n_2, r_2)$" to describe it. This denotes a network with r_1 input crossbar switches having n_1 inputs and m outputs each, so that $N = n_1r_1$; r_2 output switches of n_2 outputs and m inputs each, so that the number of outputs, often M, is n_2r_2; and m middle switches of size $r_1 \times r_2$.

Clos' work has been expanded to show

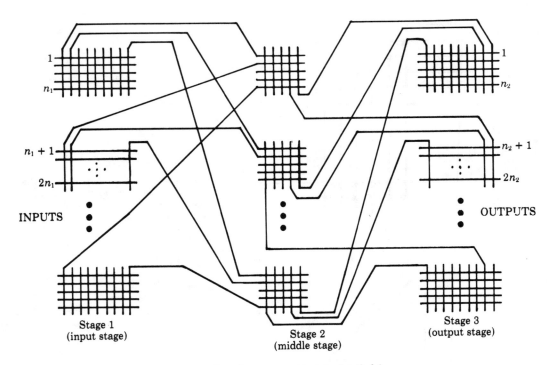

Figure 11. Clos three-stage network, detailed form.

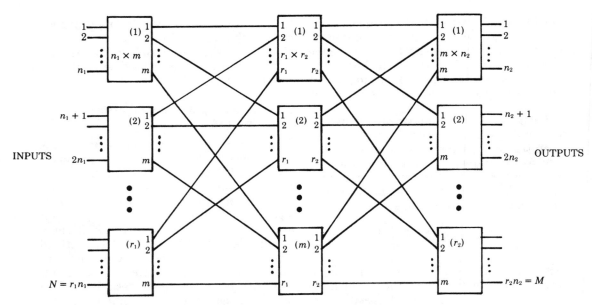

Figure 12. Clos three-stage network, block form.

that such a network used as a connector is rearrangeable if $m \geq n_2$. For $m < n_2$, the network becomes blocking. Masson and Jordan [Masson and Jordan 1972; Masson 1973] have expanded the work to include the capability for one input to be connected to any number of outputs, giving expansion capabilities, or "fan-out." In the latter case,

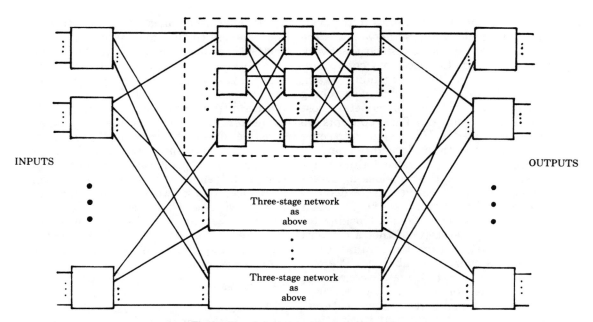

Figure 13. A five-stage Clos-type network.

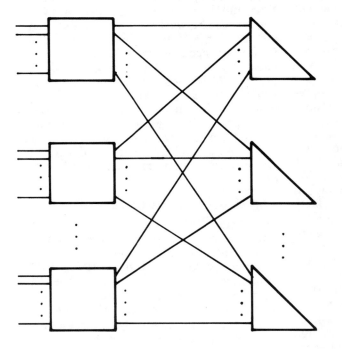

Figure 14. Single-sided Clos-type network.

the network remains strictly nonblocking if $m \geq r_2 n_2 + n_2 - 1$ and is rearrangeable if $m \geq n_2$ for $n_2 \geq r_2 n_1$, and if $m \geq r_2 n_1$ for $n_2 < r_2 n_1$. This shows that, as might be expected, for a given number of inputs per input switch, more middle switches give more possible routings from any input to any output, and reduce the likelihood of blocking. Although these are continuous functions, generally only the boundary values for the nonblocking and rearrangeable cases are of much practical importance, since the major related interest in the evaluation of circuit switching networks is the determination of the lowest cost network that will provide the necessary connection ability for a specific application. (Again, cost here is simply the number of switching elements; determination of monetary cost is much more complex.) These concepts are further expanded by Hwang [1972] to cover networks with both fan-in and fan-out.

With regard to generalized three-stage crossbar networks, the nonblocking condition has received relatively little detailed study beyond the proof that the particular specifications given yield nonblocking topologies. In general, proof that a network is nonblocking is considered sufficient to describe its connection abilities. There is little discussion in the literature concerning the control of nonblocking Clos networks. On the other hand, rearrangeable three-stage networks have received much attention for several reasons. The main appeal of rearrangeable networks is that they can realize all possible input–output permutations with far fewer switching elements than are required by nonblocking networks. Any new connection can be established in a network already carrying connections with only the possible need to rearrange one or more of the existing connections. This required rearrangement of arbitrary existing connections to establish a arbitrary additional connection has been a field of much study [Masson and Jordan 1972; Opferman and Tsao-Wu 1971; Parker 1980a; Paull 1962; Ramanujam 1973; Thompson 1976; Tsao-Wu and Opferman 1969]. Rather than attempt to give any summary of the findings in this field, which is complex and beyond the scope of an introductory overview, we refer interested readers to the references.

Opferman and Tsao-Wu [1971], in their work on rearrangeable networks, state that one switch of the network can be omitted or set to a permanent permutation (related to Waksman, below), and thus further reduce the number of crosspoints required.

Although general multistage networks have received much theoretical work, they have evidently seen little actual application in computer systems. One reported application appears in Feng's paper [Feng et al. 1979]. One problem is that, even though they are less costly than the full crossbar, they still require many switching elements. In addition, they are somewhat difficult to control, partly because of the great number of routes initially available for any required input–output connection. Rearrangeable connectors are of the most interest in many cases because of their relatively low cost. In many forms, especially when $N = M = 2^i$ (which is a good case for theoretical study), the general three-stage network can be reduced to a network of many stages of 2×2 switches [Opferman and Tsao-Wu 1971] similar to the cell-based networks discussed in Section 3.2, and are probably more easily applied in this form than in the three-stage form.

3.1.6 Beneš Networks

V. E. Beneš has done a great deal of work on the mathematics of switching systems [Beneš 1962a, 1962b, 1962c, 1964a, 1964b, 1965, 1967]. In his works, he discusses very few specific networks. He gives a very general algorithm [Beneš 1964a, 1964b, 1965] for a rearrangeable network, and discusses networks such as the example shown in Figure 15 [Beneš 1962a, 1962b, 1962c, 1965], which have taken his name in the literature [Joel 1979; Thurber 1974, 1978]. This network is composed of 2×2 switches and is often considered a crossbar network, although its topology is very similar to several cell-based networks discussed in Section 3.2.

3.1.7 Masson's Binomial Concentrator

In addition to his work on general three-stage networks, Masson has proposed a concentrator network made up of "sparse" crossbar switches [Masson 1976, 1977; Masson et al. 1979]. In a concentrator, it is only necessary to be able to connect an input to *any* unused output. Masson shows a method of using relatively few crosspoints in a two-stage arrangement to construct concentrators. A typical binomial concentrator of his design is shown in Figure 16 [Masson 1976].

3.2 Cell-Based Networks

Many networks have been developed by assuming simple switching cells, which are

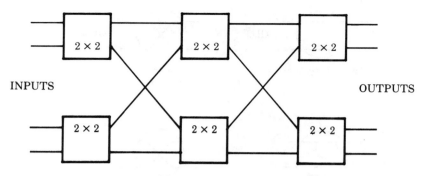

Figure 15. Beneš network.

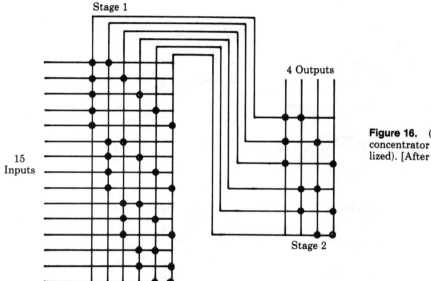

Figure 16. (15, 4, 4)ᵣ Binomial concentrator (⊕, crosspoint utilized). [After Thurber 1978.]

connected into an array to form an entire switching system for many ports. Whereas it is possible to consider the networks discussed in Section 3.1 as being composed of crossbar "cells" or "nodes," most networks considered in this section use a much simpler two-input, two-output cell. Some networks discussed use other relatively small, simple cells.

The most common two-input, two-output (2 × 2) cell is some version of the *beta element* [Joel 1968]. As an electromechanical switching device, Joel proposed the beta element as an efficient simplification of the 2 × 2 crossbar. His beta element, requiring one coil and two transfer con-

tacts, could realize the two most useful switching states, "straight-through" and "exchange," of the more complex 2 × 2 crossbar. In a more modern view, the basic beta element is a two-input, two-output switching cell, with only two states, as shown in Figure 17a. Such an element is also sometimes referred to as a "Kautz cell" [Thurber 1978], an "interchange box" [Siegel 1979a], a "permutation cell" [Gecsei 1977], and occasionally by other names, all of which have the same basic concept but differ in origin or in implementation.

In constructing switching networks, the beta element is attractive because, having only two states, it requires only one control

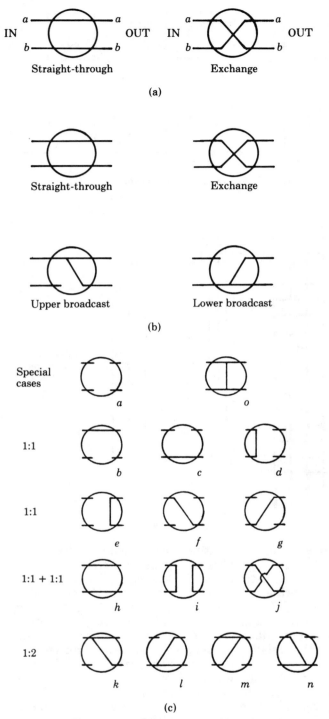

Figure 17. 2 × 2 Switching cells: (a) beta element or Kautz cell or interchange box; (b) interchange box with broadcasts; (c) general four-terminal node, in which there are 15 possible electrical states and 4 control bits required to specify all possible states. In (c) subsets allowed by common nodes are

(1) interchange box or Kautz cell or beta element—states h and j;
(2) Gecsei three-state cell—states h, i, j;
(3) 2 × 2 crossbar switch—all states except d, e, i;
(4) interchange box with broadcast—states h, j, k, l.

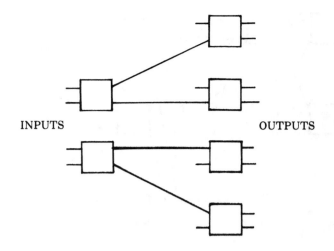

Figure 18. A 2 × 2 cell-based network for four inputs and eight outputs.

INPUTS   OUTPUTS

bit to operate a basic switching function.[4] Many networks have been developed using the beta element or a similar basic two-state cell, which could be considered a basic $(2, 2, 2)_n$ connector. The beta element is not, however, the only 2 × 2 switching cell. One simple modification, proposed for the omega network [Lawrie 1975] among others, allows two extra "forward broadcast" states, as shown in Figure 17b. This gives the cell expansion of fan-out capabilities and renders it a basic $(2, 2, 2)_n$ expander cell. These additional states, of course, require additional control bits to handle the additional capabilities. Some networks have been proposed using the simple 2 × 2 crossbar as the basic node, although in most cases the use of only the beta element's two states is assumed. Figure 17c shows all electrical states assumable by a 2 × 2 cell, and indicates those used by the most commonly found nodes.

The topologies of most basic 2 × 2 cell networks are closely related, but in some cases the actual switching capability of a given topology is dependent on the capability of the individual 2 × 2 cells used in the network. There seems to be little work discussing the capabilities of particular topol-

ogies as a function of node operation [Gecsei and Brassard 1981].

One repercussion of the use of 2 × 2 cells is that most networks using them are assumed to have $N = 2^i$ inputs and outputs, where i is an integer. Although there are methods of handling numbers of inputs and outputs that do not meet this standard, in general they do not work well, and little discussion has been given to this problem. (Some work related to this has been reported [Opferman and Tsao-Wu 1971], but it was done with respect to crossbar-based networks.) The problem is immediately apparent when one tries to envision the use of a network such as that shown in Figure 6 for, say, five, or six, or even nine inputs and outputs. In almost all cases, this requires the use of a network with inputs and outputs on some cells that are unused. One impact is that the cost in switching elements per input becomes higher than the cost figures given for specific networks in the following sections, which are based on the use of 2^i inputs and outputs. The control of the network may also be affected owing to the number of unused inputs and outputs. Unequal numbers of inputs and outputs must be accommodated in a similar manner, and the resulting network may become somewhat irregular. Figure 18 provides an illustration of the problems involved for even the simple case where the number of outputs is exactly twice the number of inputs. The main point to be made here is that most cell-based networks are

---

[4] Notice that the basic beta element does not allow the "off" condition. This would require a third state, complicating the control. In most networks, it is assumed that the devices at the inputs are capable of realizing a state in which they are disconnected from the switching network ports. Thus many types of switching elements have no "off" state.

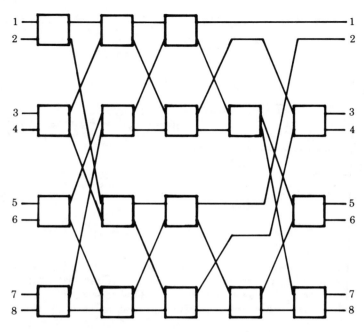

Figure 19. Waksman permutation network.

best suited for cases in which the number of inputs and outputs are equal and an integral power of two. This should be kept in mind as the specific cell-based networks are discussed.

3.2.1 General Cellular Arrays

In relatively early work, Kautz et al. [1968] discuss interconnecting "elemental" two-state permutation cells in a wide variety of patterns to give permutation or switching networks. They discuss "triangular," "diamond," "rectangular," and other arrays of the basic cell and conjecture that the "rearrangeable array" requires the minimal number of cells. Their "rearrangeable array" is essentially the same as the Waksman network discussed in the next section.

Gecsei [1977] investigated a variety of arrays using a cell that he calls the "elementary full switch." This is the two-state cell with a third state that connects both "inputs" and both "outputs" together (see Figure 17c). Many of the networks discussed are clearly single sided, and from some viewpoints the cell itself and all of the networks are single sided, since they

can connect any port to any other port. One failing, or inefficiency, in the basic concept is that a three-state cell requires two control bits and thus "wastes" one available state.

3.2.2 Waksman Permutation Network

In another relatively early paper, Waksman [1968] developed a permutation network based on the premise that any permutation of $N = 2^i$ inputs to an equal number of outputs can be reduced to a series of permutations between two inputs and two outputs. This led to a network of 2×2 permutation cells, interconnected in the pattern shown in Figure 19. Larger networks may be generated by expanding the pattern recursively, letting each input and output select between two copies of the network shown. Waksman shows that some permutations or cells which would complete the pattern are not necessary and can be omitted.

Waksman's network requires $N \log_2 N - N + 1$ cells, which, he argues, represent the lower bound on the number of cells required to realize any permutation from inputs (in-

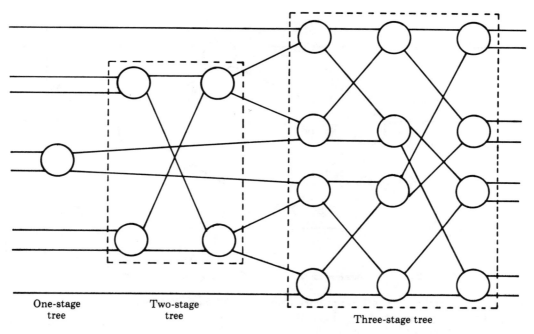

One-stage tree   Two-stage tree   Three-stage tree

Figure 20. Joel's nested tree network of beta elements. [After Joel 1968.]

tegral power of 2) to outputs. He discusses methods of setting up the network and gives an algorithm for setting the elements to achieve any desired permutation.

Applied as a circuit switch, Waksman's topology is a basic 2 × 2 cell-based rearrangeable network. The network can realize any permutation of inputs to outputs, but cannot accomplish random connections in sequence without possibly needing to rearrange existing connections. The network could be classified as an $(N, N, N)_r$ connector.

One result of the minimization of the number of cells is that the network is irregular, with different delays over different paths. The network could easily be made regular by the inclusion of the "missing" cells. The "regular" version would require $N \log_2 N - N/2$ cells.

A Clos-type rearrangeable network with an integral power of 2 inputs and outputs can be recursively reduced until it becomes a "full" (no eliminated cells) Waksman topology. Beneš devotes a great deal of discussion to this network, and it has taken his name, as discussed above. The Beneš network is "more easily rearranged" owing

to the additional switching cells, which allow more possible input–output paths than exist in the Waksman network. In fact, it is a wide-sense nonblocking network, since Beneš has developed an algorithm to operate it without blocking [Beneš 1962a, 1962b, 1962c, 1964b, 1965].

One large difference between the cell-based networks, such as Waksman's, and the basic Clos network is the delay. The basic three-stage Clos network has a delay of three for any size, whereas Waksman's and other cell-based networks have a delay that increases as $O(\log_2 N)$. The Waksman network has $2 \log_2 N - 1$ stages.

3.2.3 Joel's Networks

As examples of the use of his beta element, Joel [1968] proposed several networks. One is the Beneš network, already shown in Figure 15. He also proposed a "nested tree" type network, which is shown in Figure 20 for eight inputs and outputs. Joel's reasoning in the development of this network is rather complex, but on close examination, it can be seen that this is the same basic topology as Waksman's network. Joel also

131

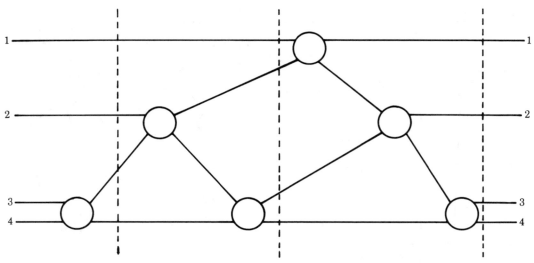

Figure 21. Joel's serial permutation network of beta elements. [After Joel 1968.]

discusses a serial permutation network that results in a more complex topology, which is shown in Figure 21.

3.2.4 Shuffle–Exchange Network

In 1968, Pease [1968] developed the concept of the "perfect shuffle" for calculating the fast Fourier transform. Stone [1971] further developed this concept. Their work was later refined into the "shuffle-exchange network" [Lang 1976; Lang and Stone 1976] shown in Figure 22. The network as proposed uses exchange boxes, which can either pass data straight through or exchange them between two inputs and two outputs (using the same concept as the 2×2 cell) and which are connected in a shuffle pattern to allow data to be permuted from box to box. Data entering at any input can be permuted to any output by repeated passes through the system.

Clearly, this is not directly applicable as a circuit switching network. The shuffle-exchange concept, however, applied as a physical series rather than as a looping algorithm, has directly or indirectly become the basis for most cell-based networks [Parker 1980b] (as will be seen in the following sections). The "full" Waksman or Beneš network can be considered a series of shuffle-exchanges in a symmetrical pattern.

3.2.5 Omega Network

In 1975, Lawrie [1975] introduced the omega network, shown in Figure 23, for array processing. Although proposed after some of the networks discussed below, the omega best illustrates the basic topology and other points, such as the relationship to the shuffle-exchange pattern, of several networks based on 2×2 cells. As shown in Figure 23a, as proposed, the network assumes $N = 2^i$ inputs and consists of $i = \log_2 N$ stages of a shuffle-exchange pattern. In Figure 23b, we redraw the network to show its similarity to what could be considered one-half of the Waksman or Beneš networks. This pattern may also be familiar as the well-known pattern of the fast Fourier transform [Pease 1968; Stone 1971]. Most cell-based networks follow or can be redrawn into these patterns.

The omega network as proposed has an efficient control algorithm. Entering from any input, the bits of the desired output number expressed in binary may be used to control the successive stages in order. If the bit is a 0, the upper output of an interchange box is taken; if it is 1, the lower output is used. This may be readily tested by using Figure 23a.

Lawrie proposed the use of four-state boxes, which allowed the broadcast states in addition to the two basic states (see Figure 17).

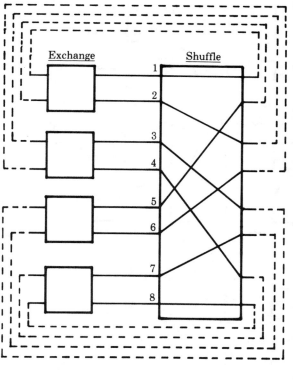

Exchange   Shuffle

1
2
3
4
5
6
7
8

Figure 22.   Shuffle–exchange network.

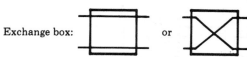

Exchange box: or

As a circuit switching topology, the omega network has a quite limited capability. For example, the network shown in Figure 23, when used as a connector, will allow only $2^{12} = 4{,}096$ different states for the 12 two-state elements as opposed to 8! = 40,320 possible permutations of 8 inputs to 8 outputs that can all be realized by a Beneš or Waksman network. Thus the network is highly blocking. With the broadcast state allowed, the network can be considered an expander, but it remains highly blocking. From a hardware standpoint, the lack of capability comes at reduced cost since the omega network requires only $N/2 \log_2 N$ (where $N = 2^i$) switching cells as opposed to the Waksman network, which requires almost twice as many.

The viability of the omega network and the many related ones depends on several factors. One is that, even though the network can realize comparatively few states, the ones that it can realize can be quite useful to array processing applications [Batcher 1976; Lawrie 1975; Pease 1977; Stone 1971]. Additionally, even though not all permutations are possible, any output can be reached from any input. Finally, the efficient algorithm proposed by Lawrie for the control of the network allows it to be set up rapidly. An individual path may be set up in the system in $O(\log_2 N)$ steps, which in some applications is done "on the fly," with each switch setting itself according to a particular bit in the destination address. The Waksman network, although it has a control algorithm, requires $O(N)$ steps in parallel to calculate the individual switch settings [Lev et al. 1981; Waksman 1968].

A complicating difference is that the Waksman algorithm sets up a full permu-

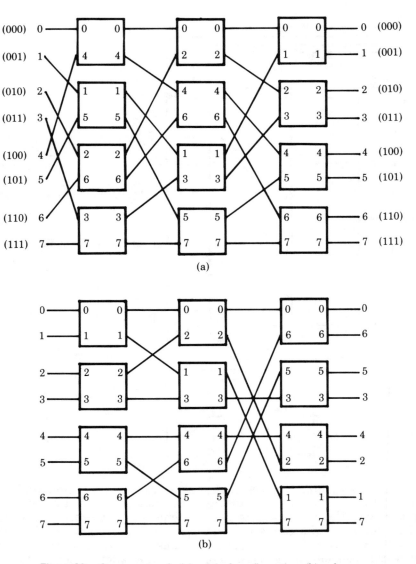

Figure 23. Omega network: (a) original configuration; (b) redrawn.

tation of inputs to outputs, whereas the very efficient Lawrie system establishes only one path. In the omega network, any single path immediately completely blocks many other connections that are still possible in the Waksman network. Determining which is more efficient depends heavily on the actual application: in some applications, few individual paths ever exist at once, and in others, full permutations are required. One compromise is to place two or more omega-type networks end to end.

Each added network increases the number of possible permutations and can still respond reasonably readily to the basic Lawrie algorithm. If $\log_2 N$ networks are connected end to end, they are capable of realizing all permutations. (There is one stage for each input–output pair, and at worst, each stage guarantees one possible match of an input to a desired output.) Such a network can be set up in $O((\log_2 N)^2)$ steps. The delay through such a network would be significantly greater than that through

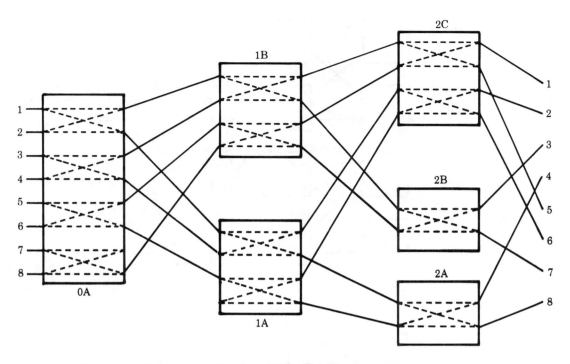

Figure 24. Flip network (with shift capabilities) for STARAN. [After Batcher 1976.]

a Waksman network, however. Again, the best solution depends on the actual application requirements.

3.2.6 Flip Network

A network very similar topologically to the omega network, and actually of earlier design, has been implemented as the "flip network" in the STARAN computing system [Batcher 1976]. This network, shown in Figure 24, is also composed of $\log_2 N$ shuffle–exchange patterns. Its major difference lies in its control efficiency. The flip network uses only the basic two-state cells and groups them or interlocks them at each stage so only a very few control bits are needed to control the network: one per block, as shown for the $N = 8$ network in Figure 24.

Although this reduces the number of possible states even below that possible in omega, the argument again is that those which are possible are highly useful in parallel processing (specifically in what are referred to as "flips" and "shifts") and any

other needs can be met by performing multiple passes through the system.

3.2.7 Binary n-Cube Network

In 1977, Pease proposed the indirect binary n-cube microprocessor array [Pease 1977]. The binary n-cube is a series of shuffle-exchanges similar to the omega and flip. The binary n-cube was proposed as a processor-to-processor communication network connecting inputs and outputs of processors, whereas omega and flip networks are oriented toward aligning data between memory and processors. The network topology, shown in Figure 25a, is identical to omega switched end for end. The binary n-cube was proposed using only two-state cells.

Pease shows that the network can execute an algorithm analogous to moving along the edges of a cube in n-space, where $N = 2^n$, as illustrated in Figure 25b for $n = 3$. This is pursued by Siegel in the cube network [Siegel 1977a, 1979b; Siegel and

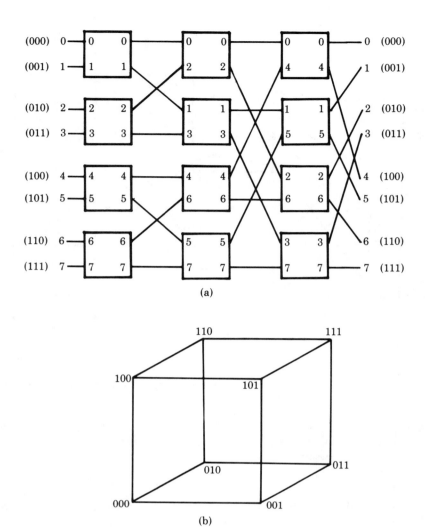

Figure 25. The cube network: (a) the network; (b) the cube transformation. [After Pease 1977.]

McMillen 1981a; Smith and Siegel 1978], related to the cube function

$$\text{Cube}_i(S_{n-1} \cdots S_1 S_0) = S_{n-1} \cdots S_{i+1} \bar{S}_i S_{i-1} \cdots S_0,$$

where $S = S_{n-1} \cdots S_1 S_0$ is the binary code for the input or output and $\bar{S}_i$ is the complement of S_i. The appropriate bits of the binary code for the output are used to control routing of data through the network in a manner similar to that of the omega network.

Just as in the omega and flip networks, the binary n-cube network is highly block-

ing, but Pease shows that the states which can exist are suitable for common multiprocessor applications. Pease discusses the possibility of a practical control scheme for the network.

3.2.8 Banyan Networks

One class of networks that is related to the cell-based networks is a group known as "banyan networks." The topology and a detailed description of the operation of a switching system using it were described by Lipovski [1970] without the banyan name.

Apex nodes (processors)

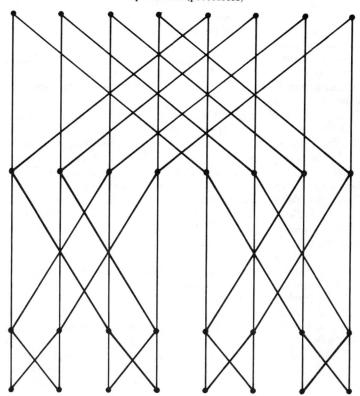

Figure 26. SW-banyan network.
[After Lipovski and Tripathi 1977.]

Base nodes (memories)

Later works by Lipovski and others [Goke and Lipovski 1973; Lipovski and Tripathi 1977; Tripathi and Lipovski 1979] give more general discussions of the banyan concept. Banyan networks are based on graph theory concepts, and as a broad class they essentially include all switching networks that have unique paths from each input to each output. The class has had a large impact on switching systems and is covered in most switching surveys.

The most commonly discussed banyan is the SW-banyan, which is being implemented as part of the TRAC system at the University of Texas, although as a packet switching system with 2 × 3 nodes [Lipovski 1979; Tripathi and Lipovski 1979]. The normal representation for the topology of an SW-banyan of 2 × 2 nodes is shown in Figure 26. In this figure, the dots represent switching cells and the lines represent connections, unlike the very similar graph of a connection network (as in Figure 7). As explained in the references, the basic banyan pattern may be used to interconnect crossbar switching elements larger than 2 × 2, making banyans as a group more general than most other cell-based topologies discussed. However, most discussions of the banyan concentrate on the SW-banyan of 2 × 2 or 2 × 3 cells.

3.2.9 General 2 × 2 Cell-Based Networks

The basic shuffle–exchange pattern for interconnecting 2 × 2 cells has been very popular in the development of networks. As was already noted with the omega network, many networks can be reduced to a readily recognizable pattern by simply redrawing the diagram of the system (as in Figure 23). As also noted, the connection capability is quite limited for these networks, which generally have the minimum number of cells to allow any input to connect to any output, but can realize only relatively few complete input–output permutations.

Although not immediately apparent, there are a large number of variations possible in the connection pattern and the method of numbering inputs and outputs, and this has impact on connection capability. This has allowed the existence of a large number of similar, differently named networks and topologies, which differ in details. Besides the omega and others mentioned above to illustrate the basic topological concept, related networks include the "delta" [Patel 1979; Dias and Jump 1980], "reverse exchange" [Wu and Feng 1979, 1980a], "baseline" [Wu and Feng 1980b], and "R-network" [Parker 1980b].

Detailed explanation and comparison of the seemingly minor topological differences between these networks, and the differences in switching capabilities that result, can be found in several papers [Parker 1980b; Patel 1981; Pradhan and Kodandapani 1980; Siegel 1977a, 1977b, 1978, 1979b, 1979c, 1980; Smith and Siegel 1978, 1979; Wu and Feng 1979, 1980a, 1980b]. Additional papers specifically address the control of such networks and/or Beneš networks [Abidi and Agrawal 1980; Chen et al. 1981; Lenfant 1978; Lev et al. 1981].

3.2.10 One-Sided Cell-Based Networks

The 2×2 basic cell may be applied to single-sided networks, but has been given only minimal discussion in the literature. Other than those in Gecsei's work [Gecsei 1977], the networks generally use a double-sided permutation network in conjunction with two-state cells to construct permutation networks or general switching networks (connectors) [Chung and Wong 1979; Marcus 1977; Osatake et al. 1973].

3.2.11 Networks with Other Basic Cells

A few networks have been proposed which use some basic switching cell other than either the crossbar switch or the 2×2 cell. Most notable of these is probably the data manipulator [Feng 1974], which uses a three-input, three-output cell connected in a pattern related to the shuffle pattern used

in 2×2 cell-based networks. The network is generally shown as in Figure 27a, but can be better compared to other topologies in this paper if represented as shown in Figure 27b.

In the original system proposed by Feng, each cell can route data from only one of the three inputs to any output at any one time, essentially acting as a data distributor. For example, the cell could not accept two inputs and route them to separate outputs, as could the beta element. This basic network topology, also referred to as the "augmented data manipulator" or ADM, opens the field of an array of 3×3 nodes, a field which has received little study compared to that of 2×2 node networks. Some papers [Siegel 1979a, 1979b, 1980; Siegel and McMillen 1981b] discuss the network, and in one [Smith and Siegel 1978] it is compared to a PM2I network, which has a three-input, three-output node with slightly different switching capabilities.

Another permuting or switching network based on a "selector cell" has been proposed by Bandyopadhyay et al. [1972], and is shown in Figure 28. Even though implementation and control of the network are discussed thoroughly, the network as a whole seems to have distinct disadvantages in that it requires a number of different sized, or nonstandard, switching cells, although each has identical basic functionality. It has a very wide range of delay through the system for different paths.

4. ANALYSIS OF SWITCHING SYSTEMS

A great deal of work has been done on the analysis of switching networks from several viewpoints, most of which is beyond the scope of this paper.

For the interested reader, some analysis of switching networks has been done from a graph-theoretic viewpoint, which is tightly related to topologies. Early work was done by Ofman [1967], who investigated topologies similar to the SW-banyan and shuffle–exchange. Thompson [1978] discusses generalized connection networks from a similar viewpoint. Pippenger has performed some specific graphical analysis of concentrators [Pippenger 1974, 1977] and connectors [Pippenger 1978].

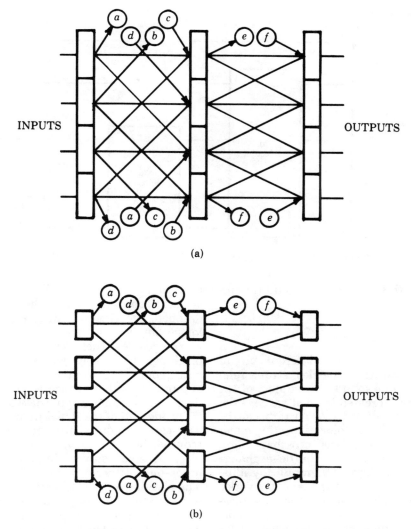

Figure 27. Data manipulator network: (a) basic diagram; (b) cell-based drawing. [After Fang 1974.]

Work has also been done on the analysis of networks from a more mathematical viewpoint. Cantor [1971] investigated blocking and nonblocking conditions and Lev et al. [1981] discuss routing in permutation networks. In other papers networks are discussed from a mathematical viewpoint [Hwang 1976; Pippenger 1976; Pradhan and Kodandapani 1979, 1980; Ramanujam 1973; Siegel 1976, 1977a; Valiant 1975].

The evaluation or comparison of various networks with respect to overall quality is a topic of much current work. This is evidenced by the many papers in which the various shuffle–exchange type networks are compared. Less work appears to have been done in comparing all possible networks, although many of the papers proposing various networks do this to some extent in defense of the network proposed. The interested reader is referred to work by

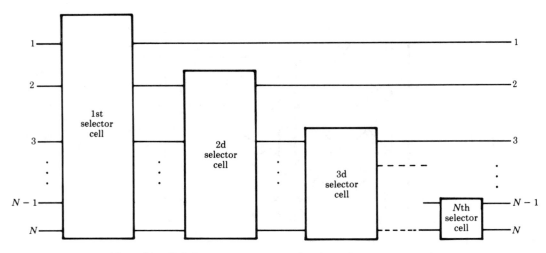

Figure 28. Cellular permuter array of Bandyopadhyay et al. [1972].

Malek and Myre [1980, 1981] and Franklin [1980].

5. CONCLUSIONS

In spite of the many different circuit switching network topologies, strong relationships can be found among most of the various types. Treated in what we consider to be descending order, a pattern can be seen linking the topologies of the major types of networks: the crossbar, Clos, Beneš, Waksman, and shuffle–exchange type networks. Figure 29 graphically illustrates this linkage of the major networks. The single-stage crossbar is the most powerful (nonblocking, delay = 1) and most costly. The nonblocking three-stage Clos network retains the switching capability of the crossbar at significantly lower cost but with greater delay. With rearrangeability allowed and inputs and outputs of equal integral powers of 2, the Clos network may be reduced to a 2 × 2 cell-based network, often called the Beneš network. This reduced network has lower cost than the nonblocking network, but can be wide-sense nonblocking, as shown by Beneš [1965]. Delay through the network can be significant for large N. If Waksman's minimization technique is employed, some cells (with the links shown in Figure 29) can be dropped, to gain even lower cost in terms of switching elements, although the non-

uniformity may add complications in areas such as control. When these cells are dropped, the Beneš routing algorithm no longer applies and the network becomes rearrangeable rather than wide-sense nonblocking. The delay varies by different paths, but all permutations of inputs to outputs can still be executed. In the shuffle–exchange omega-type network, essentially one-half of the Beneš network is used. The cost is reduced further and delay is reduced, but the network becomes highly blocking since only a limited number of input–output permutations are possible. The justification for using this kind of network is that the possible permutations are useful for specific computer applications. There are also valid arguments that this network is much easier to control than the others, partly because of its limited permutation set.

It should be emphasized that this general scheme of relationships relates only to basic topologies. Since each topoogy can be implemented in many ways, there are many more specific networks than there are topologies. Within a topology, the full capabilities of an individual network would depend heavily on the implementation, including both switching capabilities and control of the switching elements.

The basic relationships discussed above assume short-distance networks for large numbers of inputs and outputs. Cost and

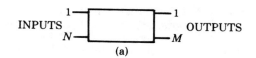

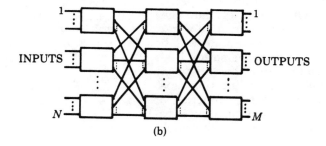

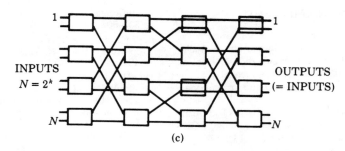

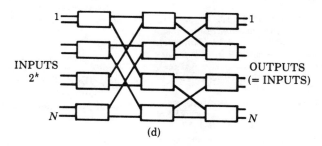

Figure 29. Linking relationship of the topologies of major network types: (a) single-stage crossbar; (b) Clos nonblocking network; (c) Clos fully decomposed rearrangeable 2×2 network or Beneš network; (d) shuffle–exchange limited permutation network. *Important criteria:* (a) nonblocking, delay = 1, cost = NM; (b) nonblocking, delay = 3, cost < NM; (c) rearrangeable, delay = $2 \log_2 N - 1$, cost = $O(N \log_2 N)$; (d) blocking, delay = $\log_2 N$, cost = $O[(N/2) \log_2 N]$.

delay relationships in particular would differ for fewer inputs and outputs.

This general scheme of relating topologies, although useful for considering some important capabilities of circuit switching networks, omits some very important ad-

ditional considerations. In particular, good systems for the control of most networks are not well documented, and the relationships among various control schemes are not clear. The ability of the 2×2 cell-based networks to handle numbers of inputs and outputs not an integral power of 2, or unequal numbers of inputs and outputs, is not well documented.

Some topologies, such as Joel's serial permutation network and the data manipulator, do not immediately fit into the system, and it may be useful to consider how they are related to the more common networks. To fully broaden the system, it will be necessary to find a firm mathematical basis for describing and relating the various topologies and possibly the connection capabilities of each topology as a function of node capabilities. This and much more remain for future research on circuit switching systems.

ACKNOWLEDGMENTS

The many useful comments from the unknown reviewers of the paper and from technical editor Rachel Rutherford and editor Bruce Weide did much to improve this paper, both in contents and readability, and their efforts are much appreciated and gratefully acknowledged.

This work was sponsored in part by the U.S. Army Ballistic Missile Defense Advanced Technology Center, Huntsville, Alabama, under contract DASG60-79-C-0052 to the University of Kentucky.

REFERENCES

ABIDI, M. A., AND AGRAWAL, D. P. 1980. On conflict-free permutations in multistage interconnection networks. *J. Digital Syst.* **4**, 2 (Summer), 115–134. Abstracted in *Proc. 1979 Int. Conf. Parallel Processing* (Detroit, Mich., Aug. 21–24), IEEE Computer Society, Long Beach, Calif., p. 159.

BANDYOPADHYAY, S., BASU, S., AND CHOUDHURY, A. K. 1972. A cellular permuter array. *IEEE Trans. Comput.* **C-21**, 10 (Oct.), 1116–1119.

BATCHER, K. E. 1968. Sorting networks and their applications. In *Proc. AFIPS Spring Joint Computer Conference*, vol. 32, Thompson, Washington, D. C., pp. 307–314.

BATCHER, K. E. 1976. The flip network in STARAN. In *Proc. 1976 Int. Conf. Parallel Processing* (Detroit, Mich., Aug. 24–27), IEEE Computer Society, Long Beach, Calif., pp. 65–71. 1976.

BENEŠ, V. E. 1962a. Heuristic remarks and mathematical problems regarding the theory of connecting systems. *Bell Syst. Tech. J.* **41**, 4 (July), 1201–1247.

BENEŠ, V. E. 1962b. Algebraic and topological properties of connecting networks. *Bell Syst. Tech. J.* **41**, 4 (July), 1249–1273.

BENEŠ, V. E. 1962c. On rearrangeable three-stage connecting networks. *Bell Syst. Tech. J.* **41**, 5 (Sept.), 1481–1492.

BENEŠ, V. E. 1964a. Optimal rearrangeable multistage connecting networks. *Bell Syst. Tech. J.* **43**, 4 (Part 2) (July), 1641–1656.

BENEŠ, V. E. 1964b. Permutation groups, complexes, and rearrangeable connecting networks. *Bell Syst. Tech. J.* **43**, 4 (Part 2) (July), 1619–1640.

BENEŠ, V. E. 1965. *Mathematical Theory of Connecting Networks and Telephone Traffic.* Academic Press, New York.

BENEŠ, V. E. 1967. Square switch distribution network employing a minimal number of crosspoints. U.S. Patent No. 3358269 (Dec. 12, 1967).

CANTOR, D. G. 1971. On nonblocking switching networks. *Networks* **1**, 4 (Winter), 367–377.

CHEN, P.-Y., LAWRIE, D. H., YEW, P.-C., AND PADUA, D. A. 1981. Interconnection networks using shuffles. *Computer* **14**, 12 (Dec.), 55–64.

CHUNG, K.-M., AND WONG, C. K. 1979. Asymptotically optimal interconnection networks from two-state cells. *IEEE Trans. Comput.* **C-28**, 7 (July), 500–505.

CLOS, C. 1953. A study of non-blocking switching networks. *Bell Syst. Tech. J.* **32**, 2 (Mar.), 406–424.

DIAS, D. M., AND JUMP, J. R. 1980. Analysis and simulation of buffered delta networks. *IEEE Trans. Comput.* **C-30**, 4 (Apr.), 273–282. Also in *Proc. Workshop Interconnection Networks* (West Lafayette, Ind., Apr. 21–22), IEEE Computer Society, Long Beach, Calif., pp. 84–92. 1980.

FENG, T.-Y. 1974. Data manipulating functions in parallel processors and their implementations, *IEEE Trans. Comput.* **C-23**, 3 (Mar.), 309–318.

FENG, T.-Y. 1981. A survey of interconnection networks. *Computer* **14**, 12 (Dec.), 12–27.

FENG, T.-Y., WU, C.-L., AND AGRAWAL, D. P. 1979. A microprocessor-controlled asynchronous circuit switching network. In *Proc. 6th Annu. Symp. Computing Architecture* (Apr. 23–25), IEEE Computing Society, Long Beach, Calif., pp. 202–215.

FRANKLIN, M. A. 1980. VLSI performance comparison of Banyan and crossbar communications networks. *IEEE Trans. Comput.* **C-30**, 4 (Apr.), 283–290. Also in *Proc. Workshop on Interconnection Networks* (West Lafayette, Ind., Apr. 21–22), IEEE Computer Society, Long Beach, Calif., 20–28.

GECSEI, J. 1977. Interconnection networks from three-state cells. *IEEE Trans. Comput.* **C-26**, 8 (Aug.), 705–711.

GECSEI, J., AND BRASSARD, J.-P. 1981. The topology of cellular partitioning networks. *IEEE Trans. Comput.* **C-30**, 2 (Feb.).

GOKE, L. R., AND LIPOVSKI, G. J. 1973. Banyan networks for partitioning multiprocessor systems. In *Proc. 1st Annu. Symp. Computing Architecture* (Gainsville, Fla., Dec. 9–11), IEEE Computer Society, Long Beach, Calif., pp. 21–28.

HWANG, F. K. 1972. Rearrangeability of multi-connection three-stage Clos networks. *Networks* **2**, 4 (Winter), 301–306.

HWANG, F. K. 1976. Balanced networks. In *Proc. 1976 Int. Conf. Communication Conference Record* (Philadelphia, June 14–16), IEEE, New York, pp. (7-13)–(7-16).

JOEL, A. E., JR. 1968. On permutation switching networks. *Bell Syst. Tech. J.* **47**, 5 (May–June), 813–822.

JOEL, A. E., JR. 1979. Circuit switching: Unique architecture and applications. *Computer* **12**, 6 (June), 10–22. Also in *Tutorial: Distributed Processor Communication Architecture*, K. J. Thurber, Ed. IEEE Comput. Soc., Long Beach, Calif., pp. 325–337.

KAUTZ, W. H., LEVITT, K. N., AND WAKSMAN, A. 1968. Cellular interconnection arrays. *IEEE Trans. Comput.* **C-17**, 5 (May), 443–451.

KUCK, D. J. 1978. *The Structure of Computers and Computations.* Wiley, New York.

LANG, T. 1976. Interconnections between processors and memory modules using the shuffle–exchange network. *IEEE Trans. Comput.* **C-25**, 5 (May), 496–503.

LANG, T., AND STONE, H. S. 1976. A shuffle–exchange network with simplified control. *IEEE Trans. Comput.* **C-25**, 1 (Jan.), 55–65.

LAWRIE, D. H. 1975. Access and alignment of data in an array processor. *IEEE Trans. Comput* **C-24**, 12 (Dec.), 1145–1155.

LENFANT, J. 1978. Parallel permutations of data: A Beneš network control algorithm for frequently used permutations. *IEEE Trans. Comput.* **C-27**, 7 (July), 637–647.

LEV, G. F., PIPPENGER, N., AND VALIANT, L. G. 1981. A fast parallel algorithm for routing in permutation networks. *IEEE Trans. Comput.* **C-30**, 2 (Feb.), 93–100.

LIPOVOSKI, G. J. 1970. The architecture of a large associative processor. In *Proc. AFIPS 1970 Spring Joint Computer Conference* (May), vol. 36, AFIPS Press, Arlington, Va., pp. 385–396.

LIPOVSKI, G. J. 1979. The architecture of the banyan switch for TRAC. Prelim. Tech. Rep. TRAC-7, Depts. Computer Science and Electrical Engineering, Univ. of Texas (Jan. 15).

LIPOVSKI, G. J., AND TRIPATHI, A. 1977. A reconfigurable varistructure array processor. *In Proc. 1977 Int. Conf. Parallel Processing* (Detroit, Mich., Aug. 23–26), IEEE Computer Society, Long Beach, Calif., pp. 165–174.

MALEK, M., AND MYRE, W. W. 1980. Figures of merit for interconnection networks. In *Proc. Workshop Interconnection Networks* (West Lafayette, Ind., Apr. 21–22), IEEE Computer Society, Long Beach, Calif., pp. 74–83.

MALEK, M., AND MYRE, W. W. 1981. A description method of interconnection networks. *IEEE Comput. Soc. Distrib. Process. Q.* **1**, 1 (Feb.), 1–6.

MARCUS, M. J. 1970a. Space–time equivalents in connecting networks. In *Proc. 1970 Int. Conf. Communication Conference Record*, pp. (35-25)–(35-31).

MARCUS, M. J. 1970b. Designs for time slot interchangers. In *Proc. 1970 Nat. Electronics Conf.* (Chicago, Ill., Dec. 7–9), vol. 26, pp. 812–817.

MARCUS, M. J. 1977. The theory of connecting networks and their complexity: A review. *Proc. IEEE* **65**, 9 (Sept.), 1263–1271.

MASSON, G. M. 1973. Upper bounds on fanout in connection networks. *IEEE Trans. Circuit Theory* **CT-20**, 3 (May), 222–229.

MASSON, G. M. 1976. On rearrangeable and nonblocking switching networks. In *Proc. 1976 Int. Conf. Communication Conference Record* (Philadelphia, Pa., June 14–16), IEEE, New York, pp. (7-1)–(7-7). Also in *Tutorial: Distributed Processor Communication Architecture*, K. J. Thurber, Ed. IEEE Computer Society, Long Beach, Calif., pp. 355–361.

MASSON, G. M. 1977. Binomial switching networks for concentration and distribution. *IEEE Trans. Commun.* **COM-25**, 9 (Sept.), 873–883.

MASSON, G. M., AND JORDAN, B. W. 1971. Realization of a class of multiple connection assignments with asymmetrical three-stage connection networks. In *Proc. 5th Annu. Princeton Conf. Information and Systems Science* (Princeton, N. J., Mar. 25–26), Princeton Univ., N. J., pp. 316–320.

MASSON, G. M., AND JORDAN, B. W. 1972. Generalized multi-stage connection networks. *Networks* **2**, 3 (Fall), 191–209.

MASSON, G. M., GINGHER, G. C., AND NAKAMURA, S. 1979. A sampler of circuit switching networks. *Computer* **12**, 6 (June), 32–48. Also in *Tutorial: Distributed Processor Communication Architecture*, K. J. Thurber, Ed. IEEE Computer Society, Long Beach, Calif., 1979, pp. 362–378.

OFMAN, J. P. 1967. A universal automaton. In *Transactions of the Moscow Mathematics Society.* American Mathematical Society, Providence, R. I., pp. 200–215.

OPFERMAN, D. C., AND TSAO-WU, N. T. 1971. On a class of rearrangeable switching networks; Part I: Control algorithms; Part II: Enumeration studies and fault diagnosis. *Bell Syst. Tech. J.* **50**, 5 (May–June), 1579–1618.

OSATAKE, T., OGAWA, T., AND HAYASHIDA, T. 1973. Optimum structure of one-sided rearrangement switching networks. *Electron. Commun. Jpn.* **56**, 1 (Jan.), 28–33.

PARKER, D. S., JR. 1980a. New points of view on three-stage rearrangeable switching networks. In *Proc. Workshop Interconnection Networks* (West Lafayette, Ind., Apr. 21–22), IEEE Computer Society, Long Beach, Calif., pp. 56–63.

PARKER, D. S., JR. 1980b. Notes on shuffle/exchange-type switching networks. *IEEE Trans. Comput.* **C-29**, 3 (Mar.), 213–222.

PATEL, J. H. 1979. Processor–memory interconnections for multiprocessors. In *Proc. 6th Annu. Symp. Computer Architecture* (Apr. 23–25), IEEE Computer Society, Long Beach, Calif., pp. 168–177.

PATEL, J. H. 1981. Performance of processor–memory interconnections for multiprocessors. *IEEE Trans. Comput.* **C-30,** 10 (Oct.), 771–780.

PAULL, M. C. 1962. Reswitching of connection networks. *Bell Syst. Tech. J.* **41,** 3 (May), 833–855.

PEASE, M. C., III 1968. An adaptation of the fast fourier transform for parallel processing. *J. ACM* **15,** 2 (Apr.), 252–264.

PEASE, M. C., III 1977. The indirect binary *n*-cube microprocessor array. *IEEE Trans. Comput.* **C-26,** 5 (May), 458–473.

PIPPENGER, N. 1974. On the complexity of strictly nonblocking connection networks. *IEEE Trans. Commun.* **COM-22,** 11 (Nov.), 1890–1892.

PIPPENGER, N. 1976. The complexity of seldom-blocking networks. In *Proc. 1976 Int. Conf. Communication Conference Record* (Philadelphia, June 14–16), IEEE, New York, pp. (7-8)–(7-12).

PIPPENGER, N. 1977. Superconconcentrators. *SIAM J. Comput.* **6,** 2 (June), 298–304.

PIPPENGER, N. 1978. Generalized connectors. *SIAM J. Comput.* **7,** 4 (Nov.), 510–514.

PRADHAN, D. K., AND KODANDAPANI, K. L. 1979. A framework for the study of permutations and applications to memory processor interconnection networks. In *Proc. 1979 Int. Conf. Parallel Processing* (Detroit, Mich., Aug. 24–27), IEEE Computer Society, Long Beach, Calif., pp. 148–158.

PRADHAN, D. K., AND KODANDAPANI, K. L. 1980. A uniform representation of single- and multistage interconnection networks used in SIMD machines. *IEEE Trans. Comput.* **C-29,** 9 (Sept.), 777–790.

RAMANUJAM, H. R. 1973. Decomposition of permutation networks. *IEEE Trans. Comput.* **C-22,** 7 (July), 639–643.

SIEGEL, H. J. 1976. Single instruction stream—Multiple data stream machine interconnection network design. In *Proc. 1976 Int. Conf. Parallel Processing* (Detroit, Mich., Aug. 24–27), pp. 273–282.

SIEGEL, H. J. 1977a. Analysis techniques for SIMD machine interconnection networks and the effects of processor address masks. *IEEE Trans. Comput.* **C-26,** 2 (Feb.), 153–161.

SIEGEL, H. J. 1977b. The universality of various types of SIMD machine interconnection networks. In *Proc. 4th Annu. Symp. Computer Architecture* (Mar. 23–25), IEEE Computer Society, Long Beach, Calif., 70–79.

SIEGEL, H. J. 1978. Partitionable SIMD computer system interconnection network universality. In *Proc. 16th Annu. Allerton Conf. Communication, Control, and Computing* (Oct.), Univ. of Illinois, Urbana-Champaign, Ill., pp. 586–595.

SIEGEL, H. J. 1979a. Interconnection networks for SIMD machines. *Computer* **12,** 6 (June), 57–65. Also in *Tutorial: Distributed Processor Communication Architecture,* K. J. Thurber, Ed. IEEE Computer Society, Long Beach, Calif., pp. 379–387.

SIEGEL, H. J. 1979b. Partitioning permutation networks: The underlying theory. In *Proc. 1979 Int. Conf. Parallel Processing* (Detroit, Mich., Aug. 21–24), IEEE Computer Society, Long Beach, Calif., pp. 175–184.

SIEGEL, H. J. 1979c. A model of SIMD machines and a comparison of various interconnection networks. *IEEE Trans. Comput.* **C-28,** 12 (Dec.), 907–917.

SIEGEL, H. J. 1980. The theory underlying the partitioning of permutation networks. *IEEE Trans. Comput.* **C-29,** 9 (Sept.), 791–801.

SIEGEL, H. J., AND McMILLEN, R. J. 1981a. The multistage cube: A versatile interconnection network. *Computer* **14,** 12 (Dec.), 65–76.

SIEGEL, H. J., AND McMILLEN, R. J. 1981b. Using the augmented data manipulator network in PASM. *Computer* **14,** 2 (Feb.), 25–33.

SIEGEL, H. J., McMILLEN, R. J., AND MUELLER, P. T., JR. 1979. A survey of interconnection methods for reconfigurable parallel processing systems. In *Proc. AFIPS 1979 Nat. Computer Conf.* (June 4–9), vol. 48, AFIPS Press, Arlington, Va., pp. 529–542.

SMITH, S. D., AND SIEGEL, H. J. 1978. Recirculating, pipelined, and multistage SIMD interconnection networks. In *Proc. 1978 Int. Conf. Parallel Processing* (Detroit, Mich., Aug. 22–25), IEEE Computer Society, Long Beach, Calif., pp. 206–214.

SMITH, S. D., AND SIEGEL, H. J. 1979. An emulator network for SIMD machine interconnection networks. In *Proc. 6th Annu. Symp. Computer Architecture* (Apr. 23–25), IEEE Computer Society, Long Beach, Calif., pp. 232–241.

STONE, H. S. 1971. Parallel processing with the perfect shuffle. *IEEE Trans. Comput.* **C-20,** 2 (Feb.), 153–161. Also in *Tutorial: Distributed Processor Communication Architecture,* K. J. Thurber, Ed. IEEE Computer Society, Long Beach, Calif., pp. 388–396.

THOMPSON, C. D. 1978. Generalized connection networks for parallel processor intercommunication. *IEEE Trans. Comput.* **C-27** (Dec.), 1119–1125.

THOMPSON, G. B. 1976. A conceptual model for crossbar switching networks. In *Proc. 1976 Int. Conf. Communication Conference Record* (Philadelphia, June 14–16), IEEE New York, pp. (7-23)–(7-25).

THURBER, K. J. 1974. Interconnection networks—A survey and assessment. In *Proc. 1974 Nat. Computer Conf.* (Chicago, May 6–10), AFIPS Press, Arlington, Va., pp. 909–919.

THURBER, K. J. 1978. Circuit switching technology: A state-of-the-art survey. *COMPCON 78 Fall* (Washington, D. C., Sept. 5–8), IEEE Computer Society, Long Beach, Calif., pp. 116–124. Also in *Tutorial: Distributed Processor Communication Architecture,* K. J. Thurber, Ed. IEEE Computer Society, 1979, pp. 338–346.

THURBER, K. J., AND MASSON, G. M. 1979. *Distributed Processor Communication Architecture.* D. C. Heath (Lexington Books), Lexington, Mass.

TRIPATHI, A. R., AND LIPOVSKI, G. J. 1979. Packet switching in banyan networks. In *Proc. 6th Annu. Symp. Computer Architecture* (Apr. 23–25), IEEE Computing Society, Long Beach, Calif., pp. 160–167.

TSAO-WU, N. T. AND OPFERMAN, D. C. 1969. On permutation algorithms for rearrangeable switching networks. In *Proc. 1969 Int. Conf. Communication Conference Record*, pp. (10-29)–(10-34).

VALIANT, L. G. 1975. On non-linear lower bounds in computational complexity. In *Proc. 7th Annu. ACM Symp. Theory of Computing*, ACM, New York, pp. 45–53.

WAKSMAN, A. 1968. A permutation network. *J. ACM 15*, 1 (Jan.), 159–163.

WU, C.-L., AND FENG, T.-Y. 1979. Fault diagnosis for a class of multistage interconnection networks. In *Proc. 1979. Int. Conf. Parallel Processing* (Detroit, Mich., Aug. 21–24), IEEE Computer Society, Long Beach, Calif., pp. 269–278.

WU, C.-L., AND FENG, T.-Y. 1980a. The reverse-exchange interconnection network. *IEEE Trans. Comput.* **C-29**, 9 (Sept.), 801–811. Also essentially in *Proc. 1979 Int. Conf. Parallel Processing* (Detroit, Mich., Aug. 21–24), IEEE Computer Society, Long Beach, Calif., pp. 160–174.

WU, C.-L., AND FENG, T.-Y. 1980b. On a class of multistage interconnection networks. *IEEE Trans. Comput.* **C-29**, 8 (Aug.), 694–702.

Received July 1981; final revision accepted June 1983.

A SURVEY OF INTERCONNECTION NETWORKS

Robert J. McMillen

Hughes Aircraft Company

ABSTRACT

This paper presents a survey of multistage interconnection networks. Applications for these networks range from telephone switching to parallel computers. Topologies of the major multistage networks that have been discussed in the literature are described and compared. An example of a control scheme is also presented. Finally five techniques for adding fault tolerance to multistage networks are described.

1. INTRODUCTION

Interconnection networks were first proposed for use in telephone exchanges to allow subscribers to talk with one another. Some decades later, researchers began to consider how networks could be incorporated into computers. The first applications dealt with the manipulation of numbers (e.g., sorting). They were later used to transfer data between various subsystems in the computer. As technology improved, component sizes shrank, and linking several computers together was explored. Many different approaches have been considered and some implemented. These include the use of busses, hierarchies of busses, direct links, single stage networks, multistage networks, and crossbars. Many of these approaches have been surveyed [3, 11, 16, 41, 46]. This survey is concerned with multistage networks only.

The interconnection networks considered for use in parallel computer systems were not suitable for use in telephone exchanges; the former were digital and the latter, primarily analog. There were also significant differences in the bandwidth requirements and set-up time. With the advent of packet switching for data transmission services, however, the applications have come full circle. Multistage interconnection networks are now being considered for use in Integrated Services Digital Networks (ISDNs) [14, 48] which are being designed to handle both voice and data packet transmissions.

2. INTERCONNECTION NETWORK APPLICATIONS

The terminology that will be used to discuss multistage interconnection networks is as fol-

lows. Network inputs or outputs are _ports_. Each _stage_ of a network is composed of _switching elements_ which will also be referred to as _nodes_. Switching elements are usually small, single-stage crossbars, however in some cases are even simpler circuits like multiplexers. Switching elements are connected to one another by _links_ and a _path_ or _connection_ consists of an alternating sequence of links and switching elements. For each network, the number of inputs (or outputs) is denoted N and $n=\log_2 N$ (if N is a power of two).

2.1. Communications

Interconnection network applications can be broadly divided into two categories: communications and computers. The use of a network in a communication system is straightforward. It must be able to establish a connection between any arbitrary pair of ports (terminations) and be able to sustain many such connections simultaneously. Two major approaches for doing this are circuit switching and packet switching.

In the circuit switching mode, once a fixed bandwidth connection is established, it is held for the duration of the conversation. It is well known that this is inefficient since conversations typically use only 40% of the available bandwidth [10] and data potentially uses much less. On the other hand, voice can be packetized and switched so that each user consumes only as much bandwidth as is needed. Packet switching has the advantage that the contents of the packet can just as easily be data as digitized voice. Consequently, researchers have considered providing both services in an integrated network [14, 48].

2.2. Computers

One of the first applications was for sorting numbers. An arbitrary list of numbers would be presented to the network, one number per input port, and the sorted list would appear at the output. Data routing applications followed in which the network moved data among various computer systems (e.g., processors and memory modules). Routing the data was similar to sorting numbers, which can be illustrated as follows. Suppose a list to be sorted has a data item associated with each number. If there are N

Reprinted from the _Proceedings of Globecom_, 1984, pages 105-113.

input and N output ports and the list presented contains one each of the of the numbers 1 through N, inclusive, then the sorting network could be used to permute the data. Permuting is a rearrangement of the data at the input such that no two items go to the same output. Hence, each number to be "sorted," acts like a routing tag. Subsequently developed routing tag schemes were generalized and simplified but motivated by this idea.

Classification of Parallel Architectures - There are a number of ways to configure a parallel system that influence the requirements for a network. Flynn [18] has proposed a taxonomy for classifying parallel architectures according to the number of concurrent instruction and data streams. The simplest system is a conventional serial processor which is classified SISD for single instruction stream - single data stream.

The first type of parallel system is classified as SIMD, single instruction stream - multiple data stream. Typically, an SIMD machine consists of a control unit, P processors, M memory modules (M is usually $\geq$ P), and an interconnection network. The control unit broadcasts instructions to all of the processors, and all active processors execute the same instruction at the same time. Each active processor executes the instruction on data in its own memory module. The interconnection network, sometimes referred to as an alignment or permutation network, provides a communications facility for the processors and memory modules.

The second type of parallel system is classified as MIMD, multiple instruction stream - multiple data stream. An MIMD machine typically consists of P processors and M memories (M $\geq$ P), where each processor can follow an independent instruction stream. As with SIMD machines, there is a multiple data stream and an interconnection network. Thus, there are N independent processors that can communicate among themselves. There may be a coordinator unit to oversee the activities of the processors.

Circuit vs. Packet Switching - Circuit switching in computers is such that a complete path is established from an input port to an output port before any information is transmitted. Networks are characterized by both, the set-up time (for establishing the path) and the transmission time (for moving information over the path). The transmission time is typically faster than the set-up time.

Packet switching in computers is a mode in which relatively small, fixed size units of information, called packets, move from stage to stage as paths between the stages become available. They do not require their entire path to be established prior to entering the network. The packets consist of a header containing routing information and some data or commands.

3. NETWORK TOPOLOGIES

3.1. Chronological History

The earliest theoretical efforts were in the context of telephone switching [8, 13]. It was eventually realized that some of that work might be applicable, with suitable modifications, to computer communication [31-32]. Also, special purpose networks for number sorting called bitonic sorters were investigated [5]. With experimentation into parallel processing or multiple computer systems such as Illiac IV [4] and C.mmp [52], interest in designing interconnection networks tailored to that application began to grow. Early work was done by Lipovski on the SW-structure [25]. That work was later refined and generalized by Goke and Lipovski who introduced a class of networks called Banyans [19]. They showed that SW-banyans are topologically equivalent to Batcher's bitonic sorter. In 1971, Stone published an influential paper on the perfect shuffle network [45]. Though presented as a single stage network, it was later extended by Lang and Stone into a multistage version [20]. Feng published work in 1974 on implementing networks for data manipulation [15]. One such network has since come to be known as the data manipulator. At that same time, Batcher was working on the Flip network used in STARAN [6], but the details were not published until 1976 [7]. Soon to follow was work by Lawrie on the Omega network [21] and by Pease on the indirect binary n-cube network [35].

In April of 1978 Siegel and Smith published a paper comparing the data manipulator, Flip, Omega, and indirect binary n-cube networks [42]. As a benchmark, they introduced the Generalized Cube network and showed that the Flip, Omega, and indirect binary n-cube networks were all topologically equivalent to it. They also relaxed some restrictions on the implementation of the data manipulator, calling their version the Augmented Data Manipulator (ADM), and proved that its capabilities were a superset of those of the Generalized Cube (and therefore all networks equivalent to the Generalized Cube). At nearly the same time, in August of 1978, Wu and Feng also published a comparison paper [49]. They introduced the baseline network as a benchmark and showed that the Flip, Omega, indirect binary n-cube, SW-banyan (with spread and fanout of two a member of the Banyan class), and the reverse (or inverse) baseline were all topologically equivalent. At the same conference where this work was presented, the Heterogeneous Element Processor (HEP) system and network were introduced [43].

The class of Delta networks was introduced by Patel [34], the reverse-exchange network was investigated by Wu and Feng [50], and properties of the inverse ADM (IADM) network have been presented by McMillen and Siegel [27]. In September 1980, Pradhan and Kodandapani published another comparison of multistage networks [36]. They defined an equivalence relation and showed the Flip, Omega, indirect binary n-cube, SW-banyan, and all of their inverses were equivalent under the defined relationship. The most recent introduction of a "new" multistage network is called the Gamma network, presented by Parker and Raghavendra [33]. It is topologically equivalent to the IADM network.

3.2. Multistage Interconnection Networks

A representative cross-section of the multistage networks will be discussed in this section, along with some of the original applications their inventors proposed. Space limitations prevent a more detailed treatment, however, the full length version [26] is available from the author.

Clos and Benes Networks - The early work on multistage interconnection networks was aimed at providing an economical telephone switch. An interesting constraint to impose on a network design is that any idle pair of input and output ports can be connected regardless of the existing connections. This is called the non-blocking property. The most obvious solution is a single-stage NxN crossbar. The drawback to this scheme is that N^2 crosspoints are required. In 1953, Clos investigated a class of multistage networks with the non-blocking property but a lower cost [13]. An example of a 36x36 three-stage network is shown in Figure 3.1. The number of crosspoints required is $6N^{3/2} - 3N = 1188$ for N=36. This compares with 1296 for a 36x36 single-stage crossbar. The larger N grows beyond 36, the greater the difference becomes.

ized by m, n, and r. Clos was able to show that for $m \geq 2n-1$, the network is strictly non-blocking [13].

Benes investigated a special case of the Clos networks with more stages, in which only 2x2 crossbars were used [9]. These networks are called rearrangeable because any idle pair of input and output ports can be connected after possibly rearranging some of the existing connections.

The Bitonic Sorter - The bitonic sorter was designed by Batcher for efficiently sorting a bitonic sequence of numbers into a monotonic sequence [5]. A bitonic sequence is the juxtaposition of two monotonic sequences, one ascending, the other descending. Thus the bitonic sorter can be used to merge two monotonic sequences (which can always be combined to form a bitonic sequence) into one. This combined with other hardware can sort an arbitrary list of numbers.

Banyan Networks - Banyan networks were originally defined in terms of their graphical representation [19]. The network has the property that there is exactly one path from any input to any output (and vice versa). A rectangular banyan is one in which the number of inputs into each node and the number of outputs from each node are equal. A network that is equivalent to the bitonic sorter is the rectangular SW-banyan with two inputs and outputs each at every node.

The Data Manipulator, ADM, IADM and Gamma Networks - Shown in Figure 3.2, Feng's data manipulator has n stages labeled from n-1 to 0, and an unlabeled output stage [15]. At stage i of the data manipulator, $0 \leq i < n$, node k is

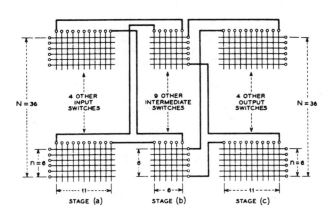

Figure 3.1. 36x36 Three-Stage Clos Network [9]

In the general three-stage Clos network, the first and last stages have r nxm crossbars and the middle stage has m rxr crossbars, where $n=N^{1/2}$. There are r·n inputs and outputs. A three-stage Clos network is completely character-

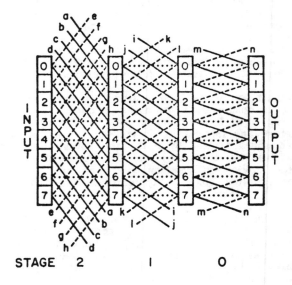

Figure 3.2. 8x8 Data Manipulator Network. Lower case letters represent end-around connections.

connected straight across to node k in the next stage, up a distance of 2^i to node $(k-2^i)$ mod N and down a distance of 2^i to node $(k+2^i)$ mod N. The "mod" operation causes some links to wrap around as shown in the figure. There are actually only two distinct data paths instead of three from each node in stage n-1 (in the figure, stage 2).

Each stage of the data manipulator receives three pairs of control signals, two each of "Straight," "Up," and "Down." The nodes in every stage are divided into two equal size groups with each group receiving one Straight, Up, and Down signal. All nodes in the same group are thus set the same way.

Feng showed that the network with this control scheme is able to perform the data manipulating functions of permuting, replicating, and spacing. Useful permutations include the shift, flip, shuffle, transpose, merge, mix, and bit reverse functions [15]. Replicating is the copying of a group of elements. Spacing is any operation that moves the data without reordering it, e.g., spreading and compressing.

The augmented data manipulator (ADM) network [42] has a topology identical to the data manipulator. The only difference is that the nodes in the ADM network can be controlled independently. The Inverse ADM (IADM) network is identical in structure to the ADM except that the stages are traversed in the opposite order. The nodes in each of these networks can connect one input to one output (or more if broadcasting) at a time. The Gamma network [33] uses 3x3 crossbars and has the same topology as the IADM. Since all of these networks use links that connect nodes distances of plus or minus 2^i, they are referred to as PM2I-type networks.

The Omega Network - Lawrie's Omega network of size N=8 is shown in Figure 3.3 [21]. Its stages are numbered from 1 to n, input to output. The links in each stage form a perfect shuffle permutation [45] which moves an item from location k to location $(2k + \lfloor 2k/N \rfloor)$ mod N. The switching elements have four states. Two items can pass straight through or be exchanged, or an item on either input can be broadcast to both outputs. Due to the way the stages are labeled, at stage i, inputs that differ in the (n-i)-th bit are compared and can be exchanged (the addresses paired can be obtained by setting all nodes in Figure 3.3 to straight and moving the input addresses on the established path throughout each stage). An Omega network can be controlled in a distributed fashion using routing tags which will be described in Section 4.

In [21] the ability of the Omega network to access vectors for processors (connected to the input) from matrices stored in memory (connected to the output) was investigated. It was shown that if a matrix is stored in memory in a skewed fashion, the Omega network provides conflict free access and alignment of rows, columns, diagonals,

backward diagonals, and N1/2xN1/2 partitions in either row or column major order. It can also produce N1/2-vector fanout and duplication functions.

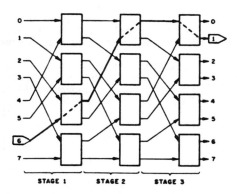

Figure 3.3. 8x8 Omega Network [21]. Path shown connects input 6 to output 1.

The Indirect Binary n-Cube Network - Pease's indirect binary n-cube network [35] has stages labeled from 1 to n, input to output. At stage i, input addresses that differ in their (i-1)-st bit can be exchanged. The switching elements have only two states, straight and exchange. The network supports the communication requirements of a large array of processors working on massive numerical problems. They include the solution of 2-D and 3-D partial differential equations, the radix-2 FFT, other signal processing algorithms, and matrix operations.

The Generalized Cube Network - The Generalized Cube network is a multistage network topology that was introduced as a standard for comparing network topologies [42]. The network has N inputs and N outputs, in Figure 3.4, N=8. There are n stages, where each stage consists of a set of N lines connected to N/2 switching elements. Each switching element is a two-input, two-output crossbar that can be set to one of the four states described for the Omega Network.

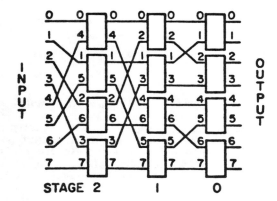

Figure 3.4. 8x8 Generalized Cube Network

It has the property that input addresses that differ in their i-th bit are paired or connected to the same switching element in stage i, $0 \leq i < n$. This property is due to constructing the network based on "cube interconnection functions" [38]. It was shown in [42] that networks topologically equivalent to the Generalized Cube also have that property and so, all are collectively referred to as cube-type networks.

There is an interesting general relationship between the Generalized Cube and Omega networks that can be illustrated by examining Figures 3.4 and 3.3. If the middle two switching elements in stage 1 of the Generalized Cube network in Figure 3.4 are swapped without disconnecting any links, the Omega network in Figure 3.3 is obtained.

3.3 The Network Family Tree

To place all these networks into perspective, a family tree for multistage interconnection networks is shown in Figure 3.5. Each network or class has a date next to it to indicate when it was first presented in the literature. Four broad categories are defined: Permutation Networks, Multiple Path Networks, Single Path Networks, and Fault Tolerant Networks. Fault tolerant networks will be discussed in Section 5, where the remainder of that family tree will be filled in.

Permutation networks are those that can connect their inputs to their outputs in any arbitrary way as long as no two inputs want the same output. For an N input, N output network, there are N! possibilities.

Multiple path networks are those that have more than one path between a given input and output (with the possible exception that there is only one path when input=output). This category includes all of the members of the Permutation Network category listed and the PM2I-type networks. The former tend to have many more paths per input/output pair than the latter. This would also include the early 4 to 8 stage telephone crossbars.

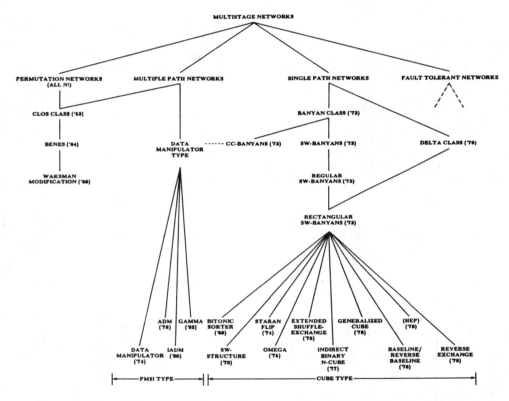

Figure 3.5. Multistage Interconnection Network Family Tree

The **single path networks** have exactly one path between every arbitrary input/output pair. The two classes listed in this category are the Banyan class and the Delta class. The Banyan class is extremely general since each switching element can have a different number of incoming and outgoing links than the other switching elements as long as some basic rules are followed. Of practical interest are the somewhat more structured subclasses called CC-banyans and SW-banyans. The Delta class is shown at this level in the tree because, in some qualitative sense, it is approximately as general as the CC-banyans and SW-banyans. The dashed line between the CC-banyans and the data manipulator type networks indicates that they are relatives.

150

In the subclass of regular SW-banyans. "Regular" implies that all the switching elements are the same. Within this class are the rectangular SW-banyans in which each node has the same number of inputs as outputs. Since the Delta class has switching elements with equal numbers of inputs and outputs, rectangular SW-banyans are also a subclass of the Delta class. Connections between switching elements in the Delta class are more general. Any of the capabilities shown for one network in the cube-type class can be built into another with suitable (often minimal) modifications.

4. CONTROL SCHEMES

4.1. Introduction

In communication and computer applications there is a distinction made in how a network is controlled. Centralized or common control is used extensively in telephone crossbars to set up connections for all users. Centralized control is often used in SIMD computers that permute or otherwise manipulate data. A control processor issues an opcode to the network control unit specifying a particular permutation (or other) configuration to be established [7, 15].

The earliest form of distributed control in communication applications was called progressive or "step-by-step" control. In the early days (cir. 1898), the physical path was the same as the logical path (telephone number). Some modern telephone exchanges have returned to using progressive control, however, now logical paths are translated into physical paths since there are many alternative connections that can be used. Distributed network control in computers can be achieved by using routing tags [20, 21, 29, 40, 47, 49]. The tag values are computed by the network users. The switching elements in each stage determine their setting by examining the tag.

As an example of how distributed control can be implemented, Lawrie's routing tag scheme for the Omega network will be described. It can be used for any cube-type network made with 2x2 switching elements. With slight modification, it can be adapted to networks made with any size cross-bar type switching elements (and an appropriate pattern of link connections between stages).

In [29] and [39] routing tag schemes are presented for controlling the ADM and IADM networks in MIMD mode and SIMD mode respectively. These schemes are equally well suited to controlling the Gamma network. Because these are all multiple path networks, the routing tag schemes are more sophisticated than those used by the cube-type networks. The MIMD mode tag scheme has the ability to perform dynamic rerouting to avoid busy or faulty switching elements when possible, without backtracking.

4.2. The Omega Routing Tag Scheme

The routing tags defined by Lawrie in [21] are called destination tags. No computation is required on the part of the network users to generate the tag. The desired destination address, D, is itself the tag. Let $d_{n-1} \ldots d_1 d_0$ be the binary representation of D. The switching element in stage i, $1 \leq i \leq n$, examines bit d_{n-i}. If $d_{n-i}=0$ the upper output is selected and if $d_{n-i}=1$ the lower output is selected. As an example, consider the path from input 6 to output 1 in an Omega network of size N=8 (n=3), as shown in Figure 3.3. D = $d_2 d_1 d_0$ = 001; in stage 1, d_2 is examined and found to be 0 so the upper output is used. Similarly, in stages 2 and 3 the upper and lower outputs are selected, respectively. If the network is bidirectional, it can be readily verified that this scheme works in reverse (from output to input).

5. FAULT TOLERANCE

5.1. Fault Detection and Diagnosis

There are two major aspects to fault tolerance: (1) detecting and diagnosing faults; and (2) avoiding known faults (if such a capability exists). The literature on fault detection and diagnosis of interconnection networks will be briefly summarized here. In the next subsection, the family tree for fault tolerant multistage networks will be completed.

General multistage network fault diagnosis is discussed in [30]. A method for diagnosing faults in the Benes network is presented in [32]. As discussed in [44], certain kinds of faults can be tolerated in all stages of the Benes network except the center stage.

Detecting and locating faults in the Baseline network are discussed in [17]. The procedures described are generally applicable to the cube-type networks. The method used is to generate test patterns that are propagated through the network. The emerging patterns are compared to precomputed, correct patterns. This requires no extra hardware in the network. Using a different approach, in [37] four methods are described for diagnosing SW-banyan networks. Extra hardware is required and it is assumed that the switching elements can diagnose themselves and set a latch if faulty. It is claimed that two of the methods can be used for any multistage network.

Most of the diagnosis studies implicitly assume the network is circuit switched (because of assuming the fault model in [22]). However, diagnosing cube type packet switching networks is addressed in [24]. Methods for diagnosing multistage networks are surveyed in [2].

5.2. Fault Tolerant Network Family Tree

The remainder of the tree headed Fault Tolerant Networks from Figure 3.5 is shown in Figure 5.1. Two kinds of fault tolerance in networks are possible: (1) correcting routing errors only (link or switching element failures that corrupt data cannot be tolerated); and (2) correcting for faulty links and switches. In synchronous systems that are typically configured in the SIMD mode, the first type of fault tolerance is the most economical to try to incorporate. In asynchronous sytems operating in the MIMD mode, the second type is used. Multiple paths are provided so that bad switching elements and links can be avoided. The telephone crossbars have this capability. Computer networks that can do this can be used in SIMD mode, but two passes of data through the network are required when a fault is present. Five ways to implement these capabilities are shown in the tree.

The earliest approach to providing SIMD mode fault tolerance was to add a double-tree repair network to a permuting network [22]. This network can be added to any of the cube type networks as well as the Benes network. It restores their fault free permuting abilities in the presence of permanent routing errors. Another approach in this category is to add an extra switching element to the Benes network [44].

To add fault tolerance to a network used in MIMD mode, one approach is to provide multiple paths between inputs and outputs. Two ways to accomplish this are (1) to add an extra stage of switching elements to the network; or (2) to add

extra links between stages. The former is done in the Extra Stage Cube network [1] and in the baseline with an extra stage [51]. In the figure, the baseline is shown with a dashed line because it requires some modification to qualify as fault tolerant. The second method for providing multiple paths is used in the modified IADM network where half-links are added [28] and in the F-network [12].

All of the approaches listed so far have drawbacks, a number of which are discussed in [23] and will be enumerated here. The SIMD mode fault model is very optimistic. It assumes that the only kind of fault that can happen is that a switching element will get stuck in one of its valid states. Not considered are invalid states, link failures, and switching element failures that alter data. The MIMD mode approaches do deal with these problems but cannot route permutation connections in one pass (two are needed). If these approaches depend on periodic diagnosis to detect faults, transient errors will go undetected. The MIMD mode networks also have more complex routing tag schemes.

The solution proposed in [23] to all these problems is the Error Correction Coded (ECC) Omega network. This is listed under both SIMD and MIMD mode in the figure. A crucial assumption required for this scheme to work is that the switching elements are bit sliced. Then different parts of one switching element can fail independently. For current technology, this is a reasonable assumption for many applications, due to the pin limitations of VLSI chips. The main drawback to this scheme is the large amount of extra hardware required since the parity bits increase the total path width significantly.

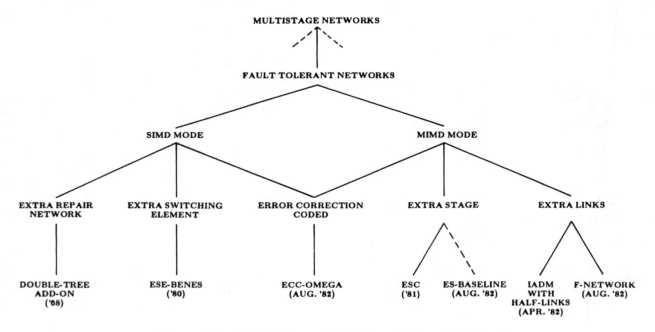

Figure 5.1. Fault Tolerant Multistage Network Family Tree

6. CONCLUSIONS

This paper has presented a broad survey of multistage interconnection networks. In addition to examining the different topologies that have been proposed, distributed control schemes and fault tolerant designs have been discussed.

REFERENCES

1. G.B. Adams III and H.J. Siegel, "The extra stage cube: a fault-tolerant interconection network for supersystems," IEEE Trans. Comp., Vol. C-31, pp. 443-454, May 1982.

2. D.P. Agrawal, "Testing and fault-tolerance of multistage interconnection networks" Computer, Vol. 15, pp. 41-53, Apr. 1982.

3. G.A. Anderson and E.D. Jensen, "Computer interconnection structures: taxonomy, characteristics, and examples," ACM Comp. Surveys, Vol. 7, pp. 197-213, Dec. 1975.

4. G.H. Barnes, etal., "The ILLIAC IV computer," IEEE Trans. Comp., Vol. C-17, pp. 746-737, Aug. 1968.

5. K.E. Batcher, "Sorting networks and their application," AFIPS 1968 Spring Joint Comp. Conf., pp. 307-314.

6. K.E. Batcher, "STARAN parallel processor system hardware," AFIPS 1974 Nat'l. Comp. Conf., May 1974, pp. 40-5-410.

7. K.E. Batcher, "The flip network in STARAN," 1976 Int'l. Conf. Parallel Proc., Aug. 1976, pp. 65-71.

8. V.E. Benes, "Optimal rearrangeable multistage connecting networks," BSTJ, Vol. 43, No. 4, Part 2, pp. 1641-1656, July 1964.

9. V.E. Benes, Mathematical Theory of Connecting Networks and Telephone Traffic, Academic Press, New York, N.Y., 1965.

10. P.T. Brady, "A statistical analysis of on-off patterns in 16 conversations," BSTJ, Vol. 47, Jan. 1968.

11. G. Broomell and J.R. Heath, "Classification categories and historical development of circuit switching topologies," ACM Comp. Surveys, Vol. 15, No. 2, pp. 95-133, June 1983.

12. L. Ciminiera and A. Serra, "A fault-tolerant connecting network for multiprocessor systems," 1982 Int'l. Conf. Parallel Proc., Aug. 1982, pp. 113-122.

13. C. Clos, "A study of non-blocking switching networks," BSTJ, Vol. 32, No. 2, pp. 406-424, March 1953.

14. M. Decina, "Progress towards user access arrangements in Integrated Services Digital Networks," IEEE Trans. Comm., Vol. 30, pp. 2117-2130, Sept. 1982.

15. T. Feng, "Data Manipulating functions in parallel processors and their implementations," IEEE Trans. Comp., Vol. C-23, pp. 309-318, Mar. 1974.

16. T. Feng, "A survey of interconnection networks," Computer, Vol. 14, pp. 12-27, Dec. 1981.

17. T. Feng and C. Wu, "Fault-diagnosis for a class of multistage interconnection networks," IEEE Trans. Comp., Vol. C-30, pp. 743-758, Oct. 1981.

18. M.J. Flynn, "Very high-speed computing systems," Proc. IEEE, Vol. 54, pp. 1901-1909, Dec. 1966.

19. L.R. Goke and G.J. Lipovski, "Banyan networks for partitioning multiprocessor systems," 1st Annual Int'l. Symp. Comp. Arch., Dec. 1973, pp. 21-28.

20. T. Lang and H.S. Stone, "A shuffle-exchange network with simplified control," IEEE Trans. Comp., Vol. C-25, pp. 55-65, Jan. 1976.

21. D.H. Lawrie, "Access and alignment of data in an array processor," IEEE Trans. Comp., Vol. C-25, pp. 1145-1155, Dec. 1975.

22. K.N. Levitt, etal., "A study of the data commutation problems in a self-repairable multiprocessor," AFIPS 1968 Spring Joint Comp. Conf., pp. 515-527.

23. J.E. Lilienkamp, D.H. Lawrie, P. Yew, "A fault tolerant interconnection network using error correcting codes," 1982 Int'l. Conf. Parallel Proc., Aug. 1982, pp. 123-125.

24. W.Y. Lim, "A test strategy for packet switching networks," 1982 Int'l. Conf. Parallel Proc., Aug. 1982, pp. 96-98.

25. G.J. Lipovski, "The architecture of a large associative processor," AFIPS 1970 Spring Joint Comp. Conf., pp. 385-396.

26. R.J. McMillen, A Study of Multistage Interconnection Networks: Design, Distributed Control, Fault Tolerance, and Performance, Ph.D. Dissertation, Purdue University, Dec. 1982, 401 p.

27. R.J. McMillen and H.J. Siegel, "MIMD machine communication using augmented data manipulator network," 7th Annual Int'l. Symp. Comp. Arch., May 1980, pp. 51-58.

28. R.J. McMillen and H.J. Siegel, "Performance and fault tolerance improvements in the inverse augmented data manipulator network," 9th Annual Int'l. Symp. Comp. Arch., Apr. 1982, pp. 63-72.

29. R.J. McMillen and H.J. Siegel, "Routing schemes for the augmented data manipulator network in an MIMD system," IEEE Trans. Comp., Vol. C-31, pp. 1202-1214, Dec. 1982.

30. J.J. Narraway and K. So., "Fault diagnosis in inter-processor switching networks," 1980 Int'l. Conf. Circuits and Comp., Oct. 1980, pp. 750-753.

31. D.C. Opferman and N.T. Tsao-Wu, "On a class of rearrangeable switching networks - Part I: control algorithm," BSTJ, Vol. 50, No. 5, pp. 1579-1600, May-June 1971.

32. D.C. Opferman and N.T. Tsao-Wu, "On a class of rearrangeable switching networks - Part II: enumeration studies and fault diagnosis," BSTJ, Vol. 50, No. 5, pp. 1601-1618, May-June 1971.

33. D.S. Parker and C.S. Raghavendra, "The Gamma network: a multiprocessor network with redundant paths," 9th Annual Int'l. Symp. Comp. Arch., Apr. 1982, pp. 73-80.

34. J.H. Patel, "Processor-memory interconnections for multiprocessors," 6th Annual Int'l. Symp. Comp. Arch., April 1979, pp. 168-177.

35. M.C. Pease, "The indirect binary n-cube microprocessor array," IEEE Trans. Comp., Vol. C-26, pp. 458-473, May 1977.

36. D.K. Pradhan and K.L. Kodandapani, "A uniform representation of single- and multistage interconnection networks used in SIMD machines," IEEE Trans. Comp., Vol. C-29, pp. 777-791, Sept. 1980.

37. B.D. Rathi and M. Malek, "Fault diagnosis of networks," Distributed Data Acquisition. Comp., and Control Symp., Dec. 1980, pp. 110-119.

38. H.J. Siegel, "Analysis techniques for SIMD machine interconnection networks and effects of processor address masks," IEEE Trans. Computers, Vol. C-26, pp. 153-161, Feb. 1977.

39. H.J. Siegel and R.J. McMillen, "Using the augmented data manipulator network in PASM," Computer, Vol. 14, pp. 25-33, Feb. 1981.

40. H.J. Siegel and R.J. McMillen, "The multistage cube: a versatile interconnection network," Computer, Vol. 14, pp. 65-76, Dec. 1981.

41. H.J. Siegel, R.J. McMillen, and P.T. Mueller, Jr., "A survey of interconnection methods for reconfigurable parallel processing systems," AFIPS 1979 Nat'l. Comp. Conf., June 1979, pp. 529-542.

42. H.J. Siegel and S.D. Smith, "Study of multistage SIMD interconnection networks," 5th Annual Int'l. Symp. Comp. Arch., Apr. 1978, pp. 223-229.

43. B.J. Smith, "A pipelined, shared resource MIMD computer," 1978 Int'l. Conf. Parallel Proc., Aug. 1978, pp. 6-8.

44. S. Sowrirajan and S.M. Reddy, "A design for fault-tolerant full connection networks," 1980 Conf. Info. Sci. and Sys., Princeton Univ., Mar. 1980, pp. 536-540.

45. H.S. Stone, "Parallel processing with the perfect shuffle," IEEE Trans. Comp., Vol. C-20, pp. 153-161, Feb 1971.

46. K.J. Thurber, "Interconnection networks - a survey and assessment," AFIPS 1974 Nat'l Comp. conf., May 1974, pp. 909-919.

47. A.R. Tripathi and G.J. Lipovski, "Packet switching in banyan networks," 6th Annual Int'l. Symp. Comp. Arch., Apr. 1979. pp. 160-167.

48. J.S. Turner and L.F. Wyatt, "A packet network architecture for integrated services," IEEE Globecom 83, pp. 45-50, Nov. 1983.

49. C. Wu and T. Feng, "Routing techniques for a class of multistage interconnection networks," 1978 Int'l. Conf. Parallel Processing, Aug. 1978, pp. 197-205.

50. C. Wu and T. Feng, "The reverse-exchange interconnection network," 1979 Int'l. Conf. Parallel Processing, Aug. 1979, pp. 160-174.

51. C. Wu, W. Lin, and M. Lin, "Distributed circuit switching starnet," 1980 Int'l. Conf. Parallel Processing, Aug. 1982, pp. 26-33.

52. W.A. Wulf and C.G. Bell, "C.mmp--a multi-miniprocessor," AFIPS 1972 Fall Joint Computer Conf., Dec. 1972, pp. 765-777.

MAILING ADDRESS

Hughes Aircraft Company, Bldg. A1, M/S 3C923, P.O. Box 9399, Long Beach, CA 90810-0399.

BIOGRAPHY

Robert J. McMillen was born in Albuquerque, NM, on July 3, 1956. He received his Ph.D. degree in electrical engineering from Purdue University in 1982. He is presently working in the Advanced Products Laboratory of Support Systems, Hughes Aircraft Company, as a Senior Staff Engineer. His responsibilities include system engineering, simulator development, and integrated circuit development for a next generation flight trainer system.

Dr. McMillen is currently on the program committees of the 12th Computer Architecture and 5th Distributed Computing Systems conferences. He is a member of the IEEE Computer Society, ACM SIGARCH, the New Mexico Academy of Science, Tau Beta Pi, and Eta Kappa Nu.

PACKET SWITCHING IN N LOG N MULTISTAGE NETWORKS

Daniel M. Dias and Manoj Kumar

IBM T. J. Watson Research Center, Yorktown Heights, NY 10598

ABSTRACT

In recent years considerable research has been done on interconnection networks for multiprocessor computer systems. This paper surveys the research on packet switched N log N Multistage Interconnection Networks used for the interconnection of multiple processors and memories. Some possible communications switching applications of these networks are briefly discussed. The emphasis is on the use of packet switching techniques to obtain high network throughput, small (but statistical) delays, simple distributed control, fault tolerance and growth capabilities.

1. INTRODUCTION

Some of the earliest work on high bandwidth Multistage Interconnection Networks (MINs) was done in the context of telephone switching networks [Clo53,Ben64]. The emphasis here was on designing non-blocking networks which use fewer cross-points than a full crossbar switch. These networks had a centralized controller for setting the crosspoints. This was adequate since the networks were circuit switched and connections were held for a relatively long period of time. Subsequently, many MINs have been studied for connecting multiple processors and memories in computer systems [Bar81, Den80a,b, Pre80, Got83, Sie81b]. The emphasis in these MINs is on the use of packet switching techniques to obtain high network throughput, small (but statistical) delays, simple distributed control, fault tolerance and growth capabilities. In these networks distributed control is essential because small packets of data are passed across the network at a rate higher than that typically handled by a centralized controller. This paper surveys research on packet switched MINs composed of O(N log N) sub-switches for multiprocessor systems and examines possible use of these networks for communications switching applications.

The early work on MINs was for strictly non-blocking networks which can connect any set of input links to distinct output links simultaneously. In addition, new connections can be made without affecting existing connections. A crossbar switch and the Clos networks [Clo53] are strictly non-blocking. Use of crossbar switches for connecting multiple processors to memories is considered in [Geo84, Wul72]. The Benes networks [Ben64,Opf71] are blocking but are rearrangeable (i.e., if a set of connections is already established, any new connection request can be satisfied by rearranging the previously set-up connections).

Single Instruction stream, Multiple Data stream (SIMD) computers have many processors, all of which execute the same instructions on different data. Thus, operations on different elements of a vector or array can be carried out simultaneously on these machines. Often, an arbitrary permutation of memory data needs to be sent to the processors for each instruction. Benes networks can be used for this purpose. However, setting up the Benes network is time consuming [Wak68] and unless the established connections are held for a long

period of time, this control overhead can be large. Efficient methods of controlling Benes networks were investigated [Lev81, Nas81, Nas82, Par80b, Cho80]. While Benes networks can provide all permutations, all of these may not be required for SIMD processing. A number of networks that provided useful permutations for different SIMD algorithms were developed such as the shuffle exchange network [Sto71,Che81], the OMEGA network [Law75], the cube network [Sie81b], the flip network [Bat76], the indirect binary n-cube [Pea77], the baseline network [Wu80a] the reverse exchange network [Wu80b], and the augmented data manipulator network [Sie81a]. These networks are blocking and not rearrangeable. Permutation capabilities of these networks are also discussed in [Par80a, Pra80, Wu81]. Surveys of the topologies of these networks can be found in [Bro83,Fen81b].

In Multiple Instruction stream Multiple Data stream (MIMD) computers, different instructions are simultaneously executed on different processors using different data. In these machines, data needs to be moved efficiently either between processors and some shared memory and/or between different processors. Permutations of data are not typically required but stochastic data streams must be supported. This environment is particularly amenable to packet switching techniques in which the processor and memory modules communicate by sending relatively small packets of information. The blocking networks used as permutation networks for SIMD machines can be used in a packet switching environment for MIMD machines. Though the networks are blocking, when used with packet switching these networks provide high throughput and small delays.

Many of the SIMD interconnection networks including the OMEGA, cube, shuffle-exchange, baseline, flip, and indirect binary n-cube networks, belong to a class of networks called **delta** networks [Pat81]. Though these networks provide different permutations, they have the same performance in a packet switching environment. This paper describes the topology and performance of delta networks and networks of related topology. The principal characteristics of these networks are:

- An (N x N) delta network consists of O(N log N) sub switches arranged in O(log N) stages.

- Control of the network is totally distributed.

- The network can be constructed modularly from smaller sub-switches.

- The networks provide high throughput and small delays in a packet switched environment.

- They can be augmented for good fault tolerance.

The topology and control of delta networks and networks of related topology are discussed in the next section. Many multistage networks do not conform to the definition of delta networks [Bat68, Sie81a,Par82]. Further details about these can be found in the listed

EH0301-2/91/0000/0155/$01.00 © 1984 IEEE

references. The performance of delta networks is outlined in section 3. Finally, section 4 discusses some potential applications of delta networks for communication switching.

2. TOPOLOGY AND ROUTING

Delta networks [Pat81] are defined as networks constructed from switches of size BxB with, with k stages, N input ports, and N output ports, where $N=B^k$. The interconnection pattern of links between two adjacent stages must be arranged so that a packet can be sent from any one of the network's inputs ports to any one of its output ports. Furthermore, a packet's movement through the network must be controlled by an k digit, base B number included in the packet, which is the destination address of the packet, in the following way: For each switch encountered by the packet as it moves from one stage to the next, the choice for the switch's output terminal which receives the packet is uniquely determined by one of the digits in the destination address. Thus the k digits in the destination address correspond the k stages of the network, and each digit controls the switches in the corresponding stage. Since there are B output terminals on each switch and each digit of the destination address has B possible values, the digits can be used to directly specify which output terminal of the switch will receive the packet. This routing technique is referred to as **digit controlled routing** or **self routing**.

Banyan networks are MINs that have unique path between any input and output port [Gok73,Tri79]. Though the digit controlled routing is desirable in MINs, it is not a requirement in the general definition of Banyan networks. Delta networks are a subclass of Banyan networks which have the digit controlled routing property.

All $B^n x B^n$ delta networks can be constructed iteratively from smaller subnetworks, using BxB switches, as follows:

- A BxB delta network consists of a single BxB switch.

- To construct a $B^n x B^n$ delta network ($n > 1$), B delta subnetworks each of size $B^{n-1} x B^{n-1}$ are stacked as shown in Fig. 1. The subnetworks are labeled 0 through B-1. A stage of B^{n-1} switches (each of size BxB) is added before the subnetworks. The B outputs of each switch are connected to a different subnetwork and labeled with the label of the subnetwork they are connected to.

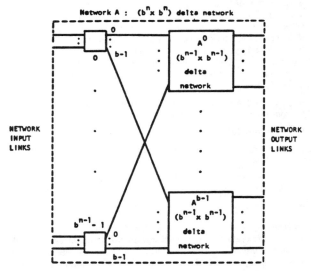

Figure 1. Construction of delta networks

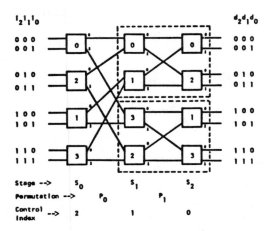

Figure 2. An (8x8) delta network

Thus, the entire network is composed of interconnected BxB switches. The outputs of this network are labelled with the concatenation of the labels of all the links traversed on a path from any network input to that output. The path between any network input/output pair is unique and the label obtained for a network output is the same for paths connecting all the different network inputs to this output. It can be readily seen that each network output gets a unique label. This label can also be viewed as a radix B number representing the output port number of the network. Figure 2 shows an 8x8 delta network constructed in this manner from 2x2 switches.

To route a request from any network input to the network output numbered L or labeled l(k-1)...l(0), the switch in stage i receiving this request ($0 \leq i < k$) forwards it to the switch output labeled l(i). When the request has travelled through a switch in each stage, it will reach the correct network output. The destination label l(i) that controls routing at stage i is called the **control digit** and the index i is referred to as the **control index**.

A **bidirectional delta network** is a delta network which is also a delta network when network input links are interpreted as network output links and vice versa. It has been shown that all bidirectional delta networks are topologically equivalent (i.e., the networks are identical to within a re-labelling of links, switches and networks ports) [Dia81b,Agr83]. Most of the SIMD N log N multistage networks listed in the previous section are bidirectional delta networks and are therefore topologically equivalent. These networks can also be constructed iteratively with a small modification to the construction procedure given above.

The BxB switches used in the network need not be full crossbars. The switches are not required to be non-blocking either, i. e., it is not necessary that the B switch inputs should be able to communicate to B distinct switch outputs simultaneously. One can use a single bus or multiple buses instead of a crossbar switch. The only requirement to be met by the BxB switches is that they must be able to connect any input to any output in the absence of any other request to the switch.

The iterative procedure presented above allows one to construct delta networks in which the number of inputs and outputs are the same and both are an integer power of the basic switch size. A more general class of NxM delta networks, where N=n(k) n(k-1) ... n(1) and M=m(k) m(k-1) ... m(1) for some integer value of k, can be constructed by slightly modifying the procedure outlined earlier as follows:

- An n(1)xm(1) network is a n(1)xm(1) switch.

- To construct a [n(i)...n(1)]x[m(i)...m(1)] network, m(i) networks of size [n(i-1)...n(1)]x[m(i-1)...m(1)] are stacked to-

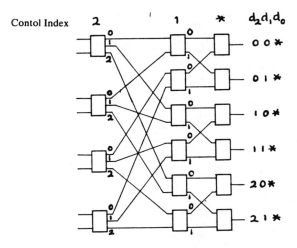

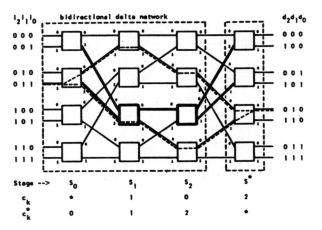

Figure 3. An (8x6) generalized delta network

Figure 4. An augmented delta network

gether and preceded by n(i-1) ...n(1) switches of size n(i)xm(i). The subnetworks are labeled 0 through m(i)-1. The outputs of each switch are connected to a different subnetwork and labeled with the label of the subnetwork it is connected to.

The network has k stages and N may be different from M. The order in which the factors of N and M are listed, affects the topology of the network. These factors need not be distinct and multiple occurrences of 1 are allowed as factors.

The labeling of the output ports is similar to the one discussed for delta networks. Now each output port label l(k).l(k-1)...l(1) can be viewed as an mixed radix number L, in which the weight of the of the digit l(i) is M/(m(k)xm(k-1)x...m(i)). Fig. 3 shows the example of an (8x6) network constructed using this method. To route a request to an output port numbered L, or labeled l(k),l(k-1),...,l(1), the switch in stage i receiving this message sends it to its output numbered l(i). The generalized shuffle networks discussed in [Bhu83] are a subclass of these general delta networks.

Any NxN delta network can also be used as a smaller N1xN2 network, where $N1 \leq N$ and $N2 \leq N$. Some switches and links in the original NxN network are not needed in the N1xN2 network and can be identified and removed as follows:

- Choose the N1 input links and the N2 output links that must be retained and remove the rest of input and output links.
- Remove all switches which have either no input links or no output links.
- Remove any link which does not have both sending and receiving switches.
- Repeat the previous two steps k times, where k is the number of stages in the network.

The above procedure will identify and remove all the switches and links which are not used by the N1xN2 network. These networks are called pruned delta networks and have been studied by Dias [Dia81b,c]. The choice for the input or output links removed affects the number of switches and links required in the network.

Finally, the networks described in this section so far provide a unique path between any input and output pair. Considering the number of switches and links involved, these networks are fault prone. Substantial research has been done for detecting and locating the faults in these networks [Fen81a, Agr82, Rat80]. It is desirable to provide multiple paths between each input and output pair to keep the network operational in the presence of few faults. One way to do so is

by adding an extra stage of switches to the network as shown in Figure 4. Delta networks with such extensions have been variously termed augmented delta networks, extra stage cube and multipath MINs and studied by [Dia81b, Ada82, Chi84, Pad83]. In an augmented delta network with one extra stage there are two paths from any input link to each output link and these paths are disjoint except for the common input and output links. Two such disjoint paths are indicated by dashed lines in Figure 4. Fault diagnosis and reliability analysis of these networks are reported in [Che84].

Augmented delta networks provide fault tolerance without affecting the fault-free network bandwidth. Both the bandwidth and the fault tolerance can be improved simultaneously by using multiple delta networks in parallel as shown in Figure 5 or by using multiple links for each switch connection as shown in Figure 6. The bandwidth of these networks has been analyzed by [Kum83a,b,Kru83]. However, the improved performance and fault tolerance is achieved at the cost of more logic.

Delta networks composed of simple switches do not allow two or more processors to access a memory module simultaneously. However, by providing adequate logic in the switches, multiple read requests and multiple write requests of certain types can be combined in the network, so that multiple processors can perform operations on a single memory location simultaneously [Got83].

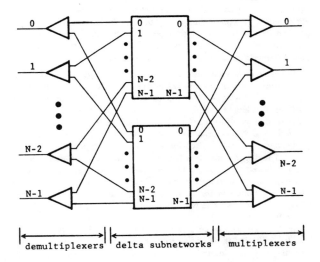

Figure 5. Two delta networks connected in parallel

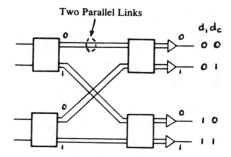

Figure 6. A (4x4) delta network with 2 parallel links

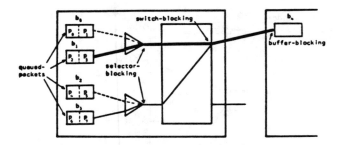

Figure 7. Blocking factors in MINs

3. NETWORK OPERATION AND PERFORMANCE

This section presents some results for the performance of delta and related networks in a **packet switching environment**. In this environment, fixed size packets arrive at the input ports of the network and are transported to the requested output ports. If the network input ports receive a stream of data bits (e.g. a stream of voice samples), it will be assumed that the stream is divided into fixed size packets. Similarly, variable length messages arriving at the network inputs are broken up into fixed size packets that are independently transported through the network.

Two principal operating modes have been considered for these networks. In an **unbuffered network** the digit controlled routing (described in Section 2) is used to establish a path from a network input port to the requested destination port; a packet of data is then passed between these ports, following which, the path is released for use by other packets. In a **buffered network** buffers are added between the stages of the MIN; packets are now passed from buffer to buffer (again using the digit controlled routing) till they reach the required output port.

These networks have been primarily studied in a synchronous mode in which the network operates in cycles synchronized by a single clock. In an unbuffered network, at the start of the network cycle, all network input ports with a packet to send simultaneously place requests (that specify the destination port tag) at the first stage of the network. At each stage of the network the requests are forwarded to the next stage along a link selected according to the control digit of the destination label. If two or more requests must pass through the same line, then one is selected and forwarded and the other requests are rejected. A request that passes successfully to the network output port causes a positive acknowledgement to be sent to the originating input port and then a packet is sent along the established path. Rejected requests cause a negative acknowledgement.

In one cycle of a synchronous buffered network, packets at the head of a buffer at a switch input select the switch output port using the digit controlled routing. If two packets select the same switch output port, one is selected and placed in the switch output buffer, while the other is retained in its current buffer. Packets which cannot be forwarded to the correct output port due to contention with other packets are said to be **switch-blocked**. If switch-blocked packets are retained in finite capacity FIFO buffers within the switch, then a packet offered to a correct output port may be rejected by the receiving switch due to insufficient buffer space. In this situation the packet is retained in the sending switch and is said to be **buffer-blocked**. Finally, packets which are not selected because they are not in the front of a FIFO buffer are said to be **queued**. Figure 7 illustrates the various blocking factors. Buffered delta networks have also been studied in an asynchronous operating mode where packet movement is coordinated by asynchronous handshaking protocols [Dia81a].

The principal measures used to evaluate and compare the performance of these networks are the throughput and delay. The **throughput** is defined informally as the average number of packets delivered by the network per clock cycle for each input module. The network **bandwidth** is defined as the maximum throughput that the network can sustain. Throughput normalized with respect the bandwidth will be denoted by NTP (i.e. NTP = throughput/bandwidth). The **delay** is defined as the average number of clock cycles between the arrival of a packet at a network input port and its delivery at the network output port. The minimum delay is the delay of a packet that has no blocking in the network. The delay normalized with respect to the minimum delay (i.e. delay/minimum delay) will be denoted by ND. Unless otherwise stated, it is assumed that the arrival of packets at the network input ports are independent and identically distributed Bernoulli processes. Packets are assumed to be equiprobably directed to the network output ports.

The performance of unbuffered delta networks and methods to improve their performance have been reported in [Pat81, Dia81a, Kum83, Lee84]. The following paragraphs summarize some of the principal results.

The bandwidth of an unbuffered delta network is shown in Figure 8 for different network sizes. This figure corresponds to an extreme load in which a packet is available at each input port in every cycle. It assumes that if a packet is rejected in a cycle (due to collision in the network) it is lost. If rejected packets are resubmitted in later network cycles, the throughput is slightly lower than than shown in Figure 8 [Dia81b]. This figure indicates that as the network size increases, the network bandwidth first falls rapidly and then flattens out. The initial rapid decrease in bandwidth is due to collisions in the network. For networks with many stages, most of the collisions occur

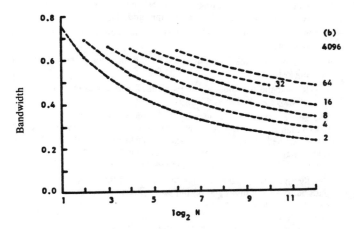

Figure 8. Bandwidth of unbuffered delta networks

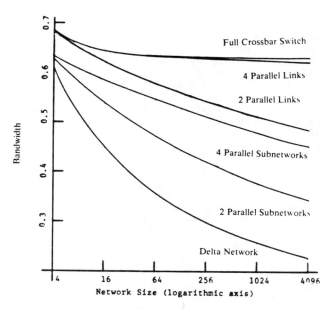

Figure 9. Bandwidth of unbuffered parallel delta networks and networks with multiple parallel links

in the first few network stages; few collisions occur in the later stages because fewer requests get through to these stages.

Some methods for improving the performance of unbuffered delta networks are presented in [Kru83, Kum83a]. One method is to parallel two delta networks (Figure 5); another is two use multiple parallel links between network stages (Figure 6). The bandwidth of these networks is shown in Figure 9. Using two parallel links between networks stages leads to a substantial improvement in bandwidth, while four parallel links provides a bandwidth close to that of a crossbar switch. Using duplicate delta networks and randomly routing packets on one or the other subnetwork provides significant bandwidth increase, but less than that obtained by the use of parallel links. Detailed analysis of these networks and implementation alternatives can be found in [Kum83a].

Studies of buffered delta networks and their variants are reported in [Dia81a;b;c, Kum83a,b, Kru83, Jen83, Chi84]. Some of their performance characteristics are presented below. Figure 10 shows the network bandwidth versus delay characteristic for synchronous buf-

fered delta networks of different sizes and buffer capacity. In this figure, switches are of size (BxB) and n is the number of stages in the network; solid lines connect performance points of the same networks with different size buffers, while dashed lines connect networks with the same total buffer capacity. The figure indicates that if the capacity of buffers between stages is increased from one to two, the bandwidth increases and the delay falls. However, if the buffer capacity is increased further, the bandwidth increases but so does the delay. The bandwidth saturates beyond about 8 buffers while delay (at maximum load) increases almost linearly with buffer size for large buffers. Thus, buffers larger than four to eight pay a large penalty in delay at high loads with marginal increase in bandwidth.

As for the unbuffered delta network, the bandwidth falls with increasing network size due to increasing congestion in the network. Comparing the normalized bandwidth of unbuffered and buffered networks (Figures 8 and 10) it is observed that unbuffered networks have a larger normalized bandwidth than a buffered network with buffer capacity of one. However, for two or more buffers, the normalized bandwidth of the buffered network is significantly larger than that of the corresponding unbuffered network. Further, the cycle time of the unbuffered delta network is O(log N), while that of the buffered network is a constant independent of the network size. This is because in one cycle of an unbuffered network path set-up signals must propagate through all the stages of the network, while packets advance only one stage at a time in a buffered network. Combining the effect of larger normalized bandwidth and smaller cycle time for the buffered network, the buffering typically leads to substantial improvement in performance [Dia81C]. However, for large packet sizes, the O(log N) set up time for the unbuffered network is dominated by the packet transfer time, making the unbuffered network more attractive.

The performance of buffered networks can also be improved by the use of parallel links between stages and by paralleling entire networks [Kru83, Kum84]. Figures 11 and 12 show the NTP and ND of a (1Kx1K) network with two parallel links and compares this with single links between stages. For a single buffer between stages, the bandwidth is about doubled and the ND reduced by about 25% by using two parallel links. Various switching strategies to improve performance are described in [Kum83b].

For fault-free conditions, augmented delta networks have the same throughput as for the delta networks with the same number of ports

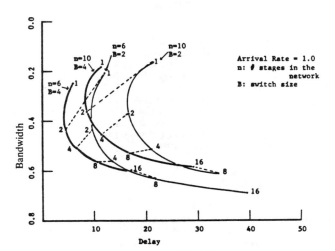

Figure 10. Bandwidth-delay characteristic for buffered delta networks

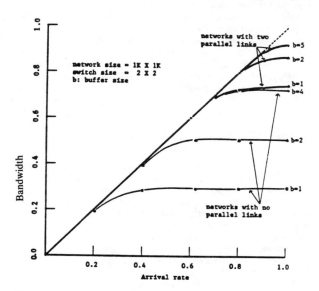

Figure 11. Throughput of buffered delta networks with parallel links

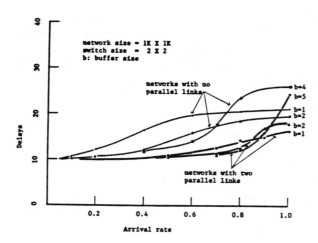

Figure 12. Delay of buffered delta networks with parallel links

if packets are passed directly through the extra stage (i.e. fixed routing at the extra stage); there is an additional cycle of delay at the extra stage. By doing load dependent routing at the extra stages, it is possible to reduce the delay [Chi84]. Detailed performance studies of these networks with faults is a topic for further investigation.

As discussed in Section 2, the topology of the non-square generalized delta networks depends on the order of the factors (n(k)x...xn(1) and m(k)x...xm(1)) of the number of network input and output ports. Further, there may be many different ways in which the port numbers can be factored. The construction of optimal unbuffered networks is studied in [Bhu83], where it is conjectured that for an (NxN) unbuffered network with N a power of 2, the most cost effective network is obtained by using as many 4x4 switches as possible. The selection of optimal pruned buffered networks is discussed in [Dia81C].

4. CONCLUSION

The previous sections have indicated that the N log N multistage networks, when used in a packet switching environment can provide high packet throughput and small delays. Further, the digit controlled routing makes the control of the network fully distributed and very simple. Adding buffers between the network stages can increase the network throughput while keeping the delay about the same or lower. While delta networks provide a unique path between input and output ports, an extra stage can be added to provide disjoint paths through the network. This extra stage retains the digit controlled routing and marginally affects fault free performance. Nonsquare networks with digit controlled routing can be constructed and optimal configurations selected. Finally, performance can be further improved by using parallel links or paralleling subnetworks. These characteristics make these networks suitable for many application where high bandwidth, low delay packet switching is required.

One possible application is to connect a number of Local Area Networks (LANs) on a campus or in a metropolitan area. Each LAN may be located in a building of a campus or complex and may generate multi-megabit per second burst traffic to other LANs. The total burst traffic between LANs, particularly for a number of nets connected in a metropolitan area, may be in the hundreds of MBPS up to the gigabit range. A delta network packet switch may be used for this purpose. However, it should be noted that the MIMD applications discussed in previous sections have assumed small packets of up to a few tens of bytes, while LAN packets may be up to several kilo bytes long. Thus, LAN packets would have to be broken up into small sub-packets and reassembled before being delivered to the destination LAN.

Delta networks can be used for more general packet switching. A packet network architecture for integrated voice and data services has been proposed in [Tur83]. Delta networks are one possible choice for use as a packet switch in such an architecture. A potential problem in using these networks to switch voice packets is the stochastic packet delay. This leads to a non-zero probability that some voice packets will be lost due to excessive delay leading to voice clipping. While the network may be engineered to make voice clipping small, this may constrain operation of the network at low to medium loads where the packet delay and differential delays are small.

These potential applications suggest several areas for further research. Most of the previous work on these networks has assumed that small fixed size packets are switched. Switching strategies and their performance for switching variable length and relatively long messages need further study. Methods for effectively handling real time and/or fixed rate traffic such as voice need to be examined. Other areas include the performance of the networks with faults and the handling of non-uniform traffic.

REFERENCES

[Ada82] G.B. Adams, III and H.J. Siegel, "The Extra Stage Cube: A Fault Tolerant Interconnection Network for Supersystems," **IEEE Trans. on Computers C-31(5)** (May 1982), pp. 443-454.

[Agr83] D.P. Agrawal, "Graph Theoretical Analysis and Design of Multistage Interconnection Network," **IEEE Trans. on Computers C-32(7)** (July 1983), pp. 637-648.

[Agr82] D.P. Agrawal, "Testing and Fault Tolerance of Multistage Interconnection Networks," **Computer 15(4)** (April 1982), pp. 41-54.

[Bar81] G.H. Barnes and S.F. Lundstrom, "Design and Validation of a Connection Network for Many-Processor Multiprocessor Systems," **Computer 14(12)** (Dec. 1981), pp. 31-41.

[Bat68] K.E. Batcher, "Sorting Networks and their Applications," **Proc. of the Spring Joint Computer Conference**, AFIPS Press (1968), pp. 307-314.

[Bat76] K.E. Batcher. "The Flip Network in Staran", **Proc. 1976 International Conference on Parallel Processing** (Aug. 1976), IEEE Computer Society, pp. 65-71.

[Ben64] V.E. Benes, "Optimal Rearrangeable Multistage Connecting Networks," **BSTJ (43)** (July 84), pp. 1641-1656.

[Bhu83] L.N. Bhuyan and D.P. Agrawal, "Design and Performance of Generalized Interconnection Networks," **IEEE Trans. on Computers C-32(12)** (Dec. 1983), pp. 1081-1090.

[Bro83] G. Broomell and J.R. Heath, "Classification Categories and Historical Development of Circuit Switching Topologies," **Computing Surveys 15(2)** (June 1983), pp. 95-133.

[Clo53] C. Clos, "A Study of Non-Blocking switching Networks", **BSTJ 32(3)** (March 53), pp. 406-424.

[Che81] P.Y. Chen, D.H. Lawrie, D.A. Padua, and P.C. Yew, "Interconnection Networks Using Shuffles," **Computer 14(12)** (Dec. 1981), pp. 55-64.

[Che84] V. Cherkassky, E. Opper and M. Malek, "Reliability and Fault Diagnosis Analysis of Fault-Tolerant Multistage Interconnection Networks," **Proc. 14th Intl. Conf. on Fault Tolerant Computing**, IEEE, Silver Spring MD, (June 1984), pp. 246-251.

[Chi84] C.Y. Chin and K. Hwang, "Connection Principles for Multipath Packet Switching Networks," **Proc. 11th Annual International Sym. on Comp. Arch.**,"ACM Sigarch (June 5-7, 1984), PP. 99-108.

[Cho80] Y.C. Chow, R.D. Dixon, Y.Y. Feng, and C.L.Wu, "Routing Techniques for Rearrangeable Interconnection Networks," Proc. Workshop on Interconnection Networks for Parallel and Distributed Processing (April 80), IEEE, pp. 64-69.

[Den80a] J.B. Dennis, G.A. Boughton, and C.K.C. Leung, "Building Blocks for Data Flow Prototypes," **Proc. 7th Annual Symposium on Computer Architecture, ACM Sigarch**, New York NY (May 6-8 1980), pp. 1-8.

[Den80b] J.B. Dennis, "Data Flow Supercomputers," **Computer 13(11)** (Nov. 1080), pp. 48-56.

[Dia81a] D.M. Dias and J.R. Jump, "Analysis and Simulation of Buffered Delta Networks," **IEEE Trans. on Computers C-30(4)** (April 1981), pp. 273-282.

[Dia81b] D.M. Dias and J.R. Jump, "Packet Switching in Delta and Related Networks," Ph. D. Dissertation, Rice University, Houston TX (May 1981)

[Dia81c] D.M. Dias and J.R. Jump, "Packet Switching Interconnection Networks for Modular Systems," **Computer 14(12)** (Dec. 1981), pp. 43-53.

[Fen81a] T. Feng and C.L. Wu, "Fault-Diagnoses for a Class of Multistage Interconnection Networks," **IEEE Trans. on Computers C-30(10)** (Oct. 81), pp. 743-758.

[Fen81b] T. Feng, A Survey of Interconnection Networks", **Computer 14(12)** (Dec.1981), pp. 12-27.

[Gok73] L.R. Goke and G.J. Lipovski, "Banyan Networks for Partitioning Multiprocessor Systems," **Proc. 1st Annual Symposium on Computer Architecture** (1973), pp. 21-28.

[Got83] A. Gottlieb, R. Grishman, C.P. Kruskal, K.P. McAullife, L. Rudolph and M.Snir, "The NYU Ultracomputer - Designing an MIMD Shared Memory Parallel Computer," **IEEE Trans. on Computers C-32(2)** (Feb. 1983), pp. 175-189.

[JEN83] Y.-C. Jenq, "Performance Analysis of a Packet Switch Based on Single-Buffered Banyan Network," **IEEE Journal on Selected Areas in Communications SAC-1(6)** (Dec. 1983), pp. 1014-1021.

[Kru83] C.P. Kruskal and M. Snir, "The Performance of Multistage Interconnection Networks for Multiprocessors," **IEEE Trans. on Computers C-32(12)** (Dec. 1983), pp. 1091-1098.

[Kum83a] M. Kumar and J.R. Jump, "Generalized Delta Networks", **Proc. 1983 International Conference on Parallel Processing**, IEEE, Silver Spring MD, (Aug. 1983), pp. 10-18.

[Kum83b] M. Kumar, "Performance Improvement in Single-Stage and Multiple-Stage Shuffle-Exchange Networks", Ph. D. Dissertation, Rice University, Houston TX (Jul. 1983).

[Kum84] M. Kumar and J.R. Jump, "Performance Enhancement in Buffered Delta Networks Using Crossbar Switches and Multiple Links," IBM Research Report RC-46070 (Jan. 1984), Yorktown Heights, New York.

[Law75] D.H. Lawrie, "Access and Alignment of Data in an Array Processor," **IEEE Trans. on Computers C-30(10)** (Dec. 1975), pp. 1145-1155.

[Lee84] M. Lee and C.L. Wu, "Performance Analysis of Circuit Switching Baseline Interconnection Networks," **Proc. 11th Annual International Sym. on Comp. Arch.,** "ACM Sigarch (June 5-7, 1984), PP. 82-90.

[Lev81] G. Lev, N. Pippenger, and L.G. Valiant, "A Fast Parallel Algorithm for Routing in Permutation Networks," **IEEE Trans. on Computers C-30(2)** (Feb. 1981), pp. 93-100.

[Nas81] D. Nassimi and S. Sahni, "Self Routing Benes Network and Parallel Permutation Algorithms," **IEEE Trans. on Computers C-30(5)** (May 1981), pp. 332-340.

[Nas82] D. Nassimi and S. Sahni, "Parallel Algorithms to Set Up the Benes Permutation Network," **IEEE Trans. on Computers C-31(2)** (Feb. 1982), pp. 148-154.

[Geo84] C. Georgiou, "Fault Tolerant Crosspoint Switching Networks," **Proc. 14th Intl. Conf. on Fault Tolerant Computing**, IEEE, Silver Spring MD, (June 1984), pp. 240-245.

[Opf71] D.C. Opferman and T.N. Tsao-Wu, "On a Class of Rearrangeable Switching Networks-Part II: Enumeration Studies and Fault Diagnosis," **BSTJ 50(5)** (May 1971), pp. 1601-1618.

[Pad83] K. Padmanabhan and and D.H. Lawrie, "A Class of Redundant Path Multistage Networks," **IEEE Trans. on Computers C-32(12)** (Dec. 1983), pp. 1099-1108.

[Par80a] D.S. Parker, Jr., "Notes on Shuffle/Exchange Type Switching Networks," **IEEE Trans. on Computers C-29(3)** (March 1980), pp. 213-222.

[Par80b] D.S. Parker, "New Points of View on Three-Stage Rearrangeable Switching Networks," Proc. Workshop on Interconnection Networks for Parallel and Distributed Processing (April 80), IEEE, pp. 56-63.

[Par82] D.S. Parker and C.S. Raghavendra, "The Gamma Network: A Multiprocessor Interconnection Network with Redundant Paths," **Proc. 9th Annual Sym. on Comp. Arch.**, ACM Sigarch, (Apr. 1982), pp. 73-80.

[Pat81] J.H. Patel, "Performance of Processor-Memory Interconnections for Multiprocessors," **IEEE Trans. on Computers C-30(10)** (Oct.1981), pp. 771-780

[Pea77] M.C. Pease, "The indirect Binary n-Cube Microprocessor Array", **IEEE Trans. on Computers C-26(5)** (May 1977), pp. 458-473.

[Pra80] D.K. Pradhan and K.L. Kodandapani, "A Uniform Representation of Single- and Multistage Interconnection Networks used in SIMD machines," **IEEE Trans on Computers C-29(9)** (Sept. 80),pp. 777-790.

[Pre80] U.V. Premkumar, R. Kapur, M. Malek, G.J. Lipovski and P. Horne, "Design and Implementation of of the Banyan Interconnection Network in TRAC," **Proc. AFIPS Conf. (49)** (1980), pp. 643-653.

[Rat80] B.D. Rathi and M. Malek, "Fault Diagnosis in Interconnection Networks," **Proc. Symp. Distributed Data Acquisition, Computing and Control** (Dec. 3-5 1980), pp. 110-119.

[Sie81a] H.J. Siegel and R.J. McMillen, "Using the Augmented Data Manipulator Network in PASM", **Computer 14(2)** (Feb. 1981), pp. 25-33.

[Sie81b] H.J. Seigel and R.J. McMillen, "The Multistage Cube: A versatile Interconnection Network", **Computer 14(12)** (Dec. 1981), pp. 65-76.

[Sto71] H.S. Stone, "Parallel Processing with the Perfect Shuffle," **IEEE Trans. on Computers C-20(2)** (Feb. 71), pp. 153-161.

[Tri79] A.R. Tripathi and G.J. Lipovaski, "Packet Switching in Banyan Networks," **Proc. 6th Annual Symposium on Computer Architecture**, ACM Sigarch, New York NY (April 1979), pp. 160-167.

[Tur83] J.S. Turner and L.F. Wyatt, "A Packet Network Architecture for Integrated Services", **Proc. Globecom 1983**, IEEE, New York NY (1983) pp. 45-50.

[Wak68] A. Waksman, "A permutation Network," **JACM 15(1)** (Jan. 68), pp. 159-163.

[Wu80a] C.L. Wu and T. Feng, "On a Class of Multistage Interconnection Networks", **IEEE Trans. on Computers C-29(8)** (Aug. 1980), pp. 694-702.

[Wu80b] C.L. Wu and T. Feng, "The Reverse Exchange Interconnection Network", **IEEE Trans. on Computers C-29(9)** (Sept. 1980), pp. 801-810.

[Wu81] C.L. Wu and T. Feng, "The Universality of Shuffle-Exchange Network", **IEEE Trans. on Computers C-30(5)** (May 1981), pp. 324-331.

[Wul72] W.A. Wulf and C.G. Bell, "C.mmp-A Multimicroprocessor" **Proc. of the Fall Joint Computer Conference**, AFIPS Press (1972), pp. 765-777.

Chapter 3:

Experimental Architectures

3.1 Introduction

One of the most significant communications issues in today's networking environment is the lack of an integrated approach for transporting information. Current telecommunications services are often provided by separate networks–voice, circuit data, packet data, telex, and private line networks. Each network may use a dedicated access, and separate signaling and addressing schemes. This method of providing services is not only inefficient but also expensive. In addition, the above-mentioned networking solutions may not be able to satisfy long-term needs, because existing and planned networks based on present technologies may not be able to support future applications efficiently. Such concerns have led to the development of the concept of an integrated information service capable of providing network resources on demand to support a broad spectrum of applications. The need for such an integrated network resource is driven by a number of trends such as future applications and the availability of high-bandwidth transmission systems.

To a large extent, demand for higher network bandwidth comes from the growing penetration and power of desktop computers. These devices now offer users such advanced features as color, audio, and full-motion graphics and video. This has changed the networking requirements of the business community. The centralized computing environment of yesteryear has been replaced by distributed computing and database systems. Increasingly, these distributed environments are linked at single sites by LANs, creating a growing demand for high-speed interconnection. The present interconnection schemes are inadequate for LANs which themselves operate at between 4 and 100 Mbps. Further evidence of user need can be seen in the emerging use of imaging as a form of information handling, the growing use of video conferencing, and the increased importance of telecommunications in education and health care. Many hospitals are experimenting with communications systems that combine voice, X-ray images, and text to improve the timely communication of radiological images and diagnostic reports to physicians treating hospital patients.

CCITT has classified such broadband communication services into *interactive* and *distributive* services. Interactive services consist of conversational, messaging, and retrieval services. Distributive services consist of broadcast services such as high-definition television and services where the user has control over the presentation. The bit rate requirements for these services range from 0.10 Kbps at the low end to 1,000 Mbps at the high end. Future advances in signal processing may significantly reduce the bandwidth requirement for certain applications without sacrificing service quality.

Optical transmission technology, offering high bandwidths approaching or exceeding 1 Gbps along with a low bit-error rate, seems to be the medium of choice for terrestrial transmission. Gradually, the fiber infrastructure is extending its reach around the world. In addition, advances in component technologies such as optoelectronics and digital signal processing are driving the component costs down as they increase performance. This trend has led to an increase in the capacity available on the public networks. In addition, fiber transmission standards such as the Synchronous Optical Network (SONET) provide the capability to support a reliable and survivable transmission infrastructure.

In such a networking environment, integration can be accomplished in a number of ways. The basic idea is to carry various types of traffic using a single network. Most initial attempts were to integrate voice and data traffic onto a common trunk. In this scheme, voice and data traffic share the transmission facility between the switching nodes, but separate switching fabrics are used to switch data and voice traffic. It is assumed that separate access networks are used to get to the switch node. It is, however, possible to provide an integrated access through a single access interface. ISDN is an example of such an integrated scheme, where different traffic types are delivered to the backbone switching fabric using a common local access line.

ISDN provides an integrated access using one or more channels through a common interface. The ISDN architecture can be thought of as consisting of a user-to-network access segment and a network segment connected by an ISDN switch. CCITT, in its definition of ISDN standards, selected a digital channel rate of 64 Kbps as a basic building block. In addition, two interfaces called the basic rate interface and the primary rate interface were defined to support bearer as well as teleservices.

The basic rate interface consists of two 64-Kbps B channels and a 16-Kbps D channel to attach end systems that require low capacity. The primary rate interface consists of $n*B$ channels ($n=23$ in North America and Japan and $n=30$ in Europe) and a D channel for signaling. In the primary rate interface, the channel rate for both the B and D channels is 64 Kbps.

The D channel provides a packet mode service to transport signaling and network control information along with packet data. High-speed interfaces such as H0 (384 Kbps), H11 (1,536 Kbps), and H12 (1,920 Kbps) to support wideband applications were also defined.

Unfortunately, ISDN is not really an integrated network. It provides integration at the physical level from the end system to the network. Within the network, the traffic is segregated into its component parts and carried on separate networks.

To meet present and future service requirements, what is needed is a network offering universal network services and capable of transporting a wide range of services in a fully integrated manner. Such a network should provide on-demand access to network resources, wideband access, uniform access protocol, application-independent network services, and integrated transport. In addition, integrating access and transport avoids a new network, which is costly to construct, administer, and maintain, for each new service. Further, an integrated access and transport network offers greater flexibility to support existing and future services with good performance and economical resource utilization. Such a unified network allows a unified network management, operation, and maintenance system to be implemented. There are a number techniques available to implement this logical universal network service, as shown in the figure below. These range from *circuit switching* at one end of the switching spectrum to *packet switching* at the other end.

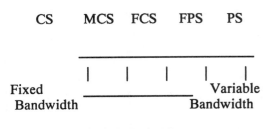

CS - Circuit Switching
MCS - Multirate Circuit Switching
FCS - Fast Circuit Switching
FPS - Fast Packet Switching
PS - Packet Switching

A network service based on circuit switching technology must provide a broad range of channel rates along with multiple switching fabrics to meet the diverse service requirements of the future services. The fixed channel structure supported by this switching technique has the advantage in that it facilitates multiplexing and cross connection of channels to the corresponding switching fabrics. On the other hand, the fixed channel structure can limit capacity sharing among various services. Consequently, blocking can occur, even when free capacity is available. In addition, the various channel structures are closely tied to the characteristics of the services. The complexity of network management is quite large compared with a packet-switched network service, because reconfiguration is needed to support service or channel changes.

Packet switching can support dynamically varying bandwidth requirements up to the maximum channel capacity. Dynamic allocation of bandwidth not only supports existing services, but can also accommodate new services. Packet switching supports adaptive bandwidth management, allowing each user to use only the amount of transport capacity needed. This type of network service decouples the trunk transmission speed from the user data rates. Suitable protocols and packetization methods achieve graceful overload performance by trading delay for throughput. In addition, a packet-switched network service supports multiple grades of service adaptively. Packet switching as supported today suffers from variable delay and limited throughput resulting from the protocol complexity. Networks based on such a protocol may not be able to support applications that require a wide range of bandwidths.

Therefore, what is needed is a switching technology as flexible as packet switching with switching performance comparable to that of circuit switching. In this regard, the most interesting development has been the acceptance of packet switching technology as the basis for a broadband integrated services digital network (BISDN) by CCITT. The objective of BISDN is to specify a user-network interface and a network that will support all the existing and future services and applications. CCITT has selected the asynchronous transfer mode (ATM), a packet mode switching and multiplexing technique, to support the various services. ATM switching is based on connection-oriented service, and it will be able to support all services—both connectionless

and connection-oriented. Another basic assumption is that a high-capacity and high-performance fiber-optic medium will form the transmission capability used to carry the traffic.

In ATM, information is organized into fixed-size blocks called *cells*. A cell consists of an information field and a header. The header size is 5 bytes and the information field size is 48 bytes. These remain constant at both the user-network and network-node interfaces. Cells are assigned on demand, depending on the source activity and available resources. The primary function of the cell header is to identify cells that belong to a connection called the *virtual channel*, a communication capability for transporting ATM cells. In addition, the ATM layer supports a communication capability called the *virtual path*, identified by a virtual path identifier (VPI). A virtual path is a group of virtual channels.

The ATM transport network common to all services is structured into an ATM layer and a physical layer. The transport functions of the ATM layer are subdivided into the virtual channel level and the virtual path level. The information field of a cell is transported transparently by the ATM layer (no error detection and correction). In addition, the ATM layer preserves the cell sequence integrity on a virtual channel connection. The ATM layer provides single-bit error correction and multiple-bit error detection on the ATM header. Additional functionalities on top of the ATM layer, such as the ATM adaptation layer, accommodate various services at the end system. The primary function of the ATM adaptation layer is to segment user data into ATM cells and to reassemble the ATM cells into user data at the destination end system. This layer can provide end-to-end error control, buffering, flow control, and multiplexing based on the application's requirements.

3.2 Overview of Papers

The papers included in this chapter describe different experimental architectures implemented in three different countries, reflecting the worldwide interest in broadband ISDN and the ATM switching technique.

The first paper, "The Prelude ATD Experiment: Assessments and Future Prospects," by Devault et al., describes a prototype network designed in France to study the feasibility of using the asynchronous time division (ATD) technique to support an integrated multirate network. A switch, a concentrator, and two home networks and terminals made up the network model. To support user applications, Prelude used 16-byte packets that included a 1-byte header. Video signals (60 Mbps), audio signals (medium quality: 64 Kbps, high quality: 786 Kbps), and digital data (64 Kbps to 16 Mbps) are shown to be easily supported by the Prelude network system.

In the second paper, "Definition of Network Options for the Belgian ATM Broadband Experiment," DePrycker describes the design issues arising in the implementation of a broadband network and derives the system architecture/options for an ATM-based network to be tested in 1992. The paper outlines the historical background of the development of the ATM techniques and fast packet switching. A fixed cell length of 32 bytes with a 2-byte header was chosen to simplify implementation. A traffic capacity of up to 600 Mbps was incorporated, allowing the use of 100- to 200-MHz CMOS technology for the switching fabric to exploit inherent parallelism. Only smaller portions of the switch are expected to use the faster emitter-coupled logic or GaAs technology. The proposal also recommends the use of a single universal low layer capable of transporting various teleservices, avoiding the use of the bearer service indicator in the ATM model.

In the third paper, "Experimental Broadband ATM Switching System," Kato et al. describe the architecture and performance of an experimental model of a broadband ISDN developed by Fujitsu. The architecture of the switching fabric was based on a multistage self-routing principle and was built from a three-stage self-routing switch module. In addition, the switching fabric included circuit switching capability, mainly to support high-quality motion video (150 Mbps). The cell format consisted of 4 bytes of header and 52 bytes of user information.

In the final paper, "Layered ATM Systems and Architectural Concepts for Subscribers' Premises Network," Vorstermans and DeVleeschouwer review the ATM technique and present a layered architectural model for an ATM-based network. For the subscribers' premises network, they propose a two-tier architecture.

The "Prelude" ATD Experiment: Assessments and Future Prospects

MICHEL DEVAULT, JEAN-YVES COCHENNEC, AND MICHEL SERVEL

Reprinted from *IEEE Journal of Selected Areas in Communications*, Vol. 6, No. 9, December 1988, pages 1528-1537.
Copyright © 1988 by The Institute of Electrical and Electronics Engineers, Inc. All rights reserved.

Abstract—The Prelude experimental model represents a complete network which can integrate any bit rate from data to moving pictures by means of the asynchronous time division (ATD) technique. In order to get a reliable and full assessment of ATD, the model is made up of a switch, a concentrator, and two home networks and terminals. The project started in 1982 and was completed in the middle of 1987. The success of the experiment proves the feasibility of ATD which is now agreed upon as a relevant solution for future ISDN. The document addresses the results derived from this experiment and analyzes the trends which can be expected for the future. First, an outline of Prelude is given in order to describe the state of the art reached within the experiment. Each element of the system is described with special emphasis on the switching matrix as its optimized and efficient implementation could lead to short term industrial developments. Then a thorough assessment of the whole system is given from which future evolutions are most likely to be derived. Finally, we conclude by giving possible ATD fields of application.

INTRODUCTION

WHEN people began to think about integrated broadband networks at the beginning of the eighties, they agreed on two major requirements: *high throughputs* and *flexibility*. High throughputs are required because of the breakthrough of video services, and flexibility is needed due to the difficulty to foresee the new service bit rates and due to the wish to transfer all the signals through one integrated network.

As soon as these two cornerstones were set up, the question was to find out the best technique to achieve the new generation of networks. In 1979, the CNET started investigating a technique mostly derived from packet switching. It was called the asynchronous time-division (ATD) technique, and in 1987 it was demonstrated in the Prelude model which implemented what could be operated in the public network in the nineties. The ATD technique belongs to a family of new transfer mode techniques now proposed all over the world, mainly in the U.S.A. [1], [3], [5], [6], [9], in Japan [8], [10], and in Europe [11]–[16]. Our main preoccupations were to investigate the technique itself and its consequences on the transfer of information, whatever the signal. With that in mind, we implemented the services that seemed the most crucial in ensuring that the critical problems be tackled. Most of the basic options of Prelude are still relevant today,

whereas some characteristics are to be reconsidered according to technology advances, or to current international trends.

It seems to be very difficult to achieve a multiservice, i.e., a multibit rate network by using circuit switched techniques: indeed, the bandwidth required for a service depends on both the coding algorithms and the technology state of the art. These two parameters are still evolving so fast, that whatever the choice for the channel bit rate, it may prove completely inefficient within a decade. Thus, the starting point of our studies was found in packet switching. Packet networks are most suitable for data transfer, simply because they have been designed for that. However, when handling signals with real-time requirements such as speech or moving pictures, some peculiarities appear. For example, one key point is the end-to-end variation of the transfer delay of the packets, and we have to reduce this variation to recover the signal properly. Besides, because of the high throughputs we need, we have to simplify the basic transfer mechanisms as much as possible regarding both multiplexing and switching, with consequences on the packet structure, on the routing mode and on the multiplexing scheme.

In terms of network design, flexibility mainly means that a complete service independence must be ensured at the interface between the terminal and the network itself. In other words, the interface has to provide the terminal with an access without any framing structure or any clock constraint. One key point is the dynamic sharing of the multiplex bandwidth, i.e., one channel transmits information when the payload of a packet has been filled up. This is an efficient way to cope with any kind of information flow including bursty sources, remembering that most of the sources are likely to be more or less bursty. Besides, our functional approach consists in making a distinction between the functions performed by the transfer network and the functions performed by the terminal applications. Consequently, some usual functions of packet networks are to be shifted to the terminal level.

These reasons lead to the following options. The information flow is split into blocks of fixed length, whatever the source. A header is added to this information block to form what is called a *cell*. The length of the cell is fixed to simplify the multiplexing and switching mechanisms. In Prelude, the length was chosen rather short: 16 octets, including one octet for the header. Reasons for that choice were numerous: the optimization of the switch fabric in

Manuscript received September 17, 1987; revised June 7, 1988.
The authors are with Centre National d'Etudes des Télécommunications (CNET), Lannion, 22301 France.
IEEE Log Number 8824395.

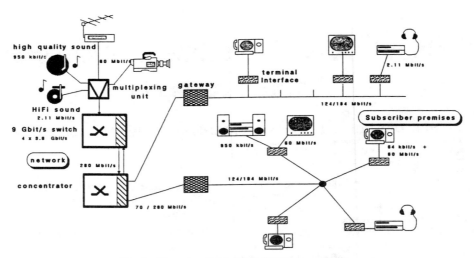

Fig. 1. General synopsis of the Prelude experiment.

terms of management, the need to take into account real-time requirements, mainly the cell assembling delay, the slightest disturbance in case of cell loss, and easy cell synchronization on the multiplexes as described below. The routing of the cells is based upon a logical virtual circuit strategy, and neither flow control nor recovery protocol is processed within the network. Depending on the signal, error recovery if any, is processed on an end-to-end basis.

It is worth noting that for the experiment, the target network was the public network, so the Prelude model implements a switch, a concentrator, two home networks and terminals (see Fig. 1).

This paper addresses an outline of Prelude concerning the terminal functions, home networks, and the switch fabric itself which is analyzed in more detail. In the last section, we assess the relevancy of the main options of the experiment, as we see them today. In particular, stress is put on the ability of ATD networks to transfer variable bit rates and on introduction strategies.

II. PERIPHERAL LAYER AND END-TO-END FUNCTIONS

Because an ATD network is comparable to a series of waiting queues, some specific problems have to be solved in a peripheral layer, i.e., between the terminal and the network itself. These problems concern both time transparency and information transparency. The corresponding functions are performed in terminal adaptors dedicated to every signal.

A. Peripheral Layer Peculiarities

1) Time Transparency: The first aspect of time transparency is related to the propagation delay of the cells within the network: it is not constant due to their buffering in the waiting queues, and it is characterized by its average value and its variation. The amplitude of this cell delay jitter is proportional to the length of the cell and inversely proportional to the bit-rate of the switched multiplexes. If the cells are short, this variation can be kept within reasonable limits: in the Prelude experiment,

we assess it to about 100 μs after a five exchange crossing (thirty stages). The problem of the cell propagation delay jitter is easily solved by using a "smoothing" buffer in the receiving terminal adaptor, the size of which is determined for each kind of signal by the bit-rate and the maximum variation of the delay. Then, waiting for a period equal to this variation before beginning to deliver the data from the buffer prevents errors due to an empty buffer.

Second, there is complete clock independence at the terminal access, i.e., there is no relationship between the transmitting terminal clock and the network access clock. However, this major feature of ATD networks has its counterpart: the actual value of this transmitting terminal clock is not given to the network. At the receiving end, the nominal value of the terminal clock frequency is used for decoding, and unavoidable frequency differences between the transmitting and the receiving clocks cause phase drifts in the receiving buffer. Therefore, a frequency synchronization may be needed in order to avoid any disturbance when retrieving the data.

2) Information Transparency: Consequences of bit errors are not specific to ATD networks but when they occur in the header, the cell may be transferred to a wrong receiving terminal ("intrusion") while it is lost for the right one. Another reason why cells may be lost is due to statistical multiplexing effects, local congestions may occur somewhere in the network leading to cell discarding through buffer overflows. In any case, we must admit that cell loss will occur in ATD networks. We tackled the problem in Prelude, and solutions were tested, specific to every signal we demonstrated. However, advanced coding algorithms, especially those related to video (conditional replenishment, entropic coding, transform coding, vector quantization) require a more thorough approach involving forward redundancy coding (see Section V-A.).

B. Coding Schemes and Terminal Adaptors

Advanced studies on coding algorithms were not the aim of the experiment because the efforts were mainly focused on the specific topics of ATD networks. So some coding schemes need a serious updating, particularly in the video field (note that the Prelude experiment began in 1982).

1) Video Signals: The video signal is component encoded following the CCIR A.601 specification through a linear PCM algorithm together with a nonlinear compression. Spreading the samples over the complete video line by using the line blanking interval yields the final bit-rate 60 Mbits/s. This codec is used both for television broadcast and for videophony, despite the high bit rate because conditional replenishment algorithms were not available when the experiment started. The video signal is framed in the transmitting terminal adaptor: 2-octet line and field synchronization patterns are inserted inside the sample flow. A video line is (3720 + 16) bit long, each field contains 288 lines and the frame frequency is 25 Hz. No forward correction is added to the signal. No information is transmitted during the field blanking, and for real-time reasons, stuffing bits are added to fill up the last incomplete cell of each field, so that it can be sent without any additional delay. It is worth noting that when a cell is lost inside the network, it is impossible to know its exact position within the video line, due to the fact that no cell numbering is used in Prelude. Two video terminal adaptors were built: one implementing a video frame memory as the receiving buffer, and one implementing a small 10 line memory. The aim was to verify whether a frame memory was of some help when implementing end-to-end functions. Therefore, two concealment techniques were tested for cell losses: replacing the disturbed line by the same one in the previous frame if the frame memory was available ("time-masking"), or if not, by the previous line in the same field ("space-masking"). Tests show that best results are obtained with the space-masking concealment technique in the case of moving pictures because with the time-masking one the replacing line is 40 ms "old," and so little trails can be seen on the pictures. Another major function of terminal receiving adaptors is frequency synchronization. In the frame memory receiver, this buffer memory is used to perform the frequency synchronization as follows. The frequency difference causes a drift between the reading pointer and the writing pointer so that they might overlap. Depending on the sign of that difference, one video frame is either skipped or repeated. Nevertheless, the crossover of the pointers has to be improved in order to cancel the cell delay jitter effects: an additional two-line memory is used to do this. As a result, visible effects are hardly seen on the monitor. In the line memory adaptor, the receiving frequency is adjusted to the transmitting one by examining the span of the field synchronization patterns in the incoming video signal (the patterns are recurrent because the coding algorithm features a constant number of bits per pixel). This feedback signal drives a PLL, yielding good results.

2) Audio Signals: For audio signals, since codecs were available, several kinds of signals were demonstrated, each requiring an adapted end-to-end protection scheme.

The "medium quality" (MQ) sound is used in Prelude for videophony: bandwidth 40 Hz–7kHz, subband and ADPCM 4 bit coding, bit-rate 64 kbits/s. No protection is added to the MQ signal: 30 samples fit exactly into one cell, and in case of cell loss, the two adjacent cells are stuck together. The results are acceptable, only little snaps can be heard in case of cell loss. For frequency synchronization, a very simple scheme is implemented: the receiving frequency adjustment is achieved by a programmable division ratio of the clock oscillator.

The "high quality" (HQ) sound is specified for DBS applications: bandwidth 30 Hz–15kHz, 14 bit linear PCM coding with near-instantaneous 10 bit compression, bit rate 2×384 kbits/s for two stereo channels. A Hamming protection algorithm is applied to the most significant bits of each sample. An interesting particularity of this signal is the indication of the silence intervals inside the signal. They are used in the receiving terminal adaptor to achieve phase slips in the buffer in order to avoid any additional frequency synchronization mechanisms.

The HiFi sound is the well-known compact-disc standard (bandwidth: 20 Hz–20kHz, 16 bit PCM coding, two stereo channels). Preliminary tests showed that bit errors are more audible than cell losses. Since listeners are highly sensitive to errors, a special protection algorithm was tested. A framing of the signal is achieved via the end-to-end forward protection. To cope with bit errors, a Hamming 6 bit protection is added to each 16 bit sample, and to cope with cell loss an interleaving process is achieved on a 220 sample basis. Then, 10 protected samples are linked together and completed by a 20 bit synchronization pattern to get a 240 bit frame to fit into two cells. For frequency synchronization, we measure the buffer filling level in the sound receiving adaptor, and its result is used to drive a PLL.

3) Data Transmission: Data transmission was implemented in three different ways. First, it was demonstrated by the signaling links. Inside the home network, data are embedded in HDLC frames and a simple one window layer 2 protocol is processed. Specific problems are further discussed with regard to the use of a *D*-like common channel. Second, a 1.2–64 kbit/s data transmission was tested between two X.25 terminals. In both cases, on transmission, the HDLC frames are segmented into ATD cells; due to the variable length of the frames, stuffing bits must be added to fill up the last cell and will be discarded at the receiving end. Finally, a high speed data transfer (up to 16 Mbits/s) is implemented between two personal computers. A specific protocol is used in order to reduce the retransmissions in case of error detections. When corrupted, the data frame (about 1000 bits including start and stop flags) can generally be recovered at reception, thanks to a seven octet error detection and correction field. A retransmission of the frame is only requested if the recovery mechanism is not sufficient. As a result a frame numbering using one octet is provided.

III. HOME NETWORKS

The Prelude home networks were designed to meet the requirements of a future customer who will have to cope

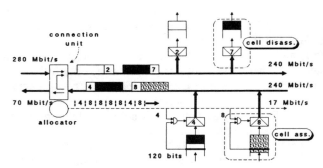

Fig. 2. Home network: basic principles.

with a lot of different terminals such as personal computers, telephones, videophones, TV sets, tape recorders, ..., spread around the house. The functional architecture was derived from the narrow-band ISDN CCITT model: gateway, passive bus distribution, single multipurpose plug, and *D*-like common signaling channel [2].

A single multipurpose plug and an automatic identification of the terminals provides a *universal connection* of terminals. In order to simplify the extensions and for reliability reasons, a passive bus-oriented structure was selected. Its length can be up to 80 m. Physically, the user access multiplexes have bit rates of 280 Mbits/s towards the user and 70 Mbits/s in the opposite direction. Two home networks were built, with the same logical bus topology but with different physical topologies: a linear and a star, both involving exactly the same line terminal equipments (see the general synopsis in Fig. 1). The bus is connected to the public network via a gateway. *Cells are built at transmission and picked out at reception in the adaptation boxes attached to each terminal.* A videophone terminal for example, requires three cell assemblers/disassemblers (CAD's) for sound, picture, and data. In-house connections are performed through the gateway.

As shown in Fig. 2, the bus is made up of three coaxial cables for reception, transmission, and allocation. Possible conflicts can occur when several cell assemblers try to obtain access to the transmission bus. They are solved by an allocator located in the gateway and programmed at call set up. The right to transmit is conveyed to all the terminals through the allocation bus. A cell assembler sends a cell under two conditions: at least 120 bits available in the queue and reception of an access right. These two conditions are time independent and therefore a cell is ready to be transmitted at any moment. Consequently, the transmission system cannot be driven by a central time-base: it is plesiochronous, i.e., each line termination located in the terminals has its own transmission clock working at the bus nominal frequency. Cells are CMI-coded and inserted into an envelope limited by start and stop bits. This allows a quick phase recovery at the beginning of each cell at the gateway. The *useful capacity* on the transmission bus shared by the terminals was fixed at 124 Mbits/s; however, in order to avoid overlapping (due to the propagation delay of 5 ns/m) between a cell transmitted by the farthest terminal followed by one transmitted by the nearest, a gap must be provided between

them, yielding a *final bit-rate of 240 Mbits/s* (before being CMI-coded). On the reception bus (using the same transmission system for internal standardization purposes), a cell disassembler picks out a cell when its header contains the right value. Note that multipoint connections to internal receivers are performed free-of-charge as in any bus topology.

As previously mentioned, the signaling channel supporting the control dialogue is derived from the narrow-band ISDN-*D* channel: it is a common channel, i.e., all the processors use cells with the same header, and the signaling data are embedded in HDLC frames. Nevertheless, it differs from the original *D*-channel by the medium access mechanism. In fact, *a CSMA-CD system can no longer be used* due to the ratio between the bit duration (4 ns) and the roundtrip delay along the bus (800 ns). As a result, an *ALOHA-like* mechanism is used alternatively and when collisions occur, the upper layers of the protocol take charge of the recovery.

IV. INTEGRATED MULTIPLEXING AND SWITCHING

A. Multiplexing

The Prelude multiplexing strategy may be described as a pseudosynchronous one. The multiplex is organized in fixed-length time slots, the size of which is equal to the cell length (see Fig. 3). In other words, the multiplex is cell interleaved, just as the synchronous time division (STD) basic multiplex is octet interleaved. Slots (i.e., cells) are not separated by a flag, but they are contiguous.

1) Framing: Framing is provided by specific "framing patterns" that are inserted into every empty slot, i.e., in every slot that is not used to transmit data ("useful cells"). The average load of a cell multiplex is never 100 percent; the rest of the capacity is used for framing, thus no additional resource is wasted. As a result, if a temporary traffic burst occurs, the whole capacity of the multiplex can be allocated to transmit that burst. This framing strategy simplifies the mechanism at the receiving end; every time that a framing pattern is recognized, a counter which delivers the cell clock is reset and then runs freely as long as no framing pattern is recognized. Due to the length of the framing pattern, the probability of framing pattern imitation is very low. Moreover, as the network selects the header values, by avoiding some of them, this prevents imitation except in the case of transmission errors. Note also that the maximum load must not exceed 80 percent in normal conditions. In addition, if a sequence of contiguous useful cells becomes too long, a framing cell is inserted by force.

Another function immediately follows framing: in order to reduce the working speed as soon as possible, a serial to parallel converter transforms the serial flow into an octet parallel flow.

2) Clock Adaptation: It is well known that clock adaptation and jitter compensation without slips raise a rather complex problem in STD networks. In packet-oriented networks, that problem becomes easier to solve because it is always possible to insert or to suppress nonsig-

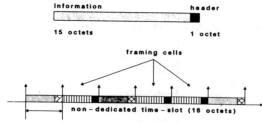

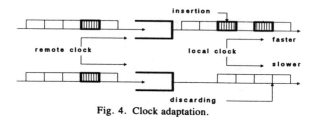

Fig. 3. Multiplexing in Prelude.

Fig. 4. Clock adaptation.

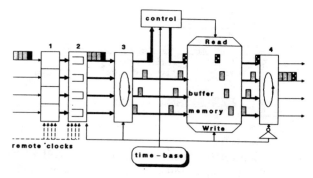

1) framing & parallelization
2) clock adaptation & phase alignment
3) super – multiplexing
4) demultiplexing & serialization

Fig. 5. Prelude switching matrix diagram.

nificant parts of the digital flow between packets. In our case, these nonsignificant parts are the slots carrying framing cells which can be removed or inserted (see Fig. 4) without any other effect upon the useful cells other than slightly modifying their propagation delay. Clock adaptation therefore only requires an asynchronous queue that is inserted between two plesiochronous or even asynchronous multiplexes. This queue is filled at the bit-rate of the incoming multiplex driven by its remote clock and emptied at the bit rate of the outgoing one, i.e., at the rate of the local clock. If the latter is slower than the remote clock, an overflow might occur but before this happens, an incoming framing cell is removed from the multiplex. On the contrary, if the local clock is faster, the queue could dry up and to safeguard against this, a framing cell is inserted into the outgoing stream. Finally, no centralized network synchronization is required to prevent information slips, which enhances safety and simplicity.

It must be pointed out that the remote and local clocks can be either plesiochronous (same nominal frequency but uncorrelated phases) or completely asynchronous (different frequencies) as long as the useful cell flow at the input does not exceed the available cell flow at the output. This feature of ATD is of great importance because it leads to total independence between the physical bit-rate of the multiplexes and the bit-rate at the inputs and at the outputs of the switching nodes, allowing the possible connection of different kinds of multiplexes to the same node.

These two functions—framing and clock adaptation—have, of course, to be performed at every switching interface, thus the switch is provided with a cell flow which has its own bit, octet, and cell clocks that can be synchronized on a switch internal clock. This is used to implement the core of the Prelude switch which works on a completely synchronous mode, at the octet clock frequency derived from the serial to parallel conversion. Subsequently, we shall call the duration of an octet "*cycle*."

B. Switching

The Prelude switch architecture (see Fig. 5) looks like a classical STD *T*-switch in so far as we can find the same functional blocks: framing, clock adaptation, phase alignment, supermultiplexing, buffer memory, control memory, and finally a demultiplexing stage. However, the main difference is that the buffer and the control memories are not cyclically operated, but operated as queues. More precisely, the buffer memory is operated as a FIFO queue and the control memory is replaced by what can actually be described as a cell switch.

1) Framing and Serial to Parallel Conversion: The first stage of the system, including the previously mentioned framing counters, is dedicated to framing and serial to parallel conversion. It operates on a per multiplex basis, i.e., it delivers on each output a cell stream transmitted octet per octet but without any phase relationship from one output to another as far as either octet or cell-rate are concerned. In other words, each parallel output still works at a rate and phase derived from the remote clock of the associated incoming multiplex.

2) Phase Alignment and Super Multiplexing: The second stage operates frame alignment by locking the octet-rate and the phase of each cell carried on all its output links on the local central clock. This is performed through the clock adaptation queues in such a way that cells present a shift of one octet (one cycle) from one link to the next (see Fig. 6). Note that *framing cells are not written in the queues*.

Due to the relationship between the cell length (16 octets in Prelude, 4 in Fig. 6) and the number of links (16 in Prelude, 4 in the figure), the effect of that shift is as follows.

1) Inside a given cycle, there is at most one header on the outputs, taken as a whole.

2) Inside cycle "*c*," a header on output "*o*" is followed inside cycle "*c* + 1" by a header on output "*o* + 1" and so on.

That phase alignment on a diagonal basis prepares cells to be supermultiplexed. The third stage is concerned with the supermultiplexing of cells. This is performed by a

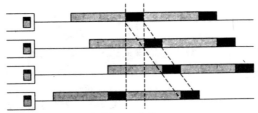

Fig. 6. Diagonal alignment.

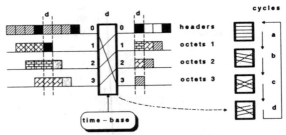

Fig. 7. Parallel-diagonal supermultiplexing.

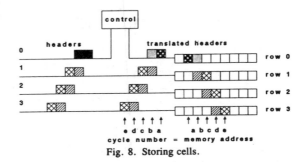

Fig. 8. Storing cells.

space-division rotative switch, which we will call a "barrel shifter" in what follows, acting on 16 input and 16 output paths simultaneously (4 × 4 in Fig. 7). Inputs ($i0$ to $i3$) are cyclically connected to outputs ($o0$ to $o3$) at the octet speed according to the following scheme:

cycle a: $i0, i1, i2, i3 \rightarrow o0, o1, o2, o3$

cycle b: $i0, i1, i2, i3 \rightarrow o1, o2, o3, o0$

cycle c: $i0, i1, i2, i3 \rightarrow o2, o3, o0, o1$

cycle d: $i0, i1, i2, i3, \rightarrow o3, o0, o1, o2$

cycle e: as cycle a and so on.

As a result,

1) parallel octets of the same row belonging to cells coming from the different inputs are gathered up on the same output path; in particular, *headers are multiplexed on the first output ($o0$)*;

2) each cell is distributed octet per octet on the output paths with a one octet offset per cycle:

octet 0(header)   on $o0$ in cycle c;

octet 1   on $o1$ in cycle $c + 1$;

octet 2   on $o2$ in cycle $c + 2$;

octet 3   on $o3$ in cycle $c + 3$.

That mechanism provides two levels of parallelization: in the first one each cell is diagonally parallelized on 16 separate paths, and in the second each path is made up of 8 wires for octet transmission. As further described, this cell organization is the key element in switching one cell every cycle at a fairly reasonable working speed.

3) Storing Cells in the Buffer Memory: A cell can be directly stored in a buffer memory as the traditional packet protocols are not relevant in ATD. That memory (made up of random access chips) is organized in rows, each of which is dedicated to an output path of the barrel shifter (see Fig. 8). It is cyclically written under the control of the time-base: during each cycle, the 16 octets present on the outputs of the barrel shifter are simultaneously stored in a column of the buffer memory. The address value is equal to the current cycle number. As a result, a given cell is diagonally stored, due to the previous diagonal arrangement of the supermultiplex.

Specific attention is given to the headers. In fact, they have to be translated because in Prelude a logical circuit identification mode was chosen: a cell is identified on a multiplex by the value of its header which is related to

this multiplex. Therefore, the internal path dedicated to headers is forwarded to the control part of the switch, which returns the translated headers. As previously seen, the latter is stored in the first row of the buffer memory. Moreover, when an empty cell occurs, corresponding to a previously discarded framing one, blank octets are written which will never be read.

4) Cell Extraction and Retransmission: At a given cycle, the control part of the switch loads the address of the first octet of the cell to be extracted in a read address register, dedicated to the first row (see below). This first octet which is the header, is transmitted towards the first octet-parallel path of the outgoing supermultiplex (see Fig. 9). At the next cycle, the same address is incremented and passed down to a second register related to the second memory row, in order to extract the second octet of the cell, and so on. Cells thus remain in the same parallel-diagonal form on the 16 outgoing supermultiplex paths. Last, a second barrel shifter working reversely reassembles the cells on each of the outgoing transmission multiplexes where they will be finally serialized.

In short,

1) the buffer memory is cyclically written and randomly read;

2) the words written or read are 128 bit long;

3) a write and a read access must be processed inside a cycle.

5) The Switching Function: The major problem is to decide very quickly what to do with an incoming cell which is going to be stored inside the buffer memory. As headers may be contiguous, just one cycle is available for taking that decision.

The cell switch control (see Fig. 10) uses a translation memory which is simultaneously addressed, at each cycle, by the incoming header and by the time-base which gives the number of the multiplex from where the header comes

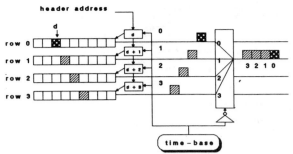

Fig. 9. Cell extraction from buffer memory.

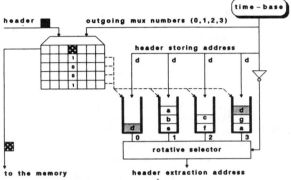

Fig. 10. The cell switch control.

(status of the incoming barrel shifter). Therefore, that translation memory contains 2^{12} words. Each of them is made up of two fields, respectively, 8 and 16 bits long (8 + 4 on Fig. 10), delivering two kinds of information for each header of each incoming multiplex:

1) the translated header, which is forwarded to the buffer memory as already mentioned;

2) a noncoded information upon the destination multiplex(es), one bit per multiplex, enabling broadcasting.

In association with this second field, a set of queues is provided. Each queue is dedicated to one of the outgoing multiplexes in order to receive the storing buffer memory address of the cells to be conveyed on that multiplex. The second field delivered by the translation memory (one bit per multiplex) enables the addition of addresses to these queues. The address that is filled into the activated queue(s) for a given cell is the address where the header of that cell is stored. This address is directly provided by the time-base counters. These queues are scanned by a rotative selector. Within the duration of a cycle, the time-base controls simultaneously this selector and the output barrel shifter in order to choose the queue and the related output multiplex. Therefore, the header buffer memory address of the cell to be conveyed on that multiplex is extracted out of the queue and stored in the address register of the first buffer memory row. It must be pointed out that broadcasting is easy due to the possibility of writing the same address at the same time in several control queues.

The control memory operates on an octet cycle basis. Besides, free cycles are available in the case of empty cells, identified by a blank header. These cycles are used by the control processor to write or erase words in order to set up or release communications.

V. Results and Likely Evolutions

A. Terminals and Home Networks

The peripheral layer is a crucial one in ATD networks for both technical and cost: it is impossible to dispense with it and we must prove that terminal adaptors are not the penalty to be payed for asynchronous transfer. As far as ''real time terminals'' are concerned, two major parameters influence adaptors: advances in coding algorithms and the state-of-the-art of technology. Their likely evolution may make it necessary to reconsider some of the functions as they were implemented in Prelude.

Advances in voice and video coding algorithms have been so great during the last decade that some coding schemes used in Prelude now appear obsolete. For example, in video coding, at least two stages have been overcome since plain PCM coding. After the first stage, the coding algorithm, whatever it may be, yields an unfixed number of bits per pixel. This is due to adapted coding tools using, for example, discrete cosine transform or variable length coding. As a result, functions such as frequency synchronization and concealment of cell loss need further investigation. The second stage takes full advantage of ATD techniques whereas the terminal access is free of any bit-rate constraint and of any clock relationship, we may then ask if it is really necessary to fix the terminal bit-rate as is usually the case when connected to a synchronous network. Of course not, because the access bottleneck disappears, and this opens up the way for variable bit-rates which feature a constant picture quality. New solutions have to be tested with particular stress on cell loss. Indeed, the cell loss ratio may be much higher because variable bit-rates mean statistical multiplexing and, therefore, we must be aware of increased queue overflows due to traffic variations within the network, even if the sources are likely to be decorrelated at least for interactive services. An effective way of solving the problem is found in priority transfer strategies. Basically, it assumes that since cells are to be lost, we would like to be able to choose these cells. In other words, any information flow can be split into several parts corresponding to priority grades. Networks procedures act in such a way so that low-priority cells are discarded first, paving the way for the others. The problem is to split the data in the most efficient manner: this is the field of our ongoing research.

The figures in Fig. 11 gives the number of gates involved in the implementation of Prelude terminal adaptors and clearly show that the use of CMOS technologies would make them credible by containing them in a handful of chips.

The whole home network should be redesigned in order to benefit from the evolution of both technology and coding techniques. Furthermore, it could be based upon new concepts, rather different than those of the narrow-band ISDN. In fact, due to the variety of home characteristics (e.g., already built or to be built) the most relevant topology could be a bus, a ring, a star, or even a combi-

VIDEOPHONE	Logical Gates	Bits of Memory
3 CADs*	13 000	4 K
Protections	13 000	56 K
TOTAL	26 000	60 K

TELEPHONE	Logical Gates	Bits of Memory
2 CADs*	9 000	3 K
Protections	1 000	4 K
TOTAL	10 000	7 K

* CAD = Cell-Assembler / Disassembler

Fig. 11. Assessment on ATD functions in the terminals.

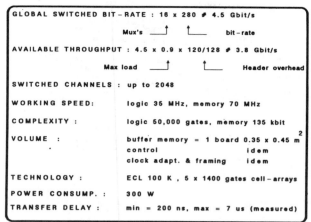

Fig. 12. Main features of Prelude matrix.

nation of all three. As far as the media are concerned, several technologies are emerging with their own field of applications. For example, plastic optical fibers could be used to link rooms together, and wireless media (diffuse infrared, millimetric microwaves) inside a single room could provide a cordfree connection of terminals. Moreover, it could be relevant in benefiting from the mains for low bit-rates. In conclusion, it may be relevant to design a S-like interface (S^*) *as independent as possible of the topology as well as of the medium technology* in order to be able to connect any terminals to any media. The results of the first studies are leading to a point-to-point interface carrying the bit rates strictly required by the connected terminal. Consequently, a *medium adaptor* may be needed between S^* and the medium.

B. Switching

The main features of a matrix are recalled in Fig. 12. At the very beginning of the experiment, ECL was the only available technology able to provide us with the high speed required. Yet, specific ECL cell arrays were implemented due to the weak degree of integration and the high power consumption of standard components. For the future, it does not seem necessary to redesign the basic principles but a few adaptations and extensions are desirable.

First of all, a header of 8 bits is obviously too short; that explains the frequently proposed values of 24–32 bits. Consequently, in order to limit the overhead, the information field could be extended to 256 bits.

Next, the storage capacity of the buffer memory is 256 cells and the length of the output control queues is 16 words. This last parameter is the most important in the case of traffic congestion. Traffic simulations issued from [17] showed that the queues should be 64 words long to provide a cell-loss probability of less than 10^{-10} with a load of 0.85.

In the case of transmission failure, the current protection parity bit of the header is not sufficient because it allows detection in most cases, but no correction. New processes are under study to cope with the actions to be undertaken on a cell with a so called wrong header.

Where broadcasting is concerned, there could be a lack of header at the call set-up because the translated header must be the same on every outgoing multiplex. There are at least two solutions. The first is to separate the control memory into two parts, one dedicated to point-to-point connections and the other to broadcasting ones, therefore, ascertaining that a header for broadcasting is simultaneously available for all the outgoing multiplexes. The other is based on a second translation of the headers. It must be remembered that a given cell is not discarded from the buffer after its transmission on a multiplex: it remains stored as long as an incoming cell is not written over it at the same address. Consequently, each time an outgoing multiplex asks for the same cell, it is possible to translate its header a second time by means of a second translation memory located at the output of the first row of the buffer memory.

Hence, the ability of ATD to benefit from variable and bursty bit-rates by using statistical mechanisms should be investigated in further studies, keeping in mind that classical flow control based upon back pressure protocols is no longer suitable.

Finally, as far as technological topics are concerned, the Prelude matrix implementation proves the feasibility and the possibility of also using such systems in trunk exchanges. However, it might prove advantageous to reduce the power dissipation and the number of interconnections in order to improve the reliability of the system. A major improvement concerns the buffer memory which was designed with single access ECL memories leading to a working frequency of 70 MHz. Fast double access CMOS memories are going to be available, providing power consumption savings and a higher degree of integration. Combined with the desirable increasing length of the cells, these technological improvements will lead to a higher throughput matrix which could be implemented on a single board convenient for trunk exchanges, concentrator units, and PABX's.

VI. CONCLUSION

The Prelude model was broadband oriented in order to prove the feasibility of ATD in the most difficult cases. The successful result of the experiment is proof of the ability of ATD to build a future universal broadband ISDN. Nevertheless, taking into account the present situation, the means by which this target is to be reached, are open to further investigation.

A first opportunity of introducing ATD is the *digitization of video distribution networks* either already built or to be built. However, as previously mentioned, we have to wait for improvements in two main areas: moving picture, coding techniques, and CMOS technology, in order to manufacture cost effective terminals. Technically speaking, this is feasible in a not too distant future (by the nineties). Furthermore, customers will be ready to purchase digital sets if the picture quality has significantly improved (see the success of the compact disc for sound).

Remaining within the consumer electronics field, ATD could be seen as a solution in providing a *universal home network*, without necessarily taking in consideration the above evolution of the public networks. This home network could initially support telecontrol and security systems and eventually evolve towards the higher bit rates required to interconnect all the customer digital terminals such as PC's, compact-disc players, video tape recorders, HiFi, TV sets, etc.

Another application concerns the *business* field. The need for flexible and fully integrated digital networks seems more and more obvious to provide a universal communication highway suitable for telephony, for communications between personal computers, between mainframe and remote CAD terminals, and so on, and of course for moving pictures. In fact, this field of application for ATD is not very different from that of the general public; for example, in each case an $S*$ interface is required and, at least in the long term, it should be the same for both. Of course, the different branches of a company must be efficiently interconnected. During a first stage, they can be ATD-linked together through the already existing digital *leased lines*. At a second stage, private *ATD cross-connects* could lead to cost savings by a better use of the above leased lines, thus providing each company with a more flexible dedicated network. Last, when an agreement will be reached on standards, *public ATD switches* could link all the companies together through a first universal multibit rate network.

Improvement of the narrow-band ISDN, which is single bit rated due to the use of 64 kbits/s circuit-switching technique, is an other area of ATD technique application. ATD can provide flexibility because as soon as a data stream is cell-assembled, the cells can be transmitted on several independent 64 kbit/s circuits. The header thus enables the recovery of the flow at the receiving end whatever the circuits used. Many applications start from such a new bearer service. For example, it becomes possible to provide videophony by a prudent sharing of $2B$ channels between picture and sound, e.g., respectively, 112 kbits/s and 16 kbits/s which is probably more suitable than 64 and 64 kbits/s. Similar services have already been proposed within the framework of CCITT [19], [20].

Whatever the means chosen to introduce ATD, it is obvious that standards are required and it is likely that a few *de facto* standards are unavoidable under commercial constraints. Nevertheless, the above ATD introduction schemes clearly show the necessity of having a consistent approach of the standards within the private field as well as within the public field.

ACKNOWLEDGMENT

We would like to acknowledge the assistance of a number of our colleagues in writing this paper: J. P. Coudreuse, P. Adam, D. Chomel, J. François, T. Houdoin, H. Le Bris, A. Lespagnol, J. P. Quinquis, Y. Rouaud, and J. C. Sapanel.

REFERENCES

[1] S. R. Amstutz, "Burst switching—An introduction," *IEEE Commun. Mag.*, Nov. 1983.
[2] M. Devault, D. Chomel, H. Le Bris, and Y. Rouaud, "From data to moving pictures: A multi-bitrate asynchronous time-division equipment at the subscriber's premises," *ISSLS '84*, Nice, France.
[3] A. Abanese, J. O. Limb, and W. D. Sincoskie, "IEEE multiservice integration with optical fiber networks," *IEEE GLOBECOM '85*, New Orleans, LA.
[4] J. Y. Cochennec, P. Adam, and T. Houdouin, "Asynchronous time-division networks: Terminal synchronization for video and sound signals," *IEEE GLOBECOM '85*, New Orleans, LA.
[5] J. S. Turner, "New directions in communications," *IEEE Commun. Mag.*, Oct. 1986.
[6] R. W. Muise, T. J. Schonfeld, and G. H. Zimmerman, "Experiments in wideband packet technology," *Int. Zurich Sem. '86*.
[7] P. Gonet, P. Adam, and J. P. Coudreuse, "Asynchronous time-division switching: The way to flexible broadband communication networks," *Int. Zurich Sem. 86*.
[8] H. Suzuki, T. Takeuchi, and T. Yamaguchi, "Very high speed and high capacity packet switching for BB-ISDN," *ICC '86*, Toronto, Ont., Canada.
[9] G. W. R. Luderer, J. J. Mansell, E. J. Messerli, R. E. Staehler, and A. K. Vaidia, "Wideband packet technology for switching system," *ISS '87*, Phoenix, AZ.
[10] K. Takami and T. Takenaka, "Architectural and functional aspects of multi-media packet switched network," *ISS '87*, Phoenix, AZ.
[11] P. Plehiers, M. Fastrez, J. Bauwens, and M. De Pryker, "Evolution towards a belgian broadband experiment," *ISS '87*, Phoenix, AZ.
[12] P. Gerke and J. F. Huber, "Fast packet switching—A principle for future system generations?" *ISS '87*, Phoenix, AZ.
[13] S. Giorcelli, C. Demichelis, G. Giandonato, and R. Melen, "Experimenting with fast packet switching techniques in first generation ISDN environment," *ISS '87*, Phoenix, AZ.
[14] F. van den Dool, "Synchronization aspects of ATD-IBC networks," *ISS '87*, Phoenix, AZ.
[15] I. D. Gallager, "A multi-service network on the Orwell protocol," *ISS '87*, Phoenix, AZ.
[16] M. Dieudonné and J. P. Quinquis, "Switching techniques review for asynchronous time-division multiplexing," *ISS '87*, Phoenix, AZ.
[17] P. Boyer, J. Boyer, J. R. Louvion, and L. Romoeuf, "Time transparency evaluation of an asynchronous time-division network," *ISS '87*, Phoenix, AZ.
[18] J. P. Coudreuse and M. Servel, "PRELUDE: An asynchronous time-division switched network," *ICC '87*, Seattle, WA.
[19] Pictel Corp., "Transmission format for $m \times 64$ kbits/s video telephony," *SG XV*, Nov. 1986.
[20] Kokusai Denshin Denwa Co. Ltd., "Transmission method for a videophone using a single 64 kbit/s channel," *COM XV*, Feb. 1987.

Michel Devault was born in 1947. He received the engineering degree from the Ecole Supérieure d'Electricité in Paris, France.

He joined the Centre National d'Etudes des Télécommunications in 1972 where he participated in the specifications of the E10 switching system. In 1979, he began studying asynchronous time-division techniques for integrated broadband networks. Since 1982, he has been involved in the Prelude experiment where he dealt with home networks. His current investigation field is the introduction strategy of ATD networks.

Jean-Yves Cochennec was born in 1948. He received the Master degree in physics from the University of Brest (France).

He joined the Centre National d'Etudes des Télécommunications in 1972 where he dealt with the implementation of the E10 switching system. In 1980, he started working on asynchronous transfer techniques. Since 1982, he has been involved in the Prelude experiment where his field was the terminal adaptation layer. His current investigation field is the implementation of variable bit rates in broadband ATD networks, and he is a member of the European Race 1041 Project.

Michel Servel was born in 1946. He received the engineering degree from the Ecole Supérieure d'Electronique de l'Ouest of Angers, France.

He joined the Centre National d'Etudes des Télécommunications in 1972 where he was involved in studies and tests of the E10 switching units. In 1979, he dealt with the design of a preliminary asynchronous transfer system, and since 1982, he has been involved in the Prelude experiment, and more specifically, in the design of the switch fabric. His current investigation field is the evolution of asynchronous switching techniques, and he is a member of the European Race 1022 Project.

Definition of Network Options for the Belgian ATM Broadband Experiment

MARTIN DE PRYCKER

Reprinted from *IEEE Journal on Selected Areas in Communications*, Vol. 6, No. 9, December 1988, pages 1538-1544. Copyright © 1988 by The Institute of Electrical and Electronics Engineers, Inc. All rights reserved.

Abstract—The definition of a new system requires the selection of a number of basic options. PTM (packet transfer mode) is a new generic concept grouping a number of similar techniques which enable very flexible switching and transmission.

This paper describes the options and parameters which have been selected for a Belgian experiment planned in the early nineties. Discussed are the following aspects: ATD (asynchronous time division) versus FPS (fast packet switching), including fixed or variable packet length, the optimal packet length, transmission and switching speed, and the number of low-layer capabilities.

I. INTRODUCTION

IN recent years, a growing interest has appeared throughout the world on broadband services with the necessary introduction of broadband networks. However, at this time it is very unclear what kind of broadband services will appear (TV, EDTV, HDTV, videophony, high speed data, · · ·), at what rate (e.g., TV between 30 and 140 Mbits, high speed data between hundreds of Kbits and tens of Mbits, · · ·), at what traffic mix and for how many and which subscribers. These uncertainties require a broadband network which is very flexible, so that it can be adapted to all expected and unexpected changes in the future requirements of this broadband network.

The classical monorate or multirate circuit switched networks are very inflexible since once the rates are defined, the networks must live with these rates for the rest of their existence (e.g., D, B, $H1$, $H4$, · · ·). Packet switching, on the other hand, is the most flexible solution since no rate must be specified in advance. However, it has problems in terms of delay and processing performance.

In 1983, an ideal mixture of both techniques (circuit and packet switching), basically offering the advantages of both techniques was introduced. In the U.S., this technique was introduced by Turner [4] and sometimes referred to as fast packet switching (FPS), in France by Coudreuse [14], and called ATD (asynchronous time division).

Rapidly, other organizations, companies as well as telecommunication administrations, became interested in these ideas. Also within the European research and de-

velopment program for broadband telecommunications (RACE), a consortium of European companies and PTT's studies these techniques.

Since 1985, also the standardization bodies, and especially CCITT SG XVIII, T1.D1, and CEPT/NA5, became interested in these techniques. Initially, they were grouped under the label of new transfer mode (NTM) by CCITT SG XVIII, in the Brasilia meeting of SG XVIII adapted to ATM (asynchronous transfer mode) and finally in the Seoul, Korea, meeting of February 1988 to PTM (packet transfer mode). Also in Seoul, it was decided to select one particular technique as the target solution, namely, the one with fixed length packets (cells) and to call this technique ATM. This may create confusion since ATM means something different depending on the time of use. We will use throughout this text the definition of Seoul. Since it resembles very much the ATD concept as defined by CNET [2], both names will be used as synonyms.

Big interest is shown in ATM, and it was agreed in Seoul by CCITT that the ultimate broadband network will be based on the ATM concepts [16]. The basic reason being its universal flexibility, described already in a large number of papers [2], [3], [4], [7]. However, a number of issues remained open to be solved in the next study period of CCITT, the major being: the cell format, the impact of voice delay, header functions and organization, load control and usage monitoring, statistical multiplexing, signaling, interface structure, interworking, · · · .

In 1984, the Research Center of Bell Telephone Mfg. Co., initiated its work on ATD. This work included conceptual studies where a number of options on system parameters for ATM have been carefully studied and decisions have been taken on these parameters, based on a detailed analysis, including simulations and analytical models, as well as on the results of the implementation of a small demonstrator [17] based on ATM.

In Belgium, an experiment is planned by 1992 based on the ATM concept performed by a consortium of Belgian telecommunication companies under guidance of the RTT [3]. A large number of options to be used in this Belgian experiment are based on the results described here.

This paper will describe the arguments which have been used to select a number of system parameters in the design of an ATD based network. The system parameters under discussion here are: fixed versus variable packet length and related to it the transmission and switching speed, the

Manuscript received September 28, 1987; revised May 4, 1988. This work was supported in part by the Instituut voor Wetenschappelijk Onderzoek in Nijverheid en Landbouw (IWONL) and the Belgian RTT (Regie voor Telegrafie en Telefonie).

The author is with Bell Telephone, B2018 Antwerp, Belgium.

IEEE Log Number 8824396.

preferred packet size, and the number of low-layer capabilities.

II. ATD and FPS

ATD and FPS are both variants of the general concept called PTM, as is shown in Fig. 1 where the genealogy of the digital switching technique is shown. The basic differences between both techniques are largely due to their historical background.

A. Asynchronous Time Division Multiplexing

The first interest in ATD was shown by the French CNET, with a number of papers describing the concepts as well as a demonstrator built in the CNET premises [2], [14].

ATD is a technique which evolves from the STD (synchronous time division multiplexing) concept, by using extra intelligence, i.e., the time slots are no longer assigned on a per call basis, but on a dynamic basis. This means that a call is no longer characterized by the position of the time slot within a "frame" but by a label designating the logical connection under consideration. It is therefore also sometimes called intelligent time division multiplexing [1]. ATD can equally well be applied both for transmission and switching.

We see that there is a great commonality between ATD and STD. This urges us to define ATD also in the layer one of the OSI reference model. Therefore, the functionality of the label is reduced to a minimum as is the case for the time information of an STD slot: the label is only used to designate the logical connection, with a possible error detection/correction on the label. The proposed packet size is also rather small (between 16 and 32 bytes information), as will be discussed later in this paper. Compare this to a slot size of 1 byte in STD.

B. Fast Packet Switching

The first description of FPS systems and prototypes were reported by the Bell Laboratories of AT&T [4]. Later, other systems and experiments were reported [5], [6] based on similar ideas.

FPS (fast packet switching) is a technique which evolves from the packet switching concept, by reducing the intelligence in the network nodes and relying on reliable links and, if required, on end-to-end protocols. By reducing the complexity of the link protocols, it becomes possible to do all protocol processing in hardware, thereby obtaining very high speed. Doing this, the disadvantages of packet switched networks for time sensitive services (e.g., voice) are removed.

Coming from the packet switching background, a large number of packet switching functionalities disappeared in the network nodes or are only performed in the edges of the network (terminals) end-to-end. However, some functions are still performed in the network itself: e.g., time stamping, and a distinction between different services (voice, · · ·) inside the network.

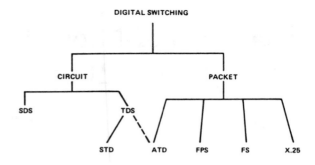

SDS : SPACE DIVISION SWITCHING
TDS : TIME DIVISION SWITCHING
STD : SYNCHRONOUS TIME DIVISION
ATD : ASYNCHRONOUS TIME DIVISION
FPS : FAST PACKET SWITCHING
FS : FRAME SWITCHING

Fig. 1. Genealogy of digital switching.

C. Conclusion

Although both techniques are very comparable, there is some difference in approach.

ATD, evolving from STD (synchronous time division), is located in the layer one of the OSI model and therefore the label functionality is reduced to an absolute minimum, only containing call identification. In contradiction, FPS, evolving from packet switching, proposes extra functionality in the label, which in the ATD solution is moved to the higher layers.

Another difference is the proposed packet length. In ATD, the proposed packet length is in the order of 16 to 32 information bytes, in FPS it is in the range of 100 bytes per packet. Again, the same historical background plays here: indeed, in byte multiplexed STD the "packet" length is only one byte.

It must also be mentioned that ATD was from the early beginning proposed to transport data, video, and voice services with a large emphasis on video. FPS was basically proposed to transport primarily high-speed data.

III. Fixed Versus Variable Cell Length

ATD, evolving from TDM is based on fixed length packets; FPS, evolving from packet switching is based on variable length frames. CCITT SGXVIII proposed in Brasilia to use the word "cell" covering both the fixed packets in ATD and the variable length frames in FPS. An important parameter to be selected is whether the cells will have a fixed or variable length.

Different factors are contributing to the possible advantages and disadvantages of both solutions. The most important are bandwidth utilization, achievable switching rates, and network performance.

A. Bandwidth Utilization

The bandwidth utilization is determined on one side by the number of effective information bits and the number of overhead bits per cell and on the other side by the size of the logical information entities to be transmitted. In [8], it is shown that the bandwidth utilization for (narrow-band) applications generating a large number of small logical bit entities (e.g., interactive data) is better if variable cell length is used instead of fixed cell length. If the application is generating mainly large logical bit entities then the gain in bandwidth utilization of variable cell length compared to fixed cell length is only very moderate (a few percent).

In broadband networks, this statement is also valid, but its implications must be carefully considered. In broadband networks, basically three different applications must be considered.

1) Video: The logical information entities which are generated in this application depend on the structuring of the video image and the coding used. Typically, for variable bit rate codecs these entities will have a size between 100 and 10 000 bits, for fixed bit rate codecs this value is in the order of a few thousand bits.

Given the bit rate necessary for video communication (videophone and video broadcast), it may be anticipated that video will by far be the largest bit rate producer in the broadband network.

2) Voice: A logical entity in a PCM voice environment is 8 bit, but these entities may fairly easily be combined into logical entities containing, e.g., 16 times 8 bits can be combined into one entity of 128 bits, thereby introducing a so-called packetization delay of 2 ms.

Even if the number of telephone subscribers will initially outnumber the video users, the total amount of traffic (bits) generated by voice applications will be much smaller than those generated by the video applications.

3) Data: Two types of data users may be considered: interactive data applications generating a large number of small cells, and bulk data transfer (e.g., file transfer and multiwindow graphic terminals) generating a large number of large cells and only a small number of very short cells (acknowledgments).

The total number of bits generated by interactive applications will only fill a very small percentage of the available bandwidth of a broadband network. For bulk data transfer applications the number of small cells will also be very small compared to the bits generated for large logical entities.

Based on these considerations it may be concluded that the percentage of small logical entities generated in a broadband network is very small. This means that the gain in bandwidth utilization which can be achieved by using variable length cells is negligible, i.e., in the order of a few percent.

B. Speed Performance

The speed performance which can be achieved in both proposals depends on the protocol which is applicable at the user-network or network-node interface (network protocol) and on the speed which can be achieved internally in the switching fabrics.

Ideally, the network protocols, on one hand, and the internal protocol of the specific switch fabric designs on the other hand, should be decoupled. While the former should be optimized for network performance, the latter is more related to the chosen switching structure, technology, access and routing mechanisms, etc. However, some simple relation between both protocols is usually required in order to achieve the maximum speed via the simplest interfaces.

1) User-Network/Network-Node Interface: In Table I, the different functions performed by the variable cell protocol and the fixed cell protocol are described.

The more complex the protocol, the lower the achievable speed performance. This is clearly explained in [9] where it is concluded that by the early 1990's, X.31 will be economically limited to primary rate operation. Frame switching, which has some additional functions compared to variable length protocols, may allow a 2–4 times speed improvement, and variable length protocols would be capable of operating at around 140 Mbits/s [9], if proper transparency techniques are used.

The complexity of the fixed length protocol is smaller than that of the variable length protocol as is shown in Table I, which may result in a higher protocol operating speed. Two practical implementations already working at this moment also show the validity of this statement. Indeed, two experiments were reported [10], [11] operating at 280 Mbits/s. It may be anticipated that a speed of around 600 Mbits/s may be achievable by the early 1990's, with economic implementation by the mid-nineties.

2) Switching Fabric Complexity: Internally, in the switching fabric, two options exist: variable cell length switching or fixed cell switching. It is clear that a large number of solutions for both options exist. However, all the variable cell length solutions will have a number of commonalities, as well as the fixed cell solutions. Some of these commonalities are described in Table II.

As can be derived from this table, the complexity of the common functions for variable cell length is higher than the one for fixed cell length. Also important is that fixed cells enable much better parallelism internally in the switches, thereby reducing to a large extent the internal speed (and thus also the power consumption) of the switches, which again will result in a more economic solution. This is also reflected in the prototypes running with fixed length cells at 280 Mbits/s against those with variable cell length.

Since the internal and external protocol may be different, it may be possible to use internally the less complex solution (fixed length) and externally the solution with optimal bandwidth utilization (variable length). However, for economical reasons, the interface between the external and internal concept should be kept as simple as possible so that no complex protocol conversion is necessary.

3) Conclusion: At this point in time it can be stated that the solution with fixed length cells is technically and economically achievable within the considered time frame

TABLE I
FUNCTIONS PERFORMED BY FIXED AND VARIABLE LENGTH CELL PROTOCOLS

Function	Variable	Fixed
1. Boundary recognition		
Flag recognition	X	
Cell synchronization		X
2. Transparency (bit stuffing)	X	
3. Error concealment		
FCS on cell	X	
Error detection on header		X
4. Discard invalid cell	X	
5. Multiplexing	X	X
6. Fill-in function		
Fill interframe time	X	
Generate empty/sync cells		X

TABLE II
COMMON FUNCTIONS IN FIXED AND VARIABLE LENGTH SWITCHING FABRICS

Function	Variable length	Fixed length
1. Queue management	bitwise allocation set of pointers	cellwise allocation top and bottom pointer
2. Switching	bit basis	cell basis allowing easy parallel operation
3. Multiplexing	variable cells	fixed time slots

and fulfills all related network requirements provided that high speeds (280–600 Mbits/s) are used. This solution permits higher processing speeds and hence larger bit rates, thus enabling larger multiplexing factors.

The minimal gain in bandwidth utilization which may be achieved by variable cell length in the broadband network protocol seems to be jeopardized by the higher complexity and the resulting lower speed (140 Mbits/s rates). It was therefore decided to build our Belgian experiment based on fixed length cells. Also, in CCITT SG XVIII it was decided to use fixed length cells [16].

IV. SWITCHING AND TRANSMISSION SPEED

When introducing a new broadband network, we must keep in mind that the network must be as open as possible to new evolutions in services, i.e., existing services with higher bandwidth requirements than today or even new services.

An important aspect which favors the use of high transmission and switching speeds is the reduction of the jitter and the delay introduced in the network. Jitter is introduced by the queues in the ATD network which are required to cope with the dynamic behavior of the network. These queues are emptied at the transmission and switching speed, and thus the queueing time of the packets is reduced when the speed is increased, thereby also reducing the jitter and the delay. It must therefore be checked for any proposed bit rate whether the CCITT Recommendations on both delay and jitter (if any) are fullfilled. The most important Recommendations for delay are G.164, G.161, and Q.507. In [10], it was shown that at speeds of about 600 Mbits/s these Recommendations could easily be met.

Also in favor of high speed is the possible multiplexing gain obtained in this case. Indeed, the higher the speed, the higher the multiplexing factor can be and thus the higher the statistical multiplexing gain.

ATD itself is already very flexible in the sense that no restriction at all is put on the bandwidth of the services transported over the network or on the combination of services transmitted or switched. The only limitation for any service in ATD is the maximum transportable and switchable bandwidth. To be as much as possible future proof, the selected speed must thus be as high as possible within the constraints that it must be economically feasible.

Knowing that HDTV will require a bandwidth in the order of 140 Mbits/s and that it should be very interesting to give any user simultaneous access to a number of TV programs, a speed of several of hundreds of Mbits/s seems required.

The current progress both in optical and electrical components and transmission will allow in the near future economical transmission and switching speeds at about 600 Mbits/s.

1) In [12], alternative economic solutions were proposed for the inexpensive transmission in the 600 Mbits/s range over a few kilometers: LED's, low cost lasers, and low cost lasers in the 800 nm range (CD lasers).

2) From [15] it can be derived that both the progress in CMOS and bipolar technology will allow the use of 600 Mbits/s switching circuit in the early nineties. Since fixed length cells will be used, parallelism in the switch will guarantee an internal reduction of the speed to CMOS achievable speeds (100–200 Mbits/s). The largest part of the switch can be built with this CMOS technology. Only a small part of the functions will run at 600 Mbits/s and thus require bipolar (ECL) or GaAs technology.

V. CELL SIZE

Another important aspect is the size of the cells. Several network aspects are in one way or another influenced by the cell size. The most important aspects are: the transmission efficiency, the packetization delay, the queueing delay, and the jitter (with associated depacketization delay). The implementation complexity is also affected by the cell size. Indeed, parallelism reduces to complexity as long as the implementation remains feasible.

A cell is composed of a header and an information field. The header size will be determined by the functionality required at the user-network interface. The information field size will influence the network performance, and some network performance parameters are only influenced by the relation between the header and information field size.

A. Header Size

We propose that the header would only consist of a virtual channel identifier (VCI) and an error detection/correction field. Additional functions should be supported implicitly in the VCI, e.g., by assigning special connection numbers to these additional functions. It must be mentioned that some additional bits may be added, e.g., to indicate the payload type.

Since a bit error in the header will multiply the bit error rate with a value equal to the information field size in bits, we think that error detection and correction is required on the header.

B. Information Field Size

Some network performance parameters are influenced by the size of the information field, independent of the header size; others are influenced by the relation between information field and header size.

1) Packetization Delay: The influence of the information field size on the packetization delay for fixed bit rate services is very obvious, e.g., a source rate of 64 Kbits/s and an information field of 256 bits will introduce 4 ms packetization delay. It is possible to reduce this packetization delay by only filling part of the cells, but this would reduce the efficiency very much (50 percent or more) and would require additional functionalities to make a distinction between full and partly filled cells at the edge of the network (terminals).

2) Queueing Delay: This parameter is influenced by the ratio between the information field size (Li) and the header size. Fig. 2 shows the maximum delay per switching stage as function of the information field length for a 2 byte header size and a link rate of 565 Mbits/s. We see that the queueing delay per stage for an effective load ($E1$, total load without the overhead introduced by the headers) of 80 percent and lower increases when this ratio increases.

3) Jitter and Associated Depacketization Delay: Services with a constant data rate require at the receiver a depacketization function which removes the jitter caused by the network on the interarrival time of the cells. This jitter is determined by statistically adding the maximum queueing delay per stage, over all stages of the network. As shown in Fig. 2, the maximum queueing delay per switching stage increases with the information field size; this relation also applies to the jitter.

4) Transmission Efficiency: It is clear that the larger the information field size is related to the header size, the header overhead will be smaller.

C. Conclusion

To select the cell size we considered the following items.

1) The transmission efficiency is influenced positively by using larger information fields.

2) The delay of the network is influenced negatively by using larger information fields increasing not only the jitter and the required smoothing buffer at the receiver but also the roundtrip delay especially in a network where a number of ATD/non-ATD conversions have to be provided.

3) At low loads (<0.80 effective load) the gain in delay, queue size and jitter of 16 information bytes is rather small compared to larger packets.

4) At higher loads (>0.85 effective load) the advantage of 32 information bytes over 16 bytes is more explicit.

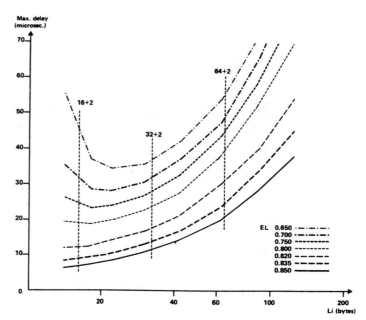

Fig. 2. Delay in function of the cell size.

5) The number of ATD/non-ATD conversions will be restricted.

Taking into account these arguments we selected an information field size of maximum 32 bytes information. This results in an ATD cell composed of 3 header bytes and 32 information bytes.

VI. A Universal Low-Layer Capability

A number of ATD proposals are proposing to introduce different low-layer capabilities to transport different teleservices. This can, e.g., be obtained by defining a special field in the ATD label, e.g., BSI (bearer service indication). This bearer service indication may be used in the network to obtain a different treatment of different applications, e.g., higher priorities for certain applications. However, we propose to have only one low-layer capability capable of transporting all different teleservices. This means that no BSI field is required in our proposal.

The basic reason for introducing this universal low-layer capability is its flexibility as a consequence of its service independence. When new services (of any type) will be introduced in the future, this low-layer capability can remain unaltered opposed to proposals where different low-layer capabilities are suggested (such as, e.g., quasisynchronous services) where possibly changes must be introduced. Indeed, multiple low-layer capabilities represent a first step away from the ideal of service independent networks since differentiation between services is made and resource pooling is reduced.

Another advantage of this universal low-layer capability is its simplicity, which has direct implications on the implementation complexity and the achievable speed both in the switching node and in the network protocol. Since only one low-layer capability must be provided in the switches, simpler implementations can be considered.

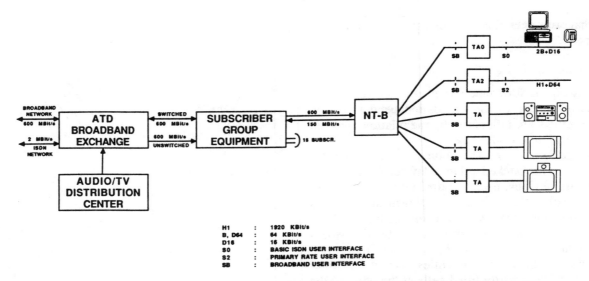

H1	:	1920 KBit/s
B, D64	:	64 KBit/s
D16	:	16 KBit/s
S0	:	BASIC ISDN USER INTERFACE
S2	:	PRIMARY RATE USER INTERFACE
SB	:	BROADBAND USER INTERFACE

ATD TARGET INTEGRATED BROADBAND ARCHITECTURE

Fig. 3. ATM network configuration.

VII. NETWORK PROTOCOLS

When a call request for a particular service is initiated by means of the signaling protocol, this request will require explicitly or implicitly a particular network performance (e.g., bit error rate, delay, · · ·). This required performance will be the basis for the acceptance or rejection of this connection based on the available sharable resources. If the requested performance can be guaranteed with the available resource, then the connection is accepted, if not it is rejected.

In order to guarantee the requested network performance certain suppliers may decide to use deterministic allocation, other suppliers may use statistic allocation; or they may internally decide to make a distinction between different support types. For the network, it is unimportant what solution is selected internally for implementation in the switching nodes, as long as the required network performance is met. None of these internal solutions should affect the network protocols and should therefore be standardized.

The connection oriented mode enables us to make the network protocol completely independent of the resource allocation and low-layer capability types. Indeed, the necessary resources in the switching fabric may be allocated (deterministically or statistically) at connection setup time, and if required, the necessary internal low-layer capability may be selected. The only thing which is important outside the switching fabric is that this particular virtual channel meets the requested performance parameters.

As soon as the call request has been accepted by the network, it will assign by means of the virtual *D*-channel protocol a virtual channel identifier (in the header of the cell), which will be used during the complete connection. There is no longer a need to refer in the cell header to the required performance parameters (e.g., by indicating that a special support type like a quasisynchronous type is re-

quired), since the virtual channel identifier contains implicitly all required information.

VIII. CONCLUSION

Based on all the arguments discussed above, the following system options have tentatively been selected as the basis of the Belgian broadband experiment.

1) Fixed cell length with minimal header functionality.

2) Short cell length, order of magnitude 32 bytes information and 2 bytes header.

3) High switching and transmission speed, in the order of 600 Mbits/s.

4) A single universal low-layer capability capable of transporting all services with an acceptable quality.

5) User-network and network-node interface independent of the implementation.

Although not all of the conclusions reached by our research group have already been agreed by everybody inside CCITT and CEPT, they will form the basis for the Belgian broadband experiment as described in [3] and shown in Fig. 3. If necessary they will be modified and adapted to the latest results of new studies and the discussions within CEPT and CCITT.

It is the goal of the Belgian experiment in 1992 to show the viability of ATM by connecting a limited number of lines and to gain experience with ATM networks.

ACKNOWLEDGMENT

I would like to thank the Advanced Switching Systems group of the Research Center of Bell Telephone Mfg. Co. for their valuable and numerous contributions to this work, and especially J. Bauwens for the numerous and interesting discussions.

REFERENCES

[1] W. Stallings, *Data and Computer Communication.* New York: MacMillan, 1985.

[2] A. Thomas, J. P. Coudreuse, and M. Servel, "ATD techniques: An experimental packet network integrating videocommunication," *ISS '84*, paper C32.2.

[3] P. Plehiers, M. Fastrez, J. Bauwens, and M. De Prycker, "Evolution towards a Belgian broadband experiment," *ISS '87*, paper B5.3.

[4] J. Turner and L. Wyatt, "A packet network architecture for integrated services," in *Proc. IEEE GLOBECOM '83*, pp. 45–50.

[5] A. Huang and S. Knauer, "Starlite: A wideband digital switch," in *Proc. IEEE GLOBECOM '84*, pp. 121–125.

[6] J. Hui, "A broadband packet switch for multi-rate services," in *Proc. ICC '87*, pp. 782–788.

[7] J. Bauwens and M. De Prycker, "Broadband experiment using asynchronous time division techniques," *Elect. Commun.*, vol. 61, no. 1, pp. 123–130, 1987.

[8] ——, "Bandwidth utilization for fixed and variable frame length," *CCITT*, SG XVIII, Contrib. D.988, Brasilia, Feb. 1987.

[9] ——, "Speed performance of frame relaying protocols for broadband ISDN," *CCITT*, SG XVIII, Contrib. D.993, Brasilia, Feb. 1987.

[10] M. De Prycker and J. Bauwens, "The ATD concept: One universal bearer service," presented at GSLB Sem. Broadband Switch., Albufeira, Portugal, Jan. 1987.

[11] P. Gonet, M. Servel, J. P. Coudreuse, "Packet-oriented techniques for integrated broadband networks: Experimenting the ATD switching concept," presented at GSLB Sem. Broadband Switch., Albufeira, Portugal, Jan. 1987.

[12] N. Cheung, "High-speed lightwave technology for future broadband integrated services digital networks," in *Proc. ICC '87*, pp. 162–165.

[13] S. Morita *et al.*, "Elastic basket-switching: application to distributed PBX," in *Proc. ICC '87*, pp. 789–793.

[14] H. Hartmann and E. Steiner, "Synchronization techniques for digital networks," *IEEE J. Select. Areas Commun.*, vol. SAC-4, pp. 506–513.

[15] H. Reiner, "Integrated circuits for broadband communication systems," *IEEE J. Select. Areas Commun.*, vol. SAC-4, pp. 480–487.

[16] ——, "Draft recommendation I.121," *CCITT*, SG XVIII, TD no. 49, Seoul, Korea, Feb. 1988.

[17] W. Verbiest, "Layered video coding model," *IEEE J. Select. Areas Commun.*, see this issue, pp. 1622–1631.

Martin De Prycker was born in Sint-Niklaas, Belgium, in 1955. He received the B.S. degree in computer science, the M.S. degree in electrical engineering, and the Ph.D. in computer science in 1978, 1979, and 1982, respectively, at the RijksUniversiteit Ghent.

Since 1982, he has been working in the Advanced Switching Systems Department of the Bell Telephone Advanced Research Center, currently as a Project Leader for the Belgian broadband experiment. Since 1985 he is also a part-time professor at the Boston University Brussels.

EXPERIMENTAL BROADBAND ATM SWITCHING SYSTEM

Yuji Kato, Toshio Shimoe, Kazuo Hajikano and Koso Murakami

FUJITSU LABORATORIES LTD.
Kawasaki, Japan

ABSTRACT

Asynchronous transfer mode (ATM) switching is considered the most appropriate for broadband ISDN because of its bandwidth flexibility and time transparency. This paper describes an ATM switching architecture based on a self-routing principle and an experimental system. It is constructed by connecting self-routing modules in a three-stage link configuration. We call this architecture multi-stage self-routing (MSSR). We confirmed that MSSR switching can deal with various media such as voice, data and video.

1. INTRODUCTION

Broadband ISDN will handle various kinds of services such as motion video, high-resolution images, and high-speed data transfer. The transmission speeds of these media will range from a few kbps to over 100 Mbps. ATM is considered the ultimate solution to broadband ISDN, and is expected to improve highway efficiency by assigning bandwidth on demand. It will increase the tolerance of traffic load variation, and offer customers an integrated access interface.

One requirement of ATM switching is to satisfy demands for differing quality of service (delay and cell-loss rate). And the quality of service should be met even under heavy traffic conditions. Our main objective is to obtain a unified access ATM switching system that does not use switching priority to control the service quality for the accessing media. We have developed a

multi-stage self-routing (MSSR) switching system to meet this requirement. This paper describes MSSR switching and a broadband ISDN prototype system.

2. MSSR SWITCHING
2.1 Configuration

The MSSR switching network is constructed by connecting self-routing switch modules (SRM) in a three-stage configuration as shown in Figure 1. Multiple routes between a first-stage SRM and a third-stage SRM enable the traffic to be routed efficiently and reduces switching delay. Furthermore, since a damaged second-stage SRM can be bypassed, high reliability is maintained.

A VCI converter (VCC) is placed at each input highway. This VCC identifies a cell by its virtual channel identifier (VCI), replaces the VCI at the input highway with the VCI of the destination output highway,

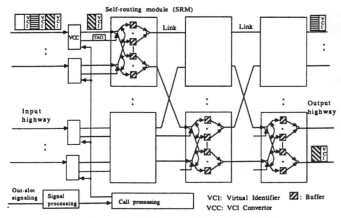

Figure 1   Multi-stage self-routing switching

EH0301-2/91/0000/0184/$01.00 © 1988 IEEE

and generates a tag containing the information about the path in the switching network. Each SRM switches cells to the outlet according to the tag without external software control. The signal processing unit deals with the signal between nodes or between a node and a subscriber.

The MSSR switching system has the following features: 1) Variable bit-rate switching based on the self-routing principle. 2) Contention control by buffer dispersion and convergence in each direction. 3) Reduced delay and real-time operation in a multi-stage configuration. 4) Building block configuration consisting of small-scale self-routing modules. 5) Suitability for LSI construction.

2.2 Operation

Figure 2 shows MSSR operation. In the call set-up phase, call processing (CP) determines an output highway, the VCI value at that output highway, and the path through the switching network. It also assigns a path tag and writes the tag and the output VCI to the VCI conversion table (VCIT) in the VCC.

In the information transfer phase, the VCC searches the VCIT using the incoming VCI, replaces this VCI with the value in the table, and transfers the cell with the tag to the switch network. Each SRM routes the cell according to its tag without external software control.

In the SRM, a cell arriving at an inlet is stored in a FIFO buffer corresponding to the outlet denoted by the tag. Outlet contentions may occur if several cells destined for the same outlet arrive simultaneously at different inlets. This is resolved by storing the cells in a buffer, as described above. The stored cells are taken out and transferred to the next SRM one at a time.

2.3 Performance

We have reported design and performance results using the MSSR switching system[1][2]. With a 90-cell buffer at the SRM outlet, the MSSR carried a 90% load with a 10 ms maximum queuing delay through 10 nodes. The end-to-end cell-loss rate was 10^{-9}. This is the same quality as that of conventional circuit switching networks. Even if the load is unbalanced, about 90% of the load can be carried if the number of internal links is increased 1.3 to 1.4 times.

Around 60 byte as a cell length is preferable from the view point of highway efficiency.

3. EXPERIMENTAL SYSTEM
3.1 System Configuration

A block diagram of the experimental ATM MSSR switch and its specifications are shown in Figure 3 and Table 1. For the initial stage of broadband ISDN, a hybrid switch configuration using both ATM and the circuit switching is adopted. The circuit switching system is used mainly for distributing high-quality full motion video because it easily enables high-speed bearer switching and is inexpensive for fixed bit rate switching. The system uses time division circuit switching

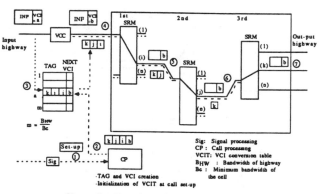

FIgure 2   Operation of multi-stage self-routing switch

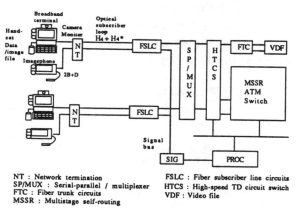

Figure 3   The experimental system

Table 1 B-ISDN experimental switching system specification

Media		Voice , Data , Image , Video
Switching system		ATM switching, Time-division switching
Switching network	ATM switch	MSSR (highway speed: 150 Mbps) cell length: 4 (header) + 52 (information) byte
	Circuit switch	TST (switching speed: 224.6 Mbps) Highway capacity: 150 Mbps x 12
	Device	CMOS, ECL, GaAs
User network interface		Layer 1 : ATM (150 Mbps x 2) Layer 2,3 : Dch protocol

for H4 rate services and ATM switching for services that are less than the H4 rate. All services will be integrated in full ATM switching by increasing the switch capacity and by adding a broadcast capability[3].

3.2 Subscriber Access Structure

The system supports a circuit switched channel (H4) used mainly for broadcast of high-quality full motion video (150 Mbps), and an ATM channel (H4*) for services with rate below 150 Mbps. The D-channel is permanently multiplexed into the H4* channel. The transmission capacity of the H4 and H4* channels is 150 Mbps.

The number of H4 and H4* channels will depend on the customer. For example, the H4* channel will probably be in demand among business customers, and the H4 channel will probably be popular among residential customers. The channel structure in our system is   H4 + H4*(D), and the channels are multiplexed in the payload of the SONET frame. Cell header error detection is needed to prevent miss-routing, but is not implemented in this system.

Figure 4 shows the cell format. Cells consist of 4 bytes of header which includes the VCI and 52 bytes of user information.

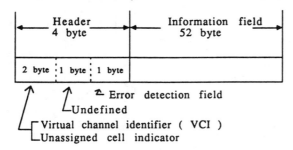

Figure 4   Cell format

Cell-synchronization is extracted from an external SONET frame.

3.3 Switching Network Configuration

The switching network is composed of an ATM switch, an H4-rate time-division circuit switch, and a line-concentration and cell-multiplexing switches which are distributed over the subscriber line circuits.

(1) ATM switch

1.8 Gbps highway from the circuit switch is de-multiplexed into twelve 150 Mbps highways. The ATM switch operates at about 18 Mbps in 8-bit parallel. We have designed with CMOS and TTL technologies.

This switch is composed of VCI convertors (VCC) and self-routing modules (SRM). The VCC converts incoming VCIs into outgoing VCIs and adds a tag containing the switch network path information. The SRM switches the cell according to the tag information.

The SRM configuration is shown in Figure 5. An SRM has eight 150 Mbps highways. Eight buffer block packages are connected with parallel bases. Cells arriving at the SRM are distributed to each buffer block package, and filtered into one by comparing the tag. The buffered cells are then led to the outlet highway cell-by-cell, avoiding cell contention from the other inlets.

Active buses reshape and retime the transmission pulses so that the MSSR can be expanded up to a 64 x 64 switch network with 8 x 8 switches in the MSSR

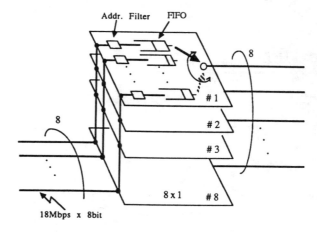

Figure 5   SRM configuration

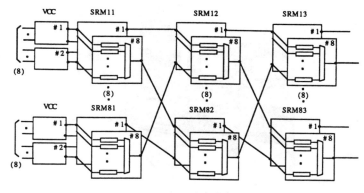

Figure 6   64 x 64 switch connection

configuration, and even more expansion is possible by enlarging SRM and increasing the number of SRMs. Figure 6 show the module connection of a 64 X 64 MSSR switch network.

(2) High speed TD circuit switch

The circuit switch is a T-S-T time division switch of 150 Mbps bearer and performs point-to-point switching, distribution of high-quality video, the separation of H4 and H4* channels, and routing of H4* channels to the ATM switch. Non-blocking distribution is achieved by duplexing the secondary T-switches. Each highway accommodates twelve H4 and H4* channels and throughput is 1.8 Gbps. The 8-bit parallel switch operates at 224 Mbps. We used GaAs, ECL, and microstrip on the backplane to achieve high-speed operation.

(3) Line concentration and cell multiplexing switch

Each fiber-subscriber line circuit (FSLC) is connected by the active buses as well as the SRM to operate at a speed of 224 Mbps, which has twelve 150 Mbps time

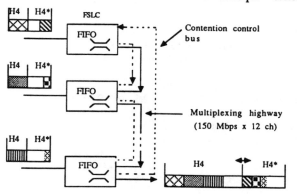

Figure 7   Subscriber line concentration

slots. H4 channels from subscribers are assigned to any time slot on a per call basis, and subscriber cells are asynchronously multiplexed into a time slot.

Figure 7 shows the line concentration scheme. It is possible to multiplex cells from the different subscribers into the same time slot for highway efficiency. In this case, multiplexing is on a per-cell basis, with tokens ensuring equal access among the different subscribers. Figure 8 shows how the token resolves the contention during cell multiplexing. The channel the FSLC uses for multiplexing is predetermined. Tokens which have a certain time slot number in the active bus are moving in a cycle among the FSLCs. If the token fined the cell to be multiplexed in the FSLC, and the time slot number matches with the predetermined channel, the tokens give it a right to cell multiplex. Multiplex switches are distributed over each FSLC.

3.4 Signaling

Subscriber signaling is carried through a dedicated virtual channel in H4*, and is used for connection set up and for selecting video programs. We use a modified D-channel protocol.

3.5 Broadband terminal

Broadband terminals consist of personal computers, TV monitors, cameras, and terminal adapters. Video and voice signal coding, signaling protocol termination,

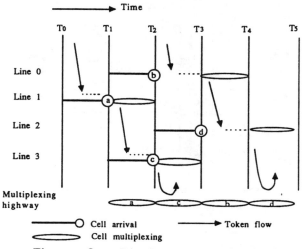

Figure 8   Distributed contention control

and cell assembly and disassembly are performed at the terminal adapter. Full motion video is coded at 75 Mbps (NTSC, non-compressed), and voice is coded at 64 kbps. Multi-media communication services with video, voice, and data are supported with a multi-window facility.

3.6 Experimental results

Figure 9 shows the experimental broadband ISDN system consisting of the ATM switching system and the high speed synchronous optical transmission system, named the "Shuttle Bus" <4>.

Figure 10 shows the experimental ATM switching system, which is composed of two cabinets for the switching system and two broadband terminals. Each cabinet is 2 m high, with 3 shelves for printed circuit boards and one shelf for power. The cabinet on the right contains the subscriber line circuits and signaling equipment. The cabinet at the left contains the switching fabric which consists of sixteen 150 Mbps ATM highways and eight 1.8 Gbps STM highways.

The power dissipation of the ATM switching fabric is 66 W per highway. That of the circuit-switching fabric is 188 W per highway.

Figure 11 shows an example of the operational waveform on an ATM highway, which shows the multiplexed cell format.

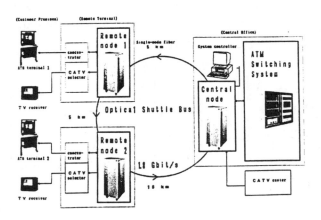

Figure 9 Experimental broadband ISDN system

Figure 10 ATM switching system

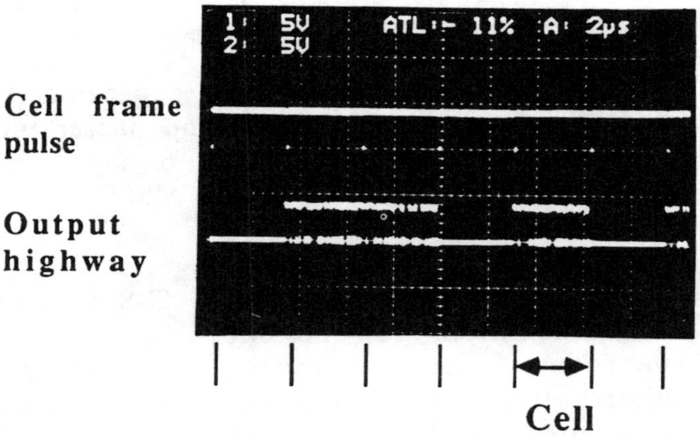

Cell frame pulse

Output highway

Cell

Figure 11 ATM switch operational wave form

4. CONCLUSION

We have developed an experimental broadband ISDN ATM switching system. We adopted an MSSR switching system for ATM and a T-S-T circuit switching system for distribution of high-quality full motion video. We confirmed that MSSR switching could deal with various media such as voice, data and video, and compiled data for developing LSIs for the MSSR switching system. The 224 Mbps TD switching network uses high-speed ECL and GaAs is obtained.

References
<1> K. Hajikano et al., "Asynchronous Transfer Mode Switching Architecture for Broadband ISDN", ICC '88, 29.3, June 1988
<2> S. Abe et al., "Traffic Design for an Asynchronous Transfer Mode Switching Network", ITC '88, 1.3A.4, June 1988
<3> T. Nishino et al., "Broadband Switching System Configuration and Access Protocol", ICC '88, 29.6, June 1988
<4> N. Fujimoto et al., "Experimental broadband drop/insert/cross-connect system", GCOM '88, 30.6, Nov. 1988

Layered ATM Systems and Architectural Concepts for Subscribers' Premises Networks

JAN P. VORSTERMANS AND ANDRÉ P. DE VLEESCHOUWER

Reprinted from *IEEE Journal on Selected Areas in Communications*, Vol. 6, No. 9, December 1988, pages 1545-1555.

Abstract—This article first reviews the essential elements of the ATM (asynchronous transfer mode) technique and presents a framework for the definition of ATM, based on the ISDN protocol reference model. A layered functional model of an ATM open system is presented and the relation with existing packet techniques is studied. The use of ATM in the subscribers' premises network (SPN) is analyzed. A two-tier SPN architecture is proposed covering both residential and business environments, in line with the modeling principles outlined in this paper. The merits of this architecture in the field of standardization are emphasized. Some implementation issues and the relation with the reference configuration are discussed.

I. Introduction

ONE of the major goals today in the world of telecommunication is the realization of a single and worldwide integrated broadband communication (IBC) network. Up to now, new networks were developed whenever a new service became relevant. This hardly seems an efficient and cost-effective way to meet emerging communication needs. The concept of IBC, based on the definition of *B*-ISDN, is to provide one network capable of handling all services, narrow-band or broadband, dialogue, or distributive. In Europe, the research program RACE was setup for this purpose. The objective of this program is the introduction of IBC taking into account the evolving ISDN and national introduction strategies, progressing to community-wide services in 1995. Since this program is aimed at introducing services, the subscribers' premises network (SPN) must certainly be considered. The SPN is the network on the subscribers' site, it can be very simple or extremely extensive for environments going from small households to large business enterprises. The subscribers' premises network may provide local switching because internal traffic must be considered (e.g., video recorder to TV set, PC to printer, intercom). Such a network, whatever the scale, is seen as one SPN if it behaves as one logical entity towards the public network.

ATM (asynchronous transfer mode) is a new and evolving transfer mode receiving broad attention. Information is divided in short, fixed length cells which are multiplexed and switched based on their header content. The

Manuscript received October 13, 1987; revised May 27, 1988. This work was supported under research grants by the European Community and the Belgian Administration RTT.

The authors are with the Applied Research Department, ATEA, B-2410 Herentals, Belgium.

IEEE Log Number 8824397.

key advantage of ATM is its flexibility. ATM is now generally seen as the ultimate solution for *B*-ISDN. Though the use of ATM in the SPN has many advantages in itself, the introduction of ATM in B-ISDN opens up a wide perspective. Indeed, the advantages of the ATM technique can be fully exploited over the whole connection. Using ATM in the SPN has major consequences on the SPN architecture, i.e., ATM is very well suited to support distributed concepts.

II. Asynchronous Transfer Mode

In CCITT, ATM is now accepted as the final transfer mode for *B*-ISDN [1]. In ATM, a new and evolving packet-oriented transfer mode [2], [3], the information to be transferred is divided in small, fixed size blocks called cells (Fig. 1). These cells are transmitted over a virtual circuit in a slotted operation with respect to the instantaneous need for information transfer. This implies that the customer applications in the terminal equipment define the actual transmitted bit rates. A virtual circuit is an end-to-end connection established and released at call connect and disconnect. A virtual circuit number is associated with this connection, this number is contained in the cell header. Since the connection is established at call setup, using setup control procedures, no further routing information is required in the cell header. Bandwidth for a virtual circuit can be varied on a dynamic basis, variable bit rate services can be accommodated. This implies that the load on the network is time dependent and that cell buffering is required. The size of the buffers can be limited by good network design and by avoiding extreme load conditions.

An ATM network must be designed to carry all services. This implies that, since in general, no network-wide synchronization is provided, other forms of service synchronizations must be looked at [4], [5]. ATM is a universal basic service, it provides the same network service to all user services. There is only one such service required (since it supports all user services) making ATM a unique basic service. The use of optical technology and VLSI in *B*-ISDN must increase the network reliability such that error detection and correction protocols on links inside the network are no longer required. However, in some cases, e.g., signaling, protocols may be required to assure good operation. These can, however, be seen as add-on capabilities on top of the ATM service. For these reasons, no processing is provided in the ATM network

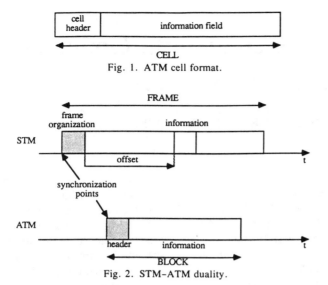

Fig. 1. ATM cell format.

Fig. 2. STM–ATM duality.

and very high bit rates are achievable. These concepts of ATM in a *B*-ISDN greatly simplify the network implementation and improve future safety.

It is said that ATM has circuit-like characteristics. Fig. 2 gives a representation of the duality between ATM and STM (synchronous transfer mode). In STM, information is divided in small fixed length words and identified by reference to a clock. Once the beginning of a new frame is detected (synchronization point) and the organization of the frame is known, the information is identified through specification of a time offset. In ATM, the information is organized to be transferred in blocks of which the recurrence depends on the instantaneous required bit rate. Blocks of information are identified by a block header. When a synchronization point is detected (beginning of a new block), the information is identified by the content of the block header. The great advantage of ATM over circuit-oriented techniques is its flexibility. This flexibility allows for dynamic bandwidth allocation, a feature that is welcomed by the end users. Integration of all services, which is greatly enjoyed by the operators, is another consequence of ATM flexibility.

ATM is a competitive technique, even on short and medium terms. It is also very well suited for use in the SPN. As ATM is introduced in B-ISDN [6], aspects concerning the SPN structure, bandwidth assignment, terminals, and terminal interfaces lead to the conclusion that it is essential to use ATM down to the terminal [7], [8]. On the long term, ATM is the best solution because of the inherent flexibility. An adaptable system can be developed using new architectural concepts (e.g., distributive control) leading to a new generation network. ATM must also support a smooth evolution to *B*-ISDN over several decades.

III. ATM OPEN SYSTEMS

A. ISDN Protocol Reference Model

The open system interconnection (OSI) basic reference model [9] is a fundamental concept for architectural modeling, still gaining strong support in a wide area of appli-

cations. Though the OSI reference model is not perfect, it is the best widely accepted model available. As a consequence, the use of OSI in *B*-ISDN and SPN is imperative. The development of ISDN showed that certain facilities are not provided in the current OSI reference model. The OSI model was oriented towards data services, in ISDN a multitude of services has to be supported. This requires facilities not needed for data services, such as the following [10]:

—out-of-band control processes
—layer service definitions for nondata-services
—associated connections (e.g., related voice and video connections, a two-way dialogue connection associated with a one-way distributive connection)

Therefore, the ISDN protocol reference model, I.320 [10], was introduced bringing together the OSI modeling principles and the ISDN requirements. It is used to model information flows including user information (information transferred between users) and control information (containing network control functions such as connection establishment and release, connection characteristics negotiation). All references in this paper to the OSI model must thus be seen in the ISDN context as specified in the protocol reference model.

A protocol block (Fig. 3) is identified which is used to describe various elements as well in the public network as in the SPN (e.g., network termination, exchange termination, terminal equipment). In a protocol block, three planes are identified as follows:

—*U*-plane for user information and associated protocols,
—*C*-plane for control information and associated protocols, and
—*M*-plane for management associated with the transfer of user and control information. Where management information exchange is required, the *U*- and *C*-planes are used to convey this information.

In each plane, OSI layering principles are applied; layers may be null or empty, i.e., not containing protocol functions. The services provided by the null layer to the layer above are then the same as the services provided by the underlying layer to the null layer. The primitives at the higher layer interface are mapped directly onto the primitives at the lower layer interface. The protocol stacks in each plane are described in an independent fashion. The protocols in the user plane are used to transfer information between user applications, the protocols in the control plane for information transfer between control systems. One application of such a control system is to control user plane connections. This implies that functions in the *C*-plane can be "called" (not unlike a subroutine) from the user plane. This nesting principle is important, e.g., for the setup of connections.

B. Modeling of ATM Systems

1) Layering: The concepts developed in the ISDN protocol reference model will be used here to describe the lower layers of open systems in an ATM network. System

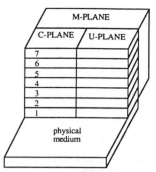

Fig. 3. ISDN protocol reference model—protocol block.

and network management are not discussed. It is recognized that these are important and very critical issues but they fall outside the scope of this paper. These modeling concepts are based on the layering principles of the OSI model. One of the fundamental ideas in the definition of layers is the strong principle of abstraction. This principle states that a layer is created where there is a need for a different level of abstraction in the handling of data [9]. The specification of the size of the layer 1 service data unit in function of serial or parallel transmission (1 or n bits) is inconsistent with this abstraction principle. Parallel or serial transmission can be seen as two different ways of providing the same service, the transmission of a group of bits. It is therefore suggested to consider that the layer one service data unit be k bits, whereby k is independent of the physical medium [11].

2) Layer 1: Layer 1 is divided into three sublayers. The lowest sublayer 1A (PHY) deals with the physical medium aspects. Sublayer 1B (MAC) controls the access on a medium shared by several systems, and the highest sublayer 1C (ATM) handles cell-based multiplexing and relaying. The ATM functions and access control functions are separated in different layers to facilitate the discussions on the ATM cell header functions and multidrop interfaces.

a) Sublayer 1A: PHY: The PHY sublayer contains all details of the transmission medium. It deals with the following:

—medium characteristics
—transmitters and receivers
—code conversions.

Techniques such as frequency and wavelength multiplexing, together with serial or parallel transmission, are functions provided by this sublayer. In B-ISDN the physical medium is optical fiber. This sublayer thus deals with the characteristics of the fiber, optical emittors and receivers, electro-optical conversions. This sublayer is responsible for the transmission of fixed size data units, without enhancement. It provides a fixed bit rate service to the above sublayer. There is no distinction in the transmission of user and control information. This results in one common sublayer protocol for both *U*- and *C*-plane where no distinction is made in the service provided to the higher sublayer. Functions performed in this sublayer include the following:

—definition of the transmission medium
—transmitter and receiver frequencies
—line coding (and scrambling), line driving (e.g., laser/LED modulation)
—line termination
—power budget, power levels
—transmission quality supervision (monitoring and testing)
—receive indication (e.g., symbol violation)
—regeneration and bypassing
—bit synchronization
—mechanical attachment
—power feeding
—activation/deactivation.

b) Sublayer 1B: MAC: Several systems can be connected on a shared medium. This results in a possible contention problem which is solved by this sublayer. The MAC sublayer can in general be described as the set of functions providing access to the medium. This sublayer provides services common for the *U*- and *C*-planes. The 1B-sublayer performs:

—definition of the medium access protocol, taking into account topology, geographical constraints and system margins
—MAC data unit synchronization
—corrupt cell indication (i.e., collision)
—QOS monitoring.

Fig. 4 gives an example of data flow in the MAC sublayer for the case where several systems are connected to a single medium. The access medium protocol indicates to the system a permission to transmit. A sequence of such transmission opportunities creates a channel on the access medium; this channel is an access channel.

c) Sublayer 1C: ATM: The ATM sublayer contains all details of the ATM technique and is unaware of the underlying medium and topology. This sublayer provides multiplexing in a flexible manner (in contrast with the PHY sublayer) and offers to layer 2 a variable bit rate service. The ATM sublayer transfers fixed size information blocks, the (1C)-SDU's. A connection between two (1C)-entities is a virtual channel. To form the (1C)-PDU, protocol control information (PCI) is added. This control information contains the virtual channel identification (VCI), error detection or forward error control on the PCI and other functions now under study (e.g., priorities). The sublayer (1C)-PDU is called a cell. A cell has a fixed size and consists of a fixed length information field, (1C)-SDU, and a fixed size header field, (1C)-PCI. The cell header format must be unique at a given interface, e.g., at all the user-network interfaces the header format must be the same, but this format could be different from the format used at the network-node interfaces. User information is not organized in fixed short length units. Data are segmented in fixed information blocks in a higher layer where a segmentation process must be provided that creates fixed information fields from the information emitted by the source. Of course, the inverse operation is needed at the destination side. The information field length must

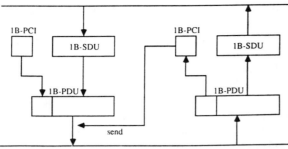

Fig. 4. MAC sublayer data flow.

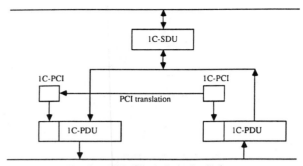

Fig. 5. ATM virtual channel relaying.

be the same everywhere throughout the network in order to avoid resegmentation. The ATM sublayer allows individual cells to be relayed through the network, based on the content of their header field, without processing. Hence, it allows cells of different connections to be multiplexed. The header indentifies the connections they belong to.

Again, no difference is seen in the handling of user and control information. This again results in one sublayer for buth U- and C-plane where no distinction is made in the service provided to the higher layer. All U- and C-plane connections make use of the virtual channels. The functions of this sublayer include the following:

—transfer of information over a (1C)-connection, a virtual channel
—virtual channel identification
—virtual channel relaying (see Fig. 5)
—multiplexing/demultiplexing by means of header information
—error control on the PCI
—synchronization and reconstruction of the cells.

3) Layer 2: This sublayer handles the enhancement of the service, if required, provided by layer 1, in order to meet the QOS (quality of service) parameters imposed through the negotiation process at connection setup time. The functions in this layer are different for the U- and C-plane. In the U-plane, this layer is transparent. Because of the high environment reliability (use of optical technology and VLSI) a low-bit error rate is foreseen. Good design (e.g., buffer lengths) and load control must result in a very low cell loss probability. Another aspect is that high throughput services do not need error control and flow control functions because of the limited lifetime of the information (i.e., television). As a result, no protection and flow control protocol on a link-by-link basis is required inside the network. Some proposals, however, include a layer two protocol in the user plane protocol stack for sensitive data services. A service can then select whether it needs this protocol or whether it wants a transparent layer two (protocol selection).

In the C-plane, an error detection/correction and flow control protocol must deal with overload situations, cell loss, problems resulting from mobile communication. Such a protocol can be derived from the ISDN LAP-D protocol. It is essential that the service to the network layer is compatible to that defined in ISDN. The level 2

functions in the C-plane include the following:
—error detection
—error correction
—flow control
—link control and identification
—multiplexing/demultiplexing
—frame structuring
—segmentation in fixed blocks (of given size).

4) Layer 3: The network layer handles routing and provides network connections. The functions in this layer are again different for the U- and C-plane.

This layer is transparent in the U-plane. User information is sent transparently over a virtual circuit that has been established at call setup time. No further routing information is required. Selection of protocols in the U-plane is for further study. Proposals exist to support semiconnectionless network services. Layer four messages are then independently sent through the network, possibly via different paths. This is accomplished by setting up a virtual circuit per layer 4 message to be sent. These "add-on" functions (and those realizing protocol selection in layer 2) are fully compatible to the ATM model outlined, and do not affect the basic ATM-oriented layer 1. These examples of variations on the basic architectural concepts are given here to demonstrate the flexibility of the ATM concept. The layer 1, containing the ATM features, always remains unchanged. It acts as transfer method that inherently includes multiplexing and switching.

In the C-plane the functions include capabilities for call establishment (e.g., routing) and release. Since B-ISDN must evolve smoothly from the existing ISDN, the layer 3 must be a compatible version of the ISDN protocol I.451, with enhancements to be studied.

C. Information Flows

Fig. 6 shows the U- and C-planes of a protocol block in an ATM environment. User information passes transparently through layers 2 and 3. It is organized in cells in the transport layer at the transmitter. The information is then transferred on basis of these cells and is passed to the higher layers transparent to layers 2 and 3 at the receiver end. There the information is reorganized in the format of the user information at the transport layer. The end-to-end performance can be upgraded by other layer 4

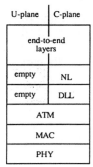

Fig. 6. ATM protocol block: *U*- and *C*-planes.

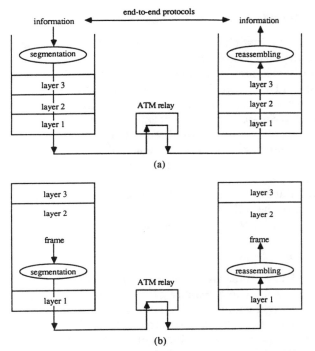

Fig. 7. Information flows. (a) User information. (b) Signaling information.

functions if necessary [Fig. 7(a)]. In the control plane, information flow is restricted to signaling (no packet data). The layer 3 protocol must be compatible to the ISDN network layer protocol (I.451) to assure a smooth evolution to B-ISDN. Control information at that layer is organized in packets or messages and passed to the layer 2 where it is organized in frames. The layer two frames are then segmented in cells and transferred over a virtual circuit [Fig. 7(b)].

IV. RELATION WITH OTHER TECHNIQUES

A. Packet Switching Techniques

X.25 is a packet switching technique based on the X.25 PLP (packet level protocol) on layer 3 and the LAP-*B* protocol in the layer 2. LAP-*B* is a datalink oriented layer 2 technique, based on HDLC. Several connections are multiplexed at layer 3 by means of the LCN (logical channel number). This means that both layers 2 and 3 must be processed in every switching node.

The *D*-channel is the common signaling channel for ISDN. Here, the LAP-*D* protocol, also based on HDLC, is used at layer 2 (I.441). This allows several datalinks to share a common channel: *s*-data for signaling of several terminals, *p*-data for packet data, and other. The datalinks are identified by an address, composed of a TEI (terminal endpoint identifier) and a SAPI (service access point identifier). Usually, there is only one logical connection per address, TEI + SAPI. The layer 2 address can be used as the identifier that discriminates between the different logical connections. If packet data is transferred over LAP-*D*, the X-25 PLP can be used at layer 3, with a fixed dummy LCN. Multiplexing is performed by means of the layer 2 address, which acts as a virtual circuit identifier. This layer 2 is then the lowest layer where packet multiplexing is possible in the circuit oriented ISDN environment.

Today no packet technique is provided in ISDN, but *B*-channels are used to convey data in a transparent way towards existing packet networks, based on X.25. In this case, there is hybrid signaling: ISDN signaling to set up the *B*-channel, and the normal procedures inside the packet network to control the packet connections; this is not a generic ISDN procedure. For some time now there is a strong tendency to adapt LAP-*D* for use inside *B*- and *H*-channels. Proposals exist to use the layer 2 address to distinguish between the logical channels. Layer 3 is suggested to be a relevant subset of X-25 PLP, but with the LCN set to a fixed dummy value. Efforts to standardize this kind of integrated packet mode inside ISDN are close to finalizing.

Two techniques are proposed for the use of packet switching at the layer 2 in the *B*-, *D*-, and *H*-channels in ISDN: frame relaying and frame switching. In order to understand the difference let us first analyze the layer 2 functions, as performed by HDLC-oriented protocols (Fig. 8). The bottom sublayer 2A provides for the addition of flags and zero bit insertion for frame synchronization. Hence, we call it the framing process of HDLC. The 2B sublayer provides for frame multiplexing, routing, and switching in every node, based on the address, identifying the different connections. This address is composed of a TEI and SAPI in case of a *D*-channel. This sublayer provides for the multiplexing process. The 2C sublayer handles error detection on the frames. Faulty frames are rejected without retransmission, this is the error detection process. The sublayer 2D finally performs all the functions that allow the error free transmission of frames in the right sequence. This is the error protection process. It also contains the flow control and the window rotation.

Between connection endpoints, or in some intermediate nodes, all protocol steps, 2A till 2D are processed, in order to provide full layer 2 service. This is called frame switching. However, in most intermediate nodes only the functions up to 2C are performed. This process is called frame relaying. Frame relaying significantly reduces the complexity of the intermediate nodes and increases the speed of the packet networks. In fact, this is possible only,

2 D	Retransmission - Flow Control Window Rotation
2 C	Error Detection
2 B	Routing - Switching Multiplexing (address)
2 A	Framing Process

Fig. 8. Functional decomposition of HDLC-oriented protocols.

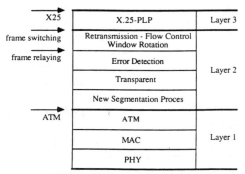

Fig. 9. ATM and HDLC-oriented packet techniques.

due to the dramatically improved transmission quality of modern digital links, compared to many existing packet networks.

B. Relation with ATM

Previous sections indicate the important role of HDLC based datalink protocols and switching systems. Supporting these protocols in an ATM environment is essential. In ATM, layer 1 provides for a packet-like variable bit rate service, with label-oriented multiplexing and switching. No higher layer protocol functions are used inside the ATM network. Hence, it is possible to transfer the $2A$ and $2B$ functions towards the ATM sublayer. Indeed, where $2B$ is the lowest possible packet multiplexing layer in STM networks, this is an inherent capability of the ATM sublayer $1C$. All connection identifiers on the HDLC and X.25 process become redundant and can be set to a fixed or dummy value. Layer 2 is linked to layer 1 by means of a segmentation process, which has to be defined. This comes instead of the HDLC framing process and will become subject of standardization. The protocol stack for the different techniques in relation with ATM is shown in Fig. 9. The HDLC layer 2 address and the X.25 layer 3 on LCN are set to a fixed dummy value. A new segmentation process has to be defined. The combination of this segmentation process and the other layer 2 functions form a new layer 2 protocol for an ATM environment.

This new protocol is directly derived from HDLC as follows:

—the $2A$ framing process is replaced by a ATM cell oriented segmentation process,

—the $2B$ process is transparent,

—the $2C$ and $2D$ control functions can be maintained or simplified considering the broadband environment.

This simpler protocol must support an identical service to layer 3, and is tailored to an ATM environment. It can make use of the ATM cells more explicitly, and must allow for operation at the high speed encountered in broadband networks. By simplifying the layer 2 process, the silicon implementation can get smaller, which allows integration inside next generation super-VLSI circuits. This helps reducing the cost of broadband networks and terminal equipment.

V. SPN Architectures in an ATM Environment

A. Introduction

In the first chapters of this paper, ATM was discussed as the final transfer mode for B-ISDN. A modeling framework was presented, allowing a clear definition of ATM and ATM related issues. The use of ATM in the SPN has of course severe impact on the SPN architectures. In this paper, an architectural concept is presented as follows:

—dealing with the SPN requirements

—allowing definition and standardization of interfaces

—based on ATM as defined in the chapter on ATM open systems.

Interfaces are proposed, independent of the implementation of the SPN, for the S and T reference points to demonstrate the value of this architectural concept.

B. SPN Requirements

Identification and evaluation of the SPN requirements are essential in the study of SPN architectures. If a generic architecture is to be defined, it must be able to deal with the most stringent requirements. It is evident that the SPN requirements are dependent on the environment. An environment is characterized by the number of users, geographical spreading, physical constraints, services needed, · · ·. Although the number of environments is virtually unlimited, some classes can be identified as follows.

—The residential environment combines the environments where a very limited amount of people, i.e., a family, use broadband services mainly for entertainment purposes. It includes small business operations.

—In a business environment, the emphasis is more towards dialogue services. A further division is made as follows:

• medium office (100 people)

• large office (1000 people).

—The factory environment class covers business environments with exceptional physical constraints.

Other environments put so specific requirements on the SPN that they cannot be grouped together like hotel, hospital or military environments.

A first type of requirements, service requirements, deal with the service mix for a specified environment, and the consequences of having to support these services. This category of requirements covers the bit rate to be supported, the information transfer characteristics such as mean and maximum delay and delay jitter, error performance, throughput, and buffer memory. The bit rate in the SPN is a very important service requirement in light of the interfaces to be defined. In the service analysis carried out in [12], a certain service mix was assumed for each environment class. Realistic service characteristics were assumed, though it is difficult to estimate the evolution of coding techniques that influence the required bit rate of a specific service. An acceptable peak rate of 34 Mbits/s was assumed for the video services, with an average rate of 10 Mbits/s. For the residential environment, a bit rate of approximately 150 Mbits/s (3 to 4 video channels) should be supported by the incoming and outgoing links of the SPN. HDTV was not included in the service analysis. For the office environment, the bit rates to be supported vary from 150 to 680 Mbits/s for the medium office, to 1 Gbit/s for the large office, depending on the need for video distribution and communications. Due to these high traffic loads, configuration constraints and redundancy requirements, a shared physical medium as a single path to the terminals is not acceptable. The total traffic must be divided over several paths, one per terminal or per terminal cluster.

A second type of requirements, structural requirements, include aspects of flexibility, reliability, physical and operating performance, and cost.

Flexibility indicates the ability of the SPN to deal with changes. Four aspects are identified as follows.

—Adaptability measures how the SPN copes with changes that do not alter the global scale of the SPN (e.g., new wiring). This requirement is very important in the terminal area, both for residential and office environments.

—Expansibility expresses how the SPN can grow, e.g., the introduction of new services increases traffic and thus the bit rate to be supported, additional terminals increasing the SPN scale.

—Mobility identifies the flexibility in moving terminals and users and can be realized through a universal terminal access method, a universal terminal interface.

—Interworking specifies how and with which other networks the SPN can interface. This is important where other large scale networks already exist (e.g., LAN).

Reliability considers the sensibility of the SPN to errors: bit errors or bursts, terminal failures, EMI problems, and human induced errors. The requirement of reliability is mainly important when a relative large number of people are affected by the error or in cases where special care must be taken to assure good operation of the SPN (e.g., fire departments).

Physical performance is concerned with an optimum use of the physical medium. It includes aspects relating to cable length, power splitters, coding efficiency and has a great influence on the overall hardware cost. The operating performance deals with installation and maintenance. In the terminal environment installation and maintenance must be very easy so that changes and reconfigurations (flexibility) can be carried out fast and cheap.

Cost is of course a major requirement. In the residential environment, low cost is very essential. In the office environments, the initial cost must be reasonable, but at the same time the incremental cost must be rather low to allow a modular build-up according to the emerging needs.

All these requirements can be met by the SPN with good design. The use of ATM in the SPN is intrinsically advantageous. The network is service independent, ATM is a straightforward and flexible technique designed to support all services. The service bit rate is no longer coupled to the network bit rate. This makes bandwidth allocation very flexible. No fixed channels are offered but flexible virtual circuits where the service determines the call characteristics. This helps to meet the service requirements and makes the SPN future safe. No bandwidth is wasted by a service that does not match a fixed channel. It also allows for further evolution in the bit rates of the services, and for the introduction of new services. An SPN based on ATM supports variable bit rate services, the quality level of which is negotiable. Taking into account the statistical variations in the bit rates of the services, statistical gain can be expected when a number of virtual circuits are multiplexed on a single connection. In the small scale SPN, the number of virtual circuits is not sufficient to guarantee a worthwhile gain. But in the large installations, statistical gain may become relevant. The distributed and dispersed architectural consequences of ATM are relevant to meet the requirements of flexibility and reliability. ATM integrates switching and multiplexing and is thus particularly well suited for dispersed switching (LAN like). No geographical constraints are imposed on the placement of the control elements because connections are realized through virtual circuits. The communication between control processes only use up relatively little bandwidth. This allows to build up a dispersed control structure. Dispersed switching and control also helps to establish an open architecture, capable of adapting to a great variety of needs in the SPN.

C. Two-Tier Architecture

1) Concepts: The SPN must be capable of carrying high traffic loads and of providing a very flexible structure at the lowest possible cost. Further constraints on the implementation of an SPN include the following:

—the need for a passive terminal interconnection network

—universal sockets

—a universal interface for a wide variety of terminals

—low initial and low incremental cost

—flexible structure, easily expansible and adaptable to emerging user needs.

There is a clear incompatibility between the need for a passive terminal interconnection network and the high traffic to be carried by that network. Such high traffic loads can only be carried by a hub switch with a point-to-point connection to every terminal, but such an SPN is in conflict with the requirements for installation flexibility. An extra terminal, not foreseen in the initial planning would require a new cable to be installed. Furthermore, it is dif-

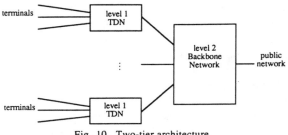

Fig. 10. Two-tier architecture.

ficult to imagine that the switch for a residential environment would be the same, or even be built in the same technology, than the one for a business environment. Clearly, there is a need for a more flexible and unified approach. Therefore, the concept of a hierarchical two-tier architecture [8], [13] is introduced (Fig. 10). The implementation of the SPN itself is not a subject for standardization. But a generic concept of SPN allowing the definition of standard interfaces, independent of implementation could be very worthwhile.

The first level is called the terminal distribution network (TDN). The TDN interconnects the terminals via a passive network to a central node and provides the interface at the S reference point, an interface common to all terminals and for all environments. Because of the passive nature the TDN is limited in geographical scope and thus connects only a limited number of terminals. The TDN can be used in the residential environment where it can be the sole level in the SPN. The TDN then connects also to the external network. But it can also be used in a business environment where in this case it connects to the second level of the two-tier architecture rather than to the public network. This second level, the backbone network (BBN), can be implemented by a number of hierarchical stages to perform switching in a flexible and reliable way. This BBN provides switching between these TDN's and the external network, if the traffic or number of terminals (or other aspects) require more than one TDN. Since the TDN is used in every environment it is candidate for high integration and volume production which must allow it to become cheap enough.

This two-tier architecture also allows a modular build-up of the SPN, which is then easily expansible. New technology can be introduced in the two levels separately, even in the different level 1 modules. From an operational point of view, terminal connection is separated from the high traffic requirements. This has also implications on the technologies which can be used at the levels. The definition of terminal and network interfaces is important for standardization, the two-tier architecture allows the definition of a passive multidrop interface towards the terminals, independent from the implementation of the SPN.

2) Terminal Distribution Network: The starting idea is that a universal terminal interface must be a high capability interface using advanced technology and allowing multidrop configurations. The first level of the architecture, the terminal distribution network (TDN), must then provide the following:

—a passive network for terminal connection with access control on the shared medium,
—a universal terminal interface at a bit rate between 150 and 300 Mbits/s [12],
—the interface to the external network which can either be the public network or the backbone network,
—possibly local switching.

Several topologies and access protocols are possible, the choice is determined by consumer-oriented parameters such as cost, simplicity, modularity, and easy terminal access. Several solutions were investigated bearing in mind the requirements, the characteristics and the consumer-oriented parameters of the TDN [14].

The proposed solution for the TDN is a passive optical tree topology with a central node [Fig. 11(a)], the interconnection network is fully passive. It has a tree-and-branch topology which makes flexible and expansible installation possible. An optical fiber is used for well known reasons of safety, high bandwidth and noise immunity. The mechanical flexibility and the low weight make a fiber easier to handle compared to a coaxial cable. The TDN is limited in speed (compared to the links in the public network) and only has to cover a limited distance (about hundred meters); cheap components and connectors can be used. The limitation in distance implies a limitation in number of terminals to be connected to the TDN. As a consequence the directional power splitters, necessary to operate the fiber as a bidirectional bus, are less critical making a cost-effective TDN possible. Of course, further study is required on the optical aspects of the TDN.

The central node (CN) controls and manages the passive interconnection network and the data flows between the terminals and between the terminal and some external network. It also realizes the interface to the external network (level two of the SPN or the public network). The fiber used to connect the CN and the terminals is operated as a bidirectional bus with a downstream direction from the CN to the terminals, and an upstream direction from the terminals to the central node. Fig. 11(b) gives a possible implementation of a simple central node where only concentration functions are performed. A more complex CN with internal switching is shown in Fig. 11(c). But essential in the two-tier concept is the realization of the same interface independent of the implementation.

The downstream bus is operated in a slotted fashion. Each slot is filled with a cell and an allocator label. This label is generated by the allocator mechanism in the central node and solves the contention problem on the upstream bus. Indeed, the allocator label addresses the next terminal that may send a cell on the upstream bus and thus acts as a bus control mechanism. Allocator labels are assigned in a cyclic fashion, thus giving a terminal an access channel with a bandwidth in multiples of the unit bandwidth (one time slot per cycle). Such a channel is not fixed. It can vary from no channel at all to a channel with the total bandwidth available, in increments equal to the unit bandwidth. Load control is inherently implemented in the access protocol since polling enforces a limitation on the negotiated bandwidth. In fact, the TDN is a logic

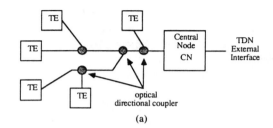

(a)

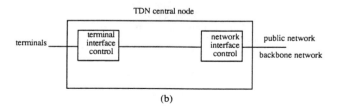

(b)

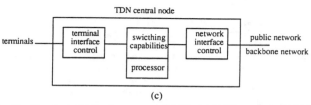

(c)

Fig. 11. Terminal distribution network (TDN). (a) TDN structure. (b) TDN simple central node. (c) TDN central node with switching capabilities.

star configuration, offering quasicircuit channels to the terminals. Such access channels are requested and assigned through a special signaling procedure similar to the TEI assignment procedures in ISDN. At start-up time, only a meta access channel and a meta signaling virtual circuit are available. The terminal uses the meta signaling circuit to request an access channel. Since all terminal have access to that meta access channel, contention is possible. Once the terminal is assigned an access channel, contention is resolved. Further procedures are then activated to request signaling and data virtual circuits. Since the terminal immediately reserves an access channel, the meta access channel is only used for the access channel assignment procedure. This contention can be solved by higher layer protocols.

This leads to an interface specification at the S reference point. For user information (U-plane) the layers 2 and 3 are transparent. The layer one is compatible to the ATM layer and allows a multidrop configuration with an access protocol based on polling. In the C-plane, layers two and three use enhanced versions of the ISDN protocols for these layers, the layer one is the same as in the U-plane. The use of a central node with the allocator mechanism reduces the complexity of the terminal interface. Constraints on technology are not too restrictive, cheap technology can be used (CMOS or BICMOS) up to about 300 Mbits/s. The hardware simplicity of the central node makes integration possible. Studies in that direction are very promising.

3) Backbone Network: The backbone level of the two-tier architecture provides the following:
—connections and interface to the TDN's and to the external network
  —flexible and reliable switching
  —high speed, bit rates of 600 Mbits/s or more
—active network with adequate redundancy for high reliability.

The complexity of this level depends on the environment. In a residential case, level 2 of the architecture can be transparent, a single TDN normally being enough to satisfy all needs for some time to come. In any case, the two-tier architecture allows expansion of the TDN. In a large office environment the backbone network can be very extensive. Of course a wide variety of environments, all with their own needs, require different complexities in the backbone network. The requirements on this network are totally different from those on the TDN. Therefore, the measures determining the choice are different. Two measures are extremely important: delay-throughput characteristics and complexity (includes a weighing of a number of other performance measures, e.g., reliability, redundancy). Studies on topologies and access control protocols indicate two preferred solutions to be selected for further study [15]. The first solution is based on a dispersed ring structure. The operation of the ring in a slotted fashion seems obvious due to the fixed short length cells in ATM, this is confirmed by numerical studies [15]. Active access control modules are preferred to realize a distributed control architecture. These modules are also required for reliability reasons. It makes special arrangements possible, such as two counterrotating rings and bypass facilities with automatic reconfiguration. The second solution takes a star switch in a centralized architecture. The switch used can be quite similar to the switches being developed in the local access area. These implementation issues, though very important, are not the main subject of this paper. The important aspect of the backbone network is the flexibility to connect TDN's and to provide trunks to the public network. The interface on these trunks are based on point-to-point optical technology with ATM as basic information carrier.

D. Reference Configuration

Recommendation I.411 [16] defines the reference points in the user access arrangement and the type of functions that can be provided between them. It is widely accepted that the ISDN reference configuration is applicable in the B-ISDN environment (Fig. 12). The functional groups can be physically integrated, e.g., NT2 and NT1 could be one piece of equipment, the same goes for TA and NT2 when an SPN provides proprietary interfaces or non-B-ISDN interfaces. In the SPN, only interfaces at the reference points will be subject for standardization. The functions defined for the NT2 can be implemented by a TDN, or by the two-tier architecture containing a backbone network (Fig. 13).

The TDN realizes two interfaces: an interface at the S reference point, towards the terminals and an interface to-

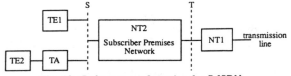

Fig. 12. Reference configuration for *B*-ISDN.

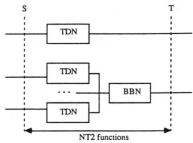

Fig. 13. NT2 implementation.

wards the external network. This external network can be the public network or the second-level backbone network. In the first case, the interface realized is the interface at the *T* reference point. In any case, the external interface of the TDN is point-to-point.

The backbone network also realizes two interfaces: an interface at the *T* reference point towards the public network and an interface towards the TDN's. This last interface must be identical to the interface realized by the TDN towards the external interface, at least if an open system architecture is wanted. This would imply that all the interfaces on the backbone network are identical to the interface at the *T* reference point. This does not preclude proprietary interfaces between TDN and backbone network, though in that case the whole installation has to be provided by one or a consortium of manufacturers.

VI. Conclusions

This article highlights the usefulness of ATM and ATM architectural concepts in the SPN. The introduction of ATM-based architects in SPN is very promising. A layered architectural model was outlined which must clarify some of the issues discussed on ATM. It can also be of help for the definition of an ATM-based network. It is assessed that the ATM technique is a layer 1 technique, but that the ATM network concepts cover the three network layers. From this model it must be clear that ATM is a unique and universal basic service, flexible enough to allow a smooth evolution towards *B*-ISDN. The relation between ATM and some other techniques is studied. A common approach to the interworking problem is given.

Starting from the SPN requirements, a two-tier architecture was proposed. Both levels were discussed in the ATM environment and solutions for the two levels presented. For the terminal distribution network, a passive optical bus structure is preferred. Two solutions for the backbone network are a slotted ring and a centralized star switch. Further study is required on this issue. The two-tier architecture is also covered in the reference configuration. The main merits of this approach is the provision of interfaces independent of the implementation of the SPN.

Acknowledgment

We wish to express our gratitude to our partners in RACE Definition Phase projects 1006 and 2023 for the discussions of a wide variety of subjects, which contributed to these conceptual ideas.

References

[1] CCITT, "Draft recommendation I.121—Broadband aspects of ISDN," CCITT, TD49 (PLEN), Seoul, Korea, Feb., 1988.
[2] P. Gonet, P. Adam, and J.-P. Coudreuse, "Asynchronous time division switching: The way to flexible broadband communication networks," in *Proc. 1986 Int. Zurich Sem. Digital Commun.*, Zurich, Switzerland, Mar., 1986, pp. 141–145.
[3] A. Thomas, J.-P. Coudreuse, and M. Servel, "Asynchronous time division techniques: An experimental packet network integrating videocommunication," presented at Proc. ISS'84, Florence, Italy, May 7–11, 1984.
[4] F. Van den Dool, "Synchronization aspects of ATM-IBC networks," presented at Proc. ISS'87, B.12.1, Phoenix, AZ, Mar. 16–20, 1987.
[5] P. Boyer, J. Boyer, J.-R. Louvion, and L. Romoeuf. "Time transparency evaluation of an asynchronous time divison network," presented at Proc. ISS'87, B.12.2, Phoenix, AZ, Mar. 16–20, 1987.
[6] P. Plehiers, M. Fastrez, J. Bauwens, and M. De Prycker, "Evolution towards a Belgian broadband experiment," presented at Proc. ISS'87, B.5.3, Phoenix, AZ, Mar. 16–20, 1987.
[7] K. Moth and S. B. Jacobsen, "Considerations on the use of ATM on the subscriber line," in *Proc. EFOC/LAN'87*, Basel, Switzerland, June 3–5, 1987, pp. 349–352.
[8] A. De Vleeschouwer, "Architectures for subscribers' premises networks using ATM," in *Proc. CEPT/GSLB Sem. Broadband Switch.*, Albufeira, Portugal, Jan. 19–20, 1987, pp. 237–246.
[9] CCITT, "Recommendation X.200—Reference model of open systems interconnection for CCITT applications," *CCITT Red Books Fascicle* VIII.5, 1984, pp. 1–53.
[10] CCITT, "Recommendation I.320—ISDN protocol reference model," *CCITT Red Books Fascicle*, III.5, 1984, pp. 79–91.
[11] RACE Definition Phase Project 1006, "Definition of the terminal environment," Fin. Consolid. Rep., June, 1987.
[12] J. Vorstermans and R. Theeuws, "Service analysis for the subscribers' premises network," presented at Proc. ISSLS'88, Boston, MA, Sept. 11–16, 1988.
[13] J. Vorstermans, "SPN architectures in an ATD environment," presented at Proc. CEPT/NA5 Sem. Subscriber Premises Network, Nürnberg, F.R. Germany, Nov. 2–4, 1987.
[14] A. De Vleeschouwer, "Terminal distribution networks using ATD," presented at Proc. CEPT/NA5 Sem. Subscriber Premises Network, Nürnberg, F.R. Germany, Nov. 2–4, 1987.
[15] RACE Definition Phase Project 2023, "Integrated broadband communications networks using asynchronous time division techniques," Fin. Rep., Dec., 1986.
[16] CCITT, "Recommendation I.411-ISDN user-network interfaces—Reference configurations," *CCITT Red Books Fascicle*, III.5, pp. 125–132, 1984.

Jan P. Vorstermans received the M.Sc.E.E. degree from the Ghent State University (Belgium), in 1983.

He stayed on at the Laboratory for Electronics and Metrology of that university as a researcher involved in the study and development of image processing architectures. In 1986, he joined the Applied Research Department at ATEA as Systems Design Engineer. There, he is responsible for the study of private broadband networks and for the performance of protocols. During 1986–1987, he was involved in several RACE activities. Since 1987, he has been a delegate to the broadband group of the CEPT. His main interests include communications networks and protocols, local area networks and modeling.

Mr. Vorstermans is a member of the Royal Flemish Society for Engineers and has published several papers.

André P. De Vleeschouwer received the M.Sc.E.E. degree from the Ghent State University (Belgium), in 1974.

After service in the Belgian Army Signals, he was involved in a variety of public and private telecommunications projects with ATEA in Belgium. From 1977 to 1985, he was Section Leader and responsible for telephone set design, later also for data terminals and transmission. Since 1985, he has been the Project Manager for the broadband research project in ATEA, and Task Leader of the SPN taskgroup of a RACE Definition Phase project on ATM in 1986, and recently of the Belgian Broadband Experiment on ATM. He participated in CEPT meetings, holds several patents, and published several papers.

Mr. De Vleeschouwer is a member of K. VIV (Royal Flemish Society for Engineers) and SMPTE (Society of Motion Picture and Television Engineers).

Chapter 4:

Switch Fabric Design and Analysis

4.1 Introduction

The fast packet switching technique offers a number of attractive features such as high throughput, better bandwidth efficiency for bursty traffic, inherent rate adaption, and an application-independent transport network. The underlying concepts of the fast packet network are the separation of signaling and data transfer capabilities, use of streamlined protocol architectures, and switching based on hardware elements as well as a self-routing switch fabric. An important element in such a network is the switch, and an interconnection subsystem is one of the essential components in a switching system. A number of interconnection techniques are available to switch information from an input port to an output port. The most common interconnection subsystems are common memory, crossbar, bus, ring, and multistage interconnection networks. Switch fabrics of future network elements should be capable of interfacing with very high speed fiber-optic transmission systems. This necessitates switch fabrics with distributed control. A self-routing interconnection network is an attractive candidate for a switch fabric.

A crossbar, as an interconnection subsystem, allows any input to be connected to any other nonbusy output at any time and with a constant delay that is basically the switching time of one switching element. The crossbar interconnection technique is characterized by its constant delay and nonblocking property. But as the number of the switch input/output ports increases, the cost increases by order N^2, where N is the number of input ports. Therefore, crossbar interconnection subsystems become impractical for systems with a large number of input/output ports. A crossbar switch fabric provides interconnection between each input-output pair using a crosspoint.

Bus interconnection architectures are simple in concept, well understood, and readily assembled from existing off-the-shelf technology. But the simplicity of the bus concept also limits its scope. The bus interconnection architecture offers a single path for a set of elements attached to it. The bandwidth available to each is inversely proportional to the total number of elements attached to the system. Also, the bus bandwidth is the product of the clock frequency and the width of the data path. An increase in system throughput can be obtained by increasing the clock frequency or the width of the data path, or both. There are, however, technological limitations beyond which the throughput cannot be increased. Any technological advances that make higher bus clock rates possible will also make faster processor speeds possible. Therefore, the ratio of the processor power to the bus bandwidth will likely remain approximately the same.

Multistage interconnection networks are an attempt to approximate the connectivity and throughput of a crossbar while reducing its cost-scaling factor from N^2 to $N \log_2 N$, where N is the number of input ports. This introduces blocking in the interconnection network and an increase in the delay on the order of $\log_2 N$.

The early work on multistage interconnection networks was done in the context of circuit-switched networks. Also, multistage interconnection networks were used to interconnect processors and memory modules in high-performance parallel computer systems. Multistage interconnection networks come in a number of flavors such as banyan networks and delta networks.

Banyan networks are part of a class of multistage interconnection networks that provide exactly one path from any input port to any output port. These switch fabrics can switch packets in parallel but their path uniqueness leads to blocking inside the switch fabric. The blocking results from a number of paths sharing common links within the switch fabric. Another form of blocking called output port blocking takes place when more than one input packet is destined for an output port. Therefore, blocking results in the reduction of the maximum throughput of the switch. The switch fabric can become congested in the presence of nonuniform traffic patterns resulting from strong communities of interest.

A number of techniques are available to avoid blocking and to increase the throughput of the switch. They are
- input buffering,
- output buffering,
- load balancing,
- buffering inside the switch fabric, and
- congestion control protocols.

Switches using pure input buffering have a throughput performance of approximately 58 percent compared to switches with pure output buffering. But providing buffering at the input of the switch results in a form of blocking called head-of-line blocking. This type of blocking takes place when the blocked packet at the head of the queue causes the subsequent packet from being delivered to the free output port, thereby reducing the switch throughput.

Output port contention can be resolved by providing buffering at the output port of the switch. The presence of buffering at the output port eliminates the head-of-line blocking present in the pure input buffering scheme. Thus, higher throughput can be obtained when the output buffering method is used. This increase in throughput is achieved at the expense of added hardware complexity.

Contention inside the switch fabric can be resolved with internal buffering, flow control between switch fabric elements, and/or operation of the switch fabric link at a higher speed than the speed of the input trunk. When a single-buffered banyan network is subject to uniform traffic of fixed-size packets, the switch throughput is about 45 percent. When the number of buffers is increased from one to four, the switch throughput is about 65 percent. This increase in the buffers increases the complexity of the switch element and also increases the delay inside the switch.

Internal blocking can be avoided if the switch fabric first sorts the packets according to their destination addresses and then switches them. The sort-banyan fabric is based on this concept. The switch fabric consists of two segments. The first segment is a sorting network that sorts the incoming packets according to their destination addresses, followed by a shuffle exchange and a banyan switching network. Output port blocking is still possible in this switch fabric when packets with identical destination addresses arrive.

The congestion resulting from nonuniform traffic distribution can be reduced by a distribution network in front of the switch network. The distribution network distributes incoming packets evenly across all the output ports. This approach provides multiple paths inside the switch fabric, thereby reducing blocking.

Yet another approach is to use a routing network in front of the switch network. The routing network also offers multiple paths to the output port. It differs from the distribution network in that a path is picked from the multiple paths and all the packets belonging to a connection follow this path until the connection is terminated.

4.2 Overview of Papers

The papers in this chapter describe various forms of multistage interconnection networks for use in the switch fabrics for broadband networks.

Newman's paper, "A Fast Packet Switch for the Integrated Services Backbone Network," describes the design and performance of a nonbuffered multistage interconnection network capable of supporting a traffic capacity of 150 Gbps. The simplified logic in this nonbuffered fast packet switch is shown to lend itself to implementation in CMOS gate array technology operating at 50 MHz, thus providing a cost-effective solution. Newman also reports on simulation studies comparing the performance of the proposed modified delta network with that of a crossbar and a pure Benes network for different switch fabric sizes, switching element sizes, multipath algorithms (searching, flooding, and random), and numbers of switch planes.

In the second paper, "Multipath Interconnection: A Technique for Reducing Congestion within Fast Packet Switching Fabrics," Anido and Seeto propose a fast packet switch configuration that includes a multipath interconnection network and an external control processor that selects a path through the switch during call setup. Although the banyan networks have the desirable properties of path uniqueness, packet sequence integrity, and minimal intracall packet delay variability, their internal structure is blocking by nature, which leads to link congestion. The connection control processor in the described scheme selects an optimal path during call setup but is not involved in any packet processing. The authors study various path-selection algorithms (least cost, binary, linear, and random) and find the binary algorithm the most cost effective. They

also report on simulation studies that predict the performance of the switch for loads with uniform, horizontal, and diagonal traffic patterns. The switch model is also used to investigate the impact of switch size and buffering on switch performance.

The third paper, "Architecture of a Packet Switch Based on Banyan Switching Network with Feedback Loops," by Uematsu and Watanabe, describes a modified banyan network with feedback connections from the output ports to the input ports. An L-input x L-output switch fabric architecture requires a 2L-input x 2L-output banyan network that provides a "banyan route" and an alternate "feedback route" for incoming packets on an input port. The topology is shown to be equivalent to a cascade banyan network with a distribution network and a routing network connected in tandem. The authors also describe an algorithm for the selection of the routing path. However, this high-throughput switch fabric architecture requires a twofold increase in the number of cells for a given input-output port count and does not preserve the self-routing feature of pure banyan networks.

"A New Self Routing Switch Driven with Input-to-Output Address Difference," by Imagawa et al., proposes a multistage self-routing switch using circulating shift registers. Both point-to-point and broadcasting connections are shown to be easily achievable. The circuit simplicity–in terms of short connection lengths and predominantly parallel connections with few crossovers–makes this architecture a potential candidate for VLSI implementation. It is estimated that a 1.6-Gbps-throughput switch fabric can be easily produced in a 100-MHz CMOS technology.

A Fast Packet Switch for the Integrated Services Backbone Network

PETER NEWMAN

Reprinted from *IEEE Journal on Selected Areas in Communications*, Vol. 6, No. 9, December 1988, pages 1468-1479.
Copyright © 1988 by The Institute of Electrical and Electronics Engineers, Inc. All rights reserved.

Abstract—With the projected growth in demand for bandwidth and telecommunications services will come the requirement for a multiservice backbone network of far greater efficiency, capacity, and flexibility than the ISDN is able to satisfy. This class of network has been termed the broadband ISDN, and the design of the switching nodes of such a network is the subject of much current research. This paper investigates one possible solution. The design and performance, for multiservice traffic, is presented of a fast packet switch based upon a nonbuffered, multistage interconnection network. It is shown that for an implementation in current CMOS technology, operating at 50 MHz, switches with a total traffic capacity of up to 150 Gbit/s may be constructed. Furthermore, if the reserved service traffic load is limited on each input port to a maximum of 80 percent of switch port saturation, then a maximum delay across the switch of the order of 100 μs may be guaranteed, for 99 percent of the reserved service traffic, regardless of the unreserved service traffic load.

I. INTRODUCTION

THE growing acceptance of the integrated services digital network (ISDN) promises increased bandwidth and new telecommunications services at reasonable cost [1]. User demand for bandwidth and services is forecast to rise rapidly following the widespread adoption of ISDN access standards. In addition, when research into multiservice metropolitan area networks (MAN's) [2] and private networks reaches the market, we may expect an acceleration in demand, possibly fueled by an increasing availability of video services. To meet this demand for increased bandwidth and for an expanding diversity of services in the backbone network, evolution towards the broadband ISDN, with the consequent requirement for new switching techniques, will become increasingly desirable [3]. The enhancements offered by a suitable switching mechanism should include: increased flexibility, high traffic capacity, enhanced bandwidth efficiency for "bursty" services, inherent rate adaption, and the service independent support of multiservice traffic.

A recent study of switching techniques appropriate to a multiservice backbone network [4] concludes in favor of fast packet switching, a statistical switching mechanism also known as asynchronous time division [5], [6]. Two problems require attention before a statistical switching mechanism may be employed in a multiservice backbone network, whether public or private. First, a fast packet switch of high maximum traffic capacity must be designed

Manuscript received October 27, 1987; revised July 7, 1988.
The author is with the Computer Laboratory, University of Cambridge, Cambridge CB2 3QG, England.
IEEE Log Number 8824386.

and implemented, in current technology, at an acceptable cost. Second, for multiservice operation, a mechanism is required to support real-time traffic across a fast packet switch at a level of service at least commensurate with that offered by the ISDN. Thus, for example, voice traffic requires a blocked calls lost service with a guaranteed maximum delay performance on each voice connection throughout the entire duration of a call.

This paper presents a simulation study of the multiservice traffic performance of a fast packet switch based upon a nonbuffered, multistage interconnection network. The design of the switch is first discussed followed by a simulation of its throughput at saturation. Then follows an investigation of the performance of the switch for multiservice traffic, in which a simple hardware mechanism is employed to offer a guaranteed maximum delay performance for real-time services such as voice. The results suggest, for example, that fast packet switches may be realized with a total switch capacity of up to 150 Gbits/s, constructed from identical switching elements, in current CMOS gate array technology operating at 50 MHz. Furthermore, if the reserved service traffic load is limited on each input port to a maximum of 80 percent of switch port saturation, then a maximum switch delay may be guaranteed of the order of 100 μs for 99 percent of all reserved service packets, while the switch is continuously loaded to saturation with multiservice traffic.

II. DESIGN OF A FAST PACKET SWITCH

Fast packet switching is a connection oriented packet switching mechanism which achieves high throughput and low delay by reducing the processing required per packet to an absolute minimum [7]–[10] and then implementing it in hardware. Routing is performed at call setup and a virtual circuit is allocated which is fixed for the duration of the call. All flow control and error recovery protocol functions are performed on an end-to-end basis. The packet length across any virtual circuit is constant and small, and the packet format is very simple: a packet header containing a priority field and a label, (to identify the virtual circuit), of fixed length, e.g., 16 bits, followed by the information component typically in the region 4–64 octets.

Two fundamental components are required to construct a fast packet switch: switching and buffering. This results in three possible classes of fast packet switch design: *input buffered*, in which the buffering precedes the switching using a nonbuffered switch fabric [13]; *output buffered*, in which the buffering follows the switching also using a nonbuffered switch fabric [11], [14]; and the *buffered switch fabric* where buffering occurs internally within the switch fabric [6], [15]–[18], [33], [34]. The decision to investigate the design of a fast packet switch based upon a nonbuffered switch fabric was taken on the basis that a nonbuffered switching element is much simpler to implement than is its buffered counterpart. This implies the possibility of implementation in gate array technology offering greater flexibility in the cost, performance, and other design parameters than that available from a dedicated VLSI solution. Furthermore, a simple design permits switching elements of greater degree to be fabricated leading to a reduction in the number of interconnections required to form a given size switch fabric compared to that of a buffered design. The long term goal of an all optical implementation of the switch fabric, or at least the switch fabric data paths, also motivated the selection of a nonbuffered switch fabric.

Pure input buffering has a performance which is approximately 58 percent that of pure output buffering [19] all other factors being equal. However, pure output buffering requires an order of magnitude, more hardware, and switch fabric interconnections than does an input buffered solution [14]. The fast packet switch to be described is thus based upon a pure input buffered switch fabric, but to improve performance, to facilitate maintenance, and to accommodate real-time traffic, a two-plane structure has been adopted which permits a limited amount of output buffering to be implemented if desired. The design may be extended to more than two switch planes in parallel but results suggest that this is unlikely to be necessary unless extremes of performance or reliability are required.

A. The Switch

The basic structure of the fast packet switch is given in Fig. 1. An incoming packet arrives in a first-in first-out (FIFO) queue. When free, the respective input port controller extracts the label from the packet at the head of the queue and uses it to reference a connection table. Each input port controller operates asynchronously, at the packet level, and independently of all other controllers. From the table it receives two components, an outgoing label and a tag. The outgoing label is used to replace the incoming label within the packet. The tag specifies the required destination output port of the switch and is attached to the front of the packet. The input port controller then initiates a setup attempt by launching the packet into the switch fabric, tag first and in bit serial form. There are two possible outcomes, either the packet will be successful and reach the desired output buffer, or it will fail. A setup attempt may fail either because it is blocked by other traffic within the switch fabric or because the requested output port is busy serving another packet. If the setup attempt fails, the switch fabric will assert a collision signal which is returned to the input port controller, along a reverse path, typically within a few bit times of emis-

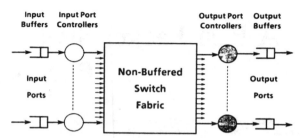

Fig. 1. Basic structure of the fast packet switch.

sion of the packet tag. On receiving the collision signal, the input port controller removes the setup attempt from the switch fabric and waits for a delay typically equivalent to 10 percent of the length of a packet. This is the retry delay and at the end of this period the input port controller begins a fresh attempt to transmit the packet. It continues to do so until it is successful or until it exceeds a limit designed to detect fault conditions.

A slightly more complex algorithm that offers an improvement in performance at high loads does not repeatedly attempt to transmit the same packet but on the failure of a setup attempt searches through the input queue and attempts to transmit the second packet. If that attempt fails the third packet on the queue is attempted and so on cyclically through the queue until a successful transmission is achieved. This overcomes the so called "head of line" blocking problem [13], [19] but care has to be taken not to get packets on the same virtual circuit out of sequence. This algorithm will be referred to as input queue by-pass [15].

A simple model of the operation of the fast packet switch may be drawn by analogy with the operation of a well-known local area network: Ethernet. Ethernet may be considered as a fast packet switch which distributes the switching function across the local area using a single shared medium switch fabric. The fast packet switch described above merely confines the switching within a box so that a multipath medium of much higher bandwidth may be implemented. The input port controller of the fast packet switch corresponds to the media access controller of Ethernet and in both cases the controller throws a packet at the switch fabric and if it is unsuccesful the switch fabric informs it immediately. The difference between the two lies in the fact that in Ethernet, a collision destroys both colliding packets therefore an exponential random backoff algorithm is required. In the fast packet switch, however, collisions are nondestructive in the sense that one of the colliding packets always survives, so a simple retransmission algorithm in sufficient.

B.. The Switch Fabric

1) The Routing Fabric: A fast packet switch requires a highly parallel structure for the switch fabric both in the number of switching elements and in the number of interconnection paths between switching elements. Also, control of the switch fabric must be distributed, with each active switching element operating independently of all others upon control information at the head of each packet. This suggests the use of a self-routing, multistage interconnection network. Such a network consists of many identical and independent switching elements, organized in stages, with the interconnection pattern of links between stages so arranged that each switching element may be controlled by the relevant digit from within a tag prefixed to the head of each packet. The tag simply contains the required destination port number of the switch. For switching elements of degree d each digit within the tag contains $\log_2(d)$ bits and the first digit controls the first stage of switches, the second digit controls the second stage, and so on. Multistage interconnection networks that display this self-routing property belong to the class of banyan networks [20] and have been called delta networks [21], and although many examples of such networks are discussed within the literature [22], they have been proven topologically equivalent [23]. An example of a single plane 64×64 delta network constructed from switching elements of degree 8 is given in Fig. 2. In general, a delta network of size N requires $\log_d(N)$ stages with N/d switching elements per stage. Each interconnection link in the delta network consists of two paths, a forward path to carry the data and a reverse path, set up in parallel with the forward path, to carry the collision signal.

While the majority of research interest has been expended upon delta networks constructed from 2×2 switching elements, our previous investigations [24] suggested that it might be possible to implement nonblocking switching elements of up to degree 16, in gate array technology. The use of switching elements of degree greater than 2 raises the problem that delta networks are only defined in sizes that are an integer power of the degree of the switching element. This would result in large increments between valid sizes of network. The proposed solution is to replicate the interconnection links between stages which permits networks to be built to any size that is an integer power of 2, from switching elements of any degree that is an integer power of 2, [25], [26]. Thus, a modified delta network of size N requires s stages where $s = \lceil \log_d(N) \rceil$, of N/d switching elements per stage and each link of the pure delta network is replicated d^s/N times.[1] Strictly speaking the modified delta network is no longer a member of the class of banyan networks and Fig. 3 illustrates a 16×16 modified delta network of switching elements of degree 8.

We now have the possibility of multiple paths existing between the same pair of input and output ports. This increases the performance and fault tolerance of the switch but requires an algorithm to select between equivalent paths. Fortunately, as there is no buffering within the switch fabric, each incident packet may be routed independently without the risk of out-of-sequence errors between packets traveling on the same virtual circuit. Two algorithms have been investigated: searching and flooding. In the searching mechanism, the input port controller

---

[1] $\lceil x \rceil$ signifies the smallest integer equal to or greater than x.

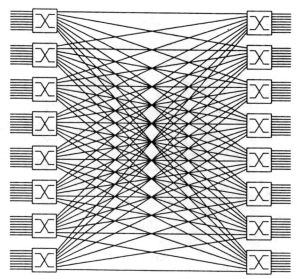

Fig. 2. A 64×64 delta network of 8×8 switching elements.

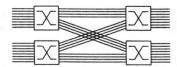

Fig. 3. A 16×16 modified delta network of 8×8 switching elements.

attempts to transmit across each of the equivalent paths in sequence until it meets with success. In the flooding method, the incoming packet is broadcast simultaneously over all paths that lead to the destination such that the destination selects one of the incident copies and all others collide and are removed immediately.

2) The Distributed Fabric: The above switch fabric performs well for traffic which has a random destination distribution but its performance can be markedly impaired for incident traffic with a worst case distribution of destinations. For some applications this may not be significant, however, for high performance switches, and in order to handle traffic sources which have an average bandwidth in excess of about 10 percent of the switch port bandwidth, extra stages of switching must be introduced to distribute the incident traffic across the routing fabric. This has been termed the distribution fabric and to distribute the incoming traffic across an entire s stage delta network requires $s - 1$ distribution stages and results in a Benes topology [28]. Fig. 4 illustrates a 64×64 Benes network of swiching elements of degree 8. Clearly, we have now introduced a large number of equivalent paths into the switch fabric and again for each incident packet we are free to select any free path independently. The simplest method of achieving this is to implement the distribution stages of the switch fabric with switching elements that select any free output at random.

C. The Two-Plane Switch Structure

It is common practice in the design of a telecommunications switch to duplicate or even replicate the switch

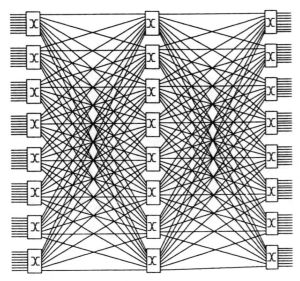

Fig. 4. A 64 × 64 Benes network of 8 × 8 switching elements.

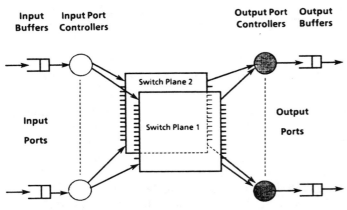

Fig. 5. A two-plane switching structure.

fabric and control hardware for reliability and ease of maintenance. If this is achieved in a load sharing manner the performance of the switch is also enhanced. The general structure of a two-plane switch is shown in Fig. 5 and may be extended to form a multiplane switch of any arbitrary number of planes. It consists of two identical switch planes, each switch plane being a complete delta network with or without a distribution fabric. The two switch planes are connected in parallel to form a load sharing arrangement [26], [27]. Once again we are introducing multiple paths and at the input port controller we may use either the searching or the flooding algorithm to select a path. Considering the output port controller: a simple implementation is only able to handle a single packet at a time and thus rejects setup attempts arriving across the free plane while it is busy serving a packet. A more complex output port controller is capable of handling two packets arriving at the same time and buffering them in a first-in first-out manner in the output buffer. Thus, a measure of output buffering may be provided at the cost of a more complex output port controller.

III. A Simulation Study of Switch Performance at Saturation

The above design of fast packet switch features a number of design parameters the effect of which, on switch performance, needs to be investigated. The simplest way to quantify the performance of a particular switch implementation is to specify the normalized average throughput of the switch when saturated with traffic with a uniform random destination distribution. A simulation model has thus been developed to investigate the throughput at saturation of the switch with respect to the design parameters summarized in Table I.

A. The Simulation Model

In order to reduce the amount of computer time required by the simulation model to reasonable proportions, the setup of a packet has been modeled as an instantaneous event. In reality, a packet will setup on a stage by stage basis, thus a packet which fails setup could itself cause blocking during its setup attempt. The effect of this simplification is to overestimate the throughput at saturation and the results of a more detailed simulation model show that the error introduced by this assumption is in general no more than about 2 percent.

In the model used to determine the throughput at saturation of the switch fabric each packet source supplies a new packet immediately upon completion of transmission of the previous packet and all output ports act as a perfect sink. Packet destinations follow a uniform random distribution and all packets are of the same length. No limit is placed upon the number of setup attempts allowed. The simulation was initialized with random time relationships between all packets and run to attain stability before measurements commenced. Simulations were run for a total of 200 000 packets minimum which yielded results with a standard deviation of about 0.8 percent of the mean for the smaller network sizes to about 0.2 percent for the larger networks. The results are normalized and presented as the throughput per port at saturation which represents the average utilization of an output port at saturation. The total traffic capacity of a fast packet switch is thus the product of the normalized throughput per port at saturation, the size of the switch, and the system clock.

B. The Crossbar Switch Fabric

First, we consider the operation of the crossbar switch fabric as it gives the ideal performance for a nonbuffered switch against which other interconnection networks may be compared. In the crossbar switch, blocking proceeds solely from the probability of multiple sources attempting to transmit to the same destination at the same time. The upper two curves of Fig. 6 show the difference between the simulator output and the analysis [21] under the assumptions of synchronous operation and blocked packets discarded. (The switch size $(N \times N)$ is expressed as $\log_2(N)$ and the curves are discrete, points being connected purely for visual convenience.) The next curve

TABLE I
SWITCH FABRIC DESIGN PARAMETERS

Parameter	Range
Switch Fabric Size	2x2 to 4096x4096
Interconnection	Crossbar
Networks	Delta
	Benes
Degree of Switching Element	2x2 to 16x16
Multiple	Searching
Path	Flooding
Algorithms	Random
Multiple Switch Planes	1 to 4
	Regular
Port	Input Queue By-Pass
Controllers	Double Buffered Output
	De-Luxe

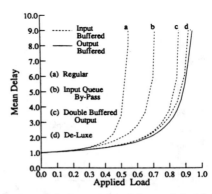

Fig. 7. Mean delay performance of crossbar switch fabrics for slotted traffic.

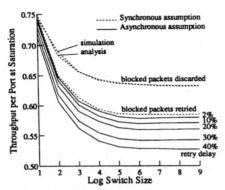

Fig. 6. Throughput at saturation for the crossbar switch.

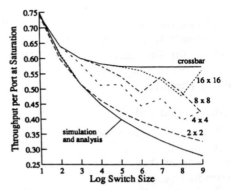

Fig. 8. Throughput at saturation for single plane, pure input buffered, flooding delta networks.

shows the effect of resubmitting blocked packets under the assumption of synchronous switch operation and its asymptote agrees with the analytical results of [19]. This is followed by a set of curves assuming asynchronous arrival of packets, with asynchronous switch operation and blocked packets retried, at different values of retry delay, expressed as a percentage of the packet length, (i.e., the emission delay of a packet).

Whilst discussing the performance of the crossbar switch fabric it is interesting to introduce a simulation study of the delay performance for slotted traffic which has been analyzed in [19]. Fig. 7 shows how input queue by-pass and the use of a two-plane output buffered crossbar switch fabric improves the average delay performance of the pure input buffered switch. For the case of a two-plane crossbar switch fabric with output buffering and input queue by-pass, (the delux model,) a performance very close to that of the pure output buffered switch may be achieved but at a much reduced cost in terms of hardware and interconnections within the switch fabric. The detailed results of the simulation model for the throughput at saturation of crossbar switch fabrics under the various design parameters are given in Appendix I.

C. The Delta Network

Fig. 8 gives the maximum throughput performance of a single plane pure input buffered delta network con-structed from switching elements of degree 2, 4, 8, and 16 using a flooding algorithm and a retry delay of 10 percent. The corresponding curve for the crossbar switch is included for comparison. The perturbations in the curves are due to the number of equivalent paths through the network with the minima indicating the pure delta network. Curves are also presented of the analysis [21] and simulation results for the 2 × 2 delta network, under the assumptions of synchronous operation and blocked packets discarded, demonstrating an agreement which renders the curves virtually coincident. Comparison to the simulation results of [32] also reveals a close agreement. In Fig. 9, the improvement in throughput obtained with a two-plane, pure input buffered delta network is shown using a routing algorithm which floods between planes but searches within a plane, commencing with a random selection from all equivalent paths within a plane. (The hardware for this hybrid mechanism is easier to implement, is more flexible, and its performance differs only marginally from that of the pure flooding case.) An investigation of multiplane, pure input buffered delta networks with more than two planes shows that, in the case of switching elements of degree 8 and 16, little is gained in increased throughput as the asymptote of crossbar network performance is approached rapidly. Further, for a two-plane, pure input buffered network, the pure searching algorithm yields a performance that is only slightly inferior to that of a flood-

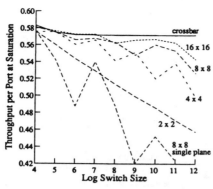

Fig. 9. Throughput at saturation for two-plane, pure input buffered delta networks.

ing mechanism, (no more than 2 percent with 8 × 8 switching elements). The detailed results of the simulation model for the throughput at saturation of delta networks with respect to the various design parameters are tabulated in Appendix II.

D. The Distribution Fabric

The performance of the Benes network as a switch fabric for a fast packet switch has been reported in [29] and for the purposes of this discussion we state the obvious that the introduction of a distribution stage into the switch fabric does not degrade its throughput performance, but rather, enhances it to approach the performance of the equivalent crossbar switch fabric. The results reported for the delta network routing fabric may thus be taken as a lower bound on performance when considering a switch fabric with distribution stages and the results for the crossbar switch fabric taken as an upper bound. Appendix III gives the throughput at saturation of the single plane pure Benes network for comparison.

IV. MULTISERVICE INTEGRATION OVER A FAST PACKET SWITCH

From the results presented of switch performance at saturation, it may be seen that switches of very high total traffic capacity may be constructed from LSI switching elements operating at conventional speeds. We now consider how to integrate multiple services, (voice, video, image, text, data, etc.) onto the structure.

A. Multiservice Traffic Requirements

We argue that all communications services may be classified into two fundamental categories according to the delay requirement they present to the network, and for lack of better terminology we will refer to them as reserved and unreserved services. A reserved service exacts an inflexible, low delay and low variance of delay requirement, whereas unreserved services are much more flexible in the range of delay that can be tolerated. The majority of reserved services derive from information based upon a physical property that changes rapidly with time, e.g., voice and video, and often contain a high degree of redundancy, thus permitting an appreciable packet

loss before any noticeable deterioration in quality is perceived. There are some reserved services, however, that are highly sensitive to error, e.g., process control, in which the delay constraint proceeds from the requirement for a high priority service, yet such services are generally of low bandwidth. Unreserved services include the bulk of data transfer, interactive, and transaction services at various priorities.

The delay constraint is not the only difference between these two basic service classifications. A reserved service requires a guaranteed bandwidth and delay performance throughout the entire duration of the connection, else the connection request must be refused. An unreserved service expects the bandwidth and delay associated with a connection to vary according to the traffic load on the network. Hence, if a minimum of these two fundamental service priorities are implemented within the hardware of the switch, a diverse range of communications services may be supported [7], [35].

B. Extensions to the Switch

In order to support the two basic services, reserved service traffic must be given priority at all input and output ports. At the input ports, the single input queue at every port of Fig. 1 is replaced by two queues, one for reserved service packets and one for unreserved service packets. A priority field is also added to the tag to distinguish the two classes of packet. The input port controller is modified so as to transmit unreserved service packets only when the reserved service packet queue is empty, and to postpone repeated setup attempts of an unsuccessful unreserved service packet on the arrival of a reserved service packet. The transmission of a successful unreserved service packet is not interrupted by the arrival of a reserved service packet. Reserved service priority at the output port is ensured by a simple mechanism implemented in hardware in each of the output port controllers. If there is competition between packets from different input ports for access to an output port, this mechanism ensures that reserved service packets are given priority.

C. Simulation Traffic Models

Two models of unreserved service traffic have been used, saturation and Poisson. In the saturation model, unreserved service traffic is generated to keep each input port continuously busy while in the Poisson model, unreserved service packets are generated according to a Poisson arrival process. Both models generate traffic with a uniform random destination distribution. Three models of reserved service traffic were investigated: Poisson, talkspurt voice, and TDM voice. In the Poisson model, reserved service packets are generated according to a Poisson arrival process with a uniform random distribution of packet destinations. In the talkspurt voice case, a superposition of individual voice sources has been modeled, on every input port of the switch, in which the on-off characteristics of speech have been used for bandwidth compression, (i.e., packet voice with silence detection.) Each voice source is assumed to exhibit two states, active, and silent, representing the talkspurts and pauses present in conversational speech [36]. In the active state each voice source generates packets at a regular rate representing 32 kbits/s voice coding, 256 bit packets with a further 32 bits overhead, and a 20 MHz system clock. No packets are generated in the silent state. The two states are modeled by an exponential distribution with means of 1.2 and 1.8 s, respectively [37], and each voice source transmits packets to a single destination which is selected at random during initialization. The TDM voice model is simply a talkspurt model with silent periods of zero duration to represent packet voice without silence detection. A random phase relationship is assumed between all voice sources.

D. Multiservice Switch Performance

The simulation results for a 64×64 fast packet switch constructed from 8×8 switching elements using a two-plane, pure input buffered delta network are now presented for various combinations of the multiservice traffic models. Investigations suggest that the major characteristics of the results are general to all sizes of fast packet switch constructed from switching elements of any degree according to any permutation of the design parameters discussed above. A good approximation to the throughput and delay performance for other sizes and designs of fast packet switch may be obtained by scaling the measurements presented for this example in proportion to the throughput at saturation of the desired switch fabric.

The measurement of delay selected for the performance of the reserved service is that of the 99th percentile of the delay distribution [38]. It is assumed that packet voice traffic may withstand a 1 percent random packet loss, for small packet sizes [39], [12], without perceptible loss of quality. Hence, our measure of guaranteed maximum delay is the delay within which 99 percent of all reserved service packets arrive at their destination. The consequence is that the accuracy of the maximum delay measurements is much lower than that of throughput as we are examining the tail of the delay distribution.

Delay is normalized to the packet length and all measurements are taken with a retry delay of 10 percent of the packet length. Applied load and throughput per port are also normalized and reflect the average utilization of input and output ports, respectively.

1) Poisson Reserved Service Traffic: Fig. 10 gives the basic result for a switch with a Poisson reserved service traffic source and a saturated unreserved service traffic source on each of the switch input ports. As the reserved service traffic load is increased, the maximum unreserved service traffic load that the switch is able to sustain falls, so as to maintain the total load on the switch reasonably constant at saturation. The reserved service throughput response, in the absence of any unreserved service traffic, is identical to that in the presence of unreserved service sources. Fig. 11 gives the corresponding maximum delay curves for reserved service traffic with and without the

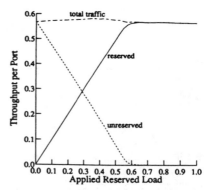

Fig. 10. Throughput performance for the Poisson reserved service + saturated unreserved service traffic model.

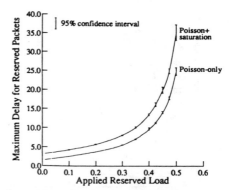

Fig. 11. Maximum reserved service packet delay for the Poisson reserved service traffic model with and without saturated unreserved service traffic.

presence of saturated unreserved service traffic. The maximum delay for reserved service traffic in the presence of saturated unreserved service traffic is approximately 50 percent greater than in the absence of unreserved service traffic. This difference is due to the probability of an incident reserved service packet finding the input node already busy serving an unreserved service packet that has achieved setup. Further, the throughput and maximum delay performance of reserved service traffic is not adversely affected by a nonuniform distribution of packet destinations for unreserved service traffic. Investigations also suggest that it is possible to operate a fast packet switch with input and output ports running at widely different mean traffic loads, as might be the case, for example, between ports connected to interswitch trunks and those connected to local area networks.

In Figs. 12 and 13, a Poisson reserved service traffic source is multiplexed with a Poisson unreserved service source at every input port of the switch. Fig. 12 shows the throughput performance of unreserved service traffic for several reserved service traffic loads. Fig. 13 shows the corresponding average delay for unreserved service traffic. Both curves saturate at a level that reflects the remaining switch bandwidth available after serving the requirements of reserved service traffic. The reserved service throughput characteristic in this case is identical to that observed with a saturated unreserved service traffic

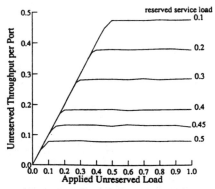

Fig. 12. Unreserved service throughput performance for the Poisson reserved service + Poisson unreserved service traffic model.

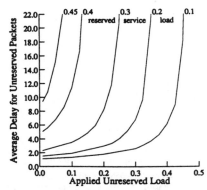

Fig. 13. Average unreserved service packet delay for the Poisson reserved service + Poisson unreserved service traffic model.

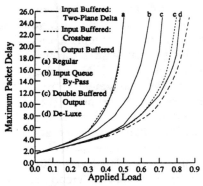

Fig. 14. Comparison of maximum delay performance of various switch designs for Poisson traffic.

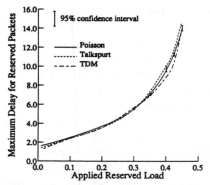

Fig. 15. A comparison of maximum reserved service packet delay for Poisson, talkspurt and TDM voice models in the absence of unreserved service traffic.

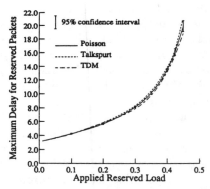

Fig. 16. A comparison of maximum reserved service packet delay for Poisson, talkspurt and TDM voice models in the presence of saturated unreserved service traffic.

source while the maximum reserved service delay is reduced in proportion to the amount that the total load on the switch falls below saturation.

To give a comparative impression of switch performance Fig. 14 shows the maximum delay performance of various designs of fast packet switch for Poisson traffic. Once again it may be seen that the performance of the pure output buffered switch [14] is only slightly greater than that of the highest performance two-plane delta design. This in turn is of slightly greater performance than a two-plane Batcher–banyan [13] as the latter is synchronous at the packet level and therefore cannot take advantage of input queue by-pass.

2) Talkspurt Voice: For the above 64 × 64 switch with Poisson traffic sources, the queue lengths at the input ports were observed to be short and to stabilize rapidly for traffic loads below about 0.45. This figure represents a load of 80 percent of saturation and is a valid conservative estimate for the upper bound of the applied reserved service traffic load for stable operation of all sizes and designs of fast packet switch. The maximum mean reserved service traffic load for any switch port may therefore be fixed at 80 percent of the saturation load for that switch. The maximum delay performance of the talkspurt and TDM voice source models, to the above value of maximum mean reserved service traffic load, is now compared to the result for the Poisson reserved service traffic model.

The maximum delay performance, in the absence of unreserved service traffic, is given by Fig. 15 and in the presence of saturated unreserved service traffic by Fig. 16. (For the talkspurt voice model an applied load of 0.45 corresponds to 625 voice sources per switch port, and to 250 voice sources per switch port for the TDM voice model.) It is evident that within the region of stable operation there is no significant difference in the guaranteed maximum delay across the switch for Poisson, talkspurt and TDM voice sources, either in the presence or absence of saturated unreserved service traffic. Furthermore, an

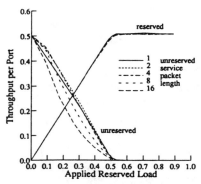

Fig. 17. Effect of unreserved service packet length on throughput performance for Poisson reserved service + saturated unreserved service traffic model.

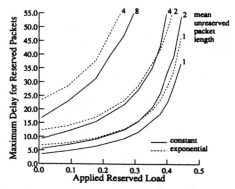

Fig. 18. Effect of unreserved service packet length, constant and exponentially distributed, on maximum reserved service packet delay.

cess, with reasonable accuracy, for applied loads below about 80 percent of saturation [41], [42].

3) Packet Length: Finally, we consider the effect of variable unreserved service packet length upon performance. In the results presented so far, we have assumed constant packet length and normalized all results to become independent of the absolute value. Now we assume that all packets consist of a header and an information component and we normalize results to the value of the information component. First, we consider the case in which reserved service packets and unreserved service packets are of different but constant length. The length of the unreserved service packet is expressed in terms of the reserved service packet information field, and all packets have a header of one eighth of the length of the reserved service packet information field. The throughput results are presented in Fig. 17 where it may be seen that the reserved service throughput performance is not unduly affected by the unreserved service packet length. However, the unreserved service throughput at saturation, for large unreserved service packet lengths, is lower than that for small packets showing that the advantage of low packet overhead is rapidly outweighed by the superior multiplexing capability of small packet sizes. An examination of the case in which the length of the information component of all unreserved service packets is given by an exponential distribution reveals similar results with a reduction in unreserved service throughput performance of between 10 to 20 percent, due to the variability in packet length. An investigation of the case in which all packet lengths follow a uniform random distribution of ±10 percent about a mean value reveals no drop in performance when compared to that of constant length. Thus, the switch is insensitive to the variation in packet length that might be introduced by a line code employing "bit-stuffing." The effect of the unreserved service packet length upon reserved service packet delay performance is given in Fig. 18. As expected, a variable length unreserved service packet exerts a greater detrimental influence than one of constant length, and the shorter the mean packet length the less the reserved service packet delay performance is affected. Hence, conventional sizes of data packet must clearly be broken down into short packets for multiplex-

observation of the interarrival times of packets generated by the talkspurt model on a single input port reveals a very close approximation to the exponential distribution [40]. Thus, the superposition of a large number of talkspurt voice sources may be modeled by a Poisson arrival proing with real-time traffic but this may not be necessary for a "data-only" environment.

V. IMPLEMENTATION

An experimental implementation of the fast packet switch has been completed in low cost 3μm HCMOS gate arrays [30], [31]. A 4×4 crossbar switching element and an experimental input port controller, with a standard 8-bit microprocessor bus interface, have been fabricated and demonstrated to operate as expected at a clock rate of 8 MHz. The throughput at saturation and delay performance of the switching element have been measured and agree with the simulation results to within 1 percent. The switching element required a total of 378 gates and the input port controller 292 gates which allows an estimate of the gate complexity of fully implemented parts to be made for crossbar switching elements of various sizes, Table II.

It is reasonable to expect an implementation in 2μm CMOS to achieve speeds of around 50 MHz without great difficulty and beyond this we observe that only the data path within the switching element is required to operate at high speed. The majority of the logic in the switching element handles packet setup and if a small increase in overhead is permitted in the packet setup time then this logic can operate at a slower speed than that within the data path. (In the current design, the data path passes through no more than two gates and a flip-flop.) We may thus consider implementation in BiCMOS, ECL, and even GaAs at speeds approaching 500 MHz and beyond without exceeding the power budget, Table III.

. For even higher speed operation the switching and data paths within the switching element may be implemented optically with the control logic in ECL or GaAs to form an electrooptic switching element [43]. Switching times down to a few nanoseconds might thus become feasible on switch ports handling several Gbits/s to yield a total switch capacity measured in Tbits/s.

213

TABLE II
ESTIMATED COMPLEXITY OF CROSSBAR SWITCHING ELEMENTS

Size	Gate Count
2x2	250
4x4	600
8x8	1900
16x16	6000
32x32	21000

TABLE III
ESTIMATED MAXIMUM BANDWIDTH PER SWITCH PORT FOR VARIOUS IMPLEMENTATION TECHNOLOGIES

Technology	Bandwidth Per Port
3 μ CMOS	10 Mbit/sec
2 μ CMOS	50 Mbit/sec
BiCMOS	250 Mbit/sec
ECL	500 Mbit/sec
GaAs	1 Gbit/sec
Photonic	>1 Gbit/sec

VI. CONCLUSIONS

The design of a fast packet switch based on a nonbuffered interconnection network has been reported and simulation results of its throughput performance at saturation discussed. The design is modular and will operate at any speed, with any device technology, including integrated optics. Maximum switch size is limited only by implementation considerations for the technology and operating speed selected. This design of fast packet switch uses fewer active elements than the equivalent crossbar switch, whilst offering a similar performance at saturation, for all sizes of switch greater than 16 × 16.

An extension to the design of the switch has been proposed in order to support multiservice traffic. Simulation results indicate that with a reserved service traffic loading of up to 80 percent of switch port saturation, the upper bound on delay for 99 percent of all incident reserved service packets is in the region of 20 packet lengths. Further, unreserved service traffic may be multiplexed with reserved service traffic, at every input port of the switch, so as to operate the switch continuously at saturation, without affecting the bounded delay performance of the reserved service. These results hold for voice traffic modeled as Poisson sources, talkspurt voice sources and TDM voice sources which yield a very similar maximum delay performance. The reserved service throughput and delay performance also appears insensitive to the arrival distribution and to the destination distribution of unreserved service traffic.

For delay-sensitive, reserved service performance, the packet length for unreserved service traffic should be short and constant. No performance impairment is introduced by a ±10 percent variation in packet length. For a single service implementation, moderately insensitive to delay, variable length packets of any reasonable maximum length may be supported.

An experimental implementation of the fast packet switch in 3μm HCMOS gate arrays has demonstrated that the switch can be implemented at low cost in conventional gate array technology and that the performance of a 4 × 4 switching element agrees closely with that predicted by the simulation model.

Work is currently in progress on the problem of supporting multicast operation across the switch, for both reserved and unreserved traffic, with a similar throughput and delay performance to that of unicast traffic. Initial results suggest that this may be achieved with the same philosophy of simple implementation in gate array technology. The much more interesting problem of how to organize, manage, control, and interface to a network of such fast packet switches is also under consideration.

Finally, by way of summary, we observe that the Cambridge fast packet switch is but: "One small chip for MAN's …"

APPENDIX I
THROUGHPUT AT SATURATION FOR CROSSBAR SWITCH FABRICS

Size	Synchronous		Asynchronous (10% retry delay)			
	Regular	Double Buffer	Regular	Queue By-Pass	Double Buffer	De-Luxe
2	.747	1.0	.742	.878	1.0	1.0
4	.650	.941	.640	.808	.931	.981
8	.613	.916	.601	.771	.897	.971
16	.594	.905	.583	.752	.883	.965
32	.586	.900	.575	.746	.876	.961
64	.585	.898	.572	.740	.872	.958
128	.585	.896	.572	.740	.872	.957
256	.584	.897	.572	.734	.872	.954
512 +	.585	.897	.572	.739	.872	.951

APPENDIX II
THROUGHPUT AT SATURATION OF DELTA NETWORKS WITH SWITCHING ELEMENTS OF DEGREE 2–16

Size	Single Plane		Two-Plane			
	Regular	Queue By-Pass	Regular	Queue By-Pass	Double Buffer	De-Luxe
2	.742	.878	.742	.878	1.0	1.0
4	.597	.791	.648	.813	.930	.979
8	.517	.711	.599	.775	.850	.950
16	.462	.642	.576	.743	.786	.916
32	.423	.583	.560	.723	.733	.879
64	.392	.536	.543	.702	.688	.842
128	.366	.496	.529	.682	.650	.807
256	.344	.463	.514	.662	.618	.773
512	.325	.435	.499	.642	.589	.741

Size	Single Plane		Two-Plane			
	Regular	Queue By-Pass	Regular	Queue By-Pass	Double Buffer	De-Luxe
4	.640	.808	.640	.808	.931	.981
8	.583	.693	.605	.767	.887	.947
16	.513	.684	.576	.752	.836	.942
32	.516	.599	.575	.725	.818	.895
64	.444	.596	.560	.724	.755	.892
128	.468	.530	.566	.697	.762	.842
256	.395	.530	.541	.696	.690	.840
512	.428	.478	.551	.668	.714	.791

Size	Single Plane		Two-Plane			
	Regular	Queue By-Pass	Regular	Queue By-Pass	Double Buffer	De-Luxe
8	.601	.771	.601	.771	.897	.971
16	.576	.652	.581	.741	.861	.927
32	.542	.649	.575	.738	.848	.924
64	.488	.647	.566	.731	.799	.922
128	.540	.560	.568	.707	.806	.865
256	.489	.559	.563	.705	.783	.864
512	.420	.558	.548	.706	.718	.863
1024	.452	.493	.560	.674	.759	.804
2048	.424	.493	.553	.674	.726	.804
4096	.371	-	.526	.674	.652	.801

Size	Single Plane		Two-Plane			
	Regular	Queue By-Pass	Regular	Queue By-Pass	Double Buffer	De-Luxe
16	.583	.752	.583	.752	.883	.965
32	.573	.635	.577	.728	.849	.916
64	.560	.632	.573	.728	.842	.913
128	.531	.632	.573	.725	.829	.913
256	.476	.630	.562	.725	.786	.912
512	.559	.544	.566	.698	.797	.851
1024	.485	.543	.567	.695	.794	.847
2048	.459	.542	.563	.695	.769	.847
4096	.409	.541	.543	-	.703	.850

Appendix III
Throughput at Saturation of the Single Plane Pure Benes Network

Degree	Size	Regular	Queue By-Pass
8	8	.601	.771
	64	.538	.648
	512	.496	.559
16	16	.583	.752
	256	.535	.632

References

[1] S. N. Pandhi, "The universal data connection," *IEEE Spectrum*, pp. 31–37, July 1987.

[2] R. W. Klessig, "Overview of metropolitan area networks," *IEEE Commun. Mag.*, vol. 24, pp. 9–15, Jan. 1986.

[3] J. S. Turner, "Design of an integrated services packet network," *IEEE J. Select. Areas Commun.*, vol. SAC-4, pp. 1373–1380, Nov. 1986.

[4] J. J. Kulzer and W. A. Montgomery, "Statistical switching architectures for future services," in *Proc. ISS '84*, Florence, Italy, May 1984, 43A1 pp. 1–6.

[5] M. Littlewood, I. D. Gallagher, and J. L. Adams, "Evolution toward an ATD multi-service network," *Brit. Telecom. Tech. J.*, vol. 5, no. 2, Apr. 1987.

[6] A. Thomas, J. P. Coudreuse, and M. Servel, "Asynchronous time-division techniques: An experimental packet network integrating video communication," in *Proc. ISS '84*, Florence, Italy, May 1984, 32C2, pp. 1–7.

[7] J. W. Forgie and A. G. Nemeth, "An efficient packetized voice/data network using statistical flow control," in *Proc. ICC '77*, 38.2, pp. 44–48.

[8] J. S. Turner and L. F. Wyatt, "A packet network architecture for integrated services," in *Proc. IEEE GLOBECOM '83*, Dec. 1983, pp. 45–50.

[9] A. G. Fraser, "DATAKIT—A modular network for synchronous and asynchronous traffic," in *Proc. ICC '79*, June 1979, pp. 20.1.1–20.1.3.

[10] P. Kirton, J. Ellershaw, and M. Littlewood, "Fast packet switching for integrated network evolution," in *Proc. ISS '87*, Mar. 1987, pp. B6.2.1.–B6.2.7.

[11] A. Huang and S. Knauer, "Starlite: A wideband digital switch," in *Proc. IEEE GLOBECOM '84*, pp. 121–125.

[12] J. G. Gruber and N. Le, "Performance requirements for integrated voice/data networks," *IEEE J. Select. Areas Commun.*, vol. SAC-1, pp. 981–1005, Dec. 1983.

[13] J. Y. Hui and E. Arthurs, "A broadband packet switch for integrated transport," *IEEE Select. Areas Commun.*, vol. SAC-5, pp. 1264–1273, Oct. 1987.

[14] Y. S. Yeh, M. G. Hluchyj, and A. S. Acompora, "The knockout switch: A simple, modular architecture for high-performance packet switching," *IEEE J. Select. Areas Commun.*, vol. SAC-5, pp. 1274–1283, Oct. 1987.

[15] R. G. Bubenik and J. S. Turner, "Performance of a broadcast packet switch," in *Proc. ICC '87*, June 1987, 31.6, pp. 1118–1122.

[16] S. Nojima *et al.*, "Integrated services packet network using bus matrix switch," *IEEE J. Select. Areas Commun.*, vol. SAC-5, pp. 1284–1292, Oct. 1987.

[17] M. De Prycker and M. De Somer, "Performance of an independent switching network with distributed control," *IEEE J. Select. Areas Commun.*, vol. SAC-5, pp. 1293–1301, Oct. 1987.

[18] G. Perucca, "Research on advanced switching techniques for the evolution to ISDN and broadband ISDN," *IEEE J. Select. Areas. Commun.*, vol. SAC-5, pp. 1356–1364, Oct. 1987.

[19] M. J. Karol, M. G. Hluchyj, and S. P. Morgan, "Input versus output queueing on a space-division packet switch," *IEEE Trans. Commun.*, vol. COM-35, pp. 1347–1356, Dec. 1987.

[20] R. J. McMillen, "A survey of interconnection networks," in *Proc. IEEE GLOBECOM '84*, 5.1.1-9, pp. 105–113.

[21] J. H. Patel, "Performance of processor to memory interconnections for multiprocessors," *IEEE Trans. Comput.*, vol. C-30, pp. 771–780, Oct. 1981.

[22] T. Feng, "A survey of interconnection networks," *IEEE Comput.*, vol. 14, pp. 12–27, Dec. 1981.

[23] C. Wu and T. Feng, "On a class of multistage interconnection networks," *IEEE Trans. Comput.*, vol. C-29, pp. 694–702, Aug. 1980.

[24] P. Newman, "Message switching: An experimental model," The GEC Hirst Res. Centre, Apr. 1983, unpublished manuscript.

[25] G. B. Adams and M. J. Siegal, "The extra stage cube: A fault tolerant interconnection network for supersystems," *IEEE Trans. Comput.*, vol. C-31, pp. 443–454, May 1982.

[26] M. Kumar and J. R. Jump "Performance of unbuffered shuffle-exchange networks," *IEEE Trans. Comput.*, vol. C-35, pp. 573–578, June 1986.

[27] C. P. Kruskal and M. Snir, "The performance of multi-stage interconnection networks for multiprocessors," *IEEE Trans. Comput.*, vol. C-32, pp. 1091–1098, Dec. 1983.

[28] V. E. Benes, "On rearrangeable three-stage connecting networks," *Bell Syst. Tech. J.*, vol. 41, no. 5, pp. 1481–1492, Sept. 1962.

[29] P. Newman, "A broad-band packet switch for multi-service communications," in *Proc. IEEE INFOCOM '88*, New Orleans, LA, Mar. 1988, 1A3, pp. 19–28.

[30] ——, "Self-routing switching element for an asynchronous time switch," Priority Pat. Appl. 8724208, Oct. 1987.

[31] ——, "Data Signal Switching Systems," UK Pat. GB 2 151 880 B, Dec. 16, 1983.

[32] D. R. Milway, "Binary routing networks," The Univ. Cambridge Comput. Lab., Tech. Rep. 101, Dec. 1986.

[33] Y. Jenq, "Performance analysis of a packet switch based on single-buffered banyan network," *IEEE J. Select. Areas Commun.*, vol. SAC-1, pp. 1014–1021, Dec. 1983.

[34] D. M. Dias and J. R. Jump, "Analysis and simulation of buffered delta networks," *IEEE Trans. Comput.*, vol. C-30, pp. 273–282, Apr. 1981.

[35] R. M. Falconer and J. L. Adams, "Orwell: A protocol for an integrated services local network," *Brit. Telecom Tech. J.*, vol. 3, no. 4, pp. 27–35, Oct. 1985.

[36] P. T. Brady, "A statistical analysis of on-off patterns in 16 conversations," *Bell Syst. Tech. J.*, vol. 47, pp. 73–91, Jan. 1968.

[37] J. N. Daigle and J. D. Langford, "Models for analysis of packet voice communications systems," *IEEE J. Select. Areas Commun.*, vol. SAC-4, pp. 847–855, Sept. 1986.

[38] J. M. Appleton and M. M. Peterson, "Traffic analysis of a token ring PBX," *IEEE Trans. Commun.*, vol. COM-34, pp. 417–422, May 1986.

[39] J. Gruber and L. Strawczynski, "Judging speech in dynamically managed voice systems," *Telesis 1983*, *II*, pp. 30–34.

[40] B. G. Kim, "Characterisation of arrival statistics of multiplexed voice packets," *IEEE J. Select. Areas Commun.*, vol. SAC-1, pp. 1133–1139, Dec. 1983.

[41] K. Sriram and W. Whitt, "Characterising superposition arrival processes in packet multiplexers for voice and data," *IEEE J. Select. Areas Commun.*, vol. SAC-4, pp. 833–846, Sept. 1986.

[42] H. Heffes and D. M. Lucantoni, "A Markov modulated characterisation of packetized voice and data traffic and related statistical multiplexer performance," *IEEE Select. Areas Commun.*, vol. SAC-4, pp. 856–867, Sept. 1986.

[43] R. W. Blackmore, W. J. Stewart, and I. Bennion, "An opto-electronic exchange for the future," in *Proc. ISS '84*, Florence, Italy, May 1984, 41A4, pp. 1–7.

Peter Newman received the B.Sc. (Tech.) degree from the University of Wales Institute of Science and Technology, Cardiff, Wales, in 1978. He is currently working toward the Ph.D degree at the Computer Laboratory, University of Cambridge.

He joined the General Electric Company Ltd. Hirst Research Centre in 1978 and worked in the Telecommunications Research Laboratory until 1983. His major contribution was in the area of high speed packet and circuit switching. In 1984, he transferred to the Computer Systems Research Laboratory where he was interested in multimedia integration across local area networks.

Mr. Newman is a member of the Institution of Electrical Engineers and a Chartered Engineer.

Multipath Interconnection: A Technique for Reducing Congestion Within Fast Packet Switching Fabrics

GARY J. ANIDO, MEMBER, IEEE, AND ANTHONY W. SEETO, MEMBER, IEEE

Reprinted from *IEEE Journal on Selected Areas in Communications*, Vol. 6, No. 9, December 1988, pages 1480-1488.

Abstract—Multistage interconnection networks such as the banyan are well suited to multiprocessor and fast packet communication systems. However, the banyan interconnection is prone to internal link congestion resulting in a blocking switch architecture. Several solutions have already been proposed, and implemented, to reduce the severity of link congestion. In general, these solutions offer packets of multiplicity of paths which tend to increase packet delay variability and allow delivery of out-of-sequence packets. This, in turn, may lead to an increase in end-to-end protocol complexity, particularly in the case of real-time services. A solution, called multipath interconnection, is proposed to overcome this difficulty. Multiple (i.e., alternate) paths are provided and one is selected at call setup time. Subsequent packets belonging to the call are constrained to follow the selected path. A number of path selection strategies are presented.

Implementation costs, in terms of relative processing and memory requirements, of each algorithm are compared. The ability of the algorithms to evenly distribute traffic across the internal links is evaluated by means of a simulation study. Queueing behavior is estimated using an approximate model of a switch path. The results indicate that the more able a path selection algorithm is to distribute switch traffic, the greater the limiting throughput. The model is also used to investigate the effect of parameters, such as switch size and buffering capacity, on switch performance.

I. INTRODUCTION

MULTISTAGE interconnection networks have found extensive use in multiprocessor applications [1]–[3] and communication systems [4], [5]. Banyan networks, and other members of a class of topologically equivalent multistage networks [6], have a number of desirable characteristics such as self-routing, suitability for VLSI implementation, and path uniqueness [7]. The latter means that connecting each inlet and outlet pair of a banyan network is a unique path. When used for fast packet switching [8], [9] this characteristic ensures that individual packets belonging to the same logical connection (or call) will follow the same route. As a result, out-of-sequence packets are prevented and intracall packet delay variability is expected to be less than would be the case if packets were routed independently through the network. This, in turn, leads to the design of simpler protocols for the synchronization of real-time services such as voice and video [10].

The path uniqueness of the banyan interconnection has the disadvantage of resulting in a blocking switch. Since a number of paths may share common links within the network, simultaneous connections of more than one inlet and outlet pair may result in conflicts in the use of these internal links [7]. In general, and for fairly uniform traffic patterns, these conflicts can be resolved by providing a small amount of buffering (typically, a single packet buffer) per switching element [11]. Such a single-buffered banyan network was analyzed in [12] under the assumptions of fixed-length packets and uniform traffic pattern. Uniform traffic strictly implies that the inlet and outlet used by each packet are independently chosen from a uniform distribution. It was found that the limiting throughput available is around 0.45. In a different study [13], a single-buffered banyan network was subjected to a somewhat nonuniform traffic pattern. In addition to a background of uniform traffic, a single point-to-point connection was established (that is, all the packets arriving on one particular inlet were directed to a given outlet). It was found that once the point-to-point traffic exceeded a throughput of about 0.45 the throughput of traffic sharing links with the point-to-point traffic fell towards zero. A simulation study of a packet switch based on the banyan network has shown that the limiting throughput can be increased from 0.45 to about 0.65 as the number of buffers per switching element is increased from 1 to 4 [14].

The use of greater packet buffering will increase not only the maximum throughput of the network, but also the delay incurred [11]. However, if the path followed by a packet through the network could be chosen so as to avoid links which are becoming congested, it may be possible to maintain high throughput and low delays. One means of doing so is to provide multiple paths within the interconnection network itself [15]. Another approach is to precede the interconnection network with a further network, the purpose of which is to establish conditions under which intercall packet conflicts are, in some sense, minimized. The use of a *sorting network* placed before an Omega (or banyan) network can be used to eliminate internal conflicts provided there are no repeated destinations amongst the input packets [4], [5]. The number of stages required by an N-input sorter, such as the Batcher network [16], requires $(\log_2 N + 1)/2$ times as many switching

Manuscript received September 1, 1987; revised June 8, 1988. This paper was presented in part at the Second Australian Teletraffic Research Seminar, University of Adelaide, Australia, November 30–December 1, 1987; and the International Zurich Seminar on Digital Communications, ETH, Zurich, Switzerland, March 8–10, 1988.

The authors are with Systems Development, Overseas Telecommunications Commission, Sydney, 2001 Australia.

IEEE Log Number 8824390.

stages than the corresponding banyan network. Instead of a presorter, the use of a randomizing, or *distribution*, network has been proposed in [8]. The purpose of this network is to distribute packets evenly throughout the succeeding banyan network. The simulation study cited earlier [14] has shown that such a distribution network is effective in reducing congestion resulting from nonuniform traffic patterns arising from *communities of interest*.

Each of the methods discussed in the preceding paragraph present individual packets with a multiplicity of paths. As mentioned earlier, this may lead to problems with the provision of real-time services owing to the increase in intracall packet delay variance due to the interaction with intercall traffic. A preferred solution may be to make the multiplicity of paths available for random selection at call setup time, yet once a path has been selected, all packets belonging to the call are constrained to follow the same path. Such an approach has been proposed in [17] as the basis for a fast packet switching exchange.

In this paper, we present a generalization of the distribution network approach. In order to avoid the problem of intracall delay variance, packets belonging to a given call are constrained to follow the same path through the switch. Path selection is at call setup and based on an algorithm designed to minimize, in some sense, the cost of the path. A number of path selection algorithms have been identified and are evaluated in this paper. In Section II, the operation of the switching network is described. The path selection algorithms are discussed in Section III together with a comparison of relative processing and memory requirements. An approximation technique using hybrid-simulation is presented in Section IV. Using this technique each path through the switch is modeled as a tandem connection of single server queues with finite buffer space. The utilization of each server, which depends on the traffic pattern and the method used to allocate paths, is determined by simulation. The results of this technique, such as the mean queue length and blocking probability as functions of mean applied load and buffering capacity, are presented in Section V.

II. THE MULTIPATH INTERCONNECTION NETWORK

A fast packet switch based on a multipath interconnection network is shown in Fig. 1. Overall switch operation is controlled by means of a *connection control processor* (CCP) which performs connection control and switch maintenance functions. The CCP, however, performs no per packet processing. *High speed trunk* (HST) lines provide either user access (via suitable network terminations) or access to other switches. Each HST is terminated by a *trunk controller* (TC) which performs link level protocol functions. Call setup packets are forwarded to the CCP which determines the routing information for that call.

The *multipath interconnection network* (MIN) in an ($N \times N$) switching fabric based on the use of self-routing baseline networks [6] with S stages of switching elements. Fig. 2 shows an example of an (8×8) MIN. The

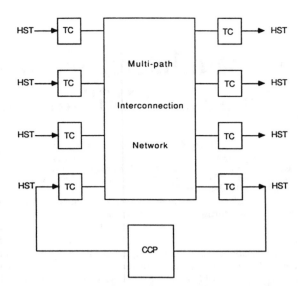

Fig. 1. Fast packet switch configuration. High speed trunk (HST) links terminate on trunk controllers (TC) which perform link level protocol functions. Multipath interconnection network (MIN) performs the switching function. Paths through the MIN are selected by the connection control processor (CCP).

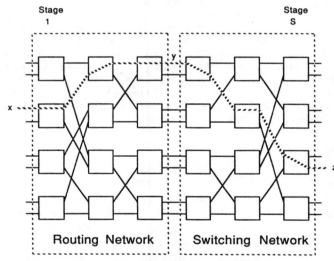

Fig. 2. An (8×8) multipath interconnection network. The dashed line illustrates a path between inlet x and outlet z via intermediate port y.

MIN consists of two baseline networks, the *routing network* (RN) and the *switching network* (SN). The addition of the RN effectively increases the number of paths available to a call from 1 to N. Outputs of the switching elements in one stage are connected to the inputs of switching elements in the next by means of *links*. Links at the input to stage 1 are called *inlets*; the links between the routing and swtiching networks are called *intermediate ports*, or simply *ports*; and links at the output of stage S are called *outlets*. So that we may consistently refer to links at the input of a stage, we regard outlets as being inputs to a null stage $S + 1$. Both networks consist of $\log_2 N$ stages of $N/2$ switching elements. Switching elements are (2×2) crossbars with buffer space for one or more packets at each input. An element at stage i will switch a packet up or down depending on the ith bit of the header. A free buffer must be available at the next stage to allow the packet to move forward. Flow between stages is controlled by means of hardware handshake lines between corresponding switching elements. In Fig. 2 the networks are shown as mirror images for conceptual simplicity; in practice, both RN and SN could be identical. This of course would simplify implementation.

The purpose of the RN is to route a packet to one of the N inlets of the SN. In its turn, the SN switches the packet to the required outlet. Since every MIN inlet has access to any of the RN outlets, each of which have access to any of the SN outlets, there exists N paths between each MIN inlet and outlet pair. In fact, the topology of the MIN is closely related to that of the Benes network which is a rearrangeable nonblocking network [7]. The CCP therefore has a choice of N possible paths when determining the route to be followed by a particular call.

III. PATH SELECTION ALGORITHM

In this section, we describe four possible path selection algorithms. References to processing and memory requirements assume a single processor CCP with random access memory. Minimum storage is taken to be $O(\log_2 N)$ per word and time requirements are estimated in terms of arithmetic and logic operations on such words.

1) Least-Cost Method: Consider the triplet (x, y, z) where x is an inlet, y is a port, and z is an outlet. Such a triplet uniquely defines a path through the MIN. Associated with the path is a set of links $R = \{(i, j)\}$ (indicated by the dashed line in Fig. 2) where each (i, j) is the jth link of stage i. In particular, $(1, j) = x$, $(S/2 + 1, j) = y$, and $(S + 1, j) = z$. Define the load matrix $\tilde{L}$ by

$$\tilde{L} = [L_{i,j}] \quad i = 1, 2, \cdots S + 1; \quad j = 1, 2, \cdots, N \tag{1}$$

where each element $L_{i,j}$ is the number of established calls on the jth link of stage i. In this method, also called the *optimal method*, the cost of a path is defined as the sum of the load on each link in the associated link set R. The CCP selects the path with the lowest cost. The method entails the following operations.

1) The set of links R associated with each of the N possible paths connecting the inlet and outlet pair is determined.
2) The cost is evaluated for each of the N possible paths.
3) The lowest cost path (with link set R_{opt}) is selected.
4) Each of the $L_{i,j}$ in R_{opt} are incremented.
5) When a call terminates each of the $L_{i,j}$, associated with that call, are decremented.

As a minimum, the least-cost method has a memory requirement of $O(N \log_2 N)$ due to the need to store the load matrix $\tilde{L}$. The processing required to determine path cost for each of the N paths is also $O(N \log_2 N)$. To this must be added the processing and/or memory needed to

obtain the link set at each path. Hence, for large N this method could be expensive to implement. On the other hand, since the complete state of the MIN is taken into account, it is expected that the method makes an optimal path selection.

2) Binary Method: Given an inlet and outlet pair (x, z), we can identify each of the N possible paths with a set of ports $P = \{y\}$. Since the intermediate port y uniquely identifies each of the path choices, it is tempting to select a path based solely on the state of these ports. Define a load array $[L_y]$ where each L_y is the number of calls applied to the yth port. Hence, for this method, we define the cost of a path through y as L_y. As in the previous method, the path with the lowest cost is selected. The method also attempts to take into account the structure of the MIN, if not its state, by using a binary search for the port with the lowest load. Implied in the method are the following steps.

1) Partition the set of intermediate ports P into two halves, P_1 containing the first $N/2$ ports and P_2 the remaining ports.
2) Calculate the sum of the link loads in each partition and select the partition with the lowest total cost.
3) Repeat the procedure until each partition contains a single port. The partition with the lesser load then identifies the selected path.

Memory requirements for the binary method are $O(N)$. Processing requirements depend primarily on the partitioning process. The set P is partitioned through at total of $\log_2 N$ steps. At the mth step, there are two partitions each containing $N/2^m$ ports whose loads must be summed and compared. Hence, the number of arithmetic and logic operations required is

$$A = 2 \sum_{m=1}^{\log_2 N} \frac{N}{2^m} = 2N \left(1 - \frac{1}{N}\right) = O(N). \tag{2}$$

Both memory and processing requirements are therefore $O(N)$ in contrast to $O(N \log_2 N)$ for the optimal method.

3) Linear Method: This method is essentially the same as the binary method with the exception that a simple linear search of the set of intermediate ports is used. Unlike either of the previous two methods, the structure of the MIN is ignored, and the sole basis for path selection is the load on the intermediate ports. As for the binary method, the memory and processing requirements are $O(N)$, although actual processing burden would be somewhat less for the linear method due to the simpler search algorithm.

4) Random Method: In this case, all state information is ignored and the path is selected by a uniformly random choice of one of the N intermediate ports. With memory and processing requirements being $O(1)$ this method is clearly the least expensive to implement and forms the basis of a number of earlier works. The distribution network proposed in [8] essentially makes a random path selection for each packet, while the randomization network of [17] establishes paths randomly for each call.

IV. MODEL AND ANALYTIC APPROACH

This section outlines the model used, and approach taken, to evaluate the relative efficacies of the path selection strategies considered in the previous section. The performance measures of interest are the expected queue length and the maximum throughput through the MIN. In the case of infinite buffering, the maximum throughput is equal to the applied load leading to an unbounded expected queue length. In the case of finite buffering, the blocking probability is also of interest. In addition to considering the different route selection strategies, we wish to study the effect of a number of switch parameters such as the number of stages in the switch and the amount of buffering space available at each stage.

A Markov chain could be used to model the system. However, the number of states involved in a Markov chain representation of even a single banyan matrix is $k^{(N \log_2 N)/2}$ where k is the number of states of a switching element [12]. Since we wish to consider, amongst other parameters, the effect of the buffering capacity per element on queueing behavior, k itself can become large. The computational effort required to handle the resulting very large state-space is clearly unfeasible. Therefore, a hybrid analytic-simulation method has been employed.

In the approach adopted to analyze the performance of the MIN, we decouple the per call path selection process from the per packet switching process. A simulation is first used to determine how, at steady state, traffic is distributed through the switch as a result of the path selection algorithms. Next, a simple analytic model of a switch path, based on a chain of finite waiting place single-server queues, is used to estimate performance metrics such as expected queue length and blocking probability.

A. The Switch Simulator

The first step of the method is to obtain moments of the distribution of the traffic intensities on the links of the MIN as a result of each path selection algorithm. This is produced by a switch simulator. Inputs to the simulator are the size of the user population, the traffic pattern, and the route selection method. Outputs from the simulator are the mean and variance of the traffic intensity on the internal links.

Traffic is modeled using a similar approach to that employed in [18]. The call arrival process is assumed to be Poisson with rate ξ and exponentially distributed call holding times with mean η^{-1}. Let n_k, $k = 1, 2, \cdots$, be the number of calls in progress at transition k, and assume that $n_0 = L$. Then the number of calls in progress at the next transition will increase to $L + 1$ with probability

$$p_{L,L+1} = \begin{cases} \dfrac{\xi}{\xi + L\eta}; & L = 0, 1, \cdots, M - 1 \\ 0, & \text{otherwise} \end{cases} \quad (3)$$

where M is the size of the user population. This is just the probability that, at the next transition of n_k, a new call arrives, given L calls currently in progress. Similarly, the

probability that, at the next transition, an existing call terminates is $p_{L,L-1} = 1 - p_{L,L+1}$. The average packet arrival·rate while $n_k = L$ is taken to be $\lambda = rL$, r^{-1} being the average packet interarrival time per call.

A newly arriving call has its input chosen, with equal probability, from one of the N inputs to the MIN. The corresponding output from the MIN depends on the nature of the traffic pattern selected. Three traffic patterns are considered as follows.

1) Uniform Pattern: The output is chosen independently, and equiprobably, from one of the N outputs.

2) Horizontal Pattern: A call arriving at input i, $i = 2, 3, \cdots N - 1$ has a corresponding output chosen, with equal probability, from $(i - 1, i, i + 1)$. Calls arriving on either the first $i = 1$ or last $i = N$ inputs have outputs chosen, equiprobably, from $(1, 2, N)$ or $(1, N - 1, N)$, respectively. This pattern is expected to result in a traffic distribution within the switch which is somewhat rougher than would be obtained with the uniform pattern.

3) Diagonal Pattern: In this case, a call arriving on input i, $i = 2, 3, \cdots N - 1$ has an output chosen, equiprobably, from $(N - i, N - i + 1, N - i + 2)$. Calls arriving on inputs $i = 1$ or $i = N$ have outputs chosen, equiprobably, from $(1, N - 1, N)$ or $(1, 2, N)$, respectively. The traffic distribution within the switch obtained by this traffic pattern would be expected to be rougher than that resulting from either of the above traffic patterns.

When a new call arrives (with inlet and outlet determined by the traffic pattern) a path selection algorithm is used to determine the appropriate path. Once the path for a call has been decided, each element of the load matrix $\tilde{L}$ associated with a link used by the path is incremented. In addition, the call is allocated a circuit number to which the path link set is bound. When a call terminates, one of the currently allocated circuit numbers is selected at random. The associated link set is then used to decrement the corresponding elements of the traffic matrix $\tilde{L}$.

The output from the switch simulator are moments of the distribution of the traffic intensities. In particular, the mean link intensity μ_λ can be written as

$$\mu_\lambda = \frac{r}{N(S + 1)} \sum_{i=1}^{S+1} \sum_{j=1}^{N} L_{i,j} \quad (4)$$

and the variance σ_λ^2 is given by

$$\sigma_\lambda^2 = \frac{r^2}{N(S + 1)} \sum_{i=1}^{S+1} \sum_{j=1}^{N} \left(L_{i,j} - \frac{\mu_\lambda}{r} \right)^2 \quad (5)$$

given that each call generates packets at an average rate r packets per second.

In Figs. 3–5, we plot normalized link intensity variance $v = \sigma_\lambda^2 / C$ as a function of normalized mean link intensity $p = \mu_\lambda / C$. Normalization is with respect to the maximum link capacity C. Fig. 3 assumes uniform traffic, Fig. 4 horizontal traffic, and Fig. 5 uses a diagonal pattern. An 8×8 MIN is assumed with link capacity equal to $20r$. Traffic is generated by a population of users each produc-

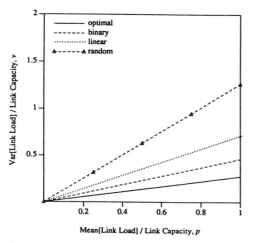

Fig. 3. Normalized variance v versus mean p of the traffic distribution through an (8×8) MIN showing the effect of each path selection algorithm with uniform traffic pattern.

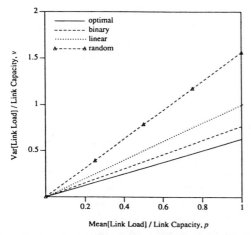

Fig. 4. Normalized variance v versus mean p of the traffic distribution through an (8×8) MIN showing the effect of each path selection algorithm with horizontal traffic pattern.

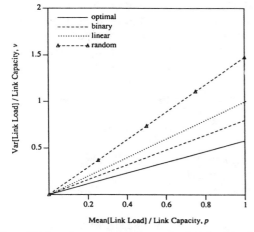

Fig. 5. Normalized variance v versus mean p of the traffic distribution through an (8×8) MIN showing the effect of each path selection algorithm with diagonal traffic pattern.

ing 0.5 Erlangs of traffic. The size of the user population is variable from 0 to 160. In each case, the four path selection algorithms are considered. Three observations can be made in the light of these results. First, for a given selection algorithm and traffic pattern σ_λ^2 is linearly proportional to μ_λ. Second, the constant of proportionality increases as the traffic pattern becomes less uniform. Third, optimal path selection results in consistently the lowest σ_λ^2. This is followed by binary and linear path selection, with random selection yielding the highest σ_λ^2. It is worth noting that, in all cases, binary is not substantially worse than optimal. In view of the $O(\log_2 N)$ reduction in memory and processing costs, the binary selection algorithm may be a useful compromise between implementation cost and variance reduction.

B. Switch Channel Model

The switch simulation discussed in Section IV-A allows us to determine how traffic is distributed across the links of the MIN as a result of the path selection algorithms described in Section III. We now wish to estimate the per packet queueing performance. Two key factors determining the queueing behavior of the MIN are the interaction between packets of different calls sharing common links, and the "back-pressure" exerted on the flow of packets by lack of free buffer space in successive stages. We wish to formulate a simple queueing model of a path through the MIN which accounts for these factors. The set of links R associated with the triplet (x, y, z) will be referred to as a *switch channel*, or simply *channel*. Traffic generated by the users of a switch channel, called *channel traffic*, enters the system at the first stage and proceeds, stage by stage, to the final stage of the channel. This procession depends on the availability of free buffer space at each stage. In addition to channel traffic, individual links of each channel will be offered traffic arising from other channels which share links with the switch channel. This traffic is said to be composed of *transit packets*.

The average arrival rate of channel traffic is λ_0, while the transit traffic sharing the link between the $(i - 1)$th stage and the ith stage has an arrival rate λ_i. Hence, the total traffic applied to this ith link has an arrival rate $\lambda_0 + \lambda_i$. The following assumptions are made to enable the formulation of a simple model of a switch channel.

1) Each packet is assumed to have a length drawn independently from a negative exponential distribution. There are essentially three reasons for this assumption: 1) the resulting Markovian service time distribution greatly simplifies the model; 2) almost without exception, previous work has dealt with the fixed length packet case (for example, [11]–[14]) with variable length packets not being investigated; 3) the assumption of exponentially distributed packets lengths is conservative in comparison to fixed length packets, and hence safe to assume.

2) The arrival process of the transit packets at each stage is independent of the arrival process of the channel packets. This assumption is based on the observation that the set of inlets with a path to one of the links of a $(2 \times$

2) element is disjoint from the corresponding set with a path to the other link of the same element [11]. The independence assumption will be good in those cases where transit packets share only one link in common with the switch channel, and will become less accurate as the number of shared links increase. However, since the path selection algorithms attempt to distribute traffic throughout the MIN, it is reasonable to suppose that sharing of links between switch channels is in some sense minimized.

3) The arrival process at stage i is assumed to be Poisson with rate $\lambda_0 + \lambda_i$ and the average service rate at each stage is μ. This assumption is motivated by the Palm-Khintchine theorem which guarantees that the sum of n independent renewal processes obeys Poisson statistics provided n is sufficiently large [19].

4) Each link within the MIN is offered traffic whose mean arrival rate is independently drawn from a normal distribution with mean μ_λ and variance σ_λ^2.

Under these assumptions each switch channel is statistically identical and can be modeled by a tandem network of $M/M/1$ queues with either finite or infinite buffering at each stage.

The buffer space at stage i is K_i. Three cases will be considered. In the first, an infinite buffer is assumed to exist at each stage. Second, finite buffers are assumed at every stage except the first, which incorporates the input FIFO of the trunk controller and is assumed infinite. Due to hardware flow control exerted between stages, no packets are lost when the buffer in any stage is full. Hence, in both of these cases, the MIN forms a pure delay system. In the third case, a finite buffer exists at each stage including the first, although it is not necessarily of the same size as the other buffers. In this case, the MIN acts as a loss system due to the nonzero blocking probability of the first stage buffer.

1) Infinite Buffers at Every Stage: With unlimited buffer capacity at each stage, the service of a packet at stage i is independent of the state of the queue at stage ($i + 1$). The number of packets that a channel packet can expect to find queued at stage i depends only on the traffic arrival rate $\lambda_0 + \lambda_i$. Hence, the total mean queue length of the switch channel can be written as

$$Q = \sum_{i=1}^{s} \frac{\rho_i}{1 - \rho_i} \qquad (6)$$

where $\rho_i = (\lambda_0 + \lambda_i)/\mu$ is the intensity at stage i. The mean queue length for the entire MIN can be obtained by averaging over all active switch channels.

2) Infinite Input Buffer, Finite Stage Buffers: In this case, a buffer of unlimited capacity is still assumed at the input to the channel (at stage 1) but each subsequent stage i is assumed to have finite capacity for K_i, $i = 2, 3 \cdots S$ packets, including the packet being served. Hence, each stage i of the channel model is modeled as a finite $M/M/1/K_i$ queue [20]. With the hardware flow control described earlier, a packet at the head of the queue in stage i cannot be served if the queue in stage $i + 1$ is blocked.

Given that the outlet of the channel is never blocked (that is, no flow control is exerted between separate MIN's), the blocking probability of stage S depends only on the combined arrival rate of channel and transit packets $\lambda_0 + \lambda_i$ and is denoted $P_b(S)$. Since the server of the previous stage will be blocked with this probability, the effective service rate available to stage $S - 1$ is $\mu(1 - P_b(S))$. The queueing performance of each stage therefore depends on the blocking probability, and hence traffic intensity, of the succeeding stage. The mean queue length and blocking probability of the ith stage can be determined iteratively from

$$\rho_i = \frac{\lambda_0 + \lambda_i}{\mu(1 - P_b(i+1))} \qquad (7a)$$

$$E[q(i)] = \frac{\rho_i}{1 - \rho_i}\left(1 - (K_i + 1)P_b(i)\right) \qquad (7b)$$

$$P_b(i) = \frac{\rho_i^{K_i}(1 - \rho_i)}{1 - \rho_i^{K_i+1}} \qquad (7c)$$

with initial condition $P_b(S + 1) = 0$ and given $K_1 = \infty$. The mean queue length of the switch channel is obtained by summing over the S stages of the channel

$$Q = \sum_{i=1}^{S} E[q(i)]. \qquad (8)$$

In both, this and the previous cases the inlet is assumed to have unlimited buffer space. We now remove this assumption and consider finite inlet buffering.

3) Finite Buffers at Every Stage: With finite buffer space provided at each inlet to the switch there is now a nonzero probability that traffic arriving for a channel will be lost. This in turn means that the traffic intensity at each subsequent stage is reduced by the probability that the first stage is full. Hence, by replacing (7a) by

$$\rho_i = \frac{\lambda_0 + \lambda_i}{\mu} \frac{1 - P_b(1)}{1 - P_b(i+1)}; \quad i = 2, 3, \cdots S \qquad (9)$$

the queueing behavior of the channel can be determined. (Note that we are assuming that the Poisson nature of the arrival process has not been destroyed.) However, (9) requires $P_b(1)$ to be known prior to its calculation by (7c). This problem can be overcome by applying an iterative procedure as follows.

1) Make an initial guess (say, zero) for the inlet blocking probability $P_b(1)$.

2) Using (9) in (7), determine $P_b(i)$ and $E[q(i)]$ for each stage i.

3) With new value of $P_b(1)$ repeat previous step.

4) Repeat previous two steps until the value obtained for the inlet blocking probability converges.

V. Results

In Fig. 6 we show the mean queue length as a function of normalized mean link load $p = \mu_\lambda/C$ with normalized

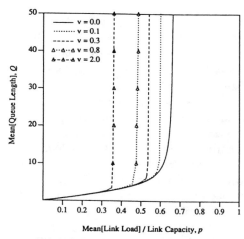

Fig. 6. Mean queue length Q of a switch channel versus mean load p with variance v as a parameter; number of stages $S = 6$, stage 1 with infinite buffer, remaining stages are singly buffered.

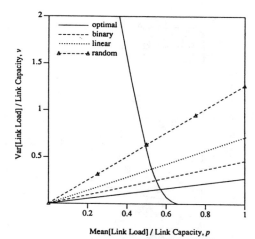

Fig. 7. Variance v versus mean link load p for uniform traffic (as in Fig. 3) superimposed on maximum throughput versus variance (derived from Fig. 6).

link load variance $v = \sigma_\lambda^2/C$ as a parameter. The MIN is assumed to have six stages. Each stage, with the exception of the first stage, has buffer space for one packet. The first stage has unlimited buffer capacity. Under light loads the queue length is independent of link variance. However, as the load increases the queue behavior becomes heavily dependent on the link variance with queue performance improving as the link variance is reduced. The lower bound on mean queue length occurs, as might be expected, when $v = 0$. We wish to relate this behavior back to the path selection algorithms. The maximum throughput as a function of link variance is plotted in Fig. 7. Superimposed on this is the variance versus mean load information earlier shown in Fig. 6 for the uniform traffic pattern case. Similar figures could be drawn for the other traffic patterns. From the results in Fig. 7, we see that maximum throughput occurs at a mean load approaching 0.5 in the case of random path selection, with throughput increasing to about 0.6 for optimal selection.

The limiting throughput in Fig. 7 occurs at a link load of about 0.65, and requires vanishingly small link variance. The path selection algorithm needed to approach this limit would doubtless make use of the rearrangeable nature of the Benes network. In other words, following each call setup or termination, paths could be reselected to jointly minimize the cost, hence load variance, of existing calls. The remaining results will be developed assuming limiting throughput conditions.

The MIN has twice as many stages as a switch using a single baseline network. We would therefore like to see if there is a penalty associated with the increased number of stages. In Fig. 8, the mean queue length is shown with the number of stages as a parameter. Again, we assume each stage is singly buffered, except stage one which has an infinite buffer. As might be expected, the maximum throughput decreases as the number of stages is increased. However, throughput rapidly reaches its limiting value with about 10 stages. This corresponds to only a switch size of (32×32). Hence, for any reasonable switch size,

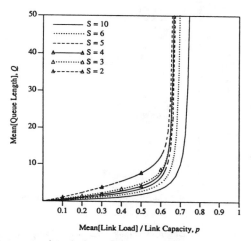

Fig. 8. Mean queue length Q versus mean load p with number of stages S as a parameter; infinite input buffer, remaining stages singly buffered.

the increase in the number of stages due to the use of a MIN has no effect on maximum throughput.

So far the MIN has been singly buffered. We now consider the effect of increasing the buffer capacity of each stage. As before, the first stage has infinite capacity, and the MIN has six stages. Mean queue length is shown in Fig. 9 with buffer size as the parameter. While increasing the buffer capacity results in increased limiting throughput, there is little incentive to using a buffer space greater than about 5. Beyond this throughput improvements become small, and to take advantage of the increased throughput significantly greater queues, and delays, are incurred. A similar conclusion was reached in [21].

The effect of finite input buffering at stage 1 is shown in Figs. 10 and 11. Fig. 10 shows the expected queue length versus mean link load with the input stage buffer size as a parameter. The MIN is assumed to have, as before, six stages, with one buffer per stage. For finite input buffer capacity the expected queue length, and hence delay, remains bounded even for high link loads. The cor-

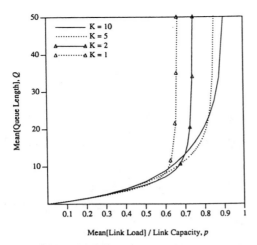

Fig. 9. Mean queue length Q versus mean load p with buffer size per stage K as a parameter; number of stages $S = 6$, infinite input buffer.

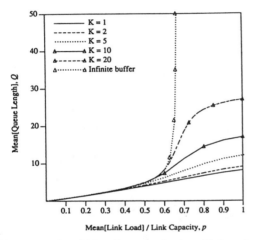

Fig. 10. Mean queue length Q versus mean load p with input buffer size K as a parameter; number of stages $S = 6$, remaining stages singly buffered.

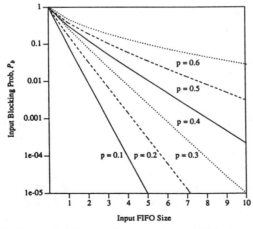

Fig. 11. Blocking probability P_b at input versus input buffer size with mean load p as a parameter.

responding blocking probability is displayed in Fig. 11. An input buffer capacity of at least 10 packets would be required for blocking probability less than, say, 10^{-3}.

VI. CONCLUSIONS

A, method has been presented which, by means of a multipath interconnection network, increases the throughput of banyan-like networks without sacrificing the property of path uniqueness. Path selection takes place at call setup time with subsequent packets belonging to the call constrained to follow the selected path. This ensures packet sequence integrity and minimal intracall packet delay variability. Four possible selection algorithms (namely, optimal, binary, linear, and random) were presented which, in some sense, minimized the cost associated with each call. Each algorithm made use of different amounts of switch state information and hence have different memory and processing requirements. The operation of the MIN under a number of traffic patterns was then simulated to determine the effect of the path selection algorithms. It was found that the ability of the path selection algorithm to distribute traffic throughout the switch is inversely related to the cost of implementing the algorithm. The binary algorithm appeared to be a useful compromise.

Next, a queueing model of a path through the MIN was formulated in order to estimate queueing behavior in terms of expected queue lengths and blocking probability. It was found that maximum throughput increases as the variance of the link load distribution decreases. The increase in throughput due to the optimal algorithm, compared to the random algorithm, was about 10 percent. This improvement increases to about 15 percent in the limiting case (corresponding to zero variance). The queueing model was also used to determine the effect of the number of stages in the MIN and the amount of buffering per stage.

The path selection algorithms considered in the current work can be described as static in the sense that, once a path is allocated to a call, the call retains that path until termination. However, the MIN can be operated as a rearrangeable nonblocking switch. Dynamic path selection algorithms, which reallocate paths to existing calls following each call setup or termination, could doubtless be formulated to take advantage of the rearrangeable nature of the MIN. The present work does not, however, address this issue.

ACKNOWLEDGMENT

The authors would like to thank the anonymous reviewers for their helpful suggestions. The authors would also like to thank the Manager Director of OTC for permission to publish this work. The views expressed are those of the authors and not necessarily the official opinion or policy of OTC.

REFERENCES

[1] L. R. Goke and G. J. Lipovski, "Banyan networks for partitioning multiprocessor systems," in *Proc. 1st Ann. Symp. Comp. Arch.*, Apr. 1973, pp. 21–28.
[2] C. Mead and L. Conway, *Introduction to VLSI Systems*. Reading, MA: Addison-Wesley, 1979.
[3] M. A. Franklin, S. A. Kahn, and M. J. Stucki, "Design issues in the

development of a modular multiprocessor communications network," in *Proc. 6th Ann. Symp. Comp. Arch.*, Apr. 1979, pp. 182–187.

[4] A. Huang and S. Knauer, "Starlite: A wideband digital switch," in *Proc. IEEE GLOBECOM '84*, Nov. 1984, pp. 5.3.1–5.3.5.

[5] C. Day, J. Giacopelli, and J. Hickey, "Applications of self-routing switches to data fiber optic networks," in *Proc. ISS'87*, vol. 3, Mar. 1987, pp. A7.3.1–A7.3.5.

[6] C.-L. Wu and T.-Y. Feng, "On a class of multistage interconnection networks," *IEEE Trans. Comput.*, vol. C-29, pp. 694–702, Aug. 1980.

[7] T.-Y. Feng, "A survey of interconnection networks," *Computer*, vol. 14, pp. 12–17, Dec. 1981.

[8] J. S. Turner, "Design of an integrated services packet network," in *Proc. 9th ACM Data Commun. Symp.*, Sept. 1985, pp. 124–133.

[9] ——, "Design of a broadcast packet switching network," presented at IEEE INFOCOM '86, Miami, FL, Apr. 1986.

[10] W. A. Montgomery, "Techniques for packet voice synchronization," *IEEE J. Select. Areas Commun.*, vol. SAC-1, pp. 1022–1028, Dec. 1983.

[11] D. M. Dias, and J. R. Jump, "Analysis and simulation of buffered delta networks," *IEEE Trans. Comput.*, vol. C-30, pp. 273–282, Apr. 1981.

[12] Y. C. Jenq, "Performance analysis of a packet switch based on single-buffered banyan network," *IEEE J. Select. Areas Commun.*, vol. SAC-1, pp. 1014–1021, Dec. 1983.

[13] L. J. Wu, "Mixing traffic in a buffered banyan network," in *Proc. 9th ACM Data Commun. Symp.*, Sept. 1985, pp. 134–139.

[14] R. G. Bubenik and J. S. Turner, "Performance of a broadcast packet switch," in *Proc. IEEE ICC'87*, June 1987, pp. 31.6.1–31.6.5.

[15] C. A. Lea, "The load-sharing banyan network," *IEEE Trans. Comput.*, vol. C-35, pp. 1025–1034, Dec. 1986.

[16] K. E. Batcher, "Sorting networks and their applications," in *Proc. AFIPS Conf., 1968 Spring Joint Comp. Conf.*, pp. 307–313.

[17] M. De Prycker and J. Bauwens, "A switching exchange for an asynchronous time division based network," in *Proc. IEEE ICC'87*, June 1987, pp. 22.3.1–22.3.8.

[18] C. J. Weinstein, M. L. Malpass, and M. J. Fischer, "Data traffic performance of an integrated, circuit- and packet-switched multiplex structure," in *Proc. IEEE ICC'79*, June 10–14, 1979, pp. 24.3.1–24.3.5.

[19] T. Suda, H. Miyahara, and T. Hasegawa, "Performance evaluation of a packetized voice system—Simulation study," *IEEE Trans. Commun.*, vol. COM-32, pp. 97–102, Jan. 1984.

[20] L. Kleinrock, *Queueing Systems, Vol. 1: Theory.* Wiley, 1975.

[21] D. M. Dias and J. R. Jump, "Packet switching interconnection networks for modular systems," *Computer*, pp. 43–53, Dec. 1981.

Gary J. Anido (M'88) received the B.E. and Ph.D. degrees from the University of N.S.W., Australia, in 1984 and 1988, respectively, in electrical engineering and computer science.

From 1984 to 1986 he was employed as a Research Engineer by Telecom Australia where he was involved in the design and development of the integrated voice and data local area network XLNET. Since 1987 he has worked under contract to the Overseas Telecommunications Commission in the area of Broadband Networks and Services. He is also a Visiting Lecturer in the Department of Electrical and Computer Engineering at the University of Wollongong. His research interests include multiservice local area networks and fast packet switching, and he is the author of a patent and a numbers of papers in these fields.

Dr. Anido is a member of the IEEE Communications Society and a member of the committee for the N.S.W. chapter of the Society.

Anthony W. Seeto (S'74–M'75) received the B.E.(Hons. I) degree from the University of Queensland, Australia, in 1975, and the Ph.D. degree from the University of N.S.W. Australia, in 1983.

He worked in AWA Ltd. (1976–1981) and in Interscan Australia P/L (1982) as a Systems Engineer in the Australian Microwave Landing System project—INTERSCAN. During this period, he also engaged in research in adaptive signal processing techniques at the University of NSW and was awarded a Ph.D. degree in 1983 for that work. From 1983 to 1986, he was a Senior Lecturer at the School of Electrical and Electronic Engineering, Nanyang Technological Institute, Singapore. He returned to Australia in 1986 and joined OTC's Systems Development Section under contract. He is now heading the Broadband Networks and Services Group in that section. He also holds a visiting academic post at the University of Technology, Sydney. His research interests include digital signal processing. ATM networks and broadband services.

Dr. Seeto is a member of IEEE Communications Society and is currently the Vice Chairman of the N.S.W. chapter of the Society.

Architecture of a Packet Switch Based on Banyan Switching Network with Feedback Loops

HITOSHI UEMATSU AND RYUICHI WATANABE

Reprinted from *IEEE Journal on Selected Areas in Communications*, Vol. 6, No. 9, December 1988, pages 1521-1527.
Copyright © 1988 by The Institute of Electrical and Electronics Engineers, Inc. All rights reserved.

Abstract—Extensive research is progressing on new communications networks which effectively use the "packetiness" of the information source. Various multistage switching networks are being proposed as the switch used at the node in these new networks. These switching networks consist of 2-input × 2-output switch cells with internal buffer memory. Of these, the banyan switching network is extremely promising. It requires the fewest switch cells for self-routing, and the cell itself can control the route within the switching network. However, to prevent congestion in this switching network, the throughput per port must be reduced when the number of ports is increased. This fact becomes an obstacle in applying this to high-speed, large-capacity communication networks.

This paper proposes and describes an original feedback banyan switching network topology and control algorithm. This feedback banyan network prevents congestion even at high throughput. It does so by establishing a feedback route for the output packets from the output port to the input port of the banyan switching network. Input accommodated logical channels at that do not encounter congestion within the switching network, are output directly from the intended output port. Other packets are returned to the input port via the feedback route, then rerouted.

I. INTRODUCTION

MANY countries are currently researching packet transport networks, based on packet multiplexing and switching, which effectively use the "packetiness" of information source signals. A large-scale high-speed packet network needs not only packet exchange nodes but packet cross-connect nodes similar to those in today's synchronous transfer networks. The packet exchange nodes set up logical channels call by call and gather packet traffic from subscribers. Consequently, their switches require frequent set up changes and may input ports to accommodate many subscribers but the traffic of each port is low. The packet cross-connect nodes set up logical channel bundles (hereafter referred to as the virtual paths) and multiplex/demultiplex virtual paths to merge different areas into a large capacity optical transmission line. Therefore, their switches require large capacity and high traffic through each port, but require less frequent set up changes [1].

Many countries are researching multistage switching networks for large capacity packet nodes. These networks consist of 2 inputs × 2 outputs switch cells connected by

Manuscript received September 11, 1987; revised June 30, 1988.
H. Uetmatsu is with the Transport Processing Laboratory, NTT Transmission Systems Laboratories, Kanagawa, 238-03 Japan.
R. Watanabe is with the Research Planning Department, NTT Transmission Systems Laboratories, Kanagawa, 238-03 Japan.
IEEE Log Number 8824394.

links. In these cells, buffer memories are placed to prevent collisions [2]. One such network is a buffered banyan network. This network, which has an equal number of input/output ports, requires the fewest switch cells. Moreover, the switch cell itself controls routing within the switching network [3]. However, the banyan network has an internal congestion problem. To prevent congestion in the switching network, the throughput per port must be reduced when the size of the switching network is increased. Thus, relatively high-speed switch cells are required to construct a high-throughput switching network.

This paper proposes a new banyan network topology and its control algorithm for packet cross-connect nodes. Setup simulation is performed to determine various characteristics. Features of the proposed feedback banyan network are explained. To reduce internal congestion of the banyan switching network, feedback loops are established which connect the switching network input and output ports. This new routing algorithm operates as follows. If input virtual paths arrive at the intended output port without encountering congestion through the switching network, they are not returned to the input port. Virtual paths which cannot avoid congestion in one path are returned to the input port via a newly established feedback loop, then rerouted.

II. FEEDBACK BANYAN SWITCHING NETWORK ARCHITECTURE

A. Properties of Banyan Switching Networks

The banyan switching network consists of switch cells with 2-input × 2-output ports, and buffer memories which prevent collisions of packets [4]. As buffered banyan network is a single-path switching network with only a one-route connection between any chosen input and output port, this switching network has the following advantages. 1) The destination port number is added to the header of the input packets. A path is selected by self-routing and the packets are delivered to the intended output ports. 2) For multistage switching networks with the same number of input and output ports, there are minimal number of switch cell stages. 3) Packets maintain sequence integrity.

However, because the switching network is a single-path network, the packet flow can concentrate at some of the links connecting switch cells. In worst case, the throughput per port which is normalized port capacity ("normalized throughput") T is represented by (1), (2)

$$T = 1/2^x \tag{1}$$

$$T = \text{int} \left\{ (\log_2 L)/2 \right\} \tag{2}$$

where L is the number of ports.

This problem has two possible solutions. Method 1) reduces the input flow to $1/2^x \{ X = \text{int} [(\log_2 L)/2] \}$ of link capacity speed. Method 2) uses a multipath switching network construction to multiply the routes connecting a pair of input and output ports [5].

As stated, method 1) reduces throughput per port according to the increased size of the switching network. Thus, it is difficult to increase throughput by increasing network size. Thus, method 1) does not resolve the inherent problem. On the other hand, if method 2) is applied, it is difficult to achieve all three advantages of the banyan switching network mentioned previously.

For example, in cascade banyan networks, which consist of distribution and routing buffered networks [6], it is difficult to maintain packet sequence integrity due to multipath queueing delay differences.

The proposed feedback banyan switching network establishes new feedback loops to connect the banyan switching network input/output ports. The input virtual paths encounter congestion are fed back to the input ports via these loops. The paths pass through the switching network twice. This algorithm prevents reduction in port throughput.

The aforementioned advantages of the banyan switching network, advantages 1) and 2) are lost by this approach.

B. Feedback Banyan Switching Network Architecture

The configuration of the proposed feedback banyan network is compared to a conventional banyan network in Fig. 1. To construct the L-input $\times$ L-output feedback banyan network, a $2L$-input $\times$ $2L$-output banyan network whose odd output ports are fed back to the same number input ports is used. Its even input and output ports are used as L-input $\times$ L-output feedback banyan network ports. Network arrangement when $L = 8$ is shown in Fig. 1.

C. Packet Sequence Integrity of Feedback Banyan Switching Networks

The packets put into the feedback banyan network pass over two types of routes. One is a direct route from input port to output port through the banyan switching network (hereafter referred to as the "banyan route"). The other is a route from the input port to output port via the feedback loop (hereafter referred to as the "feedback route"). The number of available feedback loops equals the number of output ports. Thus, many feedback routes are available for one input and output port pair. Therefore, if the same virtual path packets are passed through different routes and switched to the same output port, the resulting

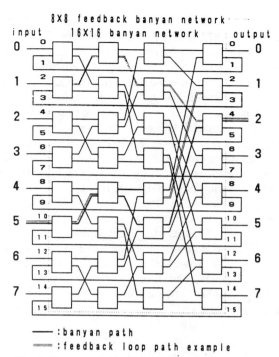

Fig. 1. Banyan switching network with feedback loops.

queueing delay differs according to the route. As a result, there is the possibility of a change in packet sequences.

To avoid this, the packets belonging to the same virtual paths should be all passed through the same route ("single-path routing"). This eliminates the packet sequence integrity problem.

D. Routing Algorithm

In the single-path routing scheme, virtual paths are selected to form effective routing as follows. First, a route for each virtual path is selected, the packets of each virtual path are sent through the designated route.

In determining this route, increasing the number of virtual paths on the banyan routes will improve the overall delay characteristics. The banyan route traverses the network only once while feedback routes traverses as many switching cells. Consequently, to decrease net flow of the network and keep delay times to minimum the usage of banyan routes must be maximized.

However, overusage of the banyan routes may cause partial internal congestion. This is because flows concentrate at some links of the switching network, as in normal banyan networks. Thus, in the proposed network, a limit value V is introduced for the flow passing through any link along the banyan route. At first, according to connection demands between the input/output ports, a route is selected from the banyan routes. If the flow exceeds the limit value V for any one of the links on the banyan route, then an alternative route is selected from the available feedback routes.

Definitive values of banyan route-flow limit values V will be discussed in Section II-E.

The feedback route is decided as follows. For the feedback route, the feedback banyan network can be transformed into a switching network where packets pass through the network only once. By some modification, the network can be transformed into a cascade banyan or Benes switching network topology as indicated in Fig. 2 [5], [6]. Therefore, the feedback routing decision is same as that in the cascade banyan or Benes switching network.

The noncongestive algorithm for cascade buffered banyan network is well known. In this algorithm, the first banyan network works as a distribution network which distributes input flow packet by packet and the last banyan network works as a routing network which transports packets to the intended output port [6].

However, the above technique cannot maintain packet sequence integrity because the packets of one virtual path flow by different routes of a Benes network, stated in Section II-C. Rearrange nonblocking circuit setup algorithms are known [7]. However, applying this to packet cross-connect switches, the moment virtual paths are rearranged, packet sequence integrity is broken. These algorithms make cascade banyan or Benes network linkages equalize completely. However, they cannot maintain packet sequence integrity. So, they are not suitable for packet cross-connect switches.

Cross-connect switches are generally used under the condition that paths or channels are multiplexed at each port. However, when there is only one path or channel at each port, jumper wires are used for cross connections. Packet flow can be distributed virtual path by path in a Benes switching network. The Benes network $B(L)$ equivalent to the feedback route is up–down and left–right symmetrical as indicated in Fig. 2. Consider the case where feedback routes are to be established between all first- and last-stage switch cells of the equivalent Benes network $B(L)$ in Fig. 2. When some virtual paths are already setup, an algorithm is needed which decides feedback routes equalize flow through each link. In this case, there are two types of feedback routes: link A equivalent sub Benes switching network $B_u(L/2) \rightarrow$ link C and link $B \rightarrow$ equivalent sub-Benes switching network $B_d(L/2) \rightarrow$ link D. However, with the previously selected banyan route and feedback route, part of the link capacity is already used in links A, B, C, and D.

Thus, routing is decided as follows to establish a new path along the feedback route from links A, B, C, and D.

1) When only one link-flow volume of links A, B, C, and D is the maximum permitted, the other feedback route is chosen.

2) When flow of two or more links among links A, B, C, and D equals the maximum permitted, the feedback route using the link with smallest flow is selected.

3) In all other cases, the feedback route through the sub-Benes switching network $B_d(L/2)$ is selected.

The above procedure completes the selection of the initial and final stage links of the equivalent Benes switching network $B(L)$.

Next, the initial and final stage links of the equivalent sub-Benes network $B_u(L/2)$ and sub-$B_d(L/2)$ are selected. For the sub-Benes network $B(L/2)$ topology,

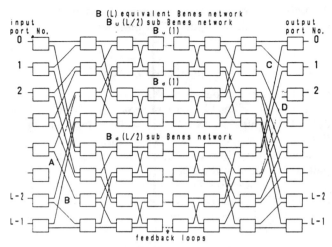

Fig. 2. Equivalent Benes switching network for feedback routes.

which is the equivalent Benes switching network $B(L)$ of one stage smaller size construction, we can apply the previously mentioned procedure for routing decisions.

The above procedure is sequential applied, to the decreasing equivalent sub-Benes networks $B(L)$. All routing decisions between the first-stage switch cells (i) and the last-stage switch cells (j) are completed by repeating this procedure down to the final equivalent sub-Benes network $B(1)$.

The definitive algorithm of this routing decision method mentioned above is shown in the Fig. 3 flow chart.

The above proposed algorithm is not rearrangeable but packet flow will not be equalized completely and may not be completely noncongestive. However, this algorithm and network reduce the internal congestion of a conventional banyan network extremely well as indicated in Fig. 4.

E. Internal Congestion of Feedback Banyan Network

In the above algorithm, the limit value V plays an important role for reducing congestion. Increasing the value V increases banyan paths in the feedback banyan network. Increase in banyan paths causes reduction of average queueing delay of the whole network. However, increasing banyan path usage may allow internal congestion in the network. Because the total flow of a link is the sum of banyan route flow and feedback route flow, it is larger or equal to banyan route flow. If the value V is larger than the maximum permissible flow per port, congestion must occur only on banyan routes. Thus, the value V must be equal to or less than the maximum permissible flow per port. Furthermore, congestion occurs with incomplete flow distribution across the equivalent Benes network. The incompleteness of flow distribution is large when the number of the multiplexed virtual paths on the ports is small and each path has large flow. In addition, because the banyan route flow overlays the feedback route in the equivalent Benes network, congestion is dependent on banyan route flow distribution. This depends on connection request patterns and network size. As mentioned

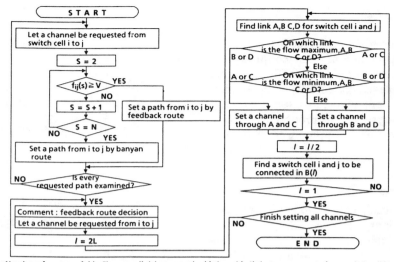

Fig. 3. Control algorithm of feedback banyan network.

L : Number of portss , $f_{ij}(s)$: Flow on a link between the (s) th and (s-1) th stage on route from switch cell i to j

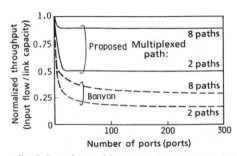

Fig. 4. Normalized throughput of banyan switching network with feedback loop.

TABLE I
TRAFFIC PATTERNS USED IN SIMULATIONS

Items	Cross type	Random type
Image of traffic pattern	switching network	switching network
Definition	Virtual paths are set up between input i and output(L - i-1). $0 \leq i \leq L-1$	Virtual paths are set up between a random pair of input and output.
Characteristic	The worst conditions for congestion	Standard traffic pattern

above, the best value of V for reducing congestion depends on the network size, connection request pattern and the number of multiplexed virtual paths.

The connection setup should be simulated until the best value of V, which occur minimizes internal congestion, is found. However, for simulations, the larger the switching network and the more connection routes between input/output ports, the longer the computation time and the greater the difficulty in obtaining results. Thus, simulation was done with the following limited parameters. The limited value V was chosen as equal to the maximum permissible flow per port, which is a required condition for banyan route flow but not the best for minimizing internal congestion for all flow patterns. The least number of multiplexed virtual paths are two on the port. For only one virtual path on a port, jumper wires are used instead of cross-connect switches. The types of connection request patterns between the switching network input/output ports have been limited to two: the cross type and the random type indicated in Table I. The reasons are as follows. From the network topology of the feedback banyan networks, conditions become worse for the banyan route when the connections between input/output ports are point-symmetrical. Conditions are considered standard when there is random connection of one input (output)

port to the intended output (input) port. Furthermore, the feedback route is also at its worst condition when the connections between input/output ports are point-symmetrical. Thus, the cross type indicated in Table I becomes the worst condition for both the banyan and feedback routes. General characteristics of the random type are rated favorably. Simulation was stopped at 10^6 connection requests. That means a congestion probability of under 10^{-6} is ensured.

III. ASSESSMENT OF PROPOSED SWITCHING NETWORK

Five factors are considered important in evaluating a switching network: 1) specified throughput, 2) the complexity of route-decision procedures, 3) the average delay, 4) number of packet buffers, and 5) the number of switch cells necessary to construct the same size switching network.

Using these five factors, the proposed feedback banyan network is compared to a conventional banyan network.

A. Normalized Throughput

As Fig. 4 indicates, to achieve an internal congestion ratio of less than 10^{-6} with conventional banyan switching networks, normalized throughput must be decreased as switching network size increases. The normalized throughput is limited to 0.17 at random type of connection at 256-input × 256-output ports.

The 256-input × 256-output ports feedback banyan network proposed here achieves a normalized throughput of 0.50 with internal congestion ratio under 10^{-6} for cross type, two paths are multiplexed per port. Though not plotted in Fig. 4, the throughput of random type is a little larger than that of cross-type in the proposed network. This throughput is achieved at nonoptimum limited value V. If the optimum value of V is used, the throughput may become larger than Fig. 4.

B. Routing-Decision Times

Routing decisions for each path within the feedback banyan switching network are made as follows. For banyan route decisions, the amount of each input flows is compared with the flow limit value V which is defined for input ports. For feedback routes, the flows of each link are compared and the largest and smallest flow links are located. Considering the total number comparisons as the routing decision complexity, the routing decision complexity of the algorithm is the sum of the decisions needed for banyan routes and feedback routes.

The simulation results are shown in Fig. 5. Though the complexity may be large or small depending on the connection pattern between the input/output ports, in both cases it is proportional with $(L \times \log_2 L)$ where L is the number of ports.

C. Delay

For a 16-input × 16-output switching network, path setup simulation was performed for the two types of input/output port connection request patterns previously mentioned, based on the following assumptions. Using the flow of each link determined by these simulation results, the average delay normalized by average packet-transmitting time $d_{i,j}$ of a Banyan path connecting input port (i) to output port (j) are derived using (4).

Similarly, the normalized average delay $d_{i,k,j}$ of a feedback path connecting input port (i) to output port (j) via feedback loop (k) is derived using (5)

$$n = \log_2 L \tag{3}$$

$$d_{i,j} = \sum_{s=1}^{n+1} f_{i,j}(s)/(1 - f_{i,j}(s)) \tag{4}$$

$$d_{i,k} = \sum_{s=1}^{n+1} f_{i,k}(s)/(1 - f_{i,k}(s)) + \sum_{s=1}^{n+1} f_{k,j}(s)/(1 - f_{k,j}(s)). \tag{5}$$

Where $f_{ij}(s), f_{ik}(s)$, and $f_{kj}(s)$ are the total flow through the link which is in the paths from input port (i) to output port (j), input port (i) to feedback loop (k), and feedback loop (k) to output port (j), respectively, between the (s)th and $(s + 1)$th stages switch cells. They are determined by the simulation and normalized link capacity.

The internal terms in (4), (5) are the values of the (s)th-

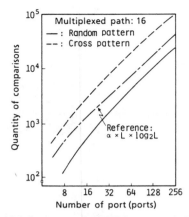

Fig. 5. Link flow comparisons of the proposed algorithm.

stage switch-cell normalized average queueing delay. The average queueing delay is found by approximating on the $M/M/1(\infty)$ model. The queueing of packets in each switch cell is found by assuming a Poisson arrival process and exponential packet length distribution.

Fig. 6 shows results obtained using this method for the proposed feedback banyan network and for conventional banyan network. Delay is small and throughput is large in the proposed feedback banyan network compared to the conventional banyan network.

D. Packet Buffers and Packet Loss Probability

Path setup simulation was performed under conditions similar to those given above.

Incorporating the $M/M/1(m)$ model with the number of packet buffers, m(packets), as the parameter, the packet-loss probability for every connection route between the input/output ports of each switch cell is found from the flow on each link in the switching network.

Next, loss probability $l_{i,j}$ between the input port (i) and the output port (j) is found by adding the loss probability of each switch cell along the routes connecting the input port (i) and output port (j) of the switching network. The relationship between the packet buffers and loss probability per switch cell is then determined by taking the average over the entire switching network. For the two connection patterns shown in Table I, normalized throughput is set of 0.2 and 0.5, respectively.

In the worst case cross-type, the feedback banyan network is advantageous, as Fig. 7 shows. Even with a normalized throughput of 0.2, the feedback banyan network requires buffer capacity of 3630 packets to ensure a packet-loss probability of 10^{-10}, while conventional banyan network requires a capacity of 4690 packets. Though considering the difference in number of switch cells (see below), it is found that the feedback banyan network requires fewer buffer memories.

E. Numbers of Switch Cells

The relation of the number of input/output ports to the number of switch cells is $L \times (\log_2 L + 1)$ for the feedback banyan network and $L \times (\log_2 L)/2$ for the banyan network where L is the number of input/output ports.

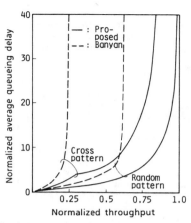

Fig. 6. Normalized average queueing delay for proposed switching network and banyan switching network.

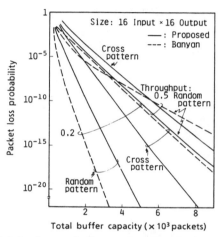

Fig. 7. Relation between buffer capacity and packet loss probability.

Thus, to construct switching networks with equal numbers of ports, the feedback banyan network requires approximately twice as many switch cells as the banyan network. Thus, the banyan networks are preferable on this point.

F. Overall Assessment

Each of the above items is summarized in Table II. Conventional banyan networks have some advantages; the self-routing capability of the switch cells makes routing decisions such as those required in feedback banyan switching networks unnecessary: fewer switch cells are used, etc. However, internal congestion appears when constructing large switching networks: the normalized throughput of the entire switching network is reduced. Banyan networks are thus suitable for low-speed, large-switching networks where frequent switching is necessary.

By contrast, the feedback banyan network proposed here requires approximately twice as many switch cells as the conventional banyan networks. It also requires routing decisions. Thus, these networks are not suitable for instant connection of logical channels or virtual paths. However, the high throughput of these networks ideally

TABLE II
COMPARISON BETWEEN SWITCHING NETWORKS

Items	The proposed feedback banyan network	The conventional banyan network
Normalized throughput	0.5 *	0.17* 1/2 ×** X = int[(log₂L)/2]
Computing complexity of route decision	Proportional to (L×log₂L)	no need
Buffer capacity ***	3630(packets)	4690(packets)
Number of switch cells	L×(log₂L + 1)	(L/2) × log₂L
Application	High-speed and small size switch	Low-speed and large size switch

L : Number of ports
* : For L = 256, virtual paths per port = 2 and congestion probability ≤ 10⁻⁶
** : Worst condition
*** : For L = 16 and virtual paths per port = 32

suits for high-speed, small-switching networks where the switching frequencies are limited.

IV. CONCLUSION

Many countries are examining switching networks consisting of 2-input × 2-output switch cells, with buffer memories which prevent packet collisions. The internal congestion problem inherent in conventional buffered banyan switching networks is that the normalized throughput must be decreased in order to ensure non-congestive connections in large switching networks. A new feedback banyan network topology and its control algorithm have been proposed to reduce internal congestion. The feedback banyan switching network is characterized by its feedback loops; one output port in the last-stage switch cell is connected to one input port in the first-stage switch cell. Virtual paths arriving at the intended output port without encountering congestion through the switching network are output directly. Other virtual paths are returned to an input port via the feedback link. They are then rerouted through the switching network. This reduces congestion and realizes connections with high specified throughput. This proposed feedback banyan network has some disadvantages compared to conventional banyan networks. These disadvantages include the approximately two-fold increase in the number of switch cells, and lack of self-routing ability. However, the network enables high throughput connection, regardless of the switching network size. As such, it is extremely effective in achieving high-speed packet cross-connect switching networks, which conventional banyan switching networks cannot achieve.

REFERENCES

[1] T. Kanada, K. Sato, and T. Tsuboi, "An ATM based transport network architecture," presented at IEEE COMSOC Int. Workshop Future Prospects Burst/Packet Multimedia Commun., Osaka, Japan, Nov. 22-24, 1987.
[2] J. S. Turner, "New directions in communications (or which way to the information age?)," *IEEE Commun. Mag.*, vol. 24, pp. 8-15, Oct. 1986.
[3] S. Giorcelli, C. Demichelis, G. Giandonato, and R. Melen, "Experimenting with fast packet switching techniques in first generation ISDN environment," presented at ISS'87, B5.4, Phoenix, AZ, Mar. 15-20, 1987.
[4] Y. C. Jenq, "Performance analysis of a packet switch based on single-buffered banyan network," *IEEE J. Select. Areas Commun.*, vol. SAC-1, pp. 1014-1021, Dec. 1983.

[5] R. J. McMillen, "A survey of interconnection network," in *Proc. IEEE GLOBECOM '84*, Atlanta, GA, Nov. 26–29, 1984, pp. 105–113.

[6] R. G. Bubenik and J. S. Turner, "Performance of a broadcast packet switch," Washington Univ. Tech. Rep., June 3, 1986.

[7] D. C. Opferman and N. T. Tsao-wu, "On a class rearrangeable switching networks part I: Control algorithm," *Bell Syst. Tech. J.*, vol. 50, no. 5, pp. 1579–1600, May–June 1971.

Hitoshi Uematsu was born in Hyogo, Japan, on July 1, 1960. He received the B.S. and M.S. degrees in communication engineering from Osaka University, Osaka, Japan, in 1983 and 1985, respectively.

He joined the Yokosuka Electrical Communication Laboratory, Nippon Telegraph and Telephone Corporation, Yokosuka, Kanagawa, Japan, in 1985, and has been engaged in research on packet cross-connect switching system.

Mr. Uematsu is a member of the Institute of Electronics, Information and Communication Engineering of Japan (EiC).

Ryuichi Watanabe was born in Hokkaido, Japan, on January 17, 1950. He received the B.S., M.S., and Ph.D. degrees in electronics engineering from Hokkaido University in 1973, 1975, and 1981, respectively.

He joined the Yokosuka Electrical Communication Laboratory, Nippon Telegraph and Telephone Public Corporation, Yokosuka, Kanagawa, Japan, in 1975, and was engaged in research into quasioptical passive devices for short-millimeter wavelengths. From 1979 to 1980 he was engaged in research on such optical devices as multiplexers for switches in optical fiber transmission systems. From 1981 to 1983, he has been engaged in the development of optical subscriber loop systems. From 1984 to 1985, he was engaged in the development of optical high speed trunk transmission systems and optical submarine transmission systems. Since 1986, he has been engaged in network architecture and the development of asynchronous transfer mode (ATM) Systems.

Dr. Watanabe is a member of the Institute of Electronics, Information and Communication Engineering of Japan (EiC).

A NEW SELF-ROUTING SWITCH DRIVEN WITH INPUT-TO-OUTPUT ADDRESS DIFFERENCE

Hitoshi Imagawa, Shigeo Urushidani and Koichi Hagishima

NTT   Communication   Switching   Laboratories
9-11   Midori-Cho   3-Chome
Musashino-shi,   Tokyo   180,   Japan

Abstract

A new self-routing switch for fast packet switching is proposed. The routing method employed in the proposed switch uses the difference between the input and output addresses as a routing header value. By this method, every information block with a routing header is easily routed to its destination line with no congestion in the switch. The switch consists of multi-stage circulated shift registers and output buffers. Multi-point connection can also be provided without additional complex hardware. This paper presents the routing algorithm, mechanism, and applications of the proposed switch.

1.   Introduction

Broadband ISDN (B-ISDN) offers various telecommunication services such as voice, data and video services. To accommodate future demand, innovations in switching technology as well as optical transmission and VLSI technologies are needed. The advanced switch must be capable of handling a wide range of bit-rates from some kilobits per second to several hundred megabits per second. Fast packet switching (FPS) [1] [2] [3] is one of the most promising technologies. Simplified protocol and hardware-oriented self-routing switches are essential for high-speed operation in FPS. This paper discusses a self-routing switch driven with a routing header.

The self-routing switch has two functions: a switching matrix and a contention-avoiding buffer. Generally, there are three types of configuration. The first implements both the above functions within a switch element. The second sets the buffer in front of the switch matrix, and the third sets the buffer behind the switch matrix. The buffered banyan switch and its topologically equivalent switches are examples of the first type[4]. These switches suffer from internal congestion because they only provide one routing path whose links are shared by other connections. Therefore, a buffer is needed within each switch element to avoid congestion on the links. This increases the deviation of switching latency. The Batcher-banyan switch[6] is an example of the second type. A sorting network checks the routing information and rearranges the information blocks in ascending order of their output line numbers. Then the information blocks are sent to a routing network. When multiple information blocks simultaneously indicate the same output line number, one information block is

assigned to the output line and the others are trapped and re-entered into the sorting network[5] [6] [7]. The input buffer size of these switches has been clearly shown to be larger than the output buffer size of the third typed switches[8] [9]. The new self-routing switch proposed here is an example of the third ones.

Self-routing switch requirements are shown in Table 1. The important features are: high-speed routing, no congestion, connection flexibility (point-to-point, multi-point), and suitability for VLSI. This paper proposes a new self-routing switch which satisfies these demands. Section 2 proposes a self-routing algorithm and a switch configuration based on input-to-output address difference. An example of the switch design, an extending method, and a broadcast connection method using the switch are discussed in section 3.

Table 1   Switch   requirements

Requirements		
Integration for various communication	Real-time	High-speed routing Small switching latency
	Small loss-rate	No congestion
	Flexbility for connection	Point-to-point Multi-point
	Broadband	High throughput
Suitablity for VLSI		
Fault tolerance		

2.   Self-Routing   Switch   Configuration

This section describes a self-routing algorithm and switch matrix configuration, and calculates the relationship between the information block loss probability and the number of buffers comprising the output buffer. Fig. 1 shows the proposed switch consisting of switch matrixes with N inputs and N outputs, and N output buffers. The information frame format is shown in Fig. 2. The information block contains header and information field. The header consists of an activity bit indicating the presence of the information field, a copy bit, an input-to-output address difference (later, called the routing header) and the source address.

EH0301-2/91/0000/0233/$01.00 © 1988 IEEE

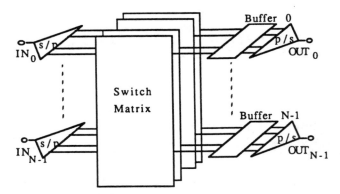

Fig. 1 N × N Switch configuration

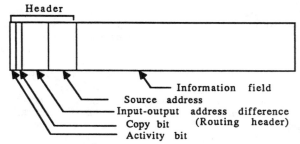

Fig. 2 Information block frame format

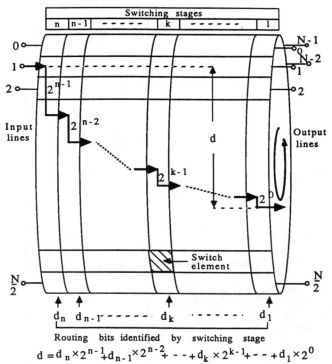

Routing bits identified by switching stage

$$d = d_n \times 2^{n-1} + d_{n-1} \times 2^{n-2} + \cdots + d_k \times 2^{k-1} + \cdots + d_1 \times 2^0$$

Fig. 3 Self-routing algorithm based on the input-to-output address difference

2.1 Self-routing algorithm based on the input-to-output address difference

A self-routing algorithm based on the input-to-output address difference is shown in Fig. 3. The switch matrix has N inputs and N outputs. The difference d between the input and output line numbers is

$$d = (Y - X) \bmod N \qquad (1)$$

where Y is the output line number and X is the input line number.

Also, using bit sequence $(d_n, d_{n-1}, \ldots, d_1)$, d can be expressed as

$$d = d_n \times 2^{n-1} + d_{n-1} \times 2^{n-2} + \cdots + d_1 \times 2^0 \qquad (2)$$

where $n = \log_2 N$.

Multiple switching stages are placed as shown in Fig. 3. Each switching stage has N switch elements. The bit sequence $(d_n, d_{n-1}, \ldots, d_1)$ is given to the routing header of each information block. In switching stage k $(1 \le k \le n)$, a switch element identifies routing bit d_k of the information block. If $d_k = 1$, the information block is shifted by 2^{k-1} in the switching stage, and then the information block is transferred to the next switching stage k-1. When d_k = 0, the information block is transferred directly to the next switching stage k-1. The same procedure is followed throughout n switching stages. As a result, the information block travels the difference d and can be routed to its assigned destination.

2.2 Basic switch matrix configuration

A basic switch matrix configuration based on the above self-routing algorithm, which is driven with the input-to-output address difference, is illustrated in Fig. 4. The switch consists of n $(= \log_2 N)$ switching stages. A kth switching stage includes switching elements E_{mk} $(0 \le m \le N-1, 1 \le k \le n)$, input links X_{mk}, output links $X_{m(k-1)}$, and inter-links Y_{mk}. Information blocks are synchronously input from lines IN_m to the switch, where they are then able to move between switching elements. Information blocks are output from each of the switching stages through an output link which is selected in accordance with the routing header. From here, the information block will ultimately be transferred to its appointed output line OUT_m.

A contention at a switch element in the kth switching stage occurs when an information block coming from the (k+1)th switching stage and another information block shifted in the kth switching stage arrive at the switch element simultaneously. The minimum output time interval of the information block in the kth switching stage must be greater than the maximum traveling time of the succeeding stages in order to avoid contention. This is expressed as

$$2^{k-1} \times T > \sum_{j=1}^{k-1} 2^{j-1} \times T \quad \text{(for } 1 \le k \le n) . \qquad (3)$$

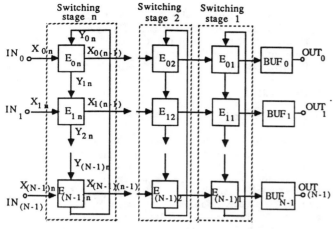

Fig. 4 Basic switch configuration

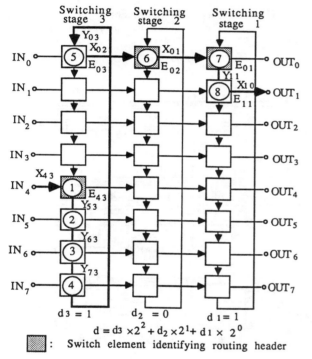

$$d = d_3 \times 2^2 + d_2 \times 2^1 + d_1 \times 2^0$$

▨ : Switch element identifying routing header

Fig. 5 Example of self-routing

Therefore, to achieve nonblocking switch, the time interval S between input information blocks in the input lines IN_m is expressed as follows:

$$S > \sum_{j=1}^{n} 2^{j-1} \times T = (2^n - 1) \times T. \qquad (4)$$

No internal congestion occurs even if two or more information blocks are destined for the same output line.

An example of self-routing is shown in Fig. 5, where N = 8 and n = 3. Consider a situation where an information block supplied to input line IN_4 (input line number $(1\ 0\ 0)_2$) is transferred to output line

OUT_1 (output line number $(0\ 0\ 1)_2$). The information block is appended with a routing header expressed in binary as $(d_3, d_2, d_1) = (1 - 4) \bmod 8 = (1\ 0\ 1)_2$, and the information block is transferred via link X_{43} to the switch element of the first switching stage. Since bit d_3 is "1", the information block is shifted from switch element E_{43} to switch element E_{03} through links Y_{53}, Y_{63}, Y_{73} and Y_{03} using four system clock pulses 4T, and then the information block is transferred on link X_{02} from switch element E_{03}. The information block is latched in switch element E_{02} of the second switching stage. Since the second bit, d_2, of the routing header corresponding to the second stage is "0", switch element E_{02} transfers the information block on output link X_{01} in the same row. Since the third bit d_1 of the header is "1", the information block is transferred from switch element E_{01} to element E_{11} during one system clock pulse T. Thus, the information block is transferred on the output line OUT_1.

2.3 Queueing buffer

In our proposed switch, a FIFO buffer is placed on every output line of the switch matrix. If information blocks independently arrive at each input, and if they are randomly assigned to outputs, and if N is very large, then the steady-state probability of the queue size is obtained by using the M/D/1 queueing model[8]. The steady-state probability P_q is a function of q and ρ, where q represents the number of information blocks in the buffer, and ρ is the average utilization rate of each input line in the switch. The information block loss probability B is

$$B = \left(1 - \sum_{k=0}^{q} P_k \right). \qquad (5)$$

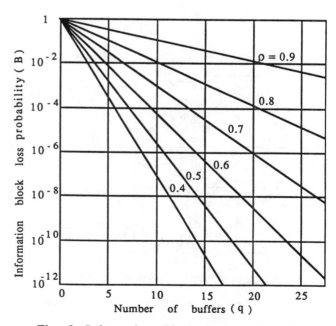

Fig. 6 Information block loss probability versus buffer size

The relationship of B versus q is shown in Fig. 6. When 20 buffers per output-line are prepared, the information block loss probabilitiy B is less than 10^{-8} for $\rho = 0.6$, and it is negligible.

The FIFO buffer is comprised of an activity bit detector, an address generator, and a random access memory. A predetermined bit position in each information block is assigned to an activity bit, indicating the presence of an information field. When this activity bit is "0", the information block is not written to the random access memory. Consequently, the buffer is used efficiently.

3.   Considerations

3.1   Switch design example

A detailed switch matrix configuration is shown in Fig. 7. The serial-parallel converter in front of the switch matrix converts the serial information block so that there is a bit for every switch matrix input. In the converter, the information block is arranged in parallel form bit by bit, and each expanded bit is input to each sub-switch element in the switch element. In the switching stage, one routing bit is sent to each switch element and cycled among all cascade-connected sub-switch elements. A variable-length information block can be easily handled by marking the front and end of the information block and identifying both marks in the switch.

This switch matrix can be easily laid out on a VLSI chip because the link lines are equal in length, and almost parallel, and have fewer crosses. Also, self-routing is accomplished with only one bit in each switching stage, regardless of the switch extension. Therefore, the simplicity of this switch makes it suitable for fast switching and VLSI technology.

A 32 × 32 switch with separated output buffers has approximately 90 kilogates. Delay time, which depends on the input-to-output address difference, is a few microseconds when the circuit is operating at 100 MHz. A throughput of 1.6 Gbits/s for a 16 × 16 switch is possible in a chip using current CMOS technology[10].

3.2   Expandability

Extending the switch matrix to a larger switch network is discussed below. The basic m × m switch matrix has $\log_2(N_{max})$ switching stages, where N_{max} is the maximum switch network using the above basic switch matrixes. An N × N ($m \le N \le N_{max}$) switch network consists of $(N/m)^2$ basic switch matrixes, N serial-parallel converters, and N parallel-serial converters, as shown in Fig. 8. A routing header, indicating the input-to-output address difference, consists of $\log_2(N_{max})$ bits, where $\log_2(N)$ are true routing bits and $[\log_2(N_{max})-\log_2(N)]$ redundant bits are unused routing bits. The redundant bits are used for extending the switch matrix.

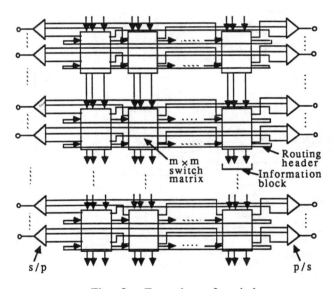

Fig. 8   Extension of switch

3.3   Broadcasting   connection

One special switch feature is its ability to handle both point-to-point and multi-point connections. It uses two methods to attain broadcasting mode in multi-point connection. In the first method, after the information blocks are copied in a copy network[5] or in random access memory in front of the switch, they are transferred to the switch. This

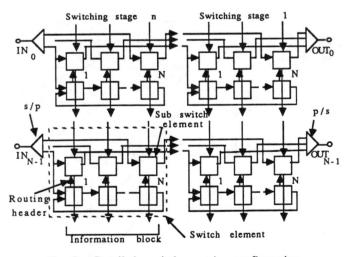

Fig. 7   Detailed switch matrix configuration

method requires large amounts of hardware to accomplish the complex copying and routing processes. In the second method, the information blocks are copied using a copy bit. Here, there is a problem with congestion between the information blocks in the switch. Therefore, it is necessary to restrict the number of input information blocks or to detain them in the switch element buffer. This section discusses the second method.

As shown in Fig. 2, the value of the copy bit determines whether or not the broadcasting mode of each switch element of each switching stage is done. The information block is transferred to both 2^{k-1} shifted switch element in the same kth switching stage and a switch element in the next switching stage, as shown in Fig. 9, and then is broadcasted. In Fig. 9, information blocks are input to each first switch element as discussed in sec. 2. 2, and they branch at the shaded switch elements. The switch has no congestion even if information blocks destined for different outputs or the same output are all simultaneously copied in the switch. This is because, as shown in equation (4), the time interval S between input information blocks is fixed, after considering that the information block may be transferred in both directions. In the outputs of the switch, it is easily to identify the information block source by the source address. It is no problem in the random traffic model if the number of buffers is designed by the M/D/1 queueing model.

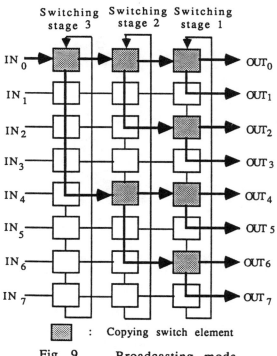

Switching   Switching   Switching
stage 3   stage 2   stage 1

▨ : Copying switch element

Fig. 9   Broadcasting mode

4. Conclusions

This paper proposed a new self-routing switch driven with input to output difference. The switch has the following features:
(1) There is no congestion in the switch matrix in either point-to-point connection or broadcasting connection.
(2) It is suitable for VLSI technology. A throughput of 1.6 Gbits/s for a 16×16 switch is possible in a chip using current CMOS technology.
(3) An output queueing buffer can easily be designed, so that the switch matrixes proposed here has deterministic ability and no internal congestion.

The switch is suited for small capacity exchange systems, interconnection networks in parallel computers, and future larger capacity exchange systems in B-ISDN.

Acknowledgments

The authors wish to thank T. Koinuma, K. Yukimatsu, and Dr. T. Aoki of NTT communication switching laboratories, for their help and encouragement.

References

(1) J. J. Kulzer and W. A. Montogomery, "Statistical switching architectures for future services," ISS' 84, pp. 43A1.1.1-6 (1984).
(2) Thomas A., Coudreuse J. P. and Servel M., "Asynchronous time-division techniques: An experimental packet network integrating videocommunication," ISS' 84, pp. 32C.2.1-7 (1984).
(3) J. S. Turner, "New direction in communications," IZS' 86, pp. A3. 1-8 (1986).
(4) Y-C JENQ, "Performance analysis of a packet switch based on single-buffered Banyan network," IEEE JSAC, vol. SAC-1, NO. 6,pp.1014-1021 Dec. (1983)
(5) A. Haung and S. Knauer, "STARLITE : A wideband digital switch," Proc. Globecom' 84, pp. 122-125 (1984).
(6) J. Y. Hui and E. Arthurs, "A broadband packet switch for integrated transport," IEEE JSAC, vol. SAC-5, NO.8, pp.1264-1273 Oct. (1987).
(7) K. E. Batcher, "Sorting networks and their application," AFIPS 1968 Spring Joint Computer Conf., pp. 307-314 (1968).
(8) M. J. Karol, M. G. Hluchyj and S. P. Morgan, "Input vs. output queueing on a space-division packet switch," Globcom' 86, pp. 659-665 (1986).
(9) Y. S. Yeh, M. G. Hluchyj and A. S. Acampara, "The knockout switch: A simple, modular architecture for high-performance packet switching," ISS' 87, pp. 801-808 B.10.2.1-8 (1987).
(10) K. Takeda, F. Ishino, Y. Ito, R. Kasai and T. Nakashima, "A single chip 80b floating point processor," ISSCC' 85, pp. 16-17 (1985).

Chapter 5:

Switch Architectures

5.1 Introduction

Any one of the switching techniques ranging from pure circuit switching to pure packet switching can be used in the design of the next-generation integrated switch. At present, there are a number of proposals for alternative switch architectures intended to switch various types of traffic in an integrated fashion. This introduction presents a brief overview of a number of switching techniques.

Switching provides an economical means to connect the source and the destination nodes by providing a path through the network. In *time-division switching*, a path of fixed bandwidth is set up during call initiation. During the data transfer phase, data are transmitted using time-division multiplexed frames. A path is formed by allocating slots in the trunks. Each of the switching nodes maps the incoming slots into the corresponding outgoing slots, using the routing information stored at the node.

As in time-division switching, *statistical time-division switching* sets up a path during call establishment. By allocating bandwidth proportional to the traffic activity, statistical time-division switching achieves higher bandwidth use than that of time-division switching. In this scheme, data from a number of connections are multiplexed statistically onto each trunk. At the intermediate switch, on the input side, data belonging to various connections are demultiplexed and stored in transit buffers allocated for that connection. On the output side, a frame is formed by collecting data from the transit buffers of all active connections, and the frame is forwarded to the next switch. The mapping between the transit buffers and the outgoing trunks is set up during the call processing phase of the connection.

In *multirate circuit switching*, bandwidth equal to one or more integer multiples of a basic rate can be allocated for a call. At call setup, the number of channels required to meet the bandwidth requested by the user is assigned by the switches.

Fast circuit switching is an attempt to allocate bandwidth dynamically to optimize bandwidth use in a bursty traffic environment. In this scheme, bandwidth is not allocated at call setup but is allocated only when data are transmitted. The bandwidth is deallocated as soon as there is no information to send, thereby allowing this bandwidth to be used by other calls. This switching method requires an end-to-end circuit setup technique with minimum delay. An inherent drawback is that an already established connection, on becoming active again, may be unable to obtain the required bandwidth, because no bandwidth was reserved for it. Thus, loss of information and an increase in delay are possible, unlike in time-division circuit switching.

Packet switching is the most dynamic form of switching, and the data are transported in the form of packets from the source to the destination. Each packet moves through the network in a store-and-forward fashion. In packet switching, bandwidth is allocated only when a packet is ready for transmission and remains allocated for the time required for the packet transfer. Two types of network services—called connection-oriented and connectionless—are available with this type of switching.

In addition to the above switching methods, a number of hybrid switching techniques have been used to integrate voice and data. An example is an integrated voice/data network where the excess bandwidth not used by circuit-switched traffic is allocated to the packet traffic (GERL78). In this integrated switch architecture, the circuit-switched traffic is processed by a switch processor module (SPM), and the packet-switched traffic is switched by a packet processing module (PPM) The SPM scans the incoming frame and switches the circuit-switched slots from the incoming trunk to the slots in the outgoing trunk based on the routing map established at call setup. The packet-switched traffic is forwarded to the PPM. After completing the protocol processing at the PPM, the data packets are placed in a queue. In this scheme, the circuit-switched traffic is handled transparently: Input slots are mapped to output slots whether each slot contains data or not. No attempt is made to reassign temporarily unused circuit-switched traffic slots to packet-switched traffic.

SENET (slotted envelope network) (COVI75) is another architecture that integrates both voice and data traffic. SENET uses a synchronous envelope to carry user traffic. User traffic is classified into two classes. Circuit-switched traffic belongs to class I, and packet-switched traffic belongs to class II. A movable boundary separates the two classes of traffic.

In this architecture, an envelope is divided into segments, and the traffic from a channel is assigned to a particular segment. If a class I channel is temporarily inactive, that bandwidth is allocated to class II traffic by dynamically adjusting the boundary. This scheme requires the envelope to be assembled completely in the switch before being transmitted.

5.2 Overview of Papers

The papers in this chapter highlight the architectural choices available for switches for the broadband networks of the future. The articles cover the two main types of packet switches:

- Those that rely on traffic statistics (statistical switches) or bandwidth limitation (knockout, bus matrix switch) to deal with fabric blocking.
- Those that use a nonblocking switch fabric (Batcher banyan).

In the first paper, "Burst Switching–An Update," Amstutz introduces the concept of switching bursty data. A burst is defined as a talk spurt or a data message. The described network architecture is noncentralized with several cooperating microprocessors connected at the edge of a transport network. A link switch, a link group, and a hub switch form the increasing levels of subgroups that can perform *self-contained* switching if the source and the destination belong to the same switch group. Each burst carries 4-byte header information that contains the burst type and the destination port address. The destination address comprises the link group, link switch within the group, and the port within the link switch identifiers, enabling switching control to be truly distributed. This paper also points out the differences and similarities of burst switching with fast packet, fast circuit, and ATM switching.

The second paper, "Starlite: A Wideband Digital Switch," attempts to define the architecture of a wideband switch that can handle bursty and continuous data of widely differing bandwidths, which are usually handled by separate circuit and packet switches. The Starlite approach relies on self-routing, nonblocking multistage interconnection networks. Bursty data (variable-length packets) and continuous data are reformatted onto fixed-length packets for routing through the network. The switch's construction out of stacks of regular routing networks is shown to allow the switch to grow modularly with increases in the number of users and the amount of per-user bandwidth, and to support a variety of interconnection types such as those that provide broadcast capability.

Wu et al., in their paper "Dynamic TDM: A Packet Approach to Broadband Networking," describe a transmission format based on the combination of conventional time-division multiplexing (TDM) and packet transmission techniques. The main theme of the paper is that even if bandwidth conservation is not the main concern (with the wide availability of fibers), dynamic time-division multiplexing (DTDM) networks can be built to provide bit-rate flexibility using a simple equipment design. The authors present a nonblocking switch architecture based on the Batcher sorting network with a banyan routing network and a slotted packet format. The DTDM concept is viewed as an extension of the asynchronous TDM (ATDM) scheme. Embedding the ATDM signals explicitly in the synchronous TDM framing structure results in a transmission format that enables slotted packets to traverse synchronous digital pipes. A 150-Mbps channel rate allowing 144,000 slots/sec. containing 130-byte slots is the expected traffic capacity of this switch.

In "A Broadband Packet Switch for Integrated Transport," Hui and Arthurs present a switch architecture to carry multiple broadband services. The switch fabric consists of a nonblocking Batcher-banyan interconnection network. The authors present a three-phase algorithm to resolve output port contention. They also discuss the issues related to the hardware implementation of the switch fabric and the results of studies on switch performance. A CMOS VLSI implementation supporting a port speed of 150 Mbps and a switch supporting 1,000-4,000 ports servicing 200,000 subscriber lines is the expected capacity of this switch.

The paper "The Knockout Switch: A Simple Modular Architecture for High Performance Packet Switching," by Yeh et al., presents a modular, self-routing, and nonblocking packet switch architecture. The knockout switch uses a fully interconnected topology, and the input packets are broadcast to all output ports. Any output port contention is resolved using the knockout technique: First the incoming packets are filtered to allow only packets addressed to that port. Then the packets are concentrated by throwing away packets that exceed the concentrator capacity. Packets are then stored in a shared buffer to be transmitted on a FIFO basis. For an arbitrarily large number of inputs, packets lost are on the order of 10^{-6} for a concentrator output of 8, amounting to a loss rate similar to that in other parts of a packet-switched network. The authors also describe the bus interface and the concentrator. An aggregate traffic rate of 50 Gbps is the capacity they expect from a switch containing 1,000 inputs, each operating at 50 Mbps.

The paper "Synchronous Composite Packet Switching–Switch Architecture for Broadband ISDN," by Takeuchi et al., describes an architecture based on a hybrid approach for integrating various services. A switch architecture consisting of a number of switch modules connected by multiple loops is described. In synchronous composite packet switching, incoming messages on multiple circuit-switched channels are assembled into composite packets and switched synchronously between switch modules maintaining time transparency and short delay time. The authors present the formats of three packet types and a transmission frame, and discuss how this switch architecture can be used in the broadband ISDN environment. They show switches with a capacity of about 4 Gbps to be possible.

Nojina et al., in "Integrated Services Packet Network Using Bus Matrix Switch," describe a packet switch that can be used for broadband ISDN. This switch architecture can switch variable-size packets and uses a matrix switching technique. The authors present the results of a performance evaluation based on a prototype model, and estimate that the maximum switch capacity will be about 2.6 Gbps in a CMOS implementation and 6.4 Gbps in emitter-coupled logic technology.

Burst Switching—An Update

Stanford R. Amstutz

Reprinted from *IEEE Communications*, Vol. 27, No. 9, September 1989, pages 50-57. Copyright © 1989 by The Institute of Electrical and Electronics Engineers, Inc. All rights reserved.

BURST SWITCHING RESEARCH IN DISPERSED control and integrated switching at GTE Laboratories is described here more fully than in previously available material [1–3].

Burst dispersed control has a non-centralized architecture employing many autonomous cooperating microprocessors connected at the edge of the transport network. It is possible to add control capacity without limit, or to compensate for control processor failure, by redistributing the processing workload among the control processors, using the transport network itself as the redistribution switch. The control reconfiguration process can be carried on even while normal call processing continues. This process is described and its timing given. The dispersed control architecture is not specific to burst switching and can work with other self-routing transports such as fast packet.

Burst transport is integrated in that voice and data are switched through the same switching fabric and transmission media. Burst switching is compared to and distinguished from fast packet, fast circuit, and ATM switching. Misunderstandings about burst transport that have appeared in the literature are corrected, to wit: burst does not immediately clip in case of channel contention; burst switches voice and data in the same way; a burst switch interfaces naturally to other types of switches. Round-trip delay performance is calculated to be less than 5 ms. The article concludes with the current status of the burst project.

Burst transport is integrated in that voice and data are switched through the same switching fabric and transmission media.

Burst Control
Dispersed Control

The original objectives for the burst control were three:

- *Unlimited capacity expandability*—We wanted to overcome the capacity limitation of a conventional centralized control, where even a tightly coupled multiprocessor has limits—its number of processors is limited by electrical constraints regarding the number of ports into shared memory and the number of ports onto an interprocessor bus.
- *Non-centralized*—We wanted a control that had no single point of failure.
- *Dispersed*—We wanted a control that could be distributed so that portions of a burst switch network could operate independently of the rest.

These objectives were met by a call processor per link switch, as was suggested in [1]. But this solution was not fully satisfactory. There was no doubt that a microprocessor had the capacity to provide call set-up and feature services for all of the ports of a link switch. The concern was that it would have more capacity (thus cost) than could be used. More important, the sizeable memory required for all the functions and features of a modern switch would have to be replicated at every link switch. Finally, a call processor failure at a link switch, while it would only disable call placement services for the ports of that switch, had no economic backup method.

An improved approach was conceived of in which the call processor was given only a port appearance, so that all of its messages to other control processors would be transferred via the network itself, as control bursts. The network certainly had the capability for this, as it had been designed to transfer data bursts between customer computers. Now there was no longer a linkage between a link switch and a call processor; a call processor could be equipped at any port.

With the new approach, the original objectives were still met, and the disadvantages of the first approach were avoided, as follows. The number of ports to be handled by a call processor became very flexible. As many or as few ports as desired could be assigned to a particular call processor; there was no longer the fixed number at a link switch to be served. The amount of replicated call processor memory in the system would be only as much as was needed to support the demand for call processing services. And most important, each call processor in the system could be backed up by transferring its load to the surviving call processors. One-for-one redundancy and special switchover networks were not required. The switching network itself could be used as the switchover network to replace failed processors with functioning processors.

There are three types of processor in the burst dispersed control:

- *Port Processor (PP)*. Every port circuit includes a PP. Its function is to convert between the external signal form and the internal burst form. Thus, for example, it converts off-hook detection or tones received to control bursts sent. It converts control bursts received to tones sent, or ringing. These are control functions.

    While the PP is considered part of the control, it also performs some transport functions, such as creating bursts from a continuous stream of digitized analog samples in an analog port, by running a voice/silence detection algorithm.

    The Z80 is used for the PP in the experimental model.

- *Call Processor (CP)*. CPs do most of the decision making in the system and they have the largest program of any of the control processors. They do call set-up and feature implementation. Each CP performs services for a subset of all of the PPs. Thus, when a PP creates an off-hook message, it sends it to its CP.

The M68000 is used for the CP in the experimental model.

- *Administrative Processor (AP)*. APs are basically database machines. Translation from directory number to equipment number and accumulation of usage information for billing are done by an AP. An AP performs services for a subset of all the system CPs.

In a smaller system, the administrative processes can run on the CPs, as is the case in the experimental model. But if the system is large enough to require separate APs, at least two would be equipped for failure protection.

Figure 1 shows the three kinds of control processors, each having a port appearance only, on the outside of the transport network.

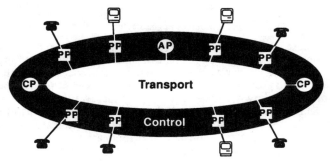

Fig. 1. Control processors outside transport.

Each PP goes to a particular CP for call processing services, and each CP goes to a particular AP for administrative processsing services. Thus we have the concept of service sets. A service set includes a processor of one type as server, and processors of another type as clients.

Figure 2 represents six service sets: two AP service sets with CPs as clients, and four CP service sets with PPs as clients. Each client has the network address of its server, so that service requests can be forwarded as control bursts. We expect that a CP will be able to serve 1,000 to 2,000 PP clients, so Figure 2 signifies nothing about service set sizes.

Figure 3 shows the message exchange for an ordinary voice call set-up and take-down. This figure requires little explanation. In Phase 11, each port receives the header that will convey bursts to the other port, thus establishing the virtual connection. In Phase 12, the connection is used to transfer voice bursts. Otherwise, each PP communicates only with its CP.

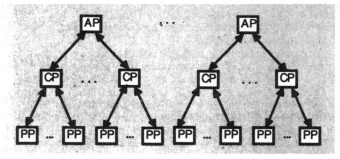

Fig. 2. Six service sets.

Control Reconfiguration

We now consider how the claimed flexibility with respect to the addition of processing capacity or the compensation for processor failure is realized. All such actions may be seen to be redefinitions of service sets. Let us first consider the addition of capacity.

A processor is overloaded if it receives more service requests than it can process with acceptable dispatch. The amount of processing requested will be related to the number

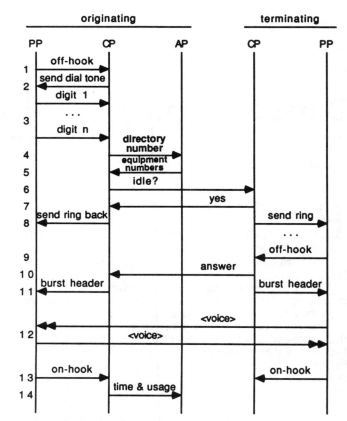

Fig. 3. Voice connection/disconnection.

of clients in its service set, and their individual activity. So the load on a processor can be reduced by transferring clients to other servers. If there are other less loaded servers in the system, clients can be transferred to them. If most or all servers need load relief, new serving processors must be added to the system to assume some of the load.

If a processor fails, its load must be assumed by other processors so that its clients continue to have a designated server.

In all cases, the desired adjustment of loading is accomplished by service set redefinition: the movement of clients from some service sets to others. The service sets of Figure 2 are shown, redefined, in Figure 4. This configuration might result if one of the CPs failed and its ports (PPs) were redistributed to other CPs. Alternatively, one can view Figure 4 as the starting point. The addition of a CP and redistribution of ports to it provides the enhanced processing configuration of Figure 2.

The transfer must be made while normal processing continues; it is not acceptable that the system shut down during the period required for control reconfiguration. Of course, the addition of reconfiguration traffic will be a significant addition to normal call handling traffic, so it is important to accomplish the reconfiguration expeditiously.

We now outline the data movements required to move a PP from one CP service set to another. (The movement of a CP from one AP service set to another is very similar, though simpler.) The movement is entirely accomplished by message exchange among control processors. However, we will not attempt to detail that message exchange; we will simply outline which processors require which kinds of information and where they get it:

- The process starts when an AP, called the initiating AP, decides to move a port, either upon the receipt of an explicit authenticated command from a craft workstation or because it has determined from background test message traffic that a failure has occurred and ports must be moved.

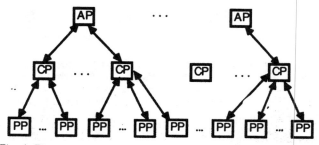

Fig. 4. Five service sets.

- All APs need to know to which service set a moved port is being added. Each receives this information from the initiating AP and updates its tables, defining all system service sets.
- The CP adding a port needs to know the port's static data—the seldom-changed class marks and features the port is entitled to use. The CP acquires this information from its serving AP, where the static data on all ports is stored.
- The CP adding a port desires to know the port's dynamic data—information describing the call status of the port. It acquires this information from the port's old CP when it can. If the old CP has failed, the information is lost and the new CP reconstructs as much of the port's call status as it can from the port's PP itself. We note that connected calls are fully reconstructed from information in the PPs themselves.
- The PP of the port being moved must know the address of its new CP. It acquires this information from the new CP itself.

All of these movements could be accomplished without difficulty were it not for the requirement that normal call handling continue while reconfiguration is under way. Because information must be moved from one processor to another, which takes time, it is possible that data may not be consistent over the whole system. Here is an example of a problem that could ensue unless care is taken.

In Phase 5 of Figure 3, the AP returns the PP address of the called port and the address of its serving CP. This CP address is used by the originating CP for the idle query of Phase 6. Suppose that a port, P, is called while it is in the process of movement from CP A to CP B. In the control reconfiguration process, the translation tables are updated with the new CP address before the new CP is informed of its new ports' identities. Thus, it would be possible for CP B to receive an idle query about port P before it knew it would be the new server for P. Knowing nothing of P, it would make no response.

This problem and others of this type are solved by locking. To continue the example, at the time P's CP entry in the translation table is changed to B, it is locked. This means that a translation request against P's directory number will be delayed (queued) until the lock is released, and the lock is released only after CP B has responded saying it now has full cognizance of and responsibility for P.

Control Reconfiguration Times

A good many messages are required to coordinate the movement of even one port. The amount of time required to move the many ports a CP might serve is an important operating parameter.

The reconfiguration software supports two kinds of reconfiguration: immediate and orderly. An immediate reconfiguration moves ports among processors as quickly as possible and is intended for compensating for processor failure, where it is important to restore service as quickly as possible. In an immediate reconfiguration, established calls are not lost but calls in the process of establishment, in digit receipt for example, will be reset.

An orderly reconfiguration puts the premium on recovery of all system data, even if it takes longer, and is intended for load balancing and the addition of processors. In an orderly reconfiguration, calls in the process of establishment are not interrupted; a port is not moved until it reaches a stable point in its process.

We have calculated how long reconfigurations take, assuming: first, that each processor can compose and send messages at 8,000 characters/s; second, that each processor can receive and process messages at 8,000 characters/s; and third, that reconfiguration traffic takes precedence over normal call handling traffic. The model calculates the total time as the sum of all the message transfer times that must be done in sequence. (The time to transfer messages sent in parallel is the time of the longest.) The times were calculated as a function of the total ports on the system and the maximum number of ports allowed per CP, R. The number of CPs on the system increases with total ports so that no CP serves more than R ports.

Figure 5 shows the times for an immediate delete of a CP: the time to move all the ports from one CP and distribute them uniformly to all the other CPs, so that at the end all the other CPs serve equal numbers of ports.

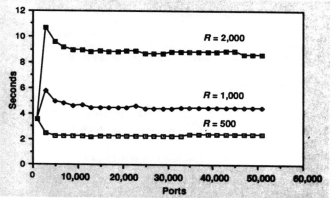

Fig. 5. Immediate delete times.

Figure 6 shows the times for an orderly add of a CP: the time to move some ports from each original CP and distribute them to the added CP, so that at the end all CPs are uniformly loaded.

We note that for both plots, the time is not primarily a function of the number of ports on the system but is proportional to the number of ports per CP. This is to be expected. The time to move ports is proportional to the number of ports moved, and that is determined by R.

The times for orderly adds are significantly longer than for immediate deletes. This is caused by two factors. First, an orderly move takes longer than an immediate move because more messages are sent. A CP acquiring a port in an orderly move will request the port's transient state information from the port's current CP. This is not done in an immediate move, since the immediate move would normally be used in case of CP failure, when a port's assigned CP would not be able to supply the information.

Second, a delete is faster than an add. In a delete, the remaining CPs work in parallel to acquire the deleted processor's ports. In an add, the added CP works alone to acquire ports from each of the other CPs.

Table I gives spot values for all four combinations of immediate/orderly and delete/add for a 50,000-port system. Here it can be seen that immediate is faster than orderly (for either delete or add), and delete is faster than add (for either immediate or orderly).

Table I also gives the lockout times. The lockout time is the interval during which critical tables are being updated and is the maximum time during which an element of a table cannot

Table I. Reconfiguration Times (s)

Ports CP	Delete			Add		
	Lockout	Immediate	Orderly	Lockout	Immediate	Orderly
500	0.3	2.4	4.6	0.2	4.5	6.6
1,000	0.4	4.5	8.9	0.4	8.7	13.0
2,000	0.8	8.8	17.6	0.7	17.1	25.5

be accessed for ordinary call handling during the reconfiguration process. The average lockout time would be half of the maximum shown.

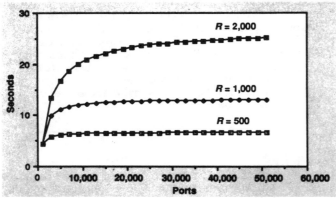

Fig. 6. Orderly add times.

Burst Dispersed Control Summary

Burst dispersed control is seen to offer the following advantages:

- Capacity expansion limited only by the number of system ports
- Ability to back up CPs and APs via the switching network itself using 1:n redundancy
- No centralized control point and thus no single point of failure
- Can operate in parts and continue to provide service even if the network is severed

The dispersed control is not limited to use with burst transport. It is useable with any self-routing transport network that provides for the delivery of information packets based on a destination address contained in the header, such that any two processors of the control can intercommunicate. Certainly a packet or ATM network supplies this capability.

Burst Transport

Architecture

Figure 7 shows that a burst switch comprises link switches and hub switches. The link switches are small switching elements designed to be remotely located from the hub switch.

Thus, the burst switch is itself a network but corresponds in this configuration to an end-office switch. Networks of burst switches and other types of switches can be formed.

Lines and trunks access a burst switch via a link switch. The experimental model link switch serves 24 ports (lines or trunks). Link switches are interconnected by links, T1 links running at 1.544 Mb/s in the experimental model. The link switches are organized into link groups, with 16 link switches per link group.

The link switches are genuine switches. A connection between ports on the same link switch is completed entirely within that link switch; a connection between ports on two link switches in the same link group is completed entirely within that link group.

The hub switch is a high-capacity ring switch designed for connections among 256 link groups. The experimental model hub switch runs at a clock rate sufficient to support 16 link groups.

While the link groups may be configured as trees or rays, the ring interconnection shown in Figure 7 is preferred because it provides alternate routing: every link switch has two ways to get to the hub, and two ways to get to every other link switch in its link group.

Burst Format

A burst consists of a header, an information field of any length, and a termination character signalling the end of the burst.

As shown in Figure 8, the 4-byte header bears the burst type, the address of the destination port, and the header check sum. The destination port address is expressed in terms of link group, link switch within link group, and port within link switch.

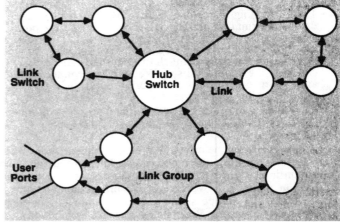

Fig. 7. Burst switch.

The information consists of the digitized samples of a talk spurt, a transmission block of data characters, or the characters of a message between control processors.

The termination character, FF_{hex} in the experimental model, is the means by which processors receiving a burst identify the end of that burst. So that this value can be used as a header or information character, it is preceeded by an escape character, $7F_{hex}$, when it does not signify end of burst.

Link Switch

Figure 9 shows the link switch in greater detail. Lines and trunks access the link switch through Port Circuits (PCs). The PC transforms the external signal form to the internal burst form. Thus, bursts are created in the PCs. When a PC determines that a burst is beginning, applying the criteria appropri-

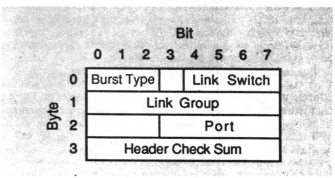

Fig. 8. Burst header format.

ate for the external signal form it handles, it prefixes the header and supplies the characters of the burst to the PC bus.

Each burst passes through a dynamically assigned buffer in Buffer Memory (BM). The Switching Processors (SPs) administer the movement of burst characters in and out of BM. A burst may arrive over a link or from a local port, and be delivered to a link or to a local port. Between bursts, the termination character is used as the fill character.

The input SP acquires a buffer from the free queue to pass the burst's characters through; deposits each character received in the buffer; interprets the destination address in the burst's header to select the appropriate output; and queues the burst on that output.

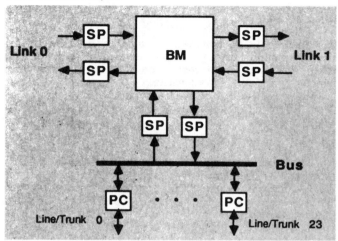

Fig. 9. Link switch detail.

The output SP finds the burst on its queue; reads the burst's characters one by one from the buffer and delivers each to its link (or port); and returns the buffer to the free queue at the end of the burst.

The SPs are specially designed processors that can execute a 64-b instruction every 100 ns. The load on the input SPs is the highest. They must be able to perform all actions associated with storing and interpreting a character each channel time.

Switching Method

Now we consider the flow of bursts through a link switch. Figure 10 shows a simplified case consisting of only one input Switching Processor (SPi) and one output Switching Processor (SPo). The queue for SPo output is represented by Q. Three bursts are in transit through three buffers in memory (B1–B3). Each burst is in one of the three major phases of burst transport.

SPi has received the first character of the first burst in its channel c1 and has stored it in buffer B1. SPi previously acquired B1 from the free queue, a queue just like Q. SPi has retained the address of B1 (b1) in its local memory associated with channel c1 so that subsequent characters of the burst will be stored in B1. On the basis of the information in the first character, SPi has determined that output should be via SPo, therefore it has enqueued the burst on Q. This means that SPi has passed the address b1 to Q.

But output of the first burst has not yet commenced. SPo, when its next idle channel occurs, will refer to Q to acquire the address of any burst awaiting output and get b1. The burst's characters will accumulate in B1 until the idle channel occurs.

SPi is inputting the characters of the second burst to B2 as SPo is outputting them. SPi inputs the characters from its channel c2 and SPo outputs the characters in its channel cX, probably a different channel number. Both processors have the buffer address b2, one for input and one for output. SPo has previously acquired b2 from Q. B2 probably has only one char-

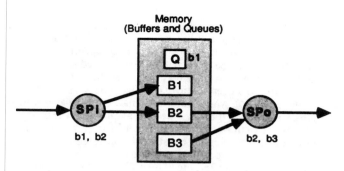

Fig. 10. Burst flow.

acter at a time in it, since the expected case is that an output channel will be acquired within one frame time from the time input begins. Since the input and output rates are equal, there is no need to accumulate characters in B2 before output starts.

SPi has detected the end of the third burst and no longer retains b3. SPo is outputting the last characters from B3. When SPo detects the end of the burst, it will return B3 to the free queue.

We note from this example that there is nearly immediate cut-through; thus, the transit delay through the switch is low. There is no accumulation of characters in a packet sense. Output can begin as soon as enough header characters have been received to permit routing, and as soon thereafter as an idle output channel occurs. In almost all cases, the first character supplies enough information to permit immediate routing. There are only two exceptions: when a burst enters the hub, the second byte must be received to determine the destination link group; and when the burst arrives at the destination link switch, the third byte must be received to determine which port the burst must be queued to (see Figure 8).

Differences in Handling Voice and Data Bursts

The description above applies equally to voice, data, and control bursts. All are switched through exactly the same switching fabric and handled in a nearly identical way. If a burst is queued for output and no output channel is free, the burst must wait until a channel becomes free. (Preemption was not included in our experimental model since the probability that a burst will need to wait more than a frame time to acquire a channel is extremely small.) While waiting, the burst's characters will accumulate in a buffer.

However, there are some differences in how voice and data ought best to be handled. Voice is more sensitive to delay than data, and data is more sensitive to loss than voice. Thus, it makes some sense to give preference to voice bursts in channel assignment (to minimize delay) and to buffer delayed data bursts (to eliminate loss). In addition, it is desirable to keep the system responsive by propagating control bursts quickly, and since they are short and occupy little channel time, they are in turn given preference over voice. Thus, a queue such as represented in Figure 10 is in fact eight queues, one for each burst type and assignment priority. An output SP will assign an idle channel to the first burst in the highest-priority non-empty queue.

We note from Figure 10 that every burst passes through a buffer; thus, a voice burst can be buffered as well. But there is a tradeoff between buffering the burst (thus contributing to delay) or clipping the burst, in the rare circumstance when a voice burst is stymied awaiting an output channel. When a burst is clipped, the accumulated characters are thrown away. Thus, there is front-end character loss but the later characters of the burst are not delayed. Beyond a certain point it is better to clip than to further delay the entire burst.

The choice we have made in the experimental model is to accumulate up to 32 voice characters (4 ms of delay at 64 kb/s) before clipping. As soon as a channel becomes available, output of the voice burst will begin on it. The whole burst is not lost unless no channel becomes free during the entire duration of the burst. So a voice burst will not be delayed by more than this amount, though it may be clipped on the front.

This ability to buffer portions of voice bursts was not appreciated in [4, p. 654], where burst was characterized as instantly clipping voice bursts if no channel was available at the moment of need. Since each burst flows through its buffer, and buffers can be chained, as much of a burst can be buffered as desired.

So bursts of all types are handled in the same way, each flowing through its buffer, etc. The differences with respect to channel assignment priority and the extent of buffering do not constitute any fundamental difference, as has been elsewhere misunderstood [5, p. 25], but are simply enhancements to improve the performance with respect to control, voice, and data.

Interfacing to Other Systems—Port Circuits

While the burst discipline is used within a burst switch (among link and hub switches), and between burst switches using burst trunks, a burst switch is entirely capable of interfacing to switches of other types. The link switch shown in Figure 9 is where outside access to a burst switch occurs, and the Port Circuit (PC) is where the conversion between the outside signal form and the burst form is done.

Figure 11 shows the essential elements of the PC for an analog telephone. All of these elements do signal conversion of their own particular sort. A burst arriving from the PC bus on the left will enter the PP, character by character. The PP strips the header and termination characters from the burst and supplies the information characters to the codec. Between bursts, the PP supplies characters of "quiet tone." The codec converts the digital sample stream to an analog signal, and the Subscriber Line Interface Circuit (SLIC) converts to 2-wire.

In the reverse direction, the PP runs the voice/silence detection algorithm on the digital sample stream from the codec. Upon detecting a silence-to-voice transition, it begins issuing a burst with header. Upon detecting a voice-to-silence transition, it ends the burst with a termination character.

PCs for other signal types are similar. The PC for an E&M trunk is much the same except that no SLIC is required. The PC for an RS-232 line is basically a PP and a Universal Asynchronous Receiver Transmitter (UART). The PC for a T1 trunk would consist of a port process for each channel, plus a multiplexer/demultiplexer. Here, a number of channel processes could run on a common processor. The PC for an X.25 trunk would consist basically of a PP and packet network interface.

There is no difficulty, conceptual or otherwise, in interfacing a burst switch to switches of other types.

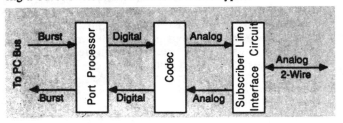

Fig. 11. Analog telephone port circuit.

Is Burst Fast Circuit Switching?

It has sometimes been suggested that burst is a form of fast circuit switching. Burst is not fast circuit switching as defined in the early 1980s [6] [7], when it was envisioned that a circuit switched connection could be broken down and reestablished with rapid signalling between talk spurts. It has rightly been observed that this procedure would be complicated and would

constitute a significant delay even with rapid signalling.

In burst, two ports are "connected" when each has received the network address of the other. The control processors, as described above, supply this address during call set-up to the PP in the form of a burst header bearing the address. When a port determines that a burst has begun, it prefixes the header to the information, and the self-routing properties of the burst transport use the header destination address to route the burst for delivery to the other port in the connection. No switching or transmission facilities are reserved until a burst occurs. And in particular, no retransmission of signalling information is required. Then burst is a form of fast circuit switching as that term is used in [8].

Is Burst Fast Packet Switching?

A burst looks like a packet, with a header bearing information that is used to route the burst to its destination. But a burst is sent in a time division channel on a link between nodes, interleaved with other bursts, whereas a packet is sent using the full link bandwidth, first one packet and then another. This small difference has significant performance consequences, as outlined below.

Delay

In a packet switch, the rate of character transmission on the links between nodes is usually greater than the rate at which the information is generated. Packets are used to convert between generation rate and transmission rate. A packet is accumulated at source rate and transmitted at link rate. The time of accumulation constitutes a delay called packetization delay. In addition, in most packet systems, a packet is fully received from a link before it is forwarded [9]. This reception delay is considerably shorter than the packetization delay, since it occurs at link rates, but it also constitutes a delay. Both of these delays are related to the packet length: the longer the packet, the longer the delay.

These delays are avoided in burst switching. Here, there is no packetization delay because the link transmission rate equals the information generation rate for voice traffic, which is presumed to continue as the dominant form of traffic even when voice compression techniques are used. As soon as a burst is found to be starting, transmission of the burst on the next link can commence. Moreover, there is no reception delay at an intermediate node. Output can begin as soon the routing on the first header byte is done.

The delay performance of burst is attractive because there are neither packetization nor reception delays. Below, we examine the computed delays for a burst switch.

Delay Variation

Traffic will ebb and flow, with the number of active users and with each active user's burst issuance frequency. As the traffic at a node increases, a burst must expect to wait longer for an available outgoing channel. So the delay experienced by one burst may vary by a few milliseconds from that experienced by another in the same connection. This delay variation translates into a variation in the time between bursts. This is of no consequence at all for data. In voice, talk spurts average about 300 ms in length. At 33% speech activity, the silence intervals between talk spurts average about 600 ms in length. If the length of the silence interval as received differs from that sent by a small percentage, it cannot be detected. Therefore, delay variation is not a problem for burst.

The situation for voice packet switching is different. So that packetization and packet reception delays are not too great, the packet length is usually constrained to be considerably less than the average length of a talk spurt, requiring most talk spurts to be sliced into many packets. But the delay experienced by one packet of a talk spurt will normally differ from the delays experienced by others. A later packet might be less delayed and outrun an earlier packet. A later packet might be more delayed than an earlier packet, creating a gap at the output. Thus, an apparatus is required, of time stamps, sequence numbering, and an additional purposeful delay at the output for the first packet (to permit later more delayed packets to catch up)—all to permit correct reconstitution of a talk spurt.

The burst approach, with an entire talk spurt or data block in one burst, avoids this complexity, and also avoids the additional destination end delay.

Header Efficiency

Each packet requires a header. The shorter the packet, the more packets and the more header characters, which increase the percentage of link bandwidth used for overhead. So a tradeoff is necessary in packet switching: short packets have less delay (desirable) but greater overhead (undesirable).

This tradeoff is avoided in burst. Here, there is no need to packetize; thus, there is no need to establish a packet length. An entire data block or talk spurt is contained in one burst, with only one header required, even for bursts that are endless (e.g., music). Thus the header overhead is significantly less in burst than in packet switching.

Bandwidth Flexibility

The burst approach is channelized. This works very well if all sources are less than or equal to the channel rate. However, multiple channels must be put together for sources that are greater than the channel rate.

In the packet approach, a packet is accumulated at the source rate, whatever that is, and transmitted at the link rate, whatever that is. Thus, a variety of source rates can be handled in a very natural way.

If the system is required to handle a great variety of rates, the packet switching approach is superior to burst. If voice traffic predominates, the burst approach is superior.

How Does Burst Compare to ATM?

Asynchronous Transfer Mode (ATM) is packet switching, so the assertions made above with respect to fast packet switching apply to ATM as well.

The main emphasis in ATM work is on broadband switching, at speeds greater than, say, 50 mb/s. In addition ATM packet sizes may be quite short—16 or 32 bytes of information [10]. Such short packet lengths mean that initial packetization delays are correspondingly short—2 or 4 ms at a voice sampling interval of 125 μs. The high speeds maintained between switches mean that the tandem packet reception delays are very short: a 36-byte packet (4 header bytes and 32 information bytes assumed) would be received in 5.76 μs at 50 mb/s. So the combination of high speeds and short packet lengths probably make the delay performance of ATM switches quite acceptable.

As noted above, short packet lengths decrease bandwidth utilization; to $32/36 = 0.89$ in the example. This is relatively high, and the 11% inefficiency is hardly of much importance given the great bandwidths available on fiber.

But if longer packets of 256 bytes or more [11] are used, the initial packetization time increases to 32 ms or more. This could create echo problems unless all phones in the network are 4-wire or have echo cancellation.

This leaves the question of delay variation, which must be compensated for. We have no information on the delay variation and compensation of ATM switches.

Computed Burst Delay

A burst will often pass through many switching nodes on the way to its destination. Even though the delay through one node is small, the cumulative delay through a series of link and hub

switches is an important performance parameter.

To determine the echo performance of a burst switch, we computed the round trip delay a burst would experience if it passed through eight link switches in the origin link group, the hub switch, and eight link switches in the destination link group; and then back again, as shown in Figure 12.

Each link switch along this path of 32 link switch and 2 hub switch transits had 24 ports generating 0.25 Erlang of traffic at 0.33 burst activity toward and from the hub.

This study showed that the burst delay had only two components: fixed delays (e.g. serial to parallel character accumula-

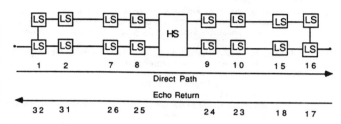

Fig. 12. Echo path.

tion time) and channel slip delays (i.e., the delay between the incoming and outgoing channel). Queueing delays (i.e., the delay a burst experiences because there is no idle channel available within a frame time) occur with negligible probability.

The total delay probability distribution then consists of the convolution of all the individual switch element channel slip delay distributions together with the sum of the individual switch fixed delays. The computed results were that the round trip delay would be < 5 ms at a probability of 0.99999. Thus, the burst delay performance is acceptable even without echo cancellers.

Burst Project Status

Four link switches and a hub switch have been built and operate satisfactorily. These are shown in the Figure 13 photograph.

Fig. 13. Burst project link switches and hub switch.

The dispersed control has been programmed to provide ordinary voice and data call set-up, as well as a few user features (e.g., line and trunk hunting, and group and directed call pick-up), sufficient to verify that a non-centralized dispersed control can handle the features likely to be required of a commercial or military switch. Finally, on-line reconfiguration, in which CPs can be added or deleted either by command or on the basis of failed background test message traffic concurrently with normal call processing, performs successfully.

References

[1] S. R. Amstutz, "Burst Switching—An Introduction," *IEEE Commun. Mag.*, Nov. 1983.

[2] S. R. Amstutz and R. D. Packard, "Dispersed Control Architecture and Specification Compilation of Burst Switching," Software Engineering for Telecommunication Switching Systems, Eindhoven, The Netherlands, Apr. 14, 1986.

[3] E. F. Haselton, S. R. Amstutz, and J. M. Lenart, "Burst-Switching Communications System," United States Patent No. 4,698,803, Oct. 6, 1987.

[4] P. O'Reilly, "Burst and Fast-Packet Switching: Performance Comparisons," Infocom '86.

[5] T. M. Chen and D. G. Messerschmitt, "Integrated Voice/Data Switching," *IEEE Commun. Mag.*, June 1988.

[6] E. A. Harrington, "Voice/Data Integration Using Circuit Switched Networks," *IEEE Trans. on Commun.*, vol. COM-28, no. 6, June 1980.

[7] P. Chen, "Use Hybrid Switches for Voice and Data," *Computer Design*, Oct. 1983.

[8] S. D. Personick and W. O. Fleckenstein, "Communications Switching—From Operators to Photonics," *Proc. of the IEEE*, vol. 75, no. 10, Oct. 1987.

[9] J. S. Turner, "Fast Packet Switch," United States Patent No. 4,491,945, Jan. 1, 1985.

[10] M. DePrycker, "Definition of Network Options for the Belgian ATM Broadband Experiment," *IEEE J. on Sel. Areas in Commun.*, Dec. 1988.

[11] K. Y. Eng, M. G. Hluchyj, Yu, and Y. S. Yeh, "A Knockout Switch for Variable-Length Packets," *Proc., IEEE Int'l. Conf. on Commun. '87*, June 1987.

[12] A. Huang and Scott Knauer, "Starlite: A Wideband Digital Switch," *Proc., IEEE GLOBECOM '84*, Dec. 1984.

Biography

Stanford R. Amstutz received his S.B. in electrical engineering from the Massachusetts Institute of Technology in 1955 and his M.S. in mathematics from Northeastern University in 1965. He is now Manager of the Distributed Systems Architecture Department at GTE Laboratories, Waltham, MA. Previously, he was Principal Investigator for Burst Switching at GTE Laboratories; Director of Software Development for the Nixdorf Computer Corporation; and Vice President of Product Development for Comshare, Inc. He holds eight patents and received the GTE Warner Achievement Award for Burst Switching Technology.

STARLITE: A WIDEBAND DIGITAL SWITCH

Alan Huang and Scott Knauer

AT&T Bell Laboratories

ABSTRACT

Starlite is a wideband digital switch intended for future visual and data communication needs. It provides arbitrary amounts of receive and transmit per user bandwidth (in fixed increments) for either bursty or continuous signals. In addition to the traditional one-to-one connection, the switch provides for simultaneous reception of information from separate sources, simultaneous transmission of different information to separate destinations and broadcast transmission to separate destinations. The switch is constructed out of self-routing, non-blocking interconnection networks. Bursty data (variable length packets) and continuous data are reformatted into small, synchronized, fixed length "switch packets" which are used to route the networks. At the switch input, unused switch packets are discarded by a *concentrator* network. A *sort-to-copy* and a *copy* network are used to provide the broadcast mechanism. A *sort-to-destination* and an *expander* network route the switch packets to their destinations. The use of multiple channels per user allows the variety of interconnections. For n user channels, the sorting networks have $O(n(\log_2 n)^2)$ elements each; the other networks have $O(n\log_2 n)$. VLSI chips implementing the networks are also described; each handles 64 5 Mbit channels.

INTRODUCTION

The Starlite switch is designed to handle a variety of communications needs, such as facsimile, electronic mail, computer communications, voice, two-way video, and anything else that can be digitized. Bursty data and continuous data of widely differing bandwidths are usually handled by separate circuit and packet switches. What is the architecture of a single switch able to handle this variety? The Starlite project explores this direction with a design that can be scaled up to accommodate the Gigabit data rates associated with optical fibers.

The basic approach relys on a self-routing, non-blocking, constant latency packet switch (with *no* store-and-forward). The switch is self-routing to avoid control bottlenecks when switch set-up can change a hundred thousand times per second, and the number of user channels, n, is arbitrarily large. The switch is non-blocking in order to avoid the hardware and control complexity needed to handle re-routing and to avoid the costs of buffering multi-megabit continuous signals. The switch is designed with constant latency so that any number of channels can be ganged together to allow users to assemble arbitrary amounts of switched bandwidth. The switch's construction out of stacks of regularly structured routing networks allows it to grow modularly with increases in number of users, amount of per user bandwidth and variety of interconnection types.

BASIC APPROACH

Figure 1 gives a capsule description of the Starlite switch architecture. At the top multiplexer various services are digitized, packetized and sent over many relatively small (5 Mbit) parallel "user channels" which are multiplexed bit-by-bit over a fiber cable to the switch. Bursty signals with variable

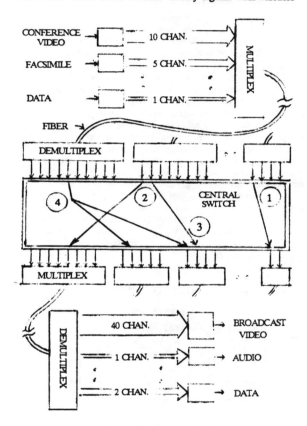

Fig. 1 Overview of Starlite System

length packets are broken up into strings of small, constant length "switch packets". Each switch packet header replicates only the routing information needed to get it through the local switch. Continuous signals are also broken up into switch packets whose headers remain the same as long as the "circuit" is connected. At the switch the user channels are demultiplexed so that each channel has its own entry point to the switch. The hardware design described later hides some of this complexity, and the illustration is for conceptual purposes.

Label (1) in Figure 1 illustrates the classic *one-to-one* connection. Since each channel is independently routed, the Starlite switch can send each user source channel to different user destinations, as illustrated at (2). Conversely, each user can simultaneously receive data from several independent sources (3). Finally, as illustrated in (4), the network has the ability to copy (at the direction of either the originator or the desired recipient) channels and route them to different users in a broadcast mode. The constant latency through the switch allows the channels to be grouped to match the bandwidths of different signals: Conference quality video - 10 channels, High definition facsimile - 5 channels, etc. By noting that the interconnections shown in (1) to (4) can be made with parallel lines, it can be seen that these can be made with variable bandwidth. The output channels are multiplexed bit-by-bit and routed on fibers back to each user. Note that users can purchase different transmit and receive bandwidths.

USING SORTING TO SWITCH

Figure 2 is a block diagram of the basic switch used in Starlite. The format of the switch packets is illustrated at the top of the figure. The first bit in the header determines whether the packet is active (activity bit = "0") or is unused, corresponding to an idle user or an idle fraction of an active user's bandwidth (activity bit = "1"). The next field in the header is the destination address - the user channel to which this particular packet of information is to be sent; this binary address is ordered most significant bit to least. The last header field gives the source address - the user channel over which the packet arrived. The n serial bit streams are fed into a self routing sorting network. Since the packets are synchronized, the header fields can be compared and the packets routed into ascending order at the output lines of the sorter as shown. The activity bits are taken as the most significant bits of the destination addresses, so that the inactive packets are to the right of the largest destination address. To complete the switching function, the relatively ordered packets must be put in an absolute order, skipping over the uncalled destinations (user channels). This is done by another routing network called the expander.

The sort function is performed by Batcher's well known bitonic sorting network (Ref 1), illustrated in Figure 3. It is constructed of small 2 line sorting elements which compare the bits of the destination address and switch to either "pass" or "exchange" the lines so that the larger number is routed to the output line indicated by the arrow. The comparison is only enabled during the destination address by a "header present" control signal supplied to the elements. After setting, the switch stays set for the duration of the packet. Note that only this control signal sets the size of the packets for the system. The figure gives an example of eight numbers being sorted by the network. The full sorter is built from a succession of "bitonic sorters" which accept a sequence that ascends, then descends, and outputs a sorted (monotonic) sequence. Figure 3 shows how the outputs of two elements (2 line sorters) are merged by a 4 element 4 line bitonic sorter, and how that in turn feeds an 8 line bitonic sorter that merges 8 lines into a sorted output. The number of elements in an n line full sorter is $n(\log_2 n)(\log_2 n + 1)/2$.

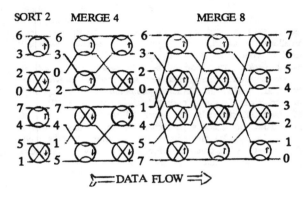

Fig. 3   Batcher Sorting Network

The interconnection pattern between the nodes in each bitonic sorter is called the "perfect shuffle" (Ref 2). This network is easily partitioned and thus a good choice for VLSI implementations. It is also the interconnection pattern used in the Expander, a version of Lawrie's Omega network (Ref 3) modified to handle inactive packets. Figure 4 illustrates the control algorithm for the nodes and gives an example of the Expander network routing the numbers in Figure 2. In this

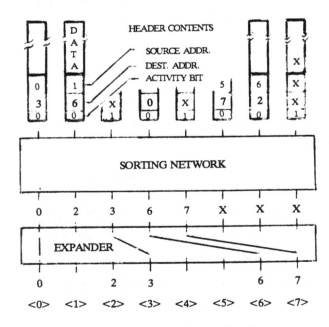

Fig. 2   Using Sorting to Switch

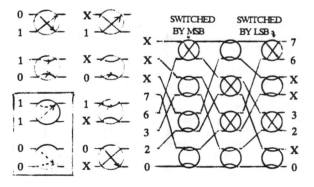

Fig. 4   Expander Network

network the first level of nodes is switched by the most significant bit, the second by the next, and so on until the last level which is switched by the least significant bit. The switching algorithm is to route the line whose destination address contains a "1" at the appropriate significant bit to the upper output and the line containing a "0" to the lower output. It can be shown that the two cases in which a conflict might arise (see box in figure) will not occur if the input to the network is sorted in ascending order and has no repeated destination addresses (user channel addresses).

CONCENTRATOR AND TRAP NETWORKS

Figure 5 illustrates two more networks. A $n\log_2 n$ element concentrator can be used to reduce the size of the $n(\log_2 n)(\log_2 n+1)/2$ sorting network by reducing n when it is likely that a significant number of users will be idle or that a significant percentage active users' bandwidth will be idle. The concentrator performs a routing that is the inverse of the expander and is indeed the same omega network turned upside down. In the concentrator the direction of the shuffles is reversed and they follow, rather than precede the switch elements at each stage. Routing for the concentrator is done by taking a running sum (sum of all bits to the left and above each position) on the inverted activity bits. This gives a set of adjacent routing addresses for active lines.

Since multiple packets addressed to the same user channel represent an overload of user bandwidth, they must be prevented by either protocols or hardware. The trap network is a hardware solution which spots repeated addresses and routes them to the rightmost outputs of the network. At this point the "repeats" can be discarded or recycled back into the sorting network in a later packet slot. The trap consists of a single stage of comparators followed by the same omega switch fabric used in the concentrator. The output of the

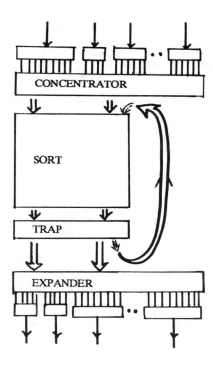

Fig. 5   Use of Trap and Concentrator

sorting network is fed to the comparators, which flag repeated addresses, but not the first instance of the address. A running sum of the flags representing either unrepeated addresses or empty packets is used to develop the routing address for the switch fabric. This puts out packets with unrepeated addresses on the left and repeats on the right; they are separated by inactive packets in the middle.

Another way of dealing with repeated user channel addresses is to switch them to unused receive channels owned by the same user. This is done by modifying the expander. In this case the trap is modified to limit the number of channels with the same user address (the more significant bits of the user channel address).

BROADCAST FUNCTION

Two extra networks are added to implement the copy function as shown in Figure 6. The input to the augmented switch consists of original source packets from users and empty "copy packets" which are input automatically by copy packet generators. These are either controlled by users transmitting data which they want broadcast or by receivers interested in getting copies of certain data (with the permission of the owner). The packet formats are shown at the top of Figure 6. Source packets have the address to which they are sent as the destination address (like the "TO" address on an envelope) and their own address or identifier listed as their source address (like the "FROM" address). A new one bit header field, the copy bit, identifies the packet as original (copy bit = "0") or empty copy packet (copy bit = "1"). Using a low bandwidth link to the input side of the switch, (typically a single signaling packet multiplexed in the user's source input) transmitters can initiate the production of empty copy packets with destination addresses for each intended recipient; they list the source address of the information that they are transmitting as the copy packet's source address. These packets have much the format of the original sources (top left of Figure 6). Receivers have the copy packet generator automatically create a stream of packets with the receiver's own address as a destination ("TO") address and the address of the data they want to copy in place of the the source ("FROM") address; this format is shown at the top right of the figure. The receiver in effect sends self-addressed empty "envelopes" into the system. Both the the empty packets and source packets are input to the modified switch.

Although not shown in Figure 6, a concentrator could be used at this stage to eliminate idle sources and unused copy packet inputs. This cuts the total number of inputs to a value less than or equal to the total number of receivers, and would be useful if there are a large number of inputs devoted to copy packets. The "sort-to-copy" network uses the source ("FROM") address of each incoming packet together with its copy bit as the least significant bit, to order the packets such that their source addresses increase from left to right. This groups copy packets together with their appropriate broadcast source packet on physically adjacent lines. Moving left to right at this sorter's output, each source packet will be followed by the number of copy packets that share the same source address.

The next stage in the broadcast switch is a $n\log_2 n$ element "copy network" which takes the information in the data field of each broadcast source packet and copies it into the empty data fields of all blank packets to the right until another source packet is encountered. This routing is accomplished using the "copy bit" in each header. The final stages of the modified network are the sort and expander stages of the original network. These route both original source packets and filled copy packets to their destination addresses.

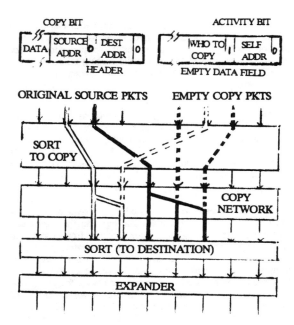

Fig. 6 Additional Hardware to Implement Broadcast Function

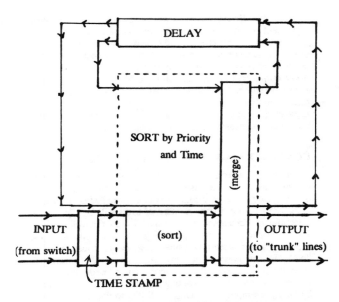

Fig. 7 Distributed Prioritized Concentrator

DISTRIBUTED PRIORITIZED CONCENTRATOR

In connecting central "star" architecture switches, there needs to be a method of allocating the "trunk lines" that connect them. A distributed, self-routing non-blocking architecture for this is shown in Figure 7. This "Distributed Prioritized Concentrator" is used to allocate lines going to one destination, such as a city. A sorting network (dashed box) takes delayed bursty data and incoming bursty and continuous data and sorts it on the basis of priority (most significant bits in the sort key) and time. Continuous data is given enough priority to be sorted to the trunk line. If no room exists, the connection is refused. Bursty data can be recycled for delay if it's priority (always less than the continuous data) is not high enough. Bursty data is "time stamped" upon entering the DPC. If it is delayed it's "time stamp" becomes relatively smaller compared to newer data arriving. Higher priorities are represented by low numbers, and by making the time stamp part of the least significant bits of the priority, delayed bursty data gains "seniority". If the DPC memory fills, the data with the lowest priority, which includes the newest bursty data, is lost. The system protocols should monitor the DPC memory to refuse new data when the load is high. The DPC efficiently packs bursty data into gaps that might occur in continuous data and also uses channels held open while being reserved for continuous data. It is possible to enhance this design so that one DPC handles multiple trunk lines to multiple cities.

HARDWARE IMPLEMENTATIONS

To test these concepts a TTL version of a Starlite switch has been built using PALs each holding two sorting or routing elements (4 lines). The networks used are: a sort-to-copy, copy, sort-to-destination, trap, and expander. The system handles 64 12 Mbit lines and is mounted in a vertical paging system with six 14" X 15" boards. This version is debugged and running.

The architecture of Starlite is really intended for VLSI implementation, and several chips have been designed. The recursive structure of the Batcher network in Figure 3 is exploited to build a a two chip set that will build any size Batcher sorting network. The "front end" chip contains as large a full sort network as will fit - this chip is area limited. It contains two 32 line full sorters (480 sorting nodes) and handles 64 5 Mbit lines. These lines are multiplexed 4:1 to conserve pins; the pins handle 20 Mbit data rates. The "back end" chip is used to build any size bitonic sorter, from 64 lines wide on up (in powers of 2). It consists of 4 levels of 64 wide perfect shuffles, and the levels can be bypassed to vary this depth. Since small shuffles can be interleaved to produce wider ones, and the chips can be stacked to any depth, it is possible to build any size bitonic sorter to merge the outputs of smaller full sorts, and thus to build any size full sorter. The "back end" chip is pin limited, and only contains 128 sorting elements. These chips were designed using the MULGA system (Ref 4), fabricated in 2.5 micron CMOS and work at the rates indicated.

The expander network has the same shuffle connections as the bitonic sort, only the algorithm used to switch the nodes is different. The expander network can thus be built to any size using chips identical to the "back end" sorting chips except for the switch nodes. By adding switching to reverse the direction and positions of the shuffle, this "Omega chip" can also serve as the concentrator and trap building blocks. This chip has been designed and is currently being fabricated. Chip designs for the running sum adders and copy network are in the layout stage.

Based on these chip designs, a 256 channel basic switch of the form shown in Figure 2 would require 36 chips (28 to sort, 8 to expand), 1024 channels require 256 (208 + 48), and 4096 (about 20 Gbits bandwidth) gets up to 1408 (1216 + 192). The corresponding counts for a broadcast system similar to Figure 6 are 80, 544, and 2944. In comparing the VLSI to the TTL PAL implementation, it turns out that two 32 line sort networks (corresponding to one front end sort chip) occupy one 14" X 15" board, draw 35 A and cost about $4000 for the board and chips. While the PALs are faster due to the bottleneck of too few pins on the VLSI and the need to multiplex, the chip only draws 20 ma and would cost under $100 in production.

SUMMARY

The Starlite project has given at least an upper bound on the practical complexity (in terms of chip count and growth as a function of channels) of a single switch architecture for both "packet" and "circuit" switching problems at high data rates. It has developed a tool box of useful techniques for solving switching problems with self-routing networks that have modular growth. Hardware implementations demonstrate that such networks are well suited to VLSI implementation. Future work looks to improving speed and reducing the "granularity" of the channel size by using denser VLSI chips. Steps of 5 Mbits per channel are a bit coarse. Wholly optical architectures are also being studied.

REFERENCES

1. K. Batcher, "Sorting Networks and their Applications," *AFIPS Conf. Proc., 1968 SJCC* pp. 307-314

2. H. Stone, "Parallel Processing with the Perfect Shuffle" *IEEE Transactions on Computers,* Vol C-20, pp. 153-161, Feb. 1971

3. D. Lawrie, "Access and Alignment of Data in an Array Processor," *IEEE Transactions on Computers,* Vol. c-24, No. 12, December 1975, pp 1145-1155

4. N. Weste, "MULGA - An Interactive Symbolic Layout System for the Design of Integrated Circuits," *BSTJ,* Volume 60, No. 6, pp. 823-857, July-Aug. 1981

DYNAMIC TDM -
A PACKET APPROACH TO BROADBAND NETWORKING

L. T. Wu, S. H. Lee and T. T. Lee

Bell Communications Research, Inc.
Morristown, NJ 07960

ABSTRACT

Packet switching techniques, have been shown to be efficient ways to provide transport for variable bandwidth or bursty traffic. However, their attractiveness in satisfying future service demands is overshadowed by the recent wide availability of single-mode fibers. One question being raised is: If transmission is getting cheaper, why do we need packet switching techniques for bandwidth conservation when these techniques introduce new network components that may not be compatible with the existing circuit switched network? To answer the above question, we show in this paper, through a flexible network transport technique (referred to here as Dynamic TDM), that packet techniques permit us to achieve bit-rate flexibility and to simplify network equipment design even if bandwidth conservation is not the major concern.

1. INTRODUCTION

Single mode fiber has become the preferred choice for interoffice trunking and feeder networks to consolidate outside plant based on the existing telephone services. The evolution toward an all fiber network in the subscriber distribution area is less clear because of the uncertainties in predicting future broadband service demands. From a network design point-of-view, these uncertainties translate into challenges that defy common network design practices. First, a broadband public switching network must be sufficiently flexible in its design to handle vastly different types of traffic, ranging from low speed data and voice to full motion video. Second, to satisfy the unknown growth pattern in future service demands, it is necessary to have a robust design that can easily be modified as the network evolves. Finally, another important design objective is the formulation of a practical migration strategy from the existing copper wires and the associated switching and transmission systems to fiber and the succeeding generations of high speed equipment. These three design criteria determine the selection of the three major components of the network architecture: network topology, transmission techniques, and switching methods.

To meet the above design objectives, it is necessary to see if there is a better set of networking techniques, other than traditional circuit switching and packet switching, capable of handling both continuous and variable rate broadband services. While traditional circuit switching can be extended to multirate (multiple-bit-rate) services by allocating multiple slots to high bandwidth applications, the approach requires larger frame buffers and relies on close synchronization between switching and transmission systems to avoid the time-slot skewing problem. On the other hand, although packet multiplexing is inherently bit-rate independent, conventional designs (e.g., X.25 networks) use sophisticated protocols to perform flow-control, error recovery and alternate routing. Thus, the substantial processing overhead and excessive packet delay render it unsuitable for most real-time applications.

More recently, several new approaches have been suggested for multirate services in a public switching network that combine the transparency of the circuit switching approach with the bandwidth flexibility of the packet switching technique. Among these, fast packet switching [1] and burst switching [2] have been identified as potentially attractive for voice and data integration. Both techniques use low functionality packet protocols that support data rates up to the full transmission speed. The main difference between them is in the header: burst switching associates one header with each activity spurt (say, up to 10,000 bit) and fast packet switching uses a header for every nominal size packet (e.g., up to 1,000 bits). Although these transport techniques have been shown to be effective for T-carrier systems, both require modifications to handle broadband applications.

A straightforward extension of the burst switching technique to allow broadband link speeds will result in the problems of time-slot skewing and larger frame buffers as mentioned before. Furthermore, when homing hundreds of such links on a central office switch, the required multirate circuit switch with the desirable throughput is a difficult design problem. The second approach, fast packet switching, is also inadequate, in its original form, for broadband applications. Fast packet switching relies on complex switching and multiplexing hardware and variable packet length to achieve a high degree of transmission efficiency. Accordingly, with T-carriers being replaced by higher bit rate optical transmission systems (requiring much shorter processing time for individual packets), the switch architecture and the packet protocol must be further simplified to avoid the potential bottleneck resulting from the processing of simultaneous packets.

In this paper, as an outgrowth of previous work on fast packet switching and burst switching, we propose a broadband packet network architecture capable of handling voice, data and video. The architecture uses a nonblocking self-routing switch together with a flexible statistical multiplexing scheme, referred to here as Dynamic Time Division Multiplexing (DTDM). In addition to the bit-rate flexibility and the transport efficiency, this approach is compatible with the existing digital multiplexing hierarchy and, as a result, lends itself to a natural migration strategy to evolve the existing public switching network into an all fiber broadband packet switching network.

The concept of DTDM is a combination of the conventional TDM and packet transmission techniques. Bit-interleaved TDM provides a simple mechanism to merge high speed signals into a higher speed time multiplexed signal for optical transmission. However, it also imposes tight synchronization requirements between incoming lines to avoid bit slips. The one-bit-at-a-time mode of operation also limits the way transmission bandwidth is shared by the incoming lines. Thus, for lower speed applications (< 150 Mbps) where VLSI technology can still be used to provide more sophisticated header processing, block interleaving (or slotted packets) appears to offer an attractive alternative as a transmission format. In other words, DTDM is a transmission format that permits synchronous multiplexing beyond a predetermined high signal rate and allows asynchronous transmission below that rate.

Section 2 exploits the advantages of designing a broadband packet switch to conform with the slotted fixed length packet format. We describe a two-level switch design by decoupling transmission formats from the internal switch protocol and show that it is possible to build any type of switching networks by associating the same nonblocking self-routing switch with appropriate channel structures. Section 3 generalizes the slotted packet concept of a self-routing switch to the network transport to define the concept of DTDM as a transmission format. Examples are then given to show the potential applications of a DTDM multiplexer.

Combining a nonblocking self-routing switch with a DTDM channel structure, we show in Section 4 a broadband packet network architecture suitable for public switching network applications. A network deployment strategy is identified that depicts DTDM as a broadband packet network embedded in a TDM based interoffice facility hubbing network. Finally, in section 5, we summarize the distinct features of the proposed approach.

2. SWITCHING AND CHANNELIZATION

The conventional classification of digital switching networks generally implies an explicit channel structure for the transmission links and a set of implicit operational characteristics for the switching systems. For instance, a digital circuit switching network typically refers to the time division multiplexing (TDM) technique that establishes a sequence of time slots for individual calls to transmit signals and a combination of space division switching and time division switching to swap time slots on different channels. A packet switching network, on the other hand, implies statistical multiplexing, the use of address header, and memory switching for routing and buffering packets.

In this section, we examine the structural properties of switching networks and show that the above dichotomy does not necessarily lead to the distinction of packet switching and circuit switching based on the use of an address header alone. More precisely, we show that, by adopting a slotted packet approach it is possible to adopt the same switch design for a circuit switching network, a packet switching network, or even a mix of both. Thus, if the header overhead is tolerable, packet transport techniques can be used to achieve flexibility and simplicity in equipment design without necessarily introducing packet blocking and variable delays.

A typical switching system can be represented as a block diagram, as in Figure 1, having three major components: Control Processor (CP), Switch Interfaces (SIs), and Interconnection Network (IN). The CP handles call processing, maintenance and administrative functions. The SIs convert transmission formats into switching formats and the IN routes information blocks from input links to output links of the switch. For the existing digital circuit switching systems used in the public network, input signals are time multiplexed and the information in a specific time slot on an incoming line is transferred, via the switching system, to a specific time slot on an outgoing line. Thus, the basic operation of the switch is to interchange the position of the time slots through SIs and to interconnect time slots using a time-multiplexed space division switch IN.

Note that, as described above, the essence of a circuit switching network is to provide an unperturbed transfer from an incoming time slot to an outgoing time slot. It does not matter whether the switching system uses crosspoint matrices, memory devices, or photonic elements. More recently, it has been shown in [3] that a circuit switching system can actually be built using a self-routing (packet) interconnection network to act like a time multiplexed switch, i.e., to provide unperturbed transfer between slots.

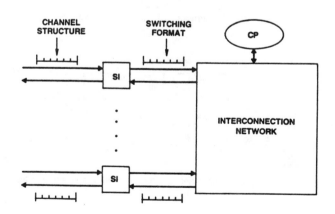

SI: SWITCH INTERFACE
CP: CONTROL PROCESSOR

FIGURE 1. COMPONENTS OF A SWITCHING SYSTEM

One possible way to construct an IN to provide circuit switched connections is to combine a Batcher sorting network with a banyan routing network, as shown in Figure 2. Based on the address headers associated with fixed size information blocks, the combined Batcher-banyan network allows packets with different destination addresses to be transmitted in parallel across the IN, without internal packet collisions. Thus, to mimic the operation of a conventional time-space-time circuit switch, one needs only to design the SIs to perform the standard time slot interchanger functions together with the ability to insert and to delete headers on the fly. By converting external TDM time slots into internal "packets," the connection pattern of the IN can be reconfigured dynamically by merely changing the headers alone (see Figure 3). In the following discussions, we will refer to this nonblocking self-routing function of a Batcher-banyan network as a self-routing time multiplexed switch or, simply, as a **self-routing TMS**.

When compared with conventional switch architectures, we find that the Batcher-banyan switch combines the advantages of both single stage and multiple stage crosspoint switches. The switch is as strictly nonblocking and virtually automatic in pathfinding as a single stage crosspoint matrix, but this simplicity is obtained with hardware complexity similar to that of a 3-stage Clos network. However, unlike a single stage crosspoint matrix, the header processing can be done in parallel by multiple SIs, allowing distributed switch control together with the benefit of easy expandability.

Note that the above construction of a self-routing TMS presents an interesting surprise: A Batcher-banyan network, even though inherently a high capacity packet switch, can also serve as a building block for a circuit switching network. Extending the observation to other types of switching networks, a self-routing TMS can be equipped with specialized SIs to provide efficient switching for a variety of switching networks. By generalizing the properties of a TSI, a variation of the Batcher-banyan switch capable of handling both continuous and bursty traffic (i.e. circuit and packet connections) has been reported in [4].

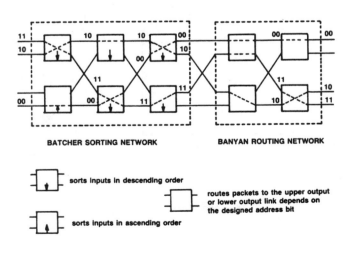

BATCHER SORTING NETWORK **BANYAN ROUTING NETWORK**

sorts inputs in descending order

routes packets to the upper output or lower output link depends on the designed address bit

sorts inputs in ascending order

FIGURE 2. A BATCHER—BANYAN INTERCONNECTION NETWORK

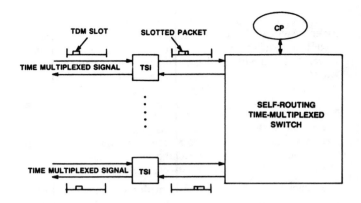

FIGURE 3. CIRCUIT SWITCHING APPLICATION OF A SELF-ROUTING TIME MULTIPLEXED SWITCH

Since the switch allows the allocation of channel bandwidth by varying the frequencies of sending packets to specified output ports, it is capable of handling multirate services even for continuous traffic.

In exploring the advantages of broadband packet transport, one should bear in mind the distinctions between the simplicity of a Batcher-banyan switch and the flexibility of a self-routing TMS. The unique properties of a self-routing TMS are derived from 1) explicit addressing and 2) fixed length packets, provided by the SIs. Explicit addressing allows information blocks to be processed in parallel to achieve high throughput and gives network equipment the freedom to participate in distributed processing. A fixed length packet format, although not essential for flexibility, simplifies equipment design by removing the need to handle contentions due to asynchronous packet arrival. Among other alternatives, it has been suggested in [5] that a nonblocking self-routing switch can also be constructed by allowing each crosspoint element of a crossbar to recognize addresses and to route packets accordingly.

In summary, the slotted packet format provides the flexibility for designing a fully integrated switch for public network applications. The switch is bit-rate independent, capable of handling both continuous and bursty traffic, and amenable to distributed switch control by remoting some header processing functions (e.g. call setups for conferencing) to appropriate places in the network. Thus, at least in the context of switching system design, we have shown that packet technique can be useful for broadband applications. In the section that follows, we will examine the validity of the above observations from the network transport point of view.

3. INTEGRATED NETWORK TRANSPORT

Similar to broadband packet switching, transmission systems for fiber-based public network must also be flexible regarding subscriber bit-rates, equipment types and access technologies (e.g., point-to-point vs. multiple access) to serve a wide variety of customer needs. Furthermore, the channel structure should be designed to be compatible with the existing digital multiplexing hierarchy to allow easy evolution of the existing network.

This section addresses issues relating to the conflict between the needs for service independence and network compatibility. We first define DTDM to be a crossover between conventional TDM and ATDM and then illustrate its applications for both local access and interoffice trunking.

3.1 Dynamic Time Division Multiplexing

Traditionally, multiplexing techniques were introduced to increase the efficiency of time-shared facilities. The public switching network uses TDM almost exclusively to carry traffic aggregated by either a pair-gain system for the feeder plant or a T-carrier system for interoffice trunking. When time slots are allocated to input terminals in proportion to their input rates, the scheme is called multirate time division multiplexing (MTDM). Another mode of digital transmission, introduced for low-speed data communications, uses channel identifiers with actual data to allow on-demand multiplexing of terminals with low channel utilization. This approach relies on timing patterns to signal the beginning and the end of an information block and permits asynchronous assignment of time slots. Hence, it is called **asynchronous** time division multiplexing or **ATDM** (see, for example, [6] and [7]) as opposed to the synchronous mode of operation for TDM.

ATDM, as a "physical" layer protocol, is inherently bit-rate flexible and allows simple timing for varying data rates. However, it has the drawback of unstable timing under heavy loading or high packet lost rates. Moreover, since a transmission system for the public network must also transmit extra information beyond that provided by the subscriber data stream (e.g. maintenance and signaling channels), the protocol hierarchy for the link and network layers can be complicated.

To resolve the above problems, we propose to extend the concept of ATDM by embedding the ATDM signals explicitly in a synchronous TDM framing structure. The modified form, referred to here as **dynamic time division multiplexing** or **DTDM,** combines the channel identifier of the ATDM with the framing pattern and control information of the TDM to form a fixed length header for each time slot. Thus, the DTDM can be viewed as a transmission format for slotted packets to traverse synchronous digital pipes. The fixed length packet envelope (vs. variable length packet format) permits us to simplify buffering in multiplexing equipment and to ease the bandwidth allocation among diverse users. Moreover, the slotting of packets makes packet detection straightforward (as opposed to bit-stuffing) and allow simple synchronization procedures (e.g. to control timing slips and to facilitate time-stamping).

DTDM, as a variation of ATDM, appears to be particularly attractive for the public switching network to support dissimilar traffic of both continuous and bursty types generated by heterogeneous users. More precisely, the slotted packet format allows us to prioritize services through the header to control packet contention and the synchronous backbone enables us to satisfy the stringent real-time constraints for continuous traffic. The need to handle continuous traffic in packet form distinguishes a broadband public packet network from a specialized data communication network (e.g. a LAN). In a public switching network, even if the point-to-point subscriber traffic is bursty, the statistical effect of multiplexing will make the aggregated traffic look almost like a smoothed stream. (This argues for a synchronous internal network with an asynchronous packet access network.)

DTDM also allows flexible transport for bursty traffic. Note that a time slot in TDM or MTDM is allocated to an input channel even in the absence of data on that channel. DTDM overcomes this inefficiency by allotting time slots only for the active channels. Thus, DTDM is a system that more efficiently uses the facility on an "instantaneous time-shared" basis as illustrated in Figure 4. In addition, the slot-by-slot allocation also eliminates the fixed channelization constraints of the TDM approach. A subscriber uses enough slots to match his bit-rate without having to live with integer multiples of the basic channel rate.

As a specific example of DTDM, let us suppose that the backbone fiber network is channelized through an 8 KHz clock to form digital pipes operating at 150 Mbps. Then, at 150 Mbps, a 125 microsecond frame will contain about 2340 bytes. Clearly, the frame size is too large to carry a single packet for most applications and needs to be further divided into smaller slots for efficient bandwidth sharing. The selection of an optimal slot size is generally a difficult engineering problem involving tradeoffs between service quality, transmission efficiency and equipment simplicity.

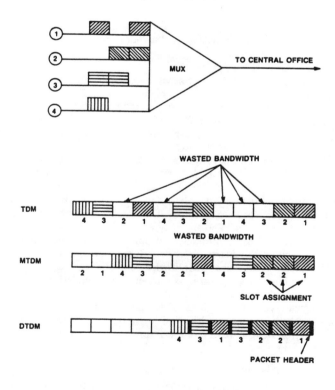

FIGURE 4. TIME DIVISION MULTIPLEXING

Without loss of generality, however, let's assume that the frame is divided into 18 slots. Then, a DTDM framing format can be represented as in Figure 5. Note that, at 18 slots per frame, each slot contains 130 bytes including the DTDM headers. The slot size is chosen here to carry up to 15 ms of PCM voice to ensure that the cross-country voice delay is within acceptable limit even if the long-haul portion of the network remains circuit switched. However, if the network traffic is predominantly entertainment video, a larger slot size, say 6 slots per frame or a 390 byte slot, could be used to simplify the processing of video packets and to reduce the header overhead.

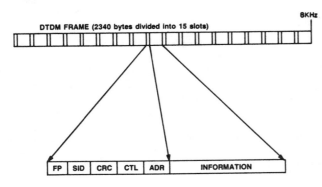

DTDM FRAME (2340 bytes divided into 15 slots)   8KHz

| FP | SID | CRC | CTL | ADR | INFORMATION |

DTDM SLOT (130 bytes with a 10 byte header)

FIGURE 5. DTDM FORMAT

For each slot of 130 bytes, we allocate the first 10 bytes to provide the following header information: 2 bytes for the framing pattern (FP), 1 byte to indicate packet type, e.g., signaling (SID), 1 byte for CRC, and the remaining 6 bytes for the address field (ADR) and other control information (CTL). The framing pattern is designed to establish the 8 KHz frame timing as well as for detecting the beginning of a slot. By combining both functions together, the multiplexing equipment can be designed to simplify the recovery of frame timing in case of a slip without having to rely on the buffering of an entire frame and tight network synchronization. The signal type field identifies whether the slot is used for continuous real-time traffic or bursty traffic, as well as the busy/idle status of the slot. Four classes of packet types are identified; control packets, packets belonging to a real-time continuous call, packets associated with a virtual circuit call, and datagrams. Depending on the signal type, the address field can be a logical channel number, routing information or physical addresses. The control field is reserved for future use (e.g. to carry time-stamps).

3.2 Applications

The major attraction of DTDM is its simplicity in providing subscriber access for varying bit-rates. As an example, Figure 6 depicts a DTDM multiplexer as a subscriber interface to a fiber loop to concentrate traffic from three different communication devices: a digital phone generating 64 Kbps PCM voice, a graphic terminal sending bursty data at 1 Mbps, and a video terminal producing a 45 Mbps constant bit stream.

At 150 Mbps, the dedicated channel yields 144,000 slots per second given a 130 byte slot size. The available time slots are shared by the three devices by giving higher priority to the video stream and allowing the voice and data terminals to contend on a first-come, first-serve basis. If the video terminal is allowed to fill up the entire data field of a slot, then, at 45 Mbps, it seizes one slot for every 3 slots passing by. Thus, the regularity of the traffic source will be maintained through the transmission link and perceived by the receiver as periodic packet arrivals. To avoid excessive end-to-end packet delay, let's assume that the voice source is allowed to accumulate up to 15 milliseconds worth of voice samples before inserting the information into a time slot. Then, a voice packet is expected to arrive every 2,160 slots. Similarly, at 1 Mbps, the graphic terminal will transmit one packet per every 150 slots.

Note that, in the above scenario, 45 Mbps is given a high priority to seize a slot to avoid queueing delay at the multiplexer. (For this particular example, the maximum queueing delay for all services is 2 slot times even without using priority!) The maximum delay from one end of the transmission link to the other is primarily the sum of the packetizing delay and the transmission delay and, for local access applications, the transmission delay is considerably less than the 125 microsecond frame time. Hence, the network transport can be viewed as a transparent transport similar to a circuit switched connection. For the voice and data packets, variations in packet delay may occur. However, since there are two free slots available for every three slots passing by, the queueing delay should be within a few slot times. Thus, packetizing delay is the dominant delay factor in determining voice quality.

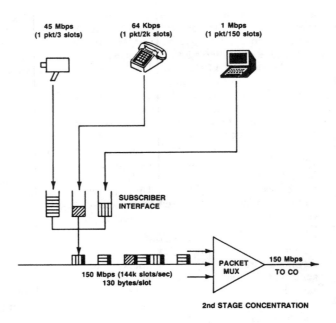

FIGURE 6. LOCAL ACCESS APPLICATIONS

260

Note that the DTDM multiplexer could have a design much simpler than a conventional TDM multiplexer. A working prototype operating at 150 Mbps has been reported in [8], consisting of no more than 8 packet buffers and simple contention resolution circuitry for each incoming line. The simplicity of the design suggests that a miniature single chip version could be possible for future applications in view of the advances in VLSI technology.

DTDM can also be used as a replacement transmission technology to carry existing interoffice traffic. More specifically, let's consider the situation that we need to multiplex three time-multiplexed signals, at DS1, DS2 and DS3 rate respectively, for a point-to-point transmission between two offices. The traditional TDM approach would have to rely on a step-by-step hierarchical approach to multiplex and to demultiplex these signals. As shown in Figure 7, the conventional multiplexing scheme would require line conditioning and synchronization circuitry at each level of the hierarchy as well as hardware for bit interleaving. (An alternative TDM approach [9] is being investigated to allow a one step multiplexing of different signal rates.)

Using a DTDM multiplexer, each of the three signals would then be fragmented and encapsulated in the time slots. Since all three signals are real-time continuous traffic, we assign the same priority to their packets and allow them to compete for available slots on a first-come, first-serve basis. Some variations in interarrival time may occur. The signal would then have to be resynchronized at the receiver end, but this can be done simply by buffering enough packets without requiring time-stamps as long as the maximum delay variation can be kept sufficiently small. The multiplexer required for this application is the same as the one used in the above example, i.e., we only need enough buffers to accumulate packets in conjunction with some hardware to perform slot interleaving. Thus, DTDM permits us to eliminate the need for step-by-step multiplexing up to 150 Mbps.

4. A BROADBAND PUBLIC SWITCHING NETWORK

A commonly held view of how to introduce packet technology into the public network is to deploy an overlay network because the existing network is optimized for circuit switched voice and is therefore incompatible with statistical switching and multiplexing techniques. Accordingly, most deployment strategies recommend constructing a skinny network for a selected set of services (e.g., WATS, low speed data, LAN interconnections, 800 services, etc.) and hope that the migration of existing and new services to the packet overlay will allow the existing network to be phased out slowly.

The main advantage of an overlay network is the quick realization of an end-to-end network for new services. However, the approach requires a large initial capital investment (as opposed to the cost of just cutting-in a new switch) and increases operational cost by having to manage multiple separate networks. Thus, an overlay network tends to prove in its economics by relying heavily on revenues generated by new services with proven market.

Based on the approach proposed in this paper, we believe that it is also possible to introduce packet techniques into the public network as a replacement technology for upgrading individual transmission and switching components. One may argue that this transitional approach requires larger capital expenses in the long run because of the need for new equipment to handle packet and circuit conversions and may prolong the time required to form an end-to-end subnetwork for new services. The above disadvantages can be partially offset by the considerably smaller operational costs and the ability to spread out capital expenditures over a longer period of time. The most significant strength of the proposed approach, however, is that it is also compatible with the overlay strategy. It is conceivable that the existing network could evolve by introducing clusters of broadband packet networks in particular areas to serve key subscribers for new services. The approach differs from the conventional overlay strategy in that each cluster would also support existing services and would be compatible with the rest of the network with respect to these services. The incentive for this strategy is to reduce operational and maintenance cost for the overall integrated transport as well as by generating new revenues from new services to compensate for the write-off of obsolete equipment. In the following discussions, we describe a possible embodiment of DTDM in the context of the public switching network.

Similar to the current network, we assume that a broadband public network is also partitioned into two parts: the local access network and the interoffice facility network. The local access network allows local dialing and traffic concentration through the local exchange. The interoffice facility network manages connectivity between switches to allow delivery of point-to-point traffic.

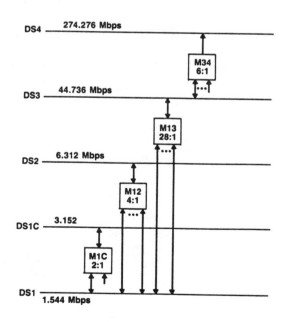

FIGURE 7. DIGITAL MULTIPLEXING HIERARCHY

As an extension of the emerging concept of interoffice fiber hubbing network [3], we postulate that the fiber facilities currently being installed for the voice trunking network would continue to be used in the future for broadband applications. However, as T-carrier technology is phased out, the facility network could also be simplified. Instead of DS3 and DS1 add/drop capability, we assume that the backbone facility network would provide real-time cross-connect functions for high speed digital pipes, say at 150 Mbps, using electronic space division or photonic switches. Regarding the local access network, we assume a star or hierarchical-star topology [10] to form a stand alone packet network. Individual local access packet networks are linked together using the fiber facilities of the backbone network, but remains operationally independent of the underlying TDM network. This approach of embedding a packet network over a facility network is referred to in the literature as "virtual networking." .

Figure 8 describes a schematic representation of a DTDM network wherein the lower speed (<150 Mbps) packet switching network coexists with other networks as a virtual network embedded in the high speed backbone network. Three classes of network equipment are illustrated: facility cross-connect systems, broadband packet switches and special service nodes. The facility cross-connect systems allocate the basic 150 Mbps digital channels between the broadband packet switch and the special service nodes. The broadband

packet switches provide the basic multirate services for the local access of the public network and the special service nodes establish private-line services for large business users and, possibly, distribute high resolution motion video to residential subscribers.

A DTDM network can also be characterized as packet switching clusters linked together by synchronous pipes. The network is divided into clusters of subscriber nodes, with each subscriber belonging to one of the nonoverlapping clusters. The synchronous links of the backbone network are managed by the facility cross-connect systems which provide fast connection setups and tear-downs on a real-time demand basis. Each individual cluster permits subscribers to communicate with each other through the packet switch. The packet switch also grooms and consolidates outgoing traffic according to the destination switch addresses. Thus, the packet switch provides the demultiplexing and multiplexing functions for the interoffice trunking network without incurring the cost for step-by-step multiplexing functions common to the current circuit switching approach.

The backbone network, as depicted in Figure 8, consists of facility cross-connect systems interconnected by fiber links transmitting at multiple gigabit speed. The links are divided into digital pipes using TDM or WDM techniques. Since the backbone network provides only basic transport for continuous bit streams, it is well suited for fiber optic

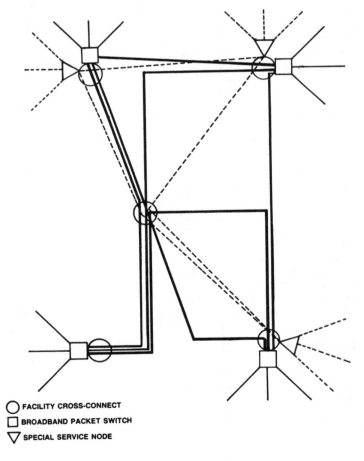

○ FACILITY CROSS-CONNECT
□ BROADBAND PACKET SWITCH
▽ SPECIAL SERVICE NODE

FIGURE 8. A DTDM BROADBAND PUBLIC NETWORK

technology and, thus, an end-to-end photonic backbone may even be possible (see [11], for example). As the backbone network being deployed, we envision the initial network topology to resemble a hubbing network, i.e., all fibers homing toward a central location for cross-connection. However, as demands for broadband services grow, direct fiber links between end offices may gradually be added. In other words, we envision that the public network to evolve from the existing T-carrier based mesh network toward fiber hubbing and finally back to a fiber based mesh network with both direct and tandem links.

There are several advantages of the two-level hierarchical networking approach including operational simplicity, facility sharing, bit-rate flexibility, and reduced overhead in network synchronization. By allowing multiple virtual networks to reside on the same backbone network, applications of different traffic characteristics can be supported with individually optimized logical topology and switching features. Furthermore, since a virtual network can be designed to be operationally independent to the backbone network and to other coexisting virtual networks, changes made to any particular network would be transparent to others.

In addition to advantages common to all virtual networks, the hierarchical realization of a DTDM network provides the bit-rate flexibility of an ATDM network, but with a simpler network synchronization scheme. The characteristics of the proposed network can be summarized as follow:

1. Uniform interface - Subscribers interact with the network at their natural rates via statistical multiplexing.

2. Bandwidth flexibility - The network acts as fixed channel allocation for continuous traffic and uses excess capacity for bursty traffic.

3. Terabit Switching - The fixed length packet format permits the construction of high throughput switches based on a self-routing TMS.

4. Easy synchronization - Network synchronization overhead is limited to higher level signals and the cost is shared by multiple users.

5. Backward Compatibility - The synchronous (8 KHz) TDM backbone provides facility sharing between DTDM and existing time-multiplexed signals.

5. SUMMARY

DTDM and slotted packet switching appear to be flexible techniques for achieving a fully integrated network for heterogeneous broadband applications. The approach allows the consolidation of outside plant and end electronics as well as the design of integrated self-routing switches to handle both continuous and bursty traffic. Moreover, by choosing the slot size to be compatible with the frame structure of the underlying TDM network, we show that it is possible to introduce packet techniques into the public switching network without necessarily constructing an overlay network. Thus, even if bandwidth conservation is not the major concern for a broadband network, statistical packet switching and multiplexing methods could still be the key to unlock the vast potential benefit of a fiber-based public switching network.

ACKNOWLEDGEMENT

The authors would like to thank P. E. White for suggesting the problem, W. D. Sincoskie and Glenn H. Estes Jr. for valuable discussions.

REFERENCES

[1] R. W. Muise, T. J. Schonfeld and G. H. Zimmerman, "Experiments in wideband packet technology," *Proc. 1986 International Zurich Seminar on Digital Communications*, pp. 135-138.

[2] S. R. Amstutz, "Burst switching - An introduction," *IEEE Communications Magazine*, November 1983, pp. 36-42.

[3] L. T. Wu and N. C. Huang, "Synchronous wideband network - An interoffice facility hubbing network," *Proc. 1986 International Zurich Seminar on Digital Communications*, pp. 33-39.

[4] Y. N. J. Hui and E. Arthurs, "A broadband packet switch for integrated transport," ICC '87, Seattle, Washington.

[5] J. J. Kulzer and W. A. Montgomery, "Statistical switching architectures for future services," *Proc. International Switching Symposium*, May 1984.

[6] W. W. Chu, "A study of asynchronous time division multiplexing for time sharing computer systems," *Proc. AFIPS*, vol. 35, pp.669-678, 1969.

[7] A. Thomas, J. P. Coudreuse and M. Servel, "Asynchronous time-division techniques: An experimental packet network integrating videocommunication," *Proc. International Switching Symposium*, May 1984.

[8] M. Wm. Beckner, T. T. Lee and S. E. Minzer, "A protocol & prototype for broadband subscriber access to ISDNs," ISS'87, to appear.

[9] R. J. Boehm, Y.-C. Ching and R. C. Sherman, "SONET (Synchronous Optical Network)," *Proc. Globecom '85*, pp. 1443-1450.

[10] L. R. Linnell, "A wide-band local access system using emerging-technology components," *IEEE Journal on Selected Areas in Communications*, Vol. SAC-4, No. 4, pp. 612-618, July 1986.

[11] M. S. Goodman, H. Kobrinski and K. W. Loh, "Application of Wavelength Division Multiplexing to Communication Network Architectures," *Proc. ICC'86*, Vol. 2, pp. 931-933, Toronto, 1986.

A Broadband Packet Switch for Integrated Transport

JOSEPH Y. HUI, MEMBER, IEEE, AND EDWARD ARTHURS

Reprinted from *IEEE Journal on Selected Areas in Communications*, Vol. SAC-5, No. 8, October 1987, pages 1264-1273. Copyright © 1987 by The Institute of Electrical and Electronics Engineers, Inc. All rights reserved.

Abstract—This paper gives a broadband (total throughput approaching 1 terabit/s) self-routing packet switch design for providing flexible multiple bit-rate broadband services for an end-to-end fiber network. The switch fabric for the slotted broadband packet switch delivers exactly one packet to each output port from one of the input ports which request packet delivery to that output port. The denied requests would try again during the next slot. We discover an effective scheme, implemented by CMOS VLSI with manageable complexity, for performing this function. First, each input port sends a request for a port destination through a Batcher sorting network, which sorts the request destinations in ascending order so that we may easily purge all but one request for the same destination. The winning request acknowledges its originating port from the output of the Batcher network, with the acknowledgment routed through a Batcher-banyan self-routing switch. The acknowledged input port then sends the full packet through the same Batcher-banyan switch without any conflict. Unacknowledged ports buffer the blocked packet for reentry in the next cycle. We also give several variations for significantly improved performance.

We then study switch performance based on some rudimentary protocols for traffic control. For the basic scheme, we analyze the throughput-delay characteristics for random traffic, modeled by random output port requests and a binomial distribution of packet arrival. We demonstrate with a buffer size of around 20 packets, we can achieve a 50 percent loading with almost no buffer overflow. Maximum throughput of switch is 58 percent. Next, we investigate the performance of the switch in the presence of periodic broadband traffic. We then apply circuit switching techniques and packet priority for high bit-rate services in our packet switch environment. We improve the throughput per port to close to 100 percent by means of parallel switch fabric, while maintaining the periodic nature of the traffic.

I. INTRODUCTION

WE believe that broadband services, supported by a broadband fiber network for end-to-end transport, shall be widely accepted [1]. One key element necessary for providing flexible services is a high capacity packet switch for interconnecting a large number of subscriber lines at the central office [2], [3]. In this paper, we design the packet switch with the following system objectives:

- support all services except entertainment video broadcast on a packet switch fabric;
- 1000–4000 ports serving a maximum of 200 000 subscriber lines;
- employ CMOS VLSI for the switch fabric and support a port speed of 150 Mbits/s.

The packet switch we design should preserve the packet sequencing for a service, have minimal variable delay, and be robust for all patterns of load on the ports. The

Manuscript received November 1, 1986; revised March 15, 1987. This paper was presented at ICC'87, Seattle, WA, June 8–11, 1987.

The authors are with Bell Communications Research, Morristown, NJ 07960-1961.

IEEE Log Number 8716300.

switch design described in this paper shall meet all these criteria.

We want to build a switch fabric performing the function of the box in Fig. 1. The fabric has input ports i and output ports j, $1 \leq i, j \leq N$. In the figure, we put a pair of parentheses around the source address i to distinguish it from the destination address j. At the beginning of a time slot, each port i may request delivery of a packet to a certain port j, which we denote by j_i. A switch fabric is internally nonblocking if it can deliver all packets to the requested output ports given that the j_i are distinct. However, an internally nonblocking switch fabric can still block at the output ports due to conflicting requests, namely, the occurrence of $j_i = j_{i'}$, for some $i \neq i'$. We call such an event an output port conflict, or external blocking, for the output port j. When this event occurs, the switch should deliver exactly one request among the requests for the output port. An example of output port conflict for output port 3 is shown in Fig. 2, in which the request from input port 1 conflicts with the request from input port 4, and loses the contention when the request from port 4 is delivered instead. Thus, a mechanism is needed for resolving output port conflict. Section II describes a simple mechanism for this purpose. If we maintain a first come first serve (FCFS) discipline for the arrival of requests at an input port, the denied request would try again during the next time slot. A buffer at each input port is required to hold incoming traffic while the request at the head of the queue is occasionally blocked.

The FCFS system may cause a phenomenon called head of the line (*HOL*) blocking, illustrated in Fig. 2. Notice that no request at the head of the line seeks delivery to output port 2, while a request for output port 2 is blocked at input port 1 because the *HOL* request at input port 1 is blocked. *HOL* blocking reduces throughput of the switch as shown in Section V, unless some mechanism is provided to arbitrate both *HOL* and non-*HOL* requests (a solution is given in Section IV), or the *HOL* requests are deliberately arranged to cause less contention (such as shifting the phase of periodic traffic destined for the same output port, described in Section VI). We shall show how these two techniques significantly improve the throughput of the switch. Furthermore, we shall show how the use of parallel switch planes can achieve up to a 100 percent efficiency in Section V.

The paper is organized as follows. Section II describes a basic 3-phase algorithm for the Batcher-banyan fabric, which resolves conflict through sorting and acknowledges the winning port before actual packet delivery. Section III

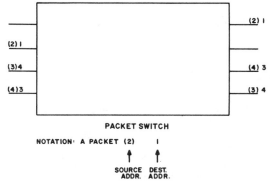

Fig. 1. A nonblocking packet switch.

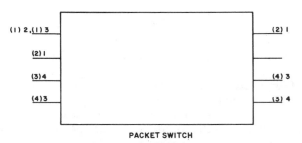

PACKET SWITCH

Fig. 2. Output conflict and head of line blocking.

describes the hardware implementation of the switching fabric. Section IV describes variations of the algorithm, some with improved performance. Section V analyzes the queue lengths and throughput of the switch assuming FCFS, uniform, and independent distribution of the destination addresses for packets, and binomial distribution for packet arrivals at each port. The analysis shows reasonable buffer requirement for loading around 50 percent for the single switch plane. Section VI demonstrates that with the use of circuit setup and priority for periodic traffic, significant improvements in throughput can be achieved, while the periodic nature of the traffic is maintained. Section VII summarizes the paper with a switch proposal.

II. A Basic 3-Phase Algorithm for the Batcher-Banyan Fabric

What sort of self-routing interconnection network may be used to deliver packets from input ports which make distinct requests for output ports? Most solutions involve a sorting network [4], [5], [7]. One implementation [7] consists of an internally nonblocking fabric which first sorts the packets according to the destination address by a sorting network (say a Batcher network [6]), subsequently routes the packets through a banyan network preceded by a shuffle exchange. It can be shown that the packets will not block within the banyan network. Unfortunately, blocking may occur if the requests are not distinct, or if we purge packets from the output of the sorter.

To resolve contention for output ports, Huang and Knauer [7] proposed for their Starlite switch the use of

reentry for packets losing the contention as shown in Fig. 3. At the output of the sorting network, a packet loses the contention if the packet on the preceding line in the sorted order has the same destination address. The packets which lose the contention are then concentrated by a concentration network, to be fed back to the front end of the Batcher network for reentry. The output of the sorting network has holes in the sequence due to purged packets. Consequently, another concentration network is required in front of the banyan network for skewing all packets to the top lines, filling all holes left behind by purged packets.

The Starlite approach has several drawbacks. First, packets can be lost due to blocking within the reentry network, and packets may be delivered out of sequence. Second, at least half of the input ports are dedicated for reentry. Third, the two extra concentration networks require more chip sets and subsystem designs. In this paper, we shall provide an alternative solution which is simpler and more flexible.

How may one arbitrate conflicting requests from a number of ports that are spatially separated? Such arbitration would require by itself an interconnection network. One solution is to bring conflicting requests together through sorting the requests. After sorting, the conflicting requests are adjacent to each other (Fig. 4), and a request decides that it wins the contention if the request above it in the sorted order is not making the same request. Thus, in the arbitration phase (Phase I) of the algorithm, each input port i sends a short request packet, which is just a source-destination pair (i, j_i). The requests are sorted in nondecreasing order according to the destination address j_i, and the request is granted only if j_i is different from the one above it in the sorted list.

However, the input port which made the request does not know the result of the arbitration. Consequently, the request packet (i, j_i) which won the arbitration must send an acknowledgment packet to input port i via an interconnection network. This process constitutes the acknowledgment phase (Phase II). By bringing a fixed connection from the kth output of the Batcher network to the kth input of the Batcher network (Fig. 5), the request packet (i, j_i) may send an acknowledgment packet from input port k, the position of the request packet in the sorted order, to input port i. The acknowledgment packets are sent to distinct output ports since each input port can send at most 1 request in Phase I. Consequently, the Batcher-banyan network described at the beginning of this section suffices for nonblocking delivery of the acknowledgments.

Input ports receiving acknowledgments for their request then transmit the full packet in the final Phase III (Fig. 6) through the same Batcher-banyan network, without conflict at the output port. Input ports which fail to receive an acknowledgment retain the packet in a buffer for retry in the next slot, when the 3-phase cycle is repeated.

III. Hardware Implementation of Switch Fabric

The Batcher-banyan network with the 3-phase control algorithm has to be partitioned into subsystems for imple-

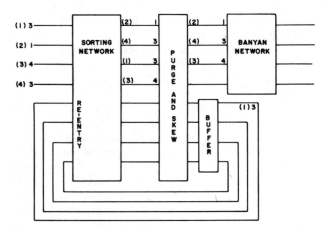

STARLITE RE-ENTRY NETWORK

Fig. 3. Starlite reentry network.

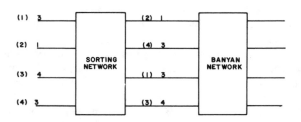

PHASE I: SEND AND RESOLVE REQUEST
- SEND SOURCE-DESTINATION PAIR THROUGH SORTING NETWORK
- SORT DESTINATION IN NON-DECREASING ORDER
- PURGE ADJACENT REQUESTS WITH SAME DESTINATION

Fig. 4. Phase I of 3-phase algorithm.

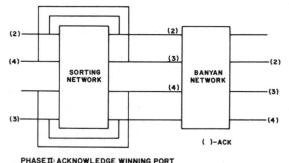

PHASE II: ACKNOWLEDGE WINNING PORT
- SEND ACK WITH DESTINATION TO PORT WINNING CONTENTION
- ROUTE ACK THROUGH BATCHER-BANYAN NETWORK

Fig. 5. Phase II of 3-phase algorithm.

mentation on CMOS VLSI chips. For high speed implementation at 150 Mbits/s, the network must be partitioned with the objective of minimizing the interconnection distance between chips, in order to avoid extensive use of power consuming ECL drivers. Realizing a way of interconnecting subnetworks in a 3-dimensional configuration, [8] proposes a compact mean of implementing the switch fabric, using ECL drivers only at

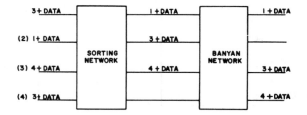

PHASE III: SEND PACKET
- ACKNOWLEDGED PORT SEND PACKET THROUGH BATCHER-BANYAN NETWORK

BUFFERS AT PORT CONTROLLER

Fig. 6. Phase III of 3-phase algorithm.

the peripheral of the fabric. The reader may read [8] concerning the network partitioning and VLSI chip design.

Given the network implementation, we shall describe the implementation of the port controller, as shown in Fig. 8. One function of the port controller is to provide the 3-phase arbitration mechanism. Besides arbitration, the port controller runs higher level protocols such as address translation, flow control and call setup procedures, etc. The implementation of these protocols is subject to further studies. We expect most of the switch cost results from implementing these protocols and switch interfaces rather than the switch fabric itself, hence concentrating switch functions at the ports helps to modularize switch cost according to number of ports needed.

The hardware implementation of the 3-phase algorithm is shown in Figs. 7, 8, and 9. Fig. 7 indicates the input/output port interface connections to the Batcher-banyan fabric. (We show a 32×32 fabric for the sake of illustration.) The port interface i transmits to input i of the Batcher network by BTI_i, and receives from outputs i and $i-1$ of the Batcher network by BTO_i and BTO_{i-1}, as well as from output i of the banyan network BNO_i. Fig. 8 shows the hardware for the port interface. The 4 : 1 multiplexer controls the phases of the algorithm through selecting one of its inputs by the MUX SELECT (MS) signal. The value of MS selects the phase of the algorithm.

In Phase I with MS = 1, a request packet, consisting of the addresses DESTINATION and SOURCE, is allowed to enter the Batcher network via BTI_i. MS is set to 1 for the duration of the request packet, which equals 2 $\log N$ bit times for a switch with N ports (All logarithms are base 2 unless stated otherwise.) Afterwards, MS = 0, consequently blocking any signal into the Batcher network. The request routes through the Batcher network, and its first bit enters another port after a time $\frac{1}{2} \log N(\log N + 1)$, the latency of the Batcher network. (See timing diagram in Fig. 9.) Just before the request enters the port, the comparison enable signal ϕ_1 clears the trigger for arbitration by its rising edge to initiate Phase II of the algorithm.

In Phase II, the two requests entering through BTO_{i-1} and BTO_i are compared by an exclusive OR gate bit-by-bit. If the requested DESTINATION's are different, the

266

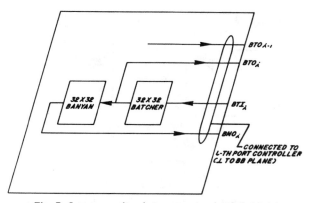

Fig. 7. Interconnections between port and switch fabric.

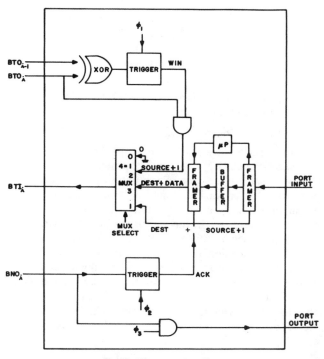

Fig. 8. The port controller.

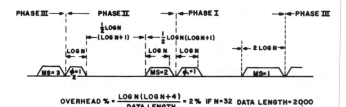

$$\text{OVERHEAD \%} = \frac{\text{LOG N (LOG N+4)}}{\text{DATA LENGTH}} = 2\% \text{ IF N=32 DATA LENGTH=2000}$$

Fig. 9. Timing diagram for 3-phase algorithm.

exclusive OR gate would give a rising edge to set the trigger WIN = 1, signaling that the request coming in on line BTO_i is granted. Thus, $\phi_1 = 1$ should last only for the duration of DESTINATION, namely, log N. Afterwards, Phase II is started by multiplexing the acknowledgment packet, which is the SOURCE part of the request packet

gated by WIN, into the network by the control MS = 2. The acknowledgment packet travels through the Batcher-banyan network for a total latency of $\frac{1}{2}$ log $N(\log N + 1)$ + log N. Afterwards, each port sets the acknowledgment enable single $\phi_2 = 1$ for a duration of log N for the purpose of receiving an acknowledgment. The rising edge of ϕ_2 clears the trigger, and the arrival of the acknowledgment would set the trigger ACK = 1.

In Phase III, a port with ACK = 1 sends the full packet by setting MS = 3 for the duration of the length of the packet. If ACK = 0, then MS = 0, and the packet is buffered for the next round of the 3-phase arbitration.

Phases I and II constitute overhead processing for the switch fabric. The timing diagram of Fig. 9 gives a total overhead of $h_t = \log N(\log N + 4)$. If the data field of the full packet has l_p bits, the switch fabric would have to be speeded up by a fraction (h_t/l_p) of the speed of the input port. For $N = 1024$ and $l_p = 1024$, the speedup required is 14 percent; consequently, the switch has to operate at 170 Mbits/s for handling a 150 Mbit/s input port. Less speedup is required if we have packets of longer duration. For packets of length 2048 bits, only 7 percent speedup at 160 Mbits/s is needed.

IV. Improvements of the Basic 3-Phase Algorithm

The 3-phase algorithm has several variations. The first variation involves no wire connection from the output of the Batcher sorting network to its corresponding input. The basic idea (Fig. 10) is to allow the requests surviving the contention of Phase I to route to their destinations through the banyan routing network, after going through a concentration network which eliminates all the holes after purging. (A Batcher network would serve that function.) Phases II and III remain the same, except that the acknowledgment is sent from the destination j_i instead. This modified 3-phase scheme can be reduced to a 2-phase scheme as follows. In Phase I, each input port with a packet sends the full packet through the network, and allows those packets winning the contention at the output of the Batcher sorting network to go all the way through the concentration network and then the banyan routing network to their destinations. In Phase II, the output ports receiving these packets acknowledge the origination of the packet by sending an acknowledgment packet to the origin. Viewed alternatively, this 2-phase scheme sends out a packet without the assurance of delivery, but retains a copy of the sent packet in a buffer, and flushes the packet only if an acknowledgment is received.

The second variation involves the designation of packet priority. Priority of packets for the contention resolution process is easily achieved by the built-in sorting function of the network. Packets of higher priority can be distinguished from those of lower priority by a priority field, appended at the end of the destination address of the request packet. The appended address is sorted by the Batcher network, consequently placing the higher priority packet at a favorable position when packets with the same request for an output port are purged.

PHASE I SEND REQUEST

Fig. 10. Modification of 3-phase algorithm.

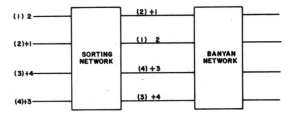

Fig. 11. Use of packet priority.

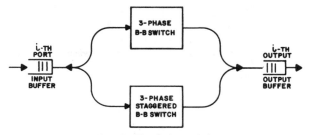

Fig. 12. The duplex switch.

This priority feature can be very profitably used for reducing *HOL* blocking. We may grant a request for the second packet in the line blocked by an *HOL* packet which lost the contention through the following process. After Phases I and II, those *HOL* packets which lost the contention would allow the second packet in the line to make requests in a second round of Phases I and II as shown in Fig. 11. These second attempts must not interfere with the requests granted during the first round. This is achieved by allowing the packets which won the first round to make requests during the second round with a higher priority, thus guaranteeing them to win in the second round. The same procedure can be repeated for a third round for the third packet in the queue. However, the diminishing return for throughput may not justify the speed overhead required for accommodating the extra phases. We shall show in Section VI how packet priority may also be used for slot assignment to avoid *HOL* blocking. Thus, packet priority allows circuit switching in a packet switch environment, in which bursty traffic can be switched by the excess capacity after circuit switching.

The third variation involves the use of 2 switch planes in parallel as shown in Fig. 12. We call this design the duplex switch. Each switch plane operates the 3-phase basic algorithm, with the second switch plane staggering by half a slot behind the first switch plane. The *HOL* packets present themselves to the first switch plane for contention resolution. Since the first 2 phases of the algorithm finish in well less than half the slot time, packets may move up the queue. The resulting *HOL* packets (blocked or fresh) would contend for the second switch plane. This process alternates between the two switch planes. A buffer is needed at the output ports for storing simultaneous arrivals from the two switch planes for the same output port. The output buffer and input buffer at the port controller may reside in the same buffer space.

More generally, we may have more than two switch planes operating in parallel, provided adjacent switch planes are staggered by a period of time longer than that required by Phases I and II of the basic algorithm. Effectively, the throughput capacity of the switch with S planes is equal to S times the single plane switch. The parallel plane switch may be used for the following advantages. First, it serves as a mean to match the speed of the incoming link when the switch fabric cannot run at the same speed as the incoming link. Second, increasing the throughput of the switch increases throughput on the output trunk beyond the 58 percent capacity for random traffic through 1 switch plane to full loading (to be derived in Section V). Third, and most important of all, this duplex switch provides redundancy of switch fabric for enhancing system reliability.

V. Switch Performance for Random Traffic

In this section, we assume that a packet arrives at each port with probability λ per slot, destined with equal probability $(1/N)$ to any one of the output ports. Each input port has a buffer for incoming packets, which are served on an FCFS basis at the begining of each slot. We are primarily interested in the throughput-delay characteristics of the switch. Since an exact performance analysis for the N-dimensional Markov chain ($N \approx 1000$) is out of the question, an approximate engineering approach will be used. The switch is modeled as N independent single server queues (one for each input port), each queue with an effective service time distribution which accounts for the *HOL* contention. More specifically, we assume for each input port that there is a probability q of winning a switch arbitration, where q is the function of λ, the traffic carried per port. The analysis is then partitioned into two parts: q is determined as a function of λ, and the individual input queues are analyzed.

There are two types of *HOL* packets at an input queue just before the arbitration phase—the blocked *HOL* packets which lost the previous arbitration, and the fresh *HOL* packets which just moved to *HOL*, which could be a new arrival in the previous slot. The fresh *HOL* packets have independent destination addresses, whereas the destination addresses of the blocked *HOL* packets are not independent because they were involved in the previous arbitration. The state of *HOL* blocking is characterized by the number of *HOL* requests for the same output port j, namely, N_j, for $1 \leq j \leq N$. The number of successful deliveries during a slot equals $\sum_{j=1}^{N} \epsilon(N_j)$, in which the indicator function $\epsilon(x) = 1$ if $x > 0$ and $\epsilon(x) = 0$ if $x \leq 0$. The throughput per port is defined as the steady-state expected value

$$T = \frac{1}{N} \sum_{j=1}^{N} E[\epsilon(N_j)] = E[\epsilon(N_j)] \qquad (1)$$

in which the last equality results from the symmetry of j. Let N_b be the total number of blocked *HOL* packets during a slot. Obviously,

$$N_b = \sum_{j=1}^{N} N_j - \sum_{j=1}^{N} \epsilon(N_j). \qquad (2)$$

We then divide (2) by N and take expectation during steady state. Substituting the result into (1), we have the throughput per port

$$T = E[N_j] - \frac{E[N_b]}{N}. \qquad (3)$$

In steady state, we have $T = \lambda$. The last two terms in (3) will be computed individually in terms of λ.

The second term in (3) can be viewed as the steady-state fraction of blocked *HOL* packets. Let ρ be the steady-state probability that a queue has a fresh *HOL* packet, given that the input port is not blocked during the previous cycle. (There are $M = N - N_b$ such ports.) Since every packet will eventually become a fresh *HOL* packet at some point, the following conservation relationship must hold:

$$E[M]\rho = N\lambda. \qquad (4)$$

Consequently, the second term in (3) can be expressed as

$$\frac{E[N_b]}{N} = 1 - \frac{E[M]}{N} = 1 - \frac{\lambda}{\rho}. \qquad (5)$$

The first term in (3) is the steady-state expected number of requests for output port j. Let N_j' be the value of N_j for the next slot, given by

$$N_j' = N_j - \epsilon(N_j) + A_j \qquad (6)$$

for which the random variable A_j is the number of fresh *HOL* arrivals for output port j. Taking expectations for both sides of (6) during steady state when $E(N_j') = E(N_j)$, we have

$$E[A_j] = E[\epsilon(N_j)] = \lambda. \qquad (7)$$

Squaring both sides of (6), we have

$$(N_j')^2 = (N_j)^2 + \epsilon(N_j) + A_j^2 - 2N_j$$
$$+ 2N_j A_j - 2\epsilon(N_j)A_j. \qquad (8)$$

We now take expectation during steady state for (8). We also notice that A_j is independent of N_j for large N. After applying the substitution in (7), we obtain the following expression:

$$E[N_j] = E[A_j] + \frac{E[A_j(A_j - 1)]}{2(1 - E[A_j])}. \qquad (9)$$

It remains to compute $E[A_j(A_j - 1)]$ for evaluating $E[N_j]$ in (9). For given M, A_j is the fresh *HOL* arrivals

from the M unblocked queues. Each of M queues may have a fresh *HOL* packet with probability ρ, and the destination address of the packet is equally distributed for each of the N output ports. Hence, each of the M input port independently contributes, with probability (ρ/N), an arrival for the total arrival A_j. Thus, the moment generating function of A_j for given M is

$$F_M(S) = E[S^{A_j}|M] = \left[1 - \frac{\rho}{N} + \frac{\rho}{N}S\right]^M. \qquad (10)$$

In equilibrium, let π_m denote the probability that $M = m$, $0 \le m \le N$. Then the moment generating function of A_j is given by

$$F(S) = \sum_{m=0}^{N} \pi_m F_m(S). \qquad (11)$$

Differentiating $F(S)$ twice with respect to S, and letting $S = 1$, we have

$$E[A_j(A_j - 1)] = \rho^2 \frac{E[M(M - 1)]}{N^2}. \qquad (12)$$

Substituting N_b in (2) into $M = N - N_b$, we can express M as the sum of N random variables X_j, $1 \le j \le N$, given by

$$X_j = 1 + \epsilon(N_j) - N_j. \qquad (13)$$

In equilibrium, the X_j are rearrangeable random variables by symmetry. As N goes to infinity, the covariance of any two X_j will go to zero. By the law of large numbers for sums of rearrangeable random variables, M/N becomes a constant (almost surely). Therefore, $N \gg 1$,

$$E[A_j(A_j - 1)] \approx \frac{E^2[M]\rho^2}{N^2} = \lambda^2, \qquad (14)$$

in which the last equality follows from (4). Thus, we obtain the second term of (3) by substituting (14) into (9):

$$E[N_j] = \lambda + \frac{\lambda^2}{2(1 - \lambda)}. \qquad (15)$$

Adding the two terms of (3) from (5) and (15), we have

$$\lambda = T = \lambda + \frac{\lambda^2}{2(1 - \lambda)} - \left(1 - \frac{\lambda}{\rho}\right) \qquad (16)$$

from which we obtain the equation

$$(2 - \rho)\lambda^2 - 2(1 + \rho)\lambda + 2\rho = 0. \qquad (17)$$

Thus, we have a relation between the input traffic rate λ and the degree of saturation ρ, which measures traffic arrival rate conditioned on a successful removal of the *HOL* packet. In particular, the maximum λ is obtained by setting $\rho = 1$, resulting in $\lambda = 2 - \sqrt{2} = 0.58$. Therefore, the maximum throughput per port is 58 percent for randon traffic. More generally, we shall use (17) to derive a relationship between λ and q, the probability that the *HOL* packet is served.

In the remainder of this section, we shall focus on a queueing model for each input port i. Let the random variable K be the number of packets in an input queue just before the arbitration phase. We shall consider a buffer of infinite size for the moment. The queue length K' for the next slot is modeled by

$$K' = K - \gamma \epsilon(K) + \alpha \tag{18}$$

in which γ is a 0–1 random variable with $E[\gamma] = q$, the probability that the HOL packet is served, and α is a 0–1 random variable with $E[\alpha] = \lambda < q$. The value of q as a function of λ can be computed as follows. The mean number of slots before a fresh HOL packet is served is $(1/q)$. After a packet is served, the mean number of slots before an arrival of a fresh HOL packet is $((1 - \rho)/\rho)$. In steady state, the sum of these two terms equals the interarrival time of packets at a switch port, namely, $(1/\lambda)$. This relationship, together with the equation of ρ in terms of λ in (14), gives

$$\frac{1}{q} = 1 + \frac{\lambda}{2(1 - \lambda)}. \tag{19}$$

From (18), the moment generating function for the steady-state probabilities p_k is given by

$$G(S) = E[S^K] = E[S^\alpha]E[S^{K - \gamma \epsilon(K)}]. \tag{20}$$

The first term on the right-hand side equal $[1 - \lambda + \lambda S]$. The second term equals

$$E_\gamma \left[\sum_{k=0}^{\infty} p_k S^{k - \gamma \epsilon(k)} \right]$$

$$= p_0 + E_\gamma \left[\sum_{k=1}^{\infty} p_k S^{k - \gamma} \right]$$

$$= p_0 [1 - E_\gamma[S^{-\gamma}]] + E_\gamma \left[\sum_{k=0}^{\infty} p_k S^{k - \gamma} \right]$$

$$= p_0 [1 - E_\gamma[S^{-\gamma}]] + G(S) E_\gamma[S^{-\gamma}]. \tag{21}$$

Substituting (21), and $E_\gamma[S^{-\gamma}] = [(1 - q) + qS^{-1}]$ into (20), we have

$$G(S) = \frac{qp_0(S - 1)(1 - \lambda + \lambda S)}{S - (1 - \lambda + \lambda S)((1 - q)S + q)}$$

$$= \frac{qp_0(1 - \lambda + \lambda S)}{((1 - \lambda)q - \lambda(1 - q)S)} \tag{22}$$

where the last equality results from cancelling the common factor $(S - 1)$. After evaluating p_0 by setting $G(1) = 1$, and substituting $\omega_\lambda = (\lambda(1 - q)/(1 - \lambda)q) = (\lambda^2/2(1 - \lambda)^2)$, we obtain finally

$$\sum_{k=0}^{\infty} p_k S^k = G(S) = (1 - \omega_\lambda) \frac{1 - \lambda + \lambda S}{1 - S\omega_\lambda}$$

$$= (1 - \omega_\lambda)(1 - \lambda + \lambda S) \sum_{k=0}^{\infty} \omega_\lambda^k S^k. \tag{23}$$

Consequently, we have

$$p_0 = (1 - \omega_\lambda)(1 - \lambda) \tag{24}$$

$$p_k = (1 - \omega_\lambda)((1 - \lambda)\omega_\lambda + \lambda) \omega_\lambda^{k-1}$$

$$= \frac{\lambda(2 - \lambda)}{2(1 - \lambda)}(1 - \omega_\lambda) \omega_\lambda^{k-1}; \quad k > 0 \tag{25}$$

from which the expected value of K is given by

$$\overline{K} = \frac{\lambda(2 - \lambda)(1 - \lambda)}{(2 - \sqrt{2} - \lambda)(2 + \sqrt{2} - \lambda)}. \tag{26}$$

Fig. 13 plots the average queue length $\overline{k}$ and average delay $D = (\overline{K}/\lambda)$ (from Little's theorem) as a function of λ.

The buffer overflow probability for a finite buffer of size B is upper bonded by the probability of $K > B$ for the case of infinite buffer size. Therefore,

$$P(\text{loss}) < P(K > B) = \frac{\lambda(2 - \lambda)}{2(1 - \lambda)} \omega_\lambda^B. \tag{27}$$

Fig. 14 plots this upper bound, which is acceptably low for λ as high as 50 percent, provided $B \approx 16$. Beyond $B = 20$, the packet loss rate is negligible.

VI. Circuit Switched Traffic in a Packet Switch Environment

Using the results of the previous section, we shall show how periodic traffic is affected by variable packet delay. We first examine a low bit-rate periodic traffic (such as voice), and then full motion video traffic. For periodic traffic occupying a significant fraction of the speed of the port, we show how we may avoid the HOL contention by shifting the relative phase of services destined for the same output port. Finally, we illustrate how we may use traffic priority for maintaining the periodicity of multirate services. We also discuss distributed circuit setup algorithms as well as how the duplex switch can be used for strict sense nonblocking circuit switching.

The use of a packet switch for circuit switching was proposed in the synchronous wideband switch [9], which is a time-space-time switch with nonblocking Batcher-banyan fabric as the space division switch. (Strictly speaking, the synchronous wideband switch provides interconnection at the 1.5 Mbits/s, or DS1 level, instead of 64 kbits/s or DS0 level. However, in this discussion, we are only interested in its time-space-time feature for illustration.) The space division switch is essentially a broadband packet switch, without any output port contention resolution mechanism. Time division switching is performed at the input and output ports of the space division switch by time slot interchangers (*TSI*), for rearranging the time slots to avoid port contention. At most half of the slots for the input and output of the space division switch are loaded for strict sense nonblocking circuit switching. In this section, we want to examine how the periodicity of mixed services would be affected if we eliminate all *TSI* in the time-space-time architecture, and

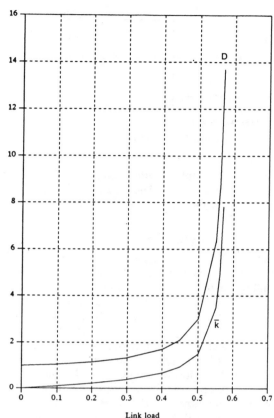

Fig. 13. Delay throughput analysis.

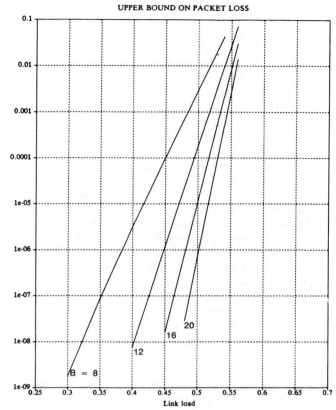

Fig. 14. Buffer overflow probability versus loading.

replace the Batcher-banyan fabric with the 3-phase packet switch described in this paper.

A signal is strictly periodic if the packet interarrival time is a constant. given flexible bit-rates and packet multiplexing of services, maintaining strict periodicity is impossible. However, for reasons discussed later, we would prefer the signal to be fairly periodic per link within the network. Thus, we need a measure of burstiness [10] of the traffic for each service, which is defined by the packet jitter ratio.

$$J = \frac{\mathrm{Var}\,[\,\text{Packet interarrival time}\,]}{\mathrm{E}\,[\,\text{Packet interarrival time}\,]}.$$

We shall see how J is affected for the different services. It should be noted that the packet loss rate derivations in the previous section also suffice for computing the probability that the packet jitter exceeds a certain tolerable duration, assuming random output addresses.

Consider loading the packet switch with voice traffic only. At a port speed of 150 Mbits/s, voice packet interarrival time for a call is 2000 slots. Subsequently, consecutive hundreds of packets on the channel belong to different calls with essentially random output address. From the analysis of the previous section, we know that we can achieve a throughput of 0.5 provided buffer size is more

than 16. Thus, we can achieve the same efficiency of the synchronous wideband switch without time slot interchanging 1000 packets. From Fig. 13, the variance of delay, which is upper bounded by the mean delay, is less than 2 slots for loading = 0.5. Therefore, J is less than $(2/2000) = 0.001$ per switch for voice. On a 150 Mbits/s link, 2 1000-bit packets has a duration of 28 μs, which is substantially less than the speech sampling interval of 125 μs or the packetization delay of 16 ms for 64 kbit/s speech. We conclude that for low bit-rate services such as speech, signal tranversing several switches and multiplexers in tandem is almost periodic. Therefore, time stamping of packets and buffering of more than one packet may not be required for playout of the signal. Thus, a packet switching network is easy to interface with the existing circuit switched network for low rate services. Furthermore, the random output port address assumption of Section V remains valid for switches in tandem.

For periodic traffic occupying a significant fraction of the capacity of a switch port, both the periodicity of traffic and the property of random output port addresses for consecutive packets are no longer preserved. Consider a 45 Mbit/s video phone service on a 150 Mbit/s packet channel. One out of three packets comes from the same service, suggesting $J \approx (2/3)$. Furthermore, a delay of 2 packets would cause 2 packets of the same service to bundle up. Without proper control of variable delay for the packet switch, this bundling continues to snowball for switches in tandem. This bundling not only requires buffering and time stamping for playout of the signal, but may also create buffer overflow problems and reduce throughput compared to the random traffic case. Thus, it is important that the periodicity of the traffic be preserved, in the sense of maintaining a small J.

There are several mechanisms for maintaining a small J for high bit-rate services. First, we may assign high priority for high bit-rate services, thus reducing their variable delay at the expense of low bit-rate services. However, the high bit-rate services still have to contend with each other. The effect of this contention on variable delay depends on the number of high bit-rate services. For example, given 2 contending services, it can be easily seen that periodicity is hardly disturbed, since each service automatically acquires a phase different from the other.

The second mechanism involves shifting the phase (packet arrival time with respect to a frame reference). Consider each input port and output port to be loaded with at most r services, and each service has a period of n packets, $n \geq r$. The Slepian–Duguid theorem states that there exists a time slot rearrangement of r packets in a frame of n slots such that there will be no output port conflict for the packet switch. Unfortunately, the computation for finding the arrangement, as well as the amount of rearrangement for adding a service to existing arrangements, becomes intractable for large n and $(r/n) > \frac{1}{2}$. For small n, the arrangement is relatively straightforward. Consider $n = 3$ and $r = 2$. It can be readily seen that adding a service to a port with at most 1 service requires no rearrangement of slot assignments. One scenario of arrangements is shown in Fig. 15. Notice the vacant slots at the input are blocking for all but one output port, hence making these slots quite useless for bursty traffic. Rearranging the 2 circuits by circuit packing algorithms would make these vacant slots less blocking. By utilizing these vacant slots at 30 percent efficiency, an output loaded with two 45 Mbit/s services can achieve an efficiency of over 70 percent.

The circuit setup and packing algorithms can be implemented in a distributed fashion. During the call setup procedure, an input port sends a packet to the destination output, requesting a multiple slot assignment within a frame. The packet also contains a vector registering the state of occupancy of the frame at the input. The output then checks the state of occupancy of the output frame, and assigns the earliest available slots common to both input and output frames. Subsequently, the output communicates the assignment (or denial) by sending a packet to the input. During call tear down, the newly vacated slots can be reassigned to services holding slots which are later than the vacated slots, thus performing packing of circuits. This reassignment has to be confirmed by the input. Circuit rearrangements involving other ports can be implemented similarly by having distributed occupancy tables and distributed computation via communication through the switch.

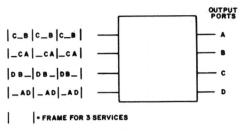

ONE 150 Mbps CHANNEL = THREE 45 Mbps VIDEO SERVICES

Fig. 15. Circuit switched traffic.

The use of the duplex switch can provide strict sense nonblocking circuit setup. The background bursty traffic can be carried by the excess capacity with low priority.

VII. Conclusion

We summarize this paper with a proposal for switch design. We advocate using an internally nonblocking switch fabric such as the Batcher-banyan network because of its ability to transport all patterns of loading with tight delay control. An unpredictable loading pattern is inevitable if we want a truly integrated transport network for all services. Such an integrated network has the advantages of being bit-rate flexible, as well as allowing easy bridging of multimedia communications. In contrast, most internally blocking networks, such as the buffered banyan network, perform rather poorly for uneven traffic [11]. To date, we have very little understanding of mixing traffic for these blocking switches in tandem.

In a hybrid transmission environment [12], a 150 Mbit/s link is framed at the speech sampling rate of 8 kHz. By dividing a frame into 18 slots, each slot can transport up to 130 bytes of information. Allowing an up to 10 byte header, each packet can carry an information payload of 120 bytes, or approximately 1000 bits. We allow high bit-rate (>1 Mbit/s) periodic traffic to take up multiple slots within a frame of 18 slots. Strict sense nonblocking slots are assigned through the duplex switch. Excess capacity assigned to a service can be used for random traffic, which has lower priority than slot assigned traffic. Services occupying a bandwidth substantially less than 1/18th of the channel may be treated as random traffic. Thus, the use of packet priority allows us to have a tight control on delay and packet loss rate of individual services. The flexible assignment of slots also allows a broad spectrum of bit-rates.

We have demonstrated switching mechanisms for a broad spectrum of bit-rates and tight control of delay for multirate services. Such switches may provide a bridge for a smooth transition from the predominantly circuit switched services at present to the highly heterogeneous traffic of future broadband services.

References

[1] J. S. Turner, "New directions in communications (or which way to the information age)," *IEEE Commun. Mag.*, vol. 24, Oct. 1986.

[2] D. R. Spears, "Broadband ISDN switching capabilities from a services perspective," this issue, pp. 1222–1230.

[3] P. Gonet, P. Adams, and J. P. Coudreuse, "Asynchronous time-division switching: The way to flexible broadband communication networks," in *Proc. IEEE 1986 Int. Zurich Seminar on Digital Commun.*, Zurich, Switzerland, Mar. 1986.

[4] A. Borodin and J. E. Hopcroft, "Routing, merging, and sorting on parallel models of computation," in *Proc. 14th Annu. ACM Symp. on Theory of Comput.*, pp. 65–71.

[5] J. D. Ullman, *Computational Aspects of VLSI*. Rockville, MD: Computer Science Press, p. 230, theorem 6.7, and p. 242, bibliographic notes, 1985.

[6] K. E. Batcher, "Sorting networks and their applications," in *Proc. 1968 Spring Joint Computer Conf.*, pp. 307–314.

[7] A. Huang and S. Knauer, "Starlite: A wideband digital switch," in *Proc. 1984 Globecom Conf.*

[8] C. M. Day, J. N. Giacopelli, and J. Hickey, "Applications of self-routing switching to LATA fiber optic networks," in *Conf. Proc. Int. Symp. Switching*, Phoenix, AZ, Mar. 1987.

[9] L. T. Wu and N. C. Huang, "Synchronous wideband network—An interoffice facility hubbing network," in *Proc. IEEE 1986 Int. Zurich Seminar on Digital Commun.*, Mar. 1986.

[10] K. Sriram and W. Whitt, "Characterizing superposition arrival processes in packet multiplexers for voice and data," *IEEE J. Select. Areas Commun.*, vol. SAC-3, Sept. 1986.

[11] L. T. Wu, "Mixing traffic in a buffered banyan network," in *Proc. 9th Data Commun. Symp.*, Whistler Mountain, B.C., Canada, Sept. 1985.

[12] L. T. Wu, S. H. Lee, and T. T. Lee, "Dynamic TDM—A packet approach to broadband networking," in *Conf. Proc. Int. Conf. Commun.*, Seattle, WA, June 1987.

Joseph Y. Hui (S'82–M'83), for a photograph and biography, see this issue, p. 1220.

Edward Arthurs received the S.B. and Sc.D. degrees from the Massachusetts Institute of Technology, Cambridge, in 1951 and 1955, respectively.

From 1955 to 1956 he was with RCA, Camden, NJ; from 1956 to 1957 with Hermes, Cambridge, MA; from 1957 to 1962 with M.I.T.; and from 1962 to 1984 with Bell Laboratories, Murray Hill, NJ. Since 1984 he has been with Bell Communications Research, Morristown, NJ, and Columbia University, New York, NY. He has been engaged in theoretical research in computer and communications science. He is coauthor of *A Computer and Communications Network Performance Primer* (Englewood Cliffs, NJ: Prentice-Hall, 1985).

The Knockout Switch: A Simple, Modular Architecture for High-Performance Packet Switching

YU-SHUAN YEH, FELLOW, IEEE, MICHAEL G. HLUCHYJ, AND ANTHONY S. ACAMPORA

Abstract—A new, high-performance packet-switching architecture, called the Knockout Switch, is proposed. The Knockout Switch uses a fully interconnected switch fabric topology (i.e., each input has a direct path to every output) so that no switch blocking occurs where packets destined for one output interfere with (i.e., block or delay) packets going to different outputs. It is only at each output of the switch that one encounters the unavoidable congestion caused by multiple packets simultaneously arriving on different inputs all destined for the same output. Taking advantage of the inevitability of lost packets in a packet-switching network, the Knockout Switch uses a novel concentrator design at each output to reduce the number of separate buffers needed to receive simultaneously arriving packets. Following the concentrator, a shared buffer architecture provides complete sharing of all buffer memory at each output and ensures that all packets are placed on the output line on a first-in first-out basis.

The Knockout Switch architecture has low latency, and is self-routing and nonblocking. Moreover, its simple interconnection topology allows for easy modular growth along with minimal disruption and easy repair for any fault. Possible applications include interconnects for multiprocessing systems, high-speed local and metropolitan area networks, and local or toll switches for integrated traffic loads.

I. INTRODUCTION

WE are currently witnessing an explosive growth in the deployment of optical fiber within buildings, across cities, states, and countries, and even between continents. In addition, continuing advances are being made in optical communications, with current transmission rates on a single fiber in the Gbit/s realm and still more than a thousand-fold increase in capacity theoretically possible. Lagging this tremendous progress in transmission, however, are the advances made in switching technology necessary to build wideband integrated communication networks. While at some time in the future such switching may be achieved through optical means, at present we are lacking suitable electronic architectures to carry us into the next decade.

To put the communication aspects of switching in proper perspective, we note that every type of switch architecture interconnecting N inputs with N outputs must perform two basic functions. First, it must route the traffic arriving on its inputs to the appropriate outputs. This may be accomplished with a single-stage, crossbarlike interconnect using N^2 simple switch elements or, at the other extreme, with an $N\log_2 N$ multistage interconnect comprised of 2×2 switch elements. Second, the switch must deal with output contention, where traffic arriving at the same time on two or more inputs may be destined for a common output. With circuit switching, output contention is prevented by using a controller that schedules the arrivals to avoid conflict. The classical time-space-time switch falls into this category [1]. Here, each input to the switch is preceded by a time slot interchange to rearrange the time sequence of the time-multiplexed traffic so that, when presented to the space switch, the data appearing at the N inputs are always destined for distinct outputs. With packet switching, however, packet arrivals to the switch are unscheduled, each containing a header bearing address information used to route the packet through the switch. Without the coordination afforded by a central scheduler, a packet switch must recognize confict among its inputs and internally store, or buffer, all but at most one of several simultaneously arriving packets destined for a common output; thereby leading to statistical delay, or latency, within the switch. Dealing with output contention in a packet switch can be more complicated than the switch fabric used to route packets to the proper outputs.

The use of high-performance packet switching for building wideband integrated communication networks has received much attention [2]–[7]. Current approaches in the design of high-performance packet switches typically employ binary switch elements appropriately interconnected and arranged to form a multistage switch, and have tended to emphasize a reduction in the number of switch elements needed to a value below that of a fully connected arrangement [8]. Unfortunately, in addition to congestion at the outputs, these element-efficient switches can also congest at each of the binary switch points, thereby requiring that additional measures be taken, such as buffering within each element, flow control between elements, and/or speed-up of the switch fabric itself. The complexity of the buffering and flow control required within the switch element far exceeds that of the basic switching mechanism used to route the inputs to the outputs. Moreover, the delay encountered within the switch fabric is greater than the unavoidable component caused by output congestion alone. Other important areas in which multistage switches have difficulties are ease of modular growth and the ability to easily locate and repair faults with minimal disruption to the operation of the

Manuscript received November 5, 1986; revised April 13, 1987. This paper was presented at the 1987 International Switching Symposium, Phoenix, AZ, March 15–20, 1987.

Y.-S. Yeh and A. S. Acampora are with AT&T Bell Laboratories, Holmdel, NJ 07733.

M. G. Hluchyj was with AT&T Bell Laboratories, Holmdel, NJ. He is now with Codex Corporation, Canton, MA 02021.

IEEE Log Number 8716301.

switch. The latter is particularly important when one considers that the hardware and software investment associated with switch maintenance often exceeds that of the switch fabric.

In this paper, we propose a new, high-performance packet-switching architecture, which we call the Knockout Switch. The switch has low latency, is self-routing, and is nonblocking. Its architecture provides for simple modular growth, fault tolerance, and easy maintenance procedures. Although based on a fully interconnected topology, the complexity of the Knockout Switch grows only linearly with N within the range of practical interest. Specifically, for $N \leq 1000$, the complexity of the switch is dominated by the buffering requirements and the input/output functions (timing recovery, address look-up and translation, etc.) present in any packet switch. Only when N is much greater than 1000 do we see the N^2 complexity of the interconnection fabric begin to dominate the overall complexity of the switch. Still, with $N = 1000$ and a data rate of 50 Mbits/s on each input line, a total switch capacity of 50 Gbits/s results. This is about 10 times the capacity of the current generation of digital central office switches. To achieve this performance, the Knockout Switch exploits one key observation: in any practical packet switching system, packet loss within the network is unavoidable (e.g., caused by bit errors on the transmission lines).

The Knockout Switch uses a fully interconnected topology to passively broadcast all input packets to all outputs. Preceding each output port is a bus interface that performs several functions. First, by means of a filtering operation, all packets not intended for the output are discarded; this operation effectively achieves the switching function in a fully distributed manner, but requires N simple filtering elements per output (hence, N^2 for the entire switch). Next, each bus interface queues the desired packets in a set of buffers that are shared, for that output, among all input lines. A simple arrangement permits the buffers to be filled in a cyclical manner and served to the output on a first-in first-out basis. Since any delay in the switch arises from output congestion only, the Knockout Switch provides the lowest latency possible in any switching arrangement [9].

The most novel aspect of the Knockout Switch, however, is its use of a knockout contention scheme in the bus interface that funnels the packets accepted from a possibly large number of packet filters (one for each of N inputs) into a far smaller number L of shared, parallel buffers, thereby vastly reducing the overall complexity of the switch. The scheme is analogous to a tournament. All inputs having packets simultaneously arriving to a particular output contend for the right to place data in the first of several parallel buffers. All losers then contend for the second buffer, and so on through buffer L; at this point, any remaining losers, having been given L attempts at victory, are simply discarded. We show that the probability of losing data in this manner is extremely small. For example, with a 90 percent load and N arbitrarily large,

$L = 8$ guarantees a lost packet rate of less than 10^{-6}, comparable with the loss expected from other sources (e.g., channel errors and buffer overflows in a packet switching network); $L = 12$ guarantees a lost packet probability of less than 10^{-10}.

For each bus interface, it is shown that the total size needed for each shared buffer is 40 packets (8 parallel buffers, each 5 packets deep) to maintain an overflow probability of 10^{-6} at an 84 percent load, again comparable to other loss mechanisms. The Knockout Switch is modular, permitting evolutionary growth from small to large configurations; is easily maintained, with faults having minimal disruption to the overall operation of the switch; and can be made fault tolerant with little additional hardware. The characteristics of the Knockout Switch make it well suited for several potential applications: 1) the interconnect for a large multiprocessing system, 2) high-capacity local and metropolitan area networking, and 3) local or toll switching.

In Section II, we describe the architecture of the Knockout Switch in greater detail. In Section III, we describe how it can grow in a modular fashion, is amenable to simple maintenance procedures, and can be made fault tolerant. We also examine in Section III the Knockout Switch implementation complexity. Finally, in Section IV we state our conclusions.

II. Knockout Switch Description

The Knockout Switch is an N-input N-output packet switch with all inputs and outputs operating at the same bit rate. Fixed-length packets arrive on the N inputs in a time-slotted fashion as shown in Fig. 1, with each packet containing the address of the output port on the switch to which it is destined. This addressing information is used by the packet switch to route each incoming packet to its appropriate output. Since the output port address for an arriving packet can be determined via a table look-up prior to the packet entering the switch fabric, the Knockout Switch has applications to both datagram and virtual circuit packet networks.

Aside from having control over the average number of packet arrivals destined for a given output, we assume no control over the specific arrival times of packets on the inputs and their associated output addresses. In other words, there is no time-slot specific scheduling that prevents two or more packets from arriving on different inputs in the same time slot destined for the same output. Hence, to avoid (or at least provide a sufficiently small probability of) lost packets, at a minimum, packet buffering must be provided in the switch to smooth fluctuations in packet arrivals destined for the same output.

A. Interconnection Fabric

The interconnection fabric for the Knockout Switch has two basic characteristics: 1) each input has a separate broadcast bus, and 2) each output has access to the packets arriving on all inputs. Fig. 2 illustrates these two characteristics where each of the N inputs is placed directly

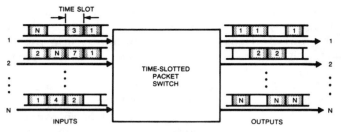

Fig. 1. Knockout Switch.

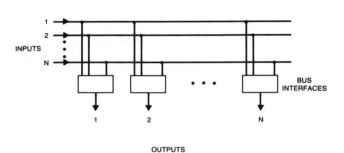

Fig. 2. Interconnection fabric.

on a separate broadcast bus and each output passively interfaces to the complete set of N buses. This simple structure provides us with several important features.

First, with each input having a direct path to every output, no switch blocking occurs where packets destined for one output interfere with (i.e., delay or block) packets going to other outputs. The only congestion in the switch takes place at the interface to each output where, as mentioned, packets can arrive simultaneously on different inputs destined for the same output. Without *a priori* scheduling of packet arrivals, this type of congestion is unavoidable, and dealing with it typically represents the greatest source of complexity within the packet switch. The focus of the Knockout Switch architecture is one of minimizing this complexity.

In addition to the above property, the switch architecture is modular: the N broadcast buses can reside on an equipment backplane with the circuitry for each of the N input/output pairs placed on a single plug-in circuit card. Hence, the switch can grow modularly from 2×2 up to $N \times N$ by adding additional circuit cards. We shall see later that the penalty of having N buses to which all outputs have to interface is not particularly significant. Most of the complexity of the bus interface is associated with the buffering of packets, which for each output does not grow with N. We show in Section III that the architecture for the bus interface also provides us with a simple way to grow the switch from $N \times N$ to $JN \times JN$, $J = 2, 3, 4 \cdots$.

Note from Fig. 2 that the bus structure has the desirable characteristic that each bus is unidirectional and driven by only one input. This allows for a higher transmission rate on the buses, with regeneration used as needed to compensate for losses, and a design more tolerant of faults compared to a shared parallel bus accessed by all inputs. In addition, the packet buffering and bus access control circuitry of the parallel bus is replaced in this architecture by, at most, an elastic buffer at each input used to synchronize the time slots from the individual input lines.

Finally, although it would require a more complex design of the packet filters than that which is described in the next section, the interconnection architecture of the Knockout Switch lends itself to broadcast and multicast features. Since every input is available at the interface to every output, arriving packets can be addressed to and received by multiple outputs.

B. Bus Interface

Fig. 3 illustrates the architecture of the bus interface associated with each output of the switch. The bus interface has three major components. At the top of the figure there are a row of N *packet filters*. Here the address of every packet broadcast on each of the N buses is examined, with packets addressed to the output allowed to pass on to the concentrator and all others blocked. The *concentrator* then achieves an N to L ($L \ll N$) concentration of the inputs lines, wherein up to L packets making it through the packet filters in each time slot will emerge at the outputs of the concentrator. These L concentrator outputs then enter a *shared buffer* composed of a shifter and L separate FIFO buffers. The shared buffer allows complete sharing of the L FIFO buffers and provides the equivalent of a single queue with L inputs and one output, operating under a first-in first-out queueing discipline. In the remainder of this section, we will expand on each of these three parts of the bus interface.

Packet Filters: Fig. 4 shows the format of the packets as they enter the packet filters from the broadcast buses. The beginning of each packet contains the address of the output on the switch for which the packet is destined, followed by a single activity bit. The destination output address contains $\log_2 N$ bits with each output having a unique address. The activity bit indicates the presence (logic 1) or absence (logic 0) of a packet in the arriving time slot and plays an important role in the operation of the concentrator.

At the start of every time slot, the path through each of the N packet filters is open, initially allowing all arriving packets to pass through to the concentrator. As the address bits for each arriving packet enter the row of N packet filters, they are compared bit-by-bit against the output address for the bus interface. If at any time the address for an arriving packet differs from that of the bus interface, the further progress of the packet to the concentrator is blocked. That is, the output of the filter is set at logic 0 for the remainder of the time slot. By the end of the output address, the filter will have either blocked the packet, and hence also set its activity bit to 0, or, if the addresses matched, allowed the packet to continue on to the concentrator. Note that even though a portion of the address bits of a blocked packet may pass through the

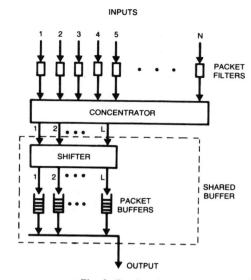

Fig. 3. Bus interface.

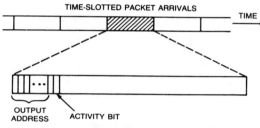

Fig. 4. Packet format.

filter and into the concentrator, these bits no longer serve any useful function and are ignored by the rest of the bus interface.

Concentrator: All packets making it through the packet filters enter the concentrator, which achieves an N to L concentration. Specifically, if there are k packets arriving in a time slot for a given output, these k packets, after passing through the concentrator, will emerge from the concentrator on outputs 1 to k, when $k \leq L$. If $k > L$, then all L outputs of the concentrator will have packets and $k - L$ packets will be dropped (i.e., lost) within the concentrator.

That packets can be dropped within the concentrator should not be of great concern. With any packet switching network, packet loss is inevitable: caused by transmission line errors (particularly in the address portion of the packet), buffer overflows, and network failures. In all cases, recovery is made possible by retransmission protocols, which, as the probability of losing packets in the network decreases, are more efficiently handled end to end rather than link by link [10], [11]. For the Knockout Switch, we implicitly assume all packet retransmissions are initiated at the endpoints of the network, so that the switch is not burdened with this task.

We must ensure, however, that the probability of losing a packet within the concentrator is no greater than that of losing a packet elsewhere in the network. If in each time

slot a packet arrives at each input independently with probability ρ, and each such packet is equally likely destined for each output, P_k, the probability of k packets arriving in a time slot all destined for a given output, has the binomial probabilities

$$P_k = \binom{N}{k}\left(\frac{\rho}{N}\right)^k \left(1 - \frac{\rho}{N}\right)^{N-k}$$
$$k = 0, 1, \cdots, N. \tag{1}$$

It follows then that the probability of a packet being dropped in a concentrator with N inputs and L outputs is given by

$$\Pr\left[\text{packet loss}\right] = \frac{1}{\rho}\sum_{k=L+1}^{N}(k - L)\binom{N}{k}$$
$$\cdot \left(\frac{\rho}{N}\right)^k \left(1 - \frac{\rho}{N}\right)^{N-k} \tag{2}$$

Taking the limit as $N \to \infty$, we obtain after some manipulation

$$\Pr\left[\text{packet loss}\right] = \left[1 - \frac{L}{\rho}\right]\left[1 - \sum_{k=0}^{L}\frac{\rho^k e^{-\rho}}{k!}\right]$$
$$+ \frac{\rho^L e^{-\rho}}{L!}. \tag{3}$$

Using (2) and (3), Fig. 5 shows for $\rho = 0.9$ (i.e., a 90 percent load) a plot of the probability of packet loss versus L, the number of outputs on the concentrator, for $N = 16, 32, 64,$ and infinity. Note that a concentrator with only eight outputs achieves a probability of lost packet less than 10^{-6} for arbitrarily large N. This is comparable to the lost packet probability resulting from transmission errors for 500-bit packets and a bit error rate of 10^{-9}. Also note from Fig. 5 that each additional output added to the concentrator beyond eight results in an order of magnitude decrease in the lost packet probability. Hence, independent of the number of inputs N, a concentrator with 12 outputs will have a lost packet probability less than 10^{-10}. Fig. 6 illustrates, for $N \to \infty$, that the required number of concentrator outputs is not particularly sensitive to the load on the switch, up to and including a load of 100 percent. It is also important to note that, assuming independent packet arrivals on each input, the simple, homogeneous model used in the analysis corresponds to the worst case, making the lost packet probability performance results shown in Figs. 5 and 6 upper bounds on any set of heterogeneous arrival statistics [12].

The basic building block of the concentrator is a simple 2×2 contention switch shown in Fig. 7(a). The two inputs contend for the "winner" output according to their activity bits. If only one input has an arriving packet (indicated by the activity bit = 1), it is routed to the winner (left) output. If both inputs have arriving packets, one input is routed to the winner output and the other input is routed to the loser output. If both inputs have no arriving packets, we do not care except that the activity bit for both should remain at logic 0 at the switch outputs.

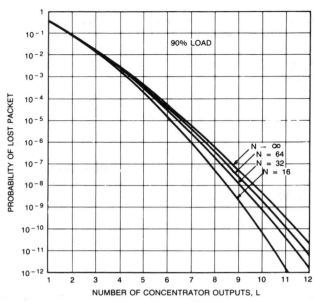

Fig. 5. Lost packet performance of concentrator.

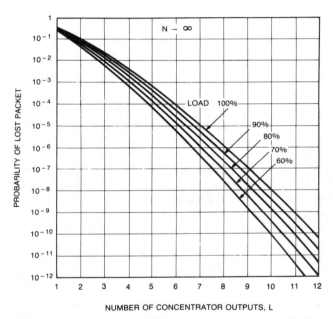

Fig. 6. Lost packet performance of concentrator for various loads.

The above requirements are met by a switch with the two states shown in Fig. 7(b). Here, the switch examines the activity bit for only the left input. If the activity bit is a 1, the left input is routed to the winner output and the right input is routed to the loser output. If the activity bit is a 0, the right input is routed to the winner output, and no path is provided through the switch for the left input. Such a switch can be realized with as few as 16 gates, and having a latency of at most one bit. Note that priority is given to the packet on the left input to the 2 × 2 switch element. To avoid this, the switch element can be designed so that it alternates between selecting the left and right inputs as winners when packets arrive to both in the

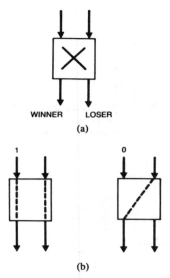

Fig. 7. (a) The 2 × 2 contention switch. (b) States of 2 × 2 contention switch.

same time slot. However, suppose the priority structure of the 2 × 2 switch element were maintained and (as described below) the concentrator were designed so that one input, say the Nth, always received lowest priority for exiting a concentrator output. The packet loss probability for this worst case input, as $N \rightarrow \infty$, is given by

$$\text{Pr [packet loss for worst case input]} = 1 - \sum_{k=0}^{L-1} \frac{\rho^k e^{-\rho}}{k!}.$$

(4)

Comparing the results of (4) to the lost packet probability averaged over all inputs, as given by (3) and shown in Fig. 6, we find that the worst case lost packet probability is about a factor of 10 greater than the average. This greater packet loss probability, however, can be easily compensated for by adding an additional output to the concentrator.

Fig. 8 shows the design of an 8-input 4-output concentrator composed of these simple 2 × 2 switch elements and single-input/single-output 1-bit delay elements (marked by ''D''). At the input to the concentrator (upper left side of Fig. 8), the N outputs from the packet filters are paired and enter a row of $N/2$ switch elements. One may view this first stage of switching as the first round of a tournament with N players, where the winner of each match emerges from the left side of the 2 × 2 switch element and the loser emerges from the right side. The $N/2$ winners from the first round advance to the second round where they compete in pairs as before using a row of $N/4$ switch elements. The winners in the second round advance to the third round, and this continues until two compete for the championship: that is, the right to exit the first output of the concentrator. Note that if there is at least one packet arriving on an input to the concentrator, a packet will exit the first output of the concentrator.

A tournament with only a single tree-structured com-

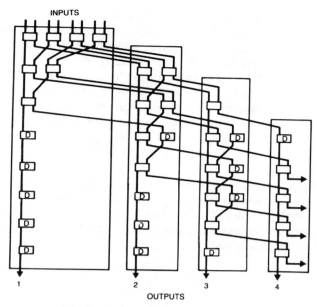

Fig. 8. The 8-input/4-output concentrator.

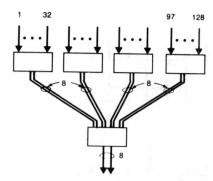

Fig. 9. The 128-to-8 concentrator constructed from 32-to-8 concentrator chips.

petition leading to a single winner is sometimes referred to as a single knockout tournament: lose one match and you are knocked out of the tournament. In a double knockout tournament, the $N - 1$ losers from the first section of competition compete in a second section, which produces a second place finisher (i.e., a second output for the concentrator) and $N - 2$ losers. As Fig. 8 illustrates, the losers from the first section can begin competing in the second section before the competition is finished in the first. Whenever there are an odd number of players in a round, one player must wait and compete in a later round in the section. In the concentrator, a simple delay element serves this function.

For a concentrator with N inputs and L outputs, there are L sections of competition, one for each output. A packet entering the concentrator is given L opportunities to exit through a concentrator output: a packet losing L times is knocked out of the competition and is discarded by the concentrator. In all cases, however, packets are only lost if more than L arrive in any one time slot. As we have seen, for $L \geq 8$, this is a low probability event.

For $N \gg L$, each section of the concentrator contains approximately N switch elements for a total concentrator complexity of $16NL$ gates. For $N = 64$ and $L = 8$, this corresponds to a relatively modest 8000 gates. Once a concentrator microcircuit is fabricated, Fig. 9 illustrates how several identical chips can be interconnected to form a larger concentrator. The approach is to select, based on an acceptable lost packet probability, the number of concentrator outputs L and then fabricate a chip with KL inputs (K an integer ≥ 2) and L outputs. A $K^j L$ input, L output concentrator can be formed by interconnecting $J + 1$ rows of KL-to-L concentrator chips in a treelike structure, with the ith row (counting from the bottom) containing K^{i-1} chips. For the example illustrated in Fig. 9, $L = 8$, $K = 4$, and $J = 2$.

Shared Buffer: The architecture of the bus interface focuses, to the extent possible, on reducing the complexity of packet buffering. This is done first by using a concentrator to reduce the number of inputs that must be buffered simultaneously. Second, through the use of a shared buffer structure, complete sharing of all packet buffer memory within the bus interface is made possible. This is accomplished while still providing a first-in first-out queueing discipline for the arriving packets and keeping the latency through the bus interface to a minimum.

Since in any given time slot up to L packets can emerge from the concentrator, the buffer within the bus interface must be capable of storing up to L packets within a single time slot. To permit high-speed low-latency operation of the Knockout Switch, the bus interface uses L separate FIFO buffers as shown in Fig. 3. A simple technique allows complete sharing of the L buffers, and at the same time provides a first-in first-out queueing discipline for all packets arriving to the output. The latter ensures fairness for access to the output and, more importantly, that successive packets arriving on an input do not get out of sequence within the switch.

As Fig. 3 shows, the L outputs from the concentrator first enter a "shifter" having L inputs and L outputs. The purpose of the shifter is to provide a circular shift of the inputs to the outputs so that the L separate buffers are filled in a cyclic fashion. This is illustrated in Fig. 10 for $L = 8$. Here, in the first time slot, five packets arrive for the output and, after passing through the concentrator, enter the first five inputs to the shifter. For this time slot, the shifter simply routes the packets straight through to the first five outputs, from which they enter buffers 1–5. In the second time slot, four packets arrive and enter the shifter on inputs 1–4. Having in the previous time slot left off by filling buffer 5, the shifter circular shifts the inputs five outputs to the right so that the arriving packets enter buffers 6, 7, 8, and 1. In the third time slot, the inputs are shifted one output to the right so that buffer 2 will receive the next arriving packet from the first output of the concentrator.

The shifter is a switch with L states: circular shift the inputs 0, 1, $\cdots$, or $L - 1$ outputs to the right. Letting S_i denote the state of the shifter and k_i the number of packets

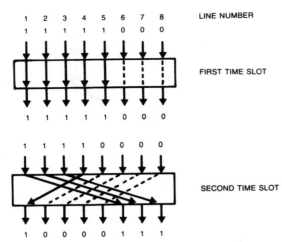

LINE NUMBER

FIRST TIME SLOT

SECOND TIME SLOT

Fig. 10. Illustration of shifter function.

exiting the concentrator in time slot i, we have

$$S_{i+1} = (S_i + k_i) \bmod L \qquad (5)$$

with $S_1 = 0$. A shifter can be constructed with $(L/2) \log_2 L$ 2×2 switch elements in the form of an omega network [13], controlled by a simple state machine obeying (5).

The flow of packets, stored in buffers $1–L$, onto the output line may be viewed as being controlled by a token. The buffer holding the token has the right to transmit one packet in the next time slot, with buffer 1 initially holding the token in the first time slot. If a buffer is empty, it will hold the token indefinitely until a new packet arrives and is transmitted. After this, the token is handed to the next buffer and wraps around in a cyclic fashion.

With the above described means for storing and removing packets for the L buffers, the shared buffer architecture has the following two characteristics.

1) Packets are stored in and removed from the L buffers in a cyclic fashion. At any time, the number of stored packets in each buffer does not differ by more than one among the L buffers. Buffer overflow only occurs when all L buffers are full. We therefore achieve the equivalent of a single buffer shared by all L outputs of the concentrator.

2) The token is held by the buffer with the highest number of stored packets, and the packet to be transmitted next is the one that has waited the longest in the bus interface. Hence, the shared buffer architecture forms the equivalent of a single-server first-in first-out queue with L inputs.

The total amount of buffering required in each bus interface depends on the assumed model for packet arrivals and the restrictions on lost packets resulting from buffer overflows. Using the homogeneous arrival model assumed previously (a packet arrives in a time slot at each input independently with probability ρ and each arriving packet is equally likely destined for each output), the probability of having k packets arrive in a time slot, all destined for the same output, has the binomial probability assignment given by (1). Since the fraction of arriving

packets lost with the concentrator is extremely small for $L \geq 8$, we make the conservative, but accurate, approximation that all arriving packets enter the shared buffer (i.e., none are lost in the concentrator). Also, so that our chosen buffer size is valid for all N, we assume a worst case situation by allowing $N \to \infty$. Under these conditions, the number of arriving packets to the shared buffer in each time slot has the Poisson probability assignment

$$\Pr\left[k \text{ arriving packets}\right] = \frac{\rho^k e^{-\rho}}{k!}. \qquad (6)$$

Since the arriving packets are fixed length, the transmission time for each packet is deterministic, equal to one time slot. Hence, for an infinite buffer, we can model the queueing process at each output as an $M/D/1$ queue [14] having mean queue size

$$\overline{Q} = \frac{\rho^2}{2(1 - \rho)} \qquad (7)$$

and steady-state probabilities[1]

$$\Pr(Q = 0) = (1 - \rho) e^\rho \qquad (8)$$

$$\Pr(Q = 1) = (1 - \rho) e^\rho (e^\rho - 1 - \rho) \qquad (9)$$

$$\vdots$$

$$\Pr(Q = n) = (1 - \rho) \sum_{j=1}^{n+1} (-1)^{n+1-j}$$
$$\cdot e^{j\rho} \left[\frac{(j\rho)^{n+1-j}}{(n+1-j)!} + \frac{(j\rho)^{n-j}}{(n-j)!} \right]$$
$$\text{for } n \geq 2 \qquad (10)$$

where the second factor in (10) is ignored for $j = n + 1$.

In Fig. 11 we plot the steady-state probability that the queue size exceeds M for loads varying from 70 to 99 percent. From Figs. 6 and 11, we can conclude that with 8 buffers, each 5 packets deep for a total of 40 packets, the probability of losing a packet within a bus interface is less than 10^{-6} for an 84 percent load on the switch. Models with more bursty arrivals (e.g., those used to model packet voice) will require a larger total buffer to achieve the same lost packet probability [15].

III. Modularity, Maintainability, Fault Tolerance, and Complexity

Often, in the design of a switch, the cost and complexity of the switch fabric play a secondary role to other factors, such as the ability to grow the switch in a modular fashion, to maintain the switch without great difficulty, and to provide the required degree of fault tolerance. In this section we describe how each of these important features can be satisfied with the Knockout Switch architecture. In addition, we show that, like most space-division

---

[1]The steady-state probabilities in Section 5.1.5 of [14] are for the *total number* of packets in an $M/D/1$ system. We are interested in *queue size*; hence, the modification to (8)–(10).

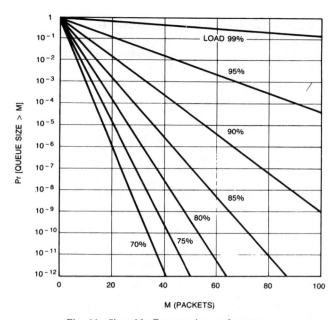

Fig. 11. Shared buffer queueing performance.

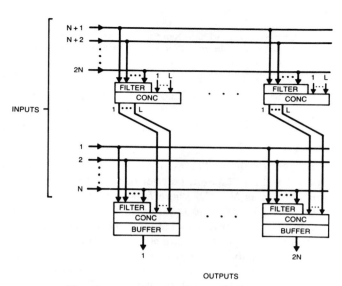

Fig. 12. Modular growth of Knockout Switch.

packet-switching architectures, the complexity of the Knockout Switch results not from the number of gates required in the design of the switch fabric, but rather from pin limitations on the circuit cards and VLSI chips.

A. Modularity

We mentioned earlier that, in addition to growing gracefully from 2×2 to $N \times N$, the Knockout Switch can grow modularly from $N \times N$ to $JN \times JN$, $J = 2$, $3, \cdots$. One way to do this is illustrated in Fig. 12 where we have provided each concentrator in the switch with L additional inputs for a total of $N + L$ inputs and L outputs. The interface for each output in a $JN \times JN$ Knockout Switch consists of J separate N-bus interfaces daisy chained together. Specifically, each of the J interfaces for one output contains a row of N packet filters and an ($N + L$)-to-L concentrator, with only the first interface (for buses 1–N) also containing the shared buffer structure with shifter and L FIFO buffers. These J individual components for each output are connected together by attaching the L outputs of the concentrator in the jth interface ($j = 2, 3, \cdots, J$) to the L extra inputs on the concentrator in the $j - 1$st interface. In effect, we have a convenient way of growing the Knockout Switch using a single ($N + L$)-to-L concentrator design and the same shared buffer (one for each output) independent of the switch size.

B. Maintainability

In the layout of the Knockout Switch, the electronics for each of the N input/output pairs can be placed on a separate interface module, with the switch then consisting of N identical interface modules interconnected by N broadcast buses (one for each input). Note that any fault on an interface module will disrupt traffic only on its input

or output and can be repaired by simply replacing the failed interface module. Even while the switch is being repaired, all other inputs and outputs can operate as usual. This is in direct contrast to a multistage switch where a failed intermediate switch point can affect multiple input/output paths and may also be difficult to locate.

C. Fault Tolerance

In many switching applications, stringent operational requirements necessitate a design that is tolerant of faults. This may be achieved by duplicating the entire switch fabric to serve as a standby in the event of a failure. With the Knockout Switch, since all interface modules are identical, a fault tolerant design can be achieved by providing only a single spare interface module attached to the N broadcast buses. This spare module could take over the operation of any one of the N interface modules if a failure occurs. Again, service would only be disrupted on the input or output attached to the failed module, and only as long as it takes to locate the fault and switch over its input and output to the spare module.

D. Complexity

As described in Section II and illustrated in Fig. 3, each bus interface in the Knockout Switch consists of a row of N packet filters, an N-to-L concentrator, and a shared buffer. The shared buffer, containing a simple shifter and L separate FIFO buffers, does not grow with the switch size N. The number of packet filters and the concentrator size, however, both grow in direct proportion to N. A packet filter can be designed with only five gates, giving us a total of $5N$ gates in the bus interface. The concentrator requires about NL simple, 16-gate 2×2 switch elements for a total of $16NL$ gates.

Table I shows, for various N, the combined packet filter and concentrator complexity of each bus interface on the switch in terms of the total number of gates and input/

TABLE I
COMBINED PACKET FILTER AND CONCENTRATOR COMPLEXITY

Switch Size, N	Gates	I/O Connections
128	17 000	136
256	34 000	264
512	68 000	520
1024	136 000	1032

output connections. The number of gates grows as $(5 + 16L)N$ and the required number of input/output connections grows as $N + L$, where $L = 8$ in Table I. At $N = 1024$, the 136 000 gate requirement in Table I is within reach of today's VLSI technology. However, the corresponding 1032-pin requirement for the VLSI chip far exceeds the capabilities that exist today, where the upper limit is in the 128–256 range. Still, the packet filters and concentrator for, say, 128 inputs can easily be integrated onto a single VLSI chip, and then several of these identical chips can be interconnected as shown in Fig. 9 to form a larger combined packet filter and concentrator. Only the chips at the top of the tree have their packet filters enabled, all others function only as concentrators. Also, as Fig. 12 illustrates, circuit card pin limitations on the backplane connector can be overcome by daisy chaining the packet filters and concentrators of multiple bus interface cards to form a larger switch. Hence, both VLSI and circuit card pin restrictions do not limit the growth of the Knockout Switch. In addition, the broadcast buses shown in Figs. 2 and 12 are unidirectional, with the packet framing clock (used to maintain time slot synchrony in the switch) traveling on a separate bus along the same path. Hence, the broadcast buses and frame clock bus can be latched and regenerated as needed to grow the switch.

IV. Conclusions

We proposed a new, high-performance packet switch architecture: the Knockout Switch. It provides direct interconnection paths from the switch inputs to the outputs, allowing us to greatly simplify the buffer design and thus achieve a more efficient switch. We observed that packet loss in any network is inevitable, whether it is caused by transmission errors or buffer overflow. By allowing the packet switch itself to introduce a small amount of additional packet loss, the concentrator required at each switch output can be reduced from $N \times N$ to $N \times 8$ for arbitrary large N. An $N \times 8$ concentrator is then designed based on knockout matches commonly used in tournaments. A new buffer sharing scheme joins the buffers associated with each of the concentrator output lines to form a simple first-in first-out buffer. The lost packet rate of the Knockout Switch can be made as small as desired and the latency is the smallest achievable by any switch. Moreover, the interconnect allows for easy modular growth, simple maintenance procedures, and a design that can be made fault tolerant. With each of $N = 1000$ inputs operating at 50 Mbits/s, possible applications for a 50 Gbit/s Knock-

out packet switch include interconnects of multiprocessors, high-speed local and metropolitan area networking, and local or toll switches for integrated traffic loads. With the advent of optical backplanes and integrated optoelectronic devices, the line speed and overall switch capacity could grow much larger.

References

[1] H. Inose, *An Introduction to Digital Integrated Communications Systems*. Tokyo, Japan: University of Tokyo Press, 1979.

[2] M. Decina and D. Vlack, Eds., Special Issue on Packet Switched Voice and Data Communication, *IEEE J. Select. Areas Commun.*, vol. SAC-1, Dec. 1983.

[3] J. S. Turner and L. F. Wyatt, "A packet network architecture for integrated services," in *GLOBECOM '84 Conf. Rec.*, Nov. 1983, pp. 45–50.

[4] J. J. Kulzer and W. A. Montgomery, "Statistical switching architectures for future services," presented at the 1984 Int. Switching Symp., Session 43, May 1984.

[5] A. Huang and S. Knauer, "STARLITE: A wideband digital switch," in *GLOBECOM '84 Conf. Rec.*, Nov. 1984, pp. 121–125.

[6] J. S. Turner, "Design of a broadcast packet network," in *Proc. IEEE INFOCOM '86*, Apr. 1986, pp. 667–675.

[7] ——, "Design of an integrated services packet network," *IEEE J. Select. Areas Commun.*, vol. SAC-4, pp. 1373–1380, Nov. 1986.

[8] D. M. Dias and M. Kumar, "Packet switching in $N \log N$ multistage networks," in *GLOBECOM '84 Conf. Rec.*, Nov. 1984, pp. 114–120.

[9] M. J. Karol, M. G. Hluchyj, and S. P. Morgan, "Input vs. output queueing in a space-division packet switch," in *GLOBECOM '86 Conf. Rec.*, Dec. 1986, pp. 659–665.

[10] D. F. Kuhl, "Error recovery protocols: Link by link vs. edge to edge," in *Proc. IEEE INFOCOM '83*, Apr. 1983, pp. 319–324.

[11] W. L. Hoberecht, "A layered network protocol for packet voice and data integration," *IEEE J. Select. Areas Commun.*, vol. SAC-1, pp. 1006–1013, Dec. 1983.

[12] W. Hoeffding, "On the distribution of the number of successes in independent trials," *Ann. Math. Statist.*, vol. 27, pp. 713–721, 1956.

[13] D. H. Lawrie, "Access and alignment of data in an array processor," *IEEE Trans. Comput.*, vol. C-24, pp. 1145–1155, Dec. 1975.

[14] D. Gross and C. M. Harris, *Fundamentals of Queueing Theory*. New York: Wiley, 1974.

[15] K. Sriram and W. Whitt, "Characterizing superposition arrival processes and the performance of multiplexers for voice and data," in *GLOBECOM '85 Conf. Rec.*, Dec. 1985.

Yu-Shuan Yeh (S'64-M'66-SM'81-F'84) was born in Wu-Kiang, Kiang-Su, China, on September 9, 1939. He received the B.S. degree in electrical engineering from the National Taiwan University, Taiwan, in 1961, and the M.S. and Ph.D. degrees from the University of California, Berkeley, in 1964 and 1966, respectively.

From 1961 to 1962 he was an Electronics Officer in the Chinese Navy. He was a Research Fellow at Harvard University, Cambridge, MA, from 1966 to 1967 doing antenna research. He joined AT&T Bell Laboratories, Holmdel, NJ, in 1967 and is currently a Supervisor in the Department of Network System Research. His research interests include switching, radio, and optical communications. He holds 20 patents.

Dr. Yeh is the recipient of two Best Paper Awards from the IEEE.

Michael G. Hluchyj was born in Erie, PA, on October 23, 1954. He received the B.S.E.E. degree in 1976 from the University of Massachusetts at Amherst and the S.M., E.E., and Ph.D. degrees in electrical engineering from the Massachusetts Institute of Technology, Cambridge, in 1978, 1978, and 1981, respectively.

From 1977 to 1981 he was a Research Assistant in the Data Communication Networks Group at the M.I.T. Laboratory for Information and Decision Systems where he investigated fundamental problems in packet radio networks and multiple access communications. In 1981 he joined the Technical Staff at Bell Laboratories where he worked on the architectural design and performance analysis of local area networks. In 1983 he moved with his department to AT&T Information Systems and continued to work on local area networks. In 1984 he transferred to the Network Systems Research Department at AT&T Bell Laboratories, performing fundamental and applied research in the area of high-performance, integrated communication networks and multiuser lightwave networks. In June 1987 he assumed his current position as Director of Networking Research and Advanced Development at Codex Corporation. His current research interests include wideband circuit and packet switching architectures, integrated voice and data networks, and local area network interconnects.

Dr. Hluchyj is active in the IEEE Communications Society and is a member of the Technical Editorial Board for the IEEE Network Magazine.

Anthony S. Acampora was born in Brooklyn, NY, on December 20, 1946. He received the B.S.E.E., M.S.E.E., and Ph.D. degrees from the Polytechnic Institute of Brooklyn in 1968, 1970, and 1973, respectively.

From 1968 through 1981 he was a member of the Technical Staff at Bell Laboratories initially working in the fields of high power microwave transmitters and radar system studies and signal processing. From 1974 to 1981 he was involved in high-capacity digital satellite systems research, including modulation and coding theory, time division multiple access methods, and efficient frequency reuse techniques. In 1981 he became Supervisor of the Data Theory Group at Bell Laboratories, working in the field of computer communications and local area networks. In January 1983 he transferred with his group to AT&T Information Systems to continue work on local area networks. In November 1983 he was appointed Head of the Radio Communications Research Department at AT&T Bell Laboratories, where his responsibilities included management of research in the areas of antennas, microwave and millimeter wavelength propagation, terrestrial radio and satellite communication systems, and multiuser radio communications. Since May 1984 he has been Head of the Network Systems Research Department responsible for research in the areas of communication systems, switching systems, and local lightwave networks.

Reprinted from *IEEE Journal on Selected Areas in Communications*, Vol. SAC-5, No. 8, October 1987, pages 1365-1376. Copyright © 1987 by The Institute of Electrical and Electronics Engineers, Inc. All rights reserved.

Synchronous Composite Packet Switching— A Switching Architecture for Broadband ISDN

TAKAO TAKEUCHI, TAKEHIKO YAMAGUCHI, MEMBER, IEEE, HIROKI NIWA, HIROSHI SUZUKI, AND SHIN-ICHIRO HAYANO

Abstract—A new switching architecture for broadband ISDN, "Synchronous Composite Packet Switching (SCPS)," is proposed and evaluated. It efficiently integrates circuit and packet switching functions on a single switching system and accommodates very high speed—up to several tens of Mbit/s—communication services, such as very high speed bursts of data, still picture, and motion video, as well as 64 kbit/s or less voice and data services.

The SCPS system comprises plural switch modules and plural very high speed synchronous loops. In the SCPS system, messages on plural circuit switched channels are assembled into quasi-packets, called "composite packets," and switched synchronously between switch modules, maintaining complete time transparency and short absolute delay time.

A system parameter design to obtain high system efficiency and appropriate system modularity is explained, and an example for a very large capacity transit switch of 4 Gbit/s throughput is presented. System implementation problems to realize the SCPS principle, such as efficient implementation of the composite packet assembling and loop transmission functions, are investigated and an experimental system constructed for circuit switching part is presented.

The most remarkable characteristic of the SCPS is that it efficiently integrates $64 \times n$ kbit/s circuit switching with packet switching. Moreover, the SCPS system retains compatibility with existing networks and the possibiliy of evolution toward a future broadband ISDN. On the basis of the above investigations and experimental system construction, the authors conclude that the SCPS is one of the most practical switching architectures for the coming broadband ISDN era.

I. INTRODUCTION

RESEARCH and development activities on the Integrated Services Digital Network (ISDN) are being continued intensively throughout the world. At the first stage, most of them were mainly devoted to the user access network integration [1] and to the 64 kbit/s and the associated submultiple-rate communication services. The user access network integration will play a key role for the ISDN introduction because of its large cost occupancy

in the entire network and integrated service offering capability to users.

After standardization for the narrowband ISDN at CCITT [2], its implementation trials have started in various places around the world. This movement surely leads to the 64 kbit/s user access integration. Then, the user will be able to access the network with a variety of communication media via "a limited set of interfaces." Nevertheless, various calls via the integrated access lines must still be discriminated at the switching office, depending on their characteristics and service categories, because individual networks, such as circuit switching network and packet switching network, or voice, data, facsimile, and videotex networks, have been constructed and operated so far. Moreover, many other useful services, not foreseen now, and broadband services will evolve in the near future. In particular, research for providing broadband services has emerged toward a future video communications society.

In the environments mentioned above, however, if the separate networks can be integrated physically into a single network with the least overhead, including circuit/packet switching integration and voice rate/broadband service integration, countless benefits and conveniences will be brought about in the communication network, such as simple network structure, unified operation and maintenance procedure, and consequent total cost reduction. Therefore, the networks themselves should be integrated to a single network, including broadband services, which is to be called a broadband ISDN.

Several alternative approaches for an ISDN switching system have been proposed [3]–[11]; two extreme examples are integration of all communication services by the packet switching technique and that by the circuit switching technique.

Generally speaking, however, the integration by the packet switching technique cannot maintain good time transparency and short absolute delay time for real-time communication services, such as voice and video services, because its queueing-based operation causes delay time variations and a large absolute delay time.

On the other hand, the integration by the circuit switching technique cannot retain flexibility and system efficiency for very high speed bursts of data and still picture services because network resources are only used sporadically by bursts of communication services.

Manuscript received November 26, 1986; revised April 20, 1987. This work was presented at the International Conference on Communications '84, Amsterdam, The Netherlands, May 14, 1984; at the International Switching Symposium '84, Florence, Italy, May 11, 1984; at the 1986 International Zurich Seminar on Digital Communications, Zurich, Switzerland, March 11, 1986; and at the International Conference on Communications '86, Montreal, P.Q., Canada, June 25, 1986.

T. Takeuchi, H. Niwa, H. Suzuki, and S.-I. Hayano are with the Communication Research Laboratory, C & C Systems Research Laboratories, NEC Corporation, 1-1, Miyazaki 4-Chome, Miyamae-ku, Kawasaki, Kanagawa 213, Japan.

T. Yamaguchi is with the 1st Switching System Engineering Development Department, Integrated Switching Development Division, NEC Corporation, 1131, Hinode, Abiko, Chiba 270-11, Japan.

IEEE Log Number 8716311.

Although much effort to progress the high speed packet switching is being done, a hybrid approach integrating circuit/packet switching seems to be a practical solution to support both forthcoming broadband communications a ʼd a huge amount of existing public switched telephone network (PSTN) services.

In order to realize the network integration mentioned above, this paper proposes a new switching system architecture, "Synchronous Composite Packet Switching (SCPS)," which enables fully integrating the circuit and packet switching functions on a single switching system. It also enables simultaneously accommodating not only the 64 kbit/s and the associated submultiple-rate services, but very high speed services up to several tens of Mbit/s, such as very high speed bursts of data, still picture, and motion video.

In this paper, the SCPS principle is described in detail, including distributed call control procedures. Next, system parameter design and system implementation problems, such as the mechanism of composite packet assembling and its transmission between switch modules over plural very high speed synchronous loops, are addressed. An experimental system for the circuit switching part, which was constructed to determine system feasibility and efficiency, is also presented. Finally, the paper evaluates the SCPS system and discusses its introduction into existing networks and evolution toward the future broadband ISDN.

II. Synchronous Composite Packet Switching

A. Switching Principle [12]

The Synchronous Composite Packet Switching (SCPS) system assembles messages on plural circuit switched channels that are communicating simultaneously into quasi-packets, and processes them in a similar manner to handle ordinary packets. The quasi-packets, however, are switched periodically and synchronously in the system to maintain complete time transparency for circuit switched channels. The details are explained in the following.

The SCPS system takes a building block system architecture to cover an extremely wide switching capacity range. The system comprises building block switch modules, each of which accommodates various kinds of circuit and packet switched channels, and an intermodule network (Fig. 1). The switching operations are accomplished by transmitting circuit and packet switched channel messages between switch modules via the common intermodule network. Intermodule signaling and command messages for system operation and control are also transmitted between modules. These three kinds of messages are transmitted in similarly structured packets, respectively, as shown below.

a) Composite Packets for Circuit Switched Calls: Each switch module assembles plural circuit switched channel messages, which are being transmitted simultaneously and have the same destination module, into a quasi-packet, called a *composite packet* [Fig. 2(a)]. They are assembled

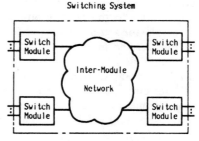

Fig. 1. Building block structure.

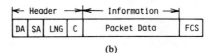

(a)

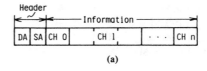

(b)

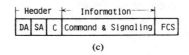

(c)

DA : Destination Module Address   LNG: Packet Length
SA : Source Module Address   C : Control Field
FCS: Frame Check Sequence

Fig. 2. Three packet structures in the SCPS system. (a) Composite packet structure for circuit switched calls. (b) Noncomposite packet structure for packet switched calls. (c) Intermodule signaling packet structure.

at regular intervals, 125 μs, for example, and transmitted synchronously between switch modules via the intermodule network.

As shown in Fig. 2(a), a composite packet comprises destination and source module addresses, *DA* and *SA* (header part), and an aggregate of circuit switched channel messages (information part). Each circuit switched channel is allotted a necessary time space in the information part, depending on its communication service speed. For example, voice channels are allotted 64 kbit/s, while video channels are allotted several or several tens of Mbit/s time space.

b) Noncomposite Packets for Conventional Packet Switched Calls: Fig. 2(b) shows the intrasystem packet structure for conventional packet switched channels. This packet is called a noncomposite packet in this paper. As illustrated in the figure, the intrasystem packet has a double-envelope structure, in which the original packet with its header is encapsulated by SCPS header and trailer.

The SCPS header is comprised of destination and source module addresses (*DA* and *SA*), packet length (*LNG*), and a control field (*C*), while the trailer is comprised of a frame check sequence (FCS). A control field and a frame check sequence play the role of intermodule packet transmission control, similar to the HDLC protocol. (Note: they are not necessary for a composite packet.)

c) Signaling Packets for Intermodule Signaling: Fig. 2(c) shows an intermodule signaling packet structure. This packet is used for intermodule signaling and control purposes, especially for circuit switched channel control and other housekeeping operations. This packet also comprises header part, signaling information part, and trailer part.

As illustrated in Fig. 2(a), (b), and (c), these three kinds of packets have almost the same structure. Thus, it is easy to handle these packets in a unified manner in a system. The noncomposite and the signaling packets differ from the composite packet only in that they do not need to be transmitted synchronously between switch modules.

B. Intermodule Network

Various intermodule network structures are conceivable to implement the SCPS system. What we have adopted is a plural-loop structure composed of plural very high speed synchronous loops having total throughput of several Gbit/s (Fig. 3). The access method to loops is similar to the so-called slotted ring method [13]. This structure is adopted for the following reasons.

• Loop structure is suitable for very high speed intermodule transmission.

• Simple topology fits autonomous and distributed switching control between switch modules. It is also excellent for intermodule cabling.

• Physically distributed configuration in transmission capacity, which is derived from the slotted ring access method, is suitable for distributed control and building block system structure.

• Logically concentrated configuration in transmission capacity enables supporting a wide range of communication speeds, because even very high speed service only occupies a part of network mass capacity.

• "Plural-loop" structure increases switching capacity flexibility and system reliability.

Electrical loop length for each loop is adjusted to the composite packet transmission interval, 125 μs, for example. Individual loops operate synchronously, having a frame structure which consists of a multiple of time baskets. The time basket is a data transmission capacity unit, which is not associated with an individual communication channel. A 125 μs frame contains 96 time baskets of 128 bits each, for example (Fig. 4). Each time basket has a basket indicator (BI), destination/source module addresses (*DA* and *SA*), sequence number (SN), information part, a parity bit (*P*), and a framing bit (*F*). The basket indicator is used to indicate four basket categories, i.e., idle, composite, noncomposite, and signaling packets.

Each switch module splits the assembled packets into time basket's information part size segments and gives sequence numbers (SN's) to them. Then, each module sends these segments to the loops, hunting for a necessary

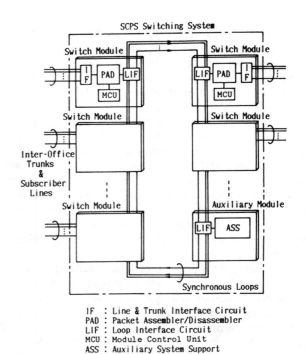

Fig. 3. Synchronous composite packet switching systems architecture.

IF : Line & Trunk Interface Circuit
PAD : Packet Assembler/Disassembler
LIF : Loop Interface Circuit
MCU : Module Control Unit
ASS : Auxiliary System Support

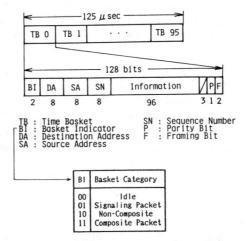

Fig. 4. Frame and time basket structure example on synchronous loops.

number of idle time baskets. At the same time, destination and source module addresses for each packet and a sequence number for each segment are moved into the associated part of each time basket. On the other hand, each module picks up split packet segments on the loops designated to it by observing the basket indicator and destination module address. The received basket becomes idle or it is reused for another packet transmission. Therefore, each packet transmission is accomplished by using idle time baskets that are not necessarily consecutive on a loop. The transmission is accomplished within one frame, particularly for composite packets. (Fig. 5).

If circuit switched calls are symmetrical bidirectional calls, frame-by-frame periodical transmission of composite packets can be guaranteed autonomously, after once

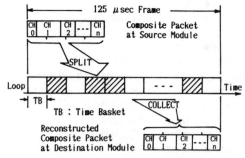

Fig. 5. Composite packet transmission over synchronous loop.

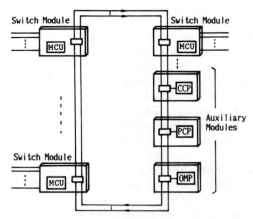

CCP : Circuit Switched Call Auxiliary Processor
PCP : Packet Switched Call Auxiliary Processor
OMP : Operation and Maintenance Processor

Fig. 6. Distributed control architecture.

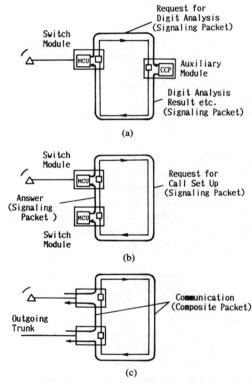

Fig. 7. Call setup procedure for circuit switched calls.

having succeeded in hunting for and finding the necessary number of idle time baskets in a frame for each composite packet. The reasons are that the same number of time baskets for module A to B packet can always be used for module B to A packet without fail, even if time basket positions for a specific composite packet may change from frame by frame. To guarantee this mechanism, however, it is necessary that each module send all composite packets prior to other category packets in every frame.

On the other hand, noncomposite packets and signaling packets can be transmitted by hunting for and finding necessary number of idle time baskets in plural frames. Sequential number (SN) in a time basket (Fig. 4) is used as a delimiter for each packet. Therefore, after transmitting composite packets in each frame, the rest of loop transmission capacity can be fully utilized by noncomposite and signaling packets. In addition, since plural physical loops are regarded as a single logical loop, higher system efficiency can be obtained by idle time basket hunting on all the loops.

C. Distributed Call Control Procedure [14]

Call-by-call control functions and O & M (operation and maintenance) functions in the SCPS system are carried out by distributed control philosophy. Fig. 6 shows an example of the SCPS distributed control architecture, which comprises individual switch modules and some auxiliary modules. These modules share control functions, as explained in the following.

Switch Modules: Line/trunk supervision, call originating/terminating control, call status control, composite packets assembling/disassembling, noncomposite packets handling, and loop transmission control.

Circuit Switched Call Auxiliary Processor (CCP) Module: Digit analysis, routing, and trunk idle/busy management for circuit switched calls.

Packet Switched Call Auxiliary Processor (PCP): The network layer protocol processing for packet switched calls, that is, routing, logical link setup, etc.

Operation and Maintenance (O & M) Module: System initialization, clock supplying, loop delay adjusting, and system supervision.

Fig. 7 shows the call set up procedure for circuit switched calls upon this configuration.

1) A switch module (source module) detects call orig-

ination and receives destination terminal number by processing D channel protocol.

2) Source module requests CCP module for destination analysis. CCP analyzes destination number, determines the route (destination module), hunts for an idle outgoing trunk, etc., and then returns the result to the source module [Fig. 7(a)].

3) Source module requests the destination module to set up a call. The destination module returns acknowledgment (or nonacknowledgment) to the source module [Fig. 7(b)].

4) Communication begins after the call setup has been completed [Fig. 7(c)].

Control procedures for packet switched calls are almost the same as those for circuit switched calls. Once a call setup is finished for a circuit or packet switched call, speech or data communication is carried out directly between source and destination switch modules.

In Fig. 6, the CCP, PCP, and O & M modules are separately illustrated. However, they are merged to a single module or to one of the switch modules, in the case of a small capacity switching system arrangement. On the other hand, fully distributed control architecture, where CCP and PCP functions are distributed to individual switch modules, is also conceivable in case that switch modules are not located neighboring upon each other.

The O & M module includes a clock oscillator, which can be synchronized to the derived clock from the prior switching office, loop associated frame aligners for adjusting the electrical loop length, and input/output devices for system operation and maintenance.

D. SCPS System Advantages

As a result of assembling messages on plural circuit switched channels into composite packets, the following advantages are brought about.

1) Circuit and packet switching functions are integrated on a single switching system.

2) Overhead loss is reduced since the composite packet header is shared by plural circuit switched channels.

3) Total absolute delay time for composite packet assembling, transmitting, and disassembling can be markedly reduced to a value comparable to the conventional time division switching system, i.e., less than 900 μs roundtrip delay (CCITT Recommendation Q507). The reason is that the message amount for each channel in a composite packet can be reduced, for example, to 1 PCM sample (8 bits) for a 64 kbit/s voice channel.

4) Complete time transparency for circuit switched channels can be maintained, since composite packet assembling and transmitting are accomplished synchronously at regular intervals, say, 125 μs.

5) Communication channels having an extremely wide range of speeds and traffic characteristics, such as voice, very high speed bursts of data, still picutre, and motion video services, can be efficiently switched in a unified manner, since the channel arrangement in a composite packet is the so-called "full availability channel arrangement" [15] in the heterogeneous traffic processing theory.

Moreover, the following advantages are derived from the system configuration described above:

6) Excellent matching with distributed control architecture; and

7) High system expandability because of the building block structure in switch modules and loops.

III. System Implementation

A. System Parameter Design [14]

How to determine the numbers of modules and loops for a specified maximum switching capacity is a significant problem for the system design. The number of modules, the frame length, and the overhead amount are selected, considering the actual module size and expansion step size, the roundtrip absolute delay in a switching system, and the system operation requirements, respectively. For example, frame length should be selected as 125 μs, to keep the roundtrip delay in the total system less than 900 μs (CCITT Recommendation Q.507).

Moreover, the time basket size must be determined optimally. A time basket is a data transmission capacity unit, which is not associated with each communication channel, as mentioned before. A time basket has a fixed length. Therefore, the time basket containing the last segment of a composite or noncomposite packet may not be fully utilized. This is a problem particularly for composite packets because their last baskets come out every frame. If time basket size is too large, the unused part in the last basket cannot be neglected. On the contrary, if the time basket size is too small, the overhead of each time basket cannot be neglected.

These design considerations are explained below. The plural physical loops are considered as a single logical loop, and the "unused part in the last time basket" problem is considered only for composite packets. A half of the last time basket is assumed to be utilized on an average. As time baskets on the loops can be used twice in each frame for bidirectional calls, total effective switching throughput C [bit/s] and total loop speed L [bit/s] are given as follows:

$$C = 2 \times (m - n/2) \times I/T \tag{1}$$

$$L = m \times (H + I)/T \tag{2}$$

where

m: total number of time baskets on the loops
$n = {}_N C_2 = N \times (N - 1)/2$
    total number of composite packets on the loops, that is, total number of last time baskets.
N: number of switch modules
I: information amount in a time basket [bits]
H: overhead amount in a time basket [bits]
T: frame length [s].

Minimum L is given under the condition that C is constant, taking the number of modules N, frame length T, and overhead amount H as given parameters. The total number of time baskets m and the time basket size I are also determined. After this calculation, the physical loop speed and the number of loops are determined from total loop speed L, taking into account transmission technology, expansion step size, reliability, etc. Finally, calculated parameters are properly rounded to convenient values.

Design examples for a transit switch, derived from the above consideration, are shown in Table I. Example 1 in the table has an effective system capacity of bidirectional 64 kbit/s $\times$ 68 000 incoming and outgoing ports equivalent, that is, total effective throughput: $C = 4.34$ Gbit/s. This example system consists of 8 optical fiber loops of

TABLE I
SYSTEM PARAMETER DESIGN EXAMPLES

Item	Example 1	Example 2
Effective System Throughput	4.34 Gb/s	1.98 Gb/s
[Module]		
Number of Switch Modules	32	32
Effective Switch Module Throughput	136 Mb/s	61.8 Mb/s
[Loop]		
Number of loops	8	16
Loop Speed	393 Mb/s	98.3 Mb/s
Frame Length	125 μsec	125 μsec
Number of Time Baskets	384/frame/loop	96/frame/loop
Time Basket Size	128 bits	128 bits
Information Part Size	96 bits	96 bits

TABLE II
SERVICE ACCOMMODATION EXAMPLE (TRANSIT SWITCH)

	Communication Speed	Activity	Number of ports
Voice	64 kbit/s	0.7 erl/port	36,000 ports
Image	1 Mbit/s	0.1 erl/port	3,600 ports
Motion Video I	6.3 Mbit/s	0.7 erl/port	360 ports
Motion Video II	32 Mbit/s	0.7 erl/port	36 ports

400 Mbit/s and 32 switch modules. Each switch module has bidirectional 64 kbit/s × 2100 ports equivalent, that is, bidirectional 136 Mbit/s throughput. Example 2 was the parameter set adopted in the experimental system, which will be presented later.

Table II shows a service accommodation example in the above transit switch of Example 1. As shown in Table II, the example system can accommodate very high speed—up to several tens of Mbit/s—communication services, such as still picture and motion video services, as well as 64 kbit/s voice rate services.

B. Implementation Studies for the Circuit Switching Part [16]

To implement the SCPS principle described above, several points in the system architecture must be investigated, such as how to assemble/disassemble composite packets, how to synchronize channel sequence change in a composite packet between source and destination modules and how to send/receive time baskets to/from plural loops.

1) Composite Packet Assembler/Disassembler: The composite packet assembler/disassembler (CPAD) consists of time switches and control memories. It is assumed here that all input lines to the switching system are multiplexed onto a single super highway (See Section IV-C). Fig. 8 shows the way of assembling circuit switched channel messages into composite packets in the time switch. In this method, the channel messages are randomly written into the time switch to arrange composite packets there, and are sequentially read out to the output highway.

The most important problem to be overcome in this method is how to originate and terminate calls. Fig. 9 shows an example of channel sequence change in a composite packet, when a communicating channel is terminated. In this example, the channels following channel 1 have to shift their positions in the composite packet. Thus, the CPAD must carry out this shifting function, while still maintaining the consistency of other still-communicating channels.

The solution for this problem is the control memory configuration illustrated in Fig. 10. In this configuration, the write address for each input channel of a time switch

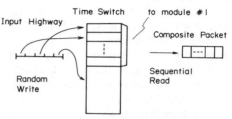

Fig. 8. Composite packet assembler configuration.

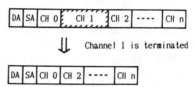

Channel 1 is terminated

Fig. 9. Channel shift by call termination.

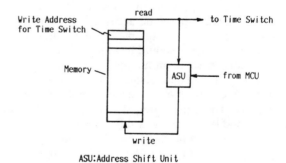

ASU: Address Shift Unit
Fig. 10. Control memory configuration.

is stored in the control memory. When a call in a composite packet is terminated, the write addresses of the channels following the terminated channel in the composite packet must be subtracted by a value equal to the terminated channel's bandwidth within one frame time The address shift unit (ASU) provides this function. Each write address from the control memory is compared to that of the terminated channel. If it is greater than the terminated channel, the address is subtracted and the result is stored in the same position as before in the control memory.

In this method, the process for the call origination is almost the same as that for the call termination. The composite packet disassembler also consists of a time switch and a control memory, whose operation mechanism is similar to that of the assembler, except that the relationship between read and write sides in Fig. 8 is reversed.

The time switch has two planes to maintain bit sequence integrity and avoid cycle slip. One plane is used for writing and the other is used for reading at every specific moment. They change places frame by frame.

2) Channel Shift Synchronization: Another significant problem is channel shift synchronization between source and destination modules. The channel shift execution time, due to call origination or termination, is reserved by negotiations between the two switch modules. Reserved time is defined by number of frames "k" from the current frame. This function is performed by channel shift time controller (CTC) in each module.

A CTC is composed of a mod-M frame counter and a cyclic buffer with M locations (Fig. 11). Each buffer location contains control memory (CM) modifying parameters, which are issued and stored by the module control unit (MCU). The buffer is accessed by the frame counter, and if any modifying parameters are contained in the location, the CTC modifies the CM according to the parameters. As a result of these processes, control memories in both modules can be changed synchronously "k" frames after detecting the change request.

Moreover, at the originating module, the address shift time for the reception side must be one frame later than that for the transmission side, as shown in Fig. 12. Hence, individual control memories must be installed for transmission and reception sides, respectively.

3) Composite Packet Transmission via Plural High Speed Loops: To achieve high efficiency, high reliability, and simple system operation, it is desirable that plural very high speed physical loops be accessed as a single ultrahigh speed logical loop. This means that individual time baskets on plural loops should be regarded as sequentially numbered time baskets on a single loop. An example is shown in Fig. 13. (These sequential numbers, however, are logical ones, and not indicated directly at individual time baskets. These are different from SN's in Fig. 4.)

This method is considerably effective because the transmission capacity can be easily managed without any complicated loop selection algorithm to obtain high transmission efficiency. Furthermore, even if a loop develops a failure, composite packets can be transmitted in the same way as usual, avoiding the failed loop channels.

This method, however, causes a new problem in that the time switch, control memory, and time basket controller that controls loop transmission of the composite packet must operate at an ultrahigh speed, which is the total sum of individual physical loop speeds. It is also inconvenient in small line capacity configurations because those circuits must operate at the same speed as that in the maximum capacity configuration, or the operation speed must vary according to the number of installed loops.

To solve this problem, the module is equipped with an individual time switch for each loop (Fig. 14). All these time switches operate simultaneously as a single logical time switch. In this configuration, input channel data are written into each time switch in parallel, and read out and

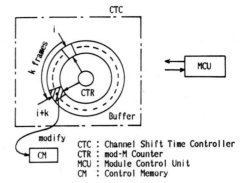

Fig. 11. Synchronization mechanism.

CTC : Channel Shift Time Controller
CTR : mod-M Counter
MCU : Module Control Unit
CM : Control Memory

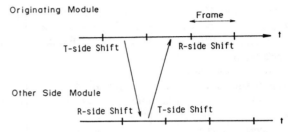

Fig. 12. Address shift coincidence at send/receive modules.

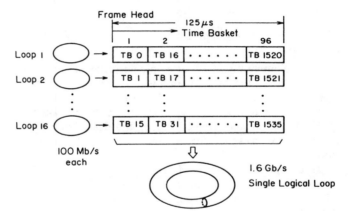

Fig. 13. A single logical loop.

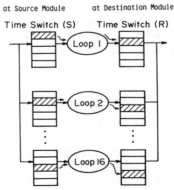

Fig. 14. Distributed time switch.

transmitted to the associated loop without any duplication of data transmission under the control of the time basket controller, as described later. On the reception side, received data on a loop are written into the associated time switch according to its source module address and sequential number marked at the transmission side. The time switches at the reception side also operate as a single logical time switch. The data received from the loops are stored in each time switch and read out by the control memory. The control memory supplies common read addresses to the individual time switches and the time switches provide stored messages and clear their positions. The output data are logically accumulated, and the resultant data are output as received data.

This method is eminently suitable for the system implementation because circuit speed is limited to the single physical loop speed, although the logical loop speed is very high. Furthermore, the loop associated part can be similarly structured, which is especially suitable for building-block architecture and LSI application.

To control the transmission and reception of composite packets in the way described above, the time basket controller is attached to each loop associated part. A block diagram of the send and receive basket controllers (SBC and RBC) is shown in Fig. 15. The SBC receives the idle/busy information from all the loop interface circuits (LIF's) every time basket, calculates the number of idle time baskets in the other loops, and determines which part of the time switch contents should be transmitted through the associated loop. Consequently, the top read address for the data to be transmitted to the loop is determined. Then, the read addresses are generated and provided to the time switch. Header information (*DA*, *SN*) for the data is also prepared and transferred to the LIF.

On the other hand, the RBC receives header information *SA* and *SN*, analyzes the data with a header analysis table, and obtains the top write address to store the received time basket data. Then, it generates the write addresses for the received data and provides them to the receive time switch. Thus, original composite packets, divided into time basket size segments and transmitted over the different loops, are reconstructed at the receive time switch.

In this way, the SBC and RBC can also operate at an individual physical loop speed, not at the total logical loop speed, which leads to easy implementation of the system.

C. An Experimental System

On the basis of the considerations described above, an experimental system, especially for the circuit switching part, has been constructed to determine the system feasibility and efficiency. The entire switch module structure is illustrated in Fig. 16. The upper half in the figure is the circuit switching part, while the lower half is the packet switching part. Example 2 in Table I was selected for the experimental system parameters.

In Fig. 16, time switches (TSW's) in the circuit switching part and packet switching unit (PSU) in the packet switching part get the channels designated to them out of the input highway. Which channels to be selected is determined at each call setup phase by using common chan-

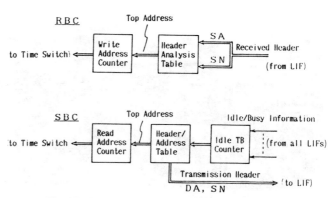

Fig. 15. Send/receive basket controller configurations.

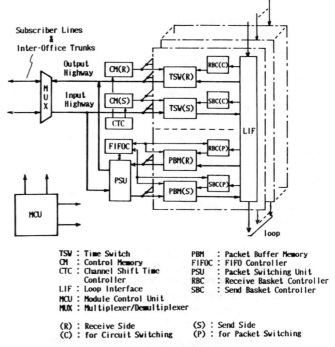

TSW : Time Switch	PBM : Packet Buffer Memory
CM : Control Memory	FIFOC : FIFO Controller
CTC : Channel Shift Time Controller	PSU : Packet Switching Unit
LIF : Loop Interface	RBC : Receive Basket Controller
MCU : Module Control Unit	SBC : Send Basket Controller
MUX : Multiplexer/Demultiplexer	

(R) : Receive Side	(S) : Send Side
(C) : for Circuit Switching	(P) : for Packet Switching

Fig. 16. Switch module configuration.

nel signaling (CCIS) or *D* channel protocol. The CCIS equipment and *D* channel handler are located at individual subscriber lines and trunk lines (see Fig. 19).

Time switches [TSW(*S*) and (*R*)], send and receive basket controller [SBC(*C*) and RBC(*C*)], and loop interface (LIF) are installed in each loop-associated part, while control memories [CM(*S*) and (*R*)] and channel shift time controller (CTC) are installed in common. The TSW's and CM's operate as a composite packet assembler/disassembler, while SBC(*C*) and RBC(*C*) work as controllers to send and receive composite packets to/from the associated loop. The loop interface (LIF) provides physical interface between the loop and the switch module. It detects the idle/busy status of each time basket, sends a composite packet segment when it is idle, receives a time basket intended for the module, and passes the received header and data to the RBC and TSW.

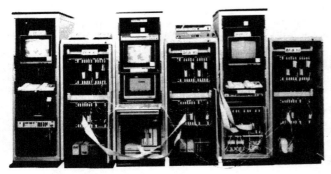

Fig. 17. Switch modules and accommodated terminals for the experimental system.

The packet switching part has almost the same structure as that of the circuit switching part. The SBC(P) and RBC(P) are similar to the SBC(C) and RBC(C), and packet buffer memory (PBM) and FIFO controller (FIFOC) correspond to the time switch and control memory in the circuit switching part, respectively, although the noncomposite packet is assembled in the packet switching unit (PSU).

The constructed experimental system comprises three switch modules and two 100 Mbit/s electrical loops. One of the modules includes operation and maintenance functions, such as electrical loop length adjustment. The experimental system accommodates three kinds of circuit switched terminals—64 kbit/s telephone, 1.024 Mbit/s (512 kbit/s × 2 channel) high quality sound program, and 8.192 Mbit/s bandcompressed motion video. It has been confirmed that the experimental system can switch these terminals successfully. It has been also confirmed that even if any one of the two loops is disconnected, ongoing communications can be continued without any particular control procedures, owing to the single logical loop mechanism.

Fig. 17 shows the appearance of the constructed three switch modules and accommodated terminals. The upper shelf in the switch module contains the loop associated parts (TSW, SBC, RBC, and LIF) for two loops. The lower shelf contains a common part, including control memories, a clock source, interface for the module control unit (MCU), and a multiplexer (MUX) of subscriber lines and interoffice trunks. This system has adopted a distributed control scheme based on the discussions in Section II-C.

IV. Evaluation and Discussion on SCPS for Broadband ISDN [17]

This section presents evaluation of the SCPS system based upon the investigations described above, and discusses its advantages and disadvantages as a broadband ISDN switching system. The introduction process of the SCPS system into existing networks and the possibility of its evolution toward a future broadband ISDN are also addressed.

A. Excellence in 64 × n kbit/s Heterogeneous Circuit Switching

The most remarkable characteristic of the SCPS is that it can efficiently handle 64 × n kbit/s heterogeneous circuit switched channels.

Essential configuration of a switching network in the SCPS is depicted in Fig. 18, which may be recognized as a time-space-space-time (TSST) switch configuration. Owing to the single logical loop mechanism, however, there is no need for complicated channel matching nor pathfinding despite bandwidth variations of individual circuit switched channels. If an idle time basket is found at any position in a frame on the loop, a time switch [TSW(S)] can send an information segment onto it, and a time switch [TSW(R)] can receive any time baskets designated to it. Therefore, if the capacity of the single logical loop is greater than the total sum of all the time switch capacities, completely nonblocking circuit switching can be obtained more efficiently than in the Clos network [18]. Since a composite packet has priority over the other category packets, this characteristic is not affected by them.

B. Circuit and Packet Switching Integration by the Hybrid Approach

The SCPS can integrate 64 × n kbit/s heterogeneous circuit switching and ordinary packet switching on a single switching system in a "hybrid" manner. The real integration is only achieved at internal loops. Nevertheless, it is a reasonable solution in order to realize network integration for a broadband ISDN, maintaining compatibility with existing networks.

Generally speaking, as long as frame-based operation is preserved for circuit switched calls, which means that information amount in a frame for each channel is strictly controlled to be constant at any place in a network, their control scheme is entirely different from that for packet switched calls. Moreover, the frame length must be short enough to reduce switching delay to a level equal to existing networks. Therefore, it is inevitable that PAD configurations for both switching categories are much different from each other. In other words, the hybrid approach is a practical and necessary solution to maintain compatibility with existing networks. Furthermore, since packet switched traffic is relatively smaller than circuit switched traffic in the hybrid configuration, packet switching unit need not have full capacity for the whole input highway.

On the other hand, if frame-based operation is abandoned, which means taking the homogeneous packet switching approach [4]–[8], a higher degree of integration becomes possible. Nevertheless, strict time transparency is hard to be preserved, and compatibility with existing networks is not maintained. Thus, serious degradation in communication qualities, such as delay and packet losses, will develop especially at the transient stage, which often disturbs to reach a final stage. Therefore, this approach should be applied to develop an overlay network to the existing one.

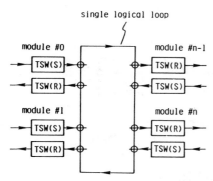

Fig. 18. Essential configuration of switching network in the SCPS.

The SCPS system taking the hybrid approach is able to maintain complete time transparency and short absolute delay time for circuit switched calls. Hence, its external interfaces can retain full compatibility with existing networks, which is one of the most valuable advantages in the SCPS architecture.

C. Interface for Interoffice Trunks and Subscriber Lines

On interoffice trunks, curcuit switched channels and packet switched channels are supposed to be physically integrated by using a movable-boundary method. Hence, all the channels on the interoffice trunks can be dynamically shared among both circuit switched and packet switched calls. For subscriber lines, basic-access channels, primary-access channels, and other high speed-access channels on the I-interface are simply multiplexed or concentrated on some highways. Consequently, degree of dynamic sharing of channels on a highway for subscriber lines is lower than that on interoffice trunks. Finally, these channels on interoffice trunks and subscriber lines are multiplexed onto a single ''super-highway,'' which is an input of the composite packet assembler (Fig. 8). Fig. 19 shows this configuration. The multiplexer for the subscriber lines in the figure can be replaced with a line concentrator.

D. Distributed Configuration

It is also possible to distribute the switching modules remotely and connect them with plural high speed loops. In this case, subscriber access line length is markedly reduced and broadband services can be easily accommodated to the switching system. Moreover, dynamic sharing of intermodule loops between composite and noncomposite packets contributes to get high system efficiency. The distributed control architecture mentioned in Section II might be changed to fully distributed configuration where auxiliary module functions are distributed to individual switch modules.

E. Broadband Switching Capability

The SCPS has the ability to handle several tens of Mbit/s video signals because plural high speed loops are accessed as a single ultrahigh speed logical loop. The

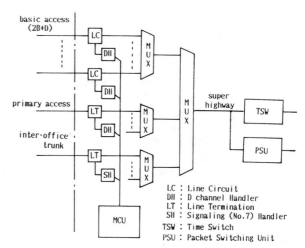

Fig. 19. External interfaces with subscriber lines and interoffice trunks.

SCPS can also provide high speed video services of more than 100 Mbit/s if necessary. However, it does not seem to be the best way to handle such broadband data (H_2-H_4: 30–150 Mbit/s) as a multiple of 64 kbit/s bearer rate. A solution for this problem will be to introduce a broadband circuit switching unit (BCSU), which is based on another bearer rate (H_2; 30–50 Mbit/s for example) for broadband services. Consequently, the value n in 64 $\times n$ kbit/s heterogeneous circuit switching should be up to around 512. In this configuration, it is possible that the BCSU accesses the dedicated loops, or shares the loops and time baskets with other narrowband communications.

F. High Speed Packet Switching

The packet switching unit (PSU) supports conventional X.25 packets, B-channel packets, and D-channel packets. However, high speed and high capacity packet switching techniques [4]–[8] can be applied to this unit. Consequently, each service can be supported by the most suitable switching technique in the hybrid approach. Furthermore, voice switching and some other switching might be shared gradually by the high speed packet switching in the future. Hence, the high speed packet switching unit (HPSU) [19] will play an important role in the system.

As a result of the above considerations, an example of SCPS architectural evolution toward broadband ISDN is shown in Fig. 20. At the first stage, SCPS is introduced into the existing network, where 64 kbit/s telephone traffic occupies a large part of network traffic [Fig. 20(a)]. Conventional packet data are processed by a packet switching unit (PSU) in each switch module.

Next, the high speed packet switching technique will be introduced. The network integration by high speed packet switching will progress further on an overlay network basis. It can share a part of the traffic, which has previously been handled by the circuit switching unit (CSU). In this stage, PSU has been replaced by the high speed packet switching unit (HPSU), which is based on the new architecture [19]. Furthermore, the broadband

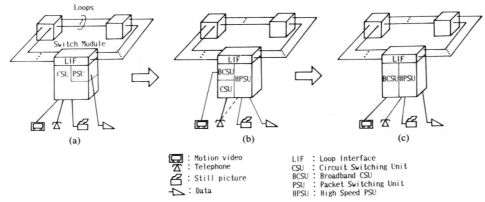

Fig. 20. SCPS architectural evolution.

circuit switching unit (BCSU) will be added to the switch module to accommodate broadband communication services [Fig. 20(b)]. Finally, conventional circuit switching traffic will be absorbed into HPSU and BCSU as shown in Fig. 20(c).

The SCPS architecture has the possibility to match requirements set forth by future technical trends as described above. Therefore, it is one of the most practical and future-proof solutions for the broadband ISDN.

V. Conclusion

This paper proposed and evaluated Synchronous Composite Packet Switching (SCPS) for the coming broadband ISDN era. The SCPS system has a building block structure consisting of plural switch modules and intermodule high speed synchronous loops. By assembling circuit switched call messages into composite packets and switching them in almost the same manner as packet switched calls, the SCPS technology provides many advantages.

An optimal system parameter design method and actual parameter design examples were also presented. When maximum system capacity and building block unit size are given, the other system parameters can be determined optimally.

Moreover, the system implementation of the SCPS was investigated and an experimental system, especially for the circuit switching part, has been constructed. In particular, composite packet assembler/disassembler configurations and efficient and highly reliable implementation of plural loops were proposed and examined.

These efforts proved that the SCPS system has the expected significant advantages. In particular, evaluation on the experimental system said that the most remarkable characteristic of the SCPS is that it efficiently integrates $64 \times n$ kbit/s (n is up to around 512) heterogenous circuit switching with packet switching. Moreover, the SCPS retains compatibility with existing networks and the possibility to evolve toward a future broadband ISDN. Therefore, the authors conclude that Synchronous Composite Packet Switching is one of the most practical solutions for a next generation switching architecture.

Acknowledgment

The authors gratefully acknowledge the encouragement and guidance given by T. Ishiguro, K. Watanabe, and F. Akashi of C & C Systems Research Laboratories; H. Goto of C & C Information Technology Research Laboratories; and A. Kitamura of the Integrated Switching Development Division, during the course of this study. The authors also appreciate the discussion and assistance given by members of the Communication Research Laboratory and Integrated Switching Development Division.

References

[1] M. Decina, "Progress towards user access arrangements in integrated services digital networks," *IEEE Trans. Commun.*, vol. COM-30, Sept. 1982.
[2] *CCITT Red Book*, Vol. 3, Fascicle 3.5, "Integrated services digital network (ISDN)," Oct. 1984.
[3] J. G. Gruber, "Performance considerations for integrated voice and data networks," *Comput. Commun.*, vol. 4, no. 3, June 1981.
[4] J. S. Turner, "A packet network architecture for integrated services," presented at the Global Telecommun. Conf. '83, Dec. 1983, Paper 2.1.
[5] J. J. Kulzer and W. A. Montgomery, "Statistical switching architectures for future services," presented at the Int. Switching Symp. '84, May 1984, Paper 43A1.
[6] J. S. Turner, "New directions in communications," presented at the 1986 Int. Zurich Seminar, Mar. 1986, Paper A3.
[7] A. Thomas, J. P. Coudreuse, and M. Servel, "Asynchronous time-division techniques: An experimental packet network integrating videocommunication," presented at the Int. Switching Symp. '84, May 1984, Paper 32C2.
[8] P. Gonet, P. Adam, and J. P. Coudreuse, "Asynchronous time-division switching: The way to flexible broadband communication networks," presented at the 1986 Int. Zurich Seminar, Mar. 1986, Paper D5.
[9] J. van Baardewijk, "An experimental all-in-one broadband switch," presented at the 1986 Int. Zurich Seminar, Mar. 1986, Paper A3.
[10] E. F. Haselton, "A PCM frame switching concept leading to burst switching network architecture," presented at the Int. Conf. Commun. '83, June 1983, Paper E6.7.
[11] S. R. Amstutz, "Burst switching—A method for dispersed and integrated voice and data switching," presented at the Int. Conf. Commun. '83, June 1983, Paper A7.8.
[12] T. Takeuchi and T. Yamaguchi, "A new switching system architecture for ISDN environment—Synchronous composite packet switching," in *Proc. Int. Conf. Commun., '84*, May 1984, pp. 38–41.
[13] J. R. Pierce, "Network for block switching of data," *Bell Syst. Tech. J.*, July–Aug. 1972.
[14] T. Takeuchi and T. Yamaguchi, "Synchronous composite packet switching for ISDN switching system architecture," presented at the Int. Switching Symp. '84, May 1984, paper 42B3.

[15] O. Enomoto *et al.*, "An analysis of mixtures of multiple band-width traffic on time division switching networks," presented at the 7th Int. Teletraffic Congr., June 1973.

[16] T. Takeuchi *et al.*, "An experimental synchronous composite packet switching system," presented at the 1986 Int. Zurich Seminar, Mar. 1986, Paper D6.

[17] H. Niwa *et al.*, "Synchronous composite packet switching for broadband ISDN," presented at the Int. Conf. Commun. '86, June 1986.

[18] C. Clos, "A study of non-blocking switching networks," *Bell Syst. Tech. J.*, vol. 32, pp. 406–424, Mar. 1953.

[19] H. Suzuki *et al.*, "Very high speed and high capacity packet switching for broadband ISDN," presented at the Int. Conf. Commun. '86, June 1986.

Hiroki Niwa was born in Nagoya, Japan, in 1953. He received the B.E. and M.E. degrees from Keio University, Yokohama, Japan, in 1977 and 1979, respectively.

After he joined NEC Corporation in 1979, he worked in developing PBX's. He has been engaged in research and development on integrated switching systems for ISDN since 1984. He is now a Supervisor at the Communication Research Laboratory, C & C Systems Research Laboratories.

Mr. Niwa is a member of the Institute of Electronics, Information, and Communication Engineers (IEICE) of Japan.

Takao Takeuchi was born in Hiroshima, Japan, in 1951. He received the B.E. and M.E. degrees in electrical engineering from the University of Tokyo, Tokyo, Japan, in 1974 and 1976, respectively.

He joined the Central Research Laboratories, NEC Corporatior., Kawasaki, Japan, in 1976, and has since been engaged in research and development of digital switching systems, and narrowband/broadband ISDN switching architectures. At present he is a Supervisor at the Communication Research Laboratory, C & C Systems Research Laboratories.

Mr. Takeuchi is a member of the Institute of Electronics, Information, and Communication Engineers (IEICE) of Japan.

Hiroshi Suzuki was born in Osaka, Japan, in 1960. He received the B.E. and M.E. degrees from Osaka University, Osaka, Japan, in 1982 and 1984, respectively.

He joined NEC Corporation in 1984. He was engaged in research and development of integrated circuit/packet switching systems. He is currently working on high speed packet switching system architectures and implementations.

Mr. Suzuki is a member of the Institute of Electronics, Information, and Communication Engineers (IEICE) of Japan.

Takehiko Yamaguchi (S'67–M'72) received the B.E., M.E., and Ph.D. degrees from the University of Tokyo, Tokyo, Japan, in 1967, 1969, and 1972, respectively.

He joined NEC Corporation in 1972, and currently is Manager of the 1st Switching Systems Engineering Department, Integrated Switching Development Division. His main interests are traffic theory, digital communication systems, and ISDN switching systems.

Dr. Yamaguchi is a member of the Institute of Electronics, Information, and Communication Engineers (IEICE) of Japan.

Shin-ichiro Hayano was born in Osaka, Japan, in 1958. He received the B.E. degree from the Nagoya Institute of Technology, Nagoya, Japan, in 1981, and the M.E. degree from the University of Tsukuba, Ibaraki, Japan, in 1983.

Since joining the NEC Corporation in 1983, he has been engaged in research and development of broadband ISDN switching systems and intraoffice optical fiber links. He is currently a Research Engineer at the Communication Research Laboratory, C & C Systems Research Laboratories.

Mr. Hayano is a member of the Institute of Electronics, Information, and Communication Engineers (IEICE) of Japan.

Integrated Services Packet Network Using Bus Matrix Switch

Reprinted from *IEEE Journal on Selected Areas in Communications*, Vol. SAC-5, No. 8, October 1987, pages 1284-1292. Copyright © 1987 by The Institute of Electrical and Electronics Engineers, Inc. All rights reserved.

SATOSHI NOJIMA, EIICHI TSUTSUI, HARUKI FUKUDA, AND
MASAMICHI HASHIMOTO, MEMBER, IEEE

Abstract—For several years, Fujitsu has been researching and developing high-speed packet switching networks, the result being an integrated multimedia information networks architecture. This architecture has already been applied to LAN systems and can now be applied to wide area corporate networks thanks to the new developments described in this paper. The most important technology—the bus matrix switch—provides a quantum leap in processing capacity up to several Gbit/s (few millions of packets every second), while keeping a very low switching delay. The performance evaluation based on our prototype models is also considered in detail. For example, a voice delay of less than 20 ms is obtained for five-hop communication. The new technologies will greatly improve for computer networks and will also enable basic and broadband ISDN.

I. INTRODUCTION

PACKET switching technology can be used to construct multimedia integrated networks. In a packet switching network, information is handled uniformly based on packet flow. Also, a packet multiplexing scheme can make efficient use of a transmission line. These features and variable length packet transmission can offer the user freedom of communication quality and bandwidth.

To fully exploit the benefits of packet switching, for the last ten years Fujitsu has been researching packetized voice technology and integrated packet networks. We have already applied these technologies to LAN systems. Recently, technologies such as optical fibers have enabled high-quality and wide bandwidth transmission. This makes it possible to construct a broadband ISDN and multimedia network. To realize these networks, we can apply high-speed packet switching technology [1], an enhanced version of packet switching. To make this technology practical, the high-speed and high-capacity packet switch and a new network controlling scheme suitable for this switch are necessary.

In this paper, we propose a new switch based on a matrix network, and discuss the realization of a multimedia integrated network using this switch. Also, we have developed an experimental system. The configuration and results obtained from this system will also be discussed. In the final section, we discuss the approach to fabricating a multimedia packet public network.

Manuscript received November 26, 1986; revised May 13, 1987. This paper was presented at the 1986 Computer Networking Symposium, Washington, DC, November 17-18, 1986.

The authors are with the Computer Network Systems Laboratory, Fujitsu Laboratories Ltd., 1015, Kamikodanaka Nakahara-ku, Kawasaki 211, Japan.

IEEE Log Number 8716302.

II. HARDWARE PACKET SWITCH

To realize a high-speed packet switching network, a high-speed high-capacity packet switch is vital. Fig. 1 shows the configuration of a packet switching node. An I-PPU (input packet processing unit) and an O-PPU (output packet processing unit) interface an input and output packet port and perform packet assembly/disassembly and interfacing functions to trunk lines and terminals. The switch is defined as an internal network which has the following properties.

1) It should be able to provide a high-speed packet transfer path between any I-PPU and O-PPU.

2) Packet transfer on each path is performed in parallel.

3) To enable expandability up to an arbitrary capacity, a building block architecture should be used.

4) The switch should be capable of handling variable length packets.

To satisfy these requirements, various internal switch structures have been proposed. These include buses [2], shuffle networks [3], loops [4], meshed networks, and matrix networks. Our new switch is based on a matrix network.

A. Bus Matrix Switch

Fig. 2 shows the architecture of our proposed switch, which we have named the BMX (bus matrix switch). The PPD (primary packet distributor) interfaces incoming packet ports and the SPD (secondary packet distributor) interfaces outgoing packet ports. Each PPD and SPD has its own packet transmission bus. These buses are arranged in a matrix. At cross points of the incoming and outgoing buses, there is an XPM (cross-point memory) with packet queueing buffers.

The PPD receives a packet from the I-PPU, determines the destination XPM and outgoing SPD by referring packet transmission header, then sends it to the XPM. The SPD scans every XPM on its bus, removes packets from the XPM, and sends them to the O-PPU.

B. Functions and Features

1) Asynchronous and Independent Operation: Because XPM's are realized with dual port buffers, their inputs and outputs can operate independently. Thus, every PPD and SPD operates independently and asynchron-

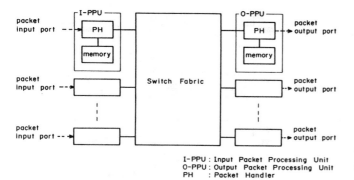

Fig. 1. Packet switching node.

I-PPU : Input Packet Processing Unit
O-PPU : Output Packet Processing Unit
PH : Packet Handler

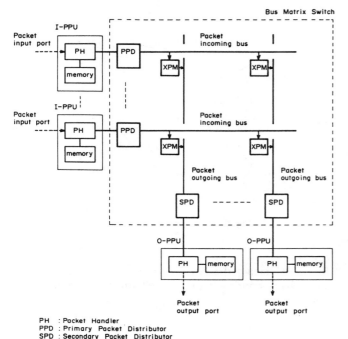

PH : Packet Handler
PPD : Primary Packet Distributor
SPD : Secondary Packet Distributor
XPM : Cross Point Memory

Fig. 2. Bus matrix switch.

ously. Thanks to this, the packet switching on each bus is realized in parallel. Also, the length of a packet can be variable.

2) Switching Using XPM: The basic XPM configuration can be realized using an FIFO (first-in first-out) buffer. If priority control is necessary, XPM can use RIFO (random-in first-out), FIRO (first-in random-out), or multiple FIFO. The XPM capacity should be determined depending on the network traffic design, which will not be described here.

3) Switching Path Definition: The switching path of a packet is defined by determining the relation between the PPD and one of the SPD's. Therefore, the PPD should have a switching definition table which performs translation from a packet transmission header to a destination SPD address. The SPD address can specify a single SPD or a group of SPD's. This group addressing enables broadcast functions to be easily achieved.

4) Switching Capacity: Using advanced C-MOS technology, a 50 ns cycle bus clock can be realized, so that if the bus width is 8 bits, the capacity of a bus will be 160 Mbits/s. Using ECL technology, the bus clock can be increased up to 10 ns, in which case the bus capacity will be 800 Mbits/s.

The number of buses can be increased as far as physical limits, such as fan-out, will allow. For C-MOS, the maximum number of buses could be increased up to 16, so that the total switch capacity could be expanded to 2.6 Gbits/s (160 Mbits/s × 16). For ECL, the fan-out limit is more critical so the maximum number will be 8. This limits the capacity to 6.4 Gbits/s (800 Mbits/s × 8).

To confirm the expandability by increasing the number of buses, we used computer simulation to evaluate the delay characteristics corresponding to the number of buses. The simulation conditions were as follows:

Bus speed: 50 Mbits/s
Transmission line: 30 Mbits/s for each bus
Packet length: 3–512 bytes, uniform distribution
Incoming packet interval: exponential distribution
    (average: 50 ms)
Packet destination: random.

Under these conditions, we measured the node transit time (from when a packet enters a queue in I-PPU to when it exits from an outgoing queue in O-PPU) and BMX transit time (from when a PPD removes a packet from an I-PPU queue to when SPD stores it in a queue in O-PPU). Fig. 3 shows the delay characteristics, where the "99 percent delay" is the delay time guaranteed for 99 percent of all transit packets. The upper graphs show the node transit time, and the lower graphs show the BMX transit time corresponding to the number of buses.

From Fig. 3, it can be seen that the switching delay is independent of the number of buses, and most of the delay occurs in the outgoing packet queue. On these conditions, the required capacity of XPM will be below 16 kbytes.

5) Hardware Size: We developed an experimental system (described in Section IV), and then estimated the amount of hardware necessary to implement the system using C-MOS gate array LSI technology. As a result, we determined that a single cross-point memory can be realized with 1000–1500 gates with 16 kbytes of memory, the PPD and SPD can be realized using 2500–3000 gates. We are now designing an LSI to provide the four XPM functions (2 × 2 matrix) in a single chip. Using this LSI, we will be able to fabricate a BMX sized 16 × 16 matrix on a single board.

C. Expandability

In the preceding section, we showed that the system can be expanded by increasing the matrix scale, but that it is limited by the hardware. Here, we discuss the switch configuration necessary to realize a broader capacity switching node. Generally, two types of requirements must be satisfied to expand the capacity.

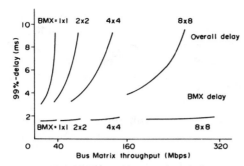

Fig. 3. Delay of bus matrix switch.

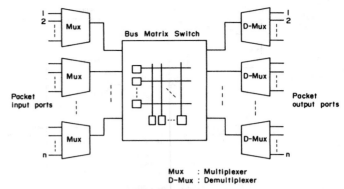

Mux : Multiplexer
D-Mux : Demultiplexer

Fig. 4. Multiplexed-type configuration.

1) Multiplexed Type: If the maximum speed of an accommodated packet port is low enough in comparison to the bus capacity, a single bus can accommodate several ports with a total bandwidth which is less than the bus capacity. Fig. 4 shows the configuration.

2) Grading Type: If the port speed is near the bus capacity, a grading scheme, commonly applied to cross-bar switches, can also be applied to the BMX switch. Using an $N \times N$ BMX switch as an element, multistage grading is possible. Using the three-stage configuration shown in Fig. 5, the speed of linking between the stages may be the same as the input/output port speed.

Type-1) configuration is suitable for a local switch which accommodates many low-speed subscriber loops. Type-2) is applicable to tandem or broadband subscriber switches.

III. ARCHITECTURE OF THE HIGH-SPEED PACKET SWITCHING NETWORK

In this section, the high-speed packet switching network, based on the BMX switch, will be described.

Recently, wide area corporate networks (WCN) have become popular in Japan because of the installing of high-speed digital transmission lines, spurred by the increasing need to connect LAN's, PBX's, and other equipments. And in the WCN, a multimedia integrated, reliable network is required. Therefore, we chose WCN as the network to consider.

An example of a WCN is the Corporate Information Network System (COINS) supplied by Fujitsu. By analyzing the various COINS networks in use, the following WCN outline was obtained.

1) Network Scale:
- Number of nodes: 2–50
- Throughput at relay node: max. 6.3 Mbits/s × 15 lines (80–90 Mbits/s)
- Number of relay nodes: 6 hops maximum
- Network topology: star topology including loop.

2) Communications:
- Services: voice, video, and data
- Bandwidth: 1200 bits/s–1.5 Mbits/s.

Most WCN's consist of local distribution networks (LDN) such as PBX's and LAN's, and back-bone networks that integrate the communication between LDN's.

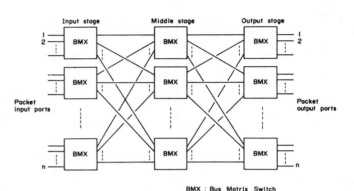

BMX : Bus Matrix Switch

Fig. 5. Grading-type configuration.

Here, we discuss the construction of a multimedia integrated back-bone network, based on high-speed packet switching technology.

A. System Configuration

Fig. 6 shows the network configuration that we propose. We call this the Network Access Processor Network (NAP-NET). This network consists of packet switching nodes (NAP-nodes) that have a BMX packet switch, and high-speed digital transmission lines (e.g., 6.3 Mbits/s).

1) Switching Node Configuration: As shown in Fig. 7, each NAP-node consists of a BMX switch (BMX), a node control processor (NCP), digital trunk interface modules (DI's), and line sets (LS's). The functions of these components are as follows.

A) Line Set and Digital Trunk Interface Module: The LS acts as the network edge. LS's perform packet assembly/disassembly, signaling, and miscellaneous functions such as delay adjustment. These functions are selected depending on the terminal characteristics. DI's interface with the trunk lines and perform serial to parallel or parallel to serial conversion and packet format conversion.

b) Node Control Processor: The NCP performs call processing and other administrative functions. It is connected to a BMX and can communicate with any LS/DI.

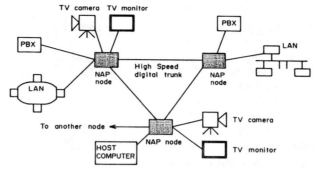

Fig. 6. Configuration of the NAP-NET.

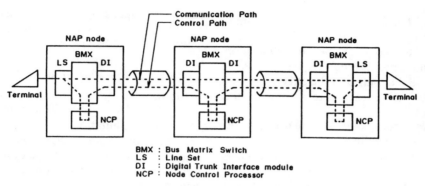

BMX : Bus Matrix Switch
LS : Line Set
DI : Digital Trunk Interface module
NCP : Node Control Processor

Fig. 7. Fundamental architecture of the NAP-NET.

B. Controlling the Multimedia Packet WCN

1) Packet Format: In the NAP-NET, all information is packetized into a uniform variable length packet format at the LS. Fig. 8 shows the packet format in the NAP-NET. This is similar to the HDLC frame structure. A packet has a TH (transmission header) field, a data field, and a 16 bit frame check sequence (FCS), all of which are encapsulated by flag sequences. The TH includes a logical channel number (LCN) which we will describe later. The flags and frame check sequence (FCS) are added at the DI, and are checked and removed at the DI of the receiving node.

2) Accommodation of Various Communication Types: The NAP-NET offers three types of communications for various media.

a) Circuit Emulation Type: For continuous communication such as voice and video, the NAP-NET offers an emulated virtual circuit using continuous packet flow. The packet length and packet generation interval is determined according to the communication.

b) High-Speed Virtual Call for Data: This type is the same as the communication realized in a conventional packet switching network. The NAP-NET can offer a conventional interface such as the X.25. The NAP-NET can offer high-speed (up to several Mbits/s) and low delay communication.

c) Datagram Communication: The NAP-NET offers various LAN interfaces, and enables packet routing and distribution for communication between LAN's so that the

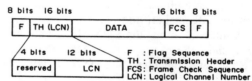

F : Flag Sequence
TH : Transmission Header
FCS: Frame Check Sequence
LCN: Logical Channel Number

Fig. 8. Packet format.

NAP-NET can realize a gateway function. This can give free inter-LAN communication.

In types b) and c), computers and terminals provide high-level protocols which guarantee flow control and process synchronization at both ends. So the network need provide only high-quality and low-delay transmission and packet sequence integrity. However, for circuit emulation, the network should offer all functions including the functions required by types b) and c). Therefore, the means of attaining this circuit emulation is discussed.

3) Communication and Control Path: The packet flow in a circuit emulation is classified into two groups. One is the control path which carries call processing and other administrative information, while the other is for actual communication. To satisfy the requirements of high-speed and low-delay times, communication flow and control flow are processed separately in the NAP-NET. The control packets terminate at the NCP's and are processed, but communication packets are processed only by the BMX's. To distinguish between control paths and communication paths in the BMX, a specific LCN (for example, number 0) is reserved for each trunk line and each LS.

4) Logical Channel Number (LCN) and Routing Table: There are various ways of defining a TH (transmission header) which indicates the switching route in a network. In the NAP-NET, to reduce the header overhead, an LCN (logical channel number) is defined as switching information. A packet's LCN is replaced at each node. Every LS and trunk line (DI) has an LCN number space, so that virtual calls are specified by the set of LCN's that are hunted at the LS at both ends of the path and at tandem trunk lines. Each node has a routing table which defines the virtual call route by translating from the input LCN to the new LCN for the next node. The contents of a routing table are updated by the NCP whenever a virtual call is set up and cleared.

5) Administration of Trunk Bandwidth: In a multimedia packet network in which various bandwidths are mixed, it is necessary to supervise the communication bandwidth of each trunk line. The call request packet in the NAP-NET has a field that indicates the effective bandwidth of the call. The NCP uses this to administer the loads on the trunk lines.

6) Virtual Call Setup Procedure: A virtual call setup sequence in NAP-NET is shown in Fig. 9. An LS receives a call setup request from a terminal, hunts an LCN, transforms the request into a CR (call request) packet, and sends it to the NCP through the control path between the NCP and the LS/DI. When the NCP receives the CR-packet, it determines the outgoing trunk line by referring to the destination address, the effective bandwidth information in the CR-packet, and the trunk line load condition. Then the NCP hunts the LCN of the outgoing trunk line and updates the routing table. Then the NCP transmits a CR packet to the adjacent node NCP using the fixed permanent virtual circuit (PVC) for a control path which is set up between each pair of adjacent NCP's (see Fig. 7).

This procedure is repeated by each node to set up a virtual call. If a hot-line connection is necessary, the PVC is set up when the network starts up.

C. Edge Function

The LS performs the edge functions and realizes various interface functions and user selectable functions. For example, delay adjustment is performed at a receiving LS for voice communication.

In the NAP-NET following delay adjustments scheme is realized. At a packet-receiving end LS, there is a packet-receiving FIFO, which absorbs packet delay fluctuation. Before the call is set up, the FIFO contains dummy samples whose length corresponds to 99 percent of the delay difference. Just after the call is set up, the first packet which arrives at the receiving end is put into the FIFO behind the last dummy sample. Each packet has a sequential number to assure time transparency.

To fully exploit high-speed packet switching, many new edge functions will be used. One of these is a variable rate coding scheme of voice and video which generates

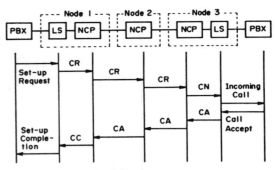

Fig. 9. Call setup sequence.

packets corresponding to the information source activity. This will offer users communication services of fixed quality, not fixed bandwidth.

D. Fault Tolerant Operation of the Network

In future networks, fault tolerant operation will be indispensable. NAP-NET offer tolerant operation using two types of reconfiguration.

1) Duplicate Path Assignment: At the time of call setup, two different paths are set up for a single communication. One is for normal operation and the other is for stand by.

2) Fast Path Selection: When a line or node fault occurs, all communication paths will be reconnected using an active part of the network.

Using method 2), packet loss is unavoidable until all paths are reconfigured. If the communication demands no packet loss when a fault occurs, a duplicate path should be assigned. To realize method 2) perfectly, a great number of routing tables for all possible faults is necessary. The reducing of routing table size will be a further study item.

IV. EXPERIMENTAL MODEL

A. Overview

To determine the exact design parameters of the NAP node, and to measure the switching characteristics more exactly, we developed an experimental model. The system configuration is shown in Fig. 10, and the specifications of the experimental model are given in Table I. The appearance is shown in Fig. 11.

The switch (BMX) is of a multiplexed type because this system is oriented to WCN. The switch configuration is shown in Fig. 12. The PPD (primary packet distributor) and SPD (secondary packet distributor) have a bus which accommodates several packet input and output ports (LS/DI). The LS's/DI's are interfaced to this bus with FIFO. The FIFO-A is for the temporary storage of incoming packets from an LS/DI, and the FIFO-C is for outgoing packets. The XPM (cross-point memory) is realized by FIFO, which in the figure is designated as FIFO-B.

In this system, the switching definition table and routing table are realized in the PPD. The PPD updates the

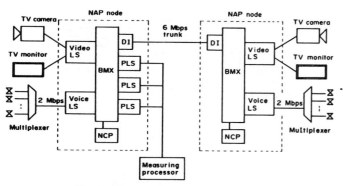

Fig. 10. Experimental NAP-NET system.

TABLE I
SPECIFICATIONS OF THE EXPERIMENTAL NAP NODE

Item	Experimental NAP node
Scale of BMX	2 x 2
Bus speed	50 Mbps (6.25 Mbps x 8 bits)
Packet length	3~512 bytes (arbitrary)
Throughput	97 Mbps (packet length= 512 bytes) 80 Mbps (packet length= 64 bytes)
FIFO capacity FIFO-A FIFO-B,C	1024 bytes 8 Kbytes (higher priority) +8 Kbytes (lower priority)
Technology	TTL-MSI, SSI 16 Kbit -SRAM

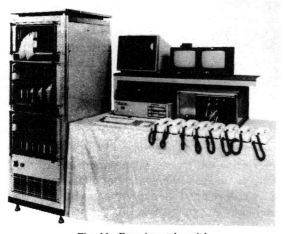

Fig. 11. Experimental model.

LCN of the packet and adds the FIFO-C address at the head of the packet.

We also developed a voice and a video line set. A voice LS accommodates a group of voice channels. At the voice channel multiplexer, 30 voice channels are time division multiplexed on a 2 Mbit/s transmission line. A video LS accommodates a channel of up to 4.7 Mbit/s video. Details of those LS's are given in Table II.

Also in the experimental system is a measuring subsystem that generates load and test packets and measures the

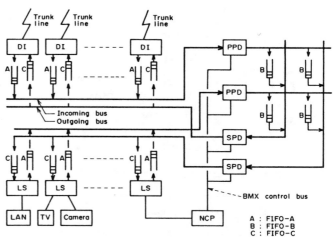

Fig. 12. Bus matrix switch (2 × 2).

A : FIFO-A
B : FIFO-B
C : FIFO-C

TABLE II
DETAILS OF LS's

Item	Voice LS	Video LS
Function	PAD for each channel	Digitize, compress and PAD
Throughput	2 Mbps (30 channels)	4.7 Mbps
Packet Length	16~128 bytes (Presettable)	16~256 bytes (Presettable)
Special Feature	Time division multiple PAD operation by hardware	PAD operation by hardware

packet delay. The measuring subsystem is composed of pseudo-LS's (PLS's) and a measuring processor. The PLS generates packets according to predefined operation table in which packet length and traffic density are defined.

As described previously, the number of buses do not affect the characteristics of the node, so we measured the characteristics using a 2 × 2 BMX.

B. Single Node Characteristics

We measured the packet delay, the packet discard rate, and the packet accumulation of each FIFO of a BMX. The measuring conditions were as follows.

1) A BMX connects with twelve 6.144 Mbit/s trunk lines.

2) The packet length is 256 bytes fixed (not including flags nor an FCS).

3) The destination address of each packet is random.

4) The packet generation interval is random.

5) The speed of each bus is 50 Mbits/s, so the total capacity of each node is 100 Mbits/s.

Fig. 13 shows the results comparing the mean and 99 percent delay. It can be seen that the switching delay is less than 6 ms even under a heavy load where the traffic density of the trunk line is 96 percent. Under these conditions, the total packet throughput of the BMX reaches 70 Mbits/s.

Fig. 14 shows the results of comparing the packet accumulation properties of an FIFO-B and an FIFO-C. It can be seen that the packet accumulation of an FIFO-B is

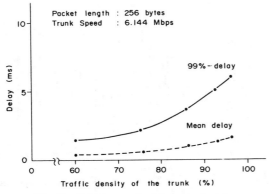

Fig. 13. Delay versus load for a single node.

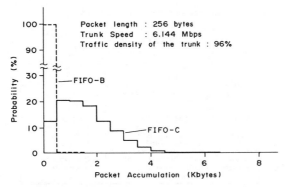

Fig. 14. Packet accumulation probability density function.

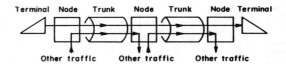

Fig. 15. Worst delay model (2 hops).

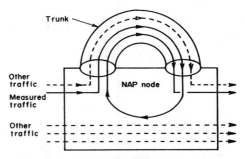

Fig. 16. Equivalent delay model as in Fig. 15.

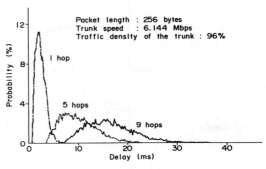

Fig. 17. Probability density function of delay corresponding to the number of hops.

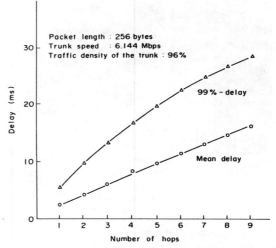

Fig. 18. Delays versus number of hops.

much smaller than that of an FIFO-C. This agrees with computer simulation results.

C. Cascade Connection

To evaluate the characteristics with cascade connections, we measured the end-to-end delay. To measure the worst case characteristics, we assumed a network model. In the NAP-NET, a packet may be delayed at tandem nodes along its route, when other packets with the same destination may be injected at the same tandem node. Therefore, we considered the model as shown in Fig. 15 as the worst case. There, packets of the other virtual calls join the route with the test packets at every trunk line entrance and leave at the next node. The model in Fig. 15 is simplified as shown in Fig. 16. The measuring conditions were the same as these described in the preceding section. The traffic density of the trunk line was fixed to 96 percent.

Experimental results of delay distribution to the number of hops are shown in Fig. 17; and mean delay and 99 percent-delay to the number of hops are shown in Fig. 18. It can be seen that the mean delay increases in proportion to the number of hops, but the 99 percent delay does not.

D. Voice Delay

This section considers voice delay properties based on the experimental results described in the previous section.

The voice delay consists of the packet assembly time, the 99 percent delay of the packet switched network, and 99 percent delay distribution. The experimental results have assumed a fixed packet length and steady traffic, so the results can generally represent the worst case when the network accommodates continuous nonbursty data such as voice and video. The network delay is propor-

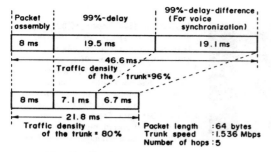

Packet assembly	99%-delay	99%-delay-difference (For voice synchronization)
8 ms	19.5 ms	19.1 ms

46.6 ms
Traffic density of the trunk = 96%

| 8 ms | 7.1 ms | 6.7 ms |

21.8 ms
Traffic density of the trunk = 80%

Packet length : 64 bytes
Trunk speed : 1.536 Mbps
Number of hops : 5

Fig. 19. Classification of the voice delay.

tional to the packet length and inversely proportional to the trunk line speed so that it can be estimated from the results. The packet assembly time can be calculated from the voice encoding speed and packet length.

The voice delay is classified as shown in Fig. 19. For 5 hops, and a packet length of 64 bytes, we confirmed an end-to-end delay time of 21.8 ms for 99 percent of packets.

From the above experimental results, it can be concluded that the delay of the NAP-NET is small enough to satisfactorily accommodate voice.

V. CONCLUSION

We proposed a new high-speed packet switch called BMX (bus matrix switch), and we showed that the packet switch can be expanded to an arbitrary limit, that it can be used not only for WCN, but also for much higher throughput networks such as future public networks. We discussed the wide area corporate network based on high-speed packet switching technology and proposed a new control scheme. And also, we demonstrated this switch's high performance by fabricating an experimental system.

To fully utilize the multimedia packet network, further study is required. For example, regarding the network architecture, we should reconsider the functions shared between the network and other equipment (e.g., computers, terminals, and local distribution networks); and for communication function, new coding schemes, such as variable rate coding for voice and video, will be important.

From now, we will be studying these problems, with the aim of realizing a public packet switching network in the future. To realize such a network, we have to consider the following.

As a first step, the proposed wide area corporate network based on high-speed packet switching technology will be realized. Using this network, terminals and communication features suitable for the network will be developed. The high-speed packet switching technology may also be applied to packet switching systems within the ISDN. The results obtained can then be used to help develop a standard. Finally, a multimedia public network based on the high-speed packet switching technology can be built.

We believe that wide area corporate networks are only the first step toward the future multimedia networks, and

that there are points still to be investigated, although most of the major problems have been solved.

ACKNOWLEDGMENT

The authors thank the reviewer for his comments. Many of our colleagues at Fujitsu Laboratories have assisted us in this work. Also, we particularly thank H. Takanashi and R. Yatsuboshi for their helpful suggestions.

REFERENCES

[1] J. S. Turner and L. F. Wyatt, "A packet network architecture for integrated services," presented at Globecom-83, Dec. 1983.
[2] A. G. Fraser, "DATAKIT-A modular network for synchronous and asynchronous traffic," in *Proc. ICC*, June 1979, pp. 20.2.1–20.1.3.
[3] J. J. Kulzer and W. A. Montgomery, "Statistical switching architectures for future services," presented at ISS-84.
[4] H. Suzuki *et al.*, "Very high speed and high capacity packet switching for broadband ISDN," in *ICC '86 Proc.*, June 1986.
[5] W. L. Hoberecht, "A layered network protocol for packet voice and data integration," *IEEE J. Select. Areas Commun.*, vol. SAC-1, Dec. 1983.
[6] T. Aoyama, T. Takahashi, H. Ueda, R. Yatsuboshi, S. Suzuki, and S. Nojima, "Packetized service integration network for dedicated voice/data subscribers," *IEEE Trans. Commun.*, vol. COM-29, Nov. 1981.
[7] T. Tsuda, S. Hattori, R. Yatsuboshi, and K. Yamauchi, "An approach to multiservice subscriber loop system using packetized voice/data terminals," *IEEE Trans. Commun.*, vol. COM-27, July 1979.
[8] A. Thomas, J. P. Coudreuse, and M. Servel, "Asynchronous time-division techniques: An experimental packet network integrating video communication," in *ISS '84 Proc.*, May 1984.

Satoshi Nojima was born in Tokyo, Japan, on March 24, 1954. He received the B.E. and M.E. degrees in electrical engineering from Waseda University, Tokyo, Japan, in 1976 and 1978, respectively.

Since 1978 he has been with Fujitsu Laboratories Ltd., Kawasaki, Japan, where he has been engaged in research and development on local communication networks, packetized voice communication, and high-speed packet communication systems.

Mr. Nojima is a member of the Institute of Electronics, Information, and Communication Engineers of Japan.

Eiichi Tsutsui was born in Kohchi, Japan, on January 8, 1955. He received the B.E. and M.E. degrees in electrical engineering from Yokohama National University, Japan, in 1978 and 1980, respectively.

In 1980 he joined Fujitsu Laboratories Ltd., Kawasaki, Japan. He has since been engaged in research of multimedia packet communication systems and local area networks.

Mr. Tsutsui is a member of the Institute of Electronics, Information, and Communication Engineers of Japan, and of the Information Processing Society of Japan.

Haruki Fukuda was born in Tokyo, Japan, on November 24, 1949. He received the B.E. degree in electrical engineering from Yokohama National University, Japan, in 1972.

In 1972 he joined Fujitsu Laboratories Ltd., Kawasaki, Japan. He has since been engaged in research of local area networks and high-speed packet communication systems.

Mr. Fukuda is a member of the Institute of Electronics, Information, and Communication Engineers of Japan.

Masamichi Hashimoto (M'82) was born in Ibaragi, Japan, on December 4, 1946. He received the B.E. degree in electrical engineering from Iwate University, Japan, in 1969.

He joined Fujitsu Ltd., Kawasaki, Japan. Since 1970 he has been with Fujitsu Laboratories Ltd., where he has been engaged in research and development on satellite and radio communication systems, fiber optic local area networks, and high-speed packet communication systems. He is currently a Section Manager of the Computer Network Systems Lab.

Mr. Hashimoto is a member of the Institute of Electronics, Information, and Communication Engineers of Japan.

Chapter 6:

Broadcast Switching Networks

6.1 Introduction

The services to be provided by broadband ISDN are classified by CCITT into *interactive services* and *distribution services*. In interactive services there is a bidirectional exchange of information between two users. Interactive services can be further subclassified into conversational services, messaging services, and retrieval services. Examples of conversational services are video conference, video telephony, and high-speed data. Messaging services are intended to support applications such as voice, video and electronic mailing, and mailbox services. The purpose of retrieval services is to let users retrieve, on demand, a wide variety of information from information centers.

Distribution services are meant to support applications in which there is a unidirectional flow of information. These services are of two types: distribution services without user presentation control and distribution services with user presentation control. In distribution services without user presentation control, the user has no control over the presentation of information by the service provider. The user can only access the information. Broadcast services for television and audio programs are examples of this type of service. On the other hand, distribution services with user presentation control let the user control the presentation. The user can control, for example, the order of the presentation. An example of this type of service is full-channel broadcast videography.

To support this diversity of applications, broadband ISDN should provide for point-to-point connections and multipoint connections. The ability to broadcast is crucial for some of these applications. Distribution services of both types cannot be supported by point-to-point networks. Of the two–circuit switching and packet switching–the adaptability of packet switching makes it the clear choice. Connection-oriented packet switching is needed for broadcast services. Users should be able to connect to and disconnect from the broadcast channel at any time. Broadcast switching networks need to be rearrangeable nonblocking networks. The switch fabric must be capable of packet replication and switching.

6.2 Overview of Papers

The importance of packet replication and switching to support such services as entertainment video and teleconferencing in the emerging broadband networks is the theme behind the papers in this chapter.

In "Design of a Broadcast Packet Switching Network," Turner describes a network architecture based on packet switching to integrate voice, data, and video communication. The switch fabric, which is the main component of the switch module, contains a copy network, broadcast and group translators, a distribution network, and a routing network. A fixed packet size of 600 bytes is used, and Turner proposes three different schemes for congestion control. The paper discusses in detail the network protocol that accommodates both connection-oriented and connectionless services, the associated packet formats, and the architecture of the switch module. Switch implementation issues and the broadcast capability of the network are also presented.

The second paper, "A New Broadcast Switching Network," by Lea, discusses the different approaches to broadcast network design and proposes a new broadcast network based on a cascading scheme for application to circuit-switched video services. Lea proposes that the multiconnection function consist of two subfunctions–consecutive spreading and routing–and that the network be built from a consecutive spreading network plus a Benes-Clos routing network. The author describes a photonic implementation of the network using directional couplers and discusses performance issues, including the number of stages, the number of crosspoints, growth, and routing. The paper also presents a self-routing version of the network and a hybrid approach to its implementation combining an optical interconnection network for data transmission and an electronic self-routing network for controlling the electrodes.

The third paper, "Nonblocking Copy Networks for Multicast Packet Switching," by Lee, discusses the disadvantages of the broadcast networks proposed by Turner (in the first paper in the chapter) and the Starlite network. Lee proposes a nonblocking, self-routing network with constant latency. A four-phase architecture consisting of a running adder network, a dummy address encoder, a broadcast banyan network, and a trunk number translator is proposed for multicast packet switches. Two problems inherent in the design of such space-division-based switches–copy network overflow and output port conflict–are discussed and solutions are proposed.

Design of a Broadcast Packet Switching Network

JONATHAN S. TURNER, MEMBER, IEEE

Abstract—Communications applications are becoming more diverse all the time. Current switching technologies, however, have been designed around single applications or small classes of applications. There is a growing need for communications systems that can handle a heterogeneous and dynamically changing mix of applications. This paper provides an overview of a system designed to meet this need. It is based on fiber optic transmission systems and high-performance packet switching and can handle applications ranging from low-speed data to voice to full-rate video. The most novel feature is a flexible multipoint connection capability suitable for broadcast and conferencing applications. The paper describes the architecture of a switching system that can be used to support this network.

I. INTRODUCTION

THIS paper proposes a design for a high-performance packet switching network that can be used to provide voice, data, and video communication on a large scale. A novel feature of the system is its flexible multipoint connection capability, which makes it suitable for a wide range of applications, including commercial television distribution and voice/video teleconferencing. The basic switching capability is provided by a high-performance packet switch called a *switch module* (SM) which terminates a set of fiber optic communications links (FOL), each operating at a speed of 100 Mbits/s. The SM is designed as a modular component that can be used to construct systems ranging from a medium-sized business system to a large packet switching exchange, capable of providing service to over 200 000 users. The paper provides a high level description of the network as a whole plus a detailed description of the design of the switch module hardware.

The proposed system was motivated by the observation that while communications applications are becoming increasingly diverse and dynamic, current systems are ill-equipped to handle a heterogeneous and changing application mix. This is because they have been designed around specific applications and tailored to those applications so closely that adding new applications is usually very difficult. The system described here was designed to handle a wide class of applications ranging from low-speed data to full-rate video with equal ease. Its most novel feature is its ability to handle multipoint connections. This feature facilitates a variety of new applications that cannot be provided effectively on conventional point-to-point networks. This research is an outgrowth of earlier work that the author and colleagues did on the design of large-scale packet switching systems for voice and data at Bell Laboratories. Details of that earlier work can be found in references [8], [10], [12], and [14]–[22]. Similar work has been done at Bolt, Beranek, and Newman [13].

This work has been influenced by two very different approaches to communications and switching. The first is

Paper approved by the Editor for Communication Switching of the IEEE Communications Society. Manuscript received October 8, 1985; revised July 2, 1987. This work was supported by the National Science Foundation under Grant DCI 8600947, and under Grants from Bell Communications Research, Italtel SIT, and NEC. This paper was presented at INFOCOM '86, Miami, FL, April 1986.

The author is with the Department of Computer Science, Washington University, St. Louis, MO 63130.

IEEE Log Number 8820344.

circuit switching as practiced in current telephone systems and the second is packet switching as practiced in computer networks. The inherent flexibility of packet switching makes is an obvious candidate for a network designed for diverse applications. On the other hand, current packet switching systems compare poorly to circuit switching systems on the basis of speed, throughput, and cost. We (and others) have observed that this relatively poor showing is not inherent to packet switching but is an artifact of current implementations which are largely software systems running on general purpose computers. Circuit switching systems, on the other hand, have always been designed using special purpose switching networks. To match the speed, throughput, and cost of circuit switching systems in a packet switched network, one must resort to the same techniques—special purpose switching networks implemented using high-speed hardware. A third major influence on this work has been the design of interconnection networks for parallel computers (see [7]). The essential component in our switching network is a buffered banyan network, tailored to the particular needs of a large-scale communications system. Some earlier work has been done on broadcast and multipoint routing of packets in computer networks (see, for example, [3], [4], [6]), but this work has little direct bearing on our research. The Prelude project described in [5] has similar goals to our work. Their system uses a switching technique called asynchronous time-division multiplexing, which is similar in spirit to ours, but differs greatly in details. Huang and Knauer [9] offer a different approach to switching high-speed broadcast channels.

The major components of the proposed network are identified in Fig. 1. Access to the network is provided through *network interfaces* (NI), which provide concentration, local switching, network protection, and accounting functions. Switching is provided by *packet switches* (PS), which are interconnected using *fiber optic links* (FOL). The PS's and NI's can be configured in a variety of sizes and configurations. A later section gives example configurations of both. FOL speeds in the neighborhood of 100 Mbits/s are achievable with current technology, and this transmission speed is used as a target for the designs presented below.

The network provides three basic communications services.

• *Point-to-Point Channels:* These are two-way channels joining pairs of users. The user requests a channel capable of providing a certain average bandwidth and the network allocates the requested bandwidth to the connection. If the requested bandwidth is not available, the connection is blocked. These connections do not provide flow control or error correction.

• *Datagrams:* These are individual packets, not associated with a preestablished connection. The network makes an effort to deliver them, but does not guarantee delivery.

• *Broadcast Channels:* Any user can set up a broadcast source that other users can then connect to. The average bandwidth of the source must be specified when it is established, and can range from a few bits per second to over a gigabit per second. There is no limit on the number of users that can receive a given broadcast signal. Thus, it is suitable for a variety of applications including commercial television and voice or video conferencing.

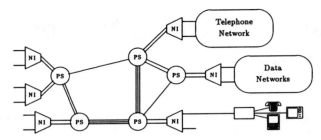

Fig. 1. Network architecture.

Fig. 2 illustrates the way broadcast channels work. Two broadcast sources are shown at the top of the figure. At various points in the network, the signals are split into multiple copies which are ultimately delivered to the appropriate users. Thus, each broadcast source and its associated connections induce a tree in the network. When a user disconnects from a broadcast channel, the corresponding branch of the tree is removed. When a user requests connection to a channel, the network attempts to find a nearby branch of the broadcast tree and connects the user at the nearest point. Thus, the broadcast trees grow and shrink as usage patterns change, but for widely distributed channels, one can expect the bulk of the activity to be concentrated near the leaves.

A special-purpose switching fabric is used to support the broadcast service. This switch fabric splits a broadcast channel into multiple parts as needed. The switching fabric also allows a set of communications links to be treated as a group with traffic distributed over all links in the group. This makes it possible for the network to provide communications channels that require more bandwidth than is available on a single FOL.

The network is predominantly connection-oriented, allowing bandwidth allocation on a per connection basis. Because the network is based on packet switching, the user consumes bandwidth only when actively sending information. This means that the instantaneous bandwidth usage varies over time. In order to permit the network to allocate bandwidth appropriately, users must specify their bandwidth requirements when a connection is established. The bandwidth specification must include average bandwidth requirements plus a measure of the burstiness of the user's application. Given such a specification, the network can allocate appropriate resources and can guarantee the availability of those resources for the duration of a connection, ensuring users adequate performance for their needs. (A detailed treatment of the bandwidth specification and allocation mechanisms is beyond the scope of this paper.)

Three mechanisms for congestion control are proposed. The primary one is connection blocking—new connections are refused if adequate bandwidth is not available. An enforcement mechanism at the NI's ensures that users do not exceed their allocated bandwidth. The second mechanism is a system of user-specified packet priorities. During periods of heavy load, low-priority packets are preferentially discarded to reduce congestion. This mechanism can be very effective if low-priority packets constitute a significant portion (say > 10 percent) of the total network traffic. The third mechanism is a congestion control field within each packet which can be modified by the network to inform the users of internal congestion. Users are expected to respond to this by temporarily slowing down their rate of transmission. Enforcement is provided at the NI's. Note that the first congestion control mechanism operates on a fairly long time scale, the second on a very short time scale, and the third on an intermediate time scale.

The system's communications protocols have been designed to facilitate the implementation of high-performance hardware protocol processors and to exploit the high-speed and low-

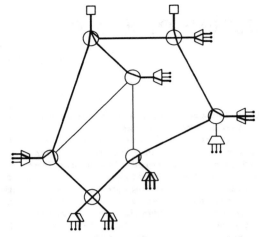

Fig. 2. Broadcast trees.

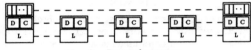

Fig. 3. Protocol structure.

error rates of the fiber optic links. The protocols are divided into three levels—the *link level*, the *network level*, and the *user level*. The interrelationships among the different levels are indicated in Fig. 3. There is a single link level protocol, two network level protocols (one for connection-based communication and one for datagrams), and a variety of user level protocols. The user level protocols are not discussed in detail here, but the intention is that these protocols would be application-dependent and built on top of one of the two network level protocols. One of these would be a telephony protocol. Another might be an internet protocol such as the DARPA IP to facilitate communication among different data networks. Still another might be a connection-oriented internet protocol.

There are essentially three packet formats of interest to the network—*connection control packets, data transfer packets,* and *datagram packets.* These packet formats are illustrated in Fig. 4. All packets are the same length; this substantially simplifies the design of high-performance protocol and switching hardware.

Aside: Fixed length packets immediately raise the question of what packet length is most appropriate. The answer is necessarily a compromise, since no one length is optimal for all applications. Applications like voice are best served by fairly short packets; a good choice is a packet length long enough for 10–20 ms worth of digitally encoded speech. Assuming 64 kbit/s coding, this translates to 640–1280 bits of speech per packet. High-speed applications like video are best served by much larger packets. Datagrams, which require a fairly high overhead per packet are also best served by long packets. Long packets also make it easier to design high-speed switching systems, since they give per packet control operations a relatively long time to take place; because data and control operations take place on very different time scales, it becomes easier to speed up the data path. Given these conflicting requirements, how does one strike a compromise? The choice proposed here is to use packets with a length of about 600 bytes. This is long enough to carry 512 bytes of data along with a generous allocation for the overhead required by common end-to-end protocols. Applications like voice would not be able to use such long packets efficiently, but would have to transmit packets that were largely empty. While this would

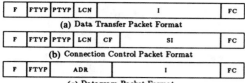

F	FTYP	PTYP	LCN	I	FC

(a) Data Transfer Packet Format

F	FTYP	PTYP	LCN	CF	SI	FC

(b) Connection Control Packet Format

F	FTYP	ADR	I	FC

(c) Datagram Packet Format

Fig. 4. Packet formats.

lead to inefficient bandwidth use for voice applications, the efficiency of the network as a whole could still be very good, since we expect that most of the packets carried by the network would belong to high-speed connections. In fact, high-speed networks such as we are proposing only make sense if this expectation is true. The decision on short versus long packets boils down to a choice between what is best in the short term versus what is best in the long term. Short packets are best for the mix of traffic that is typical today. Long packets are best for the higher speed mix of traffic expected in the future and will better enable us to speed up switching systems in order to match the speed increases in transmission. The choice proposed here is a compromise; while not optimal in the short term, it is acceptable and will become increasingly attractive as high-speed applications grow and technology pushes us to higher speed switching.

Link Level Protocol: The link level protocol provides frame delimiting, link transparency, error detection, and congestion control, but not error correction. There are two fields used by the link level protocol, the frame type field (FTYP) and the frame check field (FC). The FTYP field is one byte long and identifies the frame as a test frame, a datagram, or a frame belonging to a connection. It also has a subfield containing a user-specified priority. The network preferentially discards low-priority frames when internal network buffers are in danger of overflowing. The frame check field (FC) uses a two-byte cyclic redundancy code to detect errors in the frame. Frames are separated by flags (*F*). Bit stuffing is not used—the flag may appear within the body of the packet. Since the packet length is fixed, frame synchronization is straightforward. Frames with errors can be discarded.

Aside: Packets with errors are discarded, two approaches are possible: one is to discard packets with errors in the header and the other is to discard packets with errors in any field. The first approach can lead to a somewhat smaller packet loss rate, but is complicated by the fact that there are several different packet formats to contend with. The low-error rates typical of fiber optic transmission facilities lead to low packet loss rates even if packets are discarded on nonheader errors. This observation, together with its relative simplicity leads us to favor discarding packets if an error is detected in any field.

Network Level Protocols: There are two protocols at the network level, a datagram protocol and a connection protocol. Packets using the connection protocol contain a one byte packet type field (PTYP) and a two-byte logical channel number field (LCN). The packet type field identifies each packet as either a data packet or a control packet and contains a congestion control subfield used to inform the NI's and users of internal network congestion. The logical channel number field identifies which connection a packet belongs to. Data packets contain user information. Control packets are used to establish and control connections. Datagram packets contain an eight-byte address field which defines a hierarchical address space organized on a geographical basis, to permit routing of connections and datagrams based on local knowledge of network topology.

The remainder of the paper describes the components that make up the network. Section II describes the design of the switch modules. Section III shows how the SM's can be used to construct network interfaces and packet switches. Section

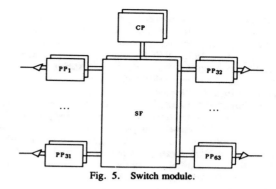

Fig. 5. Switch module.

IV offers some conclusions and outlines extensions to the SM that provide additional capabilities.

II. DESIGN OF THE SWITCH MODULE

The overall structure of the switch module (SM) is shown in Fig. 5. The SM is designed as a building block to be used in constructing larger systems.

The *switching fabric* (SF) is the heart of the SM. It has a highly parallel organization and provides multiple paths for packets to ensure good performance. The SF also replicates broadcast packets as necessary and can support distribution of packets across a set of outgoing FOL's. We have chosen to dimension the SF with 64 input and output ports; this size is large enough to provide adequate flexibility in configuring large systems, while being small enough to make high-speed synchronous operation easy to achieve and to support high reliability with a simple duplication scheme. Other choices are of course possible.

The *packet processors* (PP) perform the link level protocol functions, including the determination of how each packet is routed. This includes routing connection control packets to the connection processor and datagram packets to one of several datagram routers (not shown).

The *connection processor* (CP), is responsible for establishing connections, including both point-to-point and broadcast connections. To do this, it exchanges control packets with CP's in neighboring SM's and controls the actions of the PP's and SF by writing information in their internal control tables. It also performs a variety of administrative and maintenance functions.

The *datagram routers* (DR), which are not shown in the figure, are special-purpose devices used to route datagrams. The number of DR's can be engineered to suit the traffic. Each one occupies a port on the SF, replacing one PP. Since the vast majority of the traffic is expected to be connection-oriented, a few DR's (≤ 5) should suffice for most situations.

The duplication scheme shown in the figure is designed to ensure the highly reliable operation required in a large scale public communications network. The SM has two independent planes, each containing a CP, SF, and full complement of PP's. The normal mode of operation is for one plane to be active with the other acting as a "hot standby." A memory update channel is provided between the two CP's so that the memory of the standby CP tracks that of the active one. The standby CP executes an audit program designed to ensure that the tables in the PP's on the standby side also track those in the active side. Optical switches control which plane is connected to the FOL's. Note that the entire plane is switched as a unit. A system reconfiguration would normally cause a loss of all packets in the active switch plane before reconfiguration, but would cause no lost connections.

When a packet enters the SM, it is reformatted by the addition of three new fields, the routing field, the control field, and the source field. The *control* field (*C*) is one byte long and identifies different packet types within an SM, including various kinds of control packets. The *source* field (SRC) is one byte long and identifies the origin of the packet within the SM. The routing field (RF) is four bytes long and contains information needed to route the packet through the switch fabric. It has three subfields. The routing control subfield (RC) determines the type of routing the packet is to receive—the most important distinction is between broadcast packets and others. The interpretation of the remaining two subfields depends on the RC value. For packets belonging to point-to-point connections, the RF contains a *group number* subfield (GN) and a *logical channel number* subfield (LCN). To simplify the presentation, a complete discussion of group numbers will be deferred until later—for the present, it will suffice to identify group numbers with link numbers. The LCN subfield contains the LCN that the packet will carry when it leaves the SM. For broadcast packets, the RF contains a *fanout* subfield (FAN) and a *broadcast channel number* (BCN). The FAN subfield is equal to the number of outgoing FOL's that require a copy of this broadcast channel. The BCN is used to distinguish this channel from all other broadcast channels within the SM.

Logical channel translation is the process used to determine how to route a packet belonging to a connection through the SM. Each PP contains a *logical channel translation table* (LCXT) used for this purpose. When a packet is received by a PP, its LCN is used to index the LCXT. The selected entry is copied into the routing field of the packet. The LCXT contains 4096 entries of four bytes each. The entries in the LCXT's are maintained by the CP, which typically writes a new entry each time a new connection is established.

A. Packet Processor

The structure of the packet processor is shown in Fig. 6; it contains four packet buffers. The *receive buffer* (RCB) is used for packets arriving from the FOL and waiting to pass through the SF. The *transmit buffer* (XMB) is used for packets arriving from the SF that are to be sent out on the FOL. The *link test buffer* (LTB) and *switch test buffer* (STB) provide paths for test packets used to verify the operation of the FOL and SF, respectively. The RCB has a capacity of 16 packets, the XMB has a capacity of 32. The LTB and STB can each hold two packets. (These buffer capacities were selected based an $M/D/1$ queueing analysis and may change as a result of more detailed analyses.) The four buffers require a total of about 240 kbits of memory.

The *logical channel translation table* (LCXT) is a dual port memory that implements a table with 4096 entries of four bytes each. It can be read by the output circuit and written by the input circuit.

The *receive circuit* (RCV) converts the incoming optical signal to an electrical signal in eight-bit parallel format, synchronizes it to the local clock, discards frames containing errors, routes test packets to the LTB, and other packets to the RCB.

The *output circuit* (OUT) takes packets from the RCB, adds the routing, control, and source fields to the front of the packet as described above and sends the packets on to the switch fabric. It also processes test packets from the STB.

The *input circuit* (IN) routes data packets to the XMB, removing the routing, control, and source fields in the process. It also routes switch test packets to the STB, and updates the LCXT memory in response to LCXT write packets.

The *transmit circuit* (XMIT) takes packets from the XMB, adds the flag field and converts from the eight-bit parallel format to an optical signal. It also processes test packets from the LTB.

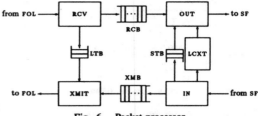

Fig. 6. Packet processor.

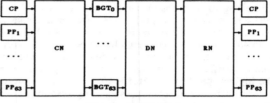

Fig. 7. Switch fabric.

B. Switch Fabric

A block diagram of the switch fabric (SF) is given in Fig. 7. It contains four major components, a *copy network*, a set of *broadcast and group translators*, a *distribution network*, and a *routing network*. The SF runs at a clock rate of 25 Mbits/s and has eight-bit wide internal data paths. This gives an effective data rate of 200 Mbits/s on the internal data paths, or two times the speed of the FOL's. An occupancy of 80 percent on the FOL's translates to a 40 percent occupancy on the internal data paths, which keeps contention and delay low. Simulation studies described in [1] show a delay of less than 100 μs for the entire switch fabric under conditions of heavy load.

When a broadcast packet having k destinations passes through the copy network (CN), it is replicated so that k copies of that packet emerge from the CN. Point-to-point packets pass through the CN without change.

The broadcast and group translators (BGT) perform two translation functions. First, for each arriving broadcast packet they determine the proper outgoing GN (group number) and LCN. Then, they translate the GN to an LN (link number). Discussion of this latter translation is postponed until later in this section. For the moment, we will assume that the GN's and LN's are identical.

The distribution and routing networks (DN, RN) move the packets to the proper outgoing PP. The RN is a straightforward binary routing network and the DN is provided to prevent internal congestion within the RN.

1) Routing Network and Distribution Network: The RN is a 64 × 64 binary routing network. Its structure is illustrated in Fig. 8 which shows a 16 × 16 version. The key property of such networks is that they are *bit addressable*—that is, the route a packet takes through the network is determined by successive bits of its destination address. The figure shows paths from two different input ports to output port 1011. Note that at the first stage the packet is routed out the lower port of the node (corresponding to the first "1" bit of the destination address), at the second stage it is routed out the upper port (corresponding to the "0" bit), and in the third and fourth stages it is routed out the lower ports. The self-routing property is shared by a variety of different interconnection patterns, including the so-called delta, shuffle-exchange, and banyan networks [7].

Buffers are provided at the inputs of each node. Each buffer is capable of holding two complete packets, implying about 20 000 bits of memory per node. (A detailed simulation study

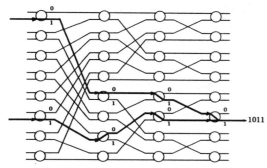

Fig. 8.   16 × 16 binary routing network.

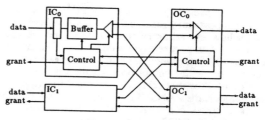

Fig. 9.   Switch node.

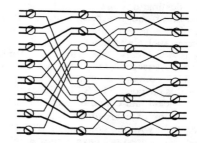

Fig. 10.   Congestion in binary routing networks.

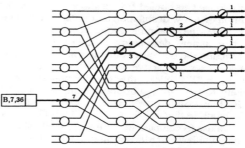

Fig. 11.   16 × 16 copy network.

has shown two buffers per input port to be sufficient to obtain the desired throughput. See [1], [2] for details.) A packet may pass through a node without being buffered at all if the desired output port is available when the packet first arrives. Indeed, in a lightly loaded network, a packet can pass through the RN with no buffering at all. In this case, it encounters a delay of just a few clock times in each node.

The data paths between nodes are eight bits wide. In addition, there is a single upstream control lead used to implement a simple flow control mechanism. This prevents loss of packets due to buffer overflows within the fabric. The entire network is operated synchronously, both on a bit basis and a packet basis—that is, all packets entering the first stage do so during the same clock cycle. The clock rate is 25 Mbits/s, giving an effective bandwidth of 200 Mbits/s for each of the paths through the network.

Fig. 9 is a more detailed picture of a single switch node. The switch node consists of two *input controllers* (IC) and two *output controllers* (OC). The IC's determine the routing required for each packet, request the appropriate OC, buffer packets if necessary, and apply upstream flow control to prevent buffer overflow. The OC's arbitrate requests for the outgoing links.

One problem with binary routing networks is that they can become congested in the presence of certain traffic patterns. This is illustrated in Fig. 10, which shows a traffic pattern corresponding to several *communities of interest*. In this pattern, all traffic entering the first four inputs is destined for the first four outputs, all traffic entering the second group of four inputs is destined for the second group of four outputs, and so forth. Note that with this pattern, only one fourth of the links joining the second and third stages are carrying traffic. Thus, if the inputs are heavily loaded, the internal links will be overloaded and traffic will back up. In a 64 × 64 network, there are eight stages and the links between the third and fourth stages can in the worst case be carrying all the traffic on just eight of the 64 links.

The DN solves this problem by evenly distributing packets it receives across all its outputs. It has the same internal structure as the RN, but its nodes ignore the destination addresses on packets and route them alternately to each of their output ports. This strategy is modified if one or both ports is unavailable. In this case, the first port to become available is used. This approach breaks up any communities of interest and makes the combination of the DN and RN robust in the face of pathological traffic patterns.

2) Copy Network and Broadcast Translation: The structure of the copy network (CN) is the same as that of the RN and DN. The CN's function is to make copies of broadcast packets as they pass through. This is illustrated in Fig. 11. The packet entering at left belongs to broadcast channel 36 and is to be sent to seven different links. At the first stage, the packet is routed out the upper port. This is an arbitrary decision—the lower link could have been used at this point. At the second

stage, the packet is sent out on both outgoing links and the fanout fields of the outgoing packets are modified. The upper packet will generate four copies and the lower one, three.

We now describe the CN routing algorithm for broadcast packets. Let BCN and FAN be the broadcast channel field and the fanout field from the packet and let sn be the stage number of the node in the switch where stages are numbered from right to left, starting with 0.

• If FAN $> 2^{sn}$ request both output links and when the links are available, simultaneously send the packet out on both.

— If BCN is even, the FAN field of the upper packet is set to $\lfloor (FAN + 1)/2 \rfloor$ and the FAN field of the lower packet is set to $\lfloor FAN/2 \rfloor$.
— If BCN is odd, the FAN field of the upper packet is set to $\lfloor FAN/2 \rfloor$ and the FAN field of the lower packet is set to $\lfloor (FAN + 1)/2 \rfloor$.

• If FAN $\leq 2^{sn}$, use the DN algorithm.

Note that this algorithm delays copying packets as long as possible. Another option is to copy the packets early. However, this approach can lead to unbounded congestion in the CN. The late copying algorithm limits the potential for congestion. See [1] for further details.

An important property of the copy algorithm is that given a particular combination of BCN and FAN, one can predict which of the FAN copies can appear at each output of the copy network. Suppose the copies of all broadcast packets were numbered from 0 to FAN − 1 with the indexes increasing as the output port numbers increase. Consider what happens when a broadcast packet is processed by the copy network. Some CN output ports may receive a copy while others do not, depending on how the arbitrary routing decisions are made. However, if an output port receives a copy, the number of that copy is completely determined. The number of the copy that can appear at a particular output port is called the *broadcast copy index* or BCI for that port. The BCI is a function of the output port number, the FAN field and the low-order bit of the BCN and is denoted bci_k (FAN, BCN). An algorithm to calculate BCI_k (FAN, BCN) appears in Fig. 12.

When the copies of a broadcast packet emerge from the CN, their final destinations are yet to be determined. The principal function of the broadcast and group translators (BGT) is to assign a new routing field to each copy of a broadcast packet in such a way that a copy is received by every PP that is supposed to receive one. This is accomplished by a simple table lookup based on the broadcast channel number (BCN). Each BGT contains a *broadcast translation table* (BTT), used for this purpose. The BTT is indexed by the BCN of the incoming packet and contains four byte entries. When BGT_k receives a copy of a broadcast packet, it replaces the packet's routing field with the contents of $BTT_k[BCN]$.

Note that two BGTs need not have identical entries in their BTTs for a given BCN, since their BCI values may be different. This point is illustrated in Fig. 13 which shows the translation process for two packets copied from a single broadcast packet.

The addition of a new destination to a broadcast channel can radically change the mapping required in the BTT's. In general, when a new destination is added to a particular broadcast channel, the CP must compute a new broadcast copy index tree as described above and use that information to update all the BTT's. Since this can happen frequently, it appears likely to present a substantial computational burden for the CP. Fortunately, there is a simple solution to the problem. It requires that for $0 \le k \le 63$, BGT_k contains a table that it can use to compute BCI_k (FAN, BCN). Since, FAN ≤ 64 and only the least significant bit of the BCN is needed to compute BCI, the table has 128 entries, each one byte long. Note that this table is static and is different for each BGT. Now, when the CP wishes to update the BTT's for a particular broadcast channel, it sends a control packet of the form shown in Fig. 14.

The CN replicates the packet so that each BGT receives a copy. When BGT_k receives its copy, it extracts FAN and BCN from the packet, then uses its lookup table to compute $j = BCI_k$ (FAN, BCN) and finally copies RF_j from the packet to BTT_k[BCN].

Using this scheme, the CP keeps a copy of the BTT__ *update* packet in its memory. To add a new destination, it increments the FAN field, adds a new RF to the end of the packet and sends it out to the SF. To remove a destination, it decrements the FAN field, removes the proper RF from the packet and sends it out.

3) Link Groups: A small addition to the mechanisms described above allows links to be collected together into *link groups*. The links in a group must have the same destination. Traffic for a link group is evenly distributed over all of its links. This allows dynamic load distribution across a set of links permitting higher utilization. It also allows the network to support channels that require more bandwidth than any one FOL can provide.

When a BGT receives a point-to-point packet, the packet contains a group number (GN) in its routing field. Each BGT contains a table, called the group translation table (GTT), which it uses to translate a GN to a link number (LN). The GTT identifies the links in each group and also contains a pointer to a particular link in the group. Every time a packet for a particular group is handled, this pointer is advanced to the next link in the group, so that each BGT distributes traffic evenly across the group. Broadcast packets are handled

```
integer function bci_k(FAN, BCN)
    integer s, f;
    Let the binary representation of k be b_{n-1} ... b_1 b_0,
        where n is number of stages in copy network.
    s := 0; f := FAN;
    p := the least significant bit of BCN.
    for i := n - 1 downto 0 →
        if f > 2^i →
            if b_i = 0 →
                f := ⌊(f + 1 - p)/2⌋;
            | b_i = 1 →
                s := s + ⌊(f + 1 - p)/2⌋;
                f := ⌊(f + p)/2⌋;
            fi;
        fi;
    rof;
    return s;
end;
```

Fig. 12. Computation of broadcast copy index.

similarly, but they must first go through the broadcast translation process described earlier. The organization of the GTT is shown in Fig. 15, which illustrates the entire translation process for a broadcast packet. Note that the GTT is a fairly static table—it changes only when groups are reconfigured, not on a per-connection basis.

One consequence of link groups is that the LCXT's for all links in the same group must be identical. Consequently, on every connection the CP must update the LCXT's for all links in the group. This can be handled easily by sending a broadcast LCXT update packet.

C. Implementation Details

This section is an an attempt to "size up" the switch modules. While the estimates given here are rough, they give a fairly accurate picture of the complexity of the SM's.

The switch nodes can best be implemented using custom integrated circuits. Each of the nodes that makes up the switch fabric consists of about 20 000 bits of memory plus some control circuitry. The memory is the dominant component, so we can estimate the size of a chip needed to implement a switch node by looking at the area required for the memory. The memory in the switch nodes is not extremely large, but it must be fast. Perhaps the simplest way to implement it is using a large dynamic shift register. A single bit of shift register memory can be implemented in 200 μm^2, assuming a 1 μm CMOS process, implying that 20 000 bits of memory will consume about 4 mm^2. This suggests that several nodes can be implemented on a single chip. Not surprisingly, pin limitations restrict the size of the chip more severely than chip area. A single node on a chip requires 36 signal pins plus additional pins for power, ground, and clock. Four nodes can be placed on a chip with about 80 pins if they are configured as a 4 × 4 switch. Actually, a better way to implement a 4 × 4 switch is to do it as a single larger node, rather than four small nodes— this structure requires half as much memory and is about twice as fast. It is convenient to have two chip types—one chip contains two independent 2 × 2 nodes and the other contains a single 4 × 4 node. Each of these can be implemented in an 80-pin package, preferably a high-density package such as a pin grid array.

The packet processors are more complex than the switch nodes. Each requires about 400 000 bits of memory if one includes all the packet buffers and the logical channel translation table. The largest item is the transmit buffer, which can be implemented as eight 128 × 150 memory arrays. The

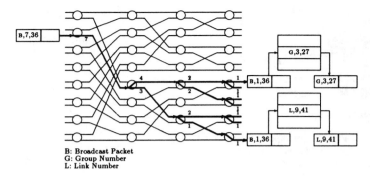

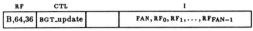

Fig. 13. Example of broadcast processing.

B: Broadcast Packet
G: Group Number
L: Link Number

RF	CTL		I
B,64,36	BGT_update		FAN, RF$_0$, RF$_1$, ..., RF$_{FAN-1}$

Fig. 14. Format of BTT update packet.

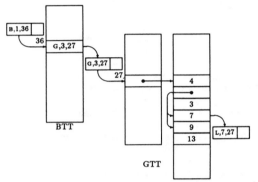

Fig. 15. Broadcast and group translation.

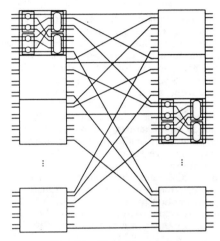

Fig. 16. 64 × 64 network.

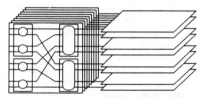

Fig. 17. Arrangement of circuit cards for a network.

cycle time of these arrays is not a critical performance factor—each access reads or writes 150 bits, leading to one access every 3 μs. Thus, fairly dense memory structures can be used. A three transistor memory cell can be implemented in about 100 μm^2 assuming a one micron technology, implying that the memory cells required by the transmit buffer will consume about 16 mm^2. The addition of the supporting circuitry required by the memory arrays could bring the total up to about 30 mm^2. The other packet buffers can be implemented in a similar fashion. Together, they contain less than two thirds the memory of the transmit buffer, but we will estimate that they consume another 30 mm^2. The LCXT contains about 150 000 bits of memory organized in nine 128 × 128 arrays. This adds about another 30 mm^2 to the chip area. If we assume (conservatively) that the memory consumes 75 percent of the chip area required to implement the PP, we arrive at an estimate of 1.2 cm^2 for the entire PP. A three chip implementation appears feasible, with the first chip containing the RCV, XMIT, and STB blocks, the second containing the RCB and XMB, and the third containing the IN, OUT, STB, and LCXT blocks.

The broadcast and group translators each contain several tables. The BCI table, used to compute a broadcast channel index contains 128 bytes, the BTT contains 4096 bytes (assuming 1024 unique BCN's in an SM), and the GTT 320 bytes. The total memory requirement is thus under 40 000 bits. Thus, two can fit comfortably on a single chip.

With these estimates in hand, we can consider how an entire SM might be put together. A convenient way to implement a 64 × 64 binary routing network is to subdivide it into 8 × 8 subnetworks as shown in Fig. 16. Each of the 8 × 8 subnets can be constructed from four 2 × 2 nodes and two 4 × 4

nodes for a total of four chips, and can be implemented on a single board with 144 signal leads. These boards can then be arranged into two orthogonally positioned ranks to give a compact implementation of the 64 × 64 network. This structure is shown in Fig. 17. What makes this arrangement particularly attractive is that the board-to-board signal paths can be kept very short.

An SM, however, contains three such networks. We can employ the same idea in the SM using the arrangement shown in Fig. 18. Here, there are four groups of eight-circuit boards each. The boards at the bottom center contain the first half of the CN. The boards at the right contain the second half of the CN, the BGT's, and the first half of the DN. The boards at the top center contain the second half of the DN and the first half of the RN. The boards at the left contain the second half of the RN and the PP's. The most densely populated boards are the ones at the left, each of which contains four switch node chips and 24 PP chips. Assuming high-density packaging in which chips are mounted in 1 in square pin grid arrays on multilayer ceramic circuit boards, this entire assembly could fit in a volume of roughly 6 in × 12 in × 20 in. A complete SM would contain two of these assemblies plus the CP's. Each CP

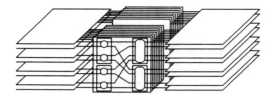

Fig. 18.   Switch module packaging.

contains a 32-bit microprocessor with perhaps 2 Mbytes of memory, an interface to the switch fabric, and a single terminal interface. Hence, it could be implemented on three- or four-circuit cards and four complete SM's could be packaged in a single 30 in wide cabinet.

III. Design of Network Components

The switch module is designed as a component that can be used to build larger systems. The decision to build large systems from small components can be criticized on the grounds that the total hardware required can be substantially larger and the performance lower than if a monolithic structure is used. There are several strong arguments in favor of the modular approach, 1) a system constructed out of small modules is easier to design and implement, 2) modular systems can grow over a large range of sizes by adding modules, and 3) manufacturing economics favor systems that contain a large number of identical components.

Modular systems are easier to design because the modules lead to a natural separation of local concerns from global ones. Modules can be defined by their external interfaces and treated as "black boxes" by their environment. Once this separation has been made explicit by a set of module specifications, the design of the individual modules is relatively straightforward. The size of the modules is a central issue. In the case of the SM's, larger modules would have been preferable for constructing larger systems. The decision to use a fairly small module was motivated by three considerations. First, it allows a single processor of moderate proportions to handle all the connection processing, avoiding a number of difficult problems that must be faced in larger systems. Second, it allows a very simple approach to reliability and maintenance—to recover from a hard failure in the active side, you simply switch to the other side. With small enough modules, failure rates are low enough that there is no need to resort to more elaborate recovery algorithms. The small size also simplifies repair in the field—once it has been determined that a particular SM side is faulty, it can be replaced as a unit. Third, hardware timing is easier to handle in a system of modest physical dimensions. The SM's are small enough that it is reasonable to operate them as synchronous systems. Asynchronous operation need only be considered at the interface between the PP's and the FOL's.

Our first example of a larger system component constructed from SM's is the network interface (NI), shown in Fig. 19. A single NI contains 15 SM's, five of which are termed back-end SM's (BSM) and 10 of which are termed front-end SM's (FSM's). Each FSM has six FOL's connecting it to each of the BSMs and each BSM has six FOLs connecting it to each of the FSM's. The NI has 330 FOL's that can be used to connect to other system components. Typically, 30 would be used to connect to the NI's host PS with the remainder going to users. The actual ratio of downstream links to upstream links is completely flexible—a ratio of about 10:1 appears reasonable, since most of the downstream bandwidth is likely to be consumed by a few popular broadcast channels.

SM's can also be used to build a packet switch (PS) as shown in Fig. 20. The PS shown can have as few as one BSM and two FSM's or as many as 30 BSM's and 60 FSM's. In the largest configuration, it terminates 1800 FOL's on its FSM's.

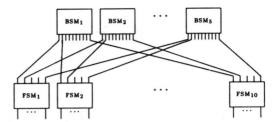

Fig. 19.   Network interface structure.

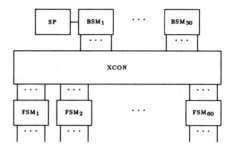

Fig. 20.   Small packet switch.

If 1500 FOL's are used to connect to NI's, such a PS can support 15 000 users. This leaves 300 FOL's for connecting to other PS's. Again, the ratio of downstream to upstream links is not limited by the architecture and can be adjusted to suit actual needs—the ratio chosen here appears reasonable. Each PS contains a supervisory processor (SP), which provides office administration and craft interface functions. The SP is a fairly large processor with its own disk subsystem and possibly other peripherals. It does no processing of individual connections, but does supervise the operation of the SM's through its control of their routing tables.

A PS must be capable of growing across a range of sizes. In the smallest configuration, it would have 30 FOL's joining each FSM to the one BSM. In the largest configuration, it would have one FOL joining each FSM to each BSM. If the SM's were joined directly to one another, growth of a PS would require manual recabling. This is avoided by the cross connect (XCON), which can be thought of as an 1800×1800 cross-bar switch. Fortunately, the XCON does not require all the functionality of a full cross bar and can implemented in much less expensive way. The structure of the XCON is illustrated in Fig. 21, which shows a small version, appropriate for SMs having just sixteen FOL's. Such an XCON can be used to construct a PS with up to eight BSM's and 16 FSM's. The XCON has a planar structure with a BSM side and an FSM side. The BSM side is organized in rows—all FOL's from a particular BSM terminate on the same row. The FSM side is organized in columns—all FOL's from a particular FSM terminate on the same column. Each of the circles at the intersection of a row and column represents a switching element. These switching elements each have four leads referred to as x^-, x^+, z^-, and z^+. The z^- and z^+ leads are arranged perpendicular to the plane of the XCON so that they can be connected to the links from the SM's. The x^- and x^+ leads are arranged in the plane of the SM in a diagonal direction. The switching elements can be switched in four different ways: (z^-, z^+), (z^-, x^+), (x^-, x^+), and (x^-, z^+). In a fully grown configuration, all the rows on the BSM side are occupied, as are all the columns on the FSM side and all the switching elements are set to (z^-, z^+) position. In this setting, all the connections pass straight through the XCON—so each BSM/FSM pair is joined by one FOL. In the smallest configuration, the top row on the BSM side is occupied, along with the first and ninth columns on the FSM side. The switching elements

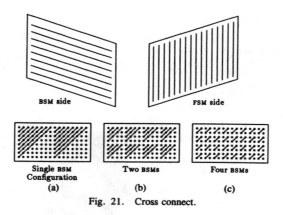

Fig. 21. Cross connect.

are set to provide eight paths connecting the BSM to each of the FSM's, as illustrated in Fig. 21(a). To illustrate, the signal arriving on the FOL connected to the switch element in the first column, fourth row is switched up and to the right by that element, which is set to (z^-, x^+), passes through two elements set to (x^-, x^+) and is then switched out the other side of the XCON by the element in the fourth column, first row, which is set to (x^-, z^+). Fig. 21(b) shows a configuration in which two BSM's are connected to four FSM's, with four FOL's joining each pair, and Fig. 21(c) shows a configuration with four BSM's and eight FSM's. The complexity of an XCON grows linearly with the number of FOL's that connect to it, making it reasonable to construct systems with fairly large XCON's.

Larger packet switches can be constructed by combining modules in a similar way. A PS capable of terminating 27 000 FOL's can be constructed from 900 FSM's and 15 of the smaller packet switches described above, which function as BSM's in that case. If 20 000 FOL's are used to connect to NI's, such a system would serve 200 000 users.

The examples in this section are illustrative of the kinds of systems one can construct using SM's as the basic building blocks and XCON's as the glue used to combine them easily. Many other variants are possible. In particular, a variety of different NI's is desirable, since not all users require the same set of services. An NI that supports say 10 Mbits/s of bandwidth to individual users can be constructed from a single SM supplemented with small front-end concentrators. Such an NI would require 16 FOL's to a PS and would support 750 users, making it substantially less expensive than the one described above.

IV. SUMMARY

It is worth noting that the network described here can support other connection types in addition to point-to-point and broadcast. First, it it possible to include an upstream capability in broadcast channels. No new mechanisms are required in the SM's. The CP must simply set the LCXT entries so that packets received from broadcast recipients get multiplexed onto the link and logical channel going to the broadcast source. Such an upstream capability can be used in a variety of ways—for example, if the channel is being used to broadcast a lecture or other presentation, the upstream capability gives listeners a way to ask questions.

The presence of an upstream capability raises new questions for congestion control. Since the data rates of all the participants add to the total upstream flow, it's not clear that one can control congestion by policies enforced at the NI's. One approach is to make upstream packets low priority, so that they get through the network only if there is sufficient bandwidth available. This may work adequately in some situations, but is not a general solution.

The system can also support a multiway broadcast, in which every packet sent by any participant is received by all others.

This is a natural sort of connection to use in providing a conference service.

In telephone networks that support conferencing, a new person is added to a conference because of an action initiated by a current participant. In a communications network that supports broadcast television, there is a need for persons not part of a connection to request that they be connected (subject to authorization procedures). Having observed this, one notices that the ability to add on to a connection from "outside" is a generally useful ability. It allows, for example, conferences that users can come and go from as they wish, rather than requiring that they be available when the conference starts. This suggests that every connection should have the potential of adding new participants either from inside or outside, with authorization options used to restrict connection from outside if that is desired.

This paper has described the design of a packet switching network that can support voice, data, and video communication on a large scale. The most novel aspect of the system is its flexible broadcast capability, which is suitable for a range of services including broadcast, television, and conferencing. The main purpose of the paper is to show that such systems are technically feasible.

ACKNOWLEDGMENT

This research is an outgrowth of earlier work that I did while at Bell Laboratories. Many individuals contributed ideas to that earlier work, which continues to influence my thinking—particularly, H. Andrews, D. Creed, M. Dècina, B. Hoberecht, W. Montgomery, and L. Wyatt. Thanks are also due to E. Nussbaum, P. White, and N. Haller of Bell Communications Research for supporting my continued work in this area and for stimulating my interest in broadcast networks.

REFERENCES

[1] R. Bubenik, "Performance evaluation of a broadcast packet switch," M.S. thesis, Comput. Sci. Dep., Washington Univ., Aug. 1985.

[2] R. Bubenik and J. S. Turner, "Performance of a broadcast packet switch," Comput. Sci. Dep., Washington Univ., Tech. Rep., WUCS-86-10, June 1986.

[3] J.-M. Chang and N. F. Maxemchuk, "Reliable broadcast protocols," *ACM Trans. Comput. Syst.*, vol. 2, no. 3, Aug. 1984.

[4] D. R. Cheriton and S. E. Deering, "Host groups: A multicast extension for datagram internetworks," in *Proc. Ninth Data Commun. Symp.*, Sept. 1985, pp. 172–179.

[5] J. P. Condreuse and M. Servel, "Asynchronous time-division techniques: An experimental packet network integrating videocommunication," presented at Proc. Int. Switching Symp., 1984.

[6] Y. K. Dalal and R. M. Metcalf, "Reverse path forwarding of broadcast packets," *Commun. ACM*, vol. 21, no. 2, pp. 1040–1047, Dec. 1978.

[7] T.-Y. Feng, "A survey of interconnection networks," *Comput.*, vol. 14, no. 12, pp. 12–30, Dec. 1983.

[8] W. L. Hoberecht, "A layered network protocol for packet voice and data integration," *IEEE J. Select. Areas Commun.*, vol. SAC-1, pp. 1006–1013, Dec. 1983.

[9] A. Huang and S. Knauer, "Starlite: A wideband digital switch," in *Proc. GLOBECOM 84*, Dec. 1984, pp. 121–125.

[10] Y.-C. Jenq, "Performance analysis of a packet switch based on a single-buffered banyan network," *IEEE J. Select. Areas Commun.*, vol. SAC-1, Dec. 1983, pp. 1014–1021.

[11] J. J. Kulzer and W. A. Montgomery, "Statistical switching architectures for future services," presented at Proc. Int. Switching Symp., May 1984.

[12] W. A. Montgomery, "Techniques for packet voice synchronization," *IEEE J. Select. Areas Commun.*, vol. SAC-1, Dec. 1983, pp. 1022–1028.

[13] R. Rettberg, C. Wyman, D. Hunt, M. Hoffman, P. Carvey, B. Hyde, W. Clark, and M. Kraley, "Development of a voice funnel system: Design report," Bolt Beranek and Newman, Rep. 4098, Aug. 1979.

[14] J. S. Turner and L. F. Wyatt, "A packet network architecture for integrated services," in *Proc. GLOBECOM 83*, Nov. 1983, pp. 45–50.

[15] J. S. Turner, "Packet load monitoring by trunk controllers," U.S. Patent 4 484 326, Nov. 20, 1984.

[16] ——,, "Packet switching loop-around network and facilities testing," U.S. Patent 4 486 877, Dec. 4, 1984.

[17] ——,, "End-to-end information memory arrangement in a line controller," U.S. Patent 4 488 288, Dec. 11, 1984.

[18] ——,, "Interface facility for a packet switching system," U.S. Patent 4 488 289, Dec. 11, 1984.

[19] ——,, "Packet error rate measurements by distributed controllers," U.S. Patent 4 490 817, Dec. 25, 1984.

[20] ——,, "Fast packet switch," U.S. Patent 4 491 945, Jan 1, 1985.

[21] ——,, "Fast packet switching system," U.S. Patent 4 494 230, Jan. 15, 1985.

[22] ——,, "Design of an integrated services packet network," *IEEE J. Select. Areas Commun.*, pp. 1373–1380, Nov. 1986.

Jonathan S. Turner (M'77) received the B.S.C.S. and B.S.E.E. degrees from Washington University, St. Louis, MO, in 1977. He received his M.S. and Ph.D. degrees in computer science from Northwestern University, Evanston, IL, in 1979 and 1981, respectively.

From 1977 to 1983 he worked for Bell Laboratories, Naperville, IL, first as a member of Technical Staff and later as a Consultant. His work there included the development of maintenance software and design of system architectures for telephone switching systems. From 1981 to 1983 he was the principal system architect for the fast packet switching project, an applied research project which established the feasibility of integrated voice and data communication using packet switching technology. He is now an Associate Professor of Computer Science at Washington University, where he continues his research on high performance communications systems. His research interests also include the study of algorithms and computational complexity, with particular interest in the probable performance of heuristic algorithms for NP-complete problems.

Dr. Turner is a member of the Association for Computing Machinery and the Society for Industrial and Applied Mathematics. He has also been awarded 11 patents for his work.

A New Broadcast Switching Network

CHIN-TAU LEA, MEMBER, IEEE

Abstract—A new rearrangeable broadcast switching network is proposed. Under the condition that the number of the outlets (customers) is much larger than the number of source inlets (video channels), the new network normally has fewer switching nodes than other approaches. It can be modularly constructed, has easy growth, and an easy path hunt. Furthermore, the new network is more suitable for the directional-coupler photonic switching technology. An implementation based on the directional-coupler photonic switching technology is presented in this paper. The numbers of crosspoints required for both electronic and photonic implementations are also compared in this paper.

I. INTRODUCTION

WITH the fast deployment of lightwave communication systems, the concept of broad-band integrated service digital network (BB-ISDN) comes closer to reality. In the future, many new services will be integrated in the BB-ISDN, and providing them offers new challenges in switching system design. One of the major services provided in a BB-ISDN is circuit-switched video. It allows customers to receive from three to four TV programs simultaneously. Providing this service requires a switching network to have a multiconnection capability, i.e., to connect any source inlet (video channel) to any or all outlets (subscribers). Furthermore, this service has to be provided cost effectively so that it can compete with the existing cable TV system. A cross-bar network can be used as such a switching network at the price of a large amount of crosspoints. Various efforts [1]–[3], [7]–[9] in the past have addressed the problem of reducing the crosspoints in a multiconnection switching network. In this paper, a new broadcast multiconnection switching network is proposed. It is based on the cascading technique (cascading two networks together). Under the condition that the number of outlets (customers) is much larger than the number of source inlets (video channels), the proposed network normally has fewer crosspoints than other approaches. It can be modularly constructed, has easy growth, an easy path hunt, and can be modified to become a totally self-routing network. In addition, it is more suitable for the directional-coupler photonic switching technology. The numbers of crosspoints required for both electronic and photonic implementations are also compared in the paper.

The rest of this paper is organized as follows. Section II discusses the problems with the conventional Clos approach to a broadcast video switch and points out the relationship between the new approach and other approaches. Section III details the design of the new network. A mathematical proof of the nonblocking nature of the new network is also given in this section. Section IV presents an implementation of the new network based on the directional-coupler photonic switching technology. This section also discusses the impact of the new technology on switching architectures. Section V compares the

numbers of crosspoints in the new network and other networks. The computing time for routing is also discussed in this section. Section VI presents a modified version of the new network which is totally self-routing. Section VII discusses the application of a broadcast network in parallel processing. Section VIII concludes the discussion.

II. VARIOUS APPROACHES FOR BROADCAST NETWORK DESIGN

A. Clos Approach and its Variations

Various rearrangeable nonblocking multiconnection networks have been proposed. Usually they are based on the Clos network [4]. The classic Clos approach reduces the number of switching elements in a nonbroadcast (i.e., one-to-one connection) network significantly. But it is not so successful for a broadcast switching network. The example given below explains why. A general three-stage Clos network, denoted by $v(m, n_1, r_1, n_2, r_2)$, is shown in Fig. 1. This network consists of r_1 rectangular $(n_1 \times m)$ input switches, $m(r_1 \times r_2)$ middle switches, and $r_2(m \times n_2)$ output switches. The total number of inlets is $n_1 r_1$, and outlets $n_2 r_2$. In a network with the restriction of one-to-one connection, i.e., no broadcast capability, it has been proved [2] that if $m \geq N = \text{Max} [\text{Min} (n_1, n_2 r_2), \text{Min} (n_2, n_1 r_1)]$, the network is rearrangeable. Fig. 2(a) is an example where the network has 2^i inlets and 2^j outlets, and $n_1 = n_2 = 2$. Based on the discussion above, the network is rearrangeable if $m \geq 2$. The size of the middle switches, as shown in Fig. 2(a), is $2^{i-1} \times 2^{j-1}$. Following the same Clos approach, we can decompose the middle switches into three-stage switches and reduce the number of crosspoints even further. When $i = j$, the result is the famous Benes-Clos network [5]. The reduction of the number of crosspoints through a Clos approach lies mainly in the continuous application of the decomposition technique. Let us apply the Clos approach to a broadcast network which has the capability to connect an inlet to any or all outlets. It has also been proved in [2] that if all input and output switches have the broadcast capability, while the middle switches do not, then the necessary and sufficient condition for a rearrangeable broadcast switching network is $m \geq \text{Max} (N_1, N_2)$ where $N_1 = \text{Min} (n_1 r_2, n_2 r_2)$ and $N_2 = \text{Min} (n_2 r_1, n_1 r_1)$. An example is given in Fig. 2(b) where the network has the same number of inlets and outlets as the network in Fig. 2(a) and $n_1 = n_2 = 2$. As shown in Fig. 2(b), the number of middle-state switches required largely exceeds that in one-to-one connection network. In order to support the broadcast capability, the number of the middle switches is increased from 2 to 2^j. Although the values of n_1, n_2, r_1, and r_2 chosen in this example may not be optimum, two observations can be made that explain why the Clos approach is not efficient for broadcast networks. First, a broadcast network requires multiple appearances of inlets on the middle-stage switches. This creates the explosion of the number of the middle-state switches. But the number of lines between the first- and the second-stage switches used by any connection pattern is at most $n_2 r_2$, which is only a small portion of the total number of input lines of the middle-stage switches. Second, the size of the first- and third-stage switches in the broadcast network is pretty odd. Further decomposition on them is impossible.

Paper approved by the Editor for Communication Switching of the IEEE Communications Society. Manuscript received April 27, 1987; revised August 21, 1987, November 12, 1987, and January 4, 1988. This paper was presented in part at ICC'87, Seattle, WA, June 1987.

The author is with the School of Electrical Engineering, Georgia Institute of Technology, Atlanta, GA 30332.

IEEE Log Number 8823011.

Reprinted from *IEEE Transactions on Communications*, Vol. 36, No. 10, October 1988, pages 1128-1137. Copyright © 1988 by The Institute of Electrical and Electronics Engineers, Inc. All rights reserved.

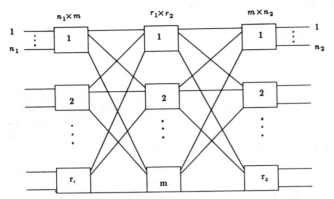

Fig. 1. A general three-stage Clos network.

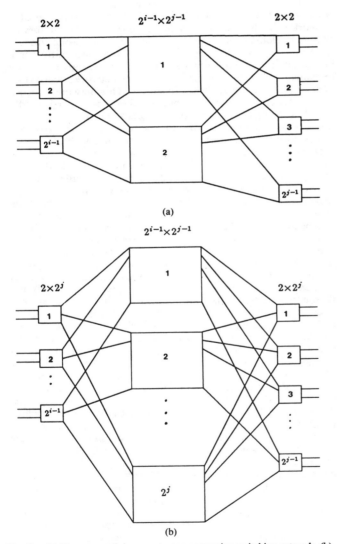

(a)

(b)

Fig. 2. (a) Clos approach in a one-to-one connection switching network. (b) Clos approach in a multiconnection switching network.

Richards and Hwang [3] proposed a novel broadcast switching network which has significantly reduced the number of crosspoints. They completely eliminated the first-state switches, and let each inlet have multiple appearances on the second-stage switches. Although eliminating the first stage also eliminates the flexibility in generating various inlet

patterns available to the second stage, a fixed inlet assignment pattern can be found that makes the network rearrangeable if the number of inlet appearances is large enough. An example is given in Fig. 3 where the number of inlets is 25 and number of appearances of every inlet is 2. The inlets are arranged into a 5-by-5 array. If we assign the inlets of the same row into one switch, and the inlets of the same column into another, the network is a rearrangeable broadcast network when $n_2 < 5$ [3]. (There is only one outlet switch shown in Fig. 3.) The relationship between the optimum size of the network, the value of n_2, and the fixed assignment pattern was discussed in detail in [3]. Although Richards and Hwang's approach reduces the number of crosspoints significantly, it shares some similar characteristics of a Clos approach. Namely, it requires each inlet have multiple appearances on the first-state switches, but not all of them are used simultaneously for any connection. The number of lines between the first- and second-stages (there are only two stages in the new approach) largely exceeds the number of active lines at any time. Also, the optimum sizes (in terms of crosspoint minimization) of the first- and second-stage switches are pretty odd. A two-stage network of this approach is locally most efficient when the number of inlets equals the square of a prime [3]. However, after the first decomposition, this condition is no longer satisfied. Another decomposition will not be exact and inefficiency can occur.

B. Cascading Scheme

Masson *et al.* [6] discussed various methods such as decomposition, staging, etc., for designing switching networks. The new broadcast network proposed in this paper is based on another scheme: cascading. The cascading scheme cascades two (or more) networks to form a new network and has been used in designing other broadcast networks [7]–[9]. In a wide sense, the cascading scheme can be considered as a function-level decomposition technique. A switching network's function is decomposed into several subfunctions, and each subfunction is accomplished by one switching network. Ofman's multiconnection network [7] (also called *generalized connection* network) consists of three networks cascaded together: hyperconcentrator, infrageneralizer, and a routing network (see Fig. 4). The hyperconcentrator concentrates all the active inlets in a consecutive sequence; the infrageneralizer generates multiple copies of each active inlet; the routing network routes the copies of each inlet to the corresponding destination. For a $N \times N$ multiconnection network, it has $O(5 \log_2 N)$ stages and $O(10N \log_2 N)$ *electronic* crosspoints. (As discussed later, the crosspoint count is different if photonic switching technology is used.) Thompson refined the hyperconcentrator and improved the result to $O(4 \log_2 N)$ stages and $O(8N \log_2 N)$ crosspoints [8] (It was also indicated in [8] that this number can be reduced to $O(7.6N \log_2 N)$ when three-way branching is used. However, to make a fair comparison, we restrict our discussion to two-way branching). In this paper, a similar cascading scheme is used for the proposed broadcast video switch. In the context of circuit-switched video, the number of outlets is normally much larger than the number of inlets. Taking advantage of this condition, we propose a new network which consists of two networks cascaded together: a consecutive spreading (CS) network and a routing network. It has $O(3N \log_2 N)$ stages and $O(6N \log_2 N)$ *electronic* crosspoints. (Here, N is the number of outlets.) If the directional coupler photonic switching technology is used, the number of crosspoints is only $O(2N \log_2 N)$. It should be noted that under the same condition (the number of outlets $\gg$ the number of inlets), the number of crosspoints of the hyperconcentrator in Fig. 4 is negligible compared to that of the other two subnetworks. Therefore, the number of electronic crosspoints in both Ofman and Thompson's broadcast networks is also reduced to $O(6N \log_2 N)$. But the

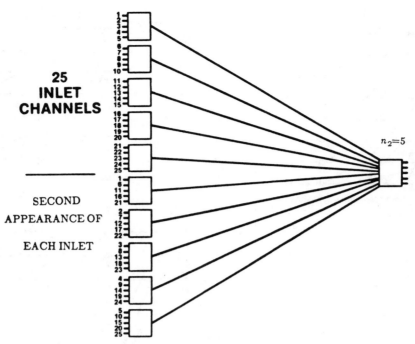

Fig. 3. An example of Richards and Hwang's broadcast network.

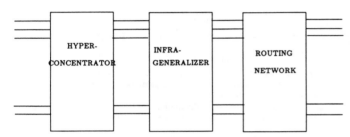

Fig. 4. Ofman's broadcast switching network.

proposed new network has two advantages. First, it has fewer stages and consequently has less signal attenuation if directional couplers are used to implement the network. Second, the setting time of the CS network is significantly less than that needed for the generalizer in Ofman's and Thompson's broadcast networks.

One final note on the difference between the Clos approach and the cascading scheme for a broadcast network: the Clos approach leads to an architecture which requires most of its nodes to have the broadcast capability. As will be discussed in Section IV, this is a significant disadvantage if the network is based on the directional-coupler photonic switching technology. But a cascading scheme leads to an architecture whose broadcast nodes reside in one subnetwork and nonbroadcast nodes in another. This architecture is more suitable for the directional-coupler photonic switching technology (see Section IV).

III. A NEW BROADCAST SWITCHING NETWORK

A connection pattern between inlets and outlets can be represented by a mapping between them. Assume that a network has n inlets and m outlets. A mapping between the inlets and the outlets can be denoted by a m-tuple $(X_0, X_1, \cdots, X_{m-1})$, where X_i represents the source inlet for outlet i. A consecutive spreading (CS) of inlets is a spreading of the first k inlets across the first l outlets, where $1 \leq k \leq n$ and $1 \leq l \leq m$, in which the order of the k inlets is preserved. For example, assume that $n = 4$, $m = 6$, and S_1, S_2, S_3, S_4 are the

four inlets. The mapping $(S_1, S_2, S_2, S_3, S_3, S_4)$ is a CS, but the mapping $(S_1, S_1, S_3, S_3, S_4, S_4)$ is not because S_2 is missing. Neither is $(S_1, S_3, S_2, S_1, S_3, S_2)$ since the order of inlets $(S_1 S_2 S_3)$ is not preserved. Another mapping $(S_1, S_2, *, S_2, S_3, S_3)$ where * stands for no subscription, is not a CS since the third outlet does not subscribe to any inlet channel.

The multiconnection function in a broadcast switching network can be decomposed into two subfunctions: *consecutive spreading* and *routing*. In consecutive spreading, copies requested for each inlet are generated. In routing, these copies are sent to their corresponding destinations. The new proposed broadcast switching network is based on this decomposition scheme. It consists of a consecutive spreading network and a routing network.

A. The Consecutive Spreading Network

The network that is nonblocking for any CS pattern is shown in Fig. 5. Hereafter, it is called the CS *network*. The topology of the CS network is a degenerate topology of a data manipulator [10]–[11]. In an n-stage CS network with a size $N \times N$ where $n = \log_2 N$, there are N nodes in each stage. Each node has two inlets (i_0, i_1) and two outlets (o_1, o_2). I_0 of the kth node in the jth stage is connected to the $(k - 2^{n-j})$th node in the previous stage and i_1 is connected to the kth node in the previous stage. If $(k - 2^{n-j}) < 0$, then i_0 is hanging open. O_0 is connected to the kth node of the next stage, and o_1 is connected to the $(k + 2^{n-j-1})$th node of the next stage. If $(k + 2^{n-j-1}) > N - 1$, then o_1 is hanging open. All the open links are not shown in Fig. 5. There are two operating states of each node: upper broadcast (UB) mode, and lower broadcast (LB) mode. In the UB(LB) mode, the information from $i_0(i_1)$ is sent to both o_0 and o_1. These two modes are depicted in Fig. 6. In the following, we are going to show that with a proper control algorithm the network in Fig. 5 is nonblocking for any CS pattern. Furthermore, the controlling algorithm can be implemented in a distributed way.

In any CS pattern, the path between outlet j and its source inlet, denoted by $s(j)$, can be represented by the binary representation of the difference between the destination address and its corresponding source address, e.g., $j - s(j)$. An example is given in Fig. 7(a) where the difference between

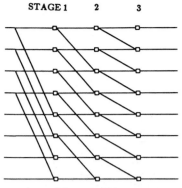

Fig. 5. The topology of the consecutive spreading (CS) network.

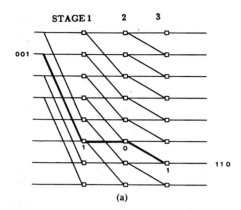

(a)

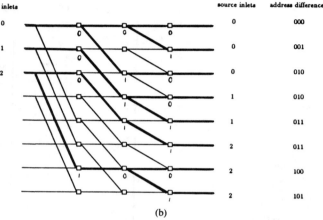

(b)

Fig. 7. (a) The path in the CS network is determined by the difference between the outlet address and the inlet address. (b) If the nodes in the CS network are controlled by the corresponding bits in the matrix of the address differences between the outlets and their corresponding source inlets, the CS network is nonblocking for any CS pattern.

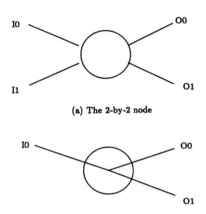

(a) The 2-by-2 node

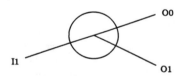

(b) The Upper Broadcast (UB) mode

(c) The Lower Broadcast (LB) mode

Fig. 6. The broadcast node has two states: UB and LB.

outlet 110 and its source inlet 001 is $110 - 001 = 101$. The path between outlet 110 and inlet 001 is indicated by heavy lines in Fig. 7(a). As can be seen, if the node in stage i along the path is controlled by the $(n - i)$th bit of the address difference in such a way that bit 1 sets the node in the UB mode and bit 0 sets the node in the LB mode, then the path between an outlet and its corresponding source inlet is set up correctly.

Assume $\{D_0, D_1, \cdots, D_{n-1}\}$ be the sequence of the address differences between all outlets and their corresponding source inlets. Let $(d_{i,n-1}, d_{i,n-2}, \cdots, d_{i,1}, d_{i,0})$ be the binary representation of D_i. Replacing D_i in the sequence with its binary representation, we have a two-dimensional array given below

$$\begin{bmatrix} d_{0,n-1} & d_{0,n-2} & \cdots & d_{0,1} & d_{0,0} \\ d_{1,n-1} & d_{1,n-2} & \cdots & d_{1,1} & d_{1,0} \\ \vdots & \vdots & \cdots & \vdots & \vdots \\ d_{n-2,n-1} & d_{n-2,n-2} & \cdots & d_{n-2,1} & d_{n-2,0} \\ d_{n-1,n-1} & d_{n-1,n-2} & \cdots & d_{n-1,1} & d_{n-1,0} \end{bmatrix} \quad (1)$$

where $d_{i,*}$ represents row i, and $d_{*,j}$ represents column j. Based on the discussion in Fig. 7(a), it is realized that if the nodes along the path between an inlet–outlet pair are controlled by $d_{i,*}$, the connection is always successful. But this control algorithm cannot be used to connect simultaneously all the inlet–outlet pairs in a CS. Many paths in a CS overlap each other. If the above algorithm is used, it implies that some nodes in the CS network may be controlled by more than one bit. This clearly causes confusion. However, one interesting property of the CS network is that if $d_{i,*}$ is used to control the nodes of *each row*, instead of the nodes along the path between outlet i and its source inlet, the CS network is still nonblocking for any CS pattern. An example is given in Fig. 7(b) where the control bit for each node in the given CS pattern is indicated. As can be seen, these control bits are identical to those in the address-difference matrix.

To give a formal proof of this property, the following lemma is introduced first. Let D_j represent the difference between outlet j and its source inlet in a CS. It is trivial to see that $D_j \geq 0$ based on the definition of a CS. Furthermore, the sequence $D_0, D_1, \cdots, D_{n-1}$ is an increasing sequence.

Lemma 1: Let $\{D_0, D_1, \cdots, D_{n-1}\}$ be the sequence of the address differences between source-destination pairs in a CS. The sequence $\{D_i\}$ is an increasing sequence.

Proof: This can be proved easily by examining the source addresses of a CS for destinations k and $k + 1$. Let $s(k)$ denote the source inlet of outlet k. According to the definition of a CS, the source address of $k + 1$ is either $s(k)$ or $s(k) + 1$. Therefore, the $D_{k+1} = k - s(k) + 1$ or $D_{k+1} = k - s(k)$. Since the difference between outlet k and its source is

320

$(k - s(k))$, i.e., $D_k = k - s(k)$, we have $D_{k+1} \geq D_k$. Hence, we conclude that D_i is an increasing sequence.

The formal proof of the nonblocking nature of the control algorithm introduced in Fig. 7(b) is given below. Two notations are introduced first to facilitate the discussion.

- $[S]_i$: bit i of the binary representation of the decimal S. For example, suppose the binary representation of S is $(s_{n-1}s_{n-2} \cdots s_1 s_0)$, then $[S]_2 = s_2$.
- $[S]_{i:j}$: $\{s_k | i \geq k \geq j\}$. For example, if $[S] = (s_{n-1}s_{n-2} \cdots s_1 s_0)$, then $[S]_{5:3} = (s_5 s_4 s_3)$.

Theorem 1: Let $\{D_0, D_1, \cdots, D_{m-1}\}$ be the address difference sequence in the CS network in Fig. 7(b), and $(d_{i,n-1} d_{i,n-2} \cdots d_{i,1} d_{i,0})$ be the binary representation of D_i. Let $d_{i,j}$ control the node i in the $(n - j)$th stage in such a way that if $d_{i,j} = 1$, the node is set to the UB mode, and if $d_{i,j} = 0$, the node is set to the LB mode. With this controlling algorithm, no blocking will occur for any CS pattern in the CS network.

Proof: Let S be the source inlet of a particular outlet y in a CS, and $(a_{n-1}, a_{n-2}, \cdots, a_1, a_0)$ be the binary representation of the address difference between outlet y and S, i.e., $D_y = (a_{n-1}, a_{n-2}, \cdots, a_1, a_0)$. According to the nature of the CS network [see Fig. 7(a)], there will be no blocking if we let a_{n-j} control the node in the jth stage along the path between outlet y and its source inlet S. In the proposed control algorithm, the node in the $(n - 1)$th stage along the path is controlled by $[D_{r_1}]_{n-1}$; the node in the $(n - 2)$th along the path is controlled by $[D_{r_2}]_{n-2}$; so forth, where $r_j = S + a_{n-1}2^{n-1} + a_{n-2}2^{n-2} + \cdots + a_{n-j}2^{n-j}$. We now prove that $[D_{r_1}]_{n-1} = a_{n-1}$, $[D_{r_2}]_{n-2} = a_{n-2}$, etc. $[D_{r_1}]_{n-1}$ is the most significant bit of the address difference between outlet r_1 and its corresponding source inlet. It has been proved in Lemma 1 that $\{D_i\}$ is an increasing sequence. Since $y \geq r_j$ where $n - 1 \geq j \geq 0$, from the definition of r_j, we have $D_y \geq D_{r_1}$. So we must have $[D_{r_1}]_{n-1} \leq a_{n-1}$. Assume that the source inlet for r_1 is T. Then we have $r_1 - T = D_{r_1}$ from the definition of D_{r_1}. We also have $r_1 - S = a_{n-1}2^{n-1}$ from the definition of r_1. If $[D_{r_1}]_{n-1} < a_{n-1}$, it implies that $S < T$, which is impossible due to the definition of a CS. Therefore, the only remaining case is $[D_{r_1}]_{n-1} = a_{n-1}$. Following similar arguments, we can prove that $[D_{r_2}]_{n-1:n-2} = a_{n-1}a_{n-2}$; $[D_{r_3}]_{n-1:n-3} = a_{n-1}a_{n-2}a_{n-3}$; so forth. Thus, we have $[D_{r_1}]_{n-1} = a_{n-1}$; $[D_{r_2}]_{n-2} = a_{n-2}$, etc. Based on the discussion above, we can see that this control algorithm is the same as using a_i to control the node in the $(n - i)$th stage along the path. Therefore, no blocking will occur for any CS pattern if the network is controlled by the proposed scheme.

To implement this algorithm in a distributed way is very easy. Just let outlet i send $d_{i,*}$ as the control tag horizontally across the network. In n-bit time, the states of all nodes will be set correctly for any CS pattern.

B. Routing Network

The routing network routes the copies of each inlet to their destinations. Therefore, any rearrangeable nonblocking network is a suitable candidate. The Benes-Clos network is the most well-known rearrangeable network which can be setup quite easily [14]. It is chosen as the routing network in the proposed broadcast switching network. The combination of the CS and the Benes-Clos routing networks is depicted in Fig. 8.

In the new broadcast switching network with a size of $N \times N$, i.e., N inlets and N outlets and N is a power of 2, not all of them can be assigned as source inlets and customer outlets. The number of source inlets denoted by P, and the number of outlets denoted by Q, much satisfy the equation $P + Q = N$. This is due to the fact that if inlet i in the CS network is requested by an outlet, then all inlets j, $j < i$, have to be broadcasted (even if they are not requested by any outlet) to satisfy the definition of a CS. Thus for inlet P, the maximum number of copies which can be created in the CS network is N

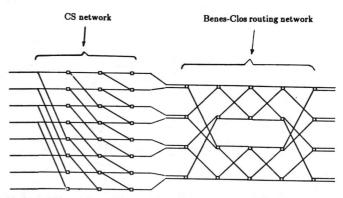

Fig. 8. The new broadcast network consists of a CS network plus a Benes-Clos routing network.

$- P$. This number imposes a limit on the number of customer outlets. An example is given in Fig. 9 where there are three source inlets and five customer outlets. Suppose all outlets request inlet 3, then inlets 1 and 2 have to be broadcasted. In a real implementation, the first P outlets can be assigned as monitor stations. Outlet i, $0 \leq i \leq P - 1$, always sends a request for the corresponding inlet i to satisfy the CS definition. The situation that any of these P monitor stations does not receive its corresponding source inlet is an indication of fault.

IV. A PHOTONIC IMPLEMENTATION BASED ON DIRECTIONAL COUPLERS

A directional coupler is an integrated optical device. It is created by diffusing titanium at high temperature into a lithium-niobate crystal to create channels in the crystal [18]. Optical signals propagate along the channel in much the same way as in optical fibers. But the evanescent field will propagate into a second, parallel channel placed about several micrometers away. The signal can be made to couple into an adjacent channel through coupling of the evanescent tail. The length required to couple light from one channel to the other is a function refractive index of the channel. The switching function of a directional coupler is based on the property that as the voltage across the two channels varies, so does the refractive index. Directional couplers are usually fabricated such that with no voltage applied, light that comes in on a particular waveguide completely couples over to the adjacent waveguide upon exiting the switch. This is shown in Fig. 10(b). When a specified voltage is applied across the electrodes, the coupling length is changed. The result is that after an optical signal is coupled into the second channel, it will couple back into the original channel. This is shown in Fig. 10(a). Based on the above description, it is realized that a directional coupler can be used like a 2-by-2 switch that has two states: the *bar* and *cross* states. At the *bar* state, the upper (lower) channel is transferred to the upper (lower) channel; at the *cross* states, the upper (lower) channel is transferred to the lower (upper) channel. An interconnection function is implemented by setting the states of each coupler properly. Many architecture issues for directional-coupler-based photonic switching systems are discussed in [12], [13], [19].

In the new network, each node in the routing network corresponds to one directional coupler. The node in the CS network has two modes: UB and LB. It can be implemented with two directional couplers and a passive combiner and splitter [see Fig. 11(a)]. The two directional couplers control the light signals from the two input links. At any instant, there is only one directional coupler is in the cross state to broadcast its signal to the two output links. The Y-branch divides the optical signal power by 2. Therefore, the number of couplers an optical signal can pass through is limited. To meet the

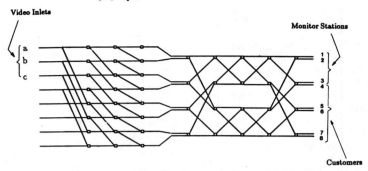

Fig. 9. In an N-by-N new broadcast network with P inlets and Q outlets, the equation $P + Q = N$ has to be satisfied. The first P outlets are reserved for monitor stations.

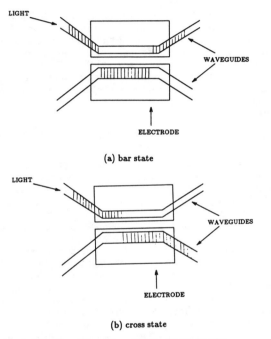

(a) bar state

(b) cross state

Fig. 10. Two states of a directional coupler.

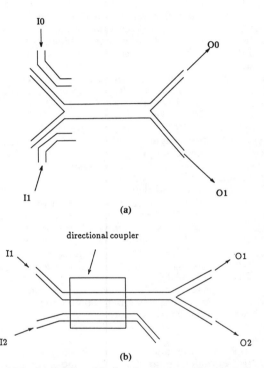

(a)

(b)

Fig. 11. (a) Implementation of the broadcast node consists of two directional couplers plus a combiner and a splitter. (b) In this implementation, only one directional coupler is needed. But two input light signals will pass through the same directional coupler. The crosstalk of this implementation is higher than that in Fig. 11(a).

signal-to-noise ratio (SNR) requirement, optical amplifiers [20]–[21] are required in a large directional-coupler-based photonic switching system. Since only one of the two directional couplers in Fig. 11(a) can be active at any time, we can replace them with only one directional. This implementation is shown in Fig. 11(b). It should be pointed out that the crosstalk in the second implementation is much higher than the first one since there is only one light signal passing through each directional coupler [23]. There is a tradeoff between the number of directional couplers and crosstalk. In this paper, we only focus on the crosspoint minimization, and the second implementation is used for comparison.

A directional has two operating modes which are totally different from those of an electronic crosspoint (its two modes are *open* and *closed*). This difference makes many architectures unsuitable for the directional-coupler photonic switching technology. As shown in Section II, the conventional Clos approach leads to an architecture requiring most of its nodes to have the broadcast capability. Similarly, in Richards and Hwang's broadcast network, the nodes in both stages must operate in both broadcast and nonbroadcast modes if they are implemented with directional couplers. While this is not an

issue in an electronic switching system, it can pose a difficult implementation problem for· a directional-coupler-based photonic switching system. For one thing, broadcast nodes will reduce the light signal power by half. Unnecessary broadcast nodes can cause a severe signal attenuation problem. Also, as shown above, a broadcast node is implemented in a different way from the nonbroadcast node. With the directional-coupler photonic switching technology, a node which has both broadcast and nonbroadcast modes is much more difficult to implement. But in the new network, the broadcast nodes and nonbroadcast nodes are separated into two different subnetworks, and can be implemented independently. The broadcast nodes are confined in the CS network, and the number of stages of the CS network is just the number required to generate the maximum number of copies for any connection. There is no redundant broadcast nodes in the new network.

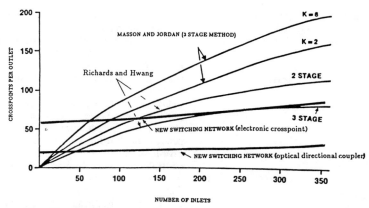

Fig. 12. Crosspoint comparison of the new network to other two approaches. The size of the new network is chosen to be 1024-by-1024.

V. PERFORMANCE

This section discusses several performance issues including the number of stages, the number of crosspoints, and the computing time for routing.

A. Number of Stages

In Masson and Jordan's or Richards and Hwang's broadcast networks, the number of stages are usually small (2 or 3). A network with few stages has a delay advantage if the crosspoints are implemented with *discrete* components. But it is not so if the large scale integration technology (such as VLSI or integrated optics) is used to implement the network. For example, if a *single* stage $N \times N$ crossbar switch is implemented on a single wafer, the number of crosspoints between an inlet-outlet pair is in the order of $O(N)$. While in the new network, a signal has to pass through only $O(3 \log_2 N)$ crosspoints. In a directional-coupler-based photonic switching system a network with fewer crosspoints between an inlet-outlet pair has less signal attenuation. It should be pointed out that a straightforward fan-out, fan-in architecture is a strictly broadcast network and has only $2 \log_2 N$ stages of switching [22]. But the number of crosspoints in this architecture is too large to make it practical for many applications.

B. Number of Crosspoints

With the advent of the VLSI technology, the number of crosspoints in a switching system is no longer as important as other criteria such as modular construction, regular topology for easy duplication, etc. But in a photonic switching system, the optical crosspoints are still expensive, and the number of crosspoints in a switching network remains to be an important design criterion. In this section, the comparison of the number of crosspoints between the new network and other broadcast networks is presented. Due to technology differences, the crosspoint counts for the electronic switching technology and for the directional-coupler photonic switching technology are not the same. For electronic switching technology, it needs four electronic crosspoints to implement each 2×2 switching element and two electronic crosspoints for each switching element in the CS network. For directional coupler technology, only one directional coupler is needed for the node in either the CS or the Benes-Clos routing network.

We first calculate the number of directional couplers required. As discussed in Section III-B, the number of source inlets denoted by P, and the number of customer outlets denoted by Q, must satisfy the equation $P + Q = N$. In general, Q is much larger than P (The number of customers is always much larger than the video channels). The comparison given below is based on the assumption that $Q \gg P$. The 2-by-

2 switching element in the Benes-Clos network can be implemented with one directional coupler. There are $2 \log_2 N - 1$ stages in the Benes-Clos routing network and each stage has $N/2$ switching nodes. Thus, there are $N/2(2 \log_2 N - 1)$ directional couplers in the Benes-Clos routing network. In the CS network, each 2×2 node with the broadcast capability can be implemented with one directional couplers [see Fig. 11(b)]. Thus, there are $N \log_2 N$ directional couplers in the CS network. In total, the number of directional couplers in the new broadcast network is $N(2 \log_2 N - 1/2)$. Since Q is the real number of customer outlets in the new network, the number of crosspoints per outlet is $N/Q(2 \log_2 N - 1/2)$. When $Q \gg P$, this number becomes approximately $(2 \log_2 Q - 1/2)$.

If electronic crosspoints are used to implement the network, the total number of electronic crosspoints is $6 N \log_2 N - 2N$. The number of electronic crosspoints per outlet becomes $N/Q(6 \log_2 N - 2)$. The results are depicted in Figs. 12–13.

The key difference between the new network and Masson and Jordan's or Richards and Huang's networks in terms of the number of crosspoints per outlet is that in Masson and Jordan's [1] or Richards and Hwang's [3] broadcast networks, it depends only on the number of *inlets*; but in the new network it depends only on the number of *outlets* under the condition that the number of outlets is much larger than the number of inlets. An example is given in Fig. 12 where the network size of the proposed network is fixed and is 1024×1024. The number of crosspoints per outlet in the proposed new network increases slightly as the number of inlets increases. On the other hand, if the number of inlets is fixed, the number of crosspoints per outlet in the other two approaches is a constant; while in the new network it increases as the number of outlets increases. An example is given in Fig. 13 where the number of inlets is fixed and equal to 128. As shown in the figure, the number of crosspoints per outlet in the new broadcast network first decreases as the network grows. It then increases and follows the equation $(2 \log_2 Q - 1/2)$ (or $6 \log_2 Q - 2$) closely when the number of outlets becomes large.

C. Growth

One advantage of the new broadcast network is its easy growth. The new network has a regular topology. An n-stage Benes-Clos network is a combination of two $(n - 1)$-stage Benes-Clos networks plus the nth stage of nodes. By adding an extra stage, the number of inlets and outlets is doubled. A n-stage CS network also has a similar structure. To make the CS network growable, all the open links, represented by the dash lines in Fig. 14, need to be kept for future growth.

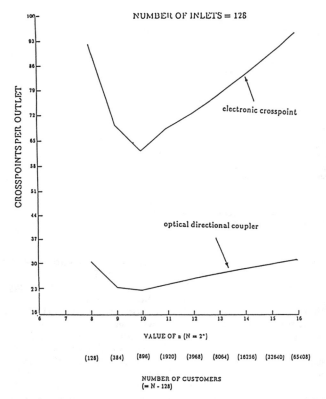

NUMBER OF INLETS = 128

CROSSPOINTS PER OUTLET

electronic crosspoint

optical directional coupler

VALUE OF a (N = 2ᵃ)

(128) (384) (896) (1920) (3968) (8064) (16256) (32640) (65408)

NUMBER OF CUSTOMERS
(= N - 128)

Fig. 13. Under the condition that the number of inlets is fixed, the number of crosspoints per outlet in the new network.

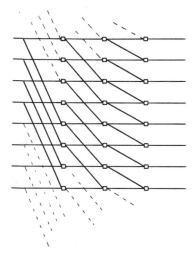

Fig. 14. The open links, indicated by the dashed lines, in the CS network are reserved for future growth.

D. Routing Algorithm

The routing algorithm in the new broadcast network is very simple. It consists of two parts: the routing in the CS network and the routing in the Benes-Clos routing network. For the CS network, the routing is straightforward. Once the number of copies requested for each source inlet is known, the matrix in (1) can be calculated. The corresponding bit in the matrix determines the path for each request. It takes a computing time of $O(N)$, N is the network size, to calculate the matrix serially. The next step is to do the routing for the Benes-Clos network. Since the destination of every copy of each inlet is known, we can determine the path for each request inside the Benes-Clos routing network. The looping algorithm [11] can do the routing for the Benes-Clos network in a time of $O(N \log_2 N)$. In total, the computing time for the new broadcast network is in the order of $O(N \log_2 N)$. Although the routing in the CS network can be done in a distributed fashion, the routing in the Benes-Clos routing network has to be calculated centrally.

One potential problem with this centralized control algorithm for a large rearrangeable network is that if incremental call setup requests are occurring at a high rate, the processing burden on the controller may be enormous. This problem can occur in almost all rearrangeable networks with a centralized controlling scheme. But most broadcast switching networks are rearrangeable nonblocking for two reasons. First, they are *nonblocking* because a video session is generally quite long. A switching network is simply not acceptable if one of its customers can get blocked for many hours. Second, they are normally *rearrangeable* nonblocking because 1) a strictly nonblocking broadcast network is too costly to implement (consider the number of crosspoints in the straightforward fan-in, fan-out approach), and 2) the setup and tearing down may not occur as often as in a voice switch (a video session is generally much longer than a voice call). But the probability of rearrangement and the number of affected connections in setting up a new call are important issues and deserve extensive future study.

In the next section a self-routing version is presented. Although it is still a rearrangeable nonblocking network, the routing is done is a totally distributed fashion (no central controller is needed). Each time slot is divided into two parts: path setup period and data transmission period. This path setup will be done at the beginning of each slot even if no new call is added. The path setup time is kept small compared to the data transmission time (this can be done by extending the length of the data field) to maintain high efficiency. The call disturbance is no longer an issue in a network with a fully distributed controlling scheme.

VI. SELF-ROUTING BROADCAST SWITCHING NETWORK

Sorting networks can be applied to switching [15]–[16]. Replacing the Benes-Clos network with a sorting network, we have a self-routing broadcast switching network. Fig. 15(a) is an implementation of the self-routing broadcast network. It consists of the CS network, a Batcher sorting network, and an array of subtractors. The direction of the data flow and the control flow are indicated in the figure. To set up the paths, the outlets send the source addresses as the routing tags into the control network. As discussed in the previous section, the equation $P + Q = N$ where P and Q are the number of source inlets and customer outlets, respectively, has to be satisfied. Similar to the network in Fig. 9, the first P outlets are designated as monitor stations. Outlet i, $0 \leq i \leq P - 1$, always requests source inlet i in any condition. The real Q customers request their own source inlets and send the source addresses as the routing tag into the sorting network. The sorting network sorts the tags in order according to the source inlet addresses. Each node in the sorting network use only one bit of the control tag, which contains the source inlet address, to set the state of the node. The node in the sorting network can be set into correct states in a m-bit time where m is the number of the stages in the sorting network. The array of subtractors in the control network subtracts the source addresses from their own addresses to form the new routing tags. Each new routing tag corresponds to a row in the matrix in (1). The new routing tag is sent horizontally into the CS network to control the nodes in the CS network. In this modified version, no central processor is needed for routing calculation. All the nodes are set up in correct states in a totally distributed way. Among the different sorting networks [17], Batcher sorting network has the advantage of easy implementation and is chosen in our

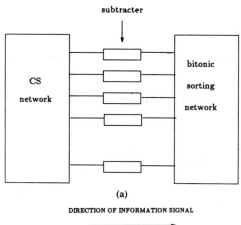

(a)

DIRECTION OF INFORMATION SIGNAL

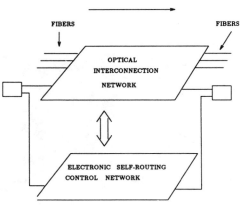

DIRECTION OF CONTROL SIGNAL

(b)

Fig. 15. (a) The self-routing broadcast network consists of an array of subtracters which generate the address difference sequence for controlling the CS network. (b) The hybrid approach consists of an optical interconnection network and an electronic self-routing control network.

design. There are $1/2(\log_2 N(\log_2 N + 1))$ stages in the Batcher sorting network. With the self-routing method, the time required to set up the path is $\log_2 N + 1/2(\log_2 N(\log_2 N + 1))$.

With the current technology, the self-routing control logic cannot be integrated on the same wafer for directional couplers. But we can combine an optical interconnection network for data transmission and an electronic self-routing control network for controlling the electrodes. Fig. 15(b) shows this hybrid approach. There is a one-to-one correspondence between a directional coupler in the optical interconnection network and an electronic switching node in the control network (the optical interconnection network does not have the array of subtractors). The total amount of directional couplers is increased to $N \log_2 N + N/4(\log_2 N (\log_2 N + 1))$. For $N = 1024$, the number of directional couplers in the self-routing broadcast network is about two times that in the nonself-routing version.

VII. APPLICATION FOR PARALLEL PROCESSING

In addition to circuit-switched video, another major application for a multiconnection network is parallel processing. Data broadcasting is a fairly common step in parallel processing [9]. The condition that the number of outlets is much larger than the number of inlets is not always true in parallel processing. The new network can become very inefficient for this

application. However, similar to Ofman's generalized connection network, we can add a concentrator in front of the CS network so that inactive inlets need not always be broadcasted to save the number of switching elements, at the cost of more stages in the network.

VIII. CONCLUSIONS

Circuit-switched video is an important service in BB-ISDN. The technical challenge is to deliver the service economically so that the majority of customers can afford it. Hopefully, it then stimulates the demand for other image related services. In this paper, a new rearrangeable broadcast switching network was presented. It has many advantages such as fewer switching elements, easy growth, an easy path hunt, and a modularly constructable topology. Of particular interest is that the new architecture is much more suitable for the directional-coupler photonic switching technology. The comparison of the number of crosspoints per outlet of the new network to that of other approaches, and the computing time for routing were also discussed. Finally, a modified version of the broadcast network was proposed in which the setting of the nodes can be done in a distributed way and only takes $O((\log_2 N)^2)$ time.

REFERENCES

[1] G. M. Masson and B. W. Jordan, "Generalized multi-stage connection networks," *Networks*, vol. 2, pp. 191–209, 1972.

[2] F. K. Hwang, "Rearrangeability of multi-connection three-stage Clos network," *Networks,* vol. 2, pp. 301–306, 1972.

[3] G. W. Richards and F. K. Hwang, "A tow-stage rearrangeable broadcast switching network," *IEEE Trans. Commun.,* vol. COM-33, pp. 1025–1035, Oct. 1985.

[4] C. Clos, "A study on non-blocking switching networks," *B.S.T.J.,* vol. 32, pp. 406–424, 1954.

[5] V. E. Benes, "On rearrangeable three-stage connecting networks," *B.S.T.J.,* vol..41, pp. 1481–1492, 1962.

[6] G. M. Masson *et al.,* "A sampler of circuit switching networks," *Computer,* pp. 32–48, June 1979.

[7] J. P. Ofman, "A universal automation," *Trans. Moscow Math. Soc.,* vol. 14, 1965 (translation published by *Amer. Math. Soc.,* Providence, RI, 1967, pp. 200–215).

[8] C. D. Thompson, "Generalized connection networks for parallel processor interconnection," *IEEE Trans. Comput.,* vol. C-27, pp. 1119–1125, Dec. 1978.

[9] D. Nassimi and S. Sahni, "Parallel permutation and sorting algorithms and a new generalized connection network," *J. Assoc. Comput. Mach.,* vol. 29, no. 3, pp. 642–667, July 1982.

[10] T.-Y. Feng, "Data manipulating functions in parallel processors and their implementations," *IEEE Trans. Comput.,* vol. C-23, pp. 309–318, Mar. 1974.

[11] R. J. McMillien and H. J. Siegel, "Routing schemes for the augmented data manipulator network in an MIMD system," *IEEE Trans. Comput.,* vol. C-31, pp. 1202–1214, Dec. 1982.

[12] C.-T. Lea, "Cross-over minimization in directional coupler based photonic switching systems," *IEEE Trans. Commun.,* to be published.

[13] W. Payne and S. Hinton, "System considerations for lithium niobate photonic switching technology" presented at 1987 Topical Meet. Photonic Switching, Reno, NV, Mar. 1987.

[14] D. C. Opferman and N. T. Tsao-Wu, "On a class of rearrangeable switching networks—Part I: Control algorithms; Part II: Enumeration studies of fault diagnosis," *B.S.T.J.,* pp. 1579–1618, 1971.

[15] K. Batcher, "Sorting networks and their applications," in *Proc. AFIPS Conf.,* 1968, pp. 307–314.

[16] A. Huang and S. Knauer, "STARLITE: A wideband digital switch," in *Proc. Globecom '84,* 1984, pp. 121–125.

[17] D. Knuth, *The Art of Computer Programming, Vol. III: Sorting and Searching.* Reading, MA: Addison-Wesley, 1973.

[18] R. V. Schmidt and R. C. Alferness, "Directional couplers switches, modulators, and filters using alternating $\Delta\beta$ techniques," *IEEE Trans. Circuits Syst.,* vol. CAS-26, pp. 1099–1108, Dec. 1979.

[19] H. S. Hinton, "A nonblocking optical interconnection network using directional couplers," presented at Proc. GLOBECOM '84, Nov. 1984.

[20] S. Kobayashi and T. Kimura, "Semiconductor optical amplifiers," *IEEE Spectrum,* pp. 26–33, May 1984.

[21] Y. Silberberg, "All-optical repeater," *Opt. Lett.,* vol. 11, no. 6, June 1986.

[22] R. A. Spanke, "Architectures for large non-blocking optical switches," *IEEE J. Quantum. Electron.,* vol. QE-22, pp. 885–889, Aug. 1986.

[23] K. Padmanabhan and A. Netravali, "Dilated networks for photonic switching," presented at 1987 Topical Meet. Photonic Switching, Reno, NV, Mar. 1987.

Chin-Tau Lea (M'82) received the B.S. and M.S. degrees from the National Taiwan University, Taiwan, Republic of China, in 1976 and 1978, respectively, and the Ph.D. degree from the University of Washington, Seattle, in 1982, all in electrical engineering.

From August 1982 to September 1985 he was with AT&T Bell Lab Laboratories, Naperville, IL, where his activities included advanced switching architecture design, protocol design, and performance evaluation. Since September 1985, he has been an Assistant Professor at Georgia Institute of Technology. His current areas of technical interests are photonic switching, interconnection networks, multiaccess protocols, and the architectures of future information networks.

Dr. Lea received the DuPont Young Faculty Award in 1986. He has also been awarded three US patents for his architecture work on fault-tolerant interconnection networks for parallel processing and high-speed switching.

Reprinted from *IEEE Journal on Selected Areas in Communications*, Vol. 6, No. 9, December 1988, pages 1455-1467.

Nonblocking Copy Networks for Multicast Packet Switching

TONY T. LEE, SENIOR MEMBER, IEEE

Abstract—In addition to handling point-to-point connections, a broadband packet network should be able to provide multipoint communications that are required by a wide range of applications such as teleconferencing, entertainment video, LAN bridging, and distributed data processing. The essential component to enhance the connection capability of a packet network is a *multicast packet switch*, capable of packet replications and switching, which is usually a serial combination of a copy network and a point-to-point switch. The copy network replicates input packets from various sources simultaneously, and then copies of broadcast packets are routed to their final destination by the switch. A nonblocking, self-routing copy network with constant latency is proposed in this paper. Packet replications are accomplished by two fundamental processes, an encoding process and a decoding process. The encoding process transforms the set of copy numbers, specified in the headers of incoming packets, into a set of monotone address intervals which form new packet headers. This process is carried out by a running adder network and a set of dummy address encoders. The decoding process performs the packet replication according to the Boolean interval splitting algorithm through the broadcast banyan network. Finally, the destinations of copies are determined by the trunk number translators. At each stage of the broadcast banyan network, the decision making is based on a two-bit header information. This yields minimum complexity in the switch nodes.

I. INTRODUCTION

THE goal of a broadband packet network is to provide flexible communication for handling multipoint connections in addition to point-to-point connections. A wide class of applications, such as teleconferencing, entertainment video, LAN bridging, and distributed data processing require multipoint communications. The essential component in the network to achieve this is a *multicast packet switch* which is capable of packet replication and switching.

A multicast packet switch is usually a serial combination of a *copy network* and a *point-to-point switch* as illustrated in Fig. 1. The copy network replicates input packets from various sources simultaneously, and then copies of broadcast packets are routed to their final destination by the point-to-point switch. Current designs of packet switch systems to support multipoint communications include the *Starlite System* proposed by A. Huang and S. Knauer [4], and the *Broadcast Packet Switch* developed by J. S. Turner [7], [8]. Both systems are parallel

Manuscript received November 1, 1987; revised June 2, 1988. This paper was presented at the 1988 International Zurich Seminar on Digital Communications, Zurich, Switzerland, ETH, March 8–10, 1988.
The author is with Bell Communications Research, Morristown, NJ 07960.
IEEE Log Number 8824385.

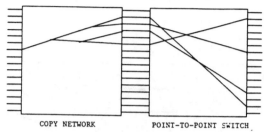

Fig. 1. A multicast packet switch consists of a copy network and a point-to-point switch.

processors with distributed control, and are constructed out of interconnected networks—regularly interconnected identical switch cells [9]. These simple structures lend themselves very well to cost-effective VLSI implementation.

The Starlite copy network is a receiver initiated system. The inputs to the network are original source packets and empty packets generated by receivers. The copy network consists of a *sorting network* [1] and a *broadcast network*. Initially, the sorting network sorts the input packets on their source addresses. At the outputs of the sorting network, the original source packet and the associated empty packets with the same source address will appear contiguously. The broadcast network then replicates the data in each source packet and inserts them into the empty data fields of subsequent empty packets until another source packet is encountered. This process is illustrated in Fig. 2. It is easy to notice that the process of packet replication in this design assumes the synchronization of the source and destinations. An "empty packet setup procedure" is also required. It is not feasible to implement this approach in a broadband packet network where packets may usually experience delay variations due to buffering, multiplexing, and switching in a multiple-hop connection.

The copy network of Turner's broadcast packet switch employs a *banyan network* and a set of *broadcast and group translators* (BGT's). The header of an input broadcast packet contains two fields, a *fan-out* indicating the number of copies requested by the packet, and a *broadcast channel number* (BCN) used by the BGT's to determine the final destination of copy packets. Fig. 3 depicts an example of this copy network, which replicates packets by splitting the fan-out. Using a table lookup method, the trunk number translation is performed by the BGT's when

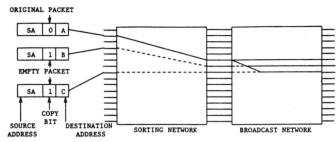

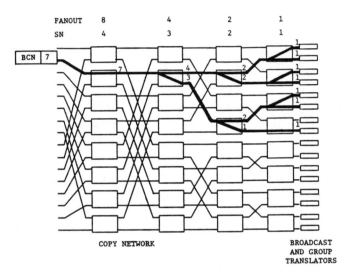

Fig. 2. Copy network of Starlite system.

Fig. 3. Copy network of broadcast packet switch.

copies appear at the outputs. Packet collisions cannot be avoided in this scheme, and it is blocking, meaning that two packets may arrive at the same node and attempt to leave on the same output link at the same time. Buffers are required for every internal node to prevent packet loss.

A nonblocking, self-routing copy network with constant latency is proposed in this paper. The basic structure consists of the following components.

• *Running Adder Network (RAN):* It generates running sums of *copy numbers*, specified in the headers of input packets.

• *Dummy Address Encoder (DAE):* It takes adjacent running sums to form a new packet header.

• *Broadcast Banyan Network (BBN):* It is a banyan network with broadcast switch nodes capable of packet replications based on two-bit header information.

• *Trunk Number Translator (TNT):* It determines the outgoing trunk number for each copy packet.

Fig. 4 illustrates the basic structure of our copy network, which can be considered as a cascaded combination of a packet header encoder and a decoder. When broadcast packets are received at the running adder network, the *number of copies* (CN's) specified in the packet headers are added up recursively. The dummy address encoders form new headers consisting of two fields: a *dummy address interval* and an *index reference* (IR). The dummy

address interval, formed by adjacent running sums, is represented by two binary numbers, namely, the *minimum* (MIN) and *maximum* (MAX). The index reference is set equal to the minimum of the address interval, used later by the trunk number translators to determine the *copy index* (CI). The broadcast banyan network replicates packets according to a *Boolean interval splitting algorithm* based on the address interval in the new header. When copies finally appear at the outputs, the TNT's compute the copy index for each copy from the output address and index reference. The *broadcast channel number* (BCN) together with copy index form a unique identifier for each copy. The TNT's then translate this identifier into a *trunk number* (TN), which is added to the packet header and used by the switch to route the packet to its final destination.

References to related articles are cited and background materials are provided in the course of the presentation to make this paper self-contained. Section II begins with a brief introduction to the self-routing and nonblocking properties of banyan networks as a point-to-point switch. Following that, these properties are generalized to broadcast banyan networks along the same principle. The basic approach to synthesize our nonblocking copy network is detailed in Section III. Broadcast networks with larger fanout than fan-in are desirable in many services applications such as entertainment video. The topology of expansion networks that preserve the self-routing and nonblocking properties of this structure are presented in Section IV. Concentration of active inputs is one of the criteria for the network being nonblocking. Two techniques, one with prioritization and the other without, are discussed in Section V. They were both developed for the Starlite switch and can also be incorporated into our copy network. Section VI is devoted to the estimation of overflow probabilities of the inputs of copy network by means of Chernoff bound. Finally, conclusions of present work and suggestions for future research are summarized in Section VII. W. H. Bechmann had proved a sufficient nonblocking condition of banyan network by introducing an elegant numbering scheme [2], which provides an insight into its topological properties. The proof is paraphrased and appended at the end of this paper.

II. BROADCAST BANYAN NETWORKS

The routing algorithm and nonblocking condition of broadcast banyan networks are discussed in this section. Based on these fundamental properties, the design of a nonblocking copy network is detailed in the next section. A banyan network is an $n \times n$ interconnection network with $N = \log_2 n$ stages. Each stage contains $n/2$ nodes and each node is a 2×2 switch element. The banyan network has been used as a point-to-point switch, based on the so-called *self-routing algorithm* in both Starlite switch (without internal buffering) and Fast Packet Switch (with internal buffering). Fig. 5 illustrates a four-stage banyan network and the self-routing algorithm. The header of each incoming packet contains an n-bit desti-

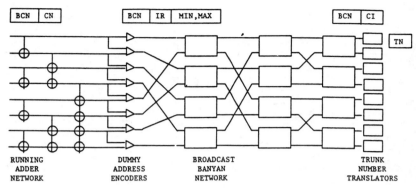

Fig. 4. The basic structure of a nonblocking copy network.

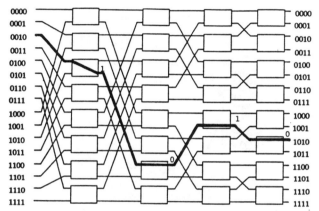

Fig. 5. A four-stage banyan network and the self-routing algorithm for point-to-point switching.

nation address, and the packet routing is done by decoding the header. That is, a node at stage k sends the packet out on link 0 (up) or link 1 (down) according to the kth bit of the header. This property is due to the topology of banyan networks: the path from any input to an output is uniquely determined by the output address.

An interconnection network is nonblocking if the packet routing is not store-and-forward, and internal buffers are not needed within the switch nodes. The banyan network itself is a blocking network. Two packets with different destination addresses may be routed through the same internal link at the same time. However, it is known that if the incoming packets with distinct destination addresses are arranged in an ascending or a descending order, then the banyan network becomes nonblocking. The Starlite switch is a nonblocking network, which is realized by combining a Batcher sorting network and a banyan network. The sorting network sorts the incoming packets on their destination addresses. The nonblocking condition of the banyan network is formally stated as follows [4].

NB1: A banyan network, as a point-to-point switch, is nonblocking if the active inputs $x_1, \cdots, x_k$ and the corresponding outputs $y_1, \cdots, y_k$ satisfy the following.

1) (Monotone): $y_1 < y_2 < \cdots < y_k$, or $y_1 > y_2 > \cdots > y_k$.

2) (Concentration): Any input between two active in-

puts is also active. That is, $x_i \leq w \leq x_j$ implies input w is active.

By induction on the stages, it is easy to see that the above statement is a sufficient nonblocking condition. A rigorous proof is given by W. H. Bechmann [see Appendix].

A. Generalized Self-Routing Algorithm

A broadcast banyan network is a banyan network with switch nodes which are capable of packet replications. A packet arriving at each broadcast switch node can be either routed to one of the output links, or it can be replicated and sent out on both links. The uncertainty of making a decision with three possibilities is $\log_2 3 = 1.585$, meaning that the minimum header information provided by a packet for each broadcast node is 2 bits. Bearing this lower bound in mind, we are going to show how to synthesize an algorithm for packet replications that minimizes the logic complexity of switch nodes.

Fig. 6 illustrates a straightforward generalization of the above self-routing algorithm to multiple addresses. A broadcast packet contains a set of arbitrary n-bit destination addresses. When a packet arrives at a node in stage k, the packet routing and replication are determined by the kth bit of all the destination addresses in the header. If they are all "0" or all "1," then the packet will be sent out on link 0 or link 1, respectively. Otherwise, the packet and its replica are sent out on both links with the following modified header information: the header of the packet sent out on link 0 or link 1, contains these addresses in the original header with their kth bit equal to 0 or 1, respectively. The set of paths from any input to a set of outputs form a (binary) tree embedded in the network, and it will be called an *input–output tree*. The modification of packet headers is performed by the node whenever the packet is replicated.

There are several problems which may arise in implementing this generalized self-routing algorithm. First, the packet header contains a variable number of addresses and the switch nodes have to read all of them. Second, the process of packet header modification depends on the entire set of addresses. This would become a burden on switch nodes. Finally, the input–output trees generated by

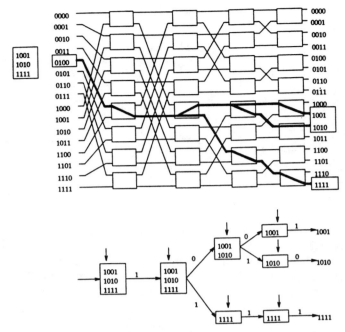

Fig. 6. An input–output tree generated by generalized self-routing algorithm.

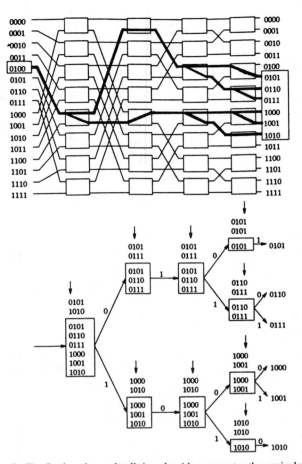

Fig. 7. The Boolean interval splitting algorithm generates the equivalent input–output tree of a packet with interval addresses.

an arbitrary set of input broadcast packets are not link-independent in general and the network is obviously blocking. All these problems are due to the irregularity of the set of destination addresses in the packet header. Fortunately, these problems can be avoided if the set of addresses is restricted to be an interval. This point is elaborated below. This interval constraint on addresses will not affect packet replication because addresses for the copy network are fictitious and are not used for switching. The switching of replicated packets is performed in the succeeding switching network based on actual addresses.

B. Boolean Interval Splitting Algorithm

An *address interval* is a set of contiguous N-bit binary numbers, which can be represented by two numbers, namely, the *minimum* and *maximum*. Suppose that a node at stage k received a packet with the header containing an address interval specified by the two binary numbers: min $(k - 1) = m_1 \cdots m_N$ and max $(k - 1) = M_1 \cdots M_N$ where the argument $(k - 1)$ denotes a node in stage $(k - 1)$ from where the packet came to stage k. Following the generalized self-routing algorithm, the direction for packet routing is described as follows.

1) If $m_k = M_k = 0$ or $m_k = M_k = 1$, then send the packet out on link 0 or 1, respectively.

2) If $m_k = 0$ and $M_k = 1$, then replicate the packet, modify the headers (according to the scheme described below) and send both packets out on both links.

It is easy to realize from these rules that $m_i = M_i$, $i = 1, \cdots, k - 1$ holds for every packet which arrives at stage k. Therefore, the event $m_k = 1$ and $M_k = 0$ is impossible, a don't care condition, due to the min–max representation of the address intervals. Fig. 7 illustrates the

Boolean interval splitting algorithm which generates the equivalent input–output tree of a broadcast packet with interval destination addresses.

The modification of a packet header is simply splitting the original address interval into two subintervals, which can be expressed by the following recursion:

$$\min (k) = \min (k - 1) = m_1 \cdots m_N,$$
$$\max (k) = M_1 \cdots M_{k-1} 01 \cdots 1 \qquad (1a)$$

for the packet sent out on link 0, and

$$\min (k) = m_1 \cdots m_{k-1} 10 \cdots 0,$$
$$\max (k) = \max (k - 1) = M_1 \cdots M_N \qquad (1b)$$

for the packet sent out on link 1 at stage k.

This modification procedure is a routine operation, meaning that it is independent of the header contents and can be considered as a built-in hardware function. Fig. 8 illustrates the switch node logic at stage k to perform this Boolean interval splitting algorithm. The decision at each node is based on a two-bit information provided by the packet header, which is the lower bound that we mentioned at the beginning of Section II-A). Thus, the logic

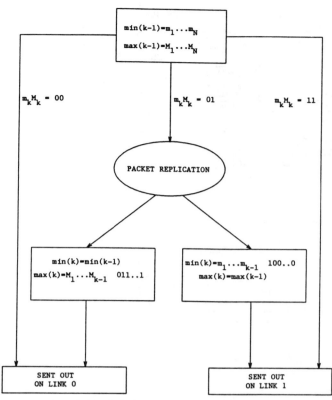

Fig. 8. The switch node logic at stage k of a broadcast banyan network.

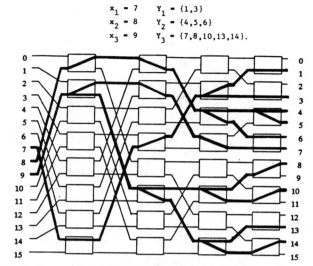

$$x_1 = 7 \qquad Y_1 = \{1,3\}$$
$$x_2 = 8 \qquad Y_2 = \{4,5,6\}$$
$$x_3 = 9 \qquad Y_3 = \{7,8,10,13,14\}.$$

Fig. 9. An example to demonstrate the nonblocking condition of broadcast banyan networks.

complexity of broadcast switch nodes reached its minimum.

C. Nonblocking Condition of Broadcast Banyan Networks

As we described earlier, a set of input broadcast packets generates a set of input–output trees embedded in the broadcast banyan network. The network is nonblocking if this set of trees is always link-independent. The following is a sufficient condition, generalized from *NB1*, which assures that the broadcast banyan network is nonblocking.

NB2: A broadcast banyan network is nonblocking if the active inputs $x_1, \cdots, x_k$ and the corresponding sets of outputs $Y_1, \cdots, Y_k$ satisfy the following:

1) (Monotone): $Y_1 < Y_2 \cdots < Y_k$, or $Y_1 > Y_2 > \cdots > Y_k$.

2) (Concentration): Same as *NB1*.

The above inequality $Y_i < Y_j$ indicates that every output address in Y_i is less than every output address in Y_j. An example with active inputs $x_1 = 7$, $x_2 = 8$, $x_3 = 9$, and corresponding outputs $Y_1 = \{1, 3\}$, $Y_2 = \{4, 5, 6\}$, $Y_3 = \{7, 8, 10, 13, 14\}$ is illustrated in Fig. 9. The proof of *NB2* is straightforward. We know from *NB1* that the set of input–output paths $\langle x_1, y_1 \rangle, \cdots, \langle x_k, y_k \rangle$ are link-independent for an arbitrary choice of output addresses $y_1 \in Y_1, \cdots, y_k \in Y_k$. It follows that the set of input–output trees embedded in the broadcast banyan network must be link-independent, and *NB2* is established.

III. The Basic Structure of Copy Networks

The basic architecture of the copy network has been described in Section I and is shown in Fig. 4. Packet replications are accomplished by two fundamental processes, an encoding process and a decoding process. The encoding process transforms the set of copy numbers, specified in the headers of incoming packets, into a set of monotone address intervals which form new packet headers. This process is carried out by a running adder network and a set of dummy address encoders. The decoding process performs the packet replication according to the Boolean interval splitting algorithm through the broadcast banyan network described in the previous section. Finally, the destinations of copies are determined by the trunk number translators.

A. Encoding Process

The running adder is a $\log_2 n$-stage network constructed from $n/2 \log_2 n$ nodes. It can be arranged as a top-down or a bottom-up recursive structure as illustrated in Fig. 10. Each node is an adder with two inputs and two outputs where the vertical line is simply a pass. The main function of the running adder network is to assign output addresses for each broadcast packet according to the requested number of copies. It first generates the running sums of copy numbers specified by input packets at each port. The dummy address encoders then form the new headers from adjacent running sums. The new header consists of two fields: one designates the dummy address interval, which is represented by two n-bit binary numbers, the minimum and maximum, and the other contains an index reference which is equal to the minimum of the address interval.

If allocation of output addresses begins with address 0, then the following sequence of dummy address intervals will be generated by a top-down running adder:

$$(0, S_0 - 1), (S_0, S_1 - 1), \cdots, (S_{n-2}, S_{n-1} - 1)$$

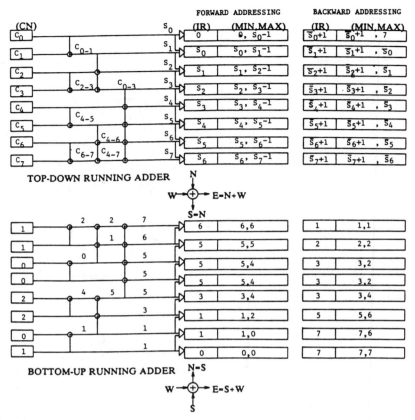

Fig. 10. Running adder networks and dummy address encoders.

where S_i is the ith running sum of the copy numbers. This sequence results in a *forward address assignment*. Suppose $\bar{S}_i$ indicates the 1's complement of S_i, which is formed by changing all 0's in S_i to 1's and 1's to 0's. Then, the sequence

$$(\bar{S}_0 + 1, n - 1), (\bar{S}_1 + 1, \bar{S}_0), \cdots, (\bar{S}_{n-1} + 1, \bar{S}_{n-2})$$

allocates addresses starting from the bottom of the outputs, a *backward address assignment*. The length of each interval is equal to the corresponding copy number in both addressing schemes. It also should be noticed that both sequences are monotone to satisfy the nonblocking condition described previously.

It is evident, as shown in Fig. 10, that the structures of top-down and bottom-up running adders are "complementary" to each other, which suggests that they can be implemented on a single chip. Although only one adder is needed at any time-slot, the other one can be a substitution. If only one running adder is used, say a top-down network, then the high numbered inputs are more likely to overflow than the lower numbered ones. Furthermore, forward or backward address assignment tends to concentrate copy packets on one side, low or high, respectively, at the outputs of the copy network. This of course may increase the potential buffer overflow at the inputs of succeeding point-to-point switch. This asymmetry can be alleviated by using *dual running adder networks*, a top-

down network with forward address assignment and a bottom-up network with backward address assignment. The two networks can be operated alternately from time-slot to time-slot. As a result, fairness for inputs and arbitration for outputs can therefore be achieved simultaneously. The adder also collects useful traffic information that is essential for flow control of broadcast channels, by counting the total number of copies at every time-slot.

B. Decoding Process

The decoding process is carried out by a broadcast banyan network and a set of trunk number translators.

The procedure of packet replication and routing within a switch node at stage k is illustrated in Fig. 11. The minimum and maximum in the packet header are bit-interleaved in the actual implementation of the address interval splitting algorithm. An *activity bit A* is posted in front of the address field to indicate that the packet is active ($A = 1$) or empty ($A = 0$). Fig. 12 illustrates the logic diagram of a switch node at stage k of a broadcast banyan network. The truth table of a *packet controller* is given in Fig. 13. It controls the procedure of packet replication and header modification. The packet routing is controlled by the *route controller* with truth table given in Fig. 14.

When packets emerge from the banyan network, the address interval in their header contains only one address at the outputs of stage N, the last stage. That is, according

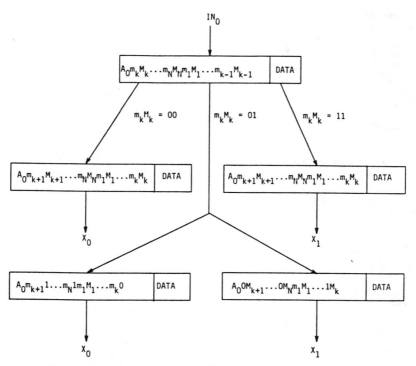

Fig. 11. The procedure of packet replication and routing within a node at stage k.

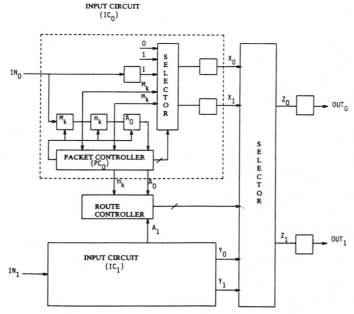

Fig. 12. The logic diagram of a broadcast switch node at stage k.

to Boolean interval splitting algorithm, we have

$$\min(N) = \max(N) = \text{output address}. \quad (2)$$

The packets belonging to the same broadcast channel should be distinguished by copy indexes. Recall that the index reference is initially set equal to the minimum of address interval. Thus, the copy index of each packet is determined by

$$\text{copy index} = \text{output address} - \text{index reference}. \quad (3)$$

This point is illustrated in Fig. 15.

The function of trunk number translators is to assign a new address to each copy of a broadcast packet. Then the copies are routed to their final destination by the succeed-

Fig. 13 Truth Table

					2(N−k) Clocks				2(k−1) Clocks								Data Length	
A_0	m_k	M_k	X_0	X_1	X_0	X_1	X_0	X_1	X_0	X_1	X_0	X_1	X_0	X_1			X_0	X_1
1	0	1	1	1	I	0	I	I	I	I	m_k	1	0	M_k			I	I
1	0	0	1	0	I	d	I	d	I	d	m_k	d	M_k	d			I	·d
1	1	1	0	1	d	I	d	I	d	I	d	m_k	d	M_k			d	I
0	d	d	0	0	d	d	d	d	d	d	d	d	d	d			d	d

d = don't care

Fig. 13. The truth table of a packet controller.

A_1	A_0	m_k	Z_0	Z_1
0	d	d	X_0	X_1
d	0	d	Y_0	Y_1
1	1	0	X_0	Y_1
1	1	1	Y_0	X_1

Fig. 14. The truth table of route controller.

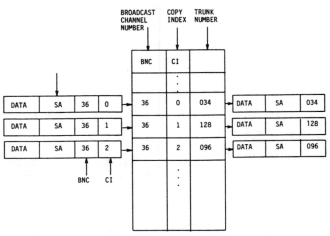

Fig. 16. Trunk number translation by table look-up.

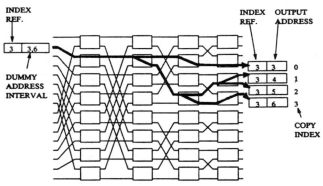

Fig. 15. Computation of copy indexes.

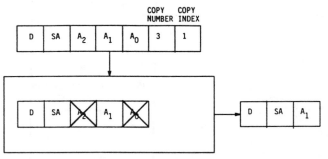

Fig. 17. Trunk number selection.

ing point-to-point switch. Trunk number assignment can be accomplished by a simple table look-up based on the identifier consisting of the broadcast channel number and the copy index associated with each packet. The trunk number translators contain a table with attributes BCN and CI as its primary key. When a TNT receives a copy of broadcast packet, it first converts the output address and IR into CI, and then replaces the BCN and CI with the corresponding trunk number in the translation table. This translation process is illustrated in Fig. 16. An alternative is trunk number selection. Suppose the BCN field contains a list of all trunk numbers belonging to the broadcast channel. The copy index can then be used as a pointer to select one of the trunk numbers in the list and erase the rest. The two parameters CN and CI are inputs to a state machine, which controls the shifting of a shift register and the loading of a buffer. This selection process is illustrated in Fig. 17. The number of copies and destinations of a broadcast channel may change frequently during a connection. Updating this information is fairly easy with both schemes.

Fig. 18 demonstrates the distributed parallel processes of packet replications described in this section.

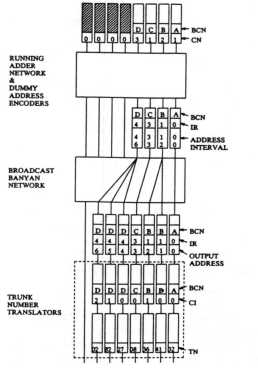

Fig. 18. The distributed parallel processes of packet replications.

IV. EXPANSION NETWORKS

The packet replications are divergent branching processes. Nonsquare networks are more appropriate in many practical applications such as entertainment video distribution and audio broadcast. An N-stage expansion network with $m = 2^M$ inputs and $n = 2^N$ outputs can be arranged as a combination of a set of 2^M $(N - M)$-stage binary trees and a set of 2^{N-M} M-stage square banyan networks. Fig. 19 illustrates an expansion network with 4 inputs and 16 outputs. The binary trees are made up of switch elements with one input and two outputs. The switch node logic is the same as the banyan network. Each input line is expanded to 2^{N-M} lines by an $(N - M)$-stage binary tree to achieve the desired expansion ratio. The remaining M stages contains square banyan networks entirely.

The interconnection of 2^M binary trees and 2^{N-M} banyan networks is similar to the link system of a multistage crossbar switch. The outputs of a binary tree can be uniquely identified by a 2-tuple $(x_1 \cdots x_M, y_1 \cdots y_{N-M})$ where $x_1 \cdots x_M$ is the top-down numbering of the binary trees and $y_1 \cdots y_{N-M}$ is the local address of each output within the binary tree. The input of a banyan network can also be identified by a 2-tuple $(a_1 \cdots a_{N-M}, b_1 \cdots b_M)$ where $a_1 \cdots a_{N-M}$ is the top-down numbering of the banyan network and $b_1 \cdots b_M$ is the local address of the input. The binary trees and banyan networks are transpose interconnected according to this numbering scheme. That is, the two ports $(x_1 \cdots x_M, y_1 \cdots y_{N-M})$ and $(a_1 \cdots a_{N-M}, b_1 \cdots b_M)$ are interconnected by a link if $(x_1 \cdots x_M, y_1 \cdots y_{N-M}) = (b_1 \cdots b_M, a_1 \cdots a_{N-M})$.

The topology of an expansion network is uniquely determined for a given fan-in m and fan-out n. The outputs of an expansion network can be concentrated and multiplexed into trunk number translators as illustrated in Fig. 20. We can use this to reduce the overflow probability of a copy network. It should be noticed that the perfect shuffle of the inputs to each banyan network is replaced by the perfect shuffle of inputs to the binary trees. It is easy to show that the $(N - M)$ stage binary trees with the transpose interconnection preserve the ordering of the inputs. Therefore, the expansion network is nonblocking if the destinations of inputs are monotone and concentrated in the manner of *NB2*.

V. INPUT CONCENTRATION

To satisfy the nonblocking condition, idle inputs between active inputs must be eliminated. Recall that the concentration criterion of *NB2* is the same as its counterpart in *NB1*, which means that the same input concentration technique for point-to-point switch can be applied to the copy network. There are two approaches proposed in the Starlite switch to concentrate the active inputs [4]. Both approaches are based on the activity bit preceding the packet header.

A. Concentration without Prioritization

The first approach uses a concentrator network consisting of a running adder network and a *reverse banyan net-*

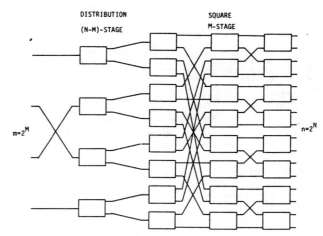

Fig. 19. An expansion network consisting of binary trees and banyan networks.

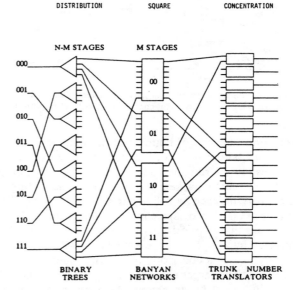

Fig. 20. An expansion network with output concentration.

work as illustrated in Fig. 21(a). The running adder computes the *routing address* based on the active bit by counting the total number of active packets above the line. The reverse banyan network is a mirror image banyan network. It ignores inactive packets and routes active packets based on the running sum of activity bits. Thus, all active packets will emerge contiguously at the outputs. For all input sets based on this addressing scheme, it is easy to see from *NB1*, that this concentrator network is also nonblocking. A copy network with this concentrator is illustrated in Fig. 21(b). Notice that fairness for inputs and arbitration for outputs are achieved by dual running adder networks in this design.

B. Concentration with Priority Sorting

The second approach involves priority sorting. The top-down (or bottom-up) running adder of the copy network allocates the output addresses to the incoming packets

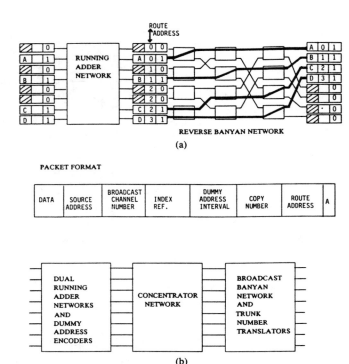

PACKET FORMAT

DATA	SOURCE ADDRESS	BROADCAST CHANNEL NUMBER	INDEX REF.	DUMMY ADDRESS INTERVAL	COPY NUMBER	ROUTE ADDRESS	A

Fig. 21. (a) An input concentrator consists of a running adder network and a reverse banyan network. (b) Input concentration without prioritization.

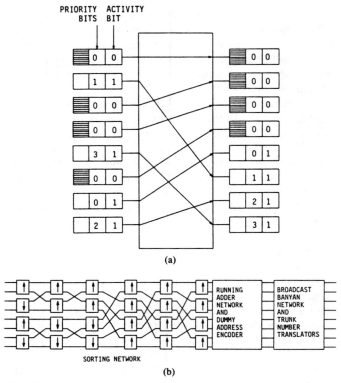

Fig. 22. (a) A sorting network sorts the priority bits and activity bit. (b) Input concentration with priority sorting.

starting from the top (or bottom) of the input lines. This top-down or bottom-up prioritization is due to the intrinsic property of the running adder. To cope with this property, the incoming packets can be sorted, according to their *priority indexes* by a Batcher sorting network. The priority index is posted adjacent to the activity bit. A lower index means higher priority. A Batcher sorting network is constructed by 2×2 sorting elements [1] which compare the binary numbers consisting of priority indexes and the activity bits. The larger number will be routed to the output line indicated by an arrow. As a result, the active packets appear monotonically with respect to their priorities at the outputs of sorting network. The sorting process is illustrated in Fig. 22(a). A copy network with priority sorting is shown in Fig. 22(b).

VI. Chernoff Bound of Overflow Probabilities

Overflow occurs when the total number of copies requested by input packets exceeds the number of output ports. This can be detected by the running adder and overflow packets can be retransmitted. A rudimentary analysis to characterize the behavior of the copy network is given in this section. To simplify the discussion, we assume the copy network is a loss system, meaning that packets are discarded when an overflow occurs.

Recall that inputs to our copy network are served in a sequentially ordered manner that can be either top-down or bottom-up. If no other fairness mechanism, such as dual running adder networks or sorting network, is provided to compensate for this biased priority scheme, then overflow probabilities are monotone increasing with respect to in-

put addresses for homogeneous input traffic. These overflow probabilities can be estimated by Chernoff bound [3], which provides a crude assessment of the copy network's performance. A complete performance analysis for waiting system, that takes retransmission of overflow packets into account, is yet to be developed.

Suppose the network under consideration has m inputs and n outputs. Input processes are assumed to be two-state Markov chains. At any given time-slot, an input line is active with probability ρ and it is idle with probability $1 - \rho$ where ρ is the *offered load* per each input line. Let Y_i be a random variable representing the number of copies requested by an incoming packet at the head of the ith input line. Since inputs are homogeneous, the Y_i are independent identically distributed random variables with density function $q(k) = P_r\{Y_i = k\}$, $k = 1, 2, \cdots$. The moment-generating function of Y_i is the expectation of e^{sY_i}, which will be denoted by $Q(s)$. Let X_i be the random variable of the number of copies generated by ith input line regardless if it is active or idle. It is obvious that

$$p(k) = P_r\{X_i = k\} = \begin{cases} 1 - \rho, & k = 0, \\ \rho q(k), & k = 1, 2, \cdots. \end{cases} \quad (4)$$

The moment-generating function of X_i is then given by

$$P(s) = E[e^{sX_i}] = \sum_k p(k)e^{sk} = (1 - \rho) + \rho Q(s).$$

$$(5)$$

Thus, the *effective offered load* for each input line is $\rho' = E[X_i] = \dot{P}(0) = \rho\dot{Q}(0) = \rho E[Y_i]$. The overflow probability of ith input line, denoted by $p_b(i)$, can be estimated by Chernoff bound as follows:

$$p_b(i) = P_r\{X_1 + \cdots + X_i > n\} \le e^{-sn}[P(s)]^i \quad (6)$$

where s is a free parameter. We may solve for the value of s that provides the tightest upper bound of $p_b(i)$ by taking the derivative of the right-hand side of (6), which is a convex function. This yields

$$i\frac{dP(s)}{ds} - nP(s) = 0. \quad (7)$$

Two distributions of Y_i are considered, the first one assumes it is a constant and the second one is geometrically distributed. The results are discussed below.

A. Constant Number of Copies

Suppose every broadcast packet requests a constant number of copies, say K. Then the moment-generating function of X_i can be expressed by

$$P(s) = (1 - \rho) + \rho e^{sK}. \quad (8)$$

The stationary value of parameter s, by solving (7), is given explicitly as

$$s = \frac{1}{K} \ln \frac{n(1 - \rho)}{(iK - n)\rho}. \quad (9)$$

Therefore, from (6), the overflow probability of ith input line is bounded by

$$p_b(i) \le \left(\frac{iK\rho}{n}\right)^{n/K}\left(\frac{iK(1 - \rho)}{iK - n}\right)^{i-(n/K)}, \quad \text{for } i > \frac{n}{K} \quad (10)$$

and it is obvious that $p_b(i) = 0$, for $i \le n/K$.

Since the X_i are statistically independent, the generating function of the random variable $X_1 + \cdots + X_i$ is the ith power of (8), yielding

$$P^i(s) = [(1 - \rho) + \rho e^{sK}]^i$$
$$= \sum_{j=0}^{i} \binom{i}{j}(1 - \rho)^{i-j}\rho^j e^{sjK}. \quad (11)$$

In this particular case, the overflow probability can also be determined exactly by

$$p_b(i) = \sum_{j=[n/K]+1}^{i} \binom{i}{j}(1 - \rho)^{i-j}\rho^j \quad (12)$$

where $[n/K]$ is the integer part of n/K. We evaluated the overflow probabilities of a 64×64 copy network as an example. The Chernoff bound is compared to the true value in Fig. 23 for fixed effective offered load $\rho' = 0.6$ with various values of K. The results show that the overflow probabilities, as a measure of the performance, are quite sensitive to the copy number K.

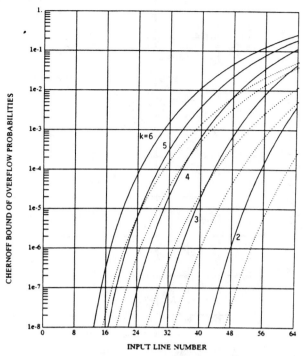

Fig. 23. Overflow probabilities of a 64×64 network with constant copy number assumption and fixed effected offered load $\rho' = 0.6$. The solid curves are Chernoff bound and the dotted curves are exact evaluation.

B. Geometric Distributed Number of Copies

Suppose the density function of the random variables Y_i is given by

$$q(k) = P_r\{Y_i = k\} = \frac{(1 - q)q^{k-1}}{1 - q^K}, \quad 1 \le k \le K \quad (13)$$

where K is the maximum allowable number of copies that can be specified in the header of a packet. The moment-generating function of Y_i is given by

$$Q(s) = \sum_{k=1}^{K} q(k)e^{sk} = \frac{e^s(1 - q)(1 - q^K e^{sK})}{(1 - q^K)(1 - qe^s)}. \quad (14)$$

The average number of copies is $\dot{Q}(0)$. This yields

$$E[Y_i] = \frac{1}{1 - q} - \frac{Kq^K}{1 - q^K}. \quad (15)$$

Similar to the previous procedure, the stationary value of parameter s can be obtained by substituting (14) into (5), and then solving (7) numerically. The Chernoff bound of overflow probabilities for both 64×64 network and 64×128 network are sketched in Fig. 24. The results demonstrate the significant improvements in performance by expanding the fan-out of a copy network.

Fig. 25 illustrates that a multicast packet switch can be arranged as a combination of a number of smaller copy networks and a large point-to-point switch. The distributed running adder in a small copy network can be re-

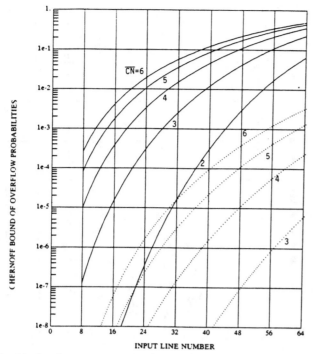

Fig. 24. Overflow probabilities with geometric distributed copy number assumption. The solid curves are results of 64 × 64 network, and the dotted curves are results of 64 × 128 network with fixed effective offered load $\rho' = 0.6$.

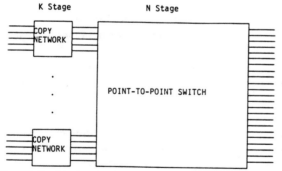

Fig. 25. Partitioning the copy network to reduce the complexity.

placed by serial adders to reduce the complexity. The results of performance analysis can be useful in the determination of the size of copy networks.

VII. CONCLUSIONS

This paper has described a design of nonblocking, self-routing copy networks for multicast packet switches. The main function of a copy network is packet replication, which is realized by cascaded encoding and decoding processes in our design. The encoding process is carried out by a running adder network and a set of dummy address encoders, which generates dummy address intervals as new headers for incoming packets. Then, a broadcast banyan network decodes these address intervals in packet headers according to a Boolean interval splitting algorithm that we have developed. At each node of the broad-

cast banyan network, the decision making is based on a two-bit header information. This yields minimum complexity of switch nodes. The address intervals of a set of inputs to the banyan network are monotone, which assures that the network is nonblocking. The computation of copy index consists of simply subtracting the index reference, initially set equal to the minimum of address interval, from the output address of each packet. The copy index and broadcast channel number combined together form a unique identifier, which is invoked by the trunk number translator to determine the packet's final destination.

The total number of copies, carried by the network, is counted by the running adder on per time-slot basis. This information provides useful traffic statistics for flow control of broadcast channels. To be flexible in many applications, we have studied nonsquare expansion network with larger fan-out than fan-in, which is an extension of the basic banyan topology and possesses the same routing properties as that of the banyan network.

There are two problems inherent in the design of a space-division-based multicast packet switch. One is the occurrence of overflows of the copy network when the total number of copy requests exceeds the number of outputs. The other is output port conflicts in the routing network when multiple packets request the same output port concurrently. The architecture of a multicast broadband packet switch is proposed in [6]. The solution to the first problem is to regulate the input packets, according to the number of copies requested and their priority, so to prevent the copy network from overflowing. The problem of output port conflicts in the routing network can be handled by a contention resolution algorithm reported in [5]. Future research should be focused on overflow control and performance analysis of the overall structure of multicast packet switches with respect to various services.

APPENDIX

Proof of NB1 [2]

In stage k of an N-stage banyan network, for $0 < k < N$, there are 2^{k-1} remaining subnetworks. Each node in stage k can be uniquely represented by two binary numbers $(a_{N-k} \cdots a_1, b_1 \cdots b_{k-1})$. Intuitively, $a_{N-k} \cdots a_1$ is the label of the node numbered from the top within the subnetwork, and $b_1 \cdots b_{k-1}$ is the label of the subnetwork numbered. Fig. 26 illustrates this numbering scheme introduced by Bechmann for a four-stage banyan network.

The node in stage $k + 1$ attached to the node $(a_{N-k} \cdots a_1, b_1 \cdots b_{k-1})$ in stage k by the output link b_k (0 or 1) is represented by $(a_{N-k-1} \cdots a_1, b_1 \cdots b_{k-1}b_k)$. Thus, the path from an input $x = a_N \cdots a_1$ to an output $y = b_1 \cdots b_N$, denoted by $\langle x, y \rangle$, consists the following sequence of nodes: $(a_{N-1} \cdots a_1, \phi)$, $(a_{N-2} \cdots a_1, b_1)$, $\cdots$, $(a_{N-k} \cdots a_1, b_1 \cdots b_{k-1})$, $\cdots$, $(\phi, b_1 \cdots b_{N-1})$. Suppose that two packets, one coming from input $x = a_N \cdots a_1$ to output $y = b_1 \cdots b_N$ and the other

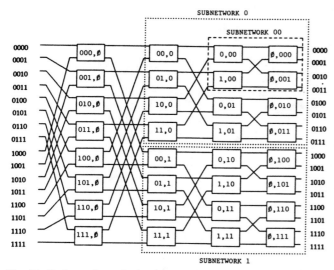

Fig. 26. Beckmann's numbering scheme of a four-stage banyan network.

coming from input $x' = a'_N \cdots a'_1$ to output $y' = b'_1 \cdots b'_N$, collide in stage k. That is, the two paths $\langle x, y \rangle$ and $\langle x', y' \rangle$ are merged at the same node $(a_{N-k} \cdots a_1, b_1 \cdots b_{k-1}) = (a'_{N-k} \cdots a'_1, b'_1 \cdots b'_{k-1})$, and shared the same output link $b_k = b'_k$. Then, we have

$$a_{N-k} \cdots a_1 = a'_{N-k} \cdots a'_1, \tag{A.1}$$

and

$$b_1 \cdots b_k = b'_1 \cdots b'_k. \tag{A.2}$$

Since inputs to banyan network are assumed to be monotone (increasing or decreasing) without gaps. It follows that the total number of active inputs between (and including) inputs x and x', $x' - x + 1$, must be less than or equal to the total number of outputs between y and y', $y' - y + 1$. We have

$$x' - x \le y' - y. \tag{A.3}$$

According to (A.1) and (A.2), we obtain

$$
\begin{aligned}
x' - x &= a'_N \cdots a'_1 - a_N \cdots a_1 \\
&= 2^{N-k}(a_N \cdots a'_{N-k+1} - a_N \cdots a_{N-k+1}) \\
&\ge 2^{N-k}, \tag{A.4}
\end{aligned}
$$

and

$$
\begin{aligned}
y' - y &= b'_1 \cdots b'_N - b_1 \cdots b_N \\
&= b'_{k+1} \cdots b'_N - b_{k+1} \cdots b_N \le 2^{N-k} - 1. \tag{A.5}
\end{aligned}
$$

But (A.3), (A.4) and (A.5) together imply $0 \le -1$, a contradiction. Therefore, the two paths $\langle x, y \rangle$ and $\langle x', y' \rangle$ must be link-independent, and the assertion in *NB1* is established.

ACKNOWLEDGMENT

I would like to thank E. Arthurs, R. P. Singh, and L. T. Wu for many discussions during the course of this work. I also want to thank W. D. Sincoskie for supporting this work, who also suggested the study of nonsquare networks. I am grateful for the comments I received from J. S. Turner, whose thoughts and work are much appreciated. I particularly want to thank H. J. Chao who helped with the logic design of broadcast switch nodes.

REFERENCES

[1] K. E. Batcher, "Sorting networks and their applications," *AFIPS Proc. Spring Joint Comput. Conf.*, 1968, pp. 307–314.
[2] W. H. Bechmann, private communication.
[3] R. G. Gallager, *Information Theory and Reliable Communication*. New York: Wiley, 1968.
[4] A Huang and S. Knauer, "Starlite: A wideband digital switch," in *Proc. IEEE GLOBECOM '84*, pp. 121–125.
[5] J. Hui and E. Arthurs, "A broadband packet switch for integrated transport," *IEEE J. Select. Areas Commun.*, vol. SAC-5, pp. 1264–1273, Oct. 1987.
[6] T. T. Lee, R. Boorstyn, and E. Arthurs, "The architecture of a multicast broadband packet switch," presented at Proc. INFOCOM '88, New Orleans, LA.
[7] J. S. Turner, "Design of a broadcast packet switching network," in *Proc. INFOCOM, '86*, pp. 667–675.
[8] ——, "Design of an integrated service packet network," *IEEE J. Select. Areas Commun.*, Nov. 1986.
[9] C. L. Wu and T. Y. Feng, "On a class of multistage interconnection networks," *IEEE Trans. Comput.*, vol. C-29, Aug. 1980.

Tony T. Lee (S'83–M'83–SM'89) received the B.S.E.E. degree from National Cheng Kung University, Taiwan, in 1971, and the M.S. and Ph.D. degrees in electrical engineering from Polytechnic Institute of New York, Brooklyn, NY, in 1976 and 1977, respectively.

He is presently with the Broadband Packet Networks Research District of the Packet Communications Research Division. He works in areas of relational database theory, protocol verification, and broadband packet switched networks. Before joining Bell Communications Research, Morristown, NJ, in 1984, he was with AT&T Bell Laboratories, Holmdel, NJ, from 1977 to 1983, where his work included teletraffic theory and engineering and switch performance analysis.

Chapter 7:

Bandwidth Allocation, Flow Control, and Congestion Control

7.1 Introduction

A major goal of an integrated broadband network is to support a diversity of services and applications. The transport needs of these services and applications vary considerably in terms of required transmission bandwidth, requirements for information buffering within the network, and acceptable levels of service quality. In addition, the various services supported will have traffic peaks that occur at different times. CCITT has defined the broadband integrated services digital network (BISDN) based on ATM to efficiently support these services. ATM is a packet-based switching and multiplexing technology and transports information using fixed-size cells, thereby avoiding the limitations and inflexibilities of circuit switching.

But congestion in a network may limit the efficient utilization of network resources and the ability to provide good service quality to all network users. In a network using packet switching technology, congestion is defined as a state in which network performance is degraded because of an excessive number of packets in transit. This performance degradation may be a reduction in throughput without any change to the traffic or an increased delay encountered by packets. Congestion may be confined to a limited region of a network, or, more seriously, it may be general throughout the network. In the extreme case, congestion may be so severe that all or almost all traffic comes to a standstill, with little or no data delivered to destinations or accepted from the sources. This is a deadly state for a network to enter, and measures should be taken to prevent it.

The basic resources in a network based on packet mode technology are the transmission bandwidth connecting the switching nodes and the buffers at the switching nodes. The limited capacity of the data links can contribute to the development of congestion. But the major cause of congestion in packet mode networks is the unavailability of data buffers in switch nodes to receive and forward arriving traffic. In a congested network, data arrive at a node and no buffers are available to store them. Then, depending on the network protocol, they must be retransmitted from the previous node or from the source node or the end system. Therefore, as the degree of congestion becomes severe, a greater proportion of the network resources are allocated for retransmission of the data. This may decrease the data delivered by the network.

Congestion can arise from several causes: a general traffic overload within the network, network bottlenecks, or faulty design of switch node software. In the packet switching world, a number of congestion control strategies are used. These strategies involve allocating resources in advance, allowing packets to be discarded when they cannot be processed, restricting the number of packets in the subnet, using flow control to avoid congestion, and choking off input when the network is overloaded.

In packet-switched networks using a connection-oriented service, a way to avoid congestion is to allocate resources such as buffers at call setup, thereby ensuring that all the resources required for connection are available. If a switch node cannot satisfy the buffer requirements for any new call, either another path is found or a busy signal is returned to the user. When a switch node preallocates and dedicates a sufficient amount of buffer for each connection passing through the node, packet switching becomes similar to circuit switching. In both cases, resources are allocated to individual connections during connection setup, without regard to the traffic activity. In both cases, congestion is avoided by allocating resources in advance, and this leads to the potentially inefficient use of resources, because resources allocated to connections not using them cannot be used by other connections.

Congestion control using the packet discarding scheme is based on the idea that switch node resources are not reserved in advance. If a packet arrives at a node where there are not enough resources, it is discarded. The responsibility of retransmitting these packets is left to the higher layer of the protocol.

A network enters the congested state when too many packets are in the network. Therefore, a direct approach to controlling congestion is to limit the number of packets in the network. The congestion control strategy using this method is called the isarithmic technique. The number of packets in the network is kept constant with a permits scheme. Permits circulate within the network, and traffic cannot be admitted at a node without a permit. This prevents the number of packets admitted into the network from exceeding the number of permits.

Some packet networks use flow control mechanisms to reduce congestion, but it is difficult to control the total amount of traffic in the network using end-to-end flow control. Although limiting the volume of traffic between the source and destination may indirectly alleviate congestion, it does so at the price of potentially reducing throughput, even when there is no threat of congestion. In the choke packet scheme, congestion control action is taken when congestion is detected. There are a number of congestion control methods based on variations of this technique. The basic idea behind such schemes is to define a set of threshold values. When a threshold value is exceeded, warning packets are sent.

In a preventive control approach, two control functions can be identified: admission control and bandwidth enforcement. Admission control is concerned with making a decision on whether a new connection can be established, and how it should be handled to satisfy the performance requirements of all active connections. Traffic enforcement is necessary for monitoring and enforcing the user traffic so it remains within the bounds determined by the admission control. Thus, compliant users do not suffer performance degradation due to violation of the negotiated bandwidth by the other users. Admission control is used during call setup and traffic enforcement is used during data transfer.

AT&T's wideband packet-switched network uses a bit dropping scheme to control congestion. In this scheme, voice traffic is packetized using a coding technique called embedded coding. The network takes advantage of this coding scheme to control congestion in the network. During congestion, the least significant bits of the voice packet are dropped, resulting in graceful degradation. The idea is that voice is redundant and the loss of a packet in a speech path may not significantly affect the quality of speech.

In ISDN, CCITT defined frame relay bearer services as part of frame mode bearer services to support high-speed data communication. In frame relay, the network node implements only the core functions of the link access protocol on the D channel by moving the complex protocols to the end system. This allows the core functions to be implemented in hardware, thereby achieving high switch node throughput (see "Packet Mode Services: From X.25 to Frame Relaying," by W.S. Lai, in Chapter 1). By reducing protocol complexity at the network nodes and supporting out-of-band signaling, a frame-relay-based network can achieve higher throughput with less delay. However, by moving protocol control functions to the end-user system, the network loses some control capabilities for effectively managing shared resources and becomes vulnerable to traffic overloads. To support guaranteed performance in such an environment, traffic controls are needed. CCITT defined cooperative protocols for congestion management to prevent the network from entering the deadly state of congestion and to recover from congestion if it occurs.

A network should use its resources as efficiently as possible and still be able to react to congestion in a controlled and fair fashion. To meet this goal, *explicit congestion notification* and *implicit congestion detection protocols*, along with committed data and a new class of uncommitted data, are specified.

As the network nodes do not have state information related to a connection, they cannot provide congestion notification to the end system. Therefore, in an implicit congestion detection protocol, the end system infers that the network has entered the congested state when one or more frames are lost. After detecting congestion, the end system then controls the input rate to the network to help reduce congestion. The problem with this method is that the frame loss may be due to other causes, such as transmission error.

In an explicit congestion control protocol, the network sends a signal to the end system to notify it about the congested state of the network. Frame relay bearer services provide two types of flow control notification mechanisms. The protocol to support congestion control is provided by layers above the core services or in a network rate-enforcing function at the user-network interface. The rate-enforcing function has access to call establishment or subscription time parameters. The committed data are identified by call setup parameters, committed burst size, and committed information rate. Committed burst size is the maximum amount of data that the network agrees to transfer under normal conditions over a well-defined measurement interval. This parameter may be used to assign extra buffer space at the network ingress node to handle bursty traffic. Committed information rate is the rate at which the network agrees to transfer data under normal conditions, measured over a defined time interval.

The frame header in the figure shows the forward explicit congestion notification (FECN), backward explicit congestion notification (BECN), and discard eligibility (DE) indicator used to support congestion control. Frames with their DE indicators on are discarded in preference to other frames when the network enters the congested

state. This maintains the committed quality of service within the network.

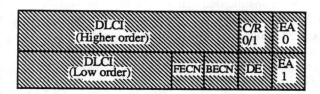

C/R	- Command /Response Indicator
EA	- Address Field Extension
DE	- Discard Eligibility Indicator
BECN	- Backward Explicit Congestion Notification
FECN	- Forward Explicit Congestion Notification
DLCI	- Data Link Connection Identifier

FECN protocol is used by the destination end system in a frame relay connection. A network resource experiencing congestion turns on the FECN bit in frames going toward the destination end system, which uses this indication to determine the state of the network. It averages the number of indicators received over a period of approximately a two-round-trip delay. (The average round-trip delay is a parameter exchanged during call setup.) An indication of congestion causes the destination end system to modify the credits allocated to the source end system. This protocol provides a delayed response to network congestion. The reduction and return to normal data transfer occur in a staged pattern.

BECN protocol is used by the source end system in a frame relay connection. The response to the arrival of a frame with BECN is immediate. The end system or rate-enforcing function reduces the data transfer rate to the committed rate. Any further reduction in data transfer is staged as a function of the number of contiguous BECNs received. Similarly, an increase in the data transfer rate is effected in stages as a function of contiguous non-BECN frames received. This protocol maintains the network at the committed rate during periods of low-level congestion. When congestion increases, the data transfer rate is reduced in stages until the congestion is cleared.

FECN protocol is an averaged approach to congestion control in which the end system evaluates the congested state of the network using information provided by the network. On the other hand, BECN protocol brings an immediate response, and the network indicates its level of congestion. These two congestion control protocols are independent of each other.

A consolidated link layer management (CLLM) message is a variation of a backward explicit congestion notification. CLLM is a group broadcast congestion indication from the congested node to the source end systems. This technique can be used when no reverse traffic is available to carry the BECN. The CLLM messages list the connection identifiers of frame relay connections that pass through the congested resource. The end systems use this information to control network congestion.

The concept of discard eligibility data was introduced to reduce the problem associated with high bandwidth utilization and arbitrary frame discard. Allowing and marking a class of uncommitted data lets the congested network determine the frames to discard. Controlling the amount of uncommitted traffic on any given connection increases network utilization with minimal committed data frames being discarded. This feature allows the network to respond quickly to the growth of congestion until FECN, BECN, and/or CLLM congestion flow controls take effect.

A major challenge in a BISDN environment is buffer and bandwidth management. Bursty traffic can lead to high cell loss rates, if the traffic flow is multiplexed indiscriminately without considering each type of traffic's effect on other traffic in networks with a minimal protocol. A complete understanding of the queuing behavior of multiplexed bursty sources is essential to achieve acceptable end-to-end cell loss rates, while making efficient use of network resources. This is an area of active research.

A set of congestion controls for a broadband network may include strategies such as admission control, buffer and queue management, traffic enforcement, and reactive control. In an ATM network, congestion may take place at the connection level, the ATM level, and the packet level. At the connection level, congestion can be controlled by blocking calls for which there are not enough resources available in the network. At the ATM level, congestion can be controlled by restricting the flow of cells on any given connection. Another way to control congestion is to limit the number of packets entering the network.

Connection admission control is a set of actions supported by an ATM network at call setup to determine whether the required network resources are available to support the connection. A connection request is accepted only if sufficient resources are available to establish the call through the whole network at its required service quality and enough resources are available to maintain the agreed service quality of existing connections. Such parameters as peak rate, average rate, burstiness, and peak duration required by the connection are supplied to the connection control function in the call setup signaling message.

The traffic enforcement system ensures that the usage of network resources by existing connections does not exceed the agreed limits negotiated at call establishment. If the negotiated limits are violated, appropriate action is taken. The traffic enforcement function is performed during the data transfer phase of a connection. It monitors the traffic volume entering the network on each active virtual path and on virtual channel connections, and the total volume of traffic on the access link.

The traffic enforcement function can invoke one or more of the following actions to prevent the network from entering the congested state: discarding the cells belonging to the connection that exceeded the negotiated parameters, forwarding congestion indication information to control the input rate, assigning a lower priority to cells belonging to the connection that exceeded negotiated parameter limits, and releasing the connection itself.

Even with call admission and traffic enforcement procedures, localized congestion may take place in a heavily loaded network. The statistical nature of variable bit rate traffic could produce temporary overloads at some internal network resources. Then additional congestion control procedures could be used to relieve local congestion in the affected segments. One such reactive technique is to refuse new connections on the basis of such measures as equipment utilization, buffer fill, alternate routing, and flow control.

Congestion occurs if the total demand exceeds the total resources. Since congestion is a dynamic problem, its relief requires dynamic reallocation of resources. Otherwise, serious performance degradation results. To invoke the congestion control schemes, mechanisms to dynamically measure demand for network resources are needed. Once identified, congestion can be relieved by dynamically reducing the demand or by dynamically increasing the resources.

The main purpose of flow control is to ensure that a source does not transmit at a rate higher than the rate that the destination can accept. This requires that there be a feedback mechanism from the receiver to the transmitter so that the receiver can, either explicitly or implicitly, indicate to the transmitter when it is ready to accept and process additional information. In a limited way, flow control may help control congestion.

7.2 Overview of Papers

The papers in this chapter cover the issues of flow control, congestion control, and resource allocation in broadband networks, which are areas of intense research within the network community.

Theoretical approaches to determining optimal routing are described in the paper "Routing of Multipoint Connections," by Waxman. Routing of multipoint connections is modeled as a graph-theoretic problem, and a static version of optimal routing is shown to be equivalent to the Steiner tree problem. Waxman measures experimental performances of different proposed approximation versions. In addition, the dynamic version of optimal routing, where endpoints are allowed to come and go, is solved using a version of the weighted greedy algorithm.

A basic requirement for integrating various types of traffic with a wide range of information transfer rates is a switch with high throughput. To achieve this high throughput, most of the switch functions are implemented in hardware. In a network based on the ATM concept, cells of fixed size are switched from the source to the destination after a connection is established. These paths are identified using a logical channel number or virtual channel identifier. In the paper "A Dynamically Controllable ATM Transport Network Based on the Virtual Path Concept," Ohta et al. introduce the concept of virtual path. The virtual path is a logical link between a pair of end nodes that can be shared by a number of virtual circuits simultaneously. Compared with the logical channel numbering scheme, this switching technique reduces the nodal processing. However, this is achieved at the expense of increased header overhead to identify the virtual path.

In "Resource Allocation for Broadband Networks," Hui addresses the techniques required to measure congestion and to allocate bandwidth in broadband networks. For measuring congestion in a broadband integrated ATM network

supporting multirate traffic, Hui proposes a new methodology based on evaluating congestion at packet, burst, and call levels. The paper also suggests methods for using these congestion measures in a multilayer bandwidth allocation algorithm. This paper contains detailed mathematical analyses of probability estimations of blocking, targeted to mathematically minded readers.

Pattavina's paper, "Multichannel Bandwidth Allocation in a Broadband Packet Switch," considers the problem of bandwidth allocation in a packet switch supporting broadband services. The author argues that a connection-oriented service with explicit resource allocation and a datagram service on top of it might not be the best scheme to use the network resources efficiently. The paper proposes a two-step bandwidth allocation scheme that is active during call setup as well as during information transfer. The implementation of the algorithm is shown to be feasible for an input buffered Batcher-banyan switch. For random traffic, large increases in throughput are obtained, and for bursty traffic the packet loss probability is decreased by several orders of magnitude.

The paper "Flow Control Schemes and Delay/Loss Tradeoff in ATM Networks," by Ohnishi et al., discusses the performance and flow control mechanisms in an asynchronous transfer mode (ATM) network supporting multimedia communications. A quasi-synchronous transfer mode (circuit switching) service is also proposed as a full-service set for ATM. With the flow control based on a call-oriented resource assignment mechanism, the call/line bit ratio is a significant parameter for line bandwidth use.

Eick's paper "Access Interface Congestion Control for Packetized Voice Transport in Wideband Networks" discusses a congestion control mechanism for an access control interface that converts end-user traffic into packets for a wideband network. The selective dynamic overload control (SDOC) scheme prevents extended access interface congestion by restricting new call traffic until congestion is relieved when calls are terminated. Assuming that the new call arrivals conform to a Poisson distribution, Eick shows that the access interface call load is the convolution of a Poisson and a binomial random variable. The model study shows that the SDOC scheme controls the congestion within 3 to 4 minutes.

In "A Congestion Control Framework for High-Speed Integrated Packetized Transport,"

Woodruff et al. consider general congestion control design approaches in the context of an integrated, packetized transport network for a public, broadband ISDN environment. Two forms of control, reactive control and preventive control, are identified. The authors argue that reactive control is not suitable for high-speed networks such as broadband ISDN, and recommend it only as a last resort during emergency situations to preserve network integrity. They propose a framework for preventive control based on access control for a connection-oriented transport. According to the authors, the framework is simple, fair, and nonservice-optimized, and it allows an explicit trade-off between transport performance and bandwidth efficiency.

Routing of Multipoint Connections

BERNARD M. WAXMAN

Reprinted from *IEEE Journal on Selected Areas in Communications*, Vol. 6, No. 9, December 1988, pages 1617-1622.

Abstract—This paper addresses the problem of routing connections in a large scale packet switched network supporting multipoint communications. We give a formal definition of several versions of the multipoint problem including both static and dynamic versions. We look at the Steiner tree problem as an example of the static problem and consider the experimental performance of two approximation algorithms for this problem. Then we consider a weighted greedy algorithm for a version of the dynamic problem which allows for endpoints to come and go during the life of a connection. One of the static algorithms serves as a reference to measure the performance of our weighted greedy algorithm in a series of experiments. In our conclusion, we consider areas for continued research including analysis of the probabilistic performance of multipoint approximation algorithms.

I. INTRODUCTION

IN a packet switched network using virtual circuits, the primary goal in routing connections is to make efficient use of the network resources. For example, one favors an algorithm which can handle the largest number of connections for a given set of network resources. In a point-to-point network, routing is often treated as a shortest path problem in a graph. Here the network is modeled as a graph $G = (V, E)$ where the nodes of a graph represent switches and the edges represent links. In addition we have two functions cap: $E \rightarrow Z_0^+$ and cost: $E \rightarrow \Re^{+1}$ which give us the bandwidth and the cost of each edge. At the time a connection is requested, a minimum cost (shortest) path connecting a pair of endpoints is selected. Of course only paths consisting of edges with sufficient unused bandwidth may be chosen.

Routing of multipoint connections may be modeled in a similar way. In the multipont problems, we wish to connect a set $D \subseteq V$. Instead of the shortest path, we are interested in the shortest subtree which contains the set D. Finding the shortest tree connecting a set of points is known as the Steiner tree problem in graphs. Of course the multipoint problem is complicated by the fact that the set D may change during the life of a connection.

II. DEFINITION: THE MULTIPOINT PROBLEM

Before proceeding, let us consider a definition for the multipoint problem. In the multipoint problem we can view the network as a graph $G = (V, E)$ with a function

cap: $E \rightarrow Z_0^+$. For simplicity we will assume that this graph is undirected. We then define a connection request c to be a pair (D, b) where

- $D \subseteq V$ is the set of nodes to be interconnected
- $b \in Z^+$ is the bandwidth required for connection c.

Finally, we define a route $R_c = (\hat{D}, \hat{E})$, for connection request $c = (D, b)$, to be a connected subgraph of G such that $D \subseteq \hat{D}$ and the bandwidth constraint

$$(\forall e \in \hat{E})(b \leq \text{cap}(e)) \tag{1}$$

is satisfied. Of course, for a given connection request c there may not be any route R_c satisfying (1).

In what we shall refer to as the *basic* multipoint problem, we are given a network $N = (G, \text{cap})$ and a connection request c. We are then asked to find a route R_c for c that is optimum. The definitions that follow represent a compromise between simplicity and complete generality. These definitions capture the essential features of this problem without being overly complex. We consider three versions which we refer to as the *static* multipoint problem, the *unitary dynamic* multipoint problem, and the *dynamic* multipoint problem.

The static version of the multipoint problem is a formalization of the *basic* problem. In this version, we assume that a function cost: $R \rightarrow \Re^+$ is defined for a network N where R is the set of all possible routes on N. The static problem is stated as follows. Given a network N and a connection request c, find a route R_c such that cost (R_c) has a minimum value among all possible routes for c. The Steiner tree problem in graphs is a specific version of the static multipoint problem which we shall discuss shortly.

In the *unitary dynamic* version of the multipoint problem, we start with a network $N = (G, \text{cap})$ and consider a sequence of requests $S = [r_0, \cdots, r_n]$ where each r_i is a pair (v, d), $v \in V$, $d \in \{add, remove\}$. Each request includes a node of the network which is to be either added to, or removed from, a connection. Thus, in the *unitary dynamic* problem we are given a network N, a bandwidth b, and a sequence of requests S. We are asked to find a sequence of routes R_i, $i \in [0, \cdots, n]$ which yields a minimum cost sequence subject to the bandwidth constraint (1). We may wish to consider several measures for the cost of a sequence such as average cost or maximum cost. In this version, once a particular set of edges has been used in a route, no reconfiguration is allowed as the algorithm proceeds. Note that if we were to allow reconfiguration, this problem would be equivalent to a sequence

Manuscript received October 30, 1987; revised June 8, 1988. This work was supported by Bell Communications Research, Italtel SIT, NEC, and National Science Foundation under Grant DCI-8600947.

The author is with Computer and Communications Research Center, Washington University, St. Louis, MO 63130.

IEEE Log Number 8824403.

[1] We use Z_0^+ ($\Re_0^+$) to stand for the positive integers (reals) including 0, and Z^+ ($\Re^+$) to stand for the positive integers (reals) excluding 0.

of static problems. Clearly, an optimal algorithm which allows reconfiguration will always do at least as well as one for which reconfiguration is not allowed.

In the *dynamic* version of the multipoint problem we again are given a sequence of requests S which are triples of the form (c, v, d), $c \in C$ where C is a set of connection identifiers. Each $r_i \in S$ is a request to add or delete a node with respect to a particular connection. In addition we are given a function $B: C \to Z^+$ which associates a bandwidth with each connection. This version comes in two flavors, one which allows reconfiguration and a second for which reconfiguration is not allowed. Finally, in the *dynamic* multipoint problem, we impose the additional constraint that the sum of the bandwidths of all routes using a given edge e may not exceed the capacity of that edge. More formally we have that for all $i \in [0, \cdots, n]$

$$(\forall e \in E) \left(\operatorname{cap}(e) \geq \sum_{R_{c_j} \in \mathcal{R}_i, e \in R_{c_j}} b_j \right) \quad (2)$$

where $\mathcal{R}_i$ is the set of all routes created by an algorithm after request i and R_{c_j} is the route for connection c_j.

Here, we measure the performance of an algorithm by the number of requests of type *add* that it can service. If $a(N, S, C, B)$ is the number of *add* requests serviced by algorithm a, we define the number of requests serviced by an optimal algorithm for a given instance (N, S, C, B) of the *dynamic* multipoint problem as

$$O(N, S, C, B) = \max_{a \in A} (a(N, S, C, B)) \quad (3)$$

where A is the collection of all *dynamic* routing algorithms which satisfy the bandwidth constraint (2). Note that the algorithms in A may or may not be allowed to use reconfiguration, depending on which version of the *dynamic* multipoint problem we consider.

III. STATIC APPROXIMATION ALGORITHMS FOR THE STEINER TREE PROBLEM

The Steiner tree problem, was not originally conceived as a problem in communication but as a problem in geometry. The original version of the Steiner tree problem derives its name from Jacob Steiner [5] and was studied even earlier by Fermat [2]. The version most closely related to multipoint routing is the Steiner tree problem in graphs, which was shown to be NP-complete by Karp in 1972 [10]. Formally, the Steiner problem in graphs can be stated as follows. Given

- a graph $G = (V, E)$
- a cost function cost: $E \to \mathcal{R}^+$, and
- a set of vertices $D \subseteq V$,

find a subtree $T = (V_T, E_T)$ of G which spans D, such that the cost $(T) = \Sigma_{e \in E_T} \operatorname{cost}(e)$ is minimized.

Since the Steiner problem is NP-complete an algorithm that finds a minimum Steiner tree will not run in polynomial time unless $P = NP$. Therefore, we would like to find approximation algorithms which have worst case performance as well as average performance that is close to

optimum. There are a number of heuristics for deriving solutions to the Steiner tree problem in graphs which run in polynomial time. At this time, none of these is known to give a solution that is better than twice optimal in the worst case [17], [16].

The approximation algorithms described here are not practical for very large networks, since they assume complete knowledge of the network and do not operate in a distributed fashion. Such algorithms are of theoretical interest and should be useful as tools to measure the performance of other routing algorithms. In addition, it is hoped that the study of such algorithms will contribute to a better understanding of the multipoint routing problem.

We have been looking at several heuristics for the Steiner tree problem in graphs. These include a heuristic which is based on minimum spanning tree algorithms, which we refer to as MST [14], [1], [7], [9], a contraction heuristic [11] and a heuristic due to V. J. Rayward-Smith [12], which we refer to as RS. Both MST and the contraction heuristic have worst case performance two times the cost of an optimal solution. In his paper, Rayward-Smith does not develop any results regarding the worst case performance of RS. We have recently proved [16] that the worst case bound for RS is also two, even though experimental studies discussed in this paper and in [13] indicate that RS does better than MST on average. In addition, analysis of probabilistic performance [15] also gives the edge to RS.

We have implemented both the MST and RS algorithms as well as an improved version of each. The improved versions result from a modification of the version of MST suggested by Kou, Markowsky, and Berman [7]. It should be noted that the improved versions do not have better worst case performance but do at least as well as the basic algorithms.

We briefly describe MST referring the reader to one of the references cited for more details.

Construct a complete graph $G' = (D, E')$ where $\operatorname{cost}_{G'}(u, v)$ is the length of the shortest path from u to v in G.

Construct a minimum spanning tree T' for G'.

Convert the edges in T' to paths in G to form a solution T_1.

Fig. 1 illustrates the application of MST to a nine node graph, with a set $D = \{a, d, e, g\}$, and with the cost of each edge indicated. The complete graph $G[D]$, a minimum spanning tree for $G[D]$, and a final solution are shown. Note that in this example the solution given by MST is not optimum.

Kou, Markowsky, and Berman take the subgraph T_1 generated by basic MST and apply a minimum spanning tree algorithm to T_1. From the tree that results they prune branches which do not contain any vertexes in D, giving their solution to the Steiner problem. What we have done is to take the nodes D_1 in T_1, rerun the basic MST on this set of nodes to produce a subgraph T_2, and then prune those branches which do not contain any vertexes from D. This version of MST will yield a tree with cost that is

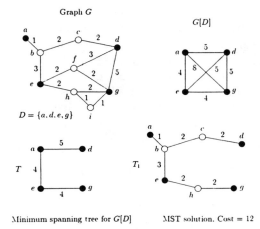

Graph G

$D = \{a.d.e.g\}$

Minimum spanning tree for $G[D]$ MST solution. Cost = 12

Fig. 1. An example of the application of MST to a graph.

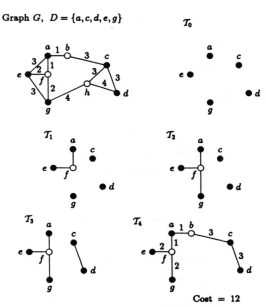

Fig. 2. The application of RS to an eight-node graph.

	f_0	f_1	f_2	f_3
a	3	3	4	4
b	4	4	4	4
c	3	3	3	4
d	3	3	3	7
e	3	3	7	7
f	2.5	2	5	5
g	3	2	7	7
h	5.5	5.5	5.5	7

Fig. 3. The values of $f_l(v)$ for $l \in [0, \cdots, 3]$.

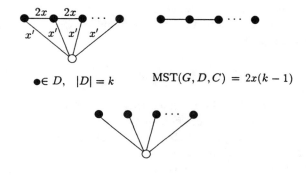

Fig. 4. MST approaches twice optimal, RS yields an optimal solution.

no greater than the cost of the solution given by the basic algorithm, but in the worst case does not do any better.

We now consider the RS algorithm proposed by Rayward-Smith. The problem with the basic MST algorithm is that it only considers the distances between nodes in the set D and does not consider the value to a final solution of nodes not in D. The nodes outside of D that are important to a solution of the Steiner tree problem will be called Steiner points. More formally, we define a node s, in a Steiner tree $T = (V_T, E_T)$ for a set D, to be a Steiner point if $s \in V_T$, $s \notin D$ and $\deg_T(s) \geq 3$. These are the nodes that RS attempts to find. Rayward-Smith defines a function $f: V \to \mathfrak{R}^+$ such that those nodes for which the function has a minimum value are the ones that should be included in a final solution. We give a brief description of RS referring the reader to [12], [13] for a more details.

Construct a collection of single node trees $\mathfrak{I}$ consisting of the nodes in D.

Repeat the following steps until $|\mathfrak{I}| = 1$.

Choose $v \in V$ such that $f(v)$ is minimum where

$$f(v) = \min_{S \subseteq \mathfrak{I}, |S| > 1} \left\{ \frac{1}{|S| - 1} \sum_{T \in S} \text{dist}(v, T) \right\}.$$

Let T_1 and T_2 be two closest trees to v.

Join T_1 and T_2 by a shortest path through v.

Informally, f is the average cost of making r joins to $r + 1$ trees through a node v where $r + 1 = |S|$.

The application of RS to an eight node graph is illustrated in Fig. 2. In this example, the repeat statement is executed four times. The collection of trees $\mathfrak{I}$ and the values for f, at each iteration, are shown in Fig. 2 and 3.

RS uses the function f to determine if a node other than those in D will be useful in solving the Steiner problem. Whenever RS chooses such a node, it has found a potential Steiner point. As can be seen in Fig. 4, we have a set of instances where MST will yield a solution whose cost is almost twice that of an optimal solution while RS will actually find an optimal solution.

To gain some insight into the probable performance of MST and RS, we implemented both the basic and the improved versions. In our experimental tests of these algo-

rithms, we also generated an optimal solution for each test case.

For the purpose of running these experiments, two simple random graph models RG1 and RG2, which have some of the characteristics of an actual network, were used. In RG1, n nodes are randomly distributed over a rectangular coordinate grid. Each node is placed at a location with integer coordinates. The Euclidean metric is then used to

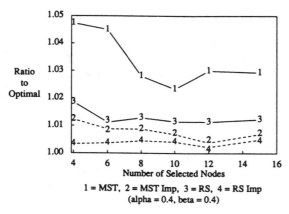

Fig. 5. Average performance of MST and RS.

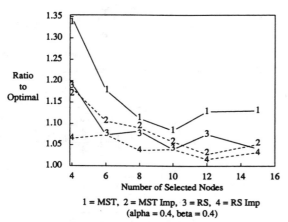

1 = MST, 2 = MST Imp, 3 = RS, 4 = RS Imp
(alpha = 0.4, beta = 0.4)

Fig. 6. Worst case performance of MST and RS.

determine the distance between each pair of nodes. In RG2, for each pair of nodes, a distance is chosen in $(0, L]$ from a uniform random distribution. For both models, edges are introduced between pairs of nodes u, v with a probability that depends on the distance between them. The edge probability is given by

$$P(\{u, v\}) = \beta \exp \frac{-d(u, v)}{L\alpha} \qquad (4)$$

where $d(u, v)$ is the distance from node u to v, L is the maximum distance between two nodes, and α and β are parameters in the range $(0, 1]$. Larger values of β result in graphs with higher edge densities, while small values of α increase the density of short edges relative to longer ones. Finally, we set the cost of an edge equal to the distance between its endpoints.

The experimental results of applying MST and RS to RG1 are illustrated in Figs. 5 and 6. For these experiments, five different 25 node random graphs were generated according to the RG1 model. Figs. 5 and 6 show the results of running 25 simulations, five for each graph, for subsets D of a given size. Each subset D contained nodes selected at random from the 25 node graphs. In each figure, the horizontal axis represents the size of the sets D, while the vertical axis indicates the ratio of the cost of a solution given by the algorithms to an optimal one. In

Fig. 5, average values are plotted for the cost of the basic and improved solutions for both algorithms. In Fig. 6, worst case results for each set of simulations is plotted. As these figures indicate both MST and RS when applied to RG1 generally yield results which are near optimum with the improved version of RS[2] giving the best performance. Our experiments also indicate that the results are essentially the same for graph model RG2 and a wide range of values for α and β.

IV. A DYNAMIC ALGORITHM FOR THE STEINER TREE PROBLEM

Let us next turn our attention to the *unitary dynamic* multipoint problem. Here individual nodes are allowed to join or leave a connection at any point during the life of that connection. This version of the multipoint problem considers the dynamic nature that we expect from real applications, although it is still a simplification of the real problem since it considers a single connection in isolation. We have run experiments with a version of the *unitary dynamic* multipoint problem using both a greedy and a weighted greedy algorithm. In the greedy algorithm, to add a node u to an existing connection, we first find a closest node v already in the connection and then connect u to v by a shortest path. We delete a node from the connection by first marking it deleted. If that node is not an internal node in the connection, we prune the branch of which it is a part. In the weighted greedy algorithm, a distinguished node o is chosen to be the owner of the connection, with the restriction that node o must remain in the connection. We add a node u to a connection by choosing a node v, in the connection, which minimizes the function

$$W(v) = (1 - \omega) d(u, v) + \omega d(v, o) \qquad (5)$$

where $0 \le \omega \le 0.5$ and $d(x, y)$ is the distance from x to y. When $\omega = 0$ the algorithm is equivalent to the basic greedy algorithm. When $\omega = 0.5$, we add a node u to the connection by the shortest path to the owner o. Fig. 7 illustrates a sequence of five events handled by the greedy algorithm. Note that events three and four are examples of a situation where the greedy algorithm is not optimal.

In our experiments we used our improved version of MST to measure the performance of the greedy algorithm. To run these experiments we used a simple probability model to determine if an event was to be an addition or deletion of a node from the connection. The function

$$P_C(k) = \frac{\alpha(n - k)}{\alpha(n - k) + (1 - \alpha)k} \qquad (6)$$

was defined for this purpose. Here, P_C is the probability that an event is the addition of a node, k is the number of nodes in set D of the current connection, n is the number of nodes in the network, and α is a parameter in $(0, 1)$. The value of α is the fraction of nodes in the connection

---

[2]Improved RS is analogous to the improved version of MST. We run RS a second time on the set of nodes generated by the initial RS solution.

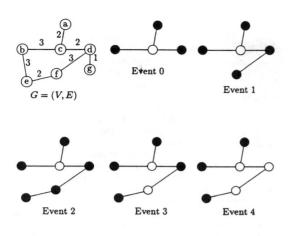

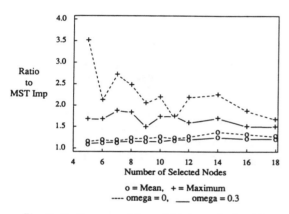

● Connected node　○ Intermediate node

Fig. 7. Greedy algorithm with sequence: $(a, b, d), f, e, \bar{f}, \bar{d}$.

Fig. 8. Performance of the weighted greedy algorithm.

at equilibrium. For example, when $k/n = \alpha$, $P_C(k) = 1/2$.

We ran several sets of experiments on 195 node graphs. Fig. 8 indicates the results of one set of experiments. For each experiment we ran a sequence of 1000 events using a fixed value for α. The horizontal axis indicates the value of αn or, in other words, the number of nodes k in the connection for which $P_C(k) = 1/2$. The vertical axis represents the ratio of the cost of the solutions produced by the greedy algorithms to the cost of the solutions produced by improved MST. The results of the worst case ratio and the average ratio for each experiment are plotted for $\omega = 0$ and $\omega = 0.3$. We can see from Fig. 8 that the greedy algorithm[3] does reasonably well for most events in comparison to MST. Those cases where the greedy algorithm does much worse than MST occur after several nodes have left the connection. In those cases several internal nodes have been deleted from D but must remain in the connection since they are part of a branch that contains other nodes in D.

This phenomenon of poor performance after several nodes leave the connection would be expected with any algorithm that does not allow for reconfiguration after each

---

[3]$\omega = 0$.

event. It is under these circumstances that the weighted algorithm shows the greatest improvement. Note that for the set of experiments shown in Fig. 8 the worst case performance improves from 3.5 to 2.0 with the weighted algorithm using $\omega = 0.3$. From our experiments it appears that a value for ω in the neighborhood of 0.3 yields the best results. There are several other techniques that might be used to smooth out the performance of the greedy algorithm. For example, we might weight the choice of a path to a node v in the connection by the distance from v to the center of the connection, or perhaps by the number of connections routed through v.

V. CONCLUSIONS

Our experiments with MST and RS indicate that they give near optimal results on our two graph models RG1 and RG2. For example, the improved version of RS gave performance within 8 percent of optimal for the entire set of experiments and within 2 percent of optimal on average. We are now looking at ways to obtain analytic results for the probable performance of Steiner tree approximation algorithms on several different random graph models [15]. Still open is the minimum value for the worst case performance of a polynomial time bounded Steiner tree algorithm, assuming $P \neq NP$.

Experimental results indicate that our weighted greedy algorithm gives reasonable results when compared to MST. Even though we got the best results with the weighting factor $\omega = 0.3$, a value for ω of 0.5 still does reasonably well. Since $\omega = 0.5$ is equivalent to shortest path routing these results indicate that extending a point-to-point routing scheme to the multipoint problem may be a reasonable thing to do.

Experimental studies of the extension of point-to-point routing as well as the extension of the weighted greedy algorithm to the general dynamic problem are worth additional consideration. These extension may then prove useful as a basis for the construction of a practical multipoint algorithm for a large network.

REFERENCES

[1] K. Bharath-Kumar and J. M. Jaffe, "Routing to multiple destinations in computer networks," *IEEE Trans. Commun.*, vol. COM-31, pp. 343–351, Mar. 1983.
[2] H. S. M. Coxeter, *Introduction to Geometry*. New York: Wiley, 1961.
[3] S. E. Dreyfus and R. A. Wagner, "The Steiner problem is graphs," *Networks*, vol. 1, no. 3, pp. 195–207, 1971.
[4] E. N. Gilbert, "Random graphs," *Ann. Math. Stat.*, vol. 30, pp. 1141–1144, 1959.
[5] E. N. Gilbert and H. O. Pollak, "Steiner minimal trees," *SIAM J. Appl. Math.*, vol. 16, no. 1, pp. 1–29, 1968.
[6] J. M. Jaffe, "Distributed multi-destination routing: The constraints of local information," *SIAM J. Comput.*, vol. 14, no. 4, pp. 875–888, Nov. 1985.
[7] L. Kou, G. Markowsky, and L. Berman, "A fast algorithm for Steiner trees," *Acta Informatica*, vol. 15, pp. 141–145, 1981.
[8] J. M. McQuillan, I. Richer, and E. C. Rosen, "The new routing algorithm for the ARPANET," *IEEE Trans. Commun.*, vol. COM-28, pp. 711–719, May 1980.
[9] H. Mehlhorn, "A faster approximation algorithm for the Steiner problem in graphs," *Inform. Process. Lett.*, vol. 27, no. 3, pp. 125–128, Mar. 1988.

[10] R. E. Miller and J. W. Thatcher, "Reducibility among combinatorial problems," in *Complexity of Computer Computations*, R. M. Karp, Ed. New York: Plenum, 1972, pp. 85–103.

[11] J. Plesnik, "A bound for the Steiner tree problem in graphs," *Math. Slovaca*, vol. 31, no. 2, pp. 155–163, 1981.

[12] V. J. Rayward-Smith, "The computation of nearly minimal Steiner trees in graphs," *Int. J. Math. Ed. Sci. Tech.*, vol. 14, no. 1, pp. 15–23, 1983.

[13] V. J. Rayward-Smith and A. Clare, "On finding Steiner vertices," *Networks*, vol. 16, pp. 283–294, 1986.

[14] H. Takahashi and A. Matsuyama, "An approximate solution for the Steiner problem in graphs," *Math. Japonic*, vol. 24, no. 6, pp. 573–577, 1980.

[15] B. M. Waxman, "Probable performance of Steiner tree algorithms," Washington Univ. Dep. Comput. Sci., St. Louis, MO, WUCS-88-4, 1988.

[16] B. M. Waxman and M. Imase, "Worst case performance of Rayward-Smith's Steiner tree heuristic," Washington Univ., Dep. Comput. Sci., St. Louis, MO, WUCS-88-13, 1988.

[17] P. Winter, "Steiner problem in networks: A survey," *Networks*, vol. 17, pp. 129–167, 1987.

Bernard M. Waxman was born in St. Louis, MO, on February 25, 1944. He received the B.S. degree in mathematics from Washington University, St. Louis, MO, in 1966, and the M.A. degree in mathematics from the University of Illinois, Urbana, IL, in 1968. More recently, he received the M.S. degree in computer science from Washington University, St. Louis, in 1985. He is currently working on a Doctorate degree in computer science at Washington University in the Advanced Communications Systems (ACS) group.

He has been investigating the multipoint routing problem as it relates to the large scale broadcast packet network being developed by ACS. Two areas of particular interest to him are the study of the probabilistic performance of Steiner tree approximation algorithms and the development of new Steiner tree algorithms with improved worst case performance. Previous to his work with ACS, he was the part owner of a professional photoprocessing laboratory.

A DYNAMICALLY CONTROLLABLE ATM TRANSPORT NETWORK BASED ON THE VIRTUAL PATH CONCEPT

Satoru OHTA, Ken-ichi SATO, and Ikuo TOKIZAWA

NTT Transmission Systems Laboratories
1-2356 Take, Yokosuka-shi, Kanagawa-ken 238-03 JAPAN

ABSTRACT

The Virtual Path concept is expected to provide an effective means in constructing an economical and efficient ATM (Asynchronous Transfer Mode) transport network. This paper describes a network based on this concept emphasizing the arrangement of the node functions and the benefits provided by Virtual Path control. Reduction of node processing and simplification of equipment are provided by the proposed function arrangement. Virtual Path control affords flexibility and high efficiency in the network. Bandwidth control of Virtual Path is evaluated and shown to effectively improve the transmission efficiency with a slight increase in node processing.

1. INTRODUCTION

Many efforts have been made to construct a more advanced communication network providing service integration and flexibility. In such an integrated services network, a wide range of information bit-rates and various types of traffic with different statistical natures have to be handled efficiently. This is made possible by dividing each information flow into short fixed-length entities, called cells, to which are attached short flow identification labels, and by transporting the entities to their destinations, in a way similar to packet multiplexing/switching [1,2,3]. The transportation mode is called the Asynchronous Transfer Mode (ATM) [4,5].

To establish an integrated service network using ATM, studies have been made on high-throughput packet switches [1,2,6], methods to assure transmission quality for various services [7,8,9], coding schemes suitable for ATM [10,11], and ATM network quality [10].

While these fundamental techniques are being developed, one of the most important subjects is how to construct the ATM transport network. To replace conventional networks, the ATM network should be economical, flexible in terms of topology and demand changes, and implemented using simple equipment. This paper proposes a technique to meet these needs. The technique is based on the concept called "Virtual Path", which is a bundle of virtual circuits. This concept reduces node processing costs, affords flexibility to the network due to its controllability, and enables implementation with simple equipment. Thus, its features meet all of challenges for future networks.

This paper presents research on an ATM network based on the Virtual Path concept. In Section 2, the Virtual Path concept and its features are defined and explained. Section 3 shows the function arrangement in the network. Finally, in Section 4, some bandwidth control algorithms for the Virtual Path are proposed, and their effects are evaluated to show the impact of bandwidth controllability.

Reprinted from the *Proceedings of Globecom*, 1988, pages 1272-1276.

2. VIRTUAL PATH CONCEPT
2.1 Labeling Schemes in ATM Network

Consider a case in which cells are transmitted from a source to a destination through some transit nodes in an ATM network, which is composed of nodes and physical links such as transmission lines. In this case, the most fundamental function required for the nodes is to identify the outgoing links, to which the cells are sent, by analyzing cell labels. To perform this function, the labels must include the information of the routes along which the cells are transported. There are several ways to represent this route information in the labels.

A typical method to include the information is to use a link-by-link logical channel number (LCN), which is a number defined for each virtual circuit in order to distinguish it from the others multiplexed in a link. In this scheme, the LCN is determined independently for each link. Thus the LCN given to a virtual circuit is different for each link along its route. Thus, when the LCN is employed as the route information, the node process for a cell is to compare the LCN of the incoming cell label with a routing table and find its outgoing link, and to update the LCN of the incoming cell label to the one on the outgoing link. This process is shown in Fig. 1.

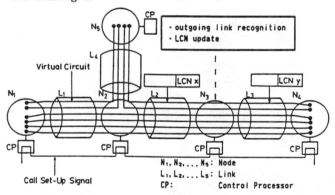

Fig. 1 Link-by-link LCN scheme.

In this scheme, the length of labels can be made short for an advantage in transmission efficiency [12]. Additionally, the routing table in the node must be renewed in every call set-up attempt because it is looked for according to the LCN determined for the call. Therefore, every node must execute this process for every call set-up.

An alternative to the link-by-link LCN scheme is a method employing the Virtual Path concept [5]. The Virtual Path is an information transport path defined as follows.

(1) A Virtual Path is a logical direct link between two nodes and can be shared by some virtual circuits simultaneously.

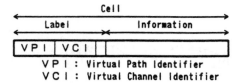

Cell

| Label | Information |

| V P I | V C I | |

V P I : Virtual Path Identifier
V C I : Virtual Channel Identifier

Fig. 2   VPI and cell structure.

(2) A pre-defined route is provided with each Virtual Path in the physical network.
(3) A Virtual Path has its own bandwidth which gives the upper limit to that of virtual circuits included in it.
(4) Virtual Paths are multiplexed in a physical link on a cell multiplexing basis.

To set a Virtual Path satisfying the above conditions, a cell label including the following two areas is required:

- Virtual Path Identifier (VPI)
  The VPI indicates to which Virtual Path a cell belongs.
- Virtual Channel Identifier (VCI)
  The VCI shows the virtual circuit which the cell belongs to.

The cell structure with the above label is shown in Fig. 2. In this scheme, the node process for the cell is described below.

(1) At a transit node through which the Virtual Path passes, the VPI in the incoming cell label is compared with a routing table and its outgoing link is found.
(2) At an end node terminating Virtual Path, the outgoing link or the destination terminal is found from the VCI.

This manner of processing for the cell is shown in Fig. 3(a). In this case, the routing table of the transit node is referred to using only the VPI. It is not necessary to rewrite the routing table at call set-up. This means that call set-up processing associated with the LCN is excluded from the transit nodes and executed only at the end nodes of the Virtual Path.

An example of a Virtual Path is shown in Fig. 4(a). In this figure, there are five nodes N_1, N_2,...,N_5 and three

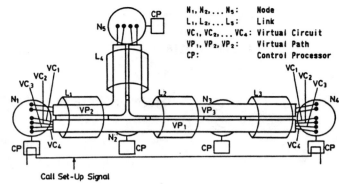

N₁, N₂,...N₅: Node
L₁, L₂,...L₅: Link
VC₁, VC₂,...VC₄: Virtual Circuit
VP₁, VP₂, VP₃: Virtual Path
CP: Control Processor

(a) Virtual Paths, virtual circuits, and links.

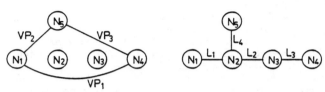

(b) Logical network for call set-up process.

(c) Physical network for VPI recognition or store-and-forward.

Fig. 4   Principle of Virtual Path.

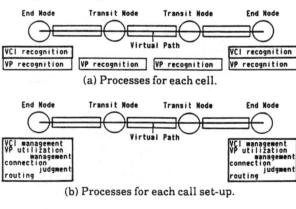

(a) Processes for each cell.

(b) Processes for each call set-up.

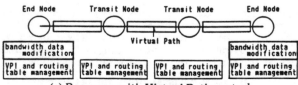

(c) Processes with Virtual Path control.

Fig. 3   Node functions in Virtual Path based network.

Virtual Paths VP_1, VP_2, and VP_3. Four virtual circuits are contained in VP_1, i.e., VC_1, VC_2, VC_3, VC_4. Cells of these virtual circuits have the same VPI from node N_1 to N_4 where VP_1 is set up. At N_2 and N_3, the outgoing link is recognized by this VPI. Therefore, these nodes are free from processing at call set-up and clearing of VC_1, VC_2, VC_3, and VC_4. As a result, the network is logically as shown in Fig. 4(b) for the call set-up process, while it is as shown in Fig. 4(c) for VPI recognition or store-and-forward process.

2.2 Virtual Path Features

The Virtual Path scheme has some advantages over the link-by-link LCN scheme. First, node processing per call set-up decreases. Hence, node processing cost is reduced. In addition, call set-up delay is short.

By contrast, label overhead is larger and thus transmission efficiency decreases with this scheme. Therefore, the Virtual Path scheme tends to give lower node cost and higher link cost than the link-by-link LCN scheme. This characteristic is valuable in constructing a low-cost ATM network because transmission links will certainly become much cheaper as high capacity optical fibers become more widely used. Additionally, the processing reduction feature enables introduction of simple equipment.

2.3 Analogy with Synchronous Transfer Mode Network

The conventional telephone network is based on a Synchronous Transfer Mode (STM) technique, which employs Time Division Multiplexing (TDM) with periodic frames and circuit switching. The analogy with the Virtual Path method is easily found here. In the existing network, two switching systems are connected by a digital path composed of several circuits, which is similar to a Virtual Path considering these four points.

- The digital path gives a logical direct link between switching systems.
- The circuits are handled in bundles, such as 384kbits/s, 1.544Mbits/s, or 6.312Mbits/s, and are not recognized by a channel at transit transmission node such as cross-

connect systems [13] or multiplexers with distributing frames.

- A pre-defined route is provided with each digital path in a physical transmission network.
- The capacity of the digital path corresponds to the bandwidth of the direct link.

In spite of these similarities, there are some essential differences between the digital path and Virtual Path depending on the way of multiplexing: the digital path is identified by its time slot position in a TDM frame while the Virtual Path is identified by cell labels. Also, the bit-rate of a digital path is always constant while that of a Virtual Path is not.

2.4 VPI Assigning Methods

VPI assigning methods are divided roughly into two categories: one is defining the VPI which explicitly exhibits the route identified globally in the network, and the other is defining the VPI with local significance, which is associated with each link, in a similar way to the link-by-link LCN scheme. Although VPI must be rewritten at transit nodes in the latter method, it has the following advantages over the former.

- VPI length is smaller because there is an upper limit to the number of Virtual Paths multiplexed on a link, and this is relatively small.
- It is easier to cope with the expansion of the network or increase in the number of Virtual Paths because VPI has local significance and an upper limit of the Virtual Path number exists as mentioned above.
- There is no need to manage VPIs in a concentrated manner.

From these features, the latter method is considered to be more suitable to large scale networks and more flexible in terms of adding nodes or links.

3. FUNCTION ARRANGEMENT

This section describes the function arrangement in a Virtual Path based network. Firstly, the function arrangement of nodes is outlined and the call set-up procedure with this method is described. Next, the benefits of controlling the Virtual Path are described, and the way to perform this is shown. Then, description on the practical equipment is presented.

3.1 Function Arrangement

The following functions needed for call set-up procedures should be placed only at the end nodes terminating the Virtual Path, since this arrangement can take advantage of its direct link nature. This helps reduce node processing and simplify equipment.

(1) LCN management
(2) Judgement to connect or refuse a call
(3) Data management of Virtual Path bandwidth utilization
(4) Selection of a Virtual Path to set-up a virtual circuit

This function arrangement is shown in Fig. 3(b). This figure shows that no special functions are needed for the call set-up process at the transit nodes.

The functions needed for Virtual Path control are,

(1) modification of the Virtual Path bandwidth data,
(2) management of VPIs and the routing tables, and
(3) determination of whether the set-up or the bandwidth increase is possible.

Among these functions, (1) is provided only at end nodes of Virtual Path while (2) must be provided at every node.

There are several alternatives for the arrangement of (3). This function can be allocated in a distributed or a concentrated fashion. The first two function arrangements are illustrated in Fig. 3(c).

3.2 Call Set-Up

The call set-up process includes traffic control (or bandwidth allocation) and routing of virtual circuits as well as LCN management as mentioned earlier. The traffic control is important to prevent congestion caused by too much input. It assures that transmission quality will not degrade in terms of queueing delay or cell loss for the virtual circuits. The routing is also significant in efficient network utilization. The routing and control procedures are executed employing the functions described earlier.

(a) Traffic control

By comparing the unused bandwidth of the Virtual Path with the new call bandwidth which the customer requests, whether or not to refuse the connection is determined. This is done by using the data of the bandwidth utilization conditions in the Virtual Paths.

(b) Routing

Some Virtual Paths with different routes are set up in advance between nodes where a logical direct link is required. Then, one of them is selected at call establishment.

3.3 Virtual Path Control

A Virtual Path is controllable in set-up, release, change of routes, and bandwidth. The benefits of the controllability include flexibility for topology changes such as node addition, link addition, or network failure, adaptability to short or long term traffic demand variations, and high utilization of transmission resources by varying bandwidths of Virtual Paths.

This controllability is achieved in the following way.

- Set-up is done by defining a VPI for each link and rewriting the routing tables of all the nodes included in the Virtual Path route.
- Release is performed by clearing the VPI and rewriting the routing tables.
- Route changes are done by executing releases and set-ups in sequence.
- Bandwidth control is carried out by modifying data of the Virtual Path bandwidth at the end nodes.

3.4 Practical Equipment

As described above, end nodes and transit nodes in the Virtual Path have very different functions. The former needs to execute more processing than the latter. Because of this, it is more economical to make separate equipment for end nodes and transit nodes. Currently, a switching system is planned for end node use, and an ATM Cross Connect System and Add Drop Multiplexer for transit node use.

The switching system is composed of a high-throughput packet switch and a high-performance processor to handle both call set-up processing and Virtual Path control. The ATM Cross Connect System also has a packet switch. However, this is a much less expensive equipment, since its control processor is much simpler because it does not do call set-up processing. This low-cost processor is also used in the Add Drop Multiplexer. The Add Drop Multiplexer is adopted in a ring topology, and has a function which allows adding or dropping cells into or from the ring.

A network configuration employing ATM Cross Connect Systems is shown in Fig. 5(a) and one employing Add Drop Multiplexers in Fig. 5(b). In large size networks, both types of equipment can be utilized.

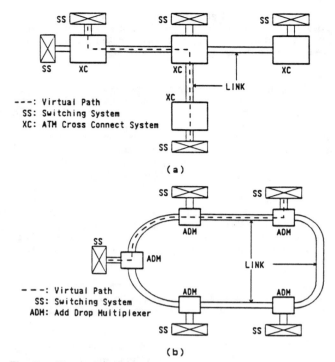

--- : Virtual Path
SS: Switching System
XC: ATM Cross Connect System

(a)

--- : Virtual Path
SS: Switching System
ADM: Add Drop Multiplexer

(b)

Fig. 5 Examples of network configurations with,
(a) ATM Cross Connect Systems, and (b) Add Drop
Multiplexers.

4. IMPACT OF VIRTUAL PATH CONTROL

The advantages of Virtual Path include controllability and processing reduction at the transit nodes. This section treats the high utilization of transmission resources allowed by controlling the bandwidth. Some control algorithms and their evaluations are described, and the bandwidth control is shown to improve transmission efficiency without losing the essential advantage of the processing reduction.

4.1 Bandwidth Control Algorithm

Bit-rates for the information transmitted through Virtual Paths vary with time due to changes in their concurrent connections. When the Virtual Path bandwidth is fixed, i.e., there is no bandwidth control, the bandwidth must be large enough to cope with the pre-defined maximum concurrent connections. Because of this, the bandwidths of few Virtual Paths in a link are usually used up, although those of the others in the same link are underutilized.

On the other hand, assume that the bandwidth is controlled so as to be small when there are few concurrent connections, and to be large for many connections. If this type of control is performed, transmission efficiency will improve, since every Virtual Path is well utilized. The fundamental algorithm to achieve this kind of control is as follows:

- Request bandwidth increase if it is insufficient for the new call arriving at the end node.
  If the increase is allowed, then increase the bandwidth and set up a virtual circuit for the call; else, keep the current bandwidth and refuse the call.
- Decrease the bandwidth if possible, according to Virtual Path utilization conditions.

This algorithm allows some variations by the unit of bandwidth changes or the range of control. By selecting these parameters, trade-offs appear between transmission

efficiency and processing quantity.

The effect of the fundamental algorithm is evaluated according to the improvement in transmission efficiency and the amount of processing increase; the advantage in reduced node processing must not be lost.

4.2 Evaluations

In this section, the algorithm is evaluated for the above characteristics. To analyze these points, the bandwidth of a call should be defined accurately as a premise. However, parameters to characterize the call bandwidth are now under study and have not yet been clarified. Thus, the following assumptions are made in this paper as a first step in evaluations.

Assumptions 1:
- The sum of the bandwidths of the Virtual Paths in a link should not exceed the link capacity. Bandwidth increase requests are rejected if they make the sum larger than the link capacity.
- A bandwidth is defined for every call. The call is refused if the sum of the call bandwidths in the Virtual Path exceeds its upper limit, i.e., the Virtual Path bandwidth.

These additional assumptions are made for simplicity.

Assumptions 2:
- Every call has the same bandwidth.
- The traffic offered to each Virtual Path is identical, and calls follow Poisson arrival pattern.

Under these assumptions, one of the links in the network is analyzed. The following values are calculated for this link.

(1) Transmission efficiency E, defined as the ratio of carried traffic and the link capacity.
(2) Processing quantity factor F, evaluated by the ratio of Virtual Path bandwidth change frequency to that of call set-up and clearing attempts.

The evaluated control methods derived from the fundamental algorithm are as follows.
Method 1 (fixed bandwidth): The Virtual Path bandwidth is always constant.
Method 2 (two-value control): The Virtual Path bandwidth is 0 or S.
Method 3 (three-value control): The Virtual Path bandwidth is 0, S, or 2S.
Here, S is the step of the bandwidth change, where the bandwidth is normalized to that of the call. In the analysis, the following parameters are defined.
a: traffic offered to Virtual Path (erl)
C: link capacity normalized to the bandwidth of a call
i_i: number of virtual circuits in the ith Virtual Path
n: number of Virtual Paths in the link
Call loss probability B for $(K+1)$-value control is calculated as follows,

$$B = \sum_{(i_1,i_2,..i_n)\in I_b} \sum \cdots \sum \frac{a^{i_1}}{i_1!} \frac{a^{i_2}}{i_2!} \cdots \frac{a^{i_n}}{i_n!} \cdot P_0,$$

where,

$$P_0 = \sum_{(i_1,i_2,..i_n)\in I} \sum \cdots \sum \frac{a^{i_1}}{i_1!} \frac{a^{i_2}}{i_2!} \cdots \frac{a^{i_n}}{i_n!}$$

$$I_b = \{(i_1,i_2,...i_n)| \ i_1 = j\cdot S \text{ and } \left\lceil\frac{i_2}{S}\right\rceil + \cdots + \left\lceil\frac{i_n}{S}\right\rceil = \left\lceil\frac{C}{S}\right\rceil - j; j = 0,1,$$

$$..K \} \cup \{(i_1,i_2,...i_n)| \ i_1 = K\cdot S \text{ and } \left\lceil\frac{i_2}{S}\right\rceil + \cdots + \left\lceil\frac{i_n}{S}\right\rceil \leq \left\lceil\frac{C}{S}\right\rceil - K\},$$

$$I = \{(i_1,i_2,...i_n)| \ \left\lceil\frac{i_1}{S}\right\rceil + \cdots + \left\lceil\frac{i_n}{S}\right\rceil \leq \left\lceil\frac{C}{S}\right\rceil\},$$

and $i_1, i_2, ...,i_n \leq K\cdot S.$

Note: the symbol $\lceil x \rceil$ denotes the smallest integer greater than or equal to x.

By using the above equation, the efficiency is evaluated for the minimum link capacity to keep the call loss below some value.

The processing quantity factor F is calculated as follows.

$$B = \sum_{(i_1,i_2,..i_n)\in I_c} \sum \cdots \sum \frac{a^{i_1}}{i_1!}\frac{a^{i_2}}{i_2!}\cdots\frac{a^{i_n}}{i_n!}\cdot P_0,$$

where,

$$I_c = \{(i_1,i_2,..i_n)|\ i_1 = j\cdot S \text{ and } \left\lceil\frac{i_2}{S}\right\rceil + \cdots + \left\lceil\frac{i_n}{S}\right\rceil \le \left\lceil\frac{C}{S}\right\rceil - j; j = 0,1, ...K\}.$$

The relation between the offered traffic and the maximum transmission efficiency of each method is shown in Fig. 6. The processing quantity factor is shown in Fig. 7 for Methods 2 and 3. These figures are obtained by setting n to 16 and $B < 10^{-3}$. The figures show that Method 3 improves the efficiency considerably, keeping additional processing to a low level. Bandwidth changes are required with a frequency below fifteen percent that of call set-up or clearing attempts for more than three erlang traffic.

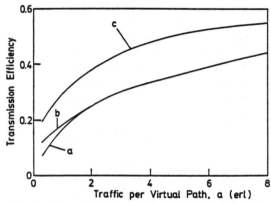

Fig. 6   Transmission efficiency for offered traffic.
a: Method 1. b: Method 2. c: Method 3.

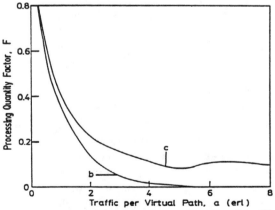

Fig. 7   Processing quantity factor for offered traffic.
b: Method 2. c: Method 3.

Fig. 8 shows trade-off between the efficiency and the processing quantity factor with the bandwidth control step. In this figure, n = 5, a = 6 (erl) and the control method is the five-value method, in which the bandwidth is 0, S, 2S, 3S, 4S. This figure shows that the efficiency has a tendency to decrease with the processing reduction. An optimal value for the bandwidth change step can be obtained employing this result when transmission/processing cost ratio is provided.

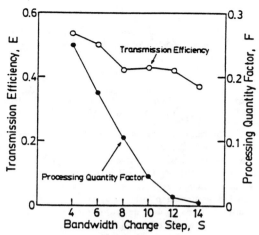

Fig. 8   Transmission efficiency and processing quantity factor for step of bandwidth change.

5. CONCLUSION

Studies were carried out on an ATM transport network based on the Virtual Path concept. The arrangement of node functions in this network were proposed. The proposed method excludes the call set-up process from transit nodes, and this leads to reduction in node processing and simplification of equipment by specialization of transit node functions. Additionally, Virtual Path control was shown to bring benefits in efficiency and flexibility. Among these benefits, the transmission efficiency improvement by the bandwidth control was evaluated to show the impact of Virtual Path control. The results showed that a simple three-value bandwidth control algorithm improves the efficiency considerably with a slight processing increase.

These studies demonstrate that the Virtual Path concept is expected to open the way to the construction of simple, economical, flexible, and high-performance networks.

REFERENCES

[1] J. J. Kulzer and W. A. Montgomery, "Statistical switching architectures for future services," ISS'84, 43A-1, May 1984.
[2] A. Thomas, J. P. Coudreuse, M. Servel, "Asynchronous time-division techniques: an experimental packet network integrating video-communication," ISS'84, 32C-2, May 1984.
[3] T. Miki, S. Kano, Y. Inoue, H. Yamaguchi, "Lightwave-based intelligent transport network," ISSLS'86, pp.47-52, Sept. 1986.
[4] T. Kanada, K. Sato, T. Tsuboi, "An ATM based transport network architecture," IEEE Int. Workshop on Future Prospects of Burst/Packetized Multimedia Commun., 2-2, Nov. 1987.
[5] I. Tokizawa, T. Kanada, K. Sato, "A new transport network architecture based on asynchronous transfer mode techniques," ISSLS'88.
[6] J. Y. Hui and E. Arthurs, "A broadband packet switch for integrated transport," IEEE J. on Selected Areas in Commun., SAC- 5, 8, pp.1264-1273, Oct. 1987.
[7] J. Cochennec, P. Adam, T. Houdoin, "Asynchronous time-division networks: terminal synchronization for video and sound signals," GLOBECOM'85, pp.791-794, Dec. 1985.
[8] W. Hoberecht, "A layered network protocol for packet voice and data integration," IEEE J. on Selected Areas in Commun., SAC-1, 6, pp.1006-1013, Dec. 1983.
[9] W. A. Montgomery, "Techniques for packet voice synchronization," IEEE J. on Selected Areas in Commun., SAC-1, 6, pp.1022-1028, Dec. 1983.
[10] K. Sato, H. Nakada, Y. Sato, "Variable rate speech coding and network delay analysis for universal transport network," Infocom'88, Mar. 1988.
[11] N. Ohta, M. Nomura, T. Fujii, "Characteristics of variable rate video coding with motion compensated DCT for burst/packetized communications," IEEE COMSOC Int. Workshop on Future Prospects of Burst/Packetized Multimedia Commun., 4-3, Nov. 1987.
[12] M. Schwartz and T. Stern, "Routing techniques used in computer communication networks," IEEE Trans. on Commun., COM-28, 4, pp.539-552, Apr. 1980.
[13] H. Ueda and I. Tokizawa, "A synchronous DS4 multiplexer with cross-connect function and its impact on the network," GLOBECOM'85, pp.547-551, Dec. 1985.

Resource Allocation for Broadband Networks

JOSEPH Y. HUI, MEMBER, IEEE

Reprinted from *IEEE Journal on Selected Areas in Communications*, Vol. 6, No. 9, December 1988, pages 1598-1608.

Abstract—The major benefit of a broadband integrated ATM network is flexible and efficient allocation of communications bandwidth for communications services. These services may have very different traffic statistics, such as average bit rate and peak bit rate. Some services are bursty, such as still picture video; or continuously varying in bit rate, such as compressed full motion video or fixed bit rate. Hence, new methodologies are needed for evaluating congestion for integrated traffic. We suggest evaluating congestion at different levels, namely, the packet level, the burst level, and the call level. Congestion is measured by the probabilities of packet blocking, burst blocking, and call blocking. We outline the methodologies for computing these blocking probabilities. We use these congestion measures for a multilayer bandwidth allocation algorithm, emulating some functions of virtual circuit setup, fast circuit switching, and fast packet switching at these levels. The analysis also sheds insight on traffic engineering issues such as appropriate link load, traffic integration, trunk group and switch sizing, and bandwidth reservation criteria for bursty services.

I. INTRODUCTION

A BROADBAND end-to-end fiber network is attractive for carrying future broadband communications services. Such services may have a broad spectrum of traffic statistics, such as average bit rate, peak bit rate, call interarrival time and holding time, and packet periodicity, etc.

To facilitate flexible sharing of the network by end users holding multirate service calls, packet (more appropriately known as cell in ATM terminologies) switching technology [1]–[4] seems promising for implementing an integrated access and transport network. Given the flexibility of packet switching, we still need an understanding of congestion and bandwidth allocation issues for enhancing network efficiency. We study in this paper congestion phenomena at the packet, burst, and call levels; and propose using multilevel bandwidth allocation.

This multilevel congestion evaluation and control is motivated by the fact that communication terminals often have traffic states characterized by these levels. Thus, bandwidth should be allocated at these levels for improving network efficiency. In Section II, we describe the resources such as time slots, trunks, and trunk groups available for allocation at switching facilities. We then examine the multilayer traffic states of terminals.

In Section III, we state more precisely what operations are needed for allocating resources at these levels. In particular, we show how traffic control at one level (say the call level) may reduce blocking at a lower level (say the burst level). At the lowest level, we assume a high throughput packet switch [1], [2], [4] for statistical allocation of the output time slots. The call level control emulates circuit switching for fixed and high bit rate services and denies calls when facility overload may cause excessive burst blocking. The burst level control emulates fast circuit switching for the individual trunks for avoiding excessive packet blocking within a trunk.

The second half of this paper studies traffic and congestion in this integrated service environment. At the packet level, we use the example of internally nonblocking packet switches [4] for analyzing blocking and throughput versus offered traffic within each trunk. Blocking occurs when a packet arrives at a full buffer, or when it suffers excessive delay. The treatment is given in Section IV. In general, the packet level loading per trunk depends on the specific packet switch design. The slot occupancy should be controlled to avoid excessive packet blocking.

The analysis in this paper focuses mostly on evaluating congestion within a trunk group at the burst level, given the mix of integrated services allowed to use the trunk group after call setup. Blocking occurs when the aggregate instantaneous bit rate of these services exceeds the capacity of the trunk group. To study this phenomenon of burst blocking, we characterize the aggregate traffic statistics within a trunk group at the burst level. We apply the theory of large deviation [5] for evaluating the probability of burst blocking. The call admission process controls the trunk group loading in the region where burst overflow is a rare event. This treatment is given in Section V.

Therefore, an arriving call which may cause excessive burst blocking for calls already setup in the trunk group should be blocked for that trunk group. Thus, a call blocking occurs for that trunk group. Other trunk groups facilitating alternative routing may then be tried. Given a method for evaluating burst blocking, we can compute the admissible region for the combinations of calls of each service type for which the burst level blocking is acceptable. Given this admissible region of call combinations, we can then calculate the probability of call blocking at the boundary of the region. This probability depends on the call arrival and holding time statistics. Trunk groups and switches should be sized for a tolerable probability of call blocking. We discuss the evaluation of call blocking and call packing in Section VI.

II. THE TERMINAL TRAFFIC PROCESS

How should communication bandwidth be allocated for a call with fluctuating bit rate in an integrated network?

Manuscript received September 15, 1987; revised June 7, 1988.

The author is with Broadband Packet Network Research, Bell Communications Research, Morristown, NJ 07960.

IEEE Log Number 8824402.

Here, communication links may be time-multiplexed for calls, and a call may choose one among many routes within the network. We show a simple network in Fig. 1 as an illustration, which consists of communications terminals and switching facilities such as multiplexers and central office switches. The switching facilities are connected by a trunk or a trunk group G. When a terminal initiates a call to another terminal, we assume that a route server (centralized or distributed) assigns one or several routes, defined by the sequence of multiplexers and switches enroute between the terminals. In Fig. 1, we have two possible routes between the terminals shown. In this paper, we assume that the candidate routes are given initially, and that only one route is used after call setup. Thus, at a switching facility such as a packet switch, we face two problems. First, which trunk groups (the choice of which determines locally the route taken across the network) should be chosen at the call level? Second, after the call is setup, which trunk may be used by a burst within a call, or should the packets of a burst be distributed evenly on all trunks of a trunk group?

Before discussing the allocation procedure, we examine the traffic process generated by communication terminals. In conventional voice telephony, a telephone is either in an on-hook or off-hook state [Fig. 2(a)]. This two-state model also suffices for other fix bit rate communications, such as fix bit rate full motion video.

When information generated by a communication terminal is packetized, we add a new state of packet transmission [Fig. 2(b)]. Given a fix rate terminal is off-hook, the terminal enters into the packet transmission state once every v packet time where v equals the channel bit rate divided by the terminal bit rate.

Some calls may communicate in alternating states of idle and burst after the call is set up [Fig. 2(c)]. Such calls include voice calls with silence removal, and graphics stations.

For some calls such as compressed full motion video, the bit rate process can be changing continuously (Fig. 4), depending on the amount of motion and the degree of granularity of the picture. For illustration, we describe two numerical examples for these traffic processes with fluctuating bit rate. These examples shall be used in the conclusion as a scenario of traffic integration.

A. Still Picture Graphics Station (Fig. 3)

The graphics station produces a burst of 10 Mbits per picture, which are sent, for example, in 10^4 packets each of 1000 bits [Fig. 3(a)] on a 100 Mbits/s multiplexed link. We assume that the whole burst is communicated through the channel in 1 s, hence the peak burst rate of the station is 10 Mbits/s [Fig. 3(b)] at 1 packet per 10 channel slots. We assume that the burst arrival process is Poisson with rate 0.1 burst/s. Hence, expected burst interarrival time is 10 s, and the average bit rate for the call duration is 1 Mbit/s [Fig. 3(c)]. We assume that a bursty call has an expected call holding time of 10^4 s, and that

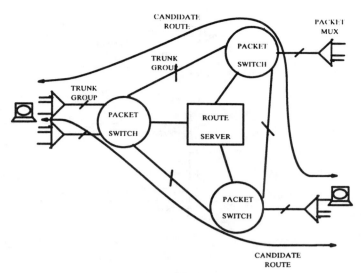

Fig. 1. Routing in broadband ATM network.

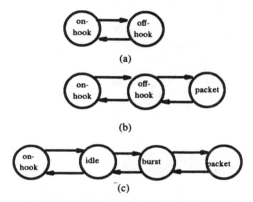

Fig. 2. Multilevel traffic states. (a) Circuit switched fix rate call. (b) Packet switched fix rate call. (c) Packet switched bursty call.

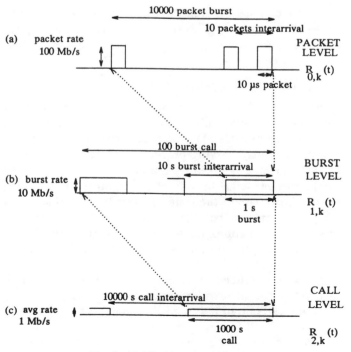

Fig. 3. Multilevel traffic processes.

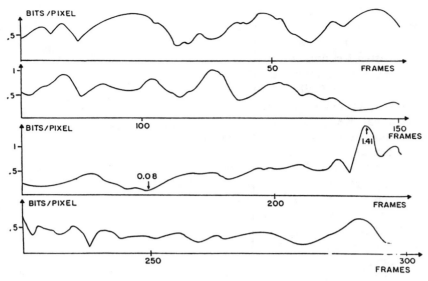

Fig. 4. Bit rate process of compressed video [15].

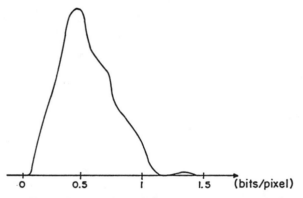

Fig. 5. Bit rate histogram of compressed video [15].

at any time, a station is in the calling process with probability 0.1. Thus, Fig. 3(a), (b), and (c) illustrate, respectively, the traffic generation process at the packet, burst, and call levels.

B. Compressed Full Motion Video (Fig. 4)

We can similarly assume that the video terminal is in the calling process with probability 0.1, and mean call holding time of 10^4 s. During a call, the bit rate (Fig. 4) is a continuously varying process [15]. Thus, the packet interarrival time is given by the channel bit rate divided by the instantaneous bit rate of the terminal. Fig. 5 [15] gives a histogram of the instantaneous bit rate. Assuming 30 picture frames per second and 250 000 pixels per frame, we shall approximate the histogram of Fig. 5 by a triangle, with peak bit rate of 10 Mbits/s and an average bit rate of 5 Mbits/s.

Blocking is defined for each level as the event of failing to allocate the bandwidth demanded. A call is blocked when the call initiation is denied. A burst is blocked when a completed call fails to deliver the burst. A packet within a burst is blocked when the packet is missing upon delivery. In the process of network design, we may designate

bounds for these blocking probabilities, say 10^{-2} for call blocking, 10^{-4} for burst blocking, and 10^{-6} for packet blocking. We shall only be concerned with computing these blocking probabilities per switching system, rather than end-to-end. We can then build an end-to-end picture given a link-by-link congestion measure.

In a network with many types of terminals, we expect that these three levels are sufficient. For high, but fixed bit rate full motion video calls, we are concerned only with the call and packet levels. The same is true for low bit rate calls such as voice with no silence removal. For compressed full motion video calls, the bit rate is far less bursty than a graphics station. Using the numerical example in the conclusion, we show that only a small number of such calls are needed to achieve the advantages of statistical averaging. Hence, the burst level allocation may not be necessary if the average bit rate for such terminals is sufficiently small compared to the channel bit rate. The burst level load statistics of these compressed video can also be captured by the derivation in later sections. For data transfer with low peak rate, its burstiness is of little significance to merit a burst level allocation. For high bit rate transfer, we may use both the call and burst levels for bandwidth allocation.

III. MULTILAYER CONTROL FOR BANDWIDTH ALLOCATION

The packet arrival process for a call k is denoted by $R_{l=0,k}(t)$ which equals u, the channel rate if a packet belonging to k is arriving at time t. When no packet for k is arriving, $R_{0,k}(t) = 0$. At a higher level l, $R_{l,k}(t)$ is defined as the average bit rate at the level of concern. (For a three-level process such as a graphics station, Fig. 3(a), (b), and (c) represent $R_{l,k}(t)$ for $l = 0, 1, 2$, respectively.) For many communications terminals, $R_{l,k}(t)$ changes only at infrequent instants, such as at burst or call arrivals. For such terminals, a pilot packet for a burst or a call may initiate a bandwidth request for the burst or the call. The pilot packet may contain the information $\Delta R_{l,k}$, the step-wise increase of bandwidth at the level of concern. The pilot packet may also contain other parameters of interest, such as an estimate of the holding time for the bandwidth increase.

The bandwidth to be allocated for these levels is illustrated in Fig. 6 for a bursty call, such as the graphics station example of the previous section. At the initiation of a call, the call requests connection to a trunk group G, or one of several candidate trunk groups in the case of alternative routing. This trunk group information, we presume, is provided by some route address server governing a service area (Fig. 1). Thus initially, the resource considered for allocation is a set of G. During call setup, a particular G is assigned.

At the arrival of a burst, a subset of the resource allocated at the call level is chosen, namely, a particular $j \in G$ may be chosen for a burst. If the capacity of the trunk group is utilized as a single entity, the burst level allocation process becomes unnecessary. Such a situation occurs when the trunk group is a single high-speed trunk, or when packets within a burst are randomly spreaded among the trunks for transmission.

At the packet level, a packet is delivered to the desire output statistically by packet switching. The packet switch may statistically assign time slots on a first-come first-serve basis.

Thus, at each level l, the bandwidth allocation algorithm choses a subset Γ_l of the communication resources Γ_{l+1} allocated at level $l + 1$. Above the call level, the bandwidth resources is initialized to Γ_{initial}, the set of candidate output trunk groups indicated by the route server.

The obvious question then is which subset of the resource allocated at the next higher level $l + 1$ should be chosen for level l. This depends on the load of the outputs. We define the offered traffic at level l to output resource Γ_l at time t by summing the traffic of all terminals k using the resource

$$W_l(t) = \sum_{k \text{ routed thru } \Gamma_l} R_{l,k}(t).$$

We define the carried traffic $W_l'(t)$ as the loading of the output resources at level l. The difference between $W_l(t)$ and $W_l'(t)$ is blocked traffic. Ideally, this should be a very small fraction of the offered traffic. We would like to obtain the sufficient statistics Σ_l for the traffic process. Take the burst level as an example, each channel j may register $\Sigma_{l=1}$ as the number of on-going bursts routing through j. If we only have bursty traffic of the type given by the graphic station example, this number is sufficient to estimate packet blocking within the interval.

When call k initiates a bandwidth request at level l, the algorithm at level l should choose a not-too-congested subset (denoted by Γ_l) of the resource allocated for k at level $l + 1$, namely, Γ_{l+1}. What do we mean by not-too-congested? Consider a burst arrival for call k, which wants to be carried by a $j \in G$ allocated at the call level. A congested j would be one which already has too many on-

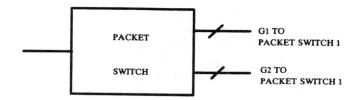

CALL LEVEL - SELECT TRUNK GROUP G1 OR G2

BURST LEVEL - SELECT TRUNK IN CHOSEN TRUNK GROUP

PACKET LEVEL - STATISTICAL ALLOCATION OF SLOT

Fig. 6. A multilevel bandwidth allocation process.

going bursts assigned to it, thus the link-load of say more than 90 percent may cause excessive output queueing at the packet level. This excessive queueing would cause blocking at the packet level. Thus at level l, the control algorithm chooses a subset of resources allocated at level $l + 1$, such that the added bandwidth to the subset may not cause excessive blocking at level $l - 1$. (For example, at call initiation, a particular G among the candidate trunk groups is chosen such that the probability of burst blocking is not excessive in the chosen G.) The algorithm tries one subset after another, and if all such subsets are congested, we say that the bandwidth request at level l is blocked. (For example, a burst is blocked if none of $j \in G$ has decent packet blocking probability when the burst is added to j.) At level l, the algorithm computes the blocking probability $P_{B,l-1}$ for a subset using the sufficient statistics Σ_l and the bandwidth request $\Delta R_{l,k}$. If the resulting blocking probability is within tolerable limits, the subset is allocated to the request, and Σ_l is updated.

Where are the routing information Γ_l and load information Σ_l held, and how are the requests and granting of requests communicated? Initially, the route address server provides a set Γ_{initial} of trunk groups for a call initiation at a switching or multiplexing system. The trunk groups individually may keep a record of calls (possibly at the output port controller of a representative trunk within the trunk group) $\Sigma_{l=2}$. A call arrival would query these representatives sequentially, by sending a request packet through the packet switch. The assigned $\Gamma_{l=2} = G$ is communicated from the representative through the packet switch to the inputs used by the call. The trunk group assigned subsequently updates Σ_2. The set Γ_2 is stored at these inputs (or the input trunk group representative) for burst routing. Next, consider the burst level $l = 1$. When a burst of a call arrives at an input, the pilot packet queries individual trunks in the assigned G. These trunks keep a record of their link load $\Sigma_{l=1}$, and if the queried output trunk can accomodate the burst without causing excessive packet blocking, the output would update the link load, and acknowledge the requesting input. Otherwise, a denial is sent. The input then remembers the output trunk assigned, and routes the packets within the burst to the assigned output. Next, consider the packet level $l = 0$. When a packet within a burst moves to the head of the input queue, the packet switch then statistically switches the packet to the desired output.

Very often, we may tolerate a delay in the resource allocation process without being considered blocked. At the packet level, conflicting input requests for the same output at a slot time may be buffered at input packet buffers [4] until the output is free. At the burst level, a burst which finds all $j \in G$ congested may buffer the entire burst in a burst buffer. The stored burst continues to search for a j with relieved load. At the call level, we may tolerate a certain delay for call completion.

We summarize the above procedures by the following three-step algorithm. At level l, the scheme allocates a subset of resources allocated at level $l + 1$, and guarantees the allocation would not cause excessive blocking at level $l - 1$ within the assigned subset.

1) *Request*—A request packet, which may contain the information Γ_{l+1} and the bandwidth request information such as $\Delta R_{k,l}$, is sent to an output which holds the load information Σ_l for the resources queried (a subset of Γ_{l+1}).

2) *Acknowledge*—The output computes $P_{B,l-1}$ using Σ_l plus $\Delta R_{k,l}$. If $P_{B,l-1}$ is acceptable, the allocated Γ_l, a subset of Γ_{l+1}, is sent to the requesting input. Subsequently, the output updates Σ_l.

3) *Send*—The requesting input stores Γ_l and routes packets through Γ_l. If the queried output failed to allocate the resource, the requesting input repeat request to other parts of Γ_{l+1}. If all of Γ_{l+1} failed to allocate the resource, the input may store packets in a buffer, if available. Otherwise, the request is considered blocked.

Note that $\Delta R_{k,l}$ can be negative when the allocated resource is relinquished, and Σ_l is updated.

This layered control protocol induces overhead processing. The burst layer may be removed if we allow packets of a burst to traverse any trunk in G. It is not clear which approach should be taken. We believe that performing a burst allocation per 10 000 packets on a link-by-link basis requires rather little processing. Burst level allocation provides better traffic control, simplified hardware, and no out-of-sequence packets. The following sections shall focus on computing blocking probabilities in networks with mixed call types.

IV. BLOCKING PROBABILITY AT THE PACKET LEVEL

The calculation of blocking probability at the packet level depends on individual packet switch design. In this section, we illustrate this calculation for internally nonblocking packet switches. The full treatment is given in [4]. With certain random traffic assumptions, the packet arrival rate λ is sufficient for computing the blocking probability, hence may be used for $\Sigma_{l=1}$.

An internally nonblocking packet switch is defined as follows. Let the switch has inputs i and outputs j, $1 \leq i$, $j \leq N$ where N is the number of inputs or outputs. Let the inputs and outputs be synchronously slotted for containing fixed length packets. At the beginning of each slot

time, each input i may request a connection to an output, say j_i. For each output j, the packet switch would deliver a packet, within the slot time, from exactly one of the conflicting i which make output requests for the same j. The packets not delivered would be buffered for contending for the next time slot. Packets are served first-come first-served (FCFS).

Assume packet arrivals at each input constitute an independent Bernoulli process with probability λ, equal for all i. Assume that the destination addresses are equally probable for all N outputs. We shall focus on the throughput performance and queueing performance for the packet switch.

A. Throughput Performance

For the moment, we assume that the buffer size B is infinite. Define the saturation factor ρ as the probability that an input has a packet in queue given the packet at the head of the queue is served in the previous slot time. The following congestion throughput relationship is proved in [4]:

$$\frac{1}{\rho} = \frac{1}{\lambda} - \frac{\lambda}{2(1 - \lambda)}. \tag{4.1}$$

By setting $\rho = 1$, maximum throughput is 58.6 percent. Thus, we need a switch speed-up of at least 2 (say with parallel fabric [4]) so that the input and output trunks may be loaded closer to 1. After the speed-up, effectively $\lambda < 0.5$. Hence, any traffic imbalance, for traffic arrival rates at inputs as well as nonuniform output addresses, would not cause excessive queueing at the inputs.

B. Input Queueing Performance

We assume that the waiting time for a packet at the head of the queue is geometrically distributed with probability q (a function of λ). The queue can then be treated as a simple discrete time process with independent arrivals, rate λ, and independent service, rate q. Relating the degree of saturation with the service and arrival rates, we have

$$\frac{1}{\lambda} = \frac{1}{q} + \frac{1 - \rho}{\rho} \tag{4.2}$$

which states that in steady state, the mean packet interarrival time should be equal to the mean number of slots before a fresh head of queue packet is served, plus the mean number of slots before an arrival of a fresh head of queue packet. Substituting ρ in (4.1) into (4.2), we can express the service rate in terms of the arrival rate

$$\frac{1}{q} = \frac{\lambda}{2(1 - \lambda)} + 1. \tag{4.3}$$

Given the service rate and the arrival rate, the queue length distribution p_l, average queue length $\bar{L}$, and buffer overflow, probability P_B can be readily solved

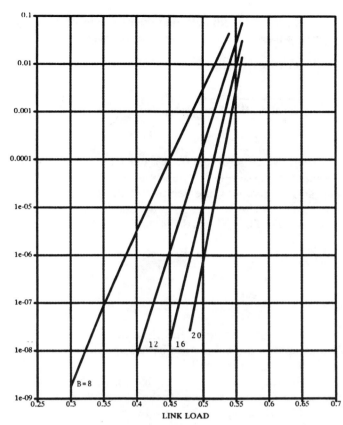

Fig. 7. Upper bound on packet loss.

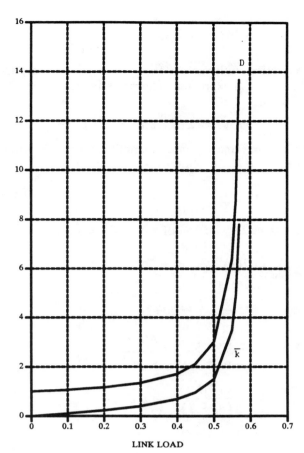

Fig. 8. Average queue length and delay.

$$p_0 = 1 - \frac{\lambda}{q}; \quad p_{l>0} = \frac{\lambda}{q}(1 - \omega)\,\omega^{l-1};$$

$$\omega = \frac{\lambda(q - 1)}{q(\lambda - 1)} \tag{4.4}$$

$$\bar{L} = \frac{\lambda(1 - \lambda)}{q - \lambda} \tag{4.5}$$

$$P_B < P(L > |B|) = \sum_{l=|B|+1}^{\infty} p_l = \frac{\lambda}{q}\,\omega^{|B|}. \tag{4.6}$$

Fig. 7 plots blocking probability versus the link load. Fig. 8 plots mean queue length and delay.

C. Output Queueing Performance

By speeding up the packet switch hence reducing the arrival rate λ, we can reduce input queueing delay and buffer overflow. The queueing effect shifts from the input to the output buffer. With a speed-up of 2 or more [4], the input queues would be short compared to the output queues. Loading on the output should be backed off to avoid excessively long output queueing delay. The output queue length distribution is given by the well-known $M/D/1$ formula [6].

V. Blocking Probability at the Burst Level

In this section, we want to characterize the sufficient statistics $\Sigma_{l=2}$ for computing the burst blocking probabil-

ity. First, we give a recursive formula for computing the tail distribution of the offered traffic W for Poisson traffic. Second, we relate the distribution for the offered traffic W with the carried traffic W'. Given this relationship, we shall focus on W later. Third, we give a tutorial on the moment generating function for W. Fourth, using the moment generating function, we apply large deviation theory to obtain good approximations for the tail distribution of W. Fifth, we generalize the computation of W to mixed Poisson traffic and continuously varying traffic such as compressed video. Numerical method is used for fast evaluation of congestion for mixed traffic types.

A. A Recursive Formula for Tail Distribution of Offered Traffic W

Let K be the set of calls assigned at the call level to the trunk group G with total bandwidth z. Let the offered load $W(t) = \Sigma_k R_k(t)$, the sum of the instantaneous burst level bit rates.

For Poisson traffic, we model $W(t)$ by jumps of different amplitudes $a_i > 0$, which arrive with Poisson rate γ_i and last for a duration b_i. (The case of a random duration can be easily accommodated by treating each duration with an associated λ.) The offered traffic $W(t)$ satisfies the following recursive relation for computing the tail distribution.

Define $Q(y) = P(W(t) < y)$, we have

$$\int_0^y x\, dQ(x) = \sum_i \gamma_i a_i b_i Q(y - a_i). \tag{5.1}$$

Differentiating with respect to y, we have the marginal distribution $q(y)$ expressed in terms of its value at smaller y as follows:

$$yq(y) = \sum_i \gamma_i a_i b_i q(y - a_i). \tag{5.2}$$

Proof: Assume n_i jumps of amplitude a_i occurs within the past period b_i. It follows then

$$Q(y) = \sum_{n_i : \Sigma n_i a_i \leq y} \prod_i \frac{(\gamma_i b_i)^{n_i}}{n_i!} e^{-\gamma_i b_i}. \tag{5.3}$$

Hence, taking expectation of x over the interval $[0, y]$

$$\int_0^y x\, dQ(x)$$

$$= \sum_{\Sigma n_i a_i \leq y} \left(\sum_i n_i a_i \prod_i \frac{(\gamma_i b_i)^{n_i}}{n_i!} e^{-\gamma_i b_i} \right)$$

$$= \sum_{\Sigma n_i a_i \leq y} \sum_j \prod_{i \neq j} \left(\frac{(\gamma_i b_i)^{n_i}}{n_i!} e^{-\gamma_i b_i} \right)$$

$$\cdot \frac{(\gamma_j b_j)^{n_j - 1}}{(n_j - 1)!} \gamma_j a_j b_j e^{-\gamma_j b_j}$$

$$= \sum_j \gamma_j a_j b_j Q(y - a_j) \qquad \square$$

$$\tag{5.4}$$

B. Relation Between Offered Traffic W and Carried Traffic W'

For the above Poisson shot noise process, we define the carried traffic process $W'(t)$ with the same arrival statistics as $W(t)$. However, a burst is blocked if its arrival causes the total load to exceed the total bandwidth z.

It is important to notice that the derivation of (5.2) is valid line by line for processes with steady state probabilities of n_i assuming a product form [6], [7] in the mean holding time b_i exponentiated by n_i; namely, $C \prod_i \cdot (\gamma_i b_i)^{n_i}/n_i!$, in which C is the normalization constant for the probabilities such that n_i satisfy $\sum_i n_i a_i \leq z$. This can be proved by the theory of reversibility [7]–[10], provided that the holding time distribution has a rational Laplace transform.

Kaufman obtained (5.2) for the lossy system W' [8]. Although (5.2) provides a method for computing $P(W(t) \leq x)$, the computation is not efficient. Instead, we extend large deviation theory for computing the tail distribution of the Poisson shot noise process W. The blocking probability for the lossy system is then obtained by the following relationship between W and W'.

For $x \leq z$, the total bandwidth of G,

$$P(W'(t) \leq x) = \frac{1}{1 - P(W(t) > z)} P(W(t) \leq x). \tag{5.5}$$

Thus, it follows immediately that the blocking probability for type i pulses for the lossy system is given by

$$P(W'(t) > z - a_i)$$

$$= \frac{1}{1 - P(W(t) > z)} \Big(P(W(t) > z - a_i)$$

$$- P(W(t) > z) \Big). \tag{5.6}$$

Proof: Since $P(W'(t) < x)$ and $P(W(t) < x)$ can be computed iteratively by the same (5.2), it follows that their ratio is given by the reciprocal of the ratio of the normalizing constants. Assume we start with an unnormalized $q(0)$. Thus, the ratio is given by

$$\frac{\int_0^\infty q(y)\, dy}{\int_0^z q(y)\, dy} = \frac{1}{1 - \left(\int_z^\infty q(y)\, dy \Big/ \int_0^\infty q(y)\, dy \right)} \tag{5.7}$$

which is the first factor on the right-hand side of (5.5). We may approximate this ratio by 1 using the assumption that $P(W(t) \geq z)$ is small. $\square$

Equation (5.5) provides the bridge for linking blocking calculations for lossy systems to the powerful central limit theory [5], which approximates the distribution density $p(W(t))$, and large deviation theory, which approximates the tail probability $P(W(t) > z)$.

C. Characteristic Functions for W

Before we describe the Gaussian and large deviation approximations for the tail distribution of W, we give a brief tutorial for the log moment generating function of $q(w) = p(W(t) = w)$ for the Poisson shot noise process, defined as

$$\mu_W(s) = \log_e \psi_W(s) = \log_e \int_0^\infty q(w) e^{sw}\, dw. \tag{5.8}$$

We rederive using (5.2) the well-known [11] result

$$\mu_W(s) = \sum_i \gamma_i b_i (e^{sa_i} - 1) \tag{5.9}$$

Proof:

$$\frac{d}{ds} \psi_W(s) = \int_0^\infty w q(w) e^{sw}\, dw$$

$$= \sum_i \int_0^\infty \gamma_i a_i b_i q(w - a_i) e^{sw}\, dw$$

$$= \sum_i \gamma_i a_i b_i e^{sa_i} \psi_W(s). \tag{5.10}$$

Thus,

$$\frac{d\psi_W(s)}{\psi_W(s)} = \sum_i \gamma_i a_i b_i e^{sa_i} ds \tag{5.11}$$

which upon integration with the proper integration constant, yields (5.9). Equation (5.9) gives an alternative way for computing $P(W(t) > y)$ via inverse Laplace transforms [12]. □

The mean and variance of $W(t)$ are given by differentiating $\psi_W(s)$

$$E(W(t)) = \psi_W'(0) = \sum_i \gamma_i a_i b_i \qquad (5.12)$$

$$\text{Var}(W(t)) = \psi_W''(0) - \psi_W'^2(0) = \sum_i \gamma_i a_i^2 b_i. \qquad (5.13)$$

D. Gaussian and Large Deviation Approximations

Knowing the mean and variance of W, $p(W(t))$ can be computed by the Gaussian approximation [5]

$$p(W = w) \approx \frac{1}{\sqrt{2\pi \text{Var}(W)}} e^{-(w - E(W))^2 / 2\text{Var}(W)}. \qquad (5.14)$$

This approximation, though can be evaluated with extreme ease, is not particularly accurate at w more than a standard deviation from the mean. We shall develope sharper estimates, at the expense of more computation.

First, we apply the Chernoff bound [13]

$$P(W(t) > y) \le e^{-(s^* y - \mu_W(s^*))} \qquad (5.15)$$

in which s^* satisfies the implicit equation for the first derivative of $\mu_W(s)$

$$y = \mu_W'(s) = \sum_i \gamma_i a_i b_i e^{sa_i}. \qquad (5.16)$$

Proof: Applying the Chebyshev's inequality for $s > 0$

$$P(W \ge y) = P(e^{sW} \ge e^{sy}) \le \min_s \frac{E[e^{sW}]}{e^{sy}}$$

$$= \min_s e^{-(sy - \mu_W(s))} = e^{-(s^* y - \mu_W(s^*))} \qquad (5.17)$$

since s^* minimizes the exponent. □

In fact, the above bound can be sharpened by the theory of large deviation [5, 13, Appendix A, ch. 5], which is concerned with the sum of a large number of random variables. We prove a new result for sharpening the above bound for inhomogeneous Poisson point processes by a factor $1/s^* \sqrt{2\pi \mu_W''(s^*)}$ such that

$$P(W > y) \approx F(s^* \sqrt{\mu_W''(s^*)}) e^{-(s^* y - \mu_W(s^*))} \qquad (5.18)$$

with

$$F(u) = e^{u^2/2} \int_u^\infty \frac{1}{\sqrt{2\pi}} e^{-v^2/2} \, dv \le \frac{1}{\sqrt{2\pi}} \frac{1}{u}. \qquad (5.19)$$

Proof: Let N_i be the number of type i bursts, of Poisson rate γ_i and amplitude a_i arriving in the past b_i. Thus,

$$p_{N_i}(n_i) = \frac{(\gamma_i b_i)^{n_i}}{n_i!} e^{-\gamma_i b_i}. \qquad (5.20)$$

Since bursts of larger amplitude contribute more significantly to $P(W > y)$, it is suggestive to skew the probabilities p_{N_i} according to a_i. Define $N_{s,i}$ as the number of

type i burst arrivals, of amplitude a_i, with Poisson arrival rate $\gamma_i e^{sa_i}$ instead of γ_i. Hence, the skewed density

$$p_{N_{s,i}}(n_{s,i}) = \frac{(\gamma_i e^{sa_i} b_i)^{n_{s,i}}}{n_{s,i}!} e^{-\gamma_i e^{sa_i} b_i}$$

$$= p_{N_i}(n_i) \, e^{sn_i a_i - \gamma_i b_i (e^{sa_i} - 1)}. \qquad (5.21)$$

Consider the probability distribution for $W = \Sigma_i N_i a_i$ and $W_s = \Sigma_i N_{s,i} a_i$, related by

$$p_W(w) = \prod_{n_i : n_i a_i = w} p_{N_i}(n_i)$$

$$= \prod_{n_{s,i} = n_i : \Sigma_i n_i a_i = w} \left[p_{N_{s,i}}(n_{s,i}) \, e^{-sn_i a_i + \gamma_i b_i (e^{sa_i} - 1)} \right]$$

$$= \left[\prod_{n_{s,i} = n_i : \Sigma_i n_i a_i = w} p_{N_{s,i}}(n_{s,i}) \right] e^{-sw + \Sigma_i \gamma_i b_i (e^{sa_i} - 1)}$$

$$= p_{W_s}(w_s = w) \, e^{-sw + \mu_W(s)} \qquad (5.22)$$

Furthermore, it can be verified easily that $E(W_s) = \mu_W'(s)$ and $\text{Var}(W_s) = \mu_W''(s)$, the derivatives of $\mu_W(s)$. The desired tail distribution is

$$P(W > y) = \int_y^\infty p_W(w) \, dw$$

$$= \int_y^\infty p_{W_s}(w_s) \, e^{-sw_s + \mu_W(s)} \, dw_s$$

$$= e^{-sy + \mu_W(s)} \int_y^\infty p_{W_s}(w_s) \, e^{-s(w_s - y)} \, dw_s. \qquad (5.23)$$

We choose $s = s^*$ which satisfies (5.16). Notice that the above integral starts from $y = \mu_W'(s^*)$, the mean of W_{s^*}. To evaluate the integral, we shall approximate $p_{W_{s^*}}$ by the Gaussian approximation of (5.14). Hence, $P(W > y)$ is approximated by

$$e^{-sy + \mu_W(s^*)} \int_y^\infty \frac{\exp\left[-\frac{(w_s - y)^2}{2\mu_W''(s^*)} - s^*(w_s - y) \right] dw_s}{\sqrt{2\pi \mu_W''(s^*)}}$$

$$= e^{-sy + \mu_W(s^*)} e^{u^2/2} \int_u^\infty \frac{1}{\sqrt{2\pi}} e^{-v^2/2} \, dv \qquad (5.24)$$

where $u = s^* \sqrt{\mu_W''(s^*)}$. Thus, we obtain (5.18). The upper bound of (5.19) results from a simple bound for the error function. □

E. Mixture of Poisson and Non-Poisson Traffic

So far, we have been restricted to traffic with Poisson burst arrivals. In fact, the estimates for $P(W > y)$ depicted in (5.18) remains true for other $R_k(t)$, such as that for compressed full motion video calls. Suppose we have the steady-state probability p_i for $R_k(t) = a_i$. Thus, $\mu_k(s)$ for call k is given by

$$\mu_k(s) = \log_e \sum_i p_i e^{sa_i}. \qquad (5.25)$$

The log moment generating function for W is given by $\mu_W(s) = \Sigma_k \mu_k(s)$. The proof of (5.14), (5.15), and (5.18) requires the use of the central limit theorem for varying components [5]. We omit the proof since it involves just some additional technicalities to the previous proof. In fact, the Poisson process is the limiting case for the superposition of renewal processes [14].

With mixed Poisson and variable rate traffic, we have a mix of the log moment generating functions of the form (5.9) and (5.25) for $\mu_W(s)$. We shall show how to compute in real-time the approximation

$$P(W > y) = \frac{1}{s^* \sqrt{2\pi \mu_W''(s^*)}} e^{-(s^* y - \mu_W(s^*))} \qquad (5.26)$$

where $\mu_W'(s^*) = y$. We expand the individual $\mu_k(s)$ by Taylor series, which can be computed beforehand for each call type. The series expansion of (5.9), which is a sum of exponentials, is straightforward. The series expansion of (5.25) is given by first expanding $\Sigma_i p_i e^{sa_i}$ as a series, and then expanding the log of the resulting series

$$\mu_k(s) = \log_e(1 + d_1 s + d_2 s^2 + \cdots)$$
$$= c_1 s + c_2 s^2 + c_3 s^3 + \cdots \text{ in which} \qquad (5.27)$$

$$c_i = d_i - \frac{1}{i} \sum_{j=1}^{i-1} j d_{i-j} c_j. \qquad (5.28)$$

Thus, with these precomputed coefficients, we may easily obtain the series expansion of $\mu_W(s)$ as well as its first two derivatives in real-time.

Equation (5.26) involves solving for s^* in an implicit equation which equates the first derivative of $\mu_W(s)$ with z. Hence, Newton's method can be applied by using the second derivative of $\mu_W(s)$. Since $\mu_W(s)$ is a positive convex function in s, the root is unique and Newton's method can be very effective. In fact, we may retain the previous values of s^* as well as $\mu_W(s^*)$ and its first two derivatives for iterations. When $\mu_W(s)$ is changed by an arriving call, we may find the perturbation to s^* via Newton's method using the previous derivatives.

We programmed the above numerical methods and found often that just 1 iterate is sufficient for updating traffic parameters for call arrivals. The values of $\mu_W(s^*)$ and its derivatives are recomputed and stored for the next iterate. The tail distribution of (5.26) is then a simple function of $\mu_W(s^*)$ and its derivatives.

VI. BLOCKING PROBABILITY AT THE CALL LEVEL

When a call arrives, we calculate the burst blocking probability at the trunk group G and accept the call only if the resulting blocking probability is acceptable. Suppose we have n types of calls, each with well defined burst level statistics. Let j denote a call type, $1 \leq j \leq n$. Let f_j be the number of calls of type j in a trunk group G. The admissible call region is defined as the n dimensional space of f_j such that within the region, the burst blocking

probability or $P(W > z)$ is acceptably small. The computation of this region depends on the approximation method used for calculating the burst blocking probability. Admissible call regions are shown in Figs. 11, 12, and 13 for two call types.

Thus, the call blocking probability for each G is given by the probability that the arriving call moves the set of n_j outside the admissible call region. This probability depends on three factors, how we compute the admissible call region, as well as the arrival and holding time statistics of each call type. Knowing these three factors, we can apply the product form solution for the probability of each point in the admissible call region. This computation, though tedious, is still tractable for trunk group sizing.

Through numerical computations in the conclusion section, we found that the admissible call region is a concave region (Fig. 11). Hence, an outer bound to the region is given by the polygon with vertexes $(0, 0, \cdots, 0)$ and $(0, 0, \cdots, 0, \bar{n}_j, 0, \cdots, 0)$ for $1 \leq j \leq n$, in which $\bar{n}_j$ is the maximum number of call j which may use the capacity z exclusively. It is also observed that the boundary of the admissible call region becomes more and more linear as z increases (Fig. 13). Hence, we conjecture that the outer bound is a good approximation for the admissible call region for large z.

Using the outer bound for approximating the admissible call region, the steady-state probabilities can be computed easily since the boundary is linear. We can use the same technique for computing burst blocking probability to compute the call blocking probability [(5.1) and (5.26)].

In the remainder of this section, we shall discuss a trunk group partitioning problem. Let G be partitioned into channels i, $1 \leq i \leq m$, each of capacity c_i. The problem of trunk group partitioning is the placement of the f_j calls for various j into the channels i, such that the maximum burst blocking probability among the i is minimized. This min–max problem uniformly reduces the blocking probability among the partitions i.

This problem is similar to bin-packing, which is known to be intractable. To make the problem more tractable, we remove the diophantine restriction. The new problem is still difficult to solve because the objective function is nonlinear. It can be shown that the optimal solution packs calls of similar types into the same partition [16].

VII. CONCLUSION

The multilevel traffic control and evaluation process has been simulated by a software package called traffic analysis, control and record-keeping system (TRACKS) on a SUN 3 work station, using the language C and X windows. The software simulates the call setup and burst allocation processes. Calls types are characterized by both the call level statistics and the burst level statistics. We use the large deviation methods of Section V for congestion evaluation during call setup.

In conclusion, we give a numerical example of traffic integration. We assume two kinds of terminals, namely,

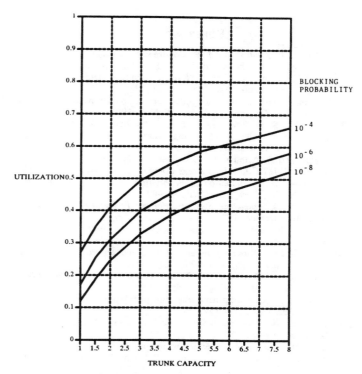

Fig. 9. Channel utilization for graphics.

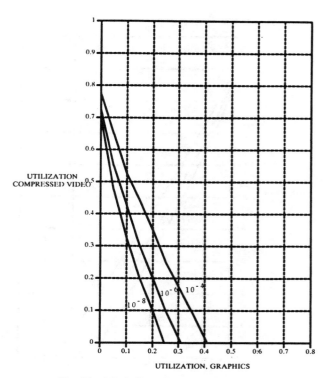

Fig. 11. Admissible call region, capacity = 2.

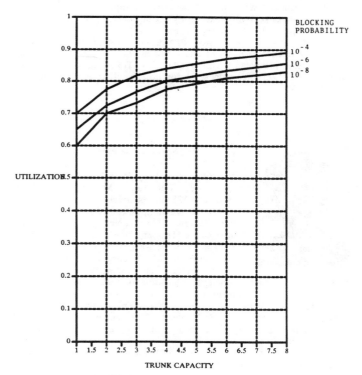

Fig. 10. Channel utilization for compressed video.

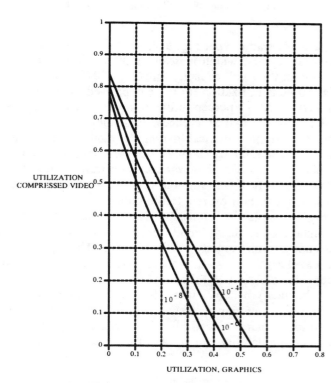

Fig. 12. Admissible call region, capacity = 4.

the still picture graphics station and compressed full motion video terminal of Section II. During the duration of a call, their respective average bit rates are 1 Mbit/s and 5 Mbits/s, whereas their peak bit rates are both 10 Mbits/s. We define trunk group utilization ρ of a type of terminal as the number of calls established times the aver-

age bit rate, divided by the total capacity. Thus, a fully utilized ($\rho = 1$) trunk group of 100 Mbits/s bandwidth can support either 100 graphics terminals or 20 compressed video terminals.

We shall consider first the effect of trunk group capacity and tolerable blocking probability on utilization, assum-

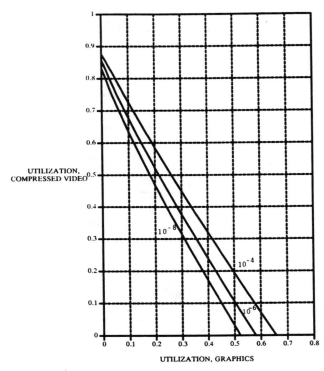

Fig. 13. Admissible call region, capacity = 8.

ing a single traffic type. Fig. 9 shows this relationship for bursty terminals, which indicates a 30 to 60 percent utilization in the trunk capacity range of 200–800 Mbits/s. Fig. 10 plots utilization for the compressed video, which indicates a 70–90 percent utilization. Comparing Figs. 9 and 10, we see two differences. First, bursty traffic benefits less from statistical averaging than variable rate traffic. Second, bursty traffic is more sensitive to tolerable blocking probability than variable rate traffic. Hence, it takes relatively fewer variable rate terminals to produce an almost constant bit rate traffic process.

Fig. 11 plots the admissible call regions for the two terminal classes for different tolerable blocking probabilities and a trunk capacity of 200 Mbits/s. The regions are concave. Figs. 12 and 13 plots the regions for trunk group capacities of 400–800 Mbits/s. It is observed that the region boundary becomes more linear as the trunk capacity increases. Assuming a linear boundary, it is straightforward to compute the call blocking probability. Such call blocking probability can be used for trunk group sizing.

Acknowledgment

The author thanks J. Amenyo of Columbia University for implementing the presentation and simulation parts of TRACKS.

References

[1] P. Gonet, P. Adams, and J. P. Coudreuse, "Asynchronous time-division switching: The way to flexible broadband communication networks," presented at IEEE 1986 Int. Zurich Sem. Digital Commun., Zurich, Switzerland, Mar. 1986.
[2] J. Turner, "Design of an integrated services packet network," *IEEE Select. Areas Commun.*, vol. SAC-4, Nov. 1986.
[3] L. T. Wu, S. H. Lee, and T. T. Lee, "Dynamic TDM—A packet approach to broadband networking," presented at Conf. Proc. ICC, Seattle, WA, June 1987.
[4] J. Hui, and E. Arthurs, "A broadband packet switch for integrated transport," *IEEE J. Select. Areas Commun.*, vol. SAC-5, Oct. 1987.
[5] W. Feller, *An Introduction to Probability Theory and its Applications, Vol. II.* New York: Wiley, 1987.
[6] L. Kleinrock, *Queueing Systems, Volume I: Theory.* New York: Wiley-Interscience, 1975.
[7] F. P. Kelly, *Reversibility and Stochastic Networks.* New York: Wiley, 1979.
[8] J. Kaufman, "Blocking in shared resource environment," *IEEE Trans. Commun.*, vol. COM-29, Apr. 1981.
[9] J. Hui, *Switching and Traffic Theory for Integrated Broadband Network.* New York: Springer-Verlag, 1989.
[10] S. S. Lam, "Queueing networks with population size constraints," *IBM J. Res. Dev.*, July 1977.
[11] E. Cinlar, *Introduction to Stochastic Processes.* Englewood Cliffs, NJ: Prentice-Hall, 1975.
[12] D. L. Jagerman, "An inversion technique for the Laplace transform with application to approximation," *Bell Syst. Tech. J.*, vol. 57, no. 3, Mar. 1978.
[13] R. G. Gallager, *Information Theory and Reliable Communication.* New York: Wiley, 1968.
[14] K. Sriram and W. Whitt, "Characterizing superposition arrival processes in packet multiplexers for voice and data," *IEEE Select. Areas Commun.*, vol. SAC-4, Sept. 1986.
[15] B. Maglaris, D. Anastassiou, P. Sen, G. Karlsson, and J. Robbins, "Performance analysis of statistical multiplexing for packet video sources," presented at GLOBECOM '87, Tokyo, Japan, Nov. 1987.
[16] J. Hui, "Resource allocation for broadband networks," Bellcore Tech. Mem.

Joseph Y. Hui (S'82-M'83) received the S.B., S.M., E.E., and Ph.D. degrees in electrical engineering from the Massachusetts Institute of Technology, Cambridge, MA, in 1981, 1981, 1982, and 1983, respectively.

After graduation, he joined Bell Communications Research, Morristown, NJ, in 1983, where he is now a Member of the Technical Staff. From 1979 to 1981, he worked at the Comsat Laboratories under the M.I.T. cooperative program. His general research interests include coding and information theory, portable telephony, and parallel processing. Currently, his research at Bell Communications Research focuses on switching, protocol, traffic, and security issues for integrated broadband networks. He is also an Adjunct Professor at the Center for Telecommunications Research of Columbia University, where he teaches a course on integrated broadband networks.

Dr. Hui is the recipient of the 1985 William Bennett Prize Paper Award given by the IEEE Communication Society. His work on multiple access communications and switching networks resulted in several patents. He served as a Guest Editor for an issue of IEEE Journal on Selected Areas in Communications on Switching Systems for Broadband Networks. He is also the author of the book, *Switching and Traffic Theory for Integrated Broadband Network*, Springer-Verlag, to be published.

Multichannel Bandwidth Allocation in a Broadband Packet Switch

ACHILLE PATTAVINA

Reprinted from *IEEE Journal on Selected Areas in Communications*, Vol. 6, No. 9, December 1988, pages 1489–1499.

Abstract—The problem of bandwidth allocation in a packet switch supporting broadband services is addressed in this paper. To reduce the performance constraints imposed by limiting a data link to a single broadband packet channel, we introduce the concept of channel group as a set of broadband packet channels that is viewed as a single data link connection by routing entities. We employ a two-step bandwidth allocation scheme. At connection setup time, a call is allocated to a channel group. At transmission time, specific channels of a group are optimally allocated to the packets destined to the group. Because of the statistical smoothing of the large number of sources served by a channel group, the traffic performance of the switch is improved. This scheme also allows super-rate switching, i.e., the support of services with peak bandwidth exceeding the capacity of a single packet channel. We show the feasibility of this scheme in a Batcher-banyan switch, by implementing in hardware the bandwidth allocation at transmission time. Performance improvements obtained by this scheme are also provided in different traffic environments.

I. Introduction

A major effort is underway to identify the main characteristics of an integrated communication network that can flexibly support current and future services. The characteristics of these services vary in terms of required bit rate, burstiness, and acceptable delay. These service requirements must be met without underutilizing switching and transmission resources.

A switch in a broadband packet network is expected to have a throughput of several tens of Gbits/s. Fast packet switching (FPS) techniques [1] are well suited as switch fabrics for this purpose. Different proposals exist based on FPS, e.g., [2]–[6]. To cope with high bit rate packet streams, the lower layer communication protocols must be simplified. Simplification is achievable because of the high speed and lower error rate qualities of fiber optic links. References [7], [8] propose simple data link layer protocols for broadband packet networks. The assumption of fixed-size packets in these protocols is well suited with the slotted switching operations of the broadband packet switches mentioned above. We will assume a slotted network environment with fixed-size packets in the following discussion.

However, high throughput switches and efficient data link protocols alone do not guarantee an effective transport system. It is crucial to be able to allocate bandwidth so that resources are efficiently utilized and congestion is kept under control. This is a difficult problem because the traffic characteristics of potential broadband services are uncertain and the capacity of the packet channel has an upper bound related to technology limits (i.e., maximum speed allowed by current technology) and to processing requirements per packet (e.g., the time required for a switch or a multiplexer to process a packet, typically on the order of a few μs). Furthermore, the peak rate of traffic sources varies by several orders of magnitude, with peak demands that may approach or exceed (e.g., for high quality video or bulk data transfer) the capacity of a single packet channel. So additional protocols are needed to support service rates that exceed the capacity of a single packet channel.

Connection-oriented service with explicit bandwidth reservation is currently believed to be the easiest mechanism for controlling resource allocation. We will assume that the basic service provided by the network is virtual circuit, on top of which a datagram service can be provided to users. However, this is a conservative strategy that might not efficiently use network resources. For example, if peak capacity is reserved for sources with long burst duration but low average bit rate (e.g., high resolution still-image sources) to ensure against packet loss due to buffer overflow, network resources can be significantly underutilized.

To solve this problem we introduce a two-step bandwidth allocation scheme, in which a data link connection between routing entities is represented by a set of packet channels, called *channel group*. At *connection setup time*, some bandwidth of a channel group is reserved for a virtual connection that is being established. At *transmission time*, slot by slot, specific channels of a group are optimally assigned to the packets to be transported over the virtual connections supported by the group. Channel assignment at transmission time is shown in this paper to be feasible in an input buffered Batcher-banyan switch with very little loss of switching capacity. Given this scheme, the peak rate in a virtual circuit is not bounded by the capacity of the packet channel, thus allowing the switch to provide super-rate switching. Nevertheless, protocol issues associated with this kind of service are still to be solved (e.g., packet sequencing).

A similar approach for allocating optimally resources in a broadband packet network has been adopted by J. Hui [9], who defines three layers of bandwidth allocation:

Manuscript received November 6, 1987; revised May 5, 1988. This work was supported by Bellcore, Morristown, NJ 07960.

The author was with Bellcore, Morristown, NJ 07960. He is now with INFOCOM Department, University "La Sapienza," 00184 Rome, Italy.

IEEE Log Number 8824391.

call level, communication burst level, and packet level. Bandwidth allocation at connection setup time or at call level express the same concept, as well as allocation at transmission time or at packet level do. The intermediate layer in [9] is used to allocate a specific channel (trunk in Hui's terminology) within a channel group (trunk group) for the whole duration of a burst from a specific user. This three-layer approach shows very interesting features (e.g., the problem of packet sequence integrity is greatly simplified), but raises questions about implementation issues that need to be answered. One of the major concerns of our proposal is to show the feasibility of our two-step bandwidth allocation scheme, by providing an implementation example.

Issues on traffic sources and communication services in a broadband packet network are discussed in Section II. Then, the problems related to the limited capacity of a broadband packet channel are examined and the multichannel bandwidth allocation scheme is described. Section III outlines the structure of a Batcher-banyan switching fabric that supports the proposed bandwidth allocation scheme, including the logical addressing of the output channels and the implications of channel failures on the bandwidth allocation scheme. The performance of the switch is evaluated in Section IV in terms of switching overhead, maximum throughput for random traffic, and packet loss probability for bursty traffic.

II. NETWORK SERVICES AND BANDWIDTH ALLOCATION

In a broadband packet network, several packet channels (also called trunks or links by others) connect two routing entities residing either in two adjacent switching nodes or in a switching node and in one of its associated user-network interfaces. In Fig. 1(a), for example, 6 packet channels connect network interface NI1 to switching node SN1 and NI2 to SN2, while 9 packet channels connect SN1 to SN2. As indicated previously, the *capacity* C of a packet channel is related to the time required for packet processing in multiplexers and switches. Thus, for a given packet size, C is basically determined by the available technology, for example, $C \leq 150$ Mbits/s with CMOS. In an all-fiber network, each fiber link would support several channels, say 16 in a single-mode fiber operating at 2.4 Gbits/s. Each channel would be switched independently in the originating switch.

A. Traffic Sources and Service Issues

The following basic issues should be considered as part of the discussion of the communication network. In a connection-oriented service, a virtual connection has to be setup before the transfer of user information takes place. In the transmission and switching facilities supporting the virtual connection, some amount of bandwidth is typically reserved for the circuit, depending on the traffic parameters of the sources using the connection.[1] In a con-

nectionless service, information is transferred in the network without any preliminary end-to-end virtual connection setup and bandwidth reservation.

As far as traffic sources are concerned, assume a source i to be characterized by two parameters: the average traffic S_i and the peak traffic P_i both expressed in bits/s. A third parameter, i.e., the activity factor α_i, can be defined as the ratio S_i/P_i between average and peak traffic. In the following discussion, unless stated otherwise, we will classify sources according to a two-level scale for each parameter, referring to these levels as high or low. For traffic source i, low-average (or low-peak) traffic means that S_i (or P_i) is much smaller than the channel capacity C and low-activity source i means $\alpha_i \ll 1$. Entertainment video or a DS-3 channel are examples of services with high-peak high-activity sources, while a 64 kbit/s PCM voice source represents a low-peak high-activity source. High resolution graphic terminals or devices for still image distributions belong to the class of high-peak low-activity sources, while interactive data terminals represent low-peak low-activity sources.

We believe that a connection-oriented service, with a suitable choice of the bandwidth reservation algorithm, allows the network to provide real-time and high-peak sources with more predictable performance, while achieving high utilization of network resources. Traditionally, connection-oriented services guarantee packet sequencing and less packet jitter than connectionless services, the latter being very important for real-time services. Low-peak nonreal-time sources can be supported by a connectionless service provided on top of a virtual circuit dedicated to these sources. Thus, a virtual circuit can carry packets of a connection-oriented call or information units of a connectionless service. Information from sources provided with a connectionless service would be transferred inside the network with a relatively low priority, thus absorbing the effect of buffer overflows due to transient congestions. Real-time sources and high-peak sources would be given a relatively high priority.

B. Unichannel Bandwidth Allocation

With the traditional *unichannel bandwidth allocation*, an amount of bandwidth W_i (bits/s) will be reserved for source i at call establishment time in one channel of each link supporting its virtual connection. W_i, which is a value satisfying the condition $S_i \leq W_i \leq P_i$, is to be selected as a tradeoff between the utilization level expected for the switching and transmission resources and the acceptable packet loss caused by buffer overflow. For high-peak low-activity sources, a high utilization level of the channels can only be obtained by accepting a significant packet loss probability.

However, resource underutilization is not only caused by high-peak sources. When sources with $\alpha_i = 1$ are allocated to a channel, thus reducing the available channel capacity to $C_r < C$, the tradeoff above between underutilization and packet loss probability applies to sources with a peak rate close to C_r, not C. For example, assume

---

[1]This information will be conveyed to the call processor during the connection setup phase.

that two sources with $P_i = 45$ Mbits/s and $\alpha_i = 1$ are allocated to a channel with $C = 150$ Mbits/s and that the residual capacity $C_r = 60$ Mbits/s is to be shared by interactive high-definition graphic sources with $P_i = 20$ Mbits/s and $\alpha_i \ll 1$. Then, packet loss probability requirements might limit acceptance to no more than three of these sources on the channel, resulting in an underutilization of the residual capacity. Underutilization increases with the decrease of the source activity factor α_i.

The channel capacity sets an upper bound on the peak rate of a virtual connection supported by the network. Services with peak rates exceeding the channel capacity can be supported only by providing specific additional protocols, which take into account the "commonality" of the different virtual circuits supporting a single source.

The expected availability of several packet channels between two switching nodes raises the problem of choosing the best policy to allocate a new virtual circuit to a specific channel. This issue becomes critical when the peak rate of the traffic source is close to the channel capacity. Indeed, an auspicious policy that uniformly books (and loads) the channels between two adjacent nodes penalizes establishment of high bandwidth connections, since the requested bandwidth is likely to exceed the residual usable capacity of each single channel.

C. Multichannel Bandwidth Allocation

A solution to the problems above could be represented by the availability of data link connections with a capacity greater than the capacity of a single packet channel. Indeed, it is believed that interswitch communication facilities with capacities on the order of 2.4 Gbits/s are likely to be required in a broadband network capable of supporting a wide variety of services. Thus, each fiber link would support several packet channels. We define a *channel group* as a set of parallel packet channels that is seen as a single data link connection between two cooperating routing entities (for interswitch or user-to-switch communication). The channel group is functionally equivalent to a set of servers sharing a single waiting list. The capacity of a channel group is the sum of the capacities of the packet channels in the group. Channels in a group need not be supported by a single transmission facility and adjacent nodes can be connected by more than one channel group. For example, the network of Fig. 1(a) is shown in Fig. 1(b), by outlining the supported channel groups. In this case, 2 groups of 3 channels connect NI1 to SN1, 3 groups of 3 channels connect SN1 to SN2, and 3 groups of 2 channels connect NI2 to SN2.

The bandwidth of a channel group is allocated in two steps, at connection setup time and at transmission time. At connection setup time, when a call request is accepted, the call processor of the switch selects an output channel group, on which some bandwidth is booked for the call. This choice is based on the booking level of the groups potentially selectable and on the traffic characteristics of the new service request. At transmission time, before packets are switched, specific channels within a group are

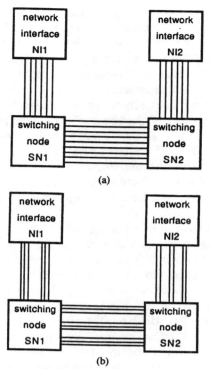

Fig. 1. Example of packet network.

assigned to the packets addressed to the group. Since the channels in a group act as servers with a shared waiting list, the number of channels within group i assigned in a slot equals the minimum between the number of packets addressed to group i in the slot and the number of channels in group i. Packets that are denied the channel allocation remain stored in the buffer at the input port. Note that this channel allocation at transmission time, which must be carried out in one slot time, requires coordination among the port controllers for each slot. Thus, a distributed implementation, possibly in hardware, of this function seems necessary.

This multichannel bandwidth allocation scheme has several advantages over the traditional approach of assigning a virtual circuit to a specific channel (unichannel bandwidth allocation scheme). "Super-rate switching" can be provided, since virtual connections with capacities greater than the channel capacity are naturally supported (sources are assigned to a channel group, not a single channel). Due to the pooling effect, a channel group can support a greater number of sources than its channels could individually. Because of the statistical smoothing of the large number of sources served by a channel group, the total instantaneous offered load on the group is not likely to vary greatly from the average offered load. Thus, assuming a constant packet loss probability, the capacity reserved for a virtual connection using multichannel bandwidth allocation can be kept closer to its average load than in the traditional unichannel bandwidth allocation. The relatively slight impact of a burst from a particular source makes assignment of a source to a particular channel group a less critical problem.

A possible disadvantage is that the sequence of packets transmitted over a virtual connection with a high-peak capacity cannot be guaranteed, since they can be assigned to different channels within a group. It is not yet clear if the packet sequence integrity or, correspondingly, a bounded packet jitter within the network is a critical requirement for a broadband packet network. If it is, specific mechanisms should be applied for the transfer of information from high-peak sources, related to the specific implementation of the channel allocation at transmission time. For lower bandwidth services, the average packet interarrival time is likely to be much greater than the waiting time jitter and hence packets should not get out of sequence.

The link group concept introduced in [3] is similar to our channel group concept. A link group is a set of links connecting two switch modules. At connection setup time, a call is allocated to a link group rather than to a specific link, thus allowing, as in our case, the network to perform super-rate switching. Nevertheless, link groups do not provide the statistical benefit given by channel groups. Indeed, at transmission time, each port controller autonomously selects a specific link within a group for the packet to be transmitted in the current slot. Each port controller offers a uniform load to the links in a group by selecting the links cyclically. Since the link selection is the result of autonomous decisions by each port controller, a link group is functionally equivalent to a set of servers, each with its own waiting list.

III. Bandwidth Allocation in a Batcher-Banyan Switch

In this section, the feasibility of the two-step bandwidth allocation scheme described in Section II is shown for an input-buffered Batcher-banyan packet switch, whose interconnection network is given in Fig. 2. It is composed of N port controllers PC_i $(i = 0, \cdots, N - 1)$, a Batcher network of $\log_2 N(\log_2 N + 1)/2$ stages of 2×2 sorters, a banyan network of $\log_2 N$ stages of 2×2 switching elements and a channel allocation network. I_i and O_i $(i = 0, \cdots, N - 1)$ are the input and output channels of the switch.

In the Batcher network, the packets offered to the input ports are sorted according to the self-routing address at the head of each packet. An ordered set of packets with nondecreasing (or nonincreasing) self-routing addresses is obtained in contiguous output ports of the Batcher network. In the banyan network, only one path exists between any input and output port, and different paths corresponding to different input and output port pairs may share internal links. However, packet collisions will not occur if the packets have different self-routing addresses and are offered as an ordered set to a contiguous set of input ports of the banyan network. Hence, a Batcher-banyan network is internally nonblocking, if the packets entering the input ports of the Batcher network are addressed to different output ports of the banyan network.

Contention between packets addressed to the same out-

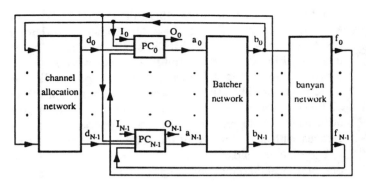

Fig. 2. Interconnection network.

put port can be resolved by a *three-phase algorithm* that, based on the algorithm described in [5], allows the handling of channel groups. In phase 1, an input port with a data packet sends a request to engage a channel of the destination channel group in the current slot. The requests are sorted in nondescending order by the Batcher network, so that the requests for the same channel group are adjacent at the output ports of the Batcher network. This sorting allows the channel allocation network to compute an index for each request that identifies a specific channel in a group (i.e., an output port of the switch). Those requests that cannot be accommodated in a channel group are given an index corresponding to a channel that does not belong to the group requested. In phase 2, the assigned index is sent back in an acknowledgment packet to the requesting port through the Batcher-banyan network. Each requesting port controller, based on the capacity of the channel group requested and the index received, determines if it won the contention for the requested group. Only the winning ports can each transmit their data packet in phase 3. A more complete description of this new three-phase algorithm is given in Section III-B.

A. Channel Addressing

A *multichannel switch*, i.e., a switch supporting channel groups and adopting the three-phase algorithm, requires the addresses of the output ports terminating the channels of the same group to be consecutive. This requirement could seriously constrain a change of the configuration of the interswitch communication facilities, e.g., following a link failure or an updating in the expected traffic patterns. For this reason, we introduce a logical addressing scheme of the output channels, which decouples the channel address from the physical address of the output port terminating the channel.

Each channel is assigned a logical address, so that a channel group is composed of channels with consecutive logical addresses, and a one-to-one correspondence exists between the channel logical address and the physical address of the port terminating the channel. The channel with the lowest logical address in a group is the *group leader*. The logical address of the group leader also represents the *group address*. A specific channel in a group is identified by a *channel offset* given by the difference between the channel logical address and the group address. Each port controller is provided with two tables, K_a and K_c. K_a maps the logical address to the physical address (i.e., the port address) of each channel. K_c specifies the maximum value, $maxoff(i)$, allowed for the channel offset in group i.

The number g of channel groups supported by the switch and the number D_i of channels in group i are such that $D_1 + D_2 + \cdots + D_g = N$. D_i, which is referred to as *capacity of channel group* i, equals $maxoff(i) + 1$. Let $D_{\max}$ be the maximum capacity allowed for a channel group and, for simplicity, N be a power of two. Let n and d denote the number of bits needed to code the logical address of a channel (or the physical address of a port) and the channel offset, respectively. Then, $n = \log_2 N$ and, according to the channel allocation procedure to be explained later, $d = 1 + \lceil \log_2 (D_{\max}) \rceil$.

B. The Three-Phase Algorithm

The three-phase algorithm of our multichannel switch is illustrated in Fig. 3. In phase 1, port controller PC_i with a data packet to be transmitted to channel group j sends a request packet REQ(j, i). This packet contains, in order of transmission, the identifier j of the *destination* channel group (i.e., the logical address of the group leader) and the physical address i of the *source* port. Packets REQ($\cdot, \cdot$) enter the channel allocation network, after being sorted by the Batcher network in a nondecreasing order of the denstinations j. Note that there is no guarantee that the number of requests for a group does not exceed the number of channels in the group. To spread the requests over all the channels of group j, the channel allocation network assigns an actual offset $actoff(j)$ to each request for group j. Let $B(j, k, l) = \{b_k, b_{k+1}, \cdots, b_{k+1}\}$ be the complete set of adjacent outlets to transmit a request for channel group j. Then the offset $actoff(j)$ assigned to request on outlet $b_h \in B(j, k, l)$ is

$$actoff(j) \begin{cases} = h - k & \text{if } h < k + D_{\max} \\ \geq D_{\max} & \text{if } h \geq k + D_{\max}. \end{cases} \quad (1)$$

Since $maxoff(j) \leq D_{\max} - 1$, each channel of group j is allocated to only one request for group j and it does not care the specific value not less than $D_{\max}$ assigned to some requests (all of these lose the contention). Then, if b_n is the port transmitting packet REQ(j, i), port d_n of the channel allocation network sends to port controller PC_n the information $actoff(j)$, meaning that port controller PC_i has been given channel $j + actoff(j)$ in group j. PC_n concurrently receives packet REQ(j, i) from b_n.

In phase 2, packet ACK($i, actoff(j)$), generated by PC_n, crosses the Batcher and the banyan networks with the self-routing address i. Field i is referred to as *source*, since it identifies the port controller originating the corresponding packet REQ(j, i). Packets ACK($\cdot, \cdot$) do not collide with each other, since each port controller cannot have sent more than one packet REQ($\cdot, \cdot$). Port controller i, receiving packet ACK($i, actoff(j)$), verifies if it has

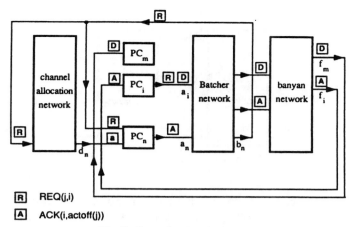

R   REQ(j,i)
A   ACK(i,actoff(j))

Fig. 3. Example of packet transfer.

been given an $actoff(j)$ corresponding to a member of channel group j.

In phase 3, if $actoff(j) \leq maxoff(j)$, port controller PC_i sends its data packet DATA($p(j + actoff(j))$) to the port whose physical address $p(x)$ is mapped by table K_a to the channel logical address x (PC_m in Fig. 3). Packets DATA($P(x)$) cross the Batcher-banyan network without collisions, since the winning requests have been assigned different output logical addresses and, hence, different physical addresses of output ports. If $actoff(j) > maxoff(j)$, the port controller waits for the start of the next slot to issue a new request for its packet, which remains stored in the input port queue.

Priority-based procedures for bounding packet jitter can be envisioned in association with the three-phase algorithm. For example, we can define a frame structure in each packet channel and allocate jitter-critical connections, e.g., high-peak circuit-emulated connections, to one or more specific slots in a channel group. Not more than D_i high-peak circuit-emulated connections will be allocated to group i in a specific slot within the frame. Then, we can bound the packet jitter in crossing a switch, by giving priority in the channel allocation to packets transported over these circuit-emulated connections. Indeed, these packets will never lose the output channel contention. Further study is required to investigate this issue, but this is outside the scope of this paper.

C. Implementation of Bandwidth Allocation at Transmission Time

While Batcher and banyan networks are well known, the channel allocation network and port controllers need to be carefully designed to carry out the channel allocation at transmission time. The channel allocation network, represented in Fig. 4, is composed of subnetworks A and B. Subnetwork A receives a set of adjacent packets with nondecreasing destination addresses and identifies the requests for the same channel group. Subnetwork B, composed of $s = \lceil \log_2 (D_{\max}) \rceil$ stages of adders, assigns an actual offset $actoff(j)$ to each packet addressed to channel group j, so that the offsets corresponding to each member of the group are assigned only once. The hard-

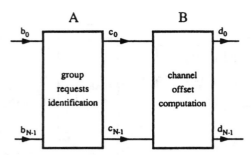

Fig. 4. Channel allocation network.

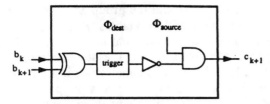

Fig. 5. Logic of subnetwork A.

ware structure of subnetworks A and B and port controller PC_k are now described. Unless specified otherwise, each AND gate is suitably enabled by a signal Φ_x, for a time equal to the transmission time of packet x or field x of a packet. For example, Φ_{ACK} and Φ_{dest} are high for the transmission time of packet ACK$(\cdot, \cdot)$ and field *destination* of packet REQ$(\cdot, \cdot)$, respectively.

1) Subnetwork A: The logic associated with port $c_{k+1}(k = 0, \cdots, N-2)$ of is given in Fig. 5. The destination address of packets REQ$(\cdot, \cdot)$ received on inputs b_k and b_{k+1} are compared bit by bit by an EX-OR gate, whose output sets the trigger by the first mismatching bits in b_k and b_{k+1}. The trigger keeps its status for a time sufficient for subnetwork B to complete the computation of the channel offset. The trigger is reset by the rising edge of signal Φ_{dest} at the start of the address comparison in subnetwork A.

Port controllers generate packet ACK$(\cdot, \cdot)$ by transmitting field *source* of the packet REQ$(\cdot, \cdot)$ being received, immediately followed by the computed *actoff*$(\cdot)$. When the first bit of packet REQ$(\cdot, \cdot)$ is transmitted on outlet b_k, it takes $n + s$ bit times to generate the first bit of *actoff*$(\cdot)$ by the channel allocation network (n bits in subnetwork A and s bits in subnetwork B), while it takes $2n$ bit times to complete the transmission of first field of packet ACK$(\cdot, \cdot)$ by a port controller. The AND gate in subnetwork A delays suitably[2] the start of computation of the channel offset in subnetwork B, so as to avoid buffering *actoff*$(\cdot)$ in the port controller. The signal on outlet c_0 is always low, independent of the input signals on $c_k(k = 1, 2, \cdots, N-1)$, as is required by subnetwork B, which always gives *actoff*$(\cdot) = 0$ to packet REQ$(\cdot, \cdot)$ transmitted on outlet b_0.

2) Subnetwork B: The channel allocation network is a running adder of $s = \lceil \log_2(D_{max}) \rceil$ stages computing the d digits of the *actoff*$(\cdot)$ assigned to each packet REQ$(\cdot, \cdot)$. Subnetwork B is represented in Fig. 6 for the case of $N = 16$ and $5 \leq D_{max} \leq 8$. The AND gate A_2 is enabled by signal Φ_1 for one bit time, so that the adder of the first stage at port c_{k+1} only receives the binary number 1 (if c_{k+1} is high) or the binary number 0 (if c_{k+1} is low). Based on the structure of subnetwork B, the output of each adder of stage $z(z = 1, 2, \cdots, s)$ is a binary stream

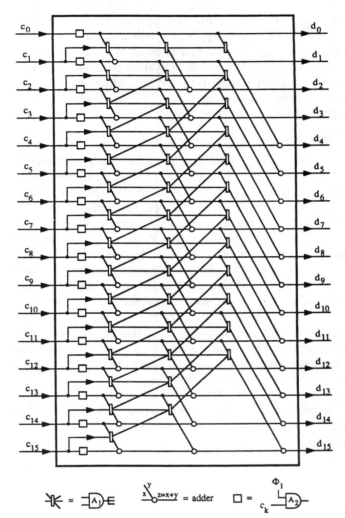

Fig. 6. Logic of subnetwork B.

smaller than or equal to 2^z with the least significant bit first. Hence, $d = 1 + s = 1 + \lceil \log_2(D_{max}) \rceil$ bits are needed to code the *actoff*$(\cdot)$ that emerges from stage s of subnetwork B. The AND gates A_1 allow an independent computation of the running sums for each requested channel group, by resetting the running sum on each inlet c_k with low signal. Indeed, subnetwork B guarantees that at least one inlet c_k with low signal separates any two sets of adjacent inlets with high signal.[3]

The binary stream on output $d_k(k = 0, \cdots, N-1)$

---

[2]The rising edge of signal Φ_{source} occurs $n-s$ bit times after the end of address comparison in the EX-OR.

[3]A set of requests for channel group j transmitted on outlets $B(j, k, l) = \{b_k, \cdots, b_{k+1}\}$ determines a low signal on inlet c_k and a high signal on inlets $c_{k+1}, \cdots, c_{k+1}$.

of subnetwork B, represents the offset $actoff(j)$ assigned to packet $REQ(j, i)$ transmitted on outlet b_k. The offset for packet $REQ(\cdot, \cdot)$ transmitted on outlet b_0 is always 0 because any other requests for the same channel group will be given an offset greater than 0. With this implementation of the channel allocation network, (1) becomes

$$actoff(j) = \begin{cases} h - k & \text{if } h < k + 2^s \\ 2^s & \text{if } h \geq k + 2^s. \end{cases} \quad (2)$$

An example of operation of the channel allocation network is given in Fig. 7 for $N = 16$ and $s = 3$. According to (2), in this case 2 out of 10 requests for the channel group 6 are given the same $actoff(j) = 8$. If $D_6 = 6$, then 4 requests for channel group 6, i.e., those transmitted by ports b_{10}–b_{13}, lose the contention, since they receive an $actoff(j) > 5$.

3) Port Controller: The hardware structure of port controller PC_k is represented in Fig. 8. At the start of phase 1, PC_k transmits a packet $REQ(j, k)$ through gates A_6 and B_2, if it has to send a data packet to channel group j. Gates A_4, A_3, and B_1 merge field *source* of packet $REQ(l, i)$ received on input b_k and $actoff(l)$ received on input d_k. The resulting packet $ACK(i, actoff(l))$ is transmitted by gates A_5 and B_2, thus starting phase 2. Packet $ACK(k, actoff(j))$, after crossing the Batcher and the banyan networks, enters port controller PC_k on input f_k. While $actoff(j)$ is being received, two different tasks are carried out as follows.

a) $actoff(j)$ and j are summed to obtain the logical address of the channel assigned to PC_k;

b) $actoff(j)$ is compared to $maxoff(j)$ to verify if the channel assigned to PC_k is a member of group j.

Task a) is a simple addition and is assumed to be performed inside the control unit. Task b) is performed by comparing $actoff(j)$ and $maxoff(j)$ bit by bit in the EX-OR gate E_1, which receives both numbers with the least significant bit first. The flip-flop D, which is cleared by the rising edge of signal Φ_{actoff}, stores the most significant bit of $actoff(j)$ that is different from the corresponding bit of $maxoff(j)$. When the comparison is over, the output Q of the flip-flop is high, if $actoff(j) \leq maxoff(j)$.

In phase 3, a high signal on Q enables PC_k to transmit packet $DATA(p(j + actoff(j)))$ by gates A_7 and B_2 to the port physical address $p(j + actoff(j))$, mapped by table K_a to the channel logical address $j + actoff(j)$. At the end of phase 3, packet $DATA(k)$, if any, is received on input port f_k and transmitted to the output channel O_k by gate A_8.

An example of the switch operations for $N = 8$ and $D_{max} = 3$ is given in Fig. 9. The logical-to-physical address mapping of the channels and the channel group capacities are shown in Fig. 9(a) and (b). Group 4 has 3 channels, group 1 has 2 channels and all the others are single-channel groups. All port controllers generate a request and packets $REQ(\cdot, \cdot)$ are offered, sorted by destination address, to the channel allocation network [see Fig. 9(c)]. The $actoff(\cdot)$ values are computed by subnet-

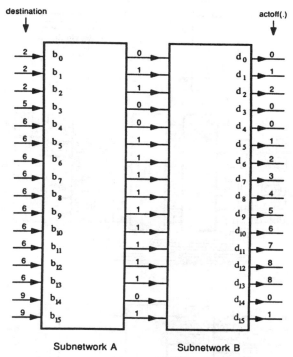

Fig. 7. Example of channel allocation.

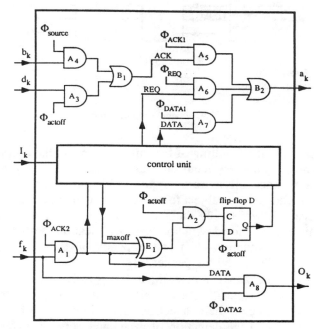

Fig. 8. Logic of port controller PC_k.

works A and B, and the port controllers generate the corresponding 8 packets $ACK(\cdot, \cdot)$. After crossing the Batcher and banyan networks, packet $ACK(i, actoff(j))$ is received by PC_i, which sends a packet $DATA(\cdot)$, if $actoff(j) \leq maxoff(j)$ [see Fig. 9(d)]. In our example, this condition is not satisfied in port controllers PC_h, $h = 3, 5, 7$.

D. Handling of Channel Failures

The multichannel bandwidth allocation scheme provides some advantages and drawbacks over unichannel al-

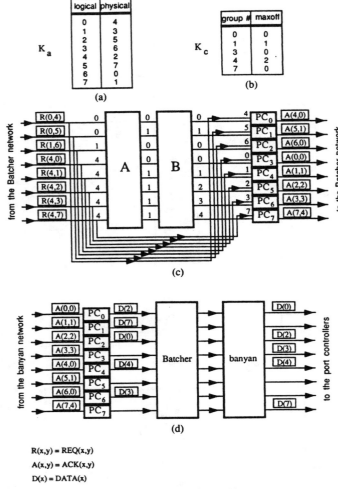

R(x,y) = REQ(x,y)
A(x,y) = ACK(x,y)
D(x) = DATA(x)

Fig. 9. Example of switching.

location in the handling of the failure events in the output channels. Between occurrence and detection of a failure in a channel, all sources supported by the group of the failed channel are subject to packet loss with multichannel allocation, while only the sources assigned to the failed channel are affected with unichannel allocation. Which of these two situations is worst depends on the kinds of sources affected by the packet loss.

With unichannel allocation, unless special procedures for circuit reallocation are available, the call processor has to tear down the connections supported by the failed channel upon the detection of a channel permanent failure. Procedures for dynamic reallocation of the connections require cooperation between the routing entities in the switches terminating the failed channel.

With multichannel allocation, a failure affecting channel group j simply reduces the capacity of group j. So, the ongoing calls on that group continue to be supported by the switch, without requiring recovery by routing entities in other switches. If the reduced capacity of channel group j is not sufficient to guarantee a given grade of service on that group, the call processor can selectively tear down some connections supported by channel group j. Upon the detection of a permanent fault in channel group j, the call processor computes the new mapping between

logical and physical addresses in group j. The new mapping and the new capacity of the only group j are broadcasted to all port controllers to update their tables K_a and K_c. Note that the updating procedure does not concern channel groups other than j.

IV. PERFORMANCE EVALUATION

A preliminary performance evaluation of the proposed bandwidth allocation scheme is given in this section. First, the processing overhead incurred by the multichannel bandwidth allocation scheme is examined and compared to unichannel bandwidth allocation. Then, the maximum throughput per port of the switch is evaluated for random traffic generated by low-peak sources. Finally, an upper bound of the packet loss probability for the switch is evaluated under simplified traffic assumptions for pure bursty traffic offered by high-peak low-activity sources.

A. Switching Overhead

The transfer of information packets takes place only in phase 3 of the three-phase algorithm, as phases 1 and 2 represent switching overhead. Let η denote the switch inefficiency, defined as the ratio between the duration T_{oh} of phases 1 and 2 and the transmission time T_{DATA} of packet DATA($\cdot$) in phase 3 (T_{oh} and T_{DATA} will be expressed in bit times). We assume that the packets received on input channels of the switch include a field to be filled with the packet self-routing address by the port controller. Then, $(1 + \eta)\,C$ bits/s is the bit rate of each digital pipe inside the switch that is required to allow a flow of C bits/s on the input and output channels.

The number of bit times it takes for a signal to cross a (sub)network will be referred to as signal latency in the (sub)network. The duration of phase 1 is given by the sum of the following latencies: $n(n + 1)/2$ in the Batcher network, n in subnetwork A. The duration of phase 2 includes the latency $n(n + 1)/2$ in the Batcher network, the latency n in the banyan network, the transmission time $n + d$ of packet ACK($\cdot$, $\cdot$) and the eventual additional time needed to complete the tasks a) and b) described in Sec III-C. Task b) is performed in d bit times and is completed at the reception time of the last bit of $actoff(\cdot)$ by the port controller. Task a) takes not more than $d + 1$ bit times. The d least significant bits of $j + actoff(j)$ are computed in d bit times, while the other $n-d$ can be computed in one bit time. So, both tasks are completed 1 bit time after the complete reception of packets ACK($\cdot$, $\cdot$) by the port controllers. We assume that the port physical address $p(j + actoff(j))$ is obtained from the channel logical address $j + actoff(j)$ in a negligible time. Hence, T_{oh} is given by $n(n + 4) + d + 1 = \log_2 N(\log_2 N + 4) + \lceil \log_2 D_{max} \rceil + 2$ for a multichannel switch, while it is given by $\log_2 N(\log_2 N + 4)$ for a unichannel switch (see [5]).

Compared to unichannel bandwidth allocation, the increase of η in our scheme is expected to be small, as the additional overhead of T_{oh} required by the channel assignment procedure is a logarithmic function of the maximum number of channels in a group. By assuming a "reasonable'' value $D_{max} = 50$, a switch with $N = 1024$ and $T_{DATA} = 1024$, we have $\eta = 0.137$ in the unichannel switch and $\eta = 0.145$ in the multichannel switch.

B. Performance for Random Traffic

In the evaluation of the throughput achievable by the multichannel switch, it is assumed that all channel groups have the same capacity $D_i = D$ and the number of ports N is significantly greater than the number of channels in a group. The input ports of the switch are uniformly loaded by traffic offered by Poisson sources, characterized by an average packet interarrival time much greater than the slot duration. Each source sends packets to a destination corresponding to a random selection of the output ports in the switch. Owing to the low-peak rate of the sources, the packet stream at each input port can be characterized by a random output port destination. In these conditions, as pointed out in [5], [10], the head of line blocking phenomenon results in a significant channel underutilization in a switch with $D = 1$, as the maximum throughput per port is 0.586.

By simulation, the maximum throughput per port ρ_{max} has been evaluated for a multichannel switch with group capacity $1 < D \le 64$ and the results are shown in Fig. 10. The maximum throughput ρ_{max} significantly increases with D and most of the advantage is obtained for small group capacities: $\rho_{max} = 0.883$ for $D = 16$ and $\rho_{max} = 0.919$ for $D = 32$. Hence, even small group capacities (e.g., 16 or 32 channels) provide a satisfactory throughput gain compared to the classical switching with single-channel groups.

C. Performance for Bursty Traffic

A simplified traffic environment is considered here to evaluate the implication of the multichannel allocation on the packet loss probability in the case of bursty traffic. Only one kind of traffic source loads the switch and the traffic pattern through the switch is such that throughput inefficiency from head of line blocking never occurs. The traffic offered to the switch is generated by a set of mutually independent high-peak low-activity sources with peak rate $P_i = C/f$ (f integer) and average load $S_i = S$. Hence, all sources have the same activity factor $\alpha_i = \alpha$. Each input port, with a buffering capability of one packet, can receive traffic from no more than one source. Each source, which is provided with a connection-oriented service, is either *active* or *silent*. A source in the active state offers the input port a periodic stream of packets, one every f slots, while no packets are generated by a source in the silent state. All the output channel groups have the same capacity $D_i = D$. The channel group selection at connection setup time is such that each of the output channel groups selected supports Dn_s virtual circuits.

To evaluate the packet loss probability B in the packet switch, we follow the approach described in [11]. B is given by the ratio between the number of packets discarded for buffer overflow at the input ports and the number of packets offered to the switch in a long time interval,

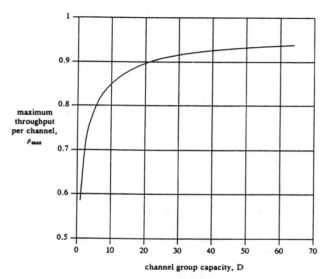

Fig. 10. Asymptotic throughput performance.

say T slots (T is much bigger than the average burst length). A packet loss occurs when more than Df packets for a specific channel group are received in f slots by the input ports. Let $\Pr(i)$ denote the probability of i active sources sharing the same output channel group. $\Pr(i)$ also represents the probability that i packets addressed to a specific channel group are offered to the switch in a time interval lasting f slots. Since α is the probability that a source is active (i.e., the probability that a source issues a packet in f slots), then $\Pr(i)$ is given by

$$\Pr(i) = \binom{Dn_s}{i} \alpha^i (1 - \alpha)^{Dn_s - i}. \tag{3}$$

The total number of packets offered to the switch in Tf slots and destined to a specific channel group is $\alpha Dn_s T$. Since up to Df packets can be transmitted by a channel group in f slots, the number of packets discarded in Tf slots is

$$\sum_{k=Df+1}^{Dn_s} \Pr(k)(k - Df) T. \tag{4}$$

Then, we easily obtain

$$B = \frac{1}{\alpha Dn_s} \sum_{k=Df+1}^{Dn_s} \binom{Dn_s}{k} \alpha^k (1 - \alpha)^{Dn_s - k} (k - Df). \tag{5}$$

The packet loss probability B has been computed as a function of the normalized offered traffic per output channel $n_s \alpha / f$ (the normalizing factor is the channel capacity C). An increasing activity factor α for $n_s = 32$ and $f = 4$ is considered in Fig. 11, while an increasing number of sources n_s for $\alpha = 1/32$ and $f = 1$ is represented in Fig. 12. These figures show that even channel groups of limited capacity (e.g., 16 channels) reduce the packet loss probability by several orders of magnitude in comparison to single-channel groups over almost the entire range of offered traffic values. In particular, we observe that a value

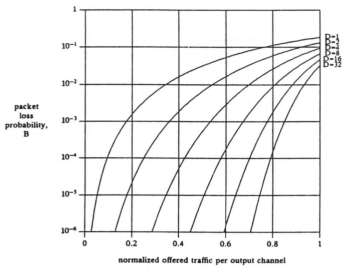

Fig. 11. Packet loss probability for bursty traffic with $n_s = 32$, $f = 4$.

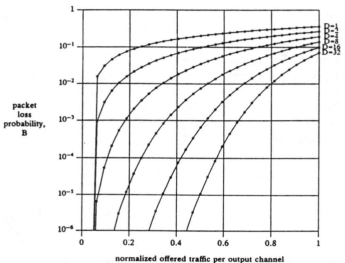

Fig. 12. Packet loss probability for bursty traffic with $\alpha_i = \alpha = 1/32$, $f = 1$.

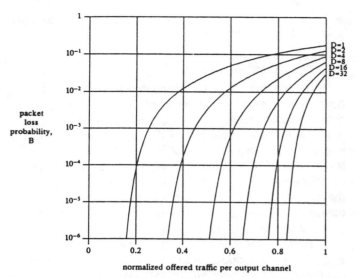

Fig. 13. Packet loss probability for bursty traffic with $n_s = 32$, $f = 4$, $M_b = 50$.

packet loss probability for the same abscissa. However, the advantage provided by multichannel bandwidth allocation is not expected to change, since the buffering capability lowers the packet loss probability for any channel group capacity. This fact can be shown by using the results obtained in [12] from an analytical model of a service system loaded by bursty traffic. Assume bursts and silences of sources to be geometrically distributed with an average burst length of 10^4 packets and a buffer of $M_b = 50$ packets at each input port. Then Fig. 13 shows the packet loss probability for $n_s = 32$ and $f = 4$. All the curves of Fig. 11 are here moved to the right and $B = 10^{-5}$ is now guaranteed by $W = 5.88S$ for $D = 1$ and by $W = 1.19S$ for $D = 32$.

V. CONCLUSIONS

A bandwidth allocation scheme has been described. It is based on the definition of data link layer connections (channel groups), whose capacity is not bounded by the capacity of a single broadband packet channel. This scheme allows more efficient use of switching and transmission resources and permits super-rate switching. The bandwidth allocation scheme has been shown to be feasible in a Batcher-banyan switch. The additional hardware required in the switch to handle channel groups negligibly increases the switch inefficiency. For random traffic, the throughput increase provided by the scheme is remarkable, even for channel groups of limited capacity. For bursty traffic, channel groups reduce the packet loss probability by several orders of magnitude.

ACKNOWLEDGMENT

The author would like to express his appreciation to N. Haller, J. Hui, M. Littlewood, S. Minzer, R. Porter, D. Sincoskie, P. White, and L. Wu for the helpful discussions and valuable suggestions. The author would also like to thank T. Lee for the use of his package for the analysis of service systems loaded by bursty traffic [12].

of $B = 10^{-5}$ can be guaranteed for the case in Fig. 11, by reserving for each source an amount of bandwidth W equal to[4] $16.7S$ with single-channel groups and to $1.4S$ for $D = 32$. For the case in Fig. 12, the same B value is guaranteed by reserving $W > 16S$ for $D = 1$ (i.e., not more than 1 source per channel can be accepted) and $W = 2.1S$ for $D = 32$. So, as one might expect, the availability of channel groups greatly reduces the underutilization of the output channels.

Note that the assumption of buffering capability limited to one packet means that B of (5) represents an upper bound for the packet loss probability, if we disregard the head of line blocking. When more than one packet can be buffered at each input port of the switch, all the curves in Figs. 11 and 12 move to the right, by providing a lower

---

[4]The "overbooking factor," W/S, is given by the inverse of the abscissa corresponding to the target value of packet loss probability.

REFERENCES

[1] J. J. Kulzer and W. A. Montgomery, "Statistical switching architecture for future services," in *Proc. Int. Switch. Symp.*, Florence, Italy, 1984, paper 43.A.1.

[2] J. P. Coudreuse and M. Servel, "Prelude: An asynchronous time-division switched network," in *Proc. Int. Conf. Commun.*, Seattle, WA, June 1987, pp. 769–773.

[3] J. Turner, "Design of a broadcast packet network," in *Proc. INFOCOM'86*, Miami, FL, Apr. 1986, pp. 667–675.

[4] A. Huang and S. Knauer, "Starlite: A wideband digital switch," in *Proc. IEEE GLOBECOM'84*, Atlanta, GA, Nov. 1984, pp. 121–125.

[5] J. Hui and E. Arthurs, "A broadband packet switch for integrated transport," *IEEE J. Select. Areas Commun.*, vol. SAC-5, pp. 1264–1273, Oct. 1987.

[6] Y. S. Yeh, M. G. Hluchyj, and A. S. Acampora, "The knockout switch: A simple, modular architecture for high-performance packet switching," in *Proc. Int. Switch. Symp.*, Phoenix, AZ, Mar. 1987, pp. 801—808.

[7] L. T. Wu, S. H. Lee, and T. T. Lee, "Dynamic TDM, a packet approach to broadband networking," in *Proc. Int. Conf. Commun.*, Seattle, WA, June 1987, pp. 1585–1592.

[8] P. Gonet, P. Adam, and J. P. Coudreuse, "Asynchronous time-division switching," in *Proc. Int. Zurich Sem. Digital Commun.*, Zurich, Switzerland, Mar. 1986, pp. 141–148.

[9] J. Y. Hui, "Resource allocation for broadband networks," *IEEE J. Select. Areas Commun.*, see this issue, pp. 1597–1607.

[10] M. J. Karol, M. G. Hluchyj, and S. P. Morgan, "Input vs. output queueing on a space-division packet switch," in *Proc. IEEE GLOBECOM'86*, Houston, TX, Dec. 1986, pp. 659–665.

[11] C. J. Weinstein, "Fractional speech loss and talker activity model for TASI and for packet-switched speech," *IEEE Trans. Commun.*, vol. COM-26, pp. 1253–1257, Aug. 1978.

[12] T. T. Lee, "Performance analysis of packet multiplexers based on dynamic time division multiplexing," Private communication.

Achille Pattavina was born in Terni, Italy, in 1954. He received the Dr. Eng. degree in electronic engineering in 1977, and specialization degree in computer and control automatic systems in 1979 from the University "La Sapienza," Rome, Italy.

From 1979 to 1983 he was a Software Analyst for CAD/CAM applications and a Professor of Communication Systems for technical officers of the Italian Air Force. Since 1983 he has been a Researcher at the INFOCOM Department, University "La Sapienza," Rome, where he has been involved since 1980 in researches on communication network synthesis, protocol specification and verification, and access protocols for local area networks. In 1987, on leave of absence from the INFOCOM Department, he spent one year at Bell Communications Research Inc., Morristown, NJ, studying protocol architectures and switching techniques for broadband packet networks.

Flow Control Schemes and Delay/Loss Tradeoff in ATM Networks

HIROKAZU OHNISHI, TADANOBU OKADA, MEMBER, IEEE, AND KIYOHIRO NOGUCHI

Reprinted from *IEEE Journal on Selected Areas in Communications*, Vol. 6, No. 9, December 1988, pages 1609-1616.
Copyright © 1988 by The Institute of Electrical and Electronics Engineers, Inc. All rights reserved.

Abstract—Broadband packet switching, called asynchronous transfer mode (ATM) in the broadband ISDN arena, is one of the more promising means to provide multimedia communications in an efficient and unified manner. Performance and flow control mechanisms, which are the features representing ATM flexibility, are discussed here. For the performance (quality of service) control, introduction of delay, and/ or loss sensitive service classes, and two mechanisms to realize these classes are proposed. It is indicated that there are possibilities to have better performance than other mechanisms, such as no-class distinction or simple priority methods. It is further suggested that this performance controllability results in the provision of multiple logical services including quasi-STM (compatible with circuit switching), by an ATM network. ATM flow control is based on a call-oriented resource allocation mechanism similar to circuit switching. This paper indicates that the concepts of call/line bit rate ratio and multiplexing degree are significant for efficient use of resources. When the network handles calls with large call/line bit rate ratios, user-specified flow control parameters at the call setup phase are important for resource assignment. The definition of two types of maximum throughput of each call and its usage for resource management are proposed.

I. INTRODUCTION

BROADBAND packet swtiching, called asychronous transfer mode (ATM) in the CCITT broadband ISDN arena, is one of the more promising means for efficient and unified support of multimedia (voice, video, and data) communications and multiple user bit rates. In accordance with current CCITT nomenclature, short (30–80 octets), fixed-length packets called cells are used as a basic information transfer vehicle.

It is well known that ATM has various flexible features, such as arbitrary, broad user bit rate support, easy network engineering, and network integration possibilities. After defining the framework and assumed conditions in Section II, Section III of this paper focuses on ATM performance controllability. ATM can provide users within a unified network with multiple services based on differing performances and capabilities.

Due to hardware switching and multimedia support, ATM flow control differs from that used in the conventional X.25-based packet networks. Call/connection oriented flow control is adopted for ATM. Within this type of flow control, basic traffic multiplexing characteristics, definition and significance of maximum call throughput, and necessary user-network interface parameter to be specified by the user are examined in Section IV.

Manuscript received December 2, 1987; revised June 11, 1988.
The authors are with Communications Switching Laboratories, Nippon Telegraph and Telephone Corporation, Tokyo 180, Japan.
IEEE Log Number 8824401.

II. ATM NETWORKS AND CHARACTERISTICS

The locations of the ATM user-network interface and its protocol structure are shown in Figs. 1 and 2. Media-dependent functions are located at the user side of the interface and are not discussed in this paper. This paper deals with some parts of layers 1–3 of the user-network interface. Services provided by the network are abstracted as much as possible to make them independent of user media, such as voice and data. (See also Section IV.)

A general comparison of network traffic controllability between conventional X.25 packet switching, ATM (fast packet switching), and circuit switching (synchronous transfer mode: STM) is shown in Table I in order to clarify the following ATM flexibility.

1) Arbitrary, broad bit rates can easily be supported.

2) Several control parameters such as quality/performance (cell delay and cell loss rate) and call blocking rates are variable and selectable by the network provider or users.

3) These parameters are variable at the time of either network engineering or after service commencement. For example, some networks may dynamically change delay/ loss performance to improve a call blocking rate when more calls than expected take place.

In order to realize these ATM control flexibilities, either per-cell or per-call based control is necessary. Although conventional X.25 packet switched networks mainly perform per-packet control, per-cell control is rather difficult in ATM. Typical per-cell and per-call control schemes in ATM are introduced in Sections III and IV, respectively.

III. PERFORMANCE CONTROL AND SERVICE CLASSES

A. Parameters to Specify Service Performance (Quality)

The main parameters which specify ATM service performance for each call are 1) network resource allocation for a call (deterministic or statistically), 2) network-wide delay characteristics (absolute value of fluctuation width and average, min/max, 99 percent value, etc.) of cells or between cells, and 3) cell loss-probability. When sufficient network resources are deterministically allocated to support maximum throughput of each call, ATM cell delay· and loss performances approach those of circuit switching. Therefore, selection and realization of cell loss and delay performances in statistical resource assignment cases are discussed in the following subsections. "Cell loss" in this section means cells discarded by the network

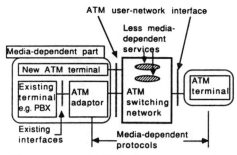

Fig. 1. Assumed ATM network configuration.

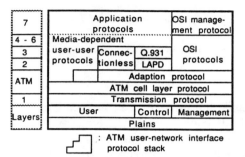

Fig. 2. ATM user-network interface protocol architecture.

TABLE I
TRAFFIC CONTROL SCHEME COMPARISON BETWEEN X.25 PACKET, ATM, AND STM SWITCHING

	PS	ATM (FPS)	STM(CS)
User throughput guarantee by the network	May not be guaranteed	High probability of success	Guaranteed
Information transfer phase involvement by the network	Large	Very small	None
User info. flow smoothing by the network	Possible by window control	No (statistical multiplexing)	None
Call blocking	No-blocking may be possible.	Exists	Exists

(switches) due to resource shortages, but does not include cells discarded due to bit errors or system malfunction.

B. Statistically Assigned Resources and Service Classes

Required performances varies depending on media. Different ones are required for the same types of different media. However, performance specification should avoid direct dependence on requirements of existing limited types of media. Service classes corresponding to various performances should be defined as abstractly as possible to cover many different types of media including open-ended capability for future media. However, service classes based on a fixed priority queueing method which is too straightforward, may not meet the requirements for existing media. Generally speaking, data media are loss sensitive, while uncompressed voice/video media are delay sensitive. Therefore, control mechanisms for a mini-

mum set consisting of the two service classes shown in Table II are discussed. For the delay sensitive class, delay fluctuation width as well as absolute delay should be small taking account of time stamp control. Cell delay in a high bit rate trunk line (e.g., 140 Mibts/s) can be so small that delay performance control may not be required. However, loss performance control is necessary in any bit rate trunk line as long as statistical multiplexing is used.

C. Performance Control Mechanism Comparison

1) Types of Mechanisms: For both delay and loss performance control, input and output mechanisms of ATM switch cell buffers, which are usually located at the exit of the ATM switch, are a dominant factor. The following discussion assumes that each output line of the switch has a buffer associated with it. Cells are first switched, then inserted into the buffer. Cells are removed from the buffer as they are transmitted on the output line. Four methods (see Table III) are listed below.

a) Variable priority method A (with loss and delay sensitive classes): The delay-sensitive class (class 1) always has priority regarding cell output from buffer to line. Regarding cell input with buffer overflow, loss-sensitive class cells always have priority, and when required for the loss sensitive class input, delay-sensitive class cells within the buffer are discarded.

b) Variable priority method B: Cell input to the buffer is identical with the above proposed method A. Cell output from the buffer is the same as the proposed method A except that when loss sensitive class cells wait for longer than a determined threshold period within the buffer, output priority is changed from the delay sensitive class to the loss sensitive class. In comparison to the above method A, this method improves the both delay and loss performance of the loss-sensitive class.

c) Fixed priority method A: Cells are either priority or normal classes. The priority class always has priority in either input or output control.

d) Fixed priority method B: Output control is identical with the above fixed method A. However, both priority and normal classes are identically handled in the input control.

The delay and loss performance comparison between the variable priority method B and the fixed priority method B is shown in Figs. 3 and 4, which also include performances of a mechanism without priority.

The traffic model for this simulation is as follows. Two independent Poisson distribution with the same average arrival rate are used for arrival process of two classes. A single server with a fixed service time (cell length/circuit bit rate) is assumed. Finite buffer whose capacity is 20 cells is shared by two classes.

These figures indicate that the variable priority methods give different loss and delay performances between classes and that the class having better performance can be reversed between loss and delay performance. This means better loss or delay performance can be given to a class under the identical resources and traffic condition if an-

TABLE II
BASIC SERVICE CLASSES FOR SUPPORTING VARIOUS MEDIA IN STATISTICAL
MULTIPLEXING

Service class	Delay quality	Loss quality
Class 1 (Delay sensitive)	**Better** than class 2	Worse than class 2
Class 2 (Loss sensitive)	Worse than class 1	**Better** than class 1

TABLE III
CONTROL MECHANISMS OF VARIOUS PERFORMANCE CONTROL METHODS

Method	Class	Cell handling in buffer I/O	
		Input	Output
No class	No class distinction (All cells are identically handled.)		
Fixed priority A	Priority / Normal	Priority class has priority.	
Fixed priority B	Priority / Normal	Identically handled	
Variable priority A	Delay sensitive	Has no priority	Has priority
	Loss sensitive	Has priority	Has no priority
Variable priority B	Delay sensitive	Identical with proposed method A	
	Loss sensitive		

* : The same as Proposed method A except that priority
is changed when the waiting time of loss-sensitive
class cells is larger than a threshold.

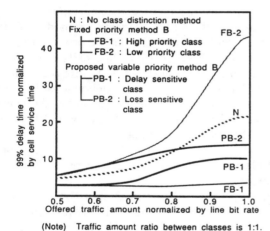

(Note) Traffic amount ratio between classes is 1:1.

Fig. 3. Delay time comparison between several performance control
methods.

other class is tolerant of that type of performance decrease.

In Fig. 3, the delay performances of both delay and loss sensitive classes are better than those of the no class distinction method in the high offered traffic part. This is due to the following reasons.

1) The loss-sensitive class can get priority when it waits longer than a threshold.

2) The total loss rate of the variable priority method *B* is larger than that of the no class distinction method, because the delay-sensitive class cells are discarded.

Though these figures are calculated by simulation assuming random arrival of cells and fixed service time, performance differences between the variable priority

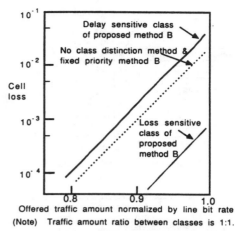

(Note) Traffic amount ratio between classes is 1:1.

Fig. 4. Cell loss probability comparison between several performance
control methods.

method and other methods can be identified. Since actual user traffic is considered to more bursty than random arrival, the variable priority methods will have a more significant effect. Traffic amount ratio between two different classes are assumed to be 1 : 1 in the simulation. When this ratio departs from this value, more resources are required for keeping identical performances. In the assumed model, common buffer is used for the two service classes. Separate buffer model is a subject for further study.

D. Possible Set of ATM Service Classes

In addition to the two basic loss and delay sensitive classes, the need for other classes (see Table IV) is discussed. Separation of the delay sensitive class into two subclasses 11 and 12 and addition of another new quasi-STM class (class 0) are required for the following reasons.

1) Classes 11 (Delay and Loss Sensitive) and 12 (Delay Sensitive): Some delay sensitive communications also include information of different loss probability requirements. For example, in compressed video or embedded ADPCM coding [1], high- and low-priority bits may be separated and delivered by different cells. Cells containing high-priority bits should be given loss probability priority. For this purpose, cells of classes 11 and 12 must keep their order at the receiver, and thus, common buffering is required for these two classes. If required, more class (e.g., 11, 12, 13) provision within the class 1 family is not so difficult issue. When sufficiently high bit rate lines are used, it may be possible to merge classes 11 and 2.

2) Quasi-STM (Class 0): Quasi-STM is an ATM network service that guarantees a certain bit rate and contiguous communications by deterministically allocating network resources to those calls. The method of transmitting cells at the user-network interface for this service must be specified. In the case of pure STM (circuit switching), user octets or bit information is periodically transferred every 125 µs. However, in the quasi-STM case, a certain

TABLE IV
POSSIBLE SERVICE CLASS SET FOR ATM

Service class		Quality		Resources
		Delay	Loss	
Quasi-STM	Class 0	Similar to STM		D**
Delay	Class 11	**Better**	**Better**	Statistically allocated
sensitive	Class 12		—*	
Loss sensitive	Class 2	—*	**Better**	
(Call) Control info.		**Better**	**Better**	D**

* : may be worse than other service classes.
** : Deterministically allocated.

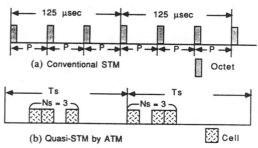

(a) Conventional STM

Octet

(b) Quasi-STM by ATM

Cell

Ts, Ns : See Fig. 8 and Sect. IV.
Fig. 5. Quasi-STM by ATM.

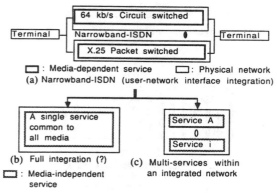

☐ : Media-dependent service   ☐ : Physical network
(a) Narrowband-ISDN (user-network interface integration)

(b) Full integration (?)

(c) Multi-services within an integrated network

☐ : Media-independent service

Fig. 6. Network and service integration.

period and the number of cells which can be transmitted in that period should be specified as shown in Fig. 5. This period and the number of cells during the period correspond to Ts and Ns described in Section IV. For a certain bit rate (BR) communications, the user can select various combination of Ts and Ns because $BR = Ns/Ts$. From the network resources allocation point of view, minimum Ts and Ns are preferable.

E. Service Integration by ATM

It is often said that ATM can integrate various trunk networks, such as 64 kbits/s circuit switched and packet switched, which narrow-band ISDN cannot. Two kinds of network integration, namely, physical network and logical service integration are shown in Fig. 6. As shown in Fig. 7, ATM can integrate physical trunk networks, however, several of the logical services based on the

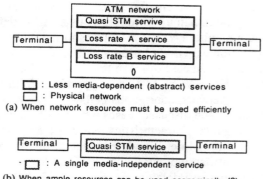

☐ : Less media-dependent (abstract) services
☐ : Physical network
(a) When network resources must be used efficiently

☐ : A single media-independent service
(b) When ample resources can be used economically (?)

Fig. 7. Network integration and service abstraction by ATM.

above-mentioned service classes may be required even for public networks. However, it must be noted that abstract services should not directly depend on existing media to support future new media.

IV. CALL ORIENTED FLOW CONTROL

A. ATM Flow Control Characteristics

In existing X.25 packet networks, flow control is performed by a so called window control mechanism using packet-by-packet acknowledge (receive-ready) signals. When congested, data packet input is stopped by the network. This mechanism can smooth resource utilization and it results in an efficient resource utilization. However, it is difficult for the network to guarantee user throughput.

Currently, it is not general ATM networks use packet-by-packet flow control in the information transfer phase due to the following reasons.

1) Some voice or video communications require guaranteed user throughput.

2) It is difficult for the high bit rate ATM network to perform window-type flow control, since windows for high bit rate services would be enormous and require processing burden.

Instead of packet-by-packet control, the ATM network allocates network resources to each call according to its throughput requirements. In this sense, ATM flow control is based on "call oriented control." Though ATM networks are based on the connection oriented flow control, connectionless communications are still possible. The connectionless mode can be supported on the top of a connection between the user and the connectionless server node within the network. Thus, the connection oriented flow control is used also for the connectionless mode. The following sections deal with basic characteristics of ATM statistical multiplexing and throughput-related parameters which should be specified by the user at the call setup phase.

B. Call/Line Bit Rate Ratio and Statistical Multiplexing

The concept of call/line bit rate ratio plays an important role in ATM flow control. This is the ratio of maximum call throughput in short duration to the bit rate of transmission line where user calls are statistically multiplexed by a cell manner. Hereafter, this ratio is abbreviated as *Rs*. Here, the maximum throughput of a call is defined as maximum number (*Ns*) of cells transmitted within a fixed short duration called *Ts* during a call as shown in Fig. 8. *Ts* and *Ns* are selected to make them as small as possible. Typically, since $Ns = 1$, *Ts* corresponds to minimum interval time between contiguous cells.

This paragraph discusses cases where calls having identical throughput characteristics, such as *Rs*, are multiplexed in a trunk line. Simulated calculation of maximum trunk line utilization ratio with various *Rs* is shown in Fig. 9 making cell delay a parameter. In this simulation, each call is assumed to have an identical bursty cell traffic pattern as shown in Fig. 10. Fig. 9 shows the following characteristics.

1) Maximum line utilization significantly depends on the *Rs* value of a call.

2) To achieve rather high line utilization ratio, *Rs* should be smaller than 0.1. This means a transmission line bit rate of 10 times larger than maximum call throughput should be used. In order to maintain service quality (performance), line utilization ratio should be controlled under the curve shown in Fig. 9.

Call blocking is not taken into account in Fig. 9. If it is taken into account, the line utilization gets smaller than that of Fig. 9.

The idea that efficient resource utilization can be realized with a large enough multiplexing degree may not always be true. Multiplexing degree is defined as the number of calls which are multiplexed on a transmission line. The line utilization ratio of the case where a small number of large *Rs* calls are multiplexed with a great number of small *Rs* calls in a line is shown in Fig. 11. This figure demonstrates that a small number of large *Rs* calls reduce the line utilization ratio even if a great number of calls are multiplexed. This is another example of the significance of *Rs*.

However, if the multiplexing degree is small, a great part of the high bit rate transmission line capacity is wasted. Therefore, both a small call/line bit rate ratio and a large multiplexing degree are important for efficient use of a line capacity as shown in Fig. 12. This can be formulated in more detail as follows.

1) A high bit rate transmission line which makes *Rs* small enough to achieve good line utilization ratio (*U*) should be used as long as it is possible.

2) Enough calls should be gathered and multiplexed to effectively utilize the transmission line to the ratio *U*.

Though a high bit rate line should be used for high bit rate communications, gathering enough traffic may be difficult. For example, traffic demands of high bit rate communications may be too small to completely fill a line. Therefore, existing traffic, such as telephony and low bit rate data could be used to fill the transmission line.

C. Another Major Throughput Related Parameter

The relationship between line utilization ratio and traffic density of a call is shown in Fig. 13. Using the same user

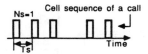
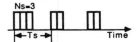
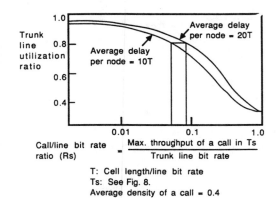

(a) Case with duration between cells
(Ts = Min.(Tsi), Ns = 1, MTs = 1/Ts)

(b) Consecutive cell transmission at a line bit rate (MTs = Min.(Nsi/Tsi))
MTs = Max. throughput of a call in short duration
Tsi = i-th short duration
Nsi = The number of cells in Tsi

Fig. 8. Definition of maximum throughput of a call in short duration.

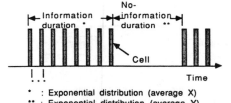

T: Cell length/line bit rate
Ts: See Fig. 8.
Average density of a call = 0.4

Fig. 9. Relation between line utilization ratio and call/line bit rate ratio (*Rs*).

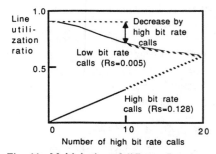

* : Exponential distribution (average X)
** : Exponential distribution (average Y)
*** : Cell interval time is assumed to be identical, and corresponds to a short duration time (Ts) for maximum throughput (MTs).

Fig. 10. Traffic model of a user call.

Fig. 11. Multiplexing of different *Rs* values.

cell transmission traffic model as that of Section B (see Fig. 10) during a call, the traffic density of a call is defined as $X/(X + Y)$. Here, *X* and *Y* are average information and no-information duration time, respectively.

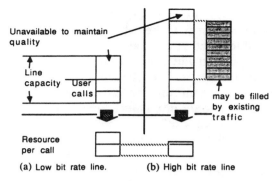

Fig. 12. Relationship between degree of multiplexing and required resources.

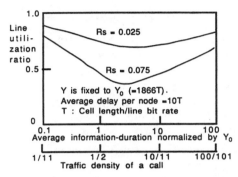

Fig. 13. Relationship between line utilization ratio and traffic density per call.

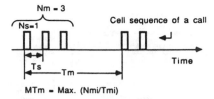

MTm: Max. throughput of call in medium duration
Tmi: i-th medium duration
Nmi: the number of cells in Tmi

Fig. 14. Definition of maximum throughput of a call in medium duration.

Strictly speaking, the horizontal axis of Fig. 13 is variable X. Since Y is fixed to $Y0$, this X corresponds to the traffic density of a call. When Rs is large, the traffic density of each call affects the line utilization, Therefore, traffic density of each call becomes another major parameter when the network handles large Rs calls. Interestingly, the line utilization ratio becomes good when the traffic density of a call is large. This is because large density means a decrease in traffic burstiness of a call.

In Fig. 13, it is assumed the traffic density of each call is constant during a call holding time. However, traffic density varies during a call holding time. Therefore, average traffic density during a call holding time is not a suitable parameter to specify call throughput characteristics. This point is discussed further and a suitable parameter is proposed in the following section.

D. Selection of Per-Call Throughput Related Parameters Specified by Users [3]

In order for the network to determine if a call can be accepted or not at the call setup phase, the network must be aware of throughput requirements or features of the call. It may be difficult, however, for a user to specify his own throughput. Therefore, suitable flow control related parameters for both the network and users must be selected.

1) Call Acceptance Decision in Small Rs Situations: When Rs is small, throughput characteristics of total multiplexed user traffic can be estimated by results of periodic measurements of line utilization or cell delay and loss quality. Therefore, little information about call throughput characteristics is required for the call acceptance decision. For example, the maximum call throughput in short duration (MTs) for a requested call is suitable for the parameter specified by the user.

2) Call Acceptance Decision When Rs is Large: When Rs is large, it is dangerous to estimate throughput characteristics of total multiplexed user traffic by results of periodic line utilization measurements. The network should reserve network resources according to the throughput requirements of each call. Therefore, these throughput requirements must be specified by user at the call setup phase. Selection of the parameters specified by the user is discussed below.

As discussed in Section IV-B, the maximum call throughput during a short period (MTs) is an important parameter in any case. MTs is a suitable parameter that the user specify at call setup time. However, MTs is ambiguous because there are various types of cell traffic pattern for an identical MTs. Therefore, Ts and Ns shown in Fig. 8 should be specified in stead of MTs. If it is possible by users, $Ns = 1$ is preferable because burstiness is minimum. However, as shown in Fig. 8, there are at least two cell transmission types. In the type a), the user can put a time interval between cells of a call. In order to cover many types of traffic, the network should allows the user to specify both a short duration time (Ts) and the number of cells (Ns) during Ts.

The pair of Ts and Ns is insufficient for exact resource determination because network resources corresponding to the maximum throughput must be allocated similarly to quasi-STM. For more efficient resource allocation, another parameter is required.

It was indicated in Section IV-C that the traffic density of each call, in addition to the maximum throughput, also plays a major role in determining available line capacity in large Rs situations. Since the situation where Rs is large is assumed here, this density should be taken into account. However, this traffic density of a call varies during a call holding time, and maximum call density during a call determines resource requirements. Therefore, maximum throughput during a medium time period (MTm), shown in Fig. 14, is proposed for the second parameter. MTm is defined as the number of cells (Nm) during a medium period (Tm) longer than Ts. At a call setup time, the

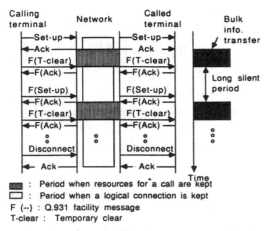

Fig. 15. Short-hold mode ATM communications.

user should specify *Nm* and *Tm* instead of *MTm*. This *MTm* specification may be constraints to users. The impact on users should be studied further. However, the user can determine the *MTm* by observing and measuring his communications in the past. During a call, the user can maintain his throughout to this *MTm* if he measures the number of cells transferred in each medium time period during a call. For example, in case of video communications, the user may be able to select one video frame interval (e.g., 1/3 ms) as *Tm*. In this case, the maximum number of cells among *Tm* intervals becomes *Nm*.

E. Short Hold Mode Communications

Very bursty communications, such as video database access where *Rs* is large, are not suitable even for ATM. These communications are characterized by the following conditions.

1) Maximum bit rate (throughput) of a call is high in comparison to the line bit rate (*Rs* is large).

2) After a period of maximum bit rate transmission, there follows a long silent period.

3) Condition 2) is repeated.

Though conventional X.25 packet switching networks delay packet input using window flow control to handle this bursty traffic efficiently, they cannot deal with high bit rate communications. In order for ATM networks to support these higher bit rates, extensive resources must be allocated even during long silent periods when *Rs* is large.

One solution to this problem is not to reserve network resources and to use variable window size end-to-end flow control protocol [2] where the window size is reduced after detected data loss by negative acknowledgment from the remote end user. A lot of cell retransmission may be required.

Another solution is a short-hold mode ATM service. As shown in Fig. 15, when an information transfer period terminates, the user sends a temporary resource clear request signal to the network, and the network releases the resources while keeping the logical call. Before the user starts next information transfer, he must ask the network to reallocate resources to the call.

Good cell loss and delay quality during the information transfer periods and efficient use of network resources can be achieved by this service, though there is some delay before restart of the information transfer period. When the network resources are insufficient to restart calls, restart must be delayed until other calls release resources. Call resetup processing causes another delay. However, this resetup processing delay can be reduced from that of the initial call setup because the number of exchanged messages, complicated message information analysis, address analysis, call reference/logical channel number hunt, and route selection can be eliminated for call resetup.

V. CONCLUSION

It is difficult for ATM to perform a full per-cell control as per-packet window flow control of conventional X.25 packet networks. However, per-cell control is possible for ATM. As one example, delay and loss performance control is discussed in Section III. In order to realize various performances for multimedia communications, the effectiveness of variable priority control mechanisms are indicated. A full service set for ATM including quasi-STM (circuit switching) is proposed.

Differing from conventional packet switching, ATM flow control is based on a call oriented resource assignment mechansim similar to circuit switching. It is demonstrated in Section IV that the call/line bit rate ratio (*Rs*) is a significant parameter for determining how much of the capacity of a transmission line can be used. The maximum throughput of a call in short duration, which constitutes *Rs*, is defined. It is further suggested that maximum traffic density during a call holding time is another major parameter in situations where *Rs* is large. For this purpose, another maximum throughput of a call during a medium period is proposed.

Based on the above-mentioned ATM traffic characteristics, a call acceptance decision scheme is also discussed. In situations where *Rs* is small, call acceptance can be determined by measuring trunk line utilization. However, in large *Rs* situations, in order for the network to allocate network resources effectively, two types of maximum throughput during both a short period and medium period should be specified by the user at the call setup phase. For extremely bursty traffic, a short hold mode service is proposed.

ACKNOWLEDGMENT

The authors wish to thank K. Manabe for his performance control simulation programs.

REFERENCES

[1] R. W. Muise, T. J. Schonfeld, and G. H. Zimmerman, III, "Experiments in wideband packet technology," presented at Zurich Sem. Digital Commun. 1986.
[2] W. Bux and D. Grillo, "Flow control in local-area networks of interconnected token rings," *IEEE Trans. Commun.*, vol. COM-33, Oct. 1985.

[3] J. S. Turner, "New directions in communications (or which way to the information age?)," *IEEE Commun. Mag.*, vol. 24, Oct. 1986; and also presented at Zurich Sem. Digital Commun., 1986.

Tadanobu Okada (M'84), Senior Research Engineer/Supervisor, received the M.S. degree in mathematics from Tohoku University in 1976.

After joining Musashino Electrical Communication Labs, NTT, in 1976, he had been involved in developmental research on DDX message communications systems and X.400 message handling systems. He is now engaged in network control for ATM networks and an active participant in CCITT SG XVIII ATM working group.

Mr. Okada is a member of IEICE of Japan.

Hirokazu Ohnishi, Senior Research Engineer/Supervisor, is engaged in R&D of network control and protocols for ATM networks and ring systems. He received the B.E. and M.E. degrees from Kyoto University in 1972 and 1974.

Since joining Musashino ECL, NTT, in 1974, he has been engaged in the development of DDX packet network. He has also worked in the fields of ISDN user-network interfaces, high-speed switched digital services using demand assignment satellite TDMA, message handling systems, and network interworking. He has submitted contributions to CCITT SG VII, XVIII, ISO TC97 SC6, and IEEE 802.6, and acted as Special Rapporteur for SG VII.

Mr. Ohnishi is a member of the IEICE and the Information Processing Society of Japan.

Kiyohiro Noguchi received the B.E. and M.E. degrees from Tohoku University, Japan, in 1982 and 1984, respectively.

Since joining the Musashino Electrical Communication Labs, NTT, Japan, in 1984, he worked on the research and development for the DDX packet switched data network. He is now engaged in the research an traffic control for ATM networks.

Mr. Noguchi is a member of the IEICE of Japan.

ACCESS INTERFACE CONGESTION CONTROLS FOR PACKETIZED VOICE TRANSPORT IN WIDEBAND NETWORKS

Stephen G. Eick

Network Management Planning Department
AT&T Bell Laboratories

The Access Interface (AI) in a wideband packet network converts all analog and digital end-user traffic into packets for high-speed transport. For voice calls, the AI detects and packetizes only during speech segments, resulting in an increased network call-carrying capacity. Each AI is engineered for a fixed number of voice, voiceband data, and data calls. During overloads, the offered traffic or composition of traffic may exceed or differ from the engineered levels, resulting in degraded transmission quality. This paper describes a modeling study investigating the performance of the the selective dynamic overload control (SDOC), a proposed AI congestion control proposed for network management. Under a range of assumptions for typical applications, SDOC will control AI congestion within 3 to 4 minutes.

1. Introduction

As shown in Figure 1, the access interface (AI) [1] in a wideband packet network (WPN) [2] converts all end-user traffic into a common, simple packet format for high-speed transport. For analog voice calls the AI detects and packetizes speech segments. By not transmitting during silence periods, it is possible to maintain transmission quality [3,4], and yet increase the capacity of the wideband network, since the activity rate for voice calls is only about 40% [5]. During overloads, the AI drops the least significant bit or bits from the voice packets, and during extreme overloads the AI drops packets, resulting in a gradual performance degradation. As a result, there is a significant increase in the voice transport capacity of wideband networks [6,7,8].

I assume that an AI is engineered for a fixed number of voice, voiceband data, and data calls. However, the offered traffic may differ from the engineered levels. During periods of network stress the AI may experience sustained overloads resulting in unsatisfactory transmission quality, unless network management controls are activated. For example, suppose a WPN facility is engineered for a fixed percentage of voiceband data calls. Since silence compression is not possible, voiceband data calls have a higher packetization rate than voice calls. If the percentage of voiceband data calls exceeds the engineered level, the aggregate number of packets will exceed the transmission link's capacity, resulting in extended performance degradation.

The Selective Dynamic Overload Control (SDOC) prevents extended AI congestion by restricting new call traffic; congestion clears as calls terminate. The focus of this paper is the time required for SDOC to control AI congestion. I assume that an AI, when operating at capacity, can carry n_0 voice calls. However,

some of the calls may have an increased activity rate or be voiceband data calls, which have increased packetization rates. In this case AI load will exceed capacity. When AI load exceeds capacity, either due to increased packetization rates for individual calls or to an increased call load, all calls are carried at degraded, potentially unacceptable quality levels. I assume that at time 0 an AI carrying n_0 voice calls experiences an instantaneous $100M_0\%$ overload. For engineered load $M_0 = 0$ and for a 50% overload $M_0 = 0.5$. Such an overload might result from multiple voice conversations simultaneously switching to voiceband data, a sudden widespread environmental noise, or from a sudden surge in new call traffic. Within seconds SDOC will restrict new call arrivals.

The AI equivalent call load is the AI packet load converted into average calls. In this paper I calculate the transient AI equivalent call load under various assumptions. In Section 2, I assume new call arrivals follow a Poisson process with a general call holding time distribution. I show that the AI call load is the convolution of a Poisson and a binomial random variable with time varying parameters. In Section 3, I assume call arrivals follow an Interrupted Poisson process [9] with exponential call holding times. I use the Runge-Kutte method to numerically solve the transient state differential equations. In Section 4, I summarize and discuss future work.

2. General Call Holding Times with Poisson Call Arrivals

Throughout this section assume that call arrivals follow a Poisson process with intensity λ and that call holding times are independently and identically distributed with distribution function H. Assume at time 0 that there are n_0 independent calls in progress whose packetization rate suddenly increases from the average rate by $100M_0\%$. (I could instead assume that the n_0 calls in progress suddenly increased to $(1 + M_0)n_0$ calls. Under these assumptions I obtained similar SDOC performance results.) Assume new calls continue to arrive at rate λ and generate packets at the original rate. Let $L(t)$ be the number of calls active at time t, so $L(0) = n_0$. The $L(t)$ active calls consist of $N(t)$ new calls which have arrived since time 0 and are still active at time t and $O(t)$ original calls which are still active at time t. The AI equivalent call load at time t is $M(t) = (1 + M_0)O(t) + N(t)$. Assume SDOC triggers instantaneously and blocks $100\alpha\%$ of the arriving calls. The intensity of the thinned call arrival process is $(1-\alpha)\lambda$.

I assume that the AI accepts all offered calls not blocked by SDOC, which implies that $O(t)$ and $N(t)$ are independent. During call overloads this assumption will not be completely true since calls also will be blocked when all provisioned trunks into the AI are occupied. However, this will result in an even lower new call arrival intensity, which will improve SDOC performance. Under the independence assumption, $N(t)$ follows a Poisson distribution with time dependent mean

$$EN(t) = n(t) = (1-\alpha)\lambda t \rho(t),$$

where

$$\rho(t) = 1 - H(t) + \int_0^t \frac{x}{t} dH(x).$$

Then $\rho(t)$ is the expected fraction of calls which arrived in (0,t) and are still active at time t.

The residual lifetimes of the original n_0 calls have identical and independent distributions with density

$$f(x) = (1-H(x))/\int_0^\infty x dH(x).$$

Each of the original n_0 calls has probability $o(t)$ of terminating after time t, where

$$o(t) = \int_t^\infty f(x) dx.$$

Then $O(t)$ has a binomial distribution with sample size n_0 and time varying probability $o(t)$. Therefore $L(t)$ is distributed as the convolution of a binomial number of original calls and a Poisson number of new calls. The AI equivalent call load mean and variance are

$$E M(t) = n_0(1+M_0)o(t) + n(t),$$

$$\text{(1)}$$

$$Var(M(t)) = n_0(1+M_0)^2 o(t)(1-o(t)) + n(t).$$

2.1 Erlangian Call Holding Times

Suppose the holding time distribution $H(t) = H(t \mid u,k)$ follows an erlang distribution with parameters $u > 0$ and $k \geq 1$, and density

$$dH(x) = (ux)^{k-1} e^{-xu} u dx / \Gamma(k) \quad \text{for } x > 0,$$

where $\Gamma(k) = (k - 1)!$ In this case

$$n(t) = (1-\alpha)\lambda t \{1 - P(tu,k) + kP(tu,k+1)/ut\},$$

$$o(t) = \{\int_t^\infty \sum_{j=1}^k \frac{(ux)^{j-1}}{\Gamma(j)} e^{-ux} dx\} u / k$$

$$= \sum_{j=1}^k \{1 - P(tu,j)\}/k,$$

where

$$P(a,m) = \int_0^a t^{m-1} e^{-t} dt / \Gamma(m)$$

is the incomplete gamma function.

Figure 2 shows $EM(t)$ curves calculated for 50% overload with Erlangian call holding times for parameters $1/u = 5$, $k = 1$ (curve 1) and $1/u = 0.05$, $k = 100$ (curve 2) with $n_o = 96$. Both sets of parameters fix the call arrival rate $\lambda = 19.2$ calls per minute and the expected call holding time at 5 minutes. The first curve with $k = 1$ is the exponential distribution with squared coefficient of variation c^2 (the variance of the arrival distribution divided by the square of its mean) of 1 and the second set with $k = 100$ has c^2 of 1/10. Figure 2 also shows the equivalent call load two standard deviation confidence bounds calculated from (1). The interval where the upper and lower two standard deviation curves cross the line $n_0 = 96$ provides an approximate 95% confidence range on when the call load will return to capacity. The curve for $k = 1$ crosses the line $n_0 = 96$ at approximately 2.5 minutes and has an approximate 95% confidence range of two minutes. The curves in Figure 2 show that decreasing c^2 increases the effectiveness of SDOC, but the improvement is small.

2.2 Hyper-Exponential Call Holding Times

Assume the call holding time distribution $H(t) = H(t \mid p,a,b,)$ is hyper-exponential with density

$$h(t) = pae^{-at} + qbe^{-bt} \quad \text{for } t > 0,$$

where $0 < p \leq 1$ and $q = 1-p$. For this distribution the square coefficient of variation,

$$c^2 = \frac{Var(T)}{E[T]^2} = \frac{2(p/a^2 + q/b^2)}{(p/a + q/b)^2} - 1 \geq 1,$$

with equality when $p = 1.0$. Then

$$o(t) = p'e^{-at} + q'e^{-bt},$$

where

$$p' = \frac{p}{p + qa/b} = 1 - q',$$

and

$$n(t) = (1-\alpha)\lambda\left\{\frac{p}{a}(1-e^{-at}) + \frac{q}{b}(1-e^{-bt})\right\}.$$

Figure 3 shows $EM(t)$ for a 50% overload with $\alpha = 0.75$, for $p = 1.0, a = 0.2$ (curve 1) and $p = 0.8, a = 0.4, b = 1/15$ (curve 2). For both curves the mean call holding time is 5 minutes, $n_0 = 96$ and $\lambda = 19.2$ calls per minute. Curve 1 with $p = 1$ is the exponential distribution considered previously with $c^2 = 1.0$. Curve 2 has $c^2 = 3.0$. Figure 3 shows that increasing the holding time coefficient of variation from 1.0 to 3.0 increases the expected time for SDOC to control congestion by approximately 1 minute. The upper and lower two standard deviation curves show that an approximate 95% confidence interval has length two minutes.

3. Interrupted Poisson Process Call Arrival Process with Exponential Call Holding Times

A possible application of Wideband Packet Technology is trunk relief. In this case an AI may receive peaked overflow traffic. Throughout this section I assume that call arrivals follow an Interrupted Poisson process (IPP) [9] and call holding times are exponential with mean holding time $1/u$. The IPP is characterized by alternating exponential length periods when calls arrive according to a Poisson process with intensity λ and waiting periods when no calls arrive. Assume the call arrival periods have mean $1/\gamma$ and the call waiting periods have mean $1/\omega$. As before, I analyze this process by assuming that the original calls, $O(t)$, and new calls, $N(t)$, are independent and consider them separately. The memoryless property of the exponential distribution implies that $O(t)$ has a binomial distribution with sample size n_0 and probability e^{-tu}. I find the distribution of $N(t)$ by numerically solving the flow transition equations.

The flow equations for the IPP are shown below. The maximum number of simultaneous calls, k, is chosen large enough so that the probability of k or more calls is negligible. The state space is two dimensional: (i,j) for $i=0,...,k$ and $j=0,1$. The first dimension, i, indicates the current number of calls in progress and the second dimension, j, indicates the call arrival status. During call waiting periods, $j=0$, and during call arrival periods, $j=1$. Assume SDOC blocks $100\alpha\%$ of the arriving calls. The thinned process is also on IPP with call arrival intensity $(1-\alpha)\lambda$. Let $\rho_0 = (\rho_{0,0}(t),\ldots,\rho_{k,0}(t))$ and $\rho_1 = (\rho_{0,1}(t),\ldots,\rho_{k,1}(t))$ be the state probability vectors at time t. Then:

$$\frac{d}{dt}\rho_{j,0}(t) = \gamma\rho_{j,1}(t) + (j+1)u\rho_{j+1,0}(t)$$

$$- (ju + \omega)\rho_{j,0}(t) \quad \text{for } 0\leq j < k,$$

$$\frac{d}{dt}\rho_{k,0}(t) = \gamma\rho_{k,1}(t) - (ku + \omega)\rho_{k,0}(t),$$

$$\frac{d}{dt}\rho_{0,1}(t) = \omega\rho_{0,0}(t) + u\rho_{1,1}(t) - (\gamma + (1-\alpha)\lambda)\rho_{0,1}(t),$$

$$\frac{d}{dt}\rho_{j,1}(t) = \omega\rho_{j,0}(t)$$

$$+ (j+1)u\rho_{j+1,1}(t) + (1-\alpha)\lambda\rho_{j-1,1}(t)$$

$$- (ju + \gamma + (1-\alpha)\lambda)\rho_{j,1}(t) \quad \text{for } 0<j<k,$$

$$\frac{d}{dt}\rho_{k,1}(t) = \omega\rho_{k,0}(t) + (1-\alpha)\lambda\rho_{k-1,1}(t)$$

$$- (ku + \gamma)\rho_{k,1}(t).$$

The appropriate boundary conditions are the probability that the process is in an on or off state initially:

$$\rho_{00}(0) = \frac{\gamma}{\gamma+\omega} \quad \text{and} \quad \rho_{01}(0) = \frac{\omega}{\gamma+\omega}.$$

Then

$$n(t) = \sum_{j=0}^{k} j(\rho_{j,0}(t) + \rho_{j,1}(t))$$

and the exponential holding time assumption implies $o(t) = e^{-tu}$. So

$$EM(t) = (1+M_o)n_o e^{-tu} + n(t) \tag{2}$$

and

$$Var(M(t)) = n_0(1+M_0)^2 e^{-tu}(1-e^{-tu})$$

$$+ \sum_{j=0}^{k} j^2(\rho_{j,0}(t)+\rho_{j,1}(t)) - n(t)^2. \tag{3}$$

I numerically evaluated (2) and (3) using the Runge-Kutta method for solving differential equations. I took $k = 100$, which was sufficiently large to minimize any approximation error for the parameter ranges which I considered.

The calculations for $n(t)$ were simplified by observing that the peakedness of the thinned arrival process is less than the original process. Suppose a call arrival process $X(t)$ has mean M, variance V, and peakedness $z = V/M$. The SDOC-thinned process then has mean $(1-\alpha)M$, variance $\alpha(1-\alpha)M + (1-\alpha)^2 V$, and peakedness $\alpha + (1-\alpha)z \leq z$ when $z \geq 1$. In extreme cases, overflow traffic might have peakedness between 5 and 10. If SDOC blocks 75% of the arrivals, then the peakedness of the thinned AI call arrival process is between 2 and 3.25. Since the peakedness of the thinned process is sharply reduced, numerical calculations for the thinned process are more stable, and the original peakedness will have less effect on the results.

Figure 4 shows $EM(t)$ for a 50% overload with $\alpha = 0.75$, $n_0 = 96$, $1/u = 5$ minute average call holding time, and IPP call arrivals with peakedness $z = 1.0$, 5.0, and 10. The $EM(t)$ is identical for each curve, but the standard deviation increases with z. The intuitive explanation is that for increased peakedness, the arrival rate is the same except the arrivals are more bursty, and so the distribution of $M(t)$ is more diffuse. However, increasing the initial peakedness from 1 to 10 increases the standard deviation of $M(t)$ by just a few seconds, which is not of practical importance. The curves in Figure 4 show that increasing the call arrival peakedness has a minimal effect on the performance of the SDOC control.

4. Summary and Conclusions

This paper describes a modeling study investigating the effectiveness of the SDOC control, proposed for AI congestion. I assume that an AI experiences a sudden overload and calculate the time for SDOC to relieve the congestion. I consider both Poisson call arrivals with various call holding time distributions and IPP call arrivals with peakedness and exponential call holding times.

For the examples I studied, if SDOC blocks 75% of the arriving calls then congestion will clear in approximately 3 minutes for an overload of 50 percent. These results are based on parametric assumptions and are probably conservative for network implementations. In actual situations SDOC may be even more effective.

The results in this paper are, however, theoretical and further study is required before they can be implemented in a wideband network. In this paper I assumed that SDOC triggered instantaneously, but in practice SDOC would trigger based on network load statistics collected over some short interval. I also assumed that when triggered, SDOC blocked a fixed fraction on the newly arriving calls. As an extension, the blocking fraction could depend on the overload level. Hysteresis could be added to the control triggering algorithm algorithm to prevent overload oscillations.

ACKNOWLEDGEMENTS

I would like to thank K. Sriram from his careful review of a previous version of this paper. His comments improved the final version.

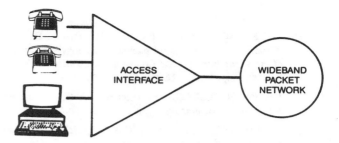

FIGURE 1. ACCESS INTERFACE IN A WIDEBAND PACKET NETWORK.

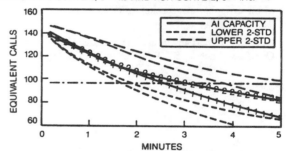

"1" = E (1/5,1)   MINUTES   "2" = E (20,100)

FIGURE 2. THIS FIGURE SHOWS EM(t) FOR A 50% OVERLOAD WITH POISSON CALL ARRIVALS, ERLANGIAN CALL HOLDING TIMES, AND SDOC BLOCKING OF 75%. TWO STANDARD DEVIATION CONFIDENCE BANDS ARE ALSO INCLUDED. FOR CURVE 1, $c^2 = 1.0$ AND FOR CURVE 2, $c^2 = 1/10$.

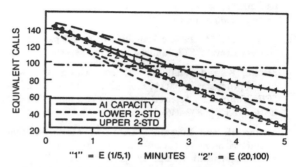

FIGURE 3. THIS FIGURE SHOWS EM(t) FOR A 50% OVERLOAD WITH POISSON CALL ARRIVALS, HYPER-EXPONENTIAL CALL HOLDING TIMES AND SDOC BLOCKING OF 75%. FOR CURVE 1, $c^2 = 1.0$ AND FOR CURVE 2, $c^2 = 3.0$.

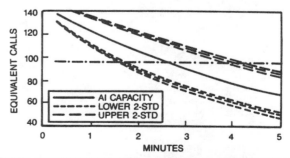

FIGURE 4. THIS FIGURE SHOWS EM(t) FOR IPP CALL ARRIVALS, WITH PEAKEDNESS z=1,5,10, EXPONENTIAL CALL HOLDING TIMES, A 50% OVERLOAD, AND SDOC BLOCKING OF 75%. THE EM(t) IS IDENTICAL FOR EACH z VALUE BUT THE UPPER AND LOWER TWO STANDARD CONFIDENCE BANDS INCREASE WITH z.

REFERENCES

[1] Muise, R.W., Schonfeld, T.J., and Zimmerman III, G.H., "Experiments in wideband packet technology," *Proceedings of the Third International Conference on Switching*, Zurich, Switzerland, March 1988, pp. 135-159.

[2] Feldman, R., "Bell Labs tests voice in packets," *Management Informations Systems Week*, July 28, 1986, pp. 28.

[3] Bowker, D.O. and Dvorak, C.A., "Speech transmission quality of wideband packet technology," *GLOBECOM 1987*.

[4] Bowker, D.O. and Armitage, C.B., "Performance issues for packetized voice communications," *National Communication Forum 1987*.

[5] Brady, P.T., "A statistical analysis of on-off patterns in 16 conversations," *BSTJ*, January 1968. pp. 73-91.

[6] Sriram, K. and Whitt, W., "Characterizing superposition arrival processes in packet multiplexers for voice and data," *IEEE Journal on Selected Areas in Commun.*, Vol. SAC-4, no. 6, 1986, pp. 833-846.

[7] Heffes, H. and Lucantoni, D.M., "A Markov-modulated characterization of packetized voice and data traffic and related statistical multiplexer performance," *IEEE Journal on Selected Areas in Commun.*, Vol. SAC-4, no. 6, 1986, pp. 856-868.

[8] Sriram, K. and Lucantoni, D.M., "Traffic smoothing effects of bit dropping in a packet voice multiplexer," *INFOCOM 88*.

[9] Kuczura, A., "The Interrupted Poisson Process as an Overflow Process," B.S.T.J., 52, No. 3, 1973, pp. 437-448.

A CONGESTION CONTROL FRAMEWORK FOR HIGH-SPEED INTEGRATED PACKETIZED TRANSPORT

G.M. Woodruff, R.G.H. Rogers, and P.S. Richards

BNR
P.O. Box 3511, Station C, Ottawa, Ontario, Canada K1Y 4H7

ABSTRACT

The advent of high-speed packet technology makes it possible to integrate the transport of services associated with a wide spectrum of traffic characteristics. This transport flexibility should be achieved with general congestion controls, which must provide adequate transport performance for all services carried in the network. This paper briefly discusses general congestion control design approaches in the context of integrated, packetized transport for a public, Broadband ISDN environment. A framework is developed for the congestion control design, applicable to Asynchronous Transfer Mode (ATM), and related performance issues and standardization requirements are identified.

1. INTRODUCTION

In today's telecommunications environment separate circuit and packet networks are operated to optimally provide low delay performance for circuit-based services and low error performance for packet-based services. The advent of cost-effective high speed packet switching and high reliability fiber-optic transmission will make it possible to integrate the transport of these traditionally segregated service types, and to support services with a spectrum of traffic characteristics ranging up to broadband bit-rates.

To this end, the Asynchronous Transfer Mode (ATM) has been proposed by the CCITT and T1 committees as the preferred transport technique for Broadband ISDN (B-ISDN). Companion papers in this conference provide an ATM architectural and applications perspective [1], and a view of protocol requirements for ATM access [2].

The major advantages of integrated broadband transport for the network access or core are one or more of the following [1]: *flexibility* to support a spectrum of service types and mixes "on demand"; *efficiency* gained by statistically multiplexing bursty traffic generated from several sources and service types; and *simplicity* by avoiding the need for multiple operations, administration, and maintenance systems.

However, the above advantages are gained at the price of additional control complexity and a performance compromise at the transport level. Congestion controls will be required to fairly allocate the shared bandwidth in a manner which satisfies the performance requirements of each service. There is a danger here of overcomplicating and optimizing the design of these transport-level congestion controls for specific service types and mixes, contrary to the underlying objectives motivating transport integration. If complex controls are required, it should be asked whether the transport integration of certain services is really desirable.

The objective of this paper is to present a framework within which this congestion control design problem can be addressed. The underlying philosophy is to "aim high" at a congestion control architecture which is simple, effective, fair, and optimized for no specific service type. The framework is applicable to connection-oriented, access-controlled high-speed packetized transport such as ATM. The protocol approach presented in [2] fits within this framework.

2. CONGESTION CONTROL APPROACH

In traditional low bit-rate packet networks, which operate over noisy transmission links, it is necessary to provide error recovery on a link-by-link basis. Protocol schemes such as windowing were designed to facilitate retransmission while retaining a high throughput. These protocols served the dual function of congestion control as they enabled a congested node to relay a throttle command to the sending nodes with the assurance that there would be sufficient buffer space to accommodate packets still in transit.

Broadband switching rates can be achieved by operating with synchronous fixed-length packets and providing "best effort" (i.e. no error recovery) packet delivery. This simplified functionality, as provided by ATM, is acceptable with low bit-error rate fiber transmission. However, congestion controls are still necessary to ensure that packet delays and loss rates, caused by buffer overflow at intermediate switching nodes, remain low. The various control approaches can be divided into two categories: reactive and preventive.

2.1 Reactive Control

Reactive control requires immediate feedback to the source nodes at the onset of congestion, instructing them to throttle their traffic flow. There are several problems with this method in high-speed networks:

- High capacity links suffer a sudden degradation in performance with increasing load, and it is therefore difficult to detect the onset of congestion with sufficient lead time to react effectively. Even if this were possible, the rapid buffer level fluctuations expected in a bursty traffic environment could unnecessarily trigger controls and lead to undesirable oscillations in the traffic flows.

- The propagation delay of approximately 5 μs/km required to send throttle feedback to the source, although insignificant for localized applications, becomes significant, in terms of number of messages still in transit (i.e. buffering requirements), for high-speed networks such as B-ISDN which may operate nation-wide.

EH0301-2/91/0000/0393/$01.00 © 1988 IEEE

- The approach of throttling all traffic sources routed through the congested point cannot be considered "fair" when a range of connection rates and performance requirements co-exist.

This type of control is therefore more suitable as a last resort for emergency situations in order to protect network integrity.

2.2 Preventive Control

The objective of preventive controls is to ensure a priori that the network traffic intensity will normally never reach the level to cause unacceptable congestion. Here, such congestion should be considered a rare event, on par with major failures in the network. This "survivability" approach is appropriate for integrated transport, which could in principle be carrying its own control traffic as well as multiple types of user traffic.

For high capacity network links, which display a sudden, rather than gradual, performance degradation with increasing load, very good delay performance can be achieved over a broad range of load. Hence, two approaches may be taken to operate in this region: either the network can be liberally over-engineered, dispelling the need for controls altogether, or the network can be operated more efficiently by controlling the traffic flow admitted into the network at the access nodes.

This latter approach, termed *access control*, is therefore best for attaining fair congestion control in an efficient manner. The framework developed in the remainder of this paper will be based on access controls.

2.3 Access Control Functions

The access controls have the function of interfacing individual users to the network. The required functions are split into three types, as depicted in Figure 1. Within the multimedia *services network*, the access controls process call requests for network resources to support a particular service need, defined by a set of service descriptors as described in [2]. An intermediate *adaptation Interface* is required at the access to convert the user traffic into the appropriate packet format (e.g. ATM fixed-length packets with short headers, called cells). Within the *transport network*, the access controls buffer and interleave packets from multiple users on outgoing links.

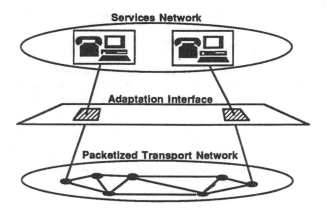

Figure 1: Access control functional framework

The effectiveness of access control depends upon whether the transport technique is connection-oriented or connectionless. ATM is connection-oriented; packets belonging to the same connection,

established at call set-up time, contain the same virtual connection identifier (VCI) in their header, and are routed at intermediate network nodes by translating the VCI into the next physical destination. Thus, packets belonging to the same connection take the same route through the network.

In contrast, connectionless packetized transport techniques route packets according to the destination address included in the header of each packet. It is possible for packets belonging to the same connection to take different routes through the network; whether they may, is dependent on network topology and the dynamic nature of routing decisions made at intermediate switching nodes.

Access control is most effective in conjunction with connection-oriented transport, since a decision to admit a connection's traffic flow can be made based on knowledge of the state of the route which the traffic would follow. Also, since connections can support a range of service types with a range of bandwidth requirements, fair bandwidth allocation -- to satisfy each service's needs -- can only be achieved by differentiating and flow-controlling the traffic of each individual connection. For these reasons, the framework developed in this paper will be based on *connection-oriented* transport.

With connection-oriented transport, the congestion controls can be divided into three levels, as presented in [3]. The first level is termed *route control,* which assigns routes through the network for each access node pair. These routes could be dynamically assigned for each connection request according to the state of the network, or could more simply be statically assigned for a longer duration, depending on the traffic environment and the degree of efficient resource utilization required. The second level is termed *admission control,* where a decision is made to accept or reject a connection request based on a description of the connection's anticipated traffic load. The third level is termed *bandwidth enforcement,* where individual connections are policed to ensure that their traffic flow admitted to the network conforms to that specified at call set-up time.

The route control objective is to make efficient use of the network's resources. The admission control and bandwidth enforcement objectives are to ensure that a satisfactory level of transport performance is provided to all users. These latter two congestion controls will be discussed in the following two sections.

3. ADMISSION CONTROL

Admission controls must decide whether to accept a new connection based on knowledge of the current network loading, the new connection's anticipated traffic characteristics, and its delay/loss performance requirements.

3.1 Transport Performance Objectives

In an integrated transport environment, a range of connection performance requirements will exist. Some delay-sensitive services, such as voice, may tolerate higher packet loss rates than others such as video. High-speed data applications may be very insensitive to delay, but require guaranteed packet delivery.

3.1.1 Universal Performance Objectives

A truely integrated transport should treat all packets equally, regardless of their associated service requirements. This philosophy can only be adopted if the integrated transport can meet a universal performance objective defined as:

- The most stringent delay requirement (mean and percentiles) for all services; and
- The most stringent cell loss probability requirement for all delay-sensitive services.

Delay-tolerant services can employ an additional layer of end-to-end error recovery which may be provided by the user, or by the transport provider as a service.

The simplest way to meet these performance requirements is to operate the links in a *non-statistical* mode, where the sum of connection peak rates on each link does not exceed the link rate, and hence there is minimal buffering. This mode of operation is appropriate for smooth, constant rate traffic sources, and also for bursty sources when bandwidth efficiency is not a concern.

If sources are predominantly bursty and bandwidth efficiency is a concern, a *statistical* mode of operation is appropriate. When statistically multiplexing cells at broadband (≥ 150 Mb/s) link rates, it will be possible to satisfy all delay requirements up to a certain maximum link utilization limit. This limit will be dependent on the traffic characteristics (e.g. peak rate, burstiness, burst lengths) of connections multiplexed on the link; each connection's peak-to-link rate ratio and burst lengths should be limited to a certain range to achieve a significant bandwidth efficiency gain, as illustrated by the simulation results in Figure 2 and discussed further in [1]. Traffic not suitable for statistical operation must be carried in the non-statistical mode.

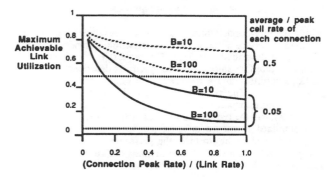

Figure 2: Maximum achievable link utilization to meet an average ATM multiplexer delay of 5 cells (connections are homogeneous with each input traffic flow modeled as a two-state Markov chain [6] [7] -- i.e. with geometrically distributed burst and idle lengths -- with an average burst length of B cells)

In the statistical mode, cell loss depends on multiplexer and switch buffer sizes - an engineerable quantity. Buffer savings can be realized by setting more stringent limits on the allowable link utilization and traffic characteristics than those determined by the delay performance requirement alone. The buffer / bandwidth efficiency tradeoff will be an important consideration when determining whether a statistical mode of operation is worthwhile.

3.1.2 Multiple Performance Classes

Greater bandwidth efficiency can also be attained through the provision of multiple performance classes. However, the gains should be carefully weighed against the additional control complexity implied - only the simplest approaches such as the following are likely candidates:

- Segregation: With multiple guaranteed performance classes, it is simplest to segregate the total bandwidth available for each class and control traffic admitted to each independently. The number of guaranteed performance classes should be kept to a minimum to retain efficiency benefits and to avoid tailoring the network to specific services.

- Priority: The simplest approach is to offer a guaranteed (high priority) and non-guaranteed (low priority) performance class. The high priority performance requirement would be the universal requirement in section 3.1.1. Note that a higher priority may be required in order to protect very loss-sensitive messages (such as network control messages) in the event of abnormal overload conditions. This high priority can be ignored by congestion controls as long as the proportion of this traffic is small.

3.2 Decision Parameters

A decision to admit a new connection is based on whether the required transport performance can be maintained. In a non-statistical mode of operation, this decision is straightforward - the sum of peak rates on a given link should not exceed the link rate. However, in a statistical mode of operation, a new virtual bandwidth measure must be defined for each connection - something between the connection's average and peak rate.

The virtual bandwidth, referred to as V (which may be a vector composed of values for each link in the route), must be derived based on a set of *traffic descriptors* for a connection's traffic flow characteristics, such as average bandwidth, peak bandwidth, burstiness, etc. This general concept of a traffic flow description is also presented in [4], and the derivation of V is addressed in [7] for the case of homogeneous connections with geometric burst and idle length distributions. The determination of an appropriate minimum set of traffic descriptors, and an effective variable V are challenging problems which must be solved if this approach is to prove viable.

Link resources can then be reserved for new connections by increasing the V-values on the route's links as appropriate, provided their allowable maximum is not exceeded. Each node could locally store the current V-values of its adjacent links; resources for a new connection could then be checked and reserved by sending "scout" packets to communicate with each node along the potential route.

Since this variable V would be used network-wide, it is important that its derivation be standardized. A standard set of traffic descriptors is therefore required to ensure the admission controls act together effectively. It will be necessary to translate service descriptors, used for call set-up [2], into these accurate traffic descriptors. This could be accomplished by defining several service classes, each of which may correspond to a range of values which the traffic descriptors may take.

4. BANDWIDTH ENFORCEMENT

Once a connection has been admitted to the network, the traffic flow of the connection must be *monitored* and *enforced* to ensure that it conforms to its allocated virtual bandwidth V, or equivalently, to its traffic descriptor values. Here, it is assumed that the flow is enforced by throwing away violating packets. It is therefore important to choose general traffic descriptors which can allow the traffic flow to be accurately monitored and enforced for all anticipated traffic types.

4.1 Bandwidth Monitoring

Traffic generated by data applications poses a problem because it is inherently unpredicable, and has traditionally been described by its statistical moments such as *average* packet length, *average* number of packets over a time interval, etc. It is possible to monitor statistical moments, given a sufficiently long time interval, but it is very difficult to enforce them since it must be decided whether a momentary burst of traffic is due to a violation, or is due to natural statistical variation.

It is therefore important to base traffic descriptors on deterministic, bounded quantities, which when violated, can immediately trigger enforcement. For example, descriptors such as maximum burst lengths, minimum inter-burst times, etc. could be used. The "leaky bucket" threshold (the maximum allowable value of a traffic flow counter which is incremented at each connection's cell arrival and decremented at the average rate) has also been suggested as a possible traffic descriptor [5][7]. An important consideration which may influence the traffic descriptor choice is the ease and accuracy with which the descriptors can be translated into V-values.

Regardless of the specific descriptors used, if the discrepancy between natural and required traffic flow is significant, there will be a need to sufficiently *shape* a statistically varying traffic flow to *match* its descriptors. Other considerations which may influence the descriptor choice are therefore the ease of implementation of a shaping function, and the gain in efficiency possible with muliplexed shaped traffic flows.

4.2 Traffic Shaping

The traffic shaping function could either be performed by the access control, or at the data source, as shown in Figure 3. If at the access control, the shaper would function as a bandwidth enforcer by *buffering*, rather than rejecting, violating packets. A sufficiently large buffer must be allocated based on estimates (from general service descriptors) of the traffic descriptors, and/or throttle feedback must be provided to the data source warning of pending buffer overflow. This mismatch between the source's and the assumed traffic characteristics is undesirable; the translation of service descriptors into traffic descriptors may result either in an overly liberal resource allocation, or in unsatisfactory performance as perceived by the user.

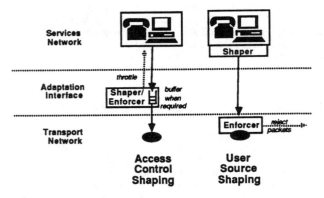

Figure 3: Access control vs. user source traffic shaping

With shaping at the data source, the shaping function may be linked directly with the source's data access arbitration method. Shaping may also be a by-product of the higher-layer transport protocol used within the data connection. For example, high throughput bulk data transfer protocols may result in a traffic flow of fixed data block sizes, which may be easily related to the standard traffic descriptors. In either case, the standard traffic descriptors could be included within the call set-up protocol, with only the bandwidth enforcement function required at the access for network protection. By generating the desired traffic characteristics at source, there is no room for doubt as to the nature of the contracted virtual bandwidth for the connection; a user who is transmitting very sporadic bursty data is given a direct means to trade-off perceived packet delay with the cost of the virtual bandwidth requested.

Note that during periods of low traffic levels, it may be desirable to boost revenue by disabling the shaping and enforcement of certain traffic descriptors. As traffic levels approach network capacity, full shaping and enforcement can be enabled to ensure that network performance objectives will be met.

5. STANDARDIZATION REQUIREMENTS

5.1 Reference Model

In summary, the congestion control framework has been developed within the context of three functional areas shown in Figure 1: the *services network* handling service requests from users, the integrated *transport network* providing the physical connectivity, and an intermediate *adaptation interface.*

Within the services network, a user requests resources for a service, specified by service descriptors contained within a standard signaling protocol as described in [2]. The request is translated by the adaptation interface into a connection request described by standard traffic descriptors and a required level of packet loss and delay performance. The request is in turn translated by the access controls at the transport network into a required performance class (if applicable), and the required amount of virtual bandwidth, described in a standard form, on specific network links. If sufficient bandwidth is available to meet the performance objectives of the transport network, the admission controls reserve the virtual bandwidth on the links, and accept the connection request.

Once a connection is set up, the adaptation interface converts the originating user's traffic stream into the transport network's packet format, shapes the traffic flow to conform to the stated traffic descriptors, and offers the traffic to the transport network. The adaptation interface also accepts traffic from the transport network, reformats, and delivers it to the destination such that it meets the service requirements. For data delivery, the adaptation interface can optionally employ end-to-end error recovery, or this function can be provided by the user.

5.2 Parameters for Standardization

The following are some of the parameters which will need to be standardized if the above framework can be applied in a public network environment:

- Access protocol with standard service descriptors [2].
- Measurable and controllable traffic flow descriptors.
- Relation between service descriptors and traffic flow descriptors.
- Transport performance objectives for each performance class.
- Virtual bandwidth measure V for a connection, and its derivation from the traffic descriptors.

With standardized traffic descriptors in place, it will be very desirable to move the packet generation and shaping function to the user, circumventing the need for the adaptation interface. At this stage, the standard traffic descriptors can be included explicitly within the standard access protocol, and the transport network need only enforce the traffic flows.

5.3 Traffic Performance Issues

This paper has raised several fundamental traffic performance issues, the most fundamental being whether the efficiency gains with a statistical mode of operation are worthwhile. If the answer is yes, then the following must be addressed before standards could be finalized:

- Can a no-priority, "universal performance" transport be operated at acceptable efficiency levels? If not, which multiple performance class scheme would suffice?
- Can a virtual bandwidth variable V be defined as an effective means to match requested and available link resources?
- How can the network performance be modeled for a heterogeneous mixture of shaped traffic flows?
- Which traffic descriptors will achieve maximum efficiency benefits with traffic shaping?
- What minimum set of traffic descriptors are necessary to characterize traffic flow for all anticipated services? Which traffic descriptor values should be assigned to known services?

6. SUMMARY

Congestion controls will be a critical component of an integrated packetized transport network. They must fairly allocate the shared bandwidth in a manner which satisfies the performance requirements of each service. This paper has developed a framework for the design of congestion controls applicable to connection-oriented, access-controlled transport such as ATM. The proposed congestion control approach is simple, fair, non service-optimized, and allows an explicit trade-off to be made between transport performance and bandwidth efficiency.

Through this framework, we can envision future integrated multimedia networks where terminals could efficiently shape their traffic flow characteristics to match those required by transport controls. We have identified some of the standards and traffic performance issues which will need to be addressed to make this view a reality.

In summary, while acknowledging the significant challenges that remain in the area of standardization, performance analysis and implementation methods, we believe that controls for a high-speed, integrated packetized transport can be developed as a viable element in the next generation of telecommunications systems.

ACKNOWLEDGEMENTS

We would like to thank our many colleagues at BNR for their helpful discussions and comments.

REFERENCES

[1] M.R. Wernik, "Architecture and Technology Considerations for Multimedia Broadband Communications", Globecom '88.

[2] M.J. Rider, "Protocols for ATM Access", Globecom '88.

[3] P.S. Richards et. al., "An Assessment of Network Control Configurations for a Packetized Multimedia Communication Network", IEEE COMSOC International Workshop on Future Prospects of Burst/Packetized Multimedia Communications, November 1987, Osaka, Japan.

[4] L. Zhang, "Designing a New Architecture for Packet Switching Communication Networks", IEEE Communications Magazine, Vol. 25, No. 9, September 1987.

[5] J. S. Turner, "New Directions in Communications (or Which Way to The Information Age?)", IEEE Communications Magazine, Vol. 24, No. 10, October 1986.

[6] A.M. Viterbi, "Approximate Analysis of Time-Synchronous Packet Networks, IEEE Journal on Selected Areas in Communications, Vol. SAC-4, No. 6, September 1986.

[7] S. Akhtar, "Congestion Control in a Fast Packet Switching Network", Master's Thesis, Washington University, Saint Louis, Missouri, December 1987.

[8] P.S. Richards et. al., "End-end Control of High-Speed Packetized Communication Networks", Eastern Communications Forum, Rye Brook, N.Y., May 1988.

Chapter 8:

Performance Modeling

8.1 Introduction

In the design and development of systems, a variety of techniques ranging from intuition to experimental evaluation may be used to comprehend the various design issues and trade-offs. While techniques based on intuition are rapid and flexible, confidence in the result is usually low because intuition is based on experience and insight, which are difficult to acquire and verify. Experimental evaluation, on the other hand, produces more accurate results but is generally expensive and inflexible–and provides insight into system behavior under only one set of assumptions.

A third approach is modeling. A model is an abstract representation of a system that omits details not essential for understanding system behavior. Analysis and simulation are the most common techniques for modeling systems. An analytical model attempts to generate equations relating system parameters to the required performance quantities. A number of simplifying assumptions are generally introduced to obtain a mathematically tractable model. Sometimes these consist of representing only a single system resource in substantial detail or using a network-of-queues representation. However, without the existence of a product-form solution, the exact analysis of a queueing network model will usually be impractical.

A simulation model can be used when analysis is not possible or detailed modeling is required to understand the system behavior. The main advantage of simulation is its great generality. The drawbacks of simulation are the costs of generating the simulation program, executing the program, and statistically analyzing program behavior.

An analytical model is generally much cheaper than a simulation model of the same system and achieves a favorable balance between accuracy and efficiency.

In a circuit-switched network, the grade of service based on blocking is used as the performance criterion. In a packet-switched network, on the other hand, delay is used as the performance criterion. Because integrated networks are still in their early evolutionary stages and are experimental, the appropriate choice of performance metrics for integrated networks is somewhat unclear. It is quite conceivable that the accepted metrics will be technology-dependent. In the packet-switched environment, the performance criteria for voice traffic may be packet delay and packet loss.

8.2 Overview of Papers

In the first paper, "Performance Analysis of a Packet Switch Based on Single-Buffered Banyan Network," Jenq describes a model for analyzing the performance of a packet switch based on an interconnection network. Using the model, Jenq computes the normalized throughput and the average delay of the banyan switch.

In "A Simulation Study of Network Delay for Packetized Voice," Gopal and Kadaba present a simulation model for studying the delay distribution and the packet loss for voice traffic across a single hop and in network environments. They investigate schemes having a 64-Kbps voice encoding with speech-activity detection and a 32-Kbps voice encoding without speech-activity detection. As long as the total offered load is below the link capacity, speech-activity detection provides better performance. As the number of active speakers increases beyond a threshold, the system performance is affected. If no speech-

activity detection is used, the mean packet delay and the variation in delay are small.

In the third paper, Bhargava et al. compare the performance of a link-by-link approach versus an end-to-end approach for error control in a high-speed network. They develop analytic and simulation performance models. The authors' research shows that an end-to-end approach is superior for the range of network parameters of practical interest.

The asynchronous transfer mode permits dynamic bandwidth allocation in a broadband ISDN environment. Fixed-size cells that consist of a header and an information field carry all the information in an ATM network. Hsing et al., in "On Cell Size and Header Error Control of Asynchronous Transfer Mode (ATM)," investigate the optimum size of the information field that maximizes resource use. Their research indicates that although an optimum value does exist, approximately the same average channel utilization is obtained for a broad range of values. The authors also investigate cell header error control and conclude that a cyclic redundancy code that can correct a single bit error is adequate.

Performance Analysis of a Packet Switch Based on Single-Buffered Banyan Network

YIH-CHYUN JENQ, MEMBER, IEEE

Reprinted from *IEEE Journal on Selected Areas in Communications*, Vol. SAC-1, No. 6, December 1983, pages 1014-1021. Copyright © 1983 by The Institute of Electrical and Electronics Engineers, Inc. All rights reserved.

Abstract — Banyan networks are being proposed for interconnecting memory and processor modules in multiprocessor systems as well as for packet switching in communication networks. This paper describes an analysis of the performance of a packet switch based on a single-buffered Banyan network. A model of a single-buffered Banyan network provides results on the throughput, delay, and internal blocking. Results of this model are combined with models of the buffer controller (finite and infinite buffers). It is shown that for balanced loads, the switching delay is low for loads below maximum throughput (about 45 percent per input link) and the blocking at the input buffer controller is low for reasonable buffer sizes.

I. Introduction

RECENTLY, there has been considerable interest in Banyan networks among computer and communication technology researchers. The Banyan network has been considered as a candidate for the interconnecting network linking a large number of memory and processor modules together in a multiprocessor system and as a switching fabric for packet-switched interconnection of computers (see, e.g., [1]–[3]).

In this paper, we present analytical results for the performance of a packet switch based on a single-buffered Banyan network. Results show reasonably low blocking probabilities and low delays for balanced internal loads. In Section II we describe the switch and its operation. In Section III we present a simple model of the switch and obtain the throughput and the internal delay of the switch. In Section IV we refine the model by removing one independence assumption used in the simple model of Section III and show that the results from two models are very close. In Section V we add a queue with unlimited waiting rooms as a front-end buffer to the switch and obtain the switching delay. The case of a finite waiting rooms front-end buffer is analyzed in Section VI. Conclusions are given in Section VII.

Manuscript received February 4, 1983; revised June 22, 1983.
The author is with Bell Laboratories, Holmdel, NJ 07733.

II. The Packet Switch

The packet switch consists of k input buffer controllers, k output buffer controllers, and a $k \times k$ square switch. A Banyan network is considered for the switching fabric because of its economy, potential VLSI implementation, and easiness of routing, A 4-stage (16×16) Banyan network is shown in Fig. 1. Each switching element (box) is a 2×2 crossbar switch with one buffer on each of its two input links. The packet routing is done by the hardware as follows. Each packet has an n-bit header in an n-stage switch. The switching element (a 2×2 crossbar) at stage 1 routes the packet up or down according to the first bit of the header ("zero" or "one" indicate up or down routing, respectively) and then removes the first bit from the header. The succeeding switching elements will perform the same routing function by removing one bit from the header and routing the packet to the next stage until the packet reaches its destination output link. It is easy to see that the header is simply the binary address of the output link at the last stage and is independent of the input link at the first stage.

If both buffers at a switching element have a packet and both packets are going to the same output link of the switching element, then a conflict occurs. In this case, we assume that one of the packets will be chosen, randomly, to go to the next stage and the other one will stay in the buffer. Of course, in order for a packet to be able to move forward, either the buffer at the next stage is empty or there is a packet in the buffer and that packet is able to move forward. Thus, the ability for a packet to move forward depends on the state of the entire portion of the network succeeding the current stage.

For simplicity, we can consider the switch operating in a clocked format. In the first portion τ_1 of a clock period $\tau = \tau_1 + \tau_2$, control signals are passed across the network from the last stage toward the first stage, so that every packet knows whether it should move forward one stage or stay in the same buffer. Then, in the second portion τ_2 of a clock period τ, packets move in accordance with control signals and the clock period ends. The whole process repeats in every clock period, and packets continue to be shifted in and out of the switch.

III. Model 1

We consider a Banyan switch with n stages. Considering the clocked operation described in the previous section, it

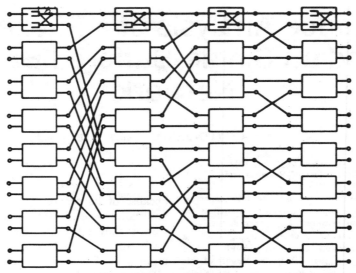

Fig. 1. A 4-stage Banyan network.

is clear that the whole switch can be modeled as a Markov chain. However, the number of the states in the chain grows exponentially with the number of stages n; the number of states in an n-stage network is $k^{n \cdot 2^n}$, where k is the number of states in a node. With such a rapidly growing number, it is almost impossible to even simulate it for large n (say $n > 10$). Therefore, some simplifying assumptions will be used.

We assume that packets arriving at each input link at stage 1 are destined uniformly (or randomly) for all output links at stage n. We also assume a uniform load for each input link at stage 1. Thus, we assume that the loads are balanced in the whole switching network. Under these assumptions, the state of a switching element (2×2 crossbar) at stage k is statistically not distinguishable from that of another switching element of the same stage. Hence, the state of a "stage" can be characterized by that of a "switching element." If we further assume that the two buffers in the same switching element are statistically independent, then the state of a "stage" can be reduced further to that of a single buffer. This last independence assumption is motivated by the fact that input packets arriving at each of the two input links of a switching element are from disjoint sets of input links at stage 1 and, hence, are independent. However, packets in the two buffers of a switching element do interfere with each other, and thus it is not clear that this independence assumption is reasonable. In this section we will assume independence to obtain a very simple model. In the next section we will remove this independence assumption to check if the removal of the independence assumption makes any noticeable difference.

We first introduce the following notation.

n = number of switching stages.

$p_0(k, t)$ = probability that a buffer of a switching element at stage k is empty at the beginning of the tth clock period.

$p_1(k, t) = 1 - p_0(k, t)$.

$q(k, t)$ = probability that a packet is ready to come to a buffer of a switching element at stage k during the tth clock period.

$r(k, t)$ = probability that a packet in a buffer of a switching element at stage k is able to move forward during the tth clock period, given that there is a packet in that buffer.

With these definitions, we can state the equations governing the relationships among these variables:

$$q(k, t) = 0.75 p_1(k-1, t) p_1(k-1, t)$$
$$+ 0.5 p_0(k-1, t) p_1(k-1, t)$$
$$+ 0.5 p_1(k-1, t) p_0(k-1, t),$$
$$k = 2, 3, \cdots, n \quad (1)$$

$$r(k, t) = [p_0(k, t) + 0.75 p_1(k, t)][p_0(k+1, t)$$
$$+ p_1(k+1, t) r(k+1, t)],$$
$$k = 1, 2, 3, \cdots, n-1 \quad (2)$$

$$r(n, t) = p_0(n, t) + 0.75 p_1(n, t)$$

$$p_0(k, t+1) = [1 - q(k, t)][p_0(k, t) + p_1(k, t) r(k, t)]$$
$$(3)$$

$$p_1(k, t+1) = 1 - p_0(k, t+1). \quad (4)$$

Equations (1)–(4) describe the dynamics of the state transition of the system. If this system has a steady state, then these quantities should converge to time-independent quantities $q(k)$, $r(k)$, $p_0(k)$, and $p_1(k)$. Then the normalized throughput S (the number of output packets per output link per clock period) and the normalized mean delay d are given by

$$S = p_1(k) r(k) \quad \text{for any } k \quad (5)$$

and

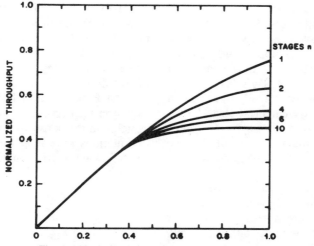

Fig. 2. Normalized throughput of Banyan networks.

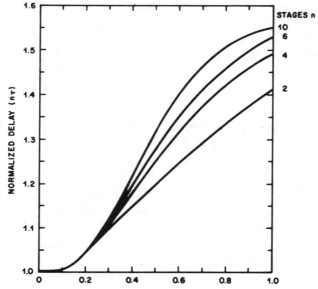

Fig. 3. Normalized internal delay of Banyan networks.

$$d = \frac{1}{n} \sum_{k=1}^{n} 1/r(k). \qquad (6)$$

The quantity $q(1)$ is determined by the load applied to the switch. In this section it is considered as an independent input parameter. The determination of $q(1)$ for a given input load is deferred to Section V.

In Fig. 2 we plot the normalized throughput S as a function of $q(1)$ for different number of stages n. This was obtained by iteratively solving for the steady state (1)–(5). There are two interesting observations. First, the throughput grows almost linearly with respect to the offered load in the range where the offered load is less than 0.4. This agrees with the intuition that under low load, all input traffic will pass through the switch with little interference. Second, with the maximum load at 1, the normalized throughput seems to converge (monotonically decreasingly) to a quantity of approximately 0.45 as the total number of

stages n in a switch increases to 10. Normalized delay d [(6)] is displayed in Fig. 3. It is interesting to see that the normalized delay also seems to converge to a quantity approximately 1.55 as n increases. A computer run is performed for $n = 20$ to check this convergence, and the result shows that the normalized delay stands at 1.55. Thus, for an n stage switch the average delay is in the range between $n\tau$ and $1.55n\tau$, depending on the load and the number of stages n.

IV. MODEL 2

In the previous section we assumed that the state of a buffer in a switching element is independent of that of the other buffer in the same switching element. In this section we present a refined model by removing this independence assumption. Numerical results show little difference from the previous model. Therefore, we conclude that the above independence assumption is a good approximation.

This model uses the following notation.

n = number of switching stages.

$p_{ij}(k, t)$ = probability that a switching element at stage k has i packets in the upper buffer and j packets in the lower buffer at the beginning of the tth clock period ($i = 0, 1; j = 0, 1$).

$q(k, t)$ = probability that a packet is ready to come to a buffer of a switching element at stage k during the tth clock period.

It is noted that "arrivals" coming to each of the two buffers in a switching element are independent of each other.

$r_{01}(k, t)$ = probability that the packet in the lower buffer of a switching element at stage k is able to move forward during the tth clock period, given that the upper buffer is empty and the lower buffer has a packet.

$r_{10}(k, t)$ = probability that the packet in the upper buffer of a switching element at stage k is

Equation (9) results from the fact that the load is assumed to be balanced and the switch is symmetric. Equation (10) results from the fact that both output links of a switching element in a Banyan network are connected either to two upper input links or to two lower input links of two different switching elements of the next stage. (See Fig. 1.) Equation (11) follows from the same reasoning. Equation (12) results from the symmetry property of the switch and the conservation law of probability.

Equations governing the state transitions are given by the following matrix equation:

$$\begin{pmatrix} p_{00}(k, t+1) \\ p_{01}(k, t+1) \\ p_{10}(k, t+1) \\ p_{11}(k, t+1) \end{pmatrix} = Q \begin{pmatrix} p_{00}(k, t) \\ p_{01}(k, t) \\ p_{10}(k, t) \\ p_{11}(k, t) \end{pmatrix} \qquad (13)$$

where

$$Q = \begin{pmatrix} \bar{q}\bar{q} & \bar{q}r_{01}\bar{q} & \bar{q}r_{10}\bar{q} & \bar{q}r_{11}^{11}\bar{q} \\ \bar{q}q & \bar{q}(r_{01}q + \bar{r}_{01}) & \bar{q}r_{10}q & \bar{q}(r_{11}^{01} + r_{11}^{11}q) \\ q\bar{q} & qr_{01}\bar{q} & \bar{q}(r_{10}q + \bar{r}_{10}) & \bar{q}(r_{11}^{10} + r_{11}^{11}q) \\ qq & q(r_{01}q + \bar{r}_{01}) & q(r_{10}q + \bar{r}_{10}) & r_{11}^{00} + r_{11}^{01}q + r_{11}^{10}q + t_{11}^{11}q^2 \end{pmatrix}$$

able to move forward during the tth clock period, given that the lower buffer is empty and the upper buffer has a packet.

$r_{11}^{ij}(k, t)$ = probability that i packets in the upper buffer and j packets in the lower buffer of a switching element at stage k are able to move forward during the tth clock period, given that both buffers have packets ($i = 0, 1, j = 0, 1$).

The equations governing the relationships among these variables are as follows:

$$q(k, t) = 0.5p_{01}(k-1, t) + 0.5p_{10}(k-1, t)$$
$$+ 0.75p_{11}(k-1, t) \qquad k = 2, 3, 4, \dots, n \quad (7)$$

$$r_{01}(k, t) = p_{00}(k+1, t)$$
$$+ 0.5p_{01}(k+1, t)[1 + r_{01}(k+1, t)]$$
$$+ 0.5p_{10}(k+1, t)[1 + r_{10}(k+1, t)]$$
$$+ p_{11}(k+1, t)[r_{11}^{11}(k+1, t) + 0.5r_{11}^{01}(k+1, t)$$
$$+ 0.5r_{11}^{10}(k+1, t)] \qquad k = 1, 2, 3, \dots, n-1 \quad (8)$$

$$r_{01}(n, t) = 1.0$$

$$r_{10}(k, t) = r_{01}(k, t), \qquad k = 1, 2, \dots, n$$

$$r_{11}^{11}(k, t) = 0.25\{[r_{01}(k+1, t)]^2 + [r_{10}(k+1, t)]^2\} \quad (9)$$
$$= 0.5[r_{01}(k+1, t)]^2 \quad (10)$$

$$r_{11}^{00}(k, t) = 0.5[1 - r_{01}(k+1, t)]$$
$$+ 0.5[1 - r_{01}(k+1, t)]^2 \quad (11)$$

$$r_{11}^{01}(k, t) = r_{11}^{10}(k, t)$$
$$= 0.5[1 - r_{11}^{11}(k, t) - r_{11}^{00}(k, t)]. \quad (12)$$

where the arguments (k, t) of the variables in the coefficient matrix are left out for the simplicity of presentation. In (13) we also use the shorthand notations $\bar{q}$, $\bar{r}_{01}$, and $\bar{r}_{10}$ to denote $(1 - q)$, $(1 - r_{01})$, and $(1 - r_{10})$, respectively.

Equations (7)–(13) govern the dynamics of the state transition of the system. If the system has a steady state, then all variables should converge to some corresponding time-independent quantities (after some iteration). The normalized throughput S and the normalized delay d are given by

$$S = 0.5\{ p_{01}(k)r_{01}(k) + p_{10}(k)r_{10}(k)$$
$$+ p_{11}(k)[r_{11}^{10}(k) + r_{11}^{01}(k) + 2r_{11}^{11}(k)]\} \quad (14)$$

and

$$d = \frac{1}{n} \sum_{k=1}^{n} 1/p(k) \quad (15)$$

where

$$p(k) = \{ p_{10}(k)r_{10}(k)$$
$$+ p_{11}(k)[r_{11}^{11}(k) + r_{11}^{10}(k)]\}/[p_{10}(k) + p_{11}(k)]. \quad (16)$$

It is noted that $p(k)$ is the probability that a packet in the upper buffer of a switching element is able to move forward, given that there is a packet in that buffer. Because of the symmetry of the problem, $p(k)$ is also applicable to the lower buffer. Therefore, (15) gives the correct expression for normalized delay d.

We have used (7)–(16) to compute the normalized throughput and average delay for varying loads and switch-

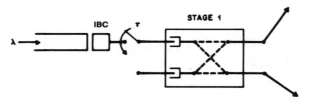

Fig. 4. Packet switch with infinite waiting rooms at its input buffer controllers.

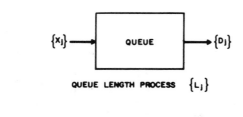

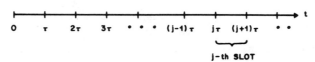

Fig. 5. A discrete time queueing model.

ing network sizes, and have found that the results of this refined model differ very little from those obtained in the previous model (differences are less than 0.1 percent). Thus, we conclude that the independence assumption is reasonable, and model 1 of Section III gives good approximate results with greatly reduce computation effort.

V. SWITCHING DELAY—IBC WITH INFINITE WAITING ROOM

In this section we study a system with input buffer controllers (IBC's) added to the front end of a Banyan switch. The model is shown in Fig. 4. Packets arrive at an input buffer controller at a rate of λ packets per clock period τ. At clock ticks, if the corresponding buffer of the switching element at stage 1 of the switching network is able to accept a packet from the input buffer controller, the IBC will place a packet into that buffer (if there is any packet in the IBC queue). Arriving packets are placed in the IBC queue according to the order of their arrivals, and they are delivered to the switching network in the first-in-first-out (FIFO) fashion. We are interested in quantifying the switching delay, which is the time a packet spends in the IBC queue and in the switch before it leaves the switch. The size of the IBC waiting room is assumed to be infinite. In the next section we will address the problem with a finite IBC waiting room.

Let us consider a discrete-time queueing model for automatic repeat request (ARQ) used in data communications as shown in Fig. 5 and previously analyzed in [4]. The time axis is slotted into uniformly spaced time epochs. The random variable X_i is the number of arriving packets

during the ith slot, D_i is the number of packets removed from the queue during the ith slot, and L_i is the queue length (number of waiting packets) at the beginning of the ith slot. Usually, a slot time is the time required to transmit a packet. If the transmission is successful, D_i is 1, otherwise, D_i is zero. The probability that a transmission is successful is p. This model can be characterized by the following two equations:

$$L_{i+1} = L_i + X_i - D_i \qquad (17)$$

and

$$D_i = \begin{cases} 0, & \text{if } L_i = 0 \\ \begin{cases} 0, & \text{with probability } 1-p, \\ 1, & \text{with probability } p, \end{cases} & \text{if } L_i > 0. \end{cases} \qquad (18)$$

If X_i is independent of i (i.e., stationary), then the steady state exists and we have $E(X) = E(D) = p \cdot \text{prob}\{L \neq 0\}$ from (17) and (18).

If we compare this model to our switching model of Fig. 4, we see that two models are mathematically equivalent and the corresponding variables and parameters are

$$E(X) = \lambda \qquad (19)$$

$$p = p_0(1) + p_1(1)r(1) \qquad (20)$$

and

$$\text{prob}\{L \neq 0\} = q(1) \qquad (21)$$

where $p_0(1)$, $p_1(1)$, $r(1)$, and $q(1)$ are steady-state variables of our Model 1 of Section III.

Now we are ready to describe an algorithmic solution to our switching delay problem.

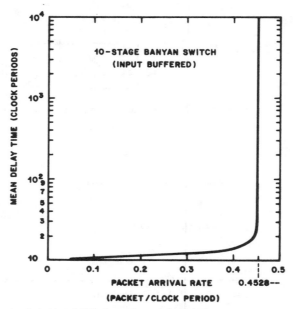

Fig. 6. Switching delay of a packet switch with 10-stage Banyan network.

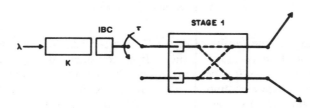

FAST PACKET SWITCH WITH FINITE WAITING ROOM

Fig. 7. Packet switch with finite waiting rooms at its input buffer controllers.

Step 1) Compute p:
a) Guess an initial $q(1)$.
b) Perform the Banyan network computation using Model 1 to obtain $p_0(1)$, $p_1(1)$, and $r(1)$.
c) Let $p = p_0(1) + p_1(1)r(1)$.
d) Let $q(1) = \lambda/p$.
e) Go to b) and iterate.
After the procedure converges, we obtain p.
Step 2) Solve the discrete-time queuing model.

With the p obtained in step 1), we then solve the well-known discrete-time queuing model.
This model has previously been solved [4], and the mean switching delay is given by

$$D_S = \left| \frac{\sigma_x^2 + \lambda(1-\lambda)}{2(p-\lambda)\lambda} - \frac{1}{2} + \sum_{k=1}^{n} 1/r(k) \right| \tau \quad (22)$$

where σ_x^2 is the variance of the arrival X.

A plot of mean switching delay for a 10-stage Banyan switch with front-end IBC queue is shown in Fig. 6. It is seen that this delay is relatively small for loads ($q(1)$) less than 0.4, and increases rapidly to infinity at about 0.45 load. Notice that 0.45 is the maximum achievable throughput for a 10-stage Banyan network as shown in Fig. 2.

VI. BLOCKING PROBABILITY—IBC WITH FINITE WAITING ROOM

In this section we study the case where the IBC waiting room is of finite size K and the quantity of interest is the blocking probability P_b. The blocking probability is the probability that an arriving packet will find the IBC waiting room is full. Blocked packets are dropped and considered to be lost.

The model is shown in Fig. 7. This model is similar to the one shown in Fig. 4, except that the IBC has only K waiting rooms. Packets arrive at the buffer controller at a rate of λ packets per clock period. Let λ' be the "carried load" of the system, that is, the average number of packets accepted by the switch; then the blocking probability P_b is given by

$$P_b = (\lambda - \lambda')/\lambda. \quad (23)$$

Therefore, if for every arrival rate λ we can compute λ', then we can obtain P_b readily by (23).

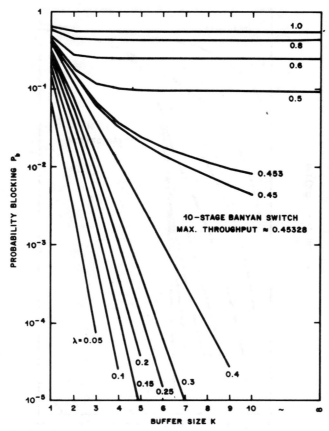

Fig. 8. Blocking probability of a packet switch with a 10-stage Banyan network.

Recall that λ' is simply the normalized throughput of the Banyan switch described in Section III. To obtain λ', we need the quantity $q(1)$. If we consider the Banyan switch as a server for the finite queue, then the service time for a customer (a packet) is $n\tau$ (recall that τ is the clock period) with probability $p(1-p)^{n-1}$ where p is the probability that an input link at the first stage of the Banyan network is able to accept a new packet at clock ticks. This variable p can be expressed in terms of the parameters of the Banyan network:

$$p = p_0(1) + p_1(1)r(1). \tag{24}$$

Because the switch operates in clocked mode, the server will not return until next clock tick if it finds an empty queue at the clock tick. Hence, assuming Poisson arrivals, the system can be modeled as an $M/G/1/K$ queue with vacation [5] (vacation time is the clock period τ). The quantity $q(1)$ is simply the probability that the server will find a nonempty queue at clock ticks. It is given by $B/(V+B)$ where B and V are, respectively, the mean busy and idle periods of a $M/G/1/K$ queue. Now we are ready to describe the algorithm for computing the blocking probability as follows.

Step 1) Guess an initial $q(1)$.
Step 2) Perform Banyan network computation using Model 1 of Section III.

Step 3) $p = p_0(1) + p_1(1)r(1)$.
Step 4) Do $M/G/1/K$ computation.
Step 5) $q(1) = B/(B+V)$.
Step 6) Go to Step 2).

After the procedure converges, the carried load λ' is given by the normalized throughput of the switch, and the blocking probability is given by (23).

In Fig. 8, we plot the blocking probability P_b against the buffer size K with load λ as a parameter for the 10-stage network. It is noted that, for $\lambda < 0.45$, P_b decreases sharply as K increases, while for $\lambda > 0.45$, the P_b curves behave dramatically differently. This is due to the fact that a 10-stage Banyan switch has the maximum achievable normalized throughput 0.45. If the input load λ is greater than 0.5, the blocking probability P_b is about $(\lambda - 0.45)/\lambda$ for $K > 3$. In Fig. 9, we plot the blocking probability P_b against the load λ with the buffer size K as a parameter. Again, we see the breakdown of the smoothness of the curves happens near $\lambda = 0.45$.

VII. Conclusions

In this paper, we analyzed the performance of the packet switch based on the single-buffered Banyan network. First, we derived a model for computing the normalized throughput and the average internal delay of the Banyan

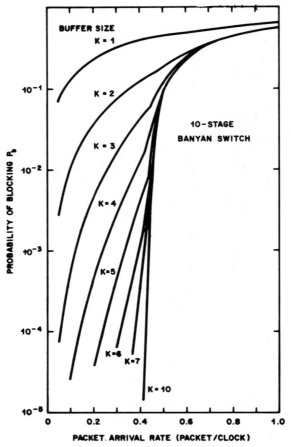

Fig. 9. Blocking probability of a packet switch with a 10-stage Banyan network.

switch. Then we used a discrete-time queuing model to compute the switch delay of the switch with infinite waiting rooms at its input buffer controllers. It was found that the switch delay is not significant until the input load approaches the maximum achievable throughput of the Banyan switch. An $M/G/1/K$ queue with vacation was used to compute the blocking probability of the packet switch with finite waiting rooms at its input buffer controllers.

REFERENCES

[1] "Interconnection Networks" Special Issue of *IEEE Computer*, Dec., 1981.
[2] D. M. Dias and R. Jump, "Analysis and simulation of buffered delta network," *IEEE Trans. Comput.*, vol. C-30, Apr. 1981.
[3] S. Cheemalavagu and M. Malek, "Analysis and simulation of Banyan interconnection networks with 2×2, 4×4, and 8×8 switching elements," in *Proc. 1982 Real-Time Syst. Symp.*, Los Angeles, CA, Dec. 1982.
[4] Y. C. Jenq, "On calculations of transient statistics of a discrete queuing system with independent general arrivals and geometric departures," *IEEE Trans. Commun.*, vol. COM-28, pp. 908–911, June 1980.
[5] T. T. Lee, "$M/G/1/N$ queue with vacation time and exhaustive service discipline," *J. Oper. Res.*, submitted for publication.

Yih-Chyun Jenq (S'74–M'77) was born in Taipei, Taiwan, on October 26, 1949. He received the B.S.E. degree from the National Taiwan University, Taipei, Taiwan, in 1971, and the M.S.E., M.A., and Ph.D. degrees in electrical engineering from Princeton University, Princeton, NJ, in 1974, 1975, and 1976, respectively.

He worked for COMSAT Laboratories, Clarksburg, MD, in the summer of 1974 and the Datalog Division of Litton Industries, Melville, NY, in the summer of 1979. From 1976 to 1980 he was Assistant Professor of Electrical Engineering at the State University of New York at Stony Brook, Stony Brook, NY. Since June 1980 he has been with Bell Laboratories, Holmdel, NJ. His current research interests are in the areas of data and computer communications, queueing analysis, and digital signal processing.

Dr. Jenq is a member of Eta Kappa Nu and Sigma Xi.

A Simulation Study of Network Delay for Packetized Voice

P. M. Gopal and Bharath Kadaba

IBM T. J. Watson Research Center, Yorktown Heights, NY 10598.

Abstract

The potential cost benefit from integrating data and voice in a single network provides a strong motivation for packetized voice transmission. In this study, simulation models are used to characterize the delay distribution of voice packets in a single hop as well a multihop network environment. The trade-off between the number of speakers that can be multiplexed using speech activity detection technique and the delay performance are quantified. This is contrasted with the performance obtained by using a lower bit-rate voice input without speech activity detection.

Introduction

Currently, packet-switching technique is widely used to carry data traffic. Voice traffic, on the other hand, is circuit-switched. This is primarily due to the differences in the characteristics of data and voice traffic. Interactive data traffic is characterized by its *burstiness*, which is well suited for packet-switching. Voice traffic has traditionally been treated as high bandwidth continuous traffic, thus suitable for circuit-switching.

For some time now, it has been recognized that packet-switching techniques are attractive for voice traffic also. Firstly, a fair amount of burstiness is present in the voice waveform, if one encodes and transmits only the talkspurts within a conversation. Secondly, significant advances have been made in lower bit-rate coding schemes and high-speed transmission media in recent years. This has made the bandwidth requirement for voice traffic much less severe. Most importantly, the potential cost saving from a single integrated network capable of switching both data and voice provides a strong motivation for packetized voice transmission. Apart from this, transmission efficiency, flexibility to accommodate various bit rates, and no synchronization requirement are some of the other significant benefits of packet-switching. Substantial investigative work has been done in this area, the most recent being [1].

There are several alternatives available for transmitting voice on a packet-switched network. One can simply packetize the input voice stream and transmit these packets in the network just like data packets. Alternatively, one can use speech activity detection (SAD) to transmit information only during periods of speaker activity. Since the activity factor of voice traffic (i.e. fraction of time a speaker

is active) is about 0.4, SAD can potentially more than double the number of simultaneous conversations sharing a single transmission facility. Bandwidth savings can also be achieved without speech activity detection, by using a lower bit-rate coding technique.

The two most important requirements for voice transmission are those of real-time delivery and output continuity. Unlike circuit-switching which offers the speakers small and fixed network delays, packet-switched transmission of voice traffic can result in the variable packet delays. As a result, additional buffering is required at the destination to smooth out this variation. The packets not arriving at the destination within a target playout time are simply dropped. In this manner, a constant delay from source to destination terminal is provided at the cost of some loss of information. Subjective evaluations have indicated that end-to-end delays of about 300 ms (echo-free) and 1-2% packet loss is tolerable [2, 3].

The performance measures for voice traffic are different from those for data traffic. Good performance in terms of mean network delays is not sufficient to ensure good output voice quality. What is also important is good performance in terms of say, the 99th percentile of delay. This allows the destination to select the target playout delay appropriately for incoming packets, with the assurance that packet loss is within tolerable limits.

Earlier work on performance evaluation of packetized voice transmission has considered only mean packet delay for voice traffic. Weinstein [4] studied the tradeoff between mean delay and packet loss when SAD is used on a single link. Suda et al [5] used a simulation model to study packetized voice, and presented results for mean delay and optimal packet size.

In this study, we use the distribution of delay and packet loss as our measures of interest for voice traffic. Specifically, the objectives of this study are threefold :

- to characterize the delay *distribution* at different levels of network load
- to evaluate the potential benefits of SAD in terms of the increase in the number of multiplexed speakers or improved delay performance for a given number of speakers
- to compare the performance achieved using SAD with that using a lower bit-rate voice input without SAD

The insights gained from this study can be summarized as follows. Speech activity detection is advantageous as long as the total offered load does not exceed the link capacities. The performance degrades very rapidly when the number of multiplexed speakers is increased beyond a certain point. If SAD is not used, the offered load never exceeds the link capacities and the performance is good throughout the range of offered loads. Moreover there are other advantages to not using SAD, e.g. no packet loss in the network, simpler playout strategy and simpler call acceptance or rejection policy.

Environment and Models

We focus on two popular voice coding schemes in this study:

1.   64 Kbps with SAD (64/SAD)
    - Coding rate used is 64 Kbps (e.g. Pulse Code Modulation)
    - Speech activity detection is used. This results in fixed sized packets being generated at regular intervals during talkspurts and no input to the network during silence periods.
2.   32 Kbps without SAD (32/NOSAD)
    - Coding rate used is 32 Kbps (e.g. Adaptive Differential PCM or Delta Modulation)
    - No speech activity detection is used. Thus a speaker generates fixed sized packets whether in talkspurt or in silence.

Both these strategies are roughly comparable in terms of bandwidth usage and have the potential of doubling the number of multiplexed speakers over the traditional circuit-switched environment with 64 Kbps speech coding (without TASI). This provides us with a basis for comparing the gain due to SAD vs. bit-rate reduction.

Both strategies require additional hardware over 64 Kbps PCM; the former for speech activity detection and the latter for more sophisticated encoding of speech. As far as output speech quality is concerned, both schemes offer somewhat lower quality than standard PCM. It is, however, not completely clear, how the quality of 64 Kbps speech with packet loss compares with 32 Kbps speech without any packet loss. For the purposes of this study, it is assumed that the two cases are comparable as far as speech quality and hardware complexity are concerned. This permits us to focus primarily on the delay performance.

Two network environments are considered in the simulation models:

- *Single Hop*: In the single hop configuration, a node can support multiple calls (the terms *call* and *speaker* are used interchangeably throughout the paper). The node is assumed to possess finite amount of buffer space and to transmit the arriving packets in a first-come first-served manner on the outgoing link. The size of the buffer serves to limit the maximum delay suffered by any packet.

- *Multi-hop*: In this case, one call travels several hops (fixed routing or permanent virtual circuits are assumed). At every hop, the interfering traffic consists of calls arriving on different links and travelling a single hop. The buffer size is assumed to be finite at each intermediate node.

The amount of interference faced by the multi-hop call is determined by the number of interfering sources at each hop and the nature of traffic generated by each source. We have assumed that the interfering calls arrive at the node over different links. Therefore, the probability of packets from several calls arriving at the node simultaneously (or within a very short interval) is higher than in the case where the calls arrive multiplexed over the same link. As far as traffic characteristics of the interfering sources is concerned, we have assumed, for the sake of simplicity, that each source corresponds to a speaker. This scenario may not give us the worst-case interference for the multi-hop call, but it certainly tends towards the worst.

The following additional assumptions are made in the simulation:

- We assume that the number of calls does not change and consider only the short-term equilibrium when the number of calls is fixed. This is not unreasonable, because the time-scale of events associated with call initiation/termination is orders of magnitude larger than that of packet generation or talkspurt/silence variation. The dynamics of call fluctuation is a nontrivial problem from the point of view of simulatio because of the vast difference in the time-scales, and is not addressed here.

- The network is assumed to be carrying only voice traffic. The results presented are expected to be a reasonable approximation even in a mixed voice and data environment, where voice traffic is given priority over data and the size of the data packets is about the same as the voice packets.

- The speakers are assumed to generate traffic independent of each other. The talkspurt and silence durations for a speaker are modelled by exponential distributions with means 1.3 s and 1.8 s respectively [6].

- It is assumed that the calls arrive at random instants. Specifically, assuming a fixed packet size, the instants at which packet arrivals take place at different speakers are taken to be uniformly distributed in an interval equal to the packet generation time.

- Link speeds of 1.544 Mbps are assumed for all the links.

- In most of the experiments, the packet size is assumed to be 1000 bits and the header size 100 bits. The question of optimal packet size is not addressed in this study.

Results

Single-hop case

We first study the *64/SAD case*. Since the activity factor of voice traffic is about 0.4, using speech activity detection reduces the offered traffic substantially. Hence we expect an improvement in the delay performance over the case without SAD. This is illustrated in Figure 1, where the mean and 99th percentiles of delay are plotted for the 64 Kbps case with and without SAD. The results[1] are shown for number of speakers up to 21, which is the maximum number for 64 Kbps without SAD and a packet size of 1000 bits (including a header of 100 bits).

- When the number of speakers is small, the delay performance is very good even without SAD. Thus the margin of improvement due to SAD is not substantial.

- However, the margin of improvement increases with the number of multiplexed speakers. For 21 speakers, the mean and the 99th percentile of delay using SAD are roughly half of those without SAD.

Since speech activity detection reduces the offered traffic due to a speaker, it is possible to put additional traffic on the link. The possibility of increasing the number of speakers sharing the link bandwidth is investigated here. Figure 2 shows the cumulative distribution of the network delay for the 64/SAD case as a function of the number of speakers, with the packet size being 1000 bits and the buffer size being 24 packets (which corresponds to a maximum delay of 15.54 ms). The following observations can be made from this figure:

- The variation of delay is negligible when the total offered traffic due to the multiplexed speakers does not exceed the link capacity.

- As the number of speakers is increased beyond 21 (which is the maximum number of speakers without SAD), there is an increase in the delay and its variance. This is due to the increased utilization of the link, and also the randomness introduced in the traffic due to speech activity detection. Even though the average offered load still remains less than the link bandwidth, there are periods when the load exceeds the bandwidth and queues build up.

The network delay and packet loss depend on the size of buffer at the outgoing link. Less buffering may lead to a large number of packets being discarded due to buffer overflow. On the other hand, large buffers can result in the network delay becoming very large. In Figure 3, we show the mean and 99th percentile of delay and the packet loss incurred due to the buffer overflow, for different values of buffer size (24 and 154 packets). We observe that:

- For a buffer size of 24 packets, the mean network delay for a single hop is of the order of a few milliseconds and the packet loss is negligible up to 32

speakers. Beyond this, the delay and packet loss increase very rapidly. This is due to the increased probability of the load due to the active speakers exceeding link capacity at any instant.

- For larger buffer size, the mean delay and the 99th percentile become substantially larger when the number of speakers is increased. On the other hand, the percentage of packets lost decreases.

- The number of multiplexed speakers depends on the constraints on maximum delay and on tolerable packet loss. For example, with the buffer size being 24 packets (maximum delay of 15.54 ms) and the tolerable packet loss being 1%, more than 43 speakers cannot be multiplexed. If the tolerable packet loss is only 0.1%, the number of speakers is limited to about 34.

- Increasing the buffer size to 154 (i.e. maximum delay of 99.74 ms) allows us to multiplex more speakers for the same constraint on packet loss. The gain is substantial at smaller values of packet loss, but only marginal for larger values of loss.

Next we look at the *32/NOSAD case*. In this case, the maximum number of speakers which can be multiplexed is 43. In Figure 4, the cumulative distribution of network delay is shown with a packet size of 1000 bits.

- We notice that the variation of delay is much smaller here than in the 64/SAD case for the same number of speakers. In fact, if one considers the output stream of packets for a particular speaker in short-term equilibrium (with no new call initiations/terminations), the delay variance for these packets is identically equal to zero.

- Furthermore, there is no packet loss in the network in this case, since there is no overallocation of the link capacity at any time.

In both 64/SAD and 32/NOSAD cases, the gain in multiplexing is less than two times the number of speakers in 64 Kbps circuit-switching. This is because of the header overhead in packet-switching.

Figure 5 shows a comparison of the mean network delays for the 64/SAD (with buffer size = 24 packets) and 32/NOSAD cases. In addition, 99th percentiles of delay for the two cases are also shown. The observations which can be made from this figure are:

- The mean delay for the 32/NOSAD case is about the same as that for the 64/SAD case with buffer size of 24 packets. As shown earlier, the delay for the 64/SAD case is larger when the buffer size is increased.

- The 99th percentile of delay for 64/SAD is smaller than that for 32/NOASAD when the number of speakers is less than 32. Beyond 32 speakers, the 99th percentile for 64/SAD increases very rapidly un-

---

[1] The 90 % confidence intervals for all the simulation experiments presented in the paper are within 10 % of the observed values.

til it is limited by the buffer size. The corresponding delay for 32/NOSAD increases gradually and is much smaller than that for 64/SAD.

Note that the packetization delay for 32/NOSAD case is twice that of 64/SAD, since we have taken the packet size to be the same in both cases. The value of this delay for the two cases is 28 and 14 ms respectively. To illustrate the case when a smaller packet size is used, figure 6 shows the cumulative distribution of network delay for the 32/NOSAD case with the packet size being 550 bits (including 100 bits header). This results in the packetization delay being the same as in the 64/SAD case, but the maximum number of speakers is limited to 39 due to the increased header overhead.

- The figure shows that the delay variation in this case is even smaller than in the 32/NOSAD case with 1000 bit packets. The corresponding mean delay is also significantly smaller.

For the NOSAD case, it is also possible to get results for the maximum delay. In Figure 7, the cumulative distribution of maximum delay is shown for different values of packet sizes. It can be seen that:

- Even though the worst-case delay can be large (a packet may have to wait for the current transmission plus the transmissions of all the other speakers), the mean of the maximum delay is quite small under uniform arrival assumption for calls. For example, when there are 43 speakers and the packet size is 1000 bits, the mean of the maximum delay is only 5 ms, though the worst-case delay could be as large as 28 ms. In fact, the probability that the maximum delay exceeds 10 ms is negligible.

Multi-hop case

Recall that in this environment, we assume fixed routing and focus on one call which travels several hops. The interference at intermediate hops is represented by calls arriving at the intermediate nodes on different links and travelling a single hop. For our experiments we assume that the multi-hop call travels 7 hops. The packet size is taken to be 1000 bits. Results were obtained with the number of interfering speakers at each hop being 23, 31, 35 and 39. For 64/SAD case, these values correspond to the utilization of each link being 0.46, 0.62, 0.69 and 0.77 respectively. The corresponding link utilizations for 32/NOSAD case are 0.55, 0.74, 0.83 and 0.92 respectively.

Figure 8 shows the distribution of network delay of the 7 hop call for the 64/SAD and 32/NOSAD cases. In the 64/SAD case, the buffer size at each hop is assumed to be 24 packets. We notice from this figure that:

- When the number of interfering speakers is small (e.g. 23), the delay distributions for both cases (64/SAD and 32/NOSAD) are very close to each other, and the variation of delay is very small.

- For 39 interfering speakers at every hop, the 64/SAD case shows a much larger variation in delay than the 32/NOSAD case. The tail of the distribution for the 64/SAD case is also much longer.

Figure 9 shows the mean and 99th percentile of delay for the 7 hop call as a function of the number of interfering speakers at each hop. It is seen that:

- The mean network delays for both 64/SAD and 32/NOSAD are about the same. The latter case shows somewhat larger delay because of slightly higher utilization of the links. The value of the mean delay for the parameters chosen is fairly small (about 15 ms for 39 interfering speakers at every hop).

- The 99th percentile of delay for 32/NOSAD case is somewhat larger than that for 64/SAD when the number of interfering speakers is small. As the number of interfering speakers increases (and the link utilizations increase correspondingly), the 99th percentile of delay for the former case becomes much smaller than that for the latter.

The packet loss due to buffer overflow at intermediate nodes for the 64/SAD case is shown in Figure 10. The corresponding values for the single hop interfering calls are also shown for comparison. The following observations can be made from this figure:

- The packet loss is quite small up to 35 interfering speakers at each hop. However, for 39 interfering speakers, the loss becomes more than 2%. To ensure that the loss is tolerable for a call travelling several hops, a more stringent constraint has to be imposed on the tolerable per hop loss. This implies that the effective gain in multiplexing due to SAD is smaller for a multi-hop environment as compared to a single hop.

It was also observed that the delay and packet loss for the 7-hop call are roughly seven times the corresponding values for the single-hop calls. This suggests that the multi-hop model considered can be approximated by a sequence of independent single-hop models, and its delay distribution can be determined using numerical convolution from the single-hop delay distribution.

For the 32/NOSAD case, the network delay for a speaker is deterministic in the absence of fluctuation in the number of calls. Also, there is no packet loss in the network. The mean delay for the multi-hop call was also observed to increase linearly with the number of hops.

Conclusions

The conclusions and insights that were obtained from this study can be grouped under different headings:

1. *Speech Activity Detection*
    - There is a significant improvement in the delay by using SAD for a given number of speakers, when the traffic due to the speakers does not exceed the capacity of the links.

- Using 64 Kbps coding rate, 1000 bit packets and SAD, up to 32 speakers can be multiplexed on a 1.544 Mbps link without any appreciable degradation in terms of delay or packet loss.
- The mean packet delay, its variation and packet loss increase rapidly as the number of speakers is increased beyond 32. The maximum number of speakers depends on the buffering delay and tolerable packet loss at each hop. For example, for a buffer size of 24 packets and 1% loss at each hop, up to 43 speakers can be multiplexed. If the constraint on loss is only 0.1% (for example on a multi-hop path), then the number of speakers is limited to about 34.

2. *No SAD*
- The mean packet delay and the variation of delay is very small if speech activity detection is not used. Furthermore there is no packet loss in the network.
- Using 32 Kbps coding rate, up to 43 speakers can be multiplexed on a 1.544 Mbps link with the packet size being 1000 bits.
- For a given number of speakers, 32 Kbps with no SAD provides comparable or better performance than 64 Kbps with SAD.
- Not using SAD has several other advantages, e.g.
  - a simpler playout strategy can be used because of smaller variation in delay
  - simpler call acceptance or rejection policy

- no packet loss in the network

Acknowledgements

Various discussions on related issues with Mon-Song Chen, Alan Baratz, Inder Gopal, Israel Cidon, Paul Green and Jeff Jaffe are gratefully acknowledged.

References

[1]   Turner, J. S.,"Design of an Integrated Services *Packet* Network", Technical report, Department of Computer Science, Washington University, St. Louis, Missouri, March 1985.

[2]   Klemmer, E. T.,"Subjective Evaluation of Transmission Delays in Telephone Conversations", Bell System Technical Journal, Vol. 46, July-August 1967.

[3]   Jayant, N.S.,"Effects of Packet Losses in Waveform-Coded Speech", ICCC 80, 275-280.

[4]   Weinstein, C. J."The Tradeoff between Delay and TASI Advantage in a Packetized Speech Multiplexer", IEEE Transactions on Communications, Vol. COM-27, No. 11, November 1979, 1716-1720.

[5]   Suda, T., Yemini, Y., Miyahara, H. and T. Hasegawa,"Performance Evaluation of a Packetized Voice System - Simulation Study", ICC 83, Boston, 1983, C5.2.1-C5.2.5.

[6]   Brady, P. T.,"A Statistical Analysis of On-Off Patterns in 16 Conversations", Bell System Technical Journal, Vol. 48, January 1968, 73-91.

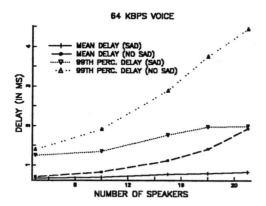

Figure 1.   Mean and 99th percentile of network delay

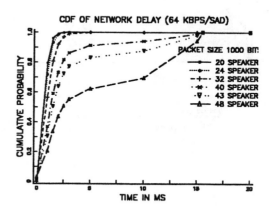

Figure 2.   CDF of network delay (buffer size = 24)

413

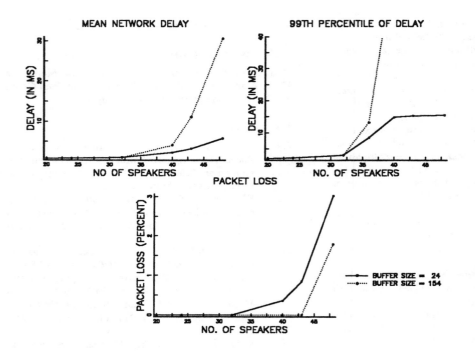

Figure 3.   Delay and packet loss (64/SAD)

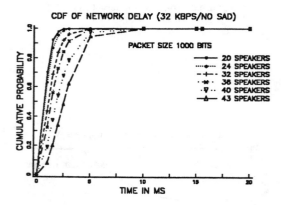

Figure 4.   CDF of network delay (32/NOSAD,
1000 bit packets)

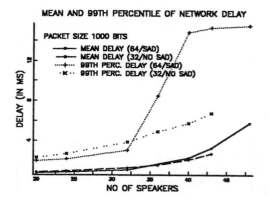

Figure 5.   Network   delay   (64/SAD   vs.
32/NOSAD)

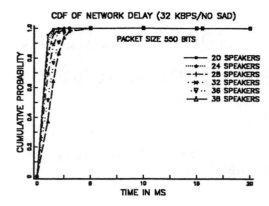

Figure 6.   CDF of network delay (32/NOSAD, 550 bit packets)

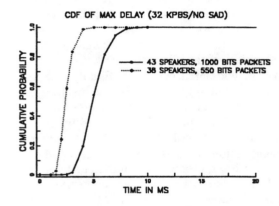

Figure 7.   CDF of maximum network delay (32/NOSAD)

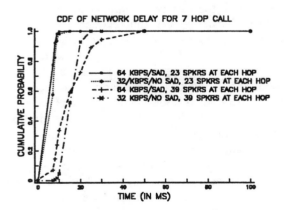

Figure 8.   CDF of network delay for 7 hop call

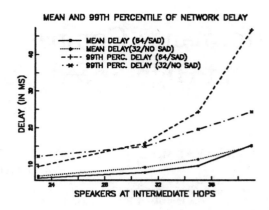

Figure 9.   Network delay for 7 hop call

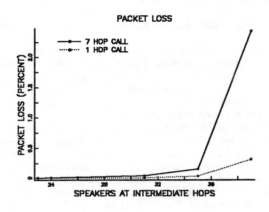

Figure 10.   Packet loss (64/SAD)

Performance Comparison of Error Control Schemes in High-Speed Computer Communication Networks

AMIT BHARGAVA, MEMBER, IEEE, JAMES F. KUROSE, MEMBER, IEEE, DON TOWSLEY, MEMBER, IEEE, AND GUY VANLEEMPUT

Reprinted from *IEEE Journal on Selected Areas in Communications*, Vol. 6, No. 9, December 1988, pages 1565-1575.

Abstract—Advances in both high-speed switching and fiber optics technology are changing some of the fundamental premises upon which traditional wide-area network packet-switching technology was first developed. It is thus appropriate to carefully reconsider design issues for protocols built upon these raw high-speed transmission and switching facilities.

We examine the performance of two different approaches for handling the loss and/or corruption of messages as they are transmitted between two end users *in a high-speed network*. In the link-by-link approach, two adjacent nodes in an end-to-end path locally detect and recover from message loss or corruption along their joining link. In the end-to-end approach, recovery is done solely on the basis of a single end-to-end protocol. We develop analytic performance models, validated with simulation, for comparing the performance of these two approaches. These models explicitly consider the effects of channel errors, propagation delays of messages and their acknowledgments, both separate and shared VC buffer pools, finite buffer capacities, buffering of unacknowledged messages, and timeout mechanisms. A result of this paper is that for the range of network parameters of practical interest, an end-to-end approach towards error control is superior to a link-by-link approach, even under assumptions that would overly favor the link-by-link approach, while at the same time requiring fewer network resources (e.g., buffers, computation time) than the link-by-link approach. The performance differences arise primarily from the increased buffer requirement of the link-by-link approach.

I. INTRODUCTION

RECENT advances in both high-speed switching [7], [18], [20] and fiber optics technology [8] make it physically possible to provide gigabit communication channels between two users in a packet-switched computer communication network. An often-cited goal of this technology is the transmission of diverse and often high bandwidth traffic, including packetsized voice, video, and data, over a single integrated communication network. From a networking standpoint, the increasing communication and switching speeds change some of the fundamental premises upon which traditional wide-area net-

work packet-switching technology was first developed. First, the ratio of transmission speed to computation speed has increased dramatically, prompting the need to shift computation-intensive protocols that once resided within the subnetwork to the "edges" of the network [19]. Second, the ratio of message propagation delay to message transmission time is also increasing, particularly in the case of wide-area networks. Finally, the use of fiber optics as a communication media is resulting in significantly lower channel error rates. Given these fundamental changes, it is appropriate to carefully reconsider the design of the protocols which lie between the raw high-speed transmission and switching facilities and the higher level application protocols in a high-speed network.

In this paper, we examine the performance of two different approaches for handling the loss and/or corruption of messages as they are transmitted between two end users *in a high-speed network* (HSN). Traditionally, this problem has been handled primarily by the data link protocol [4], which permits two *adjacent* nodes in an end-to-end path to *locally* detect and recover from message loss or corruption along their joining link. We refer to this approach as a *link-by-link* approach towards error control. Note that in this approach, a higher level end-to-end protocol is still required, for example, to recover from message loss due to node crashes. A second approach is to recover from both message loss and corruption solely on the basis of a single end-to-end protocol. In such an *end-to-end* approach, the traditional functionality of the data link protocol is mostly eliminated (although error detection, but not recovery, might still be maintained) and the burden of maintaining an error-free data path falls on the end-to-end protocol.

In this paper, we develop analytic performance models for comparing the performance of the link-by-link versus end-to-end approaches towards error control; these models are validated by simulation. A result of this paper is that for the range of network parameters of practical interest, *and end-to-end approach towards error control is superior to a link-by-link approach, even when the (higher) node processing demands of the link-by-link approach are ignored.* That is, even under assumptions that would overly favor the link-by-link approach, an end-to-end ap-

Manuscript received August 13, 1987; revised May 27, 1988. This work was supported in part by the Office of Naval Research under Grant N00014-87-K-0304, and by the National Science Foundation under Grant CER DCR 8500332. This paper was presented in part at IEEE INFOCOM '88, New Orleans, LA, March 27-31, 1988. Correspondence to G. VanLeemput may be directed to A. Bhargava.

A. Bhargava is with Codex Corporation, Mansfield, MA 02048.

J. F. Kurose and D. Towsley are with the Computer and Information Science Department, University of Massachusetts, Amherst, MA 01003.

IEEE Log Number 8824611.

proach is *still* found to provide equivalent or better performance while at the same time requiring fewer network resources (e.g., buffers, computation time) than the link-by-link approach. The performance differences arise primarily from the increased buffer requirements of the link-by-link approach. The models developed in this paper also permit *quantitative predictions* to be made for the delay, throughput, and buffering requirements of these two error control schemes in a high-speed environment.

Several previous researchers have examined the problem of end-to-end versus link-by-link error control, although *not* in the setting of a high-speed network. In [9], link-by-link and end-to-end protocols were shown to have similar performance when the number of hops between end-users was small. In all other cases, link-by-link was found to be clearly superior (exactly the opposite of our conclusions for the HSN case). We note that while the analytic models in [9] do consider the effects of finite buffer sizes, they do not consider transmission errors, propagation delays (which, as discussed above, are significant in HSN's), acknowledgment delays (which also may be relatively large due to propagation delays), or the fact that unacknowledged packets occupy buffer space at the sending station (which is also important in a HSN, again due to the large propagation delays). Nonetheless, we do follow the overall approach of [9] in developing our model of end-to-end error control. The comparison of link-by-link versus end-to-end in [11] indicated that the end-to-end approach was superior. However, we conjecture that this resulted primarily from the fact that in [11] each message was segmented in the end-to-end case, whereas a message was transmitted as a unit in the link-by-link case. Also, the effects of propagation delays, finite buffer capacities, and noninstantaneous acknowledgments (which we have found to be important in a HSN) were not considered. Finally, a model of link-by-link (only) error control was developed first in [15] and extended in [12]. This model considers the effect of blocking, propagation delays and noninstantaneous acknowledgments. Our model of each node in a link-by-link approach (only) thus closely resembles this model, although our solution technique is considerably simpler due to the structure of our virtual circuit model.

Suda [17] has recently also examined the issue of link-by-link versus end-to-end error recovery is a high-speed network. That work differs from the present work, however, in that the processing times required by an error recovery protocol were considered, while the effects of the relatively long propagation delays in the network were not considered. Finally, Brady [3] has recently studied (through simulation) a detailed implementation of the full X.25 LAPB link protocol together with a similar, but higher level, end-to-end error recovery mechanism to provide for additional error recovery. In his model, propagation delays were relatively short—on the same order as the packet transmission time.

In the following section, we describe our general approach towards modeling message flow between the two

end-users of a virtual circuit in which each VC maintains separate buffer pools at each node. In Section III, we then describe the end-to-end approach in more detail and develop a queueing-theoretic model for evaluating the performance of end-to-end error control in such a VC; the results of the analytic model are corroborated through simulation. Various modeling assumptions used in both this section and the next are also discussed. Section IV presents the corresponding model for link-by-link error control and Section V compares the performance of the two approaches for various network parameters of interest. In Section VI, we study the case in which virtual circuits routed over the same outgoing link at a node share a common buffer pool at the node; our overall conclusions for this case are identical to those drawn from the separate buffer pool model. Finally, Section VII summarizes this paper.

II. NETWORK MODEL

The network model used for comparing link-by-link versus end-to-end error control is shown in Fig. 1(a). As in [9], [11], we focus on the traffic λ_{VC} entering a virtual circuit (VC) of M links which begins at some source node (labeled 1 in Fig. 1) and terminates at some destination node (labeled $M + 1$ in Fig. 1). In Sections III, IV, and V, we study the case in which the VC is allocated a finite number of buffers K_i at all but the first link, $i = 2, \cdots, M$ and that there is enough buffer capacity at the first link to store all new packets. In Section VI, we consider the case in which *all* VC's routed over the same outgoing link share the same buffer pool at this link.

We denote the message transmission rate (capacity) of each link i along the VC by μ_0^i and denote the mean propagation delay between nodes i and $i + 1$ as $1/\mu_{prop}^i$. Note that several VC's may share the same physical link (although maintain separate buffer pools, as discussed above). For example, in Fig. 1(a) a VC of rate λ_{int} shares the link between nodes 2 and 3 with λ_{VC}. In Sections III, IV, and V, we adopt the approach of [13], [16] (also implicitly used in [9], [11]) for modeling message traffic along a virtual circuit with separate buffer pools. We thus model the effects of this interfering traffic by reducing the communication capacity seen by the λ_{VC} traffic along each outgoing link of the VC by some appropriate amount. Our model of the M-hop virtual circuit is thus shown in Fig. 1(b) where μ_{trans}^i now represents the effective (reduced) service capacity of link i on the λ_{VC} virtual circuit. In Section VI, our shared buffer pool models explicitly consider the effects of interfering traffic.

III. END-TO-END CONTROL

In this section, we develop an analytic model for studying the performance of the end-to-end error control scheme. Our model explicitly considers the effects of message propagation delays, finite buffer capacities (blocking), channel errors, and the use of a timeout mechanism in the error control scheme. We begin by describing the operation of the end-to-end error control protocol

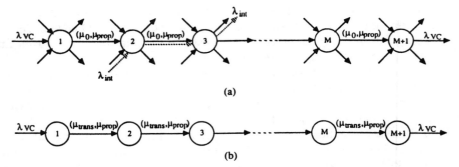

(a)

(b)

Fig. 1. Virtual circuit model.

in detail. We then develop a two-level performance model of this scheme. The lower level model, developed in Section III-B, models the interaction between two adjacent nodes along the VC. The higher level model, discussed in Section III-C, incorporates the M lower level models into a single VC-level model. The analytic results are compared to simulation results in Section III-D and various assumptions underlying the model are discussed.

A. Description of End-to-End Error Control

In an end-to-end approach towards error control, when the VC source node first transmits a message, it buffers a copy of the message until an ACK message is received from the *final* destination node on the VC. The destination station generates an ACK message only when a message from the source is received correctly. If an ACK is not received within a specified *timeout period*, the message is assumed lost or corrupted and the source node retransmits a copy of the message along the VC. In our study, we will assume (as in all previous studies [12], [15], [9], [11]) that a timeout occurs only when a message has, in fact, been lost or corrupted. We note that in a HSN where the known, fixed length, propagation delays may be significantly greater than the variable message queueing and transmission times, this assumption is even more justified than in the non-HSN case.

The function of an intermediate node in the end-to-end approach is simple; it need only check arriving messages for errors, discard any incorrectly received messages or any messages that overflow the buffer, and queue all correctly received messages for transmission on the outgoing link of the VC.

B. Link-Level Model for End-to-End Error Control

Our queueing model for studying the performance of an individual link along a VC under the end-to-end error control scheme is shown in Fig. 2. For simplicity, we assume that all nodes along the VC are homogeneous, i.e., have the same message transmission rate, propagation delay, and fixed number of buffers; this assumption can be easily relaxed at the expense only of some slightly more complicated notation.

The transmission facility for the ith outgoing link and its buffers are modeled by an $M/M/1/K$ queueing system where K represents the fixed number of buffers avail-

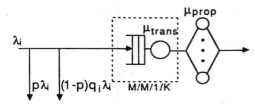

Fig. 2. Lower level queueing model for end-to-end error control.

able for this VC at all nodes other than the source node. As mentioned earlier, the source node has an infinite number of buffers. In Section III-D, we discuss the assumptions of exponential service and interarrival times implicit in the $M/M/1/K$ model. The link propagation delay is modeled by an infinite server queue in which a message has a fixed delay of $1/\mu_{\text{prop}}$. Note that a message releases its buffer immediately upon completion of its transmission. As we will see, in the link-by-link case, we must explicitly model the fact that a message continues to hold a buffer at each transmitter i while it (and its ACK message, if any) propagates across link i.

When a message arrives at node i, one of three events can occur. First, with probability p the message may be received in error, in which case it is discarded (the eventual retransmission of this message at the source node will be incorporated into the higher level model described in the following subsection). Second, a message may be received correctly but there may be no buffers available. In this case, the message is also discarded and will again be eventually requeued at the source node. In our analytic model, we assume that each arriving message *independently* finds a full buffer with probability q_i where q_i is the steady state probability that each of the K buffers at node i are in use. (Note that q_i will be different for each node along the VC, and hence, we must explicitly retain the subscript.) The use of this independence assumption, also adopted in the previously cited earlier studies of error control schemes, will also be validated by our simulation results. The third possibility is that the message is received correctly and is allocated a buffer. Last, we introduce the notation $f_i = q_i + p - q_i p$. Here, f_i is the probability that a packet is discarded by node i.

As discussed above, the arrival of messages to node i is modeled as a Poisson process with mean λ_i. Given the above assumptions and given a value of λ_i, the individual

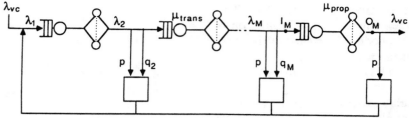

Fig. 3. VC-level model of end-to-end error control.

link-level model of end-to-end error control can now be easily solved. The average number of messages in an intermediate node i, $E[L_{1,i}]$, is simply the average queue length of the $M/M/1/K$ queue (recall that there are an infinite number of buffers at the source node). Defining

$$\rho_{1,i} = \lambda_i (1 - p)/\mu_{\text{trans}}$$

we have [1]

$$E[L_{1,i}] = \begin{cases} \rho_{1,i}/(1 - \rho_{1,i}), & i = 1, \\ \rho_{1,i}/(1 - \rho_{1,i}) - (K + 1)\rho_{1,i}^{K+1}/(1 - \rho_{1,i}^{K+1}) & i = 2, \cdots, M. \end{cases}$$

The average time a packet spends queueing for transmission plus its actual transmission time is denoted $E[W_{1,i}]$. Using Little's formula, we have

$$E[W_{1,i}] = E[L_{1,i}]/\lambda_i (1 - f_i) \qquad (1)$$

and

$$E[W_i] = E[W_{1,i}] + 1/\mu_{\text{prop}} \qquad (2)$$

where $E[W_i]$ is the average delay (queueing, transmission, and propagation) a message experiences passing through link i.

Finally, the probability q_i that all buffers at node i are full (and hence node $i - 1$ is blocked) is given by [1]

$$q_i = \Pr[L_{1,i} = K] = \frac{(1 - \rho_{1,i})\rho_{1,i}^{K+1}}{1 - \rho_{1,i}^{K+1}},$$
$$i = 2, \cdots, M. \qquad (3)$$

Equations (2) and (3) give the average delay and blocking probability at node i. Note that they are each functions of the still undetermined values, $\{\lambda_i\}$; these values will be obtained in the following section, in which the individual link-level models are combined into a single VC-level model.

C. VC-Level Model for End-to-End Error Control

Fig. 3 combines the M individual link-level models into a single model for the VC. Note that a rejected packet is removed from the VC as soon as an error occurs. As we will see, we need not explicitly consider here the amount of time that elapses between the removal of a message from the VC and its eventual retransmission at the source node. Instead, we need only consider the overall flow rate of messages back to the source node.

In order to actually solve for the blocking probabilities and average delays along the links, we must still deter-

mine $\{\lambda_i\}$. The difficulty here is that λ_i and q_i are functionally dependent on each other. Moreover, $\lambda_i \neq \lambda_j \neq \lambda_{\text{VC}}$ due to the different node blocking probabilities and the fact that a message is discarded at the point where an error is first detected. Our approach will be to first solve for the arrival rate and blocking probabilities for the last hop (M) on the VC, use these values to solve for corre-

sponding values for link $M - 1$, and then continue to solve the VC-level model working from the end of the VC back towards the front.

Clearly, the message output rate for link M at the point O_M in Fig. 3 is given by $\lambda_{\text{VC}}/(1 - p)$. By conservation of flow, the input rate to link M at the point I_M must also equal this value and thus

$$\lambda_M = \frac{\lambda_{\text{VC}}}{(1 - p)^2 (1 - q_M)}. \qquad (4)$$

Substituting this expression for λ_M into (3) yields a single equation in a single unknown (q_M) and we may thus solve for q_M immediately. Given a value for q_M, (4) gives a value for λ_M. Finally, given a value for λ_M, (2) can be solved for the average delay through link M. Once λ_M is known, q_{M-1}, λ_{M-1}, and $E[W_{M-1}]$ can be obtained using similar arguments. Working backwards along the VC, the set of unknown values, $\{q_i\}$, $\{\lambda_i\}$, and $\{E[W_i]\}$ are thus obtained.

Finally, in order to obtain the average delay along the VC, several additional quantities must be defined as follows.

p_{fail}—the probability a particular transmission of a packet from the source node will *not* successfully reach the destination node

$$p_{\text{fail}} = \sum_{i=1}^{M} f_i \prod_{j=1}^{i-1} (1 - f_j).$$

N_t—the number of times a packet must be *retransmitted* before it is successfully received at the destination station. Assuming the outcome of each packet retransmission is independent, N_t has distribution $P[N_t = k] = p_{\text{fail}}^k (1 - p_{\text{fail}})$, $k = 0, 1, \cdots$ and mean

$$E[N_t] = \sum_{i=1}^{\infty} kP[N_t = k] = \frac{p_{\text{fail}}}{1 - p_{\text{fail}}}. \qquad (5)$$

T_{ee}—the end-to-end timeout interval for the VC. The timeout interval begins when a message's *transmission* is completed at the VC source node. If the timeout interval expires and an ACK message has not been received, the message has been lost (see above) and a copy is thus retransmitted.

$E[W_{ee}]$—the average delay experienced by a packet between its initial arrival at the VC source node and its eventual successful receipt at the VC destination node, with an end-to-end error control protocol.

A packet's delay between its initial arrival and eventual successful reception at the destination results from the N_t timeout intervals during which it is unsuccessfully transmitted to the destination node plus the delays it experiences along the VC during its successful trip to the destination station. Hence,

$$E[W_{ee}] = E[N_t]\left(E[W_{1,1}] + T_{ee}\right) + \sum_{k=1}^{M} E[W_k]. \quad (6)$$

D. Discussion

In order to validate the assumptions underlying our analytic model, we compare analytic results to performance results obtained via simulation. All time-related values are given in units of the average message transmission time (i.e., in units of $1/\mu_{\text{trans}}$). We simulated a four hop virtual circuit, with a link propagation delay 25 times the message transmission time, $p = 10^{-5}$, and $T_{ee} = 224$. Given a transmission rate of 100 Mbits/s, these values would correspond to a packet size of 1000 bits, a bit error rate $= 10^{-8}$, and an internode distance $= 50$ km. In our simulations (as in the analysis), simulated timeouts occurred only when a packet was dropped at a node (either due to buffer overflow or to a transmission error).

Fig. 4 shows both simulation and analytic results plotting $E[W_{ee}]$ versus the virtual circuit arrival rate, λ_{VC}, for the cases of $K = 40$ buffers at each intermediate node. Note that the y axis values do not begin at 0. We note that as a general trend, as the number of buffers increases, the maximum throughput the protocol can support also increases. Also, for a fixed throughput, the time delay decreases (particularly at higher throughput values) as the number of buffers increases. Both these results are as expected, since a VC with fewer buffers blocks more often than one with more buffers.

In Fig. 4, simulation results are indicated as point values while analytic results are represented by continuous curves. In these results, exponentially distributed message lengths were used in both the simulation and the analysis (the appropriateness of this assumption will be examined later when the link-by-link versus end-to-end approaches are compared). Two sets of simulation results are shown in each graph. In the set of simulation results plotted using triangles, message lengths were independently generated at each node along the VC (Kleinrock's independence assumption); note that these results track the analytic results quite closely. Thus, the assumption (used in the analysis only) that each message independently

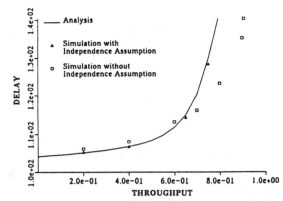

Fig. 4. Comparison of simulation and analysis: the end-to-end case.

dently finds all buffers full with probability q_i is a good one.

In the set of simulation results plotted with squares, Kleinrock's independence assumption was not used, i.e., a packet's length was preserved (after being chosen from an exponential distribution) as it passed through the VC. These simulation results show noticeably lower delay than the analytic results at high throughput values. This would indicate that the use of Kleinrock's assumption in the analysis tends to give *pessimistic* performance predictions; an observation also noted in [9]. However, as we will see, for the purposes of comparing link-by-link versus end-to-end error recovery this difference will not play a significant role, for *even using the more pessimistic analytic results derived above*, the end-to-end approach is *still* found to be superior.

IV. Link-by-Link Error Control

In this section, we develop an analytic model for studying the performance of the *link-by-link* error control scheme. The model again explicitly considers the effects of message propagation delays (of both messages and their ACK), finite buffer capacities (blocking), channel errors, and timeouts. Also, we explicitly model the fact that *a message continues to hold a buffer at each transmitter i while it (or its ACK message, if any) propagates across link i*. We begin by describing the operation of the link-by-link error control protocol. As in the end-to-end case, we then again develop a two-level performance model and compare the analytic results to those obtained via simulation.

A. Description of Link-by-Link Error Control

In the link-by-link approach, *each node* along the VC buffers a copy of each transmitted message until an ACK message is received for that message from the *next node* along the VC. A node generates an ACK message for each message it receives correctly and sends this ACK back to the preceding node on the VC. If an ACK is not received within a specified *timeout period*, the message is assumed lost or corrupted and a node retransmits a copy of the message to the next node along the VC. We again assume (as before) that a timeout occurs only when a message has, in fact, been lost or corrupted.

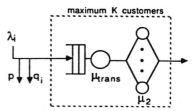

Fig. 5. Lower level queueing model for link-by-link protocol.

B. Link-Level Model for Link-by-Link Error Control

Our model for studying the performance of an *individual link* along a VC is slightly more complicated than in the end-to-end case because we must explicitly model the fact that a message continues to hold buffer space in an intermediate node while it and its ACK (if any) propagate through the channel. This behavior may be modeled using the queueing network shown in Fig. 5. For simplicity, we again assume that all nodes along the VC are homogeneous. The FCFS $M/M/1$ queue with rate $1/\mu_{\text{trans}}$ again models the transmitter at node i (although as discussed below, the number of customers waiting in this queue does *not* represent the number of messages buffered at node i). Customers in the infinite server queue represent messages for which a timeout interval has been started; each customer experiences a *fixed* delay of $1/\mu_2 = 2/\mu_{\text{prop}}$ in this queue. We may think of this as modeling the propagation delay of a message from node i to $i + 1$ plus the propagation of the ACK from $i + 1$ back to i, although the analogy is not exact since a lost or corrupted message generates no ACK. Thus, the total number of messages actually buffered at node i is given by the sum of the number of customers in the FCFS and IS queues. The fixed buffer capacity (K) and the fact that a node buffers a message while waiting for an ACK is modeled by not permitting packets to enter the node/link queueing model (i.e., enclosed by dashed lines in Fig. 5) when this sum is equal to K.

We are interested in the stationary distribution of the number of packets in the transmission queue, $L_{1,i}$ and the number of packets for which the timeout interval has started, $L_{2,i}$. Let $\pi_{i,m,n} = P[L_{1,i} = m, L_{2,i} = n]$, $0 \leq m, n \leq K$, $m + n \leq K$. Because this queueing network has been shown to have a product form queue length distribution these probabilities can be written as [15]

$$\pi_{i,m,n} = G^{-1} \rho_{1,i}^m \frac{\rho_{2,i}^n}{n!}, \quad 0 \leq m, n \leq K, \ m + n \leq K$$

(7)

where

$$\rho_{1,i} = \lambda_i (1 - p)/\mu_{\text{trans}},$$
$$\rho_{2,i} = \lambda_i (1 - p)/\mu_2.$$

Here, G^{-1} is a normalization constant given as

$$G = \frac{1 - \rho_1}{[S_K(\rho_2) - \rho_1^{K+1} S_K(\rho_2/\rho_1)]}$$

(8)

where

$$S_K(x) = \sum_{j=0}^{K} \frac{x^j}{j!}.$$

The probability q_i that all buffers in node i are full is given by

$$q_i = \sum_{m=0}^{K} \pi_{i,m,K-m}$$

(9)

and the average number of messages waiting to be transmitted along the VC is given by

$$E[L_{1,i}] = \sum_{m=0}^{K} m \sum_{n=0}^{K-m} \pi_{i,m,n}.$$

(10)

From Little's law, the amount of time a message spends actively queued waiting for transmission (i.e., as opposed to waiting for a timeout to expire) is given by

$$E[W_{1,i}] = E[L_{1,i}]/\lambda_i (1 - f_i)$$

(11)

where again $f_i = q_i + p - q_i p$ is the probability that a messge is either received in error or finds a full buffer at node i and the time spent propagating in the channel is again $1/\mu_{\text{prop}}$.

As in the case of end-to-end error control, we define $E[W_i]$ to be the average delay (queueing, transmission, and propagation) a message experiences passing through link i. Note that in general $E[W_i] \neq E[W_{1,i}] + 1/\mu_{\text{prop}}$ since the link-by-link case, a message may be retransmitted several times before it is successfully received at node $i + 1$. However, we may proceed in a manner analogous to that used in Section III-C and compute $E[W_i]$ as follows. The following definitions will be useful.

N_t^i—the number of times a packet must be *retransmitted* at node i before it is successfully received at the station $i + 1$. Under the assumption that the outcome of each packet retransmission is independent, the distribution of N_t^i is $P[N_t^i = k] = f_{i+1}^k (1 - f_{i+1})$, $k = 1, \cdots$. We thus have

$$E[N_t^i] = \frac{f_{i+1}}{1 - f_{i+1}}.$$

(12)

T_{ll}—the link timeout interval. The timeout interval again begins when a message's *transmission* is completed. In our examples, it is taken to be equal to twice the propagation delay, i.e., $T_{ll} = 2/\mu_{\text{prop}}$.

A packet's delay between its initial arrival at node i and its eventual successful reception at node $i + 1$ results from the N_t^i timeout intervals (during which it is not successfully transmitted) plus the delays it experiences during its successful transmission. Hence,

$$E[W_i] = E[N_t^i] (E[W_{1,i}] + T_{ll}) + E[W_{1,i}] + 1/\mu_{\text{prop}}.$$

(13)

Equations (13) and (9) give the average delay and blocking probability at node i. As in the end-to-end case, they are each functions of the still undetermined values, $\{\lambda_i\}$.

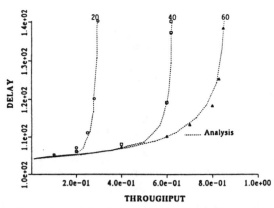

Fig. 6. Comparison of simulation and analysis: the link-by-link case.

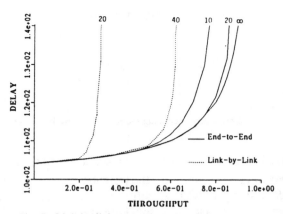

Fig. 7. Link-by-link versus end-to-end: baseline case.

C. VC-Level Model for Link-by-Link Error Control

The link-by-link VC-level model is *identical* to the end-to-end model; the only difference is that the blocking probabilities are computed using (9) rather than (3).

The model thus again consists of M link-by-link models in tandem. In order to sustain an output rate λ_{VC} of *correctly* received messages from the last link, the overall throughput rate through link M must be $\lambda_{VC}/(1 - p)$. Since a fraction $(1 - p)(1 - q)$ of the messages arriving to link M will either be blocked or in error, arrival rate of messages (first time arrivals plus retransmitted messages) to link M must also be given by (4). Once again, we can now solve for the blocking probability, arrival rate, and average delay for the last hop (M) on the VC, use these values to solve for corresponding values for link $M - 1$, and again continue to solve the VC-level model working from the end of the VC back towards the front.

A packet's delay between its initial arrival at the VC source node and its eventual successful reception at the VC destination node is simply the sum of the individual delays it experiences in each link. Thus, for the expected value of this delay, $E[W_{ll}]$, we have

$$E[W_{ll}] = \sum_{i=1}^{M} E[W_i] \quad (14)$$

D. Discussion

In order to again validate the assumptions underlying our analytic model, we compare analytic results to performance results obtained via simulation. Fig. 6 shows both simulation and analytic results plotting $E[W_{ll}]$ versus the virtual circuit arrival rate λ_{VC} for the cases of K = 20, 40, and 60 buffers at each intermediate node. All network parameters are identical to those used in Fig. 4, and the link-level timeout interval T_{ll} was taken to be 56. We again note that as a general trend, as the number of buffers increases, the maximum throughput the protocol can support also increases. We also note that the simulation results (indicated as point values) were obtained *without* Kleinrock's independence assumption. The obviously close correspondence between the analytic and the simulation results again indicates that reasonable assumptions were indeed introduced into the analytic model.

V. COMPARISON OF LINK-BY-LINK VERSUS END-TO-END

Having developed and validated our analytic models of the link-by-link and end-to-end error control protocols, we now use these models to compare the performance of these protocols for various network parameters of interest.

In Fig. 7, we compare the average end-to-end delay along the VC (i.e., the average amount of time between a message's initial arrival at the VC source node and its successful reception at the VC destination node) for the link-by-link and end-to-end protocols for the indicated intermediate-node buffer sizes. The results shown are for the network parameters given in Sections III-D and IV-D.

The results indicate that for low throughput (λ_{VC}) values, the performance of the end-to-end and link-by-link schemes is essentially identical (our numerical results indicate that link-by-link performs slightly better, but by an amount small enough so that the plotted results are indistinguishable). As the VC throughput increases, however, for the same number of buffers, the link-by-link scheme reaches saturation throughput sooner than the end-to-end scheme; consequently, higher delays are also experienced as the link-by-link throughput approaches its saturation value. This results from the fact that the link-by-link scheme continues to buffer a message as the message (and its ACK) propagate across a link; the end-to-end protocol releases the buffer immediately upon completing message transmission, effectively letting the *media* buffer a copy of the message. Thus, at higher throughputs, nodes are blocked less frequently in the end-to-end scheme, and hence the delays are lower and the maximal throughput higher.

Other interesting results are also illustrated in Fig. 7. Note that when either of the protocols is not operating near its saturation throughput, increasing the number of buffers at the intermediate nodes has little effect. Also, when the protocol is operating near saturation throughput, there is little to be gained by increasing the number of buffers beyond a certain value; in the end-to-end scheme, for example, the performance with 20 buffers at each node is quite close to the performance with an infinite number of buffers at each node.

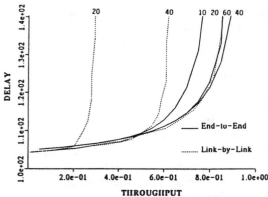

Fig. 8. Link-by-link versus end-to-end, $p = 0.001$.

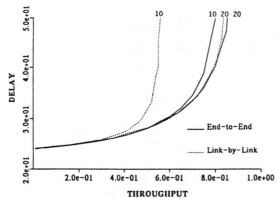

Fig. 9. Link-by-link versus end-to-end, effects of a decreased propagation delay.

In Fig. 8, the message error probability has been increased by a factor of 100; this would correspond to a bit error rate of 10^{-6}, high for a fiber optic link but reasonable for other transmission media. In this case, for a fixed number of buffers, the link-by-link scheme performs marginally better than the end-to-end scheme at lower throughput values. This results from the fact that the link-by-link protocol can recover quickly when an error occurs, i.e., after the link-by-link timeout interval has expired, while end-to-end approach must wait for the longer end-to-end timeout to occur. As the throughput increases, however, the increased buffering requirements, and hence blocking probabilities, become significant and the end-to-end scheme again eventually achieves superior performance.

In Fig. 9, the propagation delay of all links has been decreased to 5 times the message transmission time, and $p = 10^{-5}$ once again. Since the intermediate node buffers are now occupied for a shorter time, the relative performance of link-by-link improves with respect to the baseline case of Fig. 7. The blocking probabilities for the end-to-end case are unaffected, and the delay versus throughput curves for the end-to-end scheme are simply shifted down, due to the decreased propagation delays.

When the propagation delay of a single link is *increased*, the relative performance of link-by-link worsens. Fig. 10 shows the performance of the two schemes when the propagation delay of link 3 is increased by a factor of five. Again, the relative shape of the end-to-end curves remains unchanged, although the absolute value of the average delay is now higher. However, the longer link now becomes a bottleneck for the link-by-link scheme and its performance shows significant deterioration. These effects can be mitigated, of course, but only at the expense of providing additional buffers at the link with the longer propagation delay. For example, the link-by-link curves for 40 and 60 buffers would shift to the curves labeled A and B, respectively, if the number of buffers at this link were doubled.

Fig. 11 shows the effect of increasing the number of hops on the VC from 4 to 10. In order to understand the effects of increasing the number of hops, it is worthwhile reviewing the advantages and disadvantages of the end-

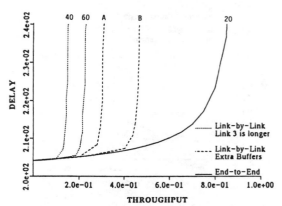

Fig. 10. Link-by-link versus end-to-end with one longer link.

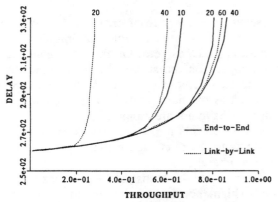

Fig. 11. Link-by-link versus end-to-end, $M = 10$ hops.

to-end and link-by-link schemes. End-to-end has the disadvantage that it recovers from an error or buffer overflow more inefficiently than link-by-link. Specifically, if either of these events occur on a link far from the source, then end-to-end "wastes" the successful transmissions over all earlier links. The problem with link-by-link is that it requires buffer storage for each packet, until it is acknowledged. Fig. 11 indicates that the relative difference in the performance of the two protocols decreases as the number of hops increases. We conjecture that, as the number of hops increases even more, the performance of link-by-link will approach and surpass that of end-to-end. This is because the inefficiency in the method by which the end-

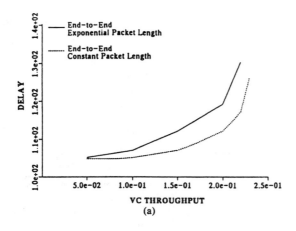

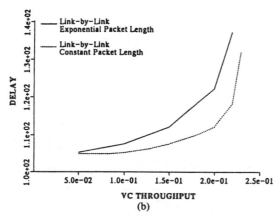

Fig. 12. Exponential versus constant message lengths.

to-end scheme recovers from errors or buffer overflows worsens to the point that the maximum sustainable throughput approaches zero. On the other hand, the constraints imposed on the sustainable throughput by finite buffer limitations is not sufficient to reduce the throughput to zero even as the number of hops approaches infinity. Although we have not attempted to determine the number of links at which this crossover occurs, we are confident that it occurs for large enough values of M so as to not be of practical interest.

Finally, we note that constant length packets are often assumed in high-speed switch architectures [7], [18], [20], although this is not always the case [6]. In order to examine the effect of constant message sizes, we again simulated a single VC, but in this case we explicitly introduced interfering traffic (at a Poison rate of $3\lambda_{VC}$) at each node along the VC, as shown in Fig. 1(a); this was done to avoid the artificial "pipelining" of constant length messages as they pass along a VC. Although the interfering traffic did not share the same buffer pools, it did utilize the outgoing link. Fig. 12 shows that in both the link-by-link and end-to-end schemes, the average delay in the case of constant length messages is uniformly lower than with exponentially distributed message lengths. Note however, that the *relative* differences in performance are almost identical in both cases. Thus, a central thesis of this paper remains unchanged, end-to-end error protocols still pro-

vide equivalent or superior performance over the range of parameters of practical interest.

VI. ERROR CONTROL WITH SHARED VC BUFFER POOLS

In previous sections, we assumed that each VC was allocated a separate pool of K buffers at each node; the effect of other VC's routed over the same outgoing link was modeled by reducing the link service capacity seen by the VC under study. An alternate buffering approach is for all VC's routed over the same outgoing link at a node to share a single common buffer pool. In this section, we examine this latter approach and its impact on the performance of link-by-link and end-to-end error control schemes. Since the models developed in this section are similar to our previous separate-buffer pool models and since the overall conclusions are identical (end-to-end is again shown to be superior to link-by-link), our discussion of the models will be somewhat abbreviated.

The modeling challenge is studying buffer sharing by multiple VC's is to avoid the complexity of explicitly modeling (in their entirety) each of the VC's sharing a buffer pool at each of the nodes along a given VC. Fortunately, in the case of homogeneous networks (of arbitrary topology) considered below in which all nodes are statistically identical (i.e., have the same traffic characteristics, buffering capacities, and service rates), this complexity can be avoided. In this case, we can study the behavior of interfering VC's along a *single* outgoing link and then extrapolate to obtain the performance of an entire VC.

We again consider a network in which each VC is M hops long, μ_0 is the service rate of the link, $1/\mu_{prop}$ is the propagation delay along a link, and K buffers are available (on a shared basis) for all VC's routed along an outgoing link. For simplicity, we assume that each node initiates and terminates a single VC, although this restriction can easily be relaxed. p is again the probability that a message is corrupted during transmission and q is the probability that a message finds all buffers occupied at a node; note that q is the same for all nodes, again due to the homogeneity condition.

An important network parameter once again is the *offered traffic* γ at each node. (We use the notation γ, instead of λ to differentiate the shared-buffer models from the separate buffer models.) Here, γ represents the rate at which messages arrive for transmission at a node and, includes both first time message transmissions as well as retransmissions. Note that as a result of the homogeneity condition, γ is now the same for *every* link in the network. In the following, we will find it useful to decompose γ into $\gamma_1, \ldots, \gamma_M$ where γ_i indicates traffic for which the link under study is the ith hop on its VC.

A. End-to-End Error Control: The Shared Buffer Approach

Our model of a single link with end-to-end control is shown in Fig. 13 (for the case $M = 4$). Note that it is

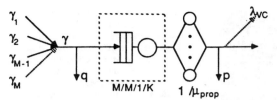

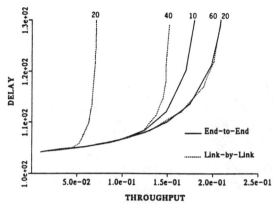

Fig. 13. Link-level model for end-to-end control with shared buffer pools.

Fig. 14. Link-by-link versus end-to-end:the shared buffer case.

similar to Fig. 2, except that the offered traffic now contains traffic from multiple VC's. In Fig. 13, λ_{VC} again represents the rate at which traffic initially enters a VC and thus also the rate at which successfully received traffic exits the network at a destination node. From flow conservation and the homogeneity condition, we have

$$\lambda_{VC} = \gamma_M (1 - f)$$

$$\gamma_i = \gamma_{i-1}(1 - f) \quad i = 2, \cdots, M$$

where $f = p + q - pq$ is the probability that a message is either blocked at, or incorrectly transmitted by a node. Note that f is the same for all nodes, again due to the homogeneity condition, Using the fact that $\gamma = \Sigma_{i=1}^{M} \gamma_i$, we have after some simple rearrangement

$$\lambda_{VC} = \gamma \frac{f(1 - f)^M}{1 - (1 - f)^M}. \tag{15}$$

Defining $\rho = \gamma / \mu_0$, we again have for the $M/M/1/K$ queue [1]

$$q = \frac{(1 - \gamma)\gamma^K}{1 - \gamma^{K+1}} \tag{16}$$

$$E[W_*] = \frac{1}{\gamma(1 - q)}$$
$$\cdot \left(\frac{\rho}{1 - \rho} - \frac{(K + 1)\rho^{K+1}}{1 - \rho^{K+1}} \right) + 1/\mu_{prop} \tag{17}$$

where $E[W_*]$ is the average delay (queueing, transmission and propagation) a message experiences passing through a link. For a given offered traffic, we can now solve (16) for q and then calculate a corresponding value of λ_{VC} using (15).

In order to incorporate the single link model into a VC-level model of end-to-end error control, we again follow the approach of Section III-C, except that p_{fail} must now be computed by

$$p_{fail} = q + (1 - q)f \sum_{i=0}^{M-1} (1 - f)^i \tag{18}$$

since a message entering the network can be blocked (and thus require later retransmission) if all the buffers at the first node are full. Using this value of p_{fail}, we can compute the total delay of a message between its initial arrival to the network and its eventual successful reception at the destination station under end-to-end error control as

$$E[W_{ee}] = E[N_t]T_{ee} + \sum_{k=1}^{M} E[W_*] \tag{19}$$

where $E[N_t]$ is the expected number of times a message is retransmitted [computed using (18) and (5)] and T_{ee} is the end-to-end timeout interval.

B. Link-by-Link Error Control: The Shared Buffer Approach

The link-level and VC-level models are similar to their counterparts described in Sections IV-B and C. The method used to solve these models is similar to the method described for solving the model for end-to-end error control with shared buffers described in the previous section. The reader is referred to [2] for details of this model.

C. Comparison of Link-by-Link Versus End-to-End to Shared Buffers

Fig. 14 shows the time delay versus throughput performance comparisons of the two error control schemes using the same network parameters as the baseline case of Fig. 7. The results indicate that in the case of shared buffers, an end-to-end again provides a performance level equivalent or superior to the link-by-link scheme. We also note that additional studies (not reported here) have shown that in the case of increased error probabilities, decreased/increased propagation delays, and increased VC hop lengths, the shared-buffer models also behave similarly to the separate-buffer models.

VII. CONCLUSION

In this paper, we have examined the performance of a link-by-link versus end-to-end approach for handling the loss and/or corruption of messages as they are transmitted between two end users in a high-speed network. Both analytic and simulation performance models were developed for comparing the performance of these two schemes. Our study has shown that for the range of network parameters of practical interest, an end-to-end approach towards error control provides equivalent or better performance while at the same time requiring fewer network resources (e.g., buffers, computation time) than the link-by-link approach. This was shown to be the case even when the (higher) node processing demands of the link-by-link approach were ignored and an overly pessimistic analytic model was used for the end-to-end approach.

Research in the area of protocols for high-speed networks has just begun. Our present work has shown that traditional wisdom, at least in the case of error control protocols, is no longer valid in a high-speed environment. We believe that the same may be true for other high-level protocols such as routing, flow control, and congestion control in a HSN. The exploration of these issues is thus a promising and challenging area for future research.

REFERENCES

[1] A. Allen, *Probability, Statistics and Queueing Theory*. New York: Academic, 1978.
[2] A. Bhargava, J. F. Kurose, D. Towsley, and G. VanLeemput, "Performance comparison of error control schemes in high speed computer communication networks," Comput. Inform. Sci. Dep., Univ. of Massachusetts, Amherst, MA, Tech. Rep. TR87-92, Sept. 1987.
[3] P. Brady, "Performance of an edge-to-edge protocol in a simulated X.25/X.75 packet network," *IEEE J. Select. Areas Commun.*, vol. 6, pp. 190–196, Jan. 1988.
[4] D. Carlson, "Bit-oriented data link control procedures," *IEEE Trans. Commun.*, vol. COM-28, pp. 455–467, Apr. 1980.
[5] K. M. Chandy, J. Howard, and D. Towsley, "Product form and local balance in queueing networks," *JACM*, vol. 24, no. 2, pp. 250–262, Apr. 1977.
[6] K. Y. Eng, M. Hluchyj, and Y. Yeh, "A knockout switch for variable-length messages," in *Proc. IEEE ICC '87*, Seattle, WA, pp. 22.6.1–22.6.5.
[7] A. Huang and S. Knauer, "Starlite: A wideband digital switch," in *Proc. IEEE GLOBCOM '84*, Atlanta, GA, pp. 121–125.
[8] Special Issue of Fiber Optic Systems for Terrestrial Applications, *IEEE J. Select. Areas Commun.*, vol. SAC-4, Dec. 1986.
[9] M. Irland and G. Pujolle, "Comparison of two packet retransmission techniques," *IEEE Trans. Inform. Theory*, vol. IT-26, pp. 92–97.
[10] H. Kobayashi, *Modeling and Analysis*. Reading, MA: Addison-Wesley, 1978.
[11] D. Kuhl, "Error recovery protocols: Link-by-link versus edge-to-edge," in *Proc. IEEE INFOCOM '83*, pp. 319–324.
[12] S. Lam, "Store and forward buffer requirements in a packet switching network," *IEEE Trans. Commun.*, vol. COM-24, pp. 394–403, Apr. 1976.
[13] M. Pennotti and M. Schwartz, "Congestion control in store and forward tandem links," *IEEE Trans. Commun.*, vol. COM-23, pp. 1434–1443, Dec. 1975.
[14] C. Samelson and W. Bulgren, "A note on product form solution for queueing networks with Poisson arrivals and general service time distributions with finite means," *JACM*, vol. 29, no. 3, pp. 830–840, July 1982.
[15] P. Schweitzer and S. Lam, "Buffer overflow in a store and forward network node," *IBM J. Res. Develop.*, pp. 542–550, Nov. 1976.
[16] M. Schwartz, "Performance analysis of the SNA virtual route pacing control," *IEEE Trans. Commun.*, vol. COM-30, pp. 172–184, Jan. 1982.
[17] T. Suda and N. Watanabe, "Evaluation of error recovery schemes for a high-speed packet switched network: Link-by-link versus edge-to-edge schemes, in *Proc. IEEE INFOCOM '88*, New Orleans, LA, Apr. 1988, pp. 722–731.
[18] J. Turner, "Design of a broadcast packet network," in *Proc. IEEE INFOCOM '86*, Miami, FL, pp. 667–675.
[19] ——, "Design of an integrated services packet network," *IEEE J. Select. Areas Commun.*, pp. 1373–1379.
[20] Y. Yeh, M. Hluchyj, and A. Acampora, "The knockout switch: A simple modular architecture for high performance packet switching," presented at Proc. Int. Switch. Symp., Phoenix, AZ.

Amit Bhargava (S'84–M'88) was born in Lucknow, India, on October 24, 1963. He received the B.Tech. degree in electrical engineering from the Indian Institute of Technology, Bombay, India, in 1985, and the M.S. degree in electrical and computer engineering from the University of Massachusetts, Amherst, MA, in 1987.

He is currently with Codex Corporation, Canton, MA. His current technical interests are in communication networks and modeling and performance analysis.

James F. Kurose (S'81–M'84) received the B.A. degree in physics from Wesleyan University, Middleton, CT, in 1978, and the M.S. and Ph.D. degrees in computer science from Columbia University, New York, NY, in 1980 and 1984, respectively.

Since 1984, he has been an Assistant Professor in the Department of Computer and Information Science at the University of Massachusetts, Amherst, where he currently leads several research efforts in the areas of computer communication networks, distributed systems, and modeling and performance evaluation.

Dr. Kurose is a member of Phi Beta Kappa, Sigma Xi, and the ACM.

Don Towsley (M'78) received the B.A. degree in physics and the Ph.D. degree in computer science from University of Texas in 1971 and 1975, respectively.

He joined the Department of Electrical and Computer Engineering at the Department of Computer and Information Sciences where he is currently an Associate Professor. His interests are in the area of computer communications, where he works on low-level and network-level protocols and on random access protocols. He also has interests in the area of modeling and analysis of distributed computer systems and parallel processing systems.

Dr. Towsley is currently an Associate Editor of *Networks* and IEEE TRANSACTIONS ON COMMUNICATIONS. He is also a member of ORSA and the ACM.

Guy VanLeemput received the Ingenieur Electricien degree from the University of Gent, Belgium, in 1985 and the M.S. degree in electrical and computer engineering from the University of Massachusetts, Amherst, MA, in 1987.

His research interests are in the areas of communication theory and communication networks. He is currently serving in the Belgian armed forces.

ON CELL SIZE AND HEADER ERROR CONTROL OF ASYNCHRONOUS TRANSFER MODE (ATM)

D. P. Hsing, F. Vakil and G. H. Estes

Bellcore, Morristown, NJ

ABSTRACT

The Asynchronous Transfer Mode (ATM) is a flexible technique that provides dynamic bandwidth allocation for the support of B-ISDN services. In an ATM network, all information is carried in fixed size cells, that consist of a header and an information field. Efforts are now underway to standardize ATM cell parameters. This article addresses two important issues related to the format of an ATM cell, the "optimum" size of the information field that maximizes the resource utilization, and the cell header error control. It is shown that from the resource utilization point of view an "optimum" size for the information field does exist, however its exact value is not critical from this perspective. The impact of cell header protection (*i.e.*, error correction and/or detection) on the cell loss of an ATM network is also investigated. It is shown that a CRC that can correct a single bit error significantly reduces the cell loss of the network and appears to be a sufficient mechanism for protection of the cell header against single bit transmission errors.

1. Introduction

Recent advances in the design of high-speed multiplexing, switching, and optical transmission systems coupled with the lure of potential new services have stimulated a great deal of interest in Integrated Services Digital Networks (ISDN) with broadband capabilities (*i.e.*, B-ISDN). Although the precise switching and transmission requirements of emerging broadband services remain to be determined,[1] it is crucial that the capabilities of Broadband ISDN (B-ISDN) be flexible.

The Asynchronous Transfer Mode (ATM) technique provides the necessary flexibility for supporting a wide spectrum of services in a B-ISDN environment. ATM is a connection-oriented technique that can be used for supporting both connection-oriented and connectionless services. In an ATM network, specific time slots are not assigned to a channel. The information is carried in cells that consist of a header and an information field. The header contains a label that uniquely identifies a channel and is used for multiplexing and routing.[2] Since an ATM interface structure consists of a set of labeled channels, it can dynamically adapt to the changes in the service demand and allocate the capacity accordingly. Hence, ATM avoids the complication associated with the multiplicity of positioned channels and enhances network integration.

Efforts are now underway to standardize ATM cell parameters. The objective of this article is to address two important issues related to the format of an ATM cell, the "optimum" length for the information field of an ATM cell that maximizes the resource utilization, and the appropriate error control strategy for cell header in an ATM network. Figure 1 shows an ATM cell.

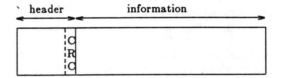

Figure 1. ATM cell

This paper is organized as follows: Section 2 discusses the problem of determining the "optimal" cell size in terms of maximizing the channel utilization rate which is the measure of network resource utilization. It is shown that from this perspective an "optimal" information field size exists that depends on the header size, the source message size, and the error characteristics of the channel. However, our results indicate that from a channel utilization point of view, the specific value of the information field size is not too critical and a broad range of values provides about the same average channel utilization. In section 3, we describe the rationale for the protection (*i.e.*, error detection and/or correction) of the cell header in an ATM network and investigate its impact on the cell loss of an ATM network. Assuming that a Cyclic Redundancy Code (CRC) is used as the protection mechanism, we obtain the cell loss as a function of the header length and CRC length, and compare it with the cell loss of an ATM network that has no header protection mechanism. These are used to determine the necessary degree of cell header protection in the ATM networks. Our results indicate that a CRC that can correct a single bit error is sufficient. Finally, section 4 summarizes our results and concludes the article with few open issues.

2. Size of the Information Field in ATM Cell

In general, implementation associated with the processing of a fixed-size cell is simpler than that of a variable-size cell. To simplify the processing and avoid potential bottlenecks, fixed-size cells are preferred for a B-ISDN environment that uses ATM techniques. Thus, a key issue is the choice of the "right" cell size. An ATM cell consists of a header and an information field and its length depends on several factors.

The length of the header depends on the required header functionalities and their implementation. *For instance*, each cell header contains route information. Since ATM is a connection-oriented technique, the route information is obtained from a logical channel number in the header. The size of this field in the header depends on the maximum number of concurrent logical channels per node in the network. On a single User-Network access link, this number is anticipated to be less than the one on a remote electronics (RE) feeder or a central office (CO) trunk. However, it is desirable to have the same header size on all interfaces, including customer access, feeder, and internal network links[2] and the LCN size should be chosen such that it imposes no unnecessary constraint on any of these interfaces. The United States Standards Committee T1 has agreed that the B-ISDN User-Network Interface (UNI) should use the same signal formats and bit rates as those in the newly created draft U.S. standard for synchronous optical networks.[3] For instance, a B-ISDN UNI will be constructed from the Concatenated Synchronous Transport Signal level 3 (STS-3c). The payload capacity available for carrying ATM cells will be the same in the access network as in the internal network. Hence the reason to retain the same size ATM cell header throughout the network. Besides carrying route information, a cell header may include additional fields such as priority and quality of service (QOS), *etc*.[4]

For a given header size, the corresponding length of the information portion is determined by a balanced trade-off among resource utilization, impact on service quality, and processing power requirement. When ATM cells are used for some circuit emulation services, e.g. basic and primary rate circuit switched services, the effect of lost cells on service quality depends on the size of the ATM cell information field. Waiting time jitter may be worse for larger cell size which will affect end-to-end cell delay and the network node buffer requirement. End-to-end cell delay has impact on the service quality of the end user. And buffer requirement may affect the implementation cost. To guarantee acceptable system performance, in ATM approaches a cell should be processed in the same amount of time it takes to be transmitted. Cell size thus has direct impact on processing power requirement which will be reflected in the implementation cost also. All these are some examples of factors that need to be investigated in choosing the ATM cell information field. However, this section only addresses the resource utilization issue. It is apparent that for a given header size, small information field makes the system to operate less efficiently, i.e., at a lower capacity utilization. A larger fraction of the system capacity is used to transport control overhead rather than information. On the other hand, if the information field is too long, the cell is more likely to be received with errors. This will cause more retransmissions for traffic types that utilize retransmission, e.g., most forms of data communications, and therefore reduce true information transfer rate. When segmentation of the source message to cells is performed, the wasted capacity of the unfilled last cell is another factor that affects system efficiency. The following analysis deals with the impact of all these factors: channel error rate, the unfilled last cell, the header size and the information field size, on the resource utilization.

2.1 Modeling and Analysis

The "optimal" value of the information field length, S_o, is the one that maximizes average channel utilization rate which is the measure of network resource utilization. The model used in our analysis assumes that:

i. A cell will be retransmitted until it is received correctly on an end-to-end basis, *i.e.*, the retransmission is performed on an end-to-end basis at the ATM layer. Note that this assumption might not be true in an ATM-based B-ISDN. Retransmission might be performed on a per source message basis at a layer higher than ATM layer. Furthermore, the retransmission of cells (or messages) for real time services is impractical. By assigning a different bit error rate, this model can still be used to analyze the effect of the channel error rate on resource utilization efficiency and obtain S_o.

ii. Source message size is geometrically distributed with parameter a[5] (source message is the higher layer information unit passed to the ATM layer).

iii. The error characteristics of the medium is that of a Binary Symmetric Channel (BSC) with parameter p, *i.e.*, bit errors occur independently with probability p.

For a given header size, S_o exists which depends on the distribution of the source message size and the error characteristics of the channel.

The key measure of the resource utilization is the average channel utilization rate that is defined as:

$$U = \bar{S}_i \; / \; (average\ actual\ number\ of\ octets\ transmitted)\ , \qquad (1)$$

where $\bar{S}_i$ is the average number of octets per source message and the average actual number of octets transmitted per source message includes all the extra octets due to unfilled last cell, cell overhead, and retransmission. The source message size, S_i, may differ from one to another and is best modeled by a random variable. Let $P[S_i = s]$ be the probability of S_i equal to s octets. In the following analysis, we assume that the size of the source message has a geometric distribution, *i.e.*,

$$P[S_i = s] = aq^{(s-1)}, \qquad s = 1,2,3,\ldots \qquad (2)$$

where $a = \dfrac{1}{\bar{S}_i}$, $\bar{S}_i$ is the average source message size in octets, and $a + q = 1$. (Information generated from one circuit connection, such as a voice connection, is considered as one single large source message.)

Since the source message is partitioned into fixed size cells, the expected number of cells per source message equals to

$$
\begin{aligned}
\bar{N}(S) &= \sum_{n=1}^{\infty} n \times P[(n-1)S < S_i \le nS] \\
&= \sum_{n=1}^{\infty} n \times \sum_{s=(n-1)S+1}^{nS} a \times q^{(s-1)} \\
&= \sum_{n=1}^{\infty} n \times (1-q^S) \times (q^S)^{(n-1)} \\
&= \frac{1}{(1-q^S)} .
\end{aligned} \qquad (3)
$$

where S is the cell information field size in octets. Let H be cell header size in octets. Thus, the total cell size is equal to $H+S$ octets. The probability of receiving a cell in error depends on the total cell size and the error characteristics of the channel. Since the bit error rate of the channel is p, the probability of an octet received in error, K, is equal to

$$K = 1 - (1-p)^8$$
$$\approx 8 \times p \quad \text{for small } p \quad , \tag{4}$$

and the error probability of a $H+S$ octets long cell, $PE(H+S)$, is equal to

$$PE(H+S) = 1 - (1-K)^{H+S}$$
$$\approx (H+S)K \quad \text{for small } K \quad . \tag{5}$$

Note that $PE(H+S)$ represents the probability of retransmitting a cell. A cell will be retransmitted until it is received correctly, therefore the average number of transmissions per cell is equal to

$$NT(S) = \sum_{i=1}^{\infty} i \times (1-PE(H+S))(PE(H+S))^{i-1}$$
$$= \frac{1}{(1-PE(H+S))} = \frac{1}{(1-(H+S)K)} \quad \text{for small } K. \tag{6}$$

Hence, the average octets transmitted per source message is equal to

$$T(S) = N(S) \times NT(S) \times (H+S)$$
$$= N(S) \times \frac{(H+S)}{(1-(H+S)K)} \quad . \tag{7}$$

Then the average channel utilization, U, is equal to

$$U = S_i / T(S) = \frac{S_i(1-(H+S)K)}{N(S)(H+S)}$$
$$= \frac{S_i(1-(H+S)K)(1-q^S)}{(H+S)} \tag{8}$$

where $q = 1 - \frac{1}{S_i}$. The S_o "optimal" size of the information field is the value of S that maximizes U. In appendix A, we prove that the maximum value of U does exist.

2.2 Numerical Results and Discussions

Using the Newton iterative method, we obtain the "optimum" value of S for different values of H, and S_i. Throughout the examples of this section, we assume that the bit error rate p is 10^{-9} and terminate the iteration when the difference between the current and previous values of S is less than unity.

Figure 2 represents the S_o versus header size for $S_i = 256$, 512 and 1000 octets. Figures 3 - 5 show channel utilization versus information field for header sizes equal to 4, 6, and 8 octets and for $S_i = 256$, 512 and 1000 octets. Table 1 includes the S_o for a selected set of H and S_i.

Table 1. S_o for Various Message Length

H	S_i		
	256	512	1000
4	44	63	89
6	54	77	108
8	62	88	124

The results of the analysis indicate that in the neighborhood of the S_o, channel utilization rate is relatively insensitive to the information field size, especially on the plus side. It is also observed from the results that the maximum channel utilization rate increases as average source message size increases or header size decreases. However, decreasing header size increases maximum channel utilization rate at a lesser rate for larger S_i. Furthermore, for a given header size, the S_o increases with the average source message size. Table 2 shows the range of values for the information field size, S, that provides no worse than 1.5% less efficiency than the maximum achievable channel utilization rate for the selected values of header size, H, and average source message size, S_i.

Table 2. Ranges of S for Various Message Length

H	S_i		
	256	512	1000
4	28 - 70	37 - 107	48 - 163
6	36 - 82	48 - 124	62 - 188
8	42 - 92	56 - 140	72 - 210+

For a given header size (which depends on the required header functionalities and their implementations), as long as the information field size is chosen within the range, the channel utilization rate remains close to the achievable maximum.

Assuming a geometric distribution for the source message size, we have derived a simple measure for resource utilization. It has been shown that a geometric distribution is a "reasonable" model for the size of source message in time sharing computer systems.[6] However, the validity of this assumption in a broadband ISDN environment remains to be proven and needs further investigation. For instance, the collected data from working LANs indicate that the source message size has a bimodal distribution.[6] [7] [8] If the information size distribution is replaced by a bimodal distribution in our analysis, the resource utilization measure will have a different form. An analytical proof of the existence of an optimal solution appears to be difficult. However a similar numerical approach can be used to show that the range of information field obtained in this paper overlaps with the range obtained if a bimodal distribution were used. One example is given in Figure 6 where two curves of utilization versus information field are presented. The dotted line corresponds to a bimodal distribution for source message size while the solid line represents the results of a geometric distribution. In this example, the bimodal distribution is modeled by two equally weighted normal distributions, where one has a mean of 64 bytes and variance of 64 bytes, and the other has a mean of 1500 bytes and variance of 400 bytes. In this case, the probability of having n cells per source message, i.e., $P[(n-1)S < S_i \leq nS]$ in equation (3), equals to

$$(\Phi[\frac{nS-\mu_1}{\sigma_1}] - \Phi[\frac{(n-1)S-\mu_1}{\sigma_1}] + \Phi[\frac{nS-\mu_2}{\sigma_2}] - \Phi[\frac{(n-1)S-\mu_2}{\sigma_2}]) / 2 \quad .$$

Here μ_1, μ_2, σ_1^2, σ_2^2 are the corresponding means and variances, and Φ is the standard normal distribution. The geometric distribution considered in this example has a mean of 782 bytes. It is currently difficult to identify the distribution of the source message in a B-ISDN environment. Suppose that the source message size has a multi-modal distribution. One *conjecture* is that the geometric distribution model used here will arrive at a "reasonable" range of cell information field, if the mean of the geometric distribution is the same as that of the multi-modal one.

3. Cell Header Error Control

In ATM networks, there are two sources of cell loss, congestion at the cell processors, and errors in the cell header, particularly in the routing information, caused by the transmission system. This section deals with the latter. If a single bit or a burst error occurs within the information field of the ATM cell, its effect will be similar to that experienced in the current circuit or packet switching networks. We *assume* that if the error event corrupts the cell header, the cell will either be lost or misdelivered to a wrong receiver.[1] Thus, the potential effect is that a single bit transmission error causes the intended receiver not to receive an entire cell, *i.e.*, experience a burst of errors. This process is known as error multiplication.[9]

To reduce error multiplication, an ATM network shall be able to *correct* and/or *detect* potential errors in a cell header.[10] From a sender's point of view, the cells whose headers are corrupted will be lost regardless of detecting the errors in their headers or not. Therefore, a sole error detection mechanism does not repair the adverse effect of lost cells on the quality of the sender's service. In the case of non-isochronous services, the *end−to−end* protocols restore the quality of service through the request for retransmission of the errored cells or errored messages that are corrupted by the loss of one or more cells. Due to the real time constraint of isochronous services, the retransmission of corrupted isochronous cells or messages (*i.e.*, messages that have lost some cells) is impractical. Some isochronous services[2] have a strict cell loss requirement, for instance, circuit emulation requires cell loss[11] of about 10^{-10}. The average BER of some fiber optic transmission system is 10^{-9}. Hence, the cell loss caused by errors in the cell header already exceeds the proposed requirement. Later, it is shown that a CRC that can correct a single bit error, reduces the cell loss of an ATM network significantly and enables the ATM network to meet the strict cell loss requirement of some isochronous services. The error correction may be either performed on the whole header or only on the logical channel number (LCN) field. The additional complexity for implementing a CRC that corrects a single bit error is negligible.

Having implemented a CRC that correct a single bit error, the network is automatically capable of detecting multiple-bit errors occurred in cell header (to some extent) and

discarding the corresponding cells. The detection capability leads to

- *Reduced "crosstalk":* It reduces the effect of the interfering traffic (*i.e.* misdelivered cells) on the actual traffic of the receiver. *For instance*, in the case of circuit emulation on the ATM networks, the receiver may perceive the effect of misdelivered cells on the circuit as a "crosstalk".

- *Increased network efficiency:* It increases the efficiency of the network because it ensures that the corrupted cells are discarded as early as possible rather than passing through the entire network. It may also reduce the retransmission rate of non-isochronous messages in the network.

3.1 The Protection Mechanism

Since the CRC implementation at high speeds is relatively simple, we *assume* that a CRC field is used to correct (or detect) errors in the cell header. The length of the CRC field is r bits and the size of the cell header (excluding the CRC) is n bits. Let us recall some properties of the Cyclic Redundancy Codes (CRC) that will be used throughout this section. Consider a r bit long CRC field whose generator polynomial is of degree r, then,[12] [13]

i. if $r \geq 2$ then all single bit errors will be detected,

ii. if the generator polynomial, $g(x)$, is divisible by $x+1$ then all odd-number bit errors will be detected,

iii. if the polynomial contains the constant term, all burst errors of length $\leq r$ will be detected,

iv. the probability of not detecting a burst error of length $> r$ is 2^{1-r},

v. given n, to detect all two bit errors (*i.e.*, correct single bit errors), the degree of the generator polynomial, r, must be chosen such that $2^r \geq n+r+1$.

We assume that the generator polynomial of the selected CRC satisfies the premises of ii.), iii.) and v.). Although the standard value of the header length is still the subject of debate and remains to be determined, different values in the range of 32-64 bits have been considered. Hence, with r check bits, n is in the range of $32-r$ to $64-r$ bits. According to v.), if $n=24$ bits, r is 5 bits, $n=32, 40, 48,$ and 56 bits requires $r=6$ check bits.

In the analysis to follow, we define that:

- P_L is the probability of an uncorrected error in the cell header. From a sender's point of view, P_L, represents the cell loss probability.
- P_E is the probability of an undetected error in the cell header, *i.e.*, it is the probability that a corrupted header passes through the protection mechanism. Note that some of these cells may be misdelivered to a wrong receiver and P_E is the upper bound of the cell misdelivery probability.

3.2 No Header Protection

First, let us obtain P_L for a network that has no protection mechanism. Since the header is n bits and the bit errors occur independently with probability p, the probability of k bit errors in the unprotected header has a Binomial

---

1. In practice, an error in the header does not always cause cell misdelivery. There may be some fields in the header (depending on the exact header format) whose error may not result in cell misdelivery.

2. Although no empirical or theoretical results are available in the literature, depending on the compression algorithm used, some video services such as HDTV may also have a strict cell loss requirement.

distribution with parameters n and p. Hence, P_L is given by:

$$P_L = 1-(1-p)^n \approx np \quad , \qquad (9)$$

where the last term approximation is due to the fact that $p \ll 1$. Note that in this case the cell loss P_L is the same as P_E, i.e., $P_E = P_L$.

3.3 With Header Protection

The probability of k bit errors in the header has a Binomial distribution with parameters $n+r$ and p. Hence, with a cyclic code that can correct t bit errors, the cell loss (i.e., P_L) of the ATM network is given by:

$$P_L = 1 - \sum_{j=0}^{t} \binom{n+r}{j} p^j (1-p)^{n+r-j} \qquad (10)$$

If the CRC is only used to detect errors (i.e., $t=0$) in the cell header, from equation (10) the cell loss, P_L, is given by:

$$P_L = 1-(1-p)^{n+r} \approx (n+r)p \quad , \; p \ll 1 \qquad (11)$$

According to equation (10), if the CRC enables the network to correct a single bit (i.e., $t=1$) error in the cell header, the cell loss of the network, P_L, is given by:

$$P_L = 1-(1-p)^{n+r}-(n+r)p(1-p)^{n+r-1} \approx (n+r)(n+r-1)p^2/2 \quad , \; p \ll 1 \qquad (12)$$

Let us also obtain the probability of undetected errors (i.e., P_E) in the cell header. Note that P_E is less than P_L. If the CRC is used only for detection, a better bound for P_E can be obtained. The probability of k bit errors in the header has a Binomial distribution with parameters $n+r$ and p, and from iv.) the probability of not detecting a burst error of length $>r$ is 2^{1-r}. Hence, the probability that a header with k errors passes through the CRC check is given by

$$\binom{n+r}{k} p^k (1-p)^{n+r-k} 2^{1-r} \leq \binom{n+r}{k} p^k 2^{1-r} \qquad (13)$$

Since we have assumed that conditions ii.), iii.), and v.) are satisfied, P_E is the probability that an even number (which is >2 by assumption v.)) of bit errors occur in the header. Hence, the probability of misdelivering a cell is given by:

$$P_E \leq \sum_{l=2}^{[(n+r)/2]} \binom{n+r}{2l} p^{2l} 2^{1-r} \approx \binom{n+r}{4} p^4 2^{1-r} \qquad (14)$$

where $[X]$ represents the largest integer less than or equal to X. The last term approximation is due to the fact that $p \ll 1$ and the higher exponents of p are negligible compared to p^4.

3.4 Parity Bit Error Check

Another error protection mechanism that has been considered for an ATM network is a single parity check. This mechanism has no correction capability, however, it detects all odd numbers of bit errors in the header. Note that this mechanism can not detect any even number of bit errors. Hence, under the independent error model, one can write

$$P_E \leq \sum_{l=1}^{[(n+1)/2]} \binom{n+1}{2l} p^{2l} \approx \binom{n+1}{2} p^2 \quad . \qquad (15)$$

3.5 Numerical Results and Discussions

Using equations (9) and (12), and varying the values of p in the range of 10^{-6} to 10^{-13}, we obtain the cell loss, P_L, for different values of n and present the logarithm of them in Figures 7-11. Note that the value of p depends on the transmission medium of the network. Without drawing hasty conclusions, from the results in Figures 7-11, one can conclude that

- the addition of a single bit forward error correction in the header results in a significant reduction in the number of lost cells in an ATM network. The reduction of the network cell loss improves the quality of senders' services.

Using equations (9), (14), and (15), and varying p in the range of 10^{-6} to 10^{-13}, we obtain P_E for different values of r and n. The results are summarized in Figures 12-16. The values of n are in the range of 24 to 56 bits, while r has taken values of 5, and 6 (for $n=24$, and $n=32-56$, respectively), 8, and 16 bits. These results indicate that:

- the use of a CRC reduces the probability of undetected errors in the header and the probability of cell misdelivery.

- as the length of CRC increases, the probability of undetected errors and cell misdelivery rate are reduced further.

4. Conclusion and Open Issues

The article has dealt with the "optimum" size of the information field of an ATM cell that maximizes the channel utilization and the cell header error control. Our analysis results indicate that an "optimum" size for the information field of an ATM cell does exist, however, its specific value is not too critical when viewed from the particular aspect of channel utilization efficiency, because a broad range of values provides about the same average channel utilization rate. What remains to be investigated in order to decide the cell size is the impact of the other factors, such as the end-to-end delay especially for time-sensitive applications (e.g. voice/video services) and the required processing resource, on the information field size.

We also conclude that cell header error control is necessary. The use of error detection in cell header reduces cell misdelivering rate, however, it does not repair the effects of lost cell on the service quality perceived by the user. With a single bit forward error correction for the cell header, the cell loss of an ATM network can be improved significantly. The reduction of cell loss improves the quality of senders' services. It appears that a single bit forward error correction is a sufficient mechanism for protecting the cell header against single bit transmission errors. A key open issue is the impact of burst errors on ATM cells and the possible needs and mechanisms for protection of ATM cells against burst errors on the medium.

Appendix A:
Existence of an Optimum Information Field Size

The value of S that maximizes U is defined to be "optimum". To prove the maximum value of U does exist [14], we shall show that

a. $\dfrac{dU}{dS}$ exists for all S > 0,

b. there exist a $S_o > 0$ such that

$$\frac{dU}{dS} > 0 \qquad \text{for all } S < S_o \quad \text{and}$$

$$\frac{dU}{dS} < 0 \qquad \text{for all } S > S_o.$$

Taking the derivative of U with respect to S, we arrive at

$$\frac{dU}{dS} = \frac{S_i\,[(q^S-1) - q^S \ln q(1-(H+S)K)(H+S)]}{(H+S)^2}$$
$$- \frac{S_i q^S\,[(1-\ln q(1-(H+S)K)(H+S)) - q^{-S}]}{(H+S)^2}. \qquad \text{(A-1)}$$

Equation (A-1) indicates that a) is true. To prove b), let us define

$$f(S) = 1 - \ln q(1-(H+S)K)(H+S)$$
$$= S^2 K \ln q - S(1-2HK)\ln q + (1-H(1-HK)\ln q), \qquad \text{(A-2)}$$

$$g(S) = q^{-S}, \qquad \text{(A-3)}$$

and

$$F(S) = f(S) - g(S). \qquad \text{(A-4)}$$

Since

$$F(0) = -\ln q(1-HK)H > 0 \qquad \text{for } K < \frac{1}{H}$$

and

$$\frac{dF}{dS} = 2SK\ln q - (1-2HK)\ln q + q^{-S}\ln q$$
$$= (-\ln q)(-2KS + 1 - 2KH - q^{-S}) < 0 \qquad \text{for } S \geq 0$$

there exists a $S_o > 0$ such that $F(S_o)$ equals zero, $F(S) > 0$ for $S < S_o$, and $F(S) < 0$ for $S > S_o$. Equation (A-1) indicates $\dfrac{dU}{dS}$ is zero if and only if $F(S)$ is zero, and $\dfrac{dU}{dS}$ and $F(S)$ always have the same sign. This completes the proof of b) and shows that the S_o does exist. Let us utilize Newton's method [15] to obtain S_o, *i.e.*, to solve the following equation equation.

$$S^2 K\ln q - S(1-2HK)\ln q + (1-H(1-HK)\ln q) - q^{-S} = 0.$$

Throughout the examples, we terminate the iteration whenever the difference between the new value and the previous value of S is less than 1 octet.

REFERENCES

1. D. R. Spears, "Broadband ISDN Switching Capabilities from a Services Perspective," *Journal on Selected Areas of Communication*, Vol. SAC-5, No. 8, October, 1987.

2. D. P. Hsing, and S. Minzer, "Preliminary Special Report on Broadband ISDN Access" *Bellcore Special Report # SR-TSY-000857*, December 1987.

3. T1X1.4/87-505R4, "American National Standard for Telecommunications Digital Hierarchy Optical Interface Rates and Formats Specifications", December 1987.

4. T1D1.1/87-492, "Some Considerations in Selecting A Cell Header Size," Bellcore, September 1987.

5. E. Fuchs and P.E. Jackson, "Estimates of Distributions of Random Variables for Certain Computer Communications Traffic Model", *Communications of the ACM*, 13 (12), December, 1970, pp. 752-757.

6. J. Shoch and J. Hupp, "Measured Performance of an Ethernet Local Network". *Communications of the ACM*, 23(12), December 1980, pp 711-721.

7. J. Shoch and J. Hupp, "Performance of the Ethernet Local Network -A Preliminary Report", *Proceedings of the IEEE COMPCON80*, February 1980, pp. 318-322.

8. D. Feldmeier, "Empirical Analysis of a Token Ring Network", *Undergraduate Thesis*, Department of Electrical Engineering and Computer Science, MIT January, 1984.

9. R. J. Schweizer, and A. K. Reilly, "Error Multiplication and its Potential Implications on ATM Header Functionality", *Document # T1S1.1/88-198*.

10. *K. A. Rao, and B. H. Bharucha, "An Analysis of the Degree of Protection for ATM Cells" Document # T1D1.1/87-604.*

11. *R. J. Schweizer, "Initial Performance Objectives for B-ISDN", Bellcore Technical Memorandum.*

12. S. Lin, and D. J. Costello Jr., *Error Control Coding: Fundamentals and Applications*, Chapter 4, Prentice Hall, 1983.

13. R. G. Gallager, *Information Theory and Reliable Communications*, John Wiley & Sons, 1968.

14. Ellis and Gulick, "*Calculus with Analytic Geometry*", Chapter 4. Harcourt, Brace, Jovanovich, Inc., 1978.

15. E. Kreyszig, "*Advanced Engineering Mathematics*", pp. 765, 4th Edition, Wiley.

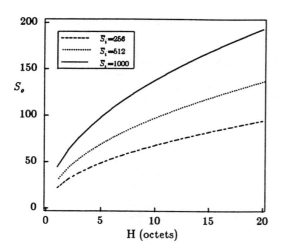

Figure 2. Optimal" Information Field vs Header Size

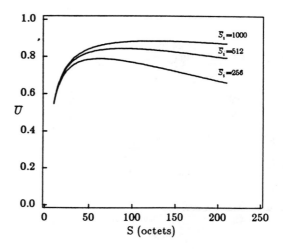

Figure 5. Utilization vs Information Field : $H=8$, $p=10^{-9}$

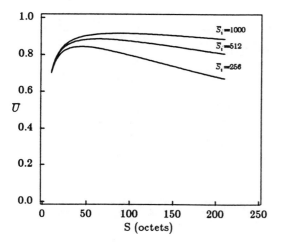

Figure 3. Utilization vs Information Field : $H=4$, $p=10^{-9}$

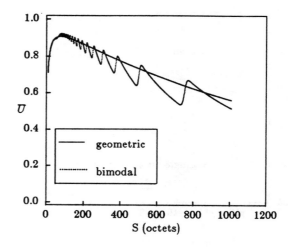

Figure 6. Utilization vs Information Field

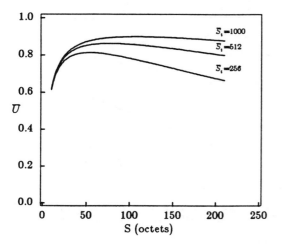

Figure 4. Utilization vs Information Field : $H=6$, $p=10^{-9}$

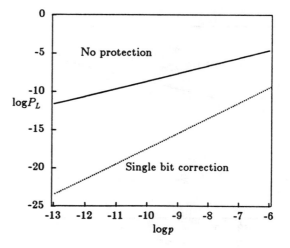

Figure 7. Cell loss vs. Bit Error Rate, $n=24$

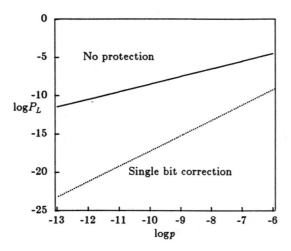

Figure 8. Cell loss vs. Bit Error Rate, $n=32$

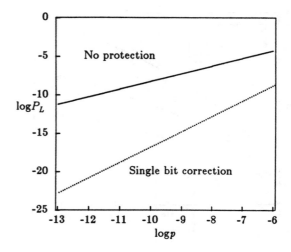

Figure 11. Cell loss vs. Bit Error Rate, $n=56$

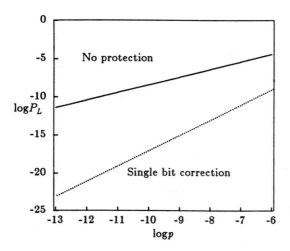

Figure 9. Cell loss vs. Bit Error Rate, $n=40$

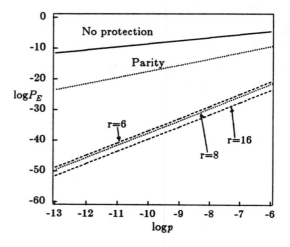

Figure 12 Probability of Undetected Error vs Bit Error Rate, $n=24$

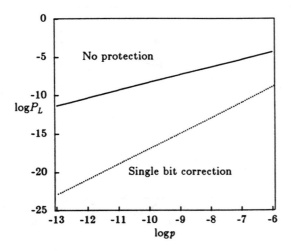

Figure 10. Cell loss vs. Bit Error Rate, $n=48$

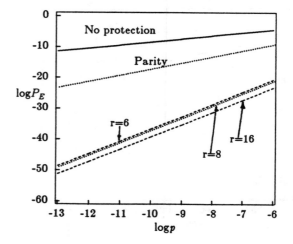

Figure 13. Probability of Undetected Error vs Bit Error Rate, $n=32$

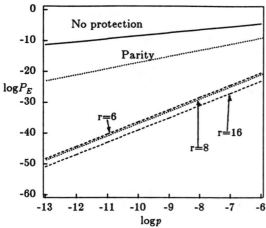

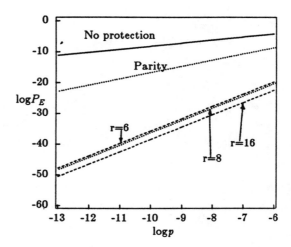

Figure 14. Probability of Undetected Error vs Bit Error Rate, $n=40$

Figure 16. Probability of Undetected Error vs Bit Error Rate, $n=56$

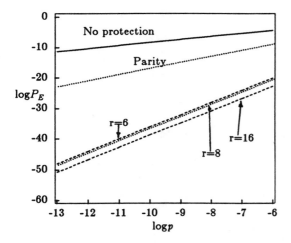

Figure 15. Probability of Undetected Error vs Bit Error Rate, $n=48$

Chapter 9:

Photonic Switching Systems

9.1 Introduction

Historically, switching technology has always followed transmission technology. During the last decade this trend has been visible in the large-scale introduction of optical fibers for transmission. Fiber-optic transmission systems are already playing an important part in long-distance communications networks. Soon they are expected to play an equally important role in the local loop. With the availability of such high-bandwidth links, the bandwidth of the optical medium is four to five orders of magnitude greater than the speed at which light can be modulated by a continuous electronic signal. Although the state of the art of high-speed electronics and electro-optics is continually advancing, their speeds will always be much lower than the bandwidth of the optical medium. Therefore, network switching elements based on electronic designs can be expected to become the bottlenecks in future networks. A way to overcome this bottleneck is to use photonic switching, where routing, switching, and other signal processing functions are performed on light signals.

Photonic devices capable of switching high-bandwidth signals can be classified as either *relational* or *logic* devices. The relational devices are also called analog or passive devices. In a relational device, switching from the input to the output is accomplished using control signals and is independent of the data signal being switched. This independence allows relational devices to support high-bit-rate channels, multiple time-multiplexed channels of different frequencies, or a large collection of dense wavelength-division channels. On the other hand, the inability of the relational devices to interpret the data bits makes them unsuitable for switching fabrics based on header information contained in the data packets. Switch fabrics such as directional couplers, optical amplifiers, and star couplers are examples of relational devices.

Logic devices allow switching to be controlled by the information contained in the data signal being switched. These devices must be able to switch as fast as or faster than the signal bit rate and to regenerate or restore the incoming signal level. The switching capacity of such devices is limited ultimately by the high-speed switching requirement. Therefore, the switch throughput that can be obtained from fabrics using logic devices will always be less than that of a relational fabric. Examples of switch fabrics based on these devices are the free-space networks based on either optical logic gates or smart pixels.

Directional couplers, optical amplifiers, and spatial light modulators are examples of switch fabrics using space channels. A space channel provides a transparent channel whose full bandwidth is available to the connected users. A directional coupler is a two-optical-input, two-optical-output device with a single control input. Depending on the control signal, the two inputs can be either cross or through connected to the output ports. Directional couplers can handle high-bit-rate information flow, but the electronic control limits their maximum reconfiguration rate.

An optical amplifier can act as an on-off switch, depending on the bias applied to the device. A switch fabric can be designed using the on/off property of the optical amplifier and an appropriate interconnection network. When the optical amplifier is turned on, the input signal is amplified, setting up a path from the input to the output. The input signal is blocked when the optical amplifier is turned off. By controlling the optical amplifiers, a path through the switch fabric can be established for the input signals.

Switching fabrics that use time channels are the time slot interchanger and those that use multiple-access techniques such as TDMA and CDMA. Switching in a time slot interchanger is achieved by interchanging the position of the time slots in a frame of a time-multiplexed information stream. The passive shared media fabrics are based on either deterministic or statistical multiple access to a passive shared medium such as a bus or star coupler. The switch fabric can be controlled using either centralized or distributed techniques. A passive shared medium such as a fiber-optic ring can be accessed using either passive taps such as fiber couplers or directional couplers operating as active taps.

A star coupler is a device with N inputs and N outputs that combines all the input channels and redistributes them equally to all the outputs. In a fixed transmitter assignment fabric, the transmitters are fixed, whereas the receivers can be adjusted to select any input. A fabric with fixed receivers and tunable transmitters is called fixed receiver assignment fabric. A code-division multiplexing fabric uses either orthogonal or pseudo-orthogonal codes to represent both the users and bits.

A switching fabric based on wavelength channels can either rearrange or reconfigure the information present in different wavelengths or share those wavelength channels through multiple-access methods. The wavelength interchanger and fabrics based on sharing wavelength channels through multiple access are examples. In a wavelength-division multiple-access fabric, the input information is used to modulate a light source that has a unique wavelength for each input. All the optical energy is combined and redistributed by a star coupler to all the output channels. On the output side, by suitably adjusting the tunable filters, an output has access to any or all of the input channels.

Even though individual photonic devices have been shown to have significant advantages over similar electronic devices in terms of speed, bandwidth, or interconnection capability, the state of the art in photonic devices has not advanced far enough to replace electronic components in the near future. Before this can happen, further progress in the size, efficiency, and reliability of photonic devices is necessary.

9.2 Overview of Papers

In "Communication Network Needs and Technologies–A Place for Photonic Switching?" Nussbaum explores the place of photonic switching in the next-generation networks. After briefly reviewing the historical trends in transmission and switching technologies, Nussbaum identifies the key components in a telecommunications network and addresses the broadband switching requirements along with the issues that need to be addressed to achieve switching in the optical domain.

The paper "Terminology for Photonic Matrix Switches," by MacDonald, presents a classification scheme for matrix switches that use photonic technology. The classification is based on three independent properties: switching principle, multiplexing scheme, and optical configuration. MacDonald presents a number of categories whose properties differ identifiably.

Hinton's paper, "Architectural Considerations for Photonic Switching Networks," reviews the photonic techniques that may be used in the design of future network elements in a communications network. This paper divides the photonic devices and systems into two classes: relational and logic. Under external control, the relational photonic devices map the input channels to the output channels. The photonic devices in the logic group perform some type or combination of Boolean logic functions. Before reviewing the optical logic devices, Hinton discusses the strengths and weaknesses of the photonic technology.

The paper "Time Division Multiplexing Using Optical Switches" describes design considerations for an integrated optical time-division multiplexing scheme. Based on high-speed optical time-division multiplexing, techniques for calculating bandwidth requirements for modulators and switches are described.

The next paper, "Optical Self-Routing Switch Using Integrated Laser Diode Optical Switch," presents an optical self-routing switch based on a Benes interconnection network. This optical self-routing switch uses integrated laser diodes. It is rearrangeable and nonblocking, and uses a synchronous multistage interconnection network for switching.

The last paper, "HYPASS: An Optoelectronic Hybrid Packet Switching System," by Arthurs et al., presents a high-performance switch architecture based on the high-throughput characteristics of multiwavelength optical interconnects together with the flexible processing power of electronic logic. Hypass is a packet switching fabric based on the wavelength-division multiple-access technique. The concept behind the Hypass switch is modulation of the tunable laser, tuned to the fixed wavelength of the designated output port. Hypass is formed from two wavelength optical networks, one for transmission and one for control. The transport network is a single-stage cross-connect with no internal collisions. The control protocol is based on an input buffered/output controlled arbitration where the output port determines when the input port is allowed to transmit. The authors also present the switch performance for random traffic.

Communication Network Needs and Technologies—
A Place for Photonic Switching?

ERIC NUSSBAUM, FELLOW, IEEE

Reprinted from *IEEE Journal on Selected Areas in* (*Invited Paper*)
Communications, Vol. 6, No. 7, August 1988, pages 1036-1043.

Abstract—Communication network architectures are driven by both
application needs and technological progress. The two dominant forces
in current network architecture research are emphasis on multimedia
"information service"-oriented network applications, including transport of voice, data, image, and video, and introduction of fiber technology, in place of the traditional copper for the transport. To date,
switching and control for such networks have been conceived of as
completely high-speed digital electronics. The role of the newly emerging photonic technologies for the switching and control functions remains uncertain. This paper projects the potential for photonic switching and control in networks by reviewing historic trends in technology
and system development, analyzing application needs as a source of
requirements on the technologies, and characterizing the limitations of
competing electronic and photonic technologies for performing these
functions. It concludes that "all optical" switching systems concepts
are in their infancy and are one to two decades away from any real
usage. Such optical structures will by then most likely involve radically
different concepts from today's electronic switches, as opposed to being
photonic analogs of today's switching elements. Deployment of more
limited application photonic subsystems in hybrid electrooptical structures are, however, distinct possibilities in earlier time frames.

I. INTRODUCTION

THE DYNAMICS of the introduction of new technologies has received extensive studies. A generic view
of the technology replacement process is shown in Fig. 1,
with a typical S-shaped introduction curve. Initial delays
in new technology introduction are often created as the
stimulus of competition extends the life of the old technologies. Parameters of the technology substitution model
[1] depend both on the capital required for deployment
and on interface compatibility with the surrounding environment. For this reason, a major system takes longer
to deploy than a new integrated circuit or subsystem (e.g.,
semiconductor chips versus core memories, electronic
calculators versus electromechanical). In comparison, a
switching system has many complex interfaces and requires massive amounts of capital for deployment. To assess the potential of optical switching, as opposed to high-
speed electronic structures, for use in the communication
networks, it is important to look at both historical trends
involving past technological developments and economic
triggers for introduction of new technology, as well as
future system requirements that the optical switching

Manuscript received February 17, 1988; revised March 18, 1988. This
paper was presented at the Topical Meeting on Photonic Switching, Lake
Tahoe, NV, March 18, 1987.
The author is with Bellcore, Morristown, NJ 07960.
IEEE Log Number 8821846.

technologies must meet to compete with their alternative
electronic technologies.

This paper reviews the capital, expense, and revenue
economic drivers for switching replacement, provides assessments of historical "basic technology invention to
system deployment intervals," and gives specific examples of future communication network needs that will
drive technology applications. These factors are then
translated into their impact on various elements of communication network structures, including access, aggregation, and control to show what the performance requirements are likely to be on future switches, thus allowing a
better determination of the probability of deploying photonic technology progress versus electronics alternatives
in this area.

II. SWITCHING TECHNOLOGY INTRODUCTION TRIGGERS

A. Traditional

A business decision to introduce a major new technology requires an economic incentive. Any combination of
value in the three key elements—capital, expense, and
revenue—of new technology systems can act as a driving
force. In the case of communications systems with their
massive costs and long deployment intervals, introduction
problems are further complicated by the need to provide
interworking between old and new technologies as the
transition takes place. Capital savings are achieved with
lower equipment first cost of new technology, both for
"new start" facilities and for replacement of aging facilities and growth of existing switches. Expense savings are
obtained by lowering annual upkeep costs through automation of labor intensive maintenance and administration
functions, and potential space and power savings in newer
more compact equipment. Revenue enhancements may be
achieved both by introducing heretofore unavailable service capabilities that a user will pay for, and through lowering the incremental cost of providing already available
services with new technology as opposed to old, thus perhaps increasing the market for them. For the past two decades, digital electronics technologies have made massive
strides in replacing electromechanical analog switching
equipment in networks, primarily with capital and expense triggers, as the learning curves (lower production
costs) of these technologies and the computerization of
the administrative processes have created extensive lever-
age to achieve these savings. Revenue potential has been

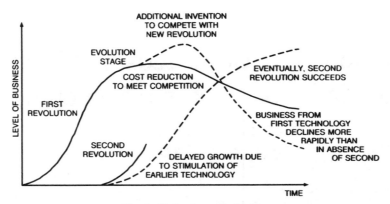

Fig. 1. Technology substitution.

more uncertain and difficult to predict than capital and expense savings and has not been factored in as extensively in the decision to introduce new switch technologies in the past.

B. Historical Trends

The time frame from the initial concept of a basic science or technology element to its deployment in a complex system deployment is usually measured in decades, with a mean of about two decades. Fig. 2 presents a brief history of switching technology changes and their scientific underpinnings. The first stored program computer-controlled switching systems, albeit using electromechanical crosspoints in the actual switching matrix, were deployed in the mid 1960's. The foundation of today's switching systems [2] goes back to the late 1930's through mid 1940's developments in three fundamental areas: pulse code modulation theory, von Neumann's fundamental work in stored program computing at Princeton, and the well-known invention of the transistor at Bell Telephone Laboratories. It was not until the mid to late 1950's that systems research experiments at Bell Telephone Laboratories, the ESSEX System [3] and the Morris ESS trial, were undertaken based on these technologies. This time lapse of more than a decade resulted from both the need for technology development and the recognition that new system concepts were required to use a new technology appropriately; that is, the transistor could not be used effectively as a one-for-one replacement for the relay. From the early system experiments to the actual deployment, almost another decade was required [4], [5], as complex problems of functionality, performance, interworking, and economics had to be addressed. By the late 1970's digital electronics technology had progressed to the point, as had system concepts, where these stored program computer-controlled switches now operated with digital PCM-based switching matrices [6], [7] as opposed to analog metallic reeds.

Looking next at the broad-band needs, the basic work in fiber and lasers took place in the late 1960's and early 1970's. Its extensive use in transmission deployment took about another ten years, and was based largely on a cost-reduction "one-for-one" replacement. Only now, in the

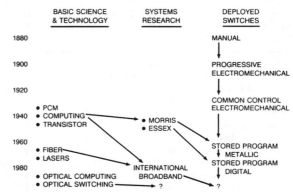

Fig. 2. A brief history of switching.

mid to late 1980's has significant systems research started throughout the world on end-to-end broad-band communication systems experiments involving transmission and switching. This builds on both fiber transmission and on progress in digital electronics, and is largely characterized by hybrid architectures employing optical transmission and electronic switching. The initial deployment of such systems concepts in the field is now believed to be viable in the early to mid 1990's time frame. Research progress to date in new materials-based nonlinear optic switching devices has been slow, and systems research in the area of optical computing and optical switching is just now getting underway. Therefore, based on historical perspectives, it seems unlikely that any form of widespread optical switching deployment is likely until well into the next century.

C. Future

Whereas the prime, but not sole, triggers for replacement of electromechanical switches by electronic switches have been capital and expense savings with only modest impact of the new revenue potential of services (e.g., cheaper touch-tone, custom calling features for customers, etc.), identifying triggers for the next generation is more difficult. Digital switches are the technology of choice today to replace the electromechanical switches that remain, but do not really prove-in as replacements for the earlier generation of stored program metallic switches, except in a toll switch environment which typically is dominated by digital (T1) trunking, or in a situation where ISDN services are provisioned.

The driving forces for broad-band networks are generally believed to be dominated by potential new revenue from various forms of voice, data, image, and video information services. Fig. 3 is a simple log plot of the information content of some broad classes of services. Today's networks are essentially voice based at somewhat less than 10^5 bits per second (64 kbits/s) and provide data services in the 1K to 10K bits/s effective bit rate range, primarily via analog modems, thus representing only a two order of magnitude range of information from sources and sinks. High-quality sound (stereo), as is found on compact discs using a linear digital code, takes in excess of 10^6 bits/s, and uncompressed NTSC television takes on the order of 10^8 bits/s (100 Mbits/s). New proposals for high-definition television increase uncompressed bandwidth by another factor of approximately five to ten. Our understanding of human vision is limited, but some extrapolations indicate that it may represent another two orders of magnitude or so beyond high-definition TV. Therefore, the future multimedia transport networks must effectively run over an application range from telemetry to high-definition TV as a minimum, representing perhaps six to eight orders of magnitude range, depending on the degree of compression of video signal that is used. As a reference point, it should be pointed out that at 64 kbits/s on an average 10 percent occupancy during the busy period, the total traffic of the current voice-based U.S. network is only 10^{12} bits/s (1 Terabit/s).

Therefore, to handle future information needs, the network requires both several orders of magnitude increase in individual switch and total network throughput and, just as important, the ability to handle a large dynamic range of signal bandwidths [8]. The high-bandwidth signals are completely dominated by video. Even delivery at home of a typical weekly periodical with half text and half photos via a communication link represents only a few hundred megabits at most, or the equivalent of a few seconds of video. Only in the area of holographic imaging research are terabit rates even discussed today for ''single channel'' links. A typical terminal display deploying ''today's'' technology using wide bandwidth communications is shown in Fig. 4. This bit mapped multiwindow display on a 1000 × 1000 pixel workstation CRT shows a combination of full-color pictures and alphanumeric text. Such a display in bit mapping requires about 10^7 bits, and in a compressed form can be done with perhaps 10^6 bits. Therefore, very quick refreshing for flipping through pages would require only several megabits per second to read material in this format.

Beyond the issue of bandwidth, communications networks of the future require a variety of different types of connections. Today's telephone communications network emphasizes point-to-point voice calls. That is basically a dialog mode with symmetric communication channels. In the future, dialogs will involve not only point-to-point but

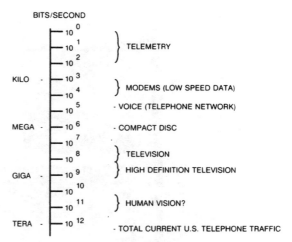

Fig. 3. Information.

also extensive point-to-multipoint (various forms of conference bridging, data broadcasting, etc.) and will deal not only with voice, but also with various forms of multimedia messaging including data, facsimile, image, etc. Second, increasing needs arise for access and retrieval of information where the communication link is between a user and some form of machine database (as is the case in Fig. 4). This may be a simple on-line text database, or may ultimately involve retrieval of video segments or access to multimedia distributed databases at various points on the network. Finally, the communications distribution function, both broadcast and narrowcast, becomes increasingly prevalent in broad-band communications. This is represented not only by the conventional NTSC or high-definition TV downstream-only broadcast, but also by various forms of video on demand where the user has interactive control over video channels. Both the access retrieval and the distribution forms of communication require asymmetric communication needs, typically with much greater downstream bandwidth than upstream bandwidth from the user. It is all these various new forms of communication, rather than cost reduction, that are likely to be the prime triggers for the next generation of switches.

III. Network Structures

Communications network topologies are made up of several key ''generic'' elements as shown in Fig. 5. The end user represents an information source or sink. That source or sink may be an individual's terminal, a computer center termination, or a TV set. The access channel provides a path from the user to the first rerouting or switching point. Paths past that switching point aggregate traffic from multiple users to higher points in the network for hierarchical routing and distribution. In addition, the user has a control or signaling and addressing scheme that allows the user to communicate with a network and give it instructions (or receive instructions from it) on how that particular communications session or call should be routed, prioritized, or generally handled.

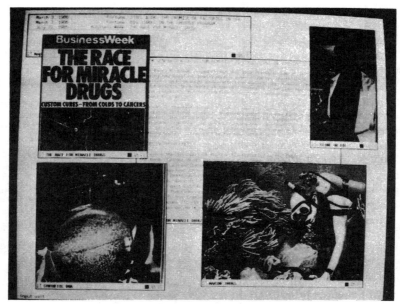

Fig. 4. Bit mapped multiwindow terminal display.

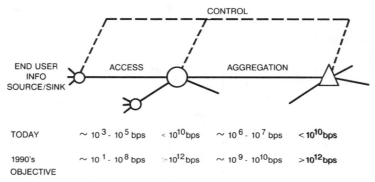

Fig. 5. Key elements of a communications network.

Examining the throughput requirements of each of these elements results in tabulation also shown in Fig. 5. This shows the increased needs based upon the applications listed in the previous section that 1990's networks must achieve. Access needs grow by 3 orders of magnitude in maximum throughput and by 5 orders of magnitude in dynamic range of throughput. The switching nodes themselves increase by 2–3 orders of magnitude in total throughput going from today's switches [9] capable of several gigabits per second throughput, to switches exceeding a terabit per second in throughput. The aggregate trunk traffic moves from today's typical few megabits per second type of throughput to a 3 order of magnitude increase as gigabit trunking becomes prevalent.

A. Network Access

We now move from this canonical network structure diagram to specific architectures and technologies under consideration to meet the broad-band needs described above which deploy a combination of fiber and electronic technologies. Fig. 6 demonstrates typical current systems experiments and configurations for broad-band access [10]. Each subscriber is provided with up to approximately 600 Mbit/s throughput capacity. The traffic for multiple subscribers is aggregated in a remote terminal where it is switched and multiplexed back via a 2.4 Gbit/s fiber to the central office. Subscribers handled include business and residence. The distances and throughput requirements are well within the bounds of capabilities of single-mode fibers using either lasers or LED transmitters [11] for the last subscriber link and laser transmitters for the feeder link. This is referred to as a double-star architecture, since both remote units and central offices have multiple outgoing routes or links. 600 Mbits/s must of course be subdivided in a manner that allows the multiple range of communication channel bandwidths to be supplied to the subscriber.

Fig. 7 shows the type of packet channel access architecture currently under prime consideration by many researchers in the field [12]–[19]. In this model, a 600 Mbit/s channel is divided into four 150 Mbit/s channels and a subscriber may be provided with anywhere from 1

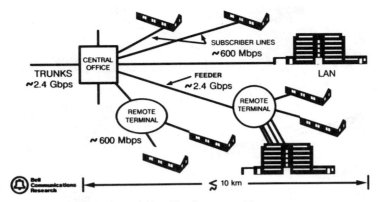

Fig. 6. Broad-band local access architectures.

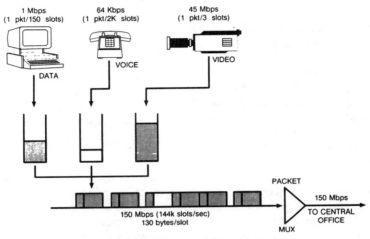

Fig. 7. Multimedia packet access.

to 4 such channels. Each channel has its traffic divided into packets or cells of fixed size (about 10^3 bits per packet) and each packet has a header or address field to identify its destination. Then, as a function of the type of communication channel bandwidth required—data, voice, video, etc.—the requisite number of packet slots is "assigned" for that channel. In this particular case, only one packet per 2000 available slots is needed for voice, but one packet per three available slots is required for 45 Mbit/s video. The job of the switching and multiplexing is then to statistically combine, reconfigure, and route these packets to their appropriate destinations. This type of a structure allows a single-fiber link and a single-switch fabric configuration to provide everything from multiple channels of conventional telephony, to high-fi audio, to a wide range of data and image needs from low-speed telemetry to very high-speed data (in excess of 10 bits/s), to various forms of video connections for NTSC and eventually high-definition TV (including both traditional one-way broadcast and moving to various forms of interactive and two-way video).

It is likely that systems deploying this type of configuration, which are now in existence as systems research

prototypes in several laboratories around the world [20], [21], will be field trialed in the early 1990's and see initial deployment by the mid 1990's.

B. Network Aggregation and Control

The interconnection of central offices to cover broader metropolitan, regional, and national areas involves the establishment of a hierarchical topology in which the multiple existing central office switches (10^4 in the United States) can be interconnected as required for a particular routing request. Today's multilevel national hierarchy has been established to minimize costs by limiting trunk routes to connections between pairs of central offices which have significant traffic flows, while providing the remaining connectivity via a tandem or higher level switched configuration. The large capacity of fiber optics allows for the creation of hubbing structures in which fewer connections for such point-to-point trunk groups are needed and more centralized reconfigurable routing through a hub can be achieved [22], [23]. Similarly, fiber optics bandwidth provides new long-term opportunities in control centralization via large databases. Today, most switching control is contained in stored programs within each of the central

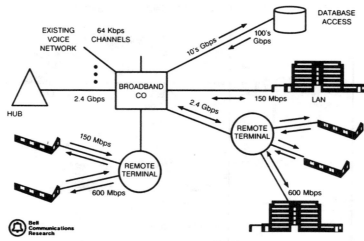

Fig. 8. A view of area broad-band networks.

office switches from which customer service offerings are provided. Greater flexibility for change and update can be obtained by the use of large database-driven structures. Such structures, in which instructions for switch reconfigurations come from interrogations and responses from central databases, would require high-throughput communication in order that inquiries on routing paths (such as "800 services") can be made on a regional and/or national basis. Here again fiber optics is likely to provide great flexibility in changing and enhancing the structure of network control.

All together these themes for broad-band access, aggregation, and control, are summarized in Fig. 8 which shows a configuration of an area broad-band network that encompasses these concepts and is currently the major thrust of extensive research and development activities in the field.

C. Broad-Band Switch Requirements

Based upon these broad gauge application needs and the various topology/hierarchy considerations associated with access, aggregation, and control, some overall guidelines can be established for the characteristics of a switch that operates effectively in such broad-band networks. The table in Fig. 9 summarizes these characteristics in terms of seven key parameters: the upper bound on bandwidth per channel, the range of bandwidth assignable directly to users, the degree of asymmetry in upstream versus downstream communication, the reconfiguration time of the switch, its connectivity in broadcast and multipoint as opposed to point-to-point applications, the total throughput, and the speed of data communications for control.

Fig. 10 then shows how this type of switch requirement is now being addressed in the laboratories around the world using hybrid electrooptical configurations. Gigabit per second fiber transported digital signals are converted to demultiplexed electrical signals on the order of 10^8 bits/s for use as inputs to switch terminations. Each of these "100 Mbit/s" bit streams arrives in a packet format with a signaling/addressing header attached and is

- High upper bandwidth bounds/channel   ($\cdot 10^8$ bps)
- Dynamically assignable bandwidth to users   ($\cdot 10^6$ range)
- Non symmetric (upstream/downstream)   ($\cdot 10^6$ range)
- Very fast switch reconfiguration   ($\cdot 10^{-5}$ sec)
- Broadcast and multipoint connectivity   (copy networks)
- Large facility hub throughput   ($\cdot 10^{12}$ bps)
- Distributed intelligence control   (high speed data comm.)

Fig. 9. Required new switch characteristics in broad-band networks.

then reconfigured in a switch with the throughput characteristics previously specified. The ability to reswitch 10^3 bit packets in approximately 10 μs for rerouting purposes requires approximately 10 ns device speeds. Control of the switch is accomplished via interpretation of the signals in the header which are processed, after reference to memory for control instructions, in order to reconfigure the switch at these speeds. In this reconfiguration then, electrical functions include multiplexing, switching, regeneration amplification, synchronization and retiming, and the signal processing and control for switch configuration. Current experimental implementations emphasize fine-line CMOS working at several hundred megabits per second. Speedups to the gigabit per second level with ECL and perhaps future fine-line CMOS, and a few gigabits per second with gallium arsenide seem possible. This switch configuration then represents the target that an optical switch must compete with on a performance and economic basis if it is to succeed in addressing this type of market.

Examining the driving forces for an optical switch, three major elements come to the fore. First, from an historical prospective, switching has always followed the format of transmission. Analog metallic transmission resulted in electromechanical metallic switches. Digital switching followed the introduction of digital transmission. If history repeats itself, so should optical switching follow optical transmission. Second, for very high-speed needs, probably in excess of 10 Gbits/s, the limitations of electronics make optical switching technology a necessity.

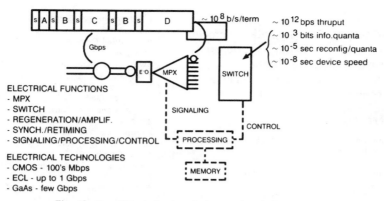

ELECTRICAL FUNCTIONS
- MPX
- SWITCH
- REGENERATION/AMPLIF.
- SYNCH./RETIMING
- SIGNALING/PROCESSING/CONTROL

ELECTRICAL TECHNOLOGIES
- CMOS - 100's Mbps
- ECL - up to 1 Gbps
- GaAs - few Gbps

Fig. 10. Broad-band electrooptical network switch structures.

And finally, one of the major costs of electronic switching as opposed to optical switching is that of *E/O* interfaces and of the demultiplexing from the high-speed fiber channels to the speeds that electronics can handle. These could be overcome by staying entirely in the optical domain which effectively means optical switching.

On the other hand, numerous research issues need to be addressed if optical switching is to fulfill these tasks. Since we are not just switching light between fibers, but rather discrete quanta of information residing in packets in each fiber, a very high-speed switch reconfiguration time is necessary. If we must convert from optical to electrical in order to extract the packet header information to control an optical switch, much has already been lost in savings, meaning that optical control of optical switches becomes a critical factor in their economic deployment. Many "analog" proposals for optical switching are plagued by crosstalk and noise buildup phenomena in large multistage configurations. Electronic digital technology has inherent in it a regeneration and amplification process, and provides means of synchronization. Optical losses must be made up for in one form or another as multiple stages of transmission and switching are chained together, in order to avoid returning to electronic format in order to recover the signal. If control is to be optical, some form of optical memory, computing, and processing will be necessary because control requires extensive logic manipulation. And lastly, all these needs must be met in a manner that economically competes with electronic alternatives. Placing these factors together over the historical prospective of time scales of change described earlier in the paper would imply that the time for the introduction of ubiquitous optical switching is measured in decades rather than years. However, it seems likely that much less demanding optical subsystems can be introduced on much shorter time frames, perhaps some forms of optical mux/demux and/or simple, low-speed, small, facility reconfiguration switches for example.

IV. SUMMARY

To achieve truly optical domain communications is most likely to require not just progress at the optical technologies (e.g., optical amplification, optical processing,

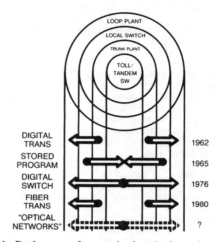

Fig. 11. Deployment of new technology in the exchange plant.

etc.) but also new system architectural concepts. New structures are likely to take advantage of the inherent huge unused bandwidth of fiber optic communications and make tradeoffs in that region, as opposed to trying to merely emulate electronic switches. Among possibilities are various very high-speed shared media configurations using code division multiple access (CDMA) technologies, and various forms of dense multiwavelength and coherent structures in which "switched" communication capabilities are built up by various "controlled broadcast" techniques as opposed to conventional rerouting and switch nodes [24]–[28]. Optical switching introduction will thus depend on progress in technology, in architectural concepts, and in new bandwidth "hogging" applications. The confluence of these elements will take time.

Finally a more recent history of deployment of new technology in the exchange plant is shown in Fig. 11 moving from digital transmission, through stored program control, digital switching, and fiber transmission. It shows both the time frame in which they were first introduced and the area of the plant which was first seeded with the technology and from which it grew inward or outward. Where and when optical domain networks will be seeded is the open question. The prognostication shown in the figure is that it is likely to start at the central toll/

tandem section of the network and work its way out. This is the highest bandwidth, slowest reconfiguration element of the plant. The local environment is the most demanding of the switch, but has the highest multiplier factor in technology usage. The date for any of these scenarios is likely to be in the next century.

REFERENCES

[1] R. C. Lenz and L. K. Vanston, *Comparisons of Technological Substitutions In Telecommunications and Other Industries*, Technology Futures, Inc., 1986.

[2] A. E. Joel, Jr., "Electronics in telephone switching systems," *Bell Syst. Tech. J.*, vol. 35, p. 91, Sept. 1956.

[3] H. E. Vaughan, "Research model for time-separation integrated communication," *Bell Syst. Tech. J.*, vol. 38, pp. 909–932, July 1959.

[4] *Bell Syst. Tech. J.*, Special Issue on No. 1 ESS, vol. 43, Sept. 1964.

[5] A. E. Spencer, Jr. and F. S. Vigilante, *Bell Syst. Tech. J.*, Special Issue on No. 2 ESS, vol. 48, Oct. 1969.

[6] A. E. Spencer, Jr. *et al.*, *Bell Syst. Tech. J.*, Special Issue on No. 4 ESS, vol. 56, Sept. 1977.

[7] *IEEE Trans. Commun.*, Special Issue on Digital Switching, vol. COM-27, p. 948, July 1979.

[8] S. B. Weinstein, "Telecommunications in the coming decades," *IEEE Spectrum*, pp. 62–67, Nov. 1987.

[9] D. L. Carney, J. I. Cochrane, L. J. Gitten, E. M. Press, and R. E. Staehler, "The 5ESS switching system architectural overview," *AT&T Tech. J.*, vol. 64, part 2, pp. 1339–1356, July–Aug. 1985.

[10] L. R. Linell, "A wide-band local access system using emerging technology components," *IEEE J. Select. Areas Commun.*, vol. SAC-4, pp. 612–618, July 1986.

[11] P. W. Shumate, "Applications of LEDs with single-mode fiber for broadband loop distribution," in *Proc. Conf. Lasers Electro-optics (CLEO)*, San Francisco, CA, June 9–13, 1986, paper THE3.

[12] J. S. Turner and L. F. Wyatt, "A packet network architecture for integrated services," in *Proc. GLOBECOM'83*, pp. 45–50.

[13] A. Thomas, J. P. Coudreuse, and M. Servel, "Asynchronous time-division techniques: An experimental packet network integrating video communication," in *Proc. Int. Switching Symp.*, Florence, Italy, May 1984.

[14] J. J. Kulzer and W. A. Montgomery, "Statistical switching architectures for future services," in *Int. Switching Symp. 1984 Proc.*, May 1984.

[15] J. S. Turner, "Design of an integrated services packet network," in *Proc. ACM Data Commun. Symp.*, Mar. 1985.

[16] ——, "Design of a broadcast packet network," in *Proc. Inforcom*, Apr. 1986.

[17] J. P. Coudreuse and M. Servel, "Prelude: An asynchronous time division switched network," in *Proc. Int. Conf. Commun.*, Seattle, WA, June 1987.

[18] L. T. Wu, T. T. Lee, and S. H. Lee, "Dynamic TDM—A packet approach to broadband networking," in *Proc. Int. Conf. Commun.*, Seattle, WA, June 7–10, 1987.

[19] T. T. Lee, R. Boorstyn, and E. Arthurs, "The architecture of a multicast broadband packet switch," in *Proc. Inforcom '88*, New Orleans, LA.

[20] Broad-band Communications Systems, *IEEE J. Select. Areas Commun.*, vol. SAC-4, July 1986.

[21] Switching Systems for Broad-band Networks, *IEEE J. Select. Areas Commun.*, vol. SAC-5, Oct. 1987.

[22] L. T. Wu and N.-C. Huang, "Synchronous wideband network—An interoffice facility hubbing network," in *Proc. Int. Zurich Seminar Digital Commun.*, Zurich, Switzerland, Mar. 11–13, 1988.

[23] M. Kerner, H. L. Lemberg, and D. M. Simmons, "An analysis of alternative architectures for the interoffice network," *IEEE J. Select. Areas Commun.*, vol. SAC-4, pp. 1404–1413, Dec. 1986.

[24] M. S. Goodman, C. A. Brackett, R. M. Bulley, C. N. Lo, H. Kobrinski, and M. P. Vecchi, "Design and demonstration of the LAMBDANET system: A multiwavelength optical network," in *Proc. GLOBECOM'87*, vol. 3, Nov. 1987.

[25] R. S. Tucker, G. Eisenstein, S. K. Korotky, G. Raybon, L. L. Buhl, J. J. Veselka, B. L. Kasper, U. Koren, and R. C. Alferness, "Optical time-division multiplexed fiber transmission using Ti : LiNbO₃ waveguide switch/modulators," in *Proc. GLOBECOM'87*, Nov. 1987.

[26] Y. Shimazu, S. Nishi, and N. Yoshikai, "Wavelength-division-multiplexing optical switch using acoustooptic deflector," *J. Lightwave Technol.*, vol. LT-5, pp. 1742–1747, Dec. 1987.

[27] D. J. Blumenthal, P. R. Prucnal, L. Thylen, and P. Granestrand, "Self-routing of 12.5 data through an 8×8 LiNbO₃ optical crossbar switch matrix with gigahertz switching speed," in *Proc. Opt. Fiber Commun. Conf., 1988 Tech. Dig. Series, Vol. 1*, Jan. 1988.

[28] S. F. Su, L. Jou, and J. Lenart, "A review on classification of optical switching systems," *IEEE Communications*, vol. 24, May 1986.

Eric Nussbaum (SM'54-M'57-SM'80-F'86) received the B.S.E.E. degree in 1955 and the M.S.E.E. degree in 1956 from Columbia University, New York, NY.

He is Assistant Vice President of the Network Systems and Services Research Laboratory at Bellcore, Morristown, NJ. He heads an organization responsible for applied research in new exchange telecommunications network architectures based on emerging silicon, photonic, and software technologies. Special emphasis is placed on network capabilities to support a full range of voice, data image, and video communications. Prior to assuming his present position on January 1, 1984, as a result of the AT&T divestiture, he held various management positions at Bell Laboratories, which he joined in 1959, primarily in fields related to electronic switching. He had responsibilities for various hardware, software, and system design aspects of several electronic switching systems, remote switching, voice storage, and advanced cellular mobile phone systems, as well as for exploratory studies of new switching networks and service capabilities. He has published numerous technical papers in several areas of telecommunications research.

Mr. Nussbaum has been active in the IEEE Communications Society and is a member of Tau Beta Pi and Eta Kappa Nu.

Terminology for Photonic Matrix Switches

R. IAN MACDONALD, SENIOR MEMBER, IEEE

Reprinted from *IEEE Journal on Selected Areas in Communications*, Vol. 6, No. 7, August 1988, pages 1141-1151.

Abstract—There are many avenues of current photonics development relevant to switching. In consequence, the properties and potentials of various technologies are difficult to assess. As an aid to such evaluations, a classification scheme is suggested here for matrix switches that employ photonic principles, generally for broad-band signal routing. The classification considers three independent properties, providing for a large number of categories whose properties differ in identifiable ways. Salient properties of each class are noted.

Matrices are classified in the following ways:
1) according to the switching principles used as
   a) optical
   b) optoelectronic;
2) according to the multiplexing system used in the switch as
   a) space division
   b) modulation division
   c) carrier division;
3) according to the optical configuration of the switched system as
   a) centrally switched
   b) optically extended
   c) distributed.

Examples of the various types are given.

I. INTRODUCTION

AS a result of the rapid development of optical communication technology, data rates at which point-to-point transmission of information is practical have passed into the gigabit/second region. The capability to control the route of very high rate data through a network lags behind. Recent developments in electronic switching range from a fully engineered, large-scale switching system for 70 Mbits/s using silicon CMOS technology [1] to an experimental hybrid FET switch array capable of bit rates past 2 Gbits/s [2]. Large-scale integration in silicon promises monolithic matrices of dimension 128 × 128 at bit rates up to 200 Mbits/s [3]. A major obstacle in electronic broad-band switching is expected to be the interconnection of the integrated matrix units, since electromagnetic cross-coupling effects are severe at high frequencies.

In spite of the difficulties involved, the ability to switch gigabit/second signals is necessary for future communication networks. Proposals for subscriber-line connections at such rates [4] illustrate a potential future need for very broad-band switching. Futhermore, very broad-band switches may have application within other apparatus: in multiple-processor computers or in microwave signal processing. Accordingly, there has been considerable effort in recent years to develop photonic techniques also for switching. Among the anticipated benefits are bit-rate independent performance and immunity to electromagnetic coupling.

Work in photonic switching has for the most part been concerned with devices. It is difficult to assess and compare various technologies in the context of system requirements. In the hope of aiding such assessments, a framework for classifying the various photonic broad-band switching technologies is offered here.

The terms "optical" and "photonic" will be used with different meanings. "Photonic" is the overall term denoting switches that direct information through an unspecified network by some method that uses light. "Optical" switches are a subclass that achieve this goal by directing light itself in an optical network. For completeness, the classification system includes multiple access methods where these provide the equivalent function to space division switching and the word "matrix" is used for all devices that provide this function.

II. ORGANIZATION OF THE CLASSIFICATION SCHEME

Previous classifications of photonic switching, e.g., [5], have been primarily concerned only with one aspect of the switches, generally the physical principles involved. The classification proposed here is organized according to three different aspects: 1) the type of multiplexing, 2) the physical basis for the switching mechanism, and 3) the optical configuration of the system. These categories are essentially independent, and therefore span a "volume" as shown in Fig. 1.

A. Multiplexing

In the broadest sense, a matrix switch is a device in which signals presented at the input ports are multiplexed in some way and distributed to provide the output ports with multiple access to them. Three main types of multiplexing are identified as follows.

1) Space Division: Space division switches employ no common paths in their internal distribution of signals. The signals are physically isolated throughout the matrix.

2) Modulation Division: This term is introduced to denote multiplexing methods in which the signals which modulate the optical carriers are multiplexed [6]. Modulation division methods cannot in principle realize the enormous potential bandwiths of optical transmission, being limited by the bandwiths of the electronic multiplexing devices. However, in practical situations it may

Manuscript received September 23, 1987; revised March 4, 1988.

The author is with the Alberta Telecommunications Research Centre, University of Alberta, Edmonton, Alta., Canada.

IEEE Log Number 8821382.

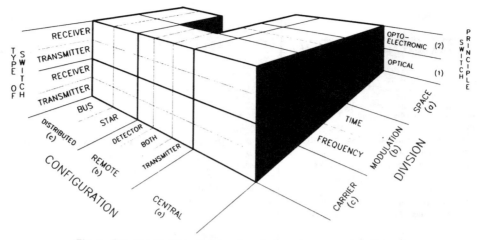

Fig. 1. The classification considers three independent properties of photonic matrices, thus spanning a "volume." The physical basis of the switching mechanism is either "optical" or "optoelectronic." The internal multiplexing is in the "carrier," the "modulation," or in "space," and the placing of the electronic elements of the system provides the categories "central," "distributed," and "remote."

be convenient to use electronic multiplexing techniques which are well developed at present.

3) Carrier Division: This category concerns the multiplexing of the optical carriers. At present, the salient example is optical frequency division, i.e., wavelength multiplexing applied to matrix switching [7]. Other carrier division multiplexing methods, for example coherence multiplex [8], may also find application in switching.

B. Switching Principle

Signals may be carried through the network in optical form for all or only part of the way. Switching can therefore take place either at an optoelectronic interface (source or detector) or in the optical or electronic path, thus defining three fundamentally different types of device. These types are illustrated in Fig. 2.

1) Optical Switching: Optical switches accept and yield optical signals. The switching action may result from alteration of the transmitted signal, for example by wavelength conversion, or from alteration of the transmission conditions, for example the state of optical waveguide switches in the transmission path.

2) Optoelectronic Switching: The switching devices accept optical signals and yield electrical signals or vice versa. When the input is optical, the term "detector" switch will be used; when the output is optical, the term is "transmitter switch."

3) Electronic Switching: The switching process accepts and produces electronic signals. In an optical communications system, this category would represent the use of optical fibers to deliver signals to and from a switching center, where they are converted to electrical signals and switched by an electronic switch. This category is excluded from the discussion here because optical signals are not employed in the internal operation of the matrix,

which is therefore not "photonic" in our definition. Nevertheless, this category may be very important. Hybrid integrated devices consisting of optical receivers, electronic switching circuitry, and optical transmitters have already been reported [9].

C. Optical Configuration

The optical signal power available to the switching matrix depends on the location of the transmitters that feed it. Likewise, the power available to the receivers of the switched signals depends on their distance from the switch. Since photonic matrix switches may contain optical power dividers and/or optoelectronic transducers, the location of the conversions of signal between optical and electrical forms is of importance to the switching function. Three system-configuration categories can be identified as shown in Fig. 3.

1) Central Configuration: The matrix is connected into the system electrically. Both optical inputs and outputs originate at the matrix.

2) Optically Extended Configuration: Either the optical inputs to the matrix originate at remote locations, or optical outputs are sent directly to remote terminals, or both.

3) Distributed Systems: For completeness, consideration is also given to systems in which there is no central matrix switch, but the equivalent function is performed by the distribution system itself. The large bandwidth of optical transmission makes it possible to multiplex many signals and deliver them to each location served. Two different topologies exist: the star network, in which signals are power-divided and distributed to all stations, and the bus network in which signals are injected onto an optical bus by the transmitting station, and extracted completely by the receiving station. Complete extraction of the de-

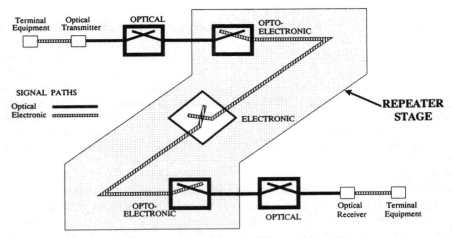

Fig. 2. Switching principle: both optical and electronic signals are present in photonic systems, and signals can be switched in either form, or switching may occur in the optoelectronic conversion between the two forms. Electronic switching is not considered here.

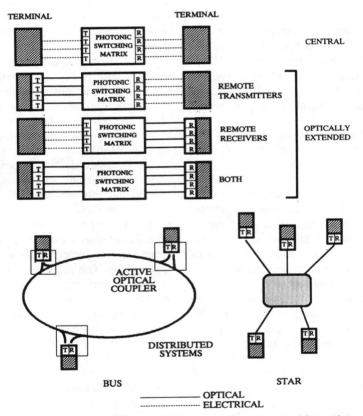

Fig. 3. System configuration: the photonic matrix may be a standalone unit connected electronically into an unspecified network ("central"), or may receive and/or produce optical signals ("extended"). Multiple access optical networks may be the equivalent of a matrix switched system ("distributed").

sired signal is the essential distinction of the bus from the star topology. Bus distribution under this definition cannot be used with optoelectronic switches because these do not influence the optical signal path and therefore cannot extract the signal completely.

III. PROPERTIES OF THE CLASSES OF PHOTONIC SWITCHING MATRICES

In the following, a short discussion of the features of matrix switches in each category of the classification scheme is given, and representative work in the literature is identified.

A. Space Division

1) Optical Switching: Matrices of this type employ switches that direct the optical carrier along a selectable path. Their performance is therefore independent of the signal except for the requirement for sufficient power at the receiver. The optical switching function is novel, and is difficult to achieve satisfactorily. Many devices are under study to perform it. Typically these are monomode integrated optical elements that depend on the perturbation of coupling between waveguides to achieve the function of a four-port "exchange bypass" [10] cell. Rearrangeable architectures such as Benes networks, or wide-sense nonblocking arrangements like Fig. 4(a) are very suitable to such switch elements. Small scale integrated optical switch matrices have been demonstrated [11]. Integration is important because losses in optical interconnections are serious. Unfortunately, the optical switches are physically long. When they are connected in tandem the integrated matrix size possible is limited to the order of ten lines by size constraints on substrate crystals. Novel parallel arrangements requiring large numbers of switching devices have been proposed to minimize loss and crosstalk in larger integrated optical exchange–bypass networks [12].

The optical equivalent of a make–break switch has also been developed, based on switchable amplifiers [13] or attenuators [14]. Crossbar architectures as shown in Fig. 4(b) are required by such devices. These have the advantage over exchange–bypass architectures that the connection paths pass through only one switch. Loss and crosstalk, therefore, do not accumulate with increasing matrix dimension and individual switch failure is less catastrophic. A second advantage is the natural provision of a "broadcast" capability which cell-based matrices only provide if the cells are multistate devices including power dividing states. Both advantages result from the division of the signal, which causes a corresponding linear decrease in delivered power with increasing matrix dimension. For small dimensions, simple exchange–bypass networks such as Fig. 4(b) deliver the most power to the receiver, but their losses accumulate multiplicatively with matrix dimension so that the crossbar architecture is advantageous for large matrices.

At present, most optical exchange–bypass cells of interest are single-mode devices, which sets limitations on the layout of the matrix because the power in two waveguides cannot be summed into a third without significant loss. This restriction weighs against optical switching architectures that require signal combiners. For example, an optical crossbar switch is not useful in combination with single-mode waveguide for broad-band connections to remote terminals.

In situations where high-speed switching and very high reliability are not required, optomechanical switches can be employed. Many such switches are now available, and relatively large matrices can be built up [15].

a) Central Configuration: Incoming electrical signals modulate optical transmitters at the input ports. The locally generated light signals are directed to detectors located at the output ports. From an external point of view, the matrix is an electronic device and its performance is influenced by optoelectronic elements as well as the optical switches. Matrices of this class closely resemble optoelectronic matrices.

b) Extended Configuration: Location of the optical transmitters or receivers remote from the switching matrix primarily adds a transmission loss to the power budget, thus making smaller the number of ports served. However, an optically transparent path is opened between subscriber and switch, or if both receivers and transmitters are remote, between subscribers. Such a channel can have extremely large bandwidth. Recently, optical switches with reduced polarization sensitivity have been demonstrated [16], so that initial requirements to maintain the polarization direction in the transmission path may be relaxed.

c) Distributed Configuration: Distributed space division would imply the distribution of each signal on a separate path, a configuration of little apparent merit.

2) Optoelectronic Switching: The elementary optoelectronic switching function is obtained in devices which convert an optical signal to an electrical one, or vice versa. In "detector-based" matrices, switching is obtained by creating, under electrical control, a sensitive "on" state and an insensitive "off" state in a detector. Photodiodes can be put into an insensitive state with forward bias, while photoconductors are insensitive when unbiased [17]. "Transmitter-based" matrices use the converse device. Laser diodes under continuous modulation are "on" when the bias current is above threshold, and "off" when it is below. The diagrams of Fig. 5(a) and (b) show the concept underlying the detector switching and transmitter switching methods.

Light signals are not actively directed through an optoelectronic matrix, but serve to distribute (or collect; in the case of transmitter switching) the signals to (from) a set of crosspoints. The matrix thus naturally takes the form of a crossbar switch. The most promising optoelectronic matrices use detector switches because arrays of detectors are considerably easier to make than arrays of transmitters. Compact integration of detectors with high isolation has been demonstrated [18].

There is no necessity to use single-mode optical transmission within an optoelectronic matrix. Since more power can be coupled into multimode waveguide, optoelectronic crossbar matrix designs for a given size may deliver greater optical power to each detector in spite of the power division required to distribute the signals [19].

Optoelectronic switching always converts between optical and electronic signals, and is best suited to single-stage switching and colocation with repeaters. Unlike op-

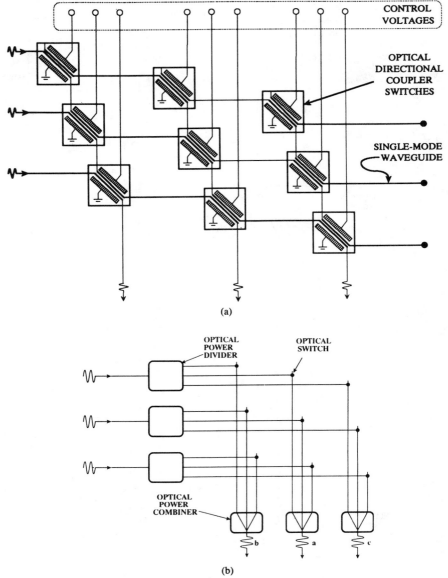

CONTROL VOLTAGES

OPTICAL DIRECTIONAL COUPLER SWITCHES

SINGLE-MODE WAVEGUIDE

(a)

OPTICAL POWER DIVIDER

OPTICAL SWITCH

OPTICAL POWER COMBINER

(b)

Fig. 4. Space division architectures using "optical" switches: (a) Typical wide-sense nonblocking matrix using optical "exchange–bypass" switches. (b) The optical "crossbar" architecture, which requires division and recombination of optical power.

tical switching, it provides access to the signal information, and is thus more suitable for the switching of packets.

a) Central Configuration: If switching is by means of detectors, incoming signals must be divided among the detectors associated with each output. The number of outputs of the matrix is the ratio of the optical power produced by the sources to the optical power requirement of the detectors, less losses. Several hundred detectors can be served by each optical signal with available optical sources and detectors [20].

For transmitter-based switching, the incoming signals must be electrically distributed to the transmitter array, and the optical signals from the crosspoints then collected into the output lines. Little work seems to have been done on this form of optoelectronic matrix, although a transmitter-switched array has been demonstrated for residue processing [21]. The transmitter-switched configuration does not divide the optical power: all the light produced by an optical source arrives at the detector except for signal combination losses. The chief disadvantage is that the number of transmitters required is high, and these are generally expensive and power-consuming items.

b) Extended Configuration: Optical power division losses in the matrix compete with losses in transmission. Arrays based on transmitter switching do no optical power

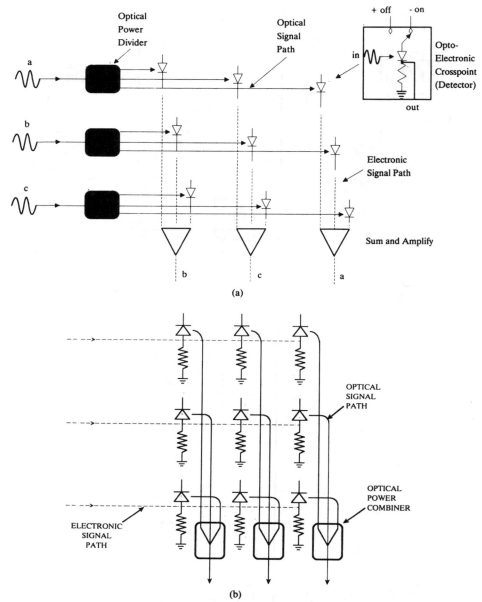

Fig. 5. Space division architectures using "optoelectronic" switches: (a) detector switched crossbar, (b) transmitter switched crossbar.

division but if monomode output paths are used, combining optical signals into the outputs leads to the same loss. Matrix dimension is thus always related to the physical size of the region served, and an optimum relation may be found between the two. With laser-driven single-mode broad-band subscriber links, considerable excesses of power may be available to cover such losses.

Arrays based on detector switches clearly cannot also have remote receivers, and similarly arrays based on transmitter switches cannot have remote transmitters. It follows that no optoelectronic space-switch can have both detectors and receivers remote.

c) Distributed Configuration: See Section III-A-1-c.

B. Modulation Division

1) Optical Switching: Purely optical techniques are not applicable to frequency division switching since there is no optical method for tuning to the modulation on an optical signal. Fast optical switches can be used to construct time division multiplexes at high rates. In a simple optical time-division multiplexing scheme using a bus, only one switch is needed for each line, as illustrated in Fig. 6(a). However, an input signal may have to pass through all the switches on the bus. Switch loss and reliability are clearly important.

In star networks with time-division multiplexing,

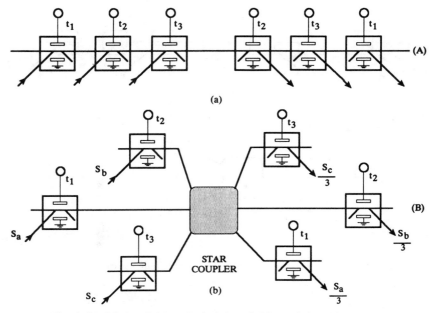

Fig. 6. Modulation division. Optical time division switch architectures:
(a) Bus. (b) Star.

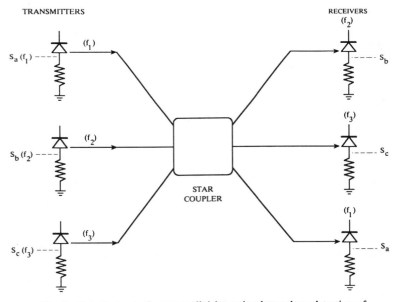

Fig. 7. Optoelectronic frequency division using heterodyne detection of
subcarriers by means of electronic IF signals applied to the detector bias.

broadcast connections are possible and no more than two switches are traversed in sequence. The cost is optical power division. A number n of signals can be passively combined and distributed to an equal number of outputs in monomode fibers without combination loss when n is a power of 2 [22].

If more than one switching stage or a resequencing of the input TDM signal is required, there is a need for optical memory. A time-division optical switching matrix using bistable laser diodes as memory elements for time-

slot interchange has been demonstrated [23]. Optical fiber delay lines are capable of long, broad-band delays, and can also be used as passive memory elements [24].

a) Central Configuration: Since fewer switches are traversed for a given matrix size, time division optical matrices may be larger than space division ones.

b) Extended Configurations: Synchronization of the demultiplexing switches at remote terminals may be a problem at the very high rates optical transmission makes otherwise attractive [25].

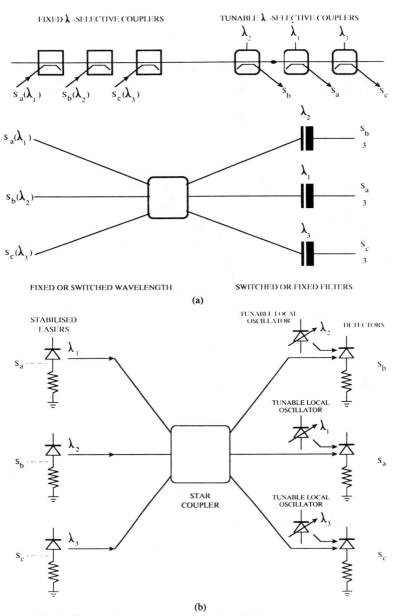

Fig. 8. Carrier division: (a) Optical wavelength division—bus and star architectures. (b) Optoelectronic wavelength division: coherent multichannel system.

c) Distributed Configuration: A time division distributed system using high-speed optical switches to gate broad-band signals on and off an optical bus has been proposed [26]. One feature of such systems is the need for optoelectronic conversion of the signal to synchronize the electronics in the stations, which cancels some of the advantages of the optical approach.

2) Optoelectronic Switching: With optoelectronic modulation division, both time and frequency domain multiplexing are possible: the transmitters and detectors can be gated for time division, or subcarrier heterodyning can be performed in the detectors [27], [28] as shown in Fig. 7. Bus architectures are not available because an optical signal cannot be completely extracted from a bus by optoelectronic methods.

There is little distinction between transmitter and detector-based optoelectronic modulation- or carrier-division switching. In detector switching, the transmitters produce a fixed multiplexed arrangement of the incoming signals and the detectors select the desired one. In transmitter switching, the detectors are assigned multiplexed channels or time slots, and the transmitters aim the incoming signals at the desired output channel.

Optical switches chopping optical signals that carry frequency multiplexed subcarriers would also achieve mixing in the detection process. This method is optoelec-

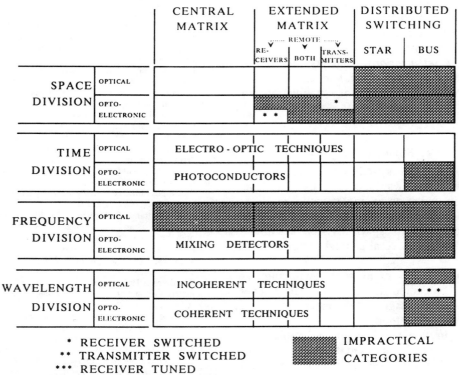

Fig. 9. Overview of the classification system. Categories noted as impractical include possible opportunities for new development.

* RECEIVER SWITCHED
** TRANSMITTER SWITCHED
*** RECEIVER TUNED

IMPRACTICAL CATEGORIES

tronic under the nomenclature described here because the switching process inherently involves detection.

a) Central Configuration: The same power division or combination losses occur in both optoelectronic matrices of this type and the corresponding optoelectronic space division matrices. However, the modulation division matrix requires fewer devices.

b) Extended Configurations: Extended time division systems have synchronization problems as noted in Section III-B-2-b. Remote-transmitter frequency division matrices seem practical, since optoelectronic heterodyne detection has been demonstrated with noise similar to direct detection with the same devices [29].

c) Distributed Configuration: The bus architecture is not available. The star is equivalent to Section III-B-2-b with both transmitters and receivers remote.

C. Carrier Division

1) Optical Switching: The only carrier division technique accessible to optical switching is that of wavelength division multiplex. This requires either wavelength-tunable sources or tunable taps or filters, plus a distribution system that is not wavelength dependent. The basic configurations of bus and star optical carrier division switches are shown in Fig. 8(a). Wide laser tuning ranges have been demonstrated [30], and tunable passive optical filters have been constructed [31]. Systems of this type have been described [32]. The star network configuration can be implemented with available technology, but tunable wavelength selective taps are requried for the bus. Optical

wavelength division methods are probably most useful in combination with another technique to give, e.g., wavelength- time-wavelength multistage matrices [5].

The distinction between optical and optoelectronic carrier division rests on the demultiplexing technique. In optical systems, passive optical filters are used. These tend to have relatively large bandwidths, and require multiplexed signals spanning relatively large optical spectral widths.

a) Central Configuration: This class resembles Section III-B-1-a.

b) Extended Configuration:,

c) Distributed Configuration: Excessively wide optical spectra composed in the multiplexing process may conflict with transmission requirements.

2) Optoelectronic Switching: This category chiefly concerns optical homodyne and heterodyne methods, usually denoted ''coherent,'' for electronic channel selection in closely spaced wavelength division multiplexes, as shown in Fig. 8(b). These techniques are the subject of much current research [33]. Matrices of this type are automatically optoelectronic, since conversion of the signal into electrical form is a consequence of the square law detection required. As with other optoelectronic matrices, multistage switching and optical retransmission are inconvenient.

a) Central Configuration: The central configuration facilitates the laser line control necessary for coherent detection, since all sources are in the same mechanical envelope, and considerably simplifies the problem of polarization control. Large matrices are potentially possible, and this category may eventually provide an attractive alternative to Section III-A-2-a.

b) Extended Configurations:

c) Distributed Configurations: Enhanced noise margin obtained by coherent transmission methods can be employed in the power distribution or combination network of the matrix, this improving the optima noted in Section III-A-2-b.

IV. Summary

This discussion of matrix switches is coarsely represented in Fig. 9, which shows the categories of the classification scheme in which no examples could be identified. Some of these categories may represent opportunity for research, others such as space-division distributed systems, are merely silly. Several general remarks can be made.

1) The very large variety of photonic switching systems that must be considered illustrates that their development is still at an early stage and the merits of various methods are not well understood.

2) Central configuration matrices are realizable in the widest variety of ways, and bus-based distributed systems in the fewest ways. This situation indicates incompatibility between photonic transmission and photonic switching. The incompatibility is a result of competition between the switch and the transmission system for signal power. Developments in optical amplifiers may alter this situation.

3) The interface between the optical and electronic form of a signal is very useful for switching. Optoelectronic switches of various kinds seem to have considerable potential if the inherent conversion of the signal is not a disadvantage, or electronic processing at the switch is an advantage.

4) Optical switches seem most suited to time division methods which minimize the number of switches needed and therefore loss. By contrast, optoelectronic switches are more suitable for space, frequency, and wavelength division matrices.

Acknowledgment

This work is based on an earlier study prepared while the author was a Guest Scientist at the Heinrich Hertz Institute (HHI), Berlin. The support of HHI and many discussions are gratefully acknowledged.

References

[1] K. D. Langer, F. Lukanek, J. Vathke, and G. Walf, "Broadband switching network and TV switching network for 70 Mb/s," in *Proc. Int. Zurich Seminar Digital Commun.*, Zurich, 1986, paper D.1.

[2] L. Ronn and P. Deichmann, "16 × 8 space switch for 2.24 Gb/s transmission," *Electron. Lett.*, vol. 23, pp. 236–237, 1987.

[3] J. E. Berthold and G. A. Hayward, "Status and future capabilities of CMOS high-speed switching arrays," in *Proc. Opt. Fibre Commun. Conf.*, New Orleans, LA, 1987, paper THI1.

[4] G. Heydt, G. Teich, and G. Walf, "1.12 Gb/s optical subscriber loop for transmission of high definition television signals," in *Proc. Int. Symp. Subscriber Loops Services* (ISSLS), 1986, pp. 289–293.

[5] M. Sakaguchi and H. Goto, "High speed optical time-division and space division switching," in *Proc. Integrated Opt. Optical Commun. Conf.* (ECOC), vol. II, Venezia, Italy, 1985, pp. 81–88.

[6] T. E. Darcie, "Subcarrier multiplexing for lightwave multiple access networks," in *Proc. Opt. Fibre Commun. Conf.* (OFC), Reno, NV, 1987, paper MI3.

[7] N. A. Olsson and W. T. Tsang, "An optical switching and routing system using frequency tunable cleaved coupled-cavity lasers," *IEEE J. Quantum Electron.*, vol. QE-20, pp. 332–334, 1984.

[8] A. D. Kersey, A. Dandridge, A. M. Yurek, and J. F. Weller, "Multiplexed fibre communications network using path-matched differential interferometry," in *Proc. Opt. Fibre Commun. Conf.* (OFC), New Orleans, LA, 1988, paper THL6.

[9] T. Iwama, Y. Oikawa, K. Yamaguchi, T. Horimatsu, M. Makiuchi, and H. Hamaguchi, "A 4 × 4 GaAs OEIC switch module," in *Proc. Opt. Fibre Commun. Conf.* (OFC), Reno, NV, 1987, paper WG3.

[10] J. E. Midwinter "Photonic switching components: Current status and future possibilities," in *Proc. Top. Meet. Photon. Switching*, Lake Tahoe, NV, 1987, paper WB1.

[11] P. Granestrand, B. Stoltz, L. Thylen, K. Bergvall, W. Doeldissen, H. Heidrich, and D. Hoffmann, "Strictly nonblocking 8 × 8 integrated optics switch matrix," *Electron. Lett.*, pp. 816–817, 1986.

[12] R. A. Spanke "Architectures for large nonblocking optical space switches," *IEEE J. Quantum Electron.*, vol. QE-22, pp. 964–970, 1986.

[13] A. Himeno, H. Terui, and M. Kobayashi, "Guided wave optical gate matrix switch," *J. Lightwave Technol.*, vol. 6, pp. 30–35, 1988.

[14] A. Ajisawa, M. Fujiwara, J. Shimizu, M. Sugimoto, M. Uchida, and Y. Ohta, "Monolithically integrated optical gate 2 × 2 matrix switch using GaAs/AlGaAs multiple quantum well structure," *Electron. Lett.*, vol. 23, pp. 1121–1122, 1987.

[15] J. le Rouzic "Optical switching in subscriber network," in *Euro. Conf. Opt. Commun.* (ECOC), *Tech. Dig.*, 1983, pp. 289–301.

[16] Y. Silberberg and P. Perlmutter, "A digital electrooptic switch," in *Proc. Top. Meet. Photon. Switching*, Lake Tahoe, 1987, paper PDP3.

[17] R. I. MacDonald and D. K. W. Lam, "Optoelectronic switch matrices: Recent developments," *Opt. Eng.*, vol. 24, pp. 220–224, 1985.

[18] D. K. W. Lam and R. I. MacDonald, "Crosstalk measurements of monolithic GaAs photoconductive detector arrays in the GHz region," *Appl. Phys. Lett.*, vol. 46, pp. 1143–1145, 1985.

[19] R. I. MacDonald and D. K. W. Lam, "Broadband matrix switches: Electro-optic or optoelectronic?" *Opt. Quantum Electron.*, vol. 18, pp. 273–277, 1986.

[20] R. I. MacDonald, "Optoelectronic switching in digital networks," *IEEE J. Select. Areas Commun.*, vol. SAC-3, pp. 336–344, 1985.

[21] A. P. Goutzoulis, D. K. Davies, E. C. Malarkey, J. C. Bradley, and P. R. Beaudet, "Residue position-coded lookup table processing," in *Proc. Top. Meet. Opt. Comput.*, Lake Tahoe, 1987, paper MD3.

[22] M. E. Marhic, "Heirarchic and combinatorial star coupler," *Opt. Lett.*, vol. 9, pp. 368–370, 1984.

[23] H. Goto "Photonic time division switching technology," in *Proc. Top. Meet. Photon. Switching*, Lake Tahoe, 1987, paper FD1.

[24] R. A. Thompson and P. P. Giordano, "Experimental photonic time-slot interchanger using optical fibers as re-entrant delay-line memories," *Opt. Fibre Commun. Conf.* (OFC), Atlanta, GA, 1986.

[25] R. S. Tucker, S. K. Korotky, G. Eisenstein, L. L. Buhl, J. J. Veselka, G. Raybon, B. L. Kasper, A. H. Gnauck, and R. C. Alferness, "16-Gb/s optical time division multiplexed transmission system experiment," in *Proc. Opt. Fibre Commun. Conf.* (OFC), New Orleans, LA, 1988, paper THB2.

[26] B. Johnson and B. Ericson, "Design of a future communications network based on optical fibre technology," in *Proc. Int. Zurich Seminar Digital Commun.*, Zurich, Switzerland, 1986, paper 13-2.

[27] R. I. MacDonald and K. O. Hill, "Avalanche optoelectronic downconverter," *Opt. Lett.*, vol. 7, pp. 83–85, 1982.

[28] A. G. Foyt and F. J. Leonberger, "InP optoelectronic mixer," *SPIE* vol. 269, pp. 109–119, 1981.

[29] D. K. W. Lam and R. I. MacDonald, "GaAs optoelectronic mixer operation at 4.5 GHz," *IEEE Trans. Electron Devices*, vol. ED-31, pp. 1766–1768, 1984.

[30] I. Mito, "Recent advances in frequency tunable DFB-DBR lasers," in *Proc. Opt. Fibre Commun. Conf.*, New Orleans, LA, 1988, paper THK1.

[31] I. P. Kaminow, P. P. Iannone, J. Stone, and L. W. Stoltz, "FDM-FSK Star network with a tunable optical filter demultiplexer," in *Euro. Conf. Opt. Commun.* (ECOC) *Tech. Dig.*, vol. III, Helsinki, Finland, 1987.

[32] R. M. Bulley, M. S. Goodman, H. Kobrinski, C. N. Lo, M. P.

Vecchi, and C. A. Brackett, "Experimental demonstration of LAMB-DANET: A multiwavelength optical network," in *Euro. Conf. Opt. Commun. (ECOC), Tech. Dig.*, Helsinki, Finland, 1987, p. 345.

[33] C. Baack, E.-J. Bachus, and G. Heydt, "Coherent multichannel techniques in future broadband communication networks," in *Euro. Conf. Opt. Commun. (ECOC), Tech. Digest*—post deadline, Helsinki, Finland, 1987.

R. Ian MacDonald (SM'87) was born in the United Kingdom in 1943. He received the B.Sc. degree in engineering physics from Queen's University, Kingston, Ont., Canada, in 1965 then joined the Northern Electric R&D Laboratory (now Bell-Northern Research) to develop neutron activation analysis methods for semiconductor dopants. Subsequent research positions involved auroral measurements at the Churchill Research Range, Fort Churchill, Manitoba, Canada, and the development of microcircuit maskmaking techniques at Microsystems International, Canada, and at SGS, Italy. After graduate work in optical communications at Carleton University, Ottawa, Ont., Canada leading to the M.Eng. and Ph.D. degrees in electrical engineering in 1972 and 1975, respectively, he joined the staff of the Communications Research Centre of the Department of Communications, Ottawa, Ont., Canada. In this position, he participated in the development of optical communications in Canada, specializing in nontransmission aspects of the technology, particularly broad-band switching. From 1985 to 1986, he was a Guest Scientist at the Heinrich Hertz Institut, Berlin, Germany. At present, he is with the Faculty of Engineering, University of Alberta, Edmonton, as Staff Professor, Alberta Telecommunications Research Centre.

Dr. MacDonald is a member of the Association of Professional Engineers of Ontario and the Optical Society of America.

Architectural Considerations for Photonic Switching Networks

H. SCOTT HINTON, MEMBER, IEEE

Reprinted from *IEEE Journal on Selected Areas in Communications*, Vol. 6, No. 7, August 1988, pages 1209-1226. (*Invited Paper*)

Abstract—This paper will review some of the photonic technologies that could become important components of future telecommunications systems. It will begin by dividing photonic devices and systems into two classes according to the function they perform. The first class, relational, will be associated with devices, which under external control, maps the input channels to the output channels. The second class, logic, requires that the devices perform some type or combination of Boolean logic functions. After the classes are defined, some of the strengths and weaknesses of the photonic domain will be presented. Relational devices and their applications will then be discussed. Finally, there will be a review of optical logic devices and their potential applications.

I. INTRODUCTION

WITHIN recent years there has been a significant amount of interest in applying the new and developing photonics technology in telecommunications switching systems [1]. As the transmission plant has converted its facilities to fiber, there is an economic interest in completing the optical path through the switching system to the terminal facilities without requiring optical-to-electrical (o/e) conversions. There are several devices that have emerged within the past few years which have the capability of meeting this goal. These devices can be arranged into two major classes according to the function they perform [2]. The first of these classes, called *relational* devices, perform the function of establishing a *relation* or a mapping between the inputs and the outputs. This *relation* is a function of the control signals to the device and is independent of the signal or data inputs. As an example, if the control signal is not enabled, the relation between the inputs and the outputs of a 2 × 2 device might be input port 1 → output port 1 and input port 2 → output port 2. When the control is enabled, the relationship might be input port 1 → output port 2 and input port 2 → output port 1. This change in the *relation* between the inputs and outputs corresponds to a change in the state of the device. Another property of this device is that the information entering and flowing through the devices cannot change or influence the current *relation* between inputs and outputs. An example of this type of device is the directional coupler as it is used in switching applications. Thus, the strength of relational devices is that they

Manuscript received October 16, 1987; revised February 24, 1988. This paper is a combination of papers presented at CLEO'87, Apr. 26–May 1, 1987, Baltimore, MD and OFC'88, Jan. 25-28, 1988, New Orleans, LA.
The author is with AT&T Bell Laboratories, Naperville, IL 60566.
IEEE Log Number 8821378.

cannot sense the presence of individual bits that are passing through them which allows them to pass high bit rates. The weakness of relational devices is that they cannot sense the presence of individual bits that are passing through them which reduces their flexibility.

The second class of devices will be referred to as *logic* devices. In these devices, the data or information carrying signal that is incident on the device controls the state of the device in such a way that some Boolean function or combination of Boolean functions is performed on the inputs. For this class of device, at least some of the devices within a total system must be able to change states or *switch* as fast or faster than the signal bit rate. This high-speed requirement for logic devices will limit the bit-rates of signals that can eventually flow through their systems to less than those that can pass through relational systems. Thus, the strength of logic devices is the added flexibility that results from their ability to sense the bits that are passing through them while their weakness is that they sense the bits that pass through them which limits the maximum bit-rate that they can handle.

This paper will begin by discussing both the strengths and limitations of the photonic technology. Some of the items to be discussed include power, speed, bandwidth, and parallelism. The next section will discuss optical relational systems with a focus on the directional coupler. The reason for this focus is that directional couplers will most likely be the initial photonic switching component to enter the marketplace. Finally, there will be a discussion of optical logic systems. These systems are still primarily in the research labs and not ready for development.

II. STRENGTHS AND LIMITATIONS OF THE PHOTONIC TECHNOLOGY

Prior to discussing either photonic devices or their applications it is important to understand both their potential and limitations. This section has the purpose of discussing the strengths and weaknesses of the photonics switching technology. It will begin by discussing the power, speed, and bandwidth limitations of photonic devices. Next it will focus on parallelism and how it can be used in photonic systems. Finally, there will be a brief discussion on the size of future devices.

A. Power, Speed, and Bandwidth

There are two speed limitations that must be considered in the design of photonic switching systems. The first of these limitations is the time required to *switch* or change the state of a device. *Switching*, in this case, refers to the changing of the present state of a device to an alternate state, as opposed to the "switching" that is analogous to an interconnection network reconfiguration. In the normal operating regions of most devices, a fixed amount of energy, the switching energy, is required to make them change states. This switching energy can be used to establish a relationship between both the switching speed and the power required to change the state of the device. Since the power required to switch the device is equal to the switching energy divided by the switching time, then a shorter switching time will require more power. As an example, for a photonic device with an area of 100 μm^2 and a switching energy of 1 fJ/μm^2 to change states in 1 ps requires 100 mW of power instead of the 100 μW that would be required if the device were to switch at 1 ns. Thus, for high-power signals the device will change states rapidly, while low-power signals yield a slow switching response.

Some approximate limits on the possible *switching* times of a given device, whether optical or electrical, are illustrated in Fig. 1 [3]. In this figure the time required to *switch* the state of a device is on the abscissa while the power/bit required to *switch* the state of a device is on the ordinate. The region of spontaneous switching is the result of a background thermal energy that is present in a device. If the switching energy for the device is too low, the background thermal energy will cause the device to change states spontaneously. To prevent these random transitions in the state of a device, the switching energy required by the device must be much larger than the background thermal energy. To be able to differentiate statistically between two states, this figure assumes that each bit should be composed of at least 1000 photons [4]. Thus, the total energy of 1000 photons sets the approximate boundary for this region of spontaneous switching. For a wavelength of 850 nm, this implies a minimum switching energy on the order of 0.2 fJ.

For the thermal transfer region, Smith assumed that for continuous operation the thermal energy present in the device cannot be removed any faster than 100 W/cm^2(1 μW/μm^2). There has been some work done to indicate that this value could be as large as 1000 W/cm^2 [5]. This region also assumes that there will be no more than an increase of 20°C in the temperature of the device. Devices can be operated in this region using a pulsed rather than continuous mode of operation. Thus, high-energy pulses can be used if sufficient time is allowed between pulses to allow the absorbed energy to be removed from the devices.

The cloud represents the performance capabilities of current electronic devices. This figure illustrates that optical devices will not be able to *switch* states orders of magnitude faster than electronic devices when the system is in the continuous rather than the pulsed mode of operation. There are, however, other considerations in the use of optical computing or photonic switching devices than

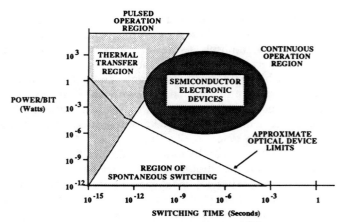

Fig. 1. Fundamental switching limits at 850 nm.

how fast a single device can change states. Assume that several physically small devices need to be interconnected so that the state information of one device can be used to control the state of another device. To communicate this information, there needs to be some type of interconnection with a large bandwidth that will allow short pulses to travel between the separated devices. Fortunately, the optical domain can support the bandwidth necessary to allow bit rates in excess of 100 Gbits/s, which will allow high-speed communication between these individual *switching* devices. In the electrical domain, the communications bandwidth between two or more devices is limited by the resistance, capacitance, and inductance of the path between the different devices. Therefore, even though photonic devices cannot *switch* orders of magnitude faster than their electronic counterparts, the communications capability or transmission bandwidth present in the optical domain should allow higher speed *systems* than are possible in the electrical domain.

The second speed limitation, which applies only to relational devices, will be referred to as the transmission bandwidth. After a relational device has been put into a particular state, it acts like a transmission line to any data entering its inputs. This input data cannot change the state of the device, thus the signal bit-rates passing through a relational device are not limited by the constraints outlined in Fig. 1. For most relational devices, this transmission bandwidth should be able to support bit-rates in excess of 100 Gbits/s.

In summary, networks composed of relational devices will have their signal bit rates limited by the transmission bandwidth and their reconfiguration rates limited by the switching time of the devices, while switching networks based on optical logic will have both their signal bit-rates and reconfiguration rates limited by the switching time of their devices.

B. Parallelism

Another method of increasing the capacity of a system, in addition to operating at higher speeds, is to operate on information in parallel instead of in series. In pursuing this parallelism, attention has recently been placed on free-space optics. These types of systems normally are composed of multiple two-dimensional (2-D) arrays of optical devices that are interconnected through either bulk optics or holography. Fig. 2 shows the optical interconnection between two 2-D arrays of optical elements. The interconnection in this case is a simple lens system. The optical elements, which will be referred to as pixels, could be optical NOR gates, optical light valves, etc. The number of pixels that can be interconnected in this manner is limited by the resolution of the optical interconnection system. Even relatively inexpensive optical imaging systems exhibit resolutions on the order of 10 μm over a 1 mm field. This provides access to 100 $\times$ 100 or 10^4 pixels. If each pixel can be equated to a pin-out, then for a 2-D array there can be greater than 10^4 pin-outs. The maximum number of pixels or pin-outs that can be supported by a lens or any optical system is referred to as its space–bandwidth product [6] (SBWP) or the degree of freedom of the system. Satellite imaging systems have been made that have an SBWP of 10^8 pixels.

Fig. 3, on the other hand, illustrates an optical interconnect based on holograms. Holography offers the promise of extremely high SBWP ($>10^{10}$). This will be discussed in more detail in Section IV-B-2.

There is no reason to limit the pixels of these 2-D arrays to all-optical logic gates; they could also be a mixture of electronic and optical devices (smart pixels) as shown in Fig. 4. This mixture of electronic and optical devices is designed to take advantage of the strengths in both the electrical and optical domain. The optical devices include detectors to convert the signals from the previous 2-D array to electronic form and modulators (surface emitting lasers or LED's) to enable the results of the electronically processed information to be transferred to the next stage of 2-D arrays. The electronics does the intelligent processing on the data. Since the electronics is localized with short interconnection lengths, the speed of this electronic island should be fast. The applications of smart pixels will be discussed later in Section IV-A-2.

To take advantage of the parallelism inherent in free-space optics, a device should have the capability of driving other devices (fan-out) in addition to being controlled by more than one device (fan-in). Since fan-out corresponds to a division of the energy emitted from a device output to the inputs of other devices, the output energy must be significantly larger than the energy required by the input to the subsequent device. Another factor that affects the fan-in is the contrast ratio, the contrast ratio being the ratio of the transmitted intensity of both states of a device. For low contrast ratios, unwanted noise is present in the system, thus reducing the ability of a device to sample the input correctly.

C. Device Size

The minimum size of an optical switching device cannot be reached below a volume of $(\lambda/n)^3$ [3] where n is

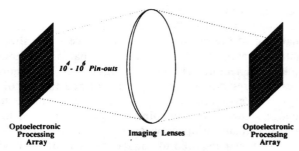

Fig. 2. Optical parallel interconnections.

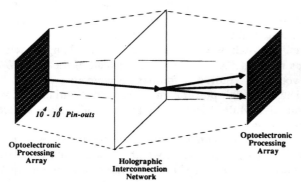

Fig. 3. Holographic interconnects.

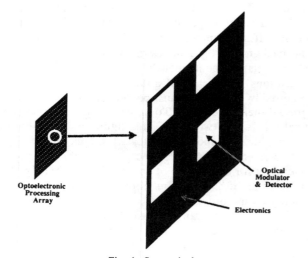

Fig. 4. Smart pixels.

the index of refraction of the device. For the case of a 2-D array of devices, assuming that these devices are fabricated such that there is a separation of λ/n between the devices, then approximately 25 million devices per square centimeter could be possible at $\lambda/n = 1$ μm. As the device switching speed increases, the thermal transfer capacity of the devices will most likely prevent such a large number of devices from ever being realized. As an example, if there are 25 million devices per square centimeter that require 1 pJ to switch, and it is desired to switch these devices in 100 ps, then the total array would require 250 kW/cm^2 of optical power if all of the devices were to switch at the same time.

III. Optical Relational System

In this section, several photonic switching systems that are based on relational devices will be reviewed. The first type of system to be discussed is based on spatial light modulators (SLM) which are 2-D arrays of devices, each of which has the capability of modulating the light that is incident upon it. The second example of a relational system is based on wavelength-division switching. Finally, the systems based on Ti : LiNbO$_3$ directional couplers will be discussed.

A. Spatial Light Modulators

An SLM is a two-dimensional array of optical modulators [7]. Each of these modulators is independent of the others and has the capability of modulating the incident light. For the applications described in this paper, the modulators will be assumed to be digital, in that they possess two states: transparent to be the incoming light (on) and opaque to the incoming light (off). These arrays are electrically controlled such that an electrically enabled pixel will be transparent while a disabled pixel will block the incident light. An SLM that is currently available in the marketplace is based on the magnetooptic effect [8]. Some other SLM's include the liquid crystal light valves (LCLV) [9], PLZT modulators [10], deformable mirrors [11], and GaAs multiple-quantum well (MQW) modulators [12].

One example of how an SLM can be used as a photonic switch is shown in Fig. 5 [13], [14]. In this figure, the fiber inputs are horizontally aligned as a row of inputs. The inputs are aligned to associate each fiber with a unique column of the SLM. A lens system is used to spread these inputs vertically so that the light emitted from each input is spread over all the elements' of the SLM's associated column. The appropriate pixels of the SLM are enabled before the data pass through the system. The one or more enabled pixels in each column allow the incident light to be transmitted through the device while the remaining pixels block the incident light. The output column of fibers acccepts the light that is passed through the SLM. An important restriction for this type of structure is that only one pixel on each row can be enabled at any time. The relational nature of this structure is evident in that each row of the SLM acts like an $N \times 1$ switch where N is the number of pixels per row. The total structure is topologically equivalent to a nonblocking crossbar interconnection network. The weakness of these systems is the losses that occur in the spreading and collection of the light going to and from the SLM can be greater than $1/N^2$.

As with all relational structures, high signal bit-rates pass through the switch with the speed limitation being the system reconfiguration time.

B. Wavelength-Division Switching Systems

Another type of relational architecture that has received a considerable amount of attention is wavelength-division switching [15]. This is schematically shown in Fig. 6. In this figure, the entering information is used to modulate a light source that has a unique wavelength for each input. All the optical energy is combined and then split so it can be distributed to all the output channels. The tunable filter on each ouput is adjusted such that it only allows the wavelength associated with the desired input channel to pass to the detector. Thus, by varying the tunable filter, an output has access to any or all of the input channels. Obviously, another method of detecting the appropriate wavelength would be through the use of coherent detection techniques [16]. This wavelength-division concept can be generalized to include other orthogonal basis functions as the carrier instead of wavelength [17], [18].

C. Directional Couplers

A directional coupler is a device that has two optical inputs, two optical outputs, and one control input. The control input is electrical and has the capability of putting the device in the *bar* state, input port 1 → output port 1 and input port 2 → output port 2, or the *cross* state, input port 1 → output port 2 and input port 2 → output port 1 [19]. The most advanced implementations of these devices have occurred using the Ti : LiNbO$_3$ technology [20]. The strength of directional couplers is their ability to control extremely high bit rate information. They are limited by several factors: 1) the electronics required to control them limits their maximum reconfiguration rate, 2) the long length of each directional coupler prevents large scale integration, and 3) the losses and crosstalk associated with each device limit the maximum size of a possible network unless some type of signal regeneration is included at critical points within the system [21]. A modest number of these devices have been integrated onto a single substrate to create larger photonic interconnection networks such as an 8 × 8 crossbar interconnection network [22]. As another example, a 4 × 4 crossbar interconnection network composed of 16 integrated directional couplers, all having crosstalk less than −35 dB with an average fiber-to-fiber insertion loss of less than 5.2 dB, has been fabricated [23]. The following sections will discuss some of the practical considerations that need to be addressed in the design of a photonic switching system based on directional couplers. Several potential applications will then be discussed.

1) Practical Considerations: There are several practical issues that need to be considered when designing a system based on Ti : LiNbO$_3$ directional couplers. These issues include the required parallel development of polarization maintaining (PM) fiber, optical amplifiers, and the packaging required to make the devices reliable and easy to use.

a) Polarization Maintaining Fiber: To minimize the required drive voltages, directional couplers have been optimized to operate with a single linear polarization. This requirement reduces the required switching voltage from approximately 50 V to the 10–15 V range. These lower voltages are desired to allow high-speed switching of the directional couplers. One problem with using single po-

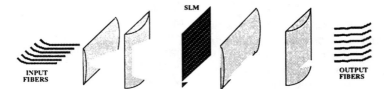

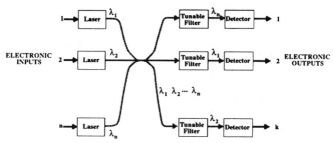

Fig. 5. Optical crossbar interconnection network.

Fig. 6. Wavelength-division switching system.

larization devices is that as light propagates through standard single-mode fiber, the state of polarization can be changed from a linearly polarized wave to a wave having an elliptical polarization. Another complicating factor is that this change in the state of polarization does not remain constant over time. To solve this problem, a PM fiber is required for the interconnection from all the laser sources to the Ti:LiNbO$_3$ substrates and for any substrate-to-substrate interconnection. In addition to PM fiber, there must also be PM fiber connectors. These connectors become important when networks involving the interconnection of multiple LiNbO$_3$ substrates are required.

b) Optical Amplifiers: The eventual size of a switching fabric composed of directional couplers is limited by either the losses through the system or the individual crosstalk terms which degrade the system signal-to-noise ratio (SNR) below an acceptable value [24]. To avoid this problem, thresholding optical amplifiers can be inserted at critical points in the network to both boost the strength of the signal and remove accumulated noise. Unfortunately, most of the optical amplifiers under investigation are linear amplifiers [25], these amplifiers have the disadvantage of amplifying the low-level noise, signals as well as the desired signals. It is desired to have an amplifier that will not amplify low-level signals associated with the logic level "0" but will amplify signals above the threshold of the logic level of a "1." The characteristics of a thresholding optical amplifier [26] are shown in Fig. 7. In the ideal case, there would be no amplification for signals below a given intensity level. Once an intensity threshold has been surpassed, a large gain is desired until a saturated or maximum value of output intensity is obtained [27]. Such a device could both amplify a signal and improve the SNR of the signals passing between substrates of a dimensionally large relational photonic switching system.

OPTICAL AMPLIFIERS

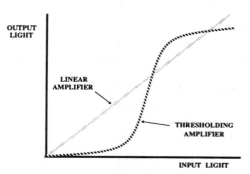

Fig. 7. Optical amplifier characteristic curves.

A component required in conjunction with optical amplifiers is an optical isolator. In addition to isolating lasers from reflections that can occur at connectors, splices, and other interfaces, optical isolators are required to prevent optical amplifiers from lasing. This lasing action can occur since the two requirements for a laser are met: optical gain and positive feedback. The optical amplifier provides the gain and the connectors, splices, or other reflecting interfaces on both sides of the optical amplifier provide the positive feedback required. For high-gain optical amplifiers, this lasing action can occur with small interface reflectivities. By placing an isolator between the unwanted reflections and the optical amplifier, the positive feedback is reduced, thus preventing the undesired lasing.

c) Packaging: To develop both a marketable and reliable system, devices have to be packaged in a useful and reliable manner. This is one of the most overlooked aspects in the development of this technology. At the current time, most of the devices are packaged in their own separate package. For large systems, this could involve an enormous amount of physical space just to house all the individual components. High-speed packaging is another unresolved issue. This subject will require a large amount of attention before this technology can successfully enter the marketplace.

2) Applications of Directional Couplers: Perhaps the key event required to drive the Ti : LiNbO$_3$ directional coupler technology into the marketplace within the next three to five years is a *good* application. Therefore, several potential applications will be outlined in this section. The first application to be discussed utilizes the directional couplers in a space switching environment. Second, the signal formats required to allow these relational devices to operate in a time-division application will be outlined. Finally, there will be brief examples of both time-multiplexed and packet switching.

a) Space Switching: This type of switching will most likely be the first application of the Ti : LiNbO$_3$ directional couplers. It requires long hold times with moderate reconfiguration rates. Once a path has been set up, high-speed data, multiplexed speech, or video can be transferred through the fabric.

The implementation of a large space switch requires the interconnection of many smaller photonic switches that are used as building blocks. These building blocks will most likely have dimensions less than 16×16 because of the large size of directional couplers and the large bending radii required in the integrated waveguides. Two examples of topologies for these building blocks are the crossbar interconnection network [21] and the broadcast network proposed by Spanke [28]. For point-to-point networks, the interconnection of these building blocks to construct a larger switching system can be done with Clos, Benes, banyan, omega, or shuffle networks. If video information is to be a main component of the system traffic, then a broadcast environment becomes important. A good topology for a broadcast network is a Richards network [29]. An example of a space-switching experiment is shown in Fig. 8 [30]. In this system, each terminal has the capability of transmitting and receiving one of two wavelengths. Initially a path is set up between two terminals through the folded optical switching network. This network is composed of switching arrays made using Ti : LiNbO$_3$ directional couplers. After the path is set up, an input signal is sent to terminal *B* on a light source of wavelength λ_1. Terminal *B* will then create the return path to terminal *A* using the same physical path but modulating the information onto the light source of wavelength λ_2.

A good application of a directional coupler is a protection switch. In this environment, the only time the switch will need to be reconfigured is when a failure occurs in an existing path. Thus, high bit rates can be passed through the switch with moderate reconfiguration rate requirements. This application matches the capabilities of the directional coupler.

b) Signal Formats: As high bit rate transmission systems are being developed, there is pressure to build switching systems that will switch these complex signals. Transmission systems are designed such that the high bit-rate signal only passes through a few elements as it is

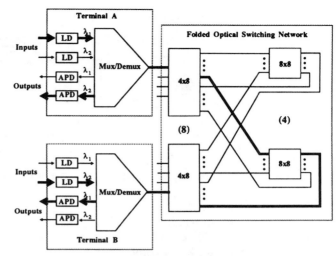

Fig. 8. 32 line optical space switch [30].

generated or processed. These few elements allow all the controlling electronics to be placed close together. In a switching system, the signal must pass through a large number of switching elements which are normally spread over a large physical space; this makes switching difficult at high bit-rates. As bit-rates continue to increase, there will need to be a compromise between the complexity of transmission and the switching systems [31]. One solution to this problem is to change the signal format from a bit-multiplexed to a block-multiplexed format. By changing this format, transmission systems will become more complex because buffers will be required at all the inputs to the multiplexers. Such a new format will simplify switching systems by allowing them to operate directly on the information rather than having to demultiplex the information and then switch the lower bit-rate signals. An example of such a proposed format is SYNTRAN [32]. The concept of this block-multiplexed format is good for future photonic applications but the DS3 bit rate (45 Mbits/s) it proposes is too slow for the photonic domain. These slower bit rates can be handled easily in the electronic domain. This concept does have merit if it is extended to higher bit rates by using individual 125 μs frames of the 45 Mbit/s DS3 information as the basic block of data for the system. As an example, in the FT series G transmission systems, a single 125 μs frame of the 1.7 Gbit/s data stream contains 36 frames of DS3 information. The bits in these frames, plus overhead bits, are interleaved and mixed so that individual DS3 frames cannot be extracted from the stream unless that stream is at least partially demultiplexed down to DS3 channels. By requiring that the DS3 frames be block multiplexed onto the high-speed channel with a small gap between them, individual extraction and insertion of DS3 frames should be possible. The characteristics of a directional coupler (slow reconfiguration rates, but high signal bit rates) make it an ideal device to be used with a block-multiplexed format.

c) Time-Multiplexed Switching: As the bit rates

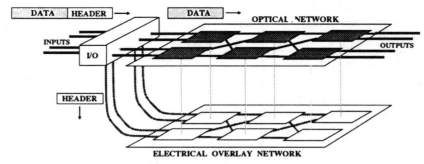

Fig. 9. Directional coupler-based packet switching system [63].

passing through these photonic switches increase, there approaches a point where no single information source can occupy all the bandwidth without some type of time-division multiplexing. As an example, uncompressed digital NTSC television signals require approximately 100 Mbits/s while digital high-definition television will be about 600 Mbits/s. Thus, if the transmission system operates at 1.7 Gbits/s then at least 16 NTSC channels or two high-definition channels can be time-multiplexed together. To build such a system will require a time-slot interchanger, and an elastic store. The elastic store will also be required to remove both frequency and phase jitter from the system inputs. Proposed elastic stores and time-slot interchangers have been demonstrated using fiber delay lines as the memory [33] or bistable laser diodes as bit memories [26].

d) Packet Switching Systems: The problem in implementing a packet switching system based on directional couplers is that a header has to be read to determine the final destination of the packet. This can be done by stripping off a portion of the optical energy and monitoring the header (trailer) electronically [63] as illustrated in Fig. 9. If necessary, the header (trailer) could be at a lower bit rate than the packet data thus allowing the slower electronics to respond to the controlling information. If the header bit rate is at the same bit rate as the data, then the eventual bit-rate upper limit would be governed by the speed of the electronics. It would be possible to have headers encoded at slower bit rates than the data at the cost of reducing the total throughput of the switching system. The cost for such a system is the large amount of electronics that is surrounding the small number of photonic devices. Thus, to be economically justifiable, large packet lengths are required.

IV. Optical Logic Systems

There are certain applications that are not well suited for relational devices. One such application requires the ability to both sense and respond to individual bits of information. A packet switch is a good example of this requirement. A packet entering a network requires a system of devices that can read and understand the header and then reconfigure the network to allow the packet to pass to its desired destination. The ability to interact and sense the individual bits in a stream of information is one of the strengths of optical logic devices [34].

To make a useful system, these devices need to be interconnected, as in current electronic switching systems, to create a large interconnection network. The purpose of the following sections is to discuss several of the optical logic devices that have been fabricated in addition to describing two methods of optically interconnecting these devices. The section on optical logic systems will begin by discussing several optical logic devices. Next, two possible methods of optically interconnecting 2-D arrays of logic devices will be discussed. Finally, there will be a brief discussion of a packet switching system that could be based on these devices.

A. Digital Optical Logic Devices

Digital optical logic devices can be further classified according to the effective number of ports, where a port is either a physical input or output. The discussion on optical logic devices will begin by reviewing several two-port devices and the constraints they impose upon a switching system architecture. Multiport devices and their associated system issues will then be discussed.

1) Two-Port Devices: Two-port devices can be divided into two different types. The first, which will be referred to as combinatorial, are devices whose output is a function of present inputs. These devices include any device that simulates a Boolean logic function. The characteristic curves of these devices are shown in Fig. 10. Part (a) illustrates the characteristic curve of an inverter. Normally, there is a bias beam, separate from the signal beam, which provides the optical energy to be modulated by the device. When the energy from the signal beam is added to the energy of the bias beam, this combined energy is enough to exceed the nonlinear threshold of the device. As the device changes states, the output level goes from a high value or a ''one'' to the lower value which is equated to a ''zero.'' By reducing the energy in the bias beam and adding more signal beam energy through the use of multiple signal beams, an optical NOR gate can be formed. Part (b) is the characteristic curve of a thresholding optical amplifier. By reducing the bias beam and adding more inputs, this device could be an optical AND gate.

The second type of device is sequential in nature. These

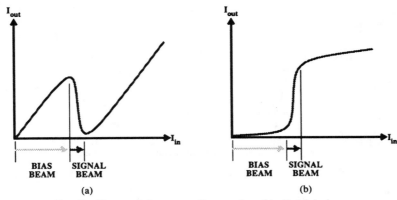

Fig. 10. Characteristics curves of two-port combinational devices.

devices could be smart pixels with one optical input, one optical output, and some high-speed electronics configured as a sequential finite state machine. Perhaps the simplest example of these structures is a bistable device [35]. Since these devices have memory, their outputs are a function of the present inputs and their present state (previous inputs). Another example of this type of two-port device is an optical regenerator. When an optical signal enters the device, the information is both amplified and retimed prior to outputting the reconstructed signal.

a) Proposed Devices: There are two optical logic devices that have received a considerable amount of attention in the past few years. The first of these devices, the self-electrooptic effect device (SEED)[36], is an electrooptical device that requires both optical and electrical energy. The second device, referred to as a nonlinear Fabry–Perot etalon (NLFP) is an all-optical device [34] in that all the energy required to switch the device is supplied optically. Both of these devices can be fabricated into 2-D arrays of devices that are both optically enabled and controlled. The purpose of the 2-D arrays is to provide the opportunity to exploit the parallelism present in the optical domain.

A functional diagram of the SEED is shown in Fig. 11. A p-i-n diode with a multiple quantum well (MQW) material in the intrinsic region is connected in series with a resistor to form the SEED structure. The characteristic curve for the device is shown in part (a) of Fig. 10. A bias beam is required to provide the energy that will be modulated by the signal beam. This modulation of the energy source by the signal beam provides the differential gain that is required in a digital logic system. With low signal beam intensities, the SEED is virtually transparent. This allows nearly all the energy present in the bias beam to pass through the device to become the output beam. When the signal beam is added with the bias beam, the combined energy is enough to force the SEED structure to become an absorber, thus reducing the total amount of energy in the output beam. This characteristic curve is equivalent to the characteristic curve of an optical NOR gate. SEED structures have been integrated into two-dimensional arrays [36]. It is conceivable that within 10

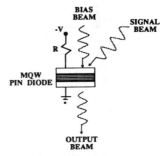

Fig. 11. The self-electrooptic effect device.

years, arrays of 10 000–100 000 individual SEED's per square centimeter will be possible. Currently, 4 fJ/μm^2 of optical energy and 16 fJ/μm^2 of electrical energy are required to change the state of the SEED.

A second possible device for the optical logic domain is a nonlinear Fabry–Perot etalon (NLFP). This device exploits reflection instead of absorption to control or modulate a bias beam. This is shown in Fig. 12. Part (a) of this figure illustrates that without a signal beam, the bias beam passes through the NLFP. With an incident signal beam, part (b) of the figure, the NLFP changes states which forces the bias beam to be reflected instead of transmitted. This device, like the previously described SEED, is operating as a NOR gate. The NLFP can also be designed such that it can operate as an AND, OR, NAND, or XOR gate [34]. These devices, like SEED's, can also be integrated into large arrays [37]. The optical switching energy for NLFP's has been measured to be less than 40 fJ/μm^2 [38] with achieved switching times less than 100 ps [39].

At the current time, there are two types of NLFP etalon structures being studied. The first is based on a thermal nonlinearity which makes it slow, on the order of microseconds [34]. The hope for these devices is that large arrays can be fabricated to exploit the strength of parallelism. The other type of device, referred to as an optical logic etalon (OLE) is a pulsed device that requires two separate wavelengths, one for the bias and one for the signal [40]. The pulsed operation is illustrated in Fig. 13. The two inputs, the data and clock, are separated in both

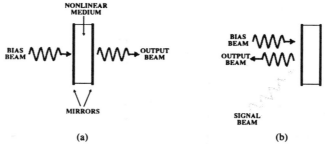

Fig. 12. The nonlinear Fabry–Perot etalon optical gate.

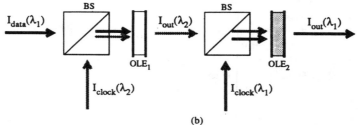

(a)

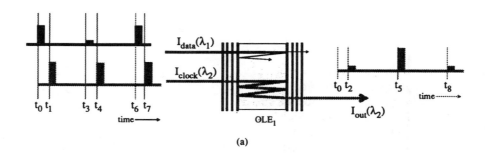

(b)

Fig. 13. OLE operation. (a) Input and output signals for OLE operation.
(b) Complementary pair of OLE's.

time and wavelength. The input data $I_{data}(\lambda_1)$ is incident on the OLE a short period of time prior to the clock input $I_{clock}(\lambda_2)$. In part (a) of the figure, $I_{data}(\lambda_1)$ is incident upon the device at t_0 while $I_{clock}(\lambda_2)$ arrives at t_1. Ideally, $I_{data}(\lambda_1)$ is chosen such that λ_1 corresponds to an absorption peak of the nonlinear material in the OLE, bulk GaAs. Thus, nearly all of the energy associated with the incoming data is absorbed by the material. This absorbed energy alters the effective index of refraction in the GaAs intercavity material [41]. This change in the index of refraction shifts the resonant peak of the Fabry–Perot etalon by changing the optical path length of the cavity. Through the proper choice of initial detuning, several potential optical logic gates can be implemented. In this figure, the initial detuning was set so that the OLE would behave like an optical NOR gate. If the material relaxation time of the intercavity material is long compared to the pulse duration, the state of the device will not change significantly prior to the arrival of the clock pulse $I_{clock}(\lambda_2)$. When the clock arrives, the OLE will be in either a transmissive or reflective state depending on the value of $I_{data}(\lambda_1)$. In this figure, if there was a logical "one" on the data input,

then the OLE would be reflective forcing the output to be a logical "zero." On the other hand, if the input data was a logical "zero," the device would be transmissive allowing the majority of the clock energy through the OLE creating a logical "one" output. Since λ_2 is not near the absorptive peak of the intercavity material, clock signals much larger than the data signals can provide an effective gain through the device. This separation of the data and clock inputs in either time or wavelength has converted a two-port device into an effective three-port device.

Part (b) of this figures shows what is required to integrate these devices into larger systems. Because the input wavelength of an OLE is different from the output wavelength, the devices are not cascadable. Therefore, to implement any type of system will require a complementary device that will change the output wavelength of the two devices back to the original input wavelength. This part of the figure shows how, through the use of polarization beamsplitters, the two devices can be interconnected. Unfortunately, at the current time a complementary pair has not been demonstrated.

b) System Considerations: When any of the two-port

468

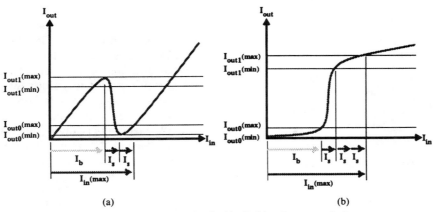

Fig. 14. Input/output levels of critically biased two-port devices.

devices operate with a bias beam as shown in Fig. 14, there are strict requirements on the bias beam's stability [42]–[44]. If the bias beam varies in intensity, it could spontaneously exceed the nonlinear threshold forcing the device to change states. Also, if the intensity of the bias beam decreases significantly, it could prevent the device from changing states when a signal beam is incident upon it. In addition to the bias beam, the point of nonlinearity in the characteristic curve must also remain constant. This required stability puts tight restrictions on thermal control for most devices. As an example, for any structure using a Fabry–Perot cavity, temperature variations will cause changes in the location of the resonant peaks of the cavity. This in turn alters the initial detuning of the device which can alter the characteristic curve of the device. Such variation could cause erroneous device behavior. Also, any device that is based on a material resonance could also suffer detrimental effects with temperature variations.

Another problem associated with the critically biased devices that have been previously described is the size of the fan-in and fan-out of the device. The fan-in is the maximum number of inputs (signal, not clock beams) that can be used as inputs to a given device and can be represented, as shown in Fig. 14, by fan-in = $(I_{in}(\max) - I_b)/I_s$. The fan-out, which is the maximum number of input signals that an output signal can be decomposed into, is represented by fan-out = $(I_{out1}(\min) - I_{out0}(\max))/I_s$. These two equations imply that for a large fan-out and fan-in, which is desirable, a large sharp nonlinearity is required. The problem with a sharp nonlinearity is the strict requirement on the stability of the bias beam. Because of these stability requirements, it is unlikely that for critically biased devices, fan-out greater than five will be practical.

Another area of concern with respect to fan-in occurs when the light in a system has a large coherence length and a single polarization. In this situation, both constructive and destructive interference could occur when two or more beams are incident on the same physical input. As the devices become smaller, which will be required to re-

duce the switching energy, it will be possible for this interference to prevent a device from operating properly. To avoid this problem, the fan-in could be limited to two inputs with each input being associated with a different polarization. Another alternative is to separate all inputs in time as in the case of the OLE.

2) Multiport Devices: The devices discussed so far can be classified as two-port devices in that they have one input and one output. By adding more inputs and outputs, many of the problems associated with two-port devices can be removed. As an example, by adding more physical input ports, the problem of low device fan-in can be resolved even though the fan-in per port is small (the desired fan-in per port is two, one for each polarization). Also, the effects of critical biasing can be removed by separating the signal input from the bias input thus creating a three-port device.

a) Proposed Devices: There are several devices that have been proposed that could be classified as "multiport devices." The first device to be discussed will be a three-port device in which the signal input is separated from the bias input. The second device is a four-port device referred to as the *symmetric SEED*. This device consists of two electrically interconnected MQW p-i-n diodes. Finally, a five-port device that performs an exchange–bypass operation will be discussed.

An example of a three-port device is shown in Fig. 15 [45]. In this figure, an MQW modulator is electrically connected to a phototransistor [46]. The objective of this device is twofold: 1) increased sensitivity for the input signal (optical gain) and 2) isolation between the input and output signal. With an input signal present on the phototransistor, a photocurrent is created which is roughly proportional to the input signal power. The optical bias beam also gives rise to a photocurrent through the modulator which is proportional to the power absorbed. These two currents must be equal or a charge will build up on the modulator affecting the absorption such that the two currents will equalize. If the operating wavelength is set just below the band edge, then with no input signal, the majority of the voltage will be dropped across the pho-

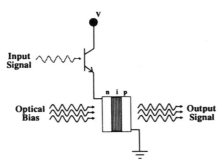

Fig. 15. Photonic three-port device.

totransistor which puts the modulator in the transmissive state. When an input signal is present, the phototransistor will turn on putting the majority of the voltage across the modulator. With this applied voltage, the band edge of the MQW material will shift, putting the modulator in the absorptive state. Thus, a lower power input signal has the capability of modulating the higher power optical source providing gain. The modulation effect provides both gain between the input and output signal and a logical inversion of the input data. A noninverting three-port device has also been demonstrated [47].

The symmetric SEED (S-SEED) is a four-port device with two inputs and two outputs as shown in part (a) of Fig. 16 [48]. This device is the result of electrically interconnecting two MQW p-i-n diodes in series. When the diodes are connected in this fashion, they become complementary, in that, when one of the diodes is "on" the other will be "off." Thus, one of the diodes will be in the absorbing state while the other is in the transmissive state. This is illustrated in the characteristic curves shown in parts (b) and (c). Perhaps the greatest strength of these devices is that changing states is a function of the *ratio* of the two input powers and not the function of the absolute intensity of the input beams. This can be seen in the characteristic curves of part (b) and (c). The optically bistable loop is centered around the point where the two inputs, P_{in1} and P_{in2}, are equal. From these figures, it can be seen that the device will remain in its current state until that ratio exceeds 1.3 or is less than 0.7. The importance of this is that the allowable noise on the signal inputs can be much greater than for the case of a critically biased device such as a SEED.

The S-SEED can be configured such that it can operate as an *S-R* latch. This is illustrated in Fig. 17. In part (a) of this figure, the inputs are separated into an *S* (set) input, and *R* (reset) input, and a clock input, where the clock has approximately the same intensity for both inputs. The *S* and *R* inputs are also separated in time from the clock inputs as shown in part (b) of this figure. Note that this clocking scheme is similar to that used by the OLE. The *S* or *R* inputs are used to set the state of the device. When the *S* input is illuminated, the S-SEED will enter a state where the upper MQW p-i-n diode will be transmissive while the lower diode will be absorptive. When the *R* input occurs, the opposite condition will occur. Since the

energy required to change the state of the devices is a function of the ratio of the *S* and *R* inputs, then when only one of those two inputs occurs, low switching intensities should be able to change the device's state. After the device has been put in its proper state, the clock beams are incident on both inputs. Since the two clock beams are roughly equivalent, the ratio between the incident beams should be close to one which will prevent the device from changing states. This higher energy clock pulse will be used to transmit the state of the device to the next stage of the system. Since the *S* or *R* inputs are low-intensity pulses and the clock is a high-intensity pulse, a large differential gain may be achieved.

Another example of a multiport device is an exchange-bypass node that has been proposed by Midwinter [49]. This device tries to capitalize on the advanced device fabrication and large scale integration capabilities of the electronic domain and the effective large number of pin-outs possible in the optical domain. Each bypass-exchange node is composed of three optical inputs, one electrical input (it could be optical), and two optical outputs as shown in Fig. 18. The optical inputs and outputs are to be implemented using MQW p-i-n diodes. By sensing the current, these diodes are detectors, while applying a time-varying voltage allows them to modulate light incident upon them. All of the processing on the data is done in the electrical domain. Since all of the interconnections are local, high-speed operation should be possible. The device is a logic device implementation of the directional coupler. If I_c is a logical "one" and the clock is a logical "one," then the inputs I_a and I_b will effectively *bypass* the node being directed to outputs O_a and O_b. On the other hand, if I_c is a logical "zero," then the node will *exchange* the inputs sending I_a to O_b and I_b to O_a with an asserted clock signal. Note that one of the main advantages of this device over a directional coupler is that there is gain and the thresholding nonlinearity of optical logic gates which will reduce the signal-to-noise requirements of the nodes. Eventually, large numbers of these nodes or "smart pixels" can be integrated into two-dimensional arrays to take advantage of the large pin-out capability available in the optical domain. Another example of a multiport device is the 4×4 OEIC switch implemented by Iwama *et al.* [50]. This device, or collection of devices, is composed of an array of optical detectors that receive the optical information. This array is then electrically connected to an electrical GaAs 4×4 switch. The output of this switch then drives four laser diodes that have been integrated onto a single substrate. This device has operated at 560 Mbits/s.

When integrating large numbers of these multiport devices into arrays, thermal problems could eventually limit their maximum size. This limitation occurs because both the electrical and optical devices dissipate power, and as the bit rates increase so does the required power.

3) Device Capabilities: In Fig. 19, the two strengths of photonics, bandwidth (data capacity) and parallelism (connectivity), are each assigned to an axis of a graph.

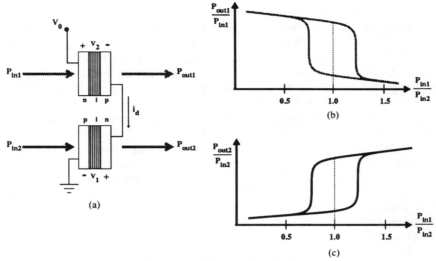

Fig. 16. Symmetric SEED operating characteristics.

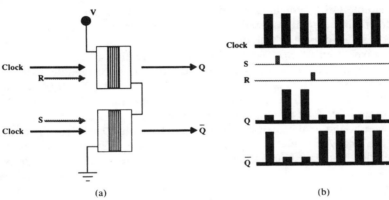

Fig. 17. Symmetric SEED as an *S–R* latch.

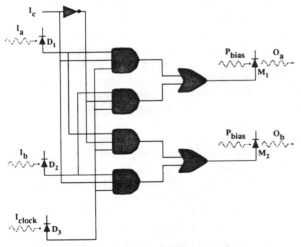

Fig. 18. Exchange–bypass smart pixel.

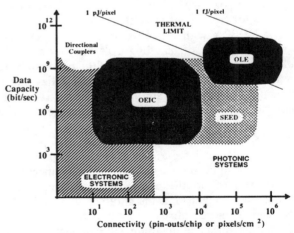

Fig. 19. Photonic device capabilities.

The ordinate, which is the bandwidth component, is labeled as data capacity and has the units of bits per second. The abscissa represents parallelism and is listed as connectivity with the units of either pin-outs per chip or pix-

els per square centimeter. The capabilities of current electronic systems are located in the lower left hand corner of the graph. The upper limit of electronic systems data capacity is approximately 10 Gbits/s with the maximum number of pin-outs of approximately 500. For the direc-

471

tional coupler, which is a relational device, the upper data capacity limit is in excess of 100 Gbits/s while the number of pin-outs is less than 100. Pin-outs in this case refers to the number of optical fiber inputs and outputs that can be connected to a single $LiNbO_3$ substrate. The lower boundary on the area designated for directional couplers was chosen at a data capacity of approximately 1 Gbit/s where high-speed electronics has been successfully demonstrated. Optoelectronic integrated circuits (OEIC) are devices that combine both the optical and electronic domains; thus, the data capacity limit will be the same as for the electronic domain while the connectivity can approach an effective pin-outs/chip of 10^4. A good example of an OEIC is a smart pixel. The lower boundary was chosen at a data capacity of 64 kbits/s (a single digital voice channel requires 64 kbits/s). The left boundary was chosen to be approximately 10. This boundary includes the possibility of simple linear arrays of optoelectronic regenerators. The self-electrooptic effect device (SEED) is a specific example of a simple OEIC which has the potential for a large number of pin-outs/chip. Finally, the optical logic etalon (OLE) is an all-optical device that has the potential of fabricating pixels (optical NOR gates) with diameters on the order of 1 μm. This small device size could potentially allow more than 10^6 pixels per cm^2 to be fabricated. The left boundary was set in excess of 10^4 pixels since OLE devices need to be small to maintain low switching energies. The lower bound on data capacity was chosen in the Gbits/s region to accommodate the pulsed mode of operation although this boundary could extend to the 64 kbit/s level.

The thermal limit region of this figure illustrates the maximum energy required to change the state of each pixel to maintain thermal stability. This is based on the assumption that 100 W/cm^2 can be removed from an array of devices for continuous operation. As an example, if an array of NLFP's with a pixel density of 10^6/cm^2 is to have the capability of handling 10^{12} bits per second, it would require the switching energy of the pixels to be less than 1 fJ. For the case of OEIC's, this implies that the combination of electrical and optical energy required to change the state of the pixel be less than the thermal limit.

B. Device-to-Device Interconnection

Once two-dimensional arrays of optical logic gates are available, it will become necessary to interconnect the individual devices on the arrays. This interconnection, which can be thought of as the photonic wires of the network, can be accomplished through the use of either space-variant or space-invariant interconnects. A space-variant network has the property that each input into the network can be redirected (connected) to any or all of the outputs and is dependent on its spatial location. This type of interconnect provides a different interconnection pattern for each spatially separated input. Alternatively, a space-invariant network can interconnect an input to any or all of the outputs, but is independent of its spatial location. These interconnects provide a single pattern for all inputs regardless of their spatial location. This spatial independence implies that the output pattern created by the inputs will be the same only shifted in space. These concepts are illustrated in Fig. 20. In part (a) the space-variant interconnect shows how the optical energy from two spatially separated inputs impinging upon the interconnection network redistribute the energy differently creating different input/output connection patterns. In part (b), the energy from the two inputs is redistributed in the same manner only shifted in spatial dimensions.

The following sections will begin by discussing free-space interconnection networks based on bulk optics and then outline some of the constraints of holographic interconnection networks.

1) Bulk Optics: The first type of free-space interconnection network can be implemented with bulk optics (lenses, prisms, mirrors, etc.). The simplest example of this type of an interconnection network is an optical imaging system composed of conventional lenses (see Fig. 2). Such a system is space-invariant and can be used to transfer the information present on the outputs of one array to the inputs of a second (or the same) array, effectively creating a large pin-out capability. As an example, the minimum resolvable spot size of a lensing system is given by $a = 1.22 \lambda (f/\#)$ where a is the diameter of the minimum resolvable spot size, λ is the wavelength of the light, and $f/\#$ is the f number of the lens (f/D where $f =$ focal length, $D =$ clear aperature of lens). This implies that for a lens system with $f/\# = 8$ and $\lambda = 850$ nm, the minimum resolvable spot size is 8.3 μm. Assuming that the image to be supported by the lens system is square, the SBWP is given by $(F/a)^2$ where F is the size of the unabberated field in one direction. From our previous example, if $F = 1$ cm then the SBWP $= 1.45 \times 10^6$ pixels or pin-outs. In order to maximize the SBWP of a lens system, the $f/\#$ must be kept small.

Another type of interconnect that is used in several types of multistage networks is the perfect shuffle, [51]–[53]. An example of a perfect shuffle interconnect that can be implemented with bulk optics is shown in Fig. 21 [54]. Part (a) of this figure illustrates the permutation performed by a perfect shuffle network. Part (b) shows how an optical perfect shuffle can be implemented with a beamsplitter, a lens, and two mirrors. Each of the inputs passed through the beamsplitter where their power is divided and directed to mirrors M_1 and M_2. The optical beams incident on M_2 will be shifted upward, pass again through the beamsplitter, and on to the lens where a spatial magnification of the row of inputs takes place. Thus, the lower four rays of information (5–8) will be shifted and magnified imaged onto the pixels associated with the output plane. On the other hand, the light rays reflecting from M_1 will be shifted downward and then magnified. The top four rays (1–4) controlled by M_1 will then be shuffled between the rays imaged by M_2. In this way, the process of perfect shuffling is accomplished by splitting, shifting, and then magnifying the rows of inputs. It should also be pointed out that there are other methods of creat-

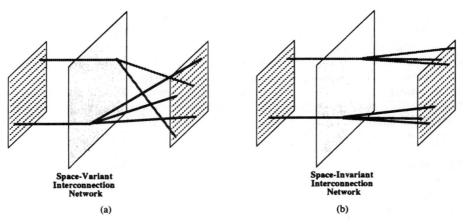

Fig. 20. Space-variant versus space-invariant networks.

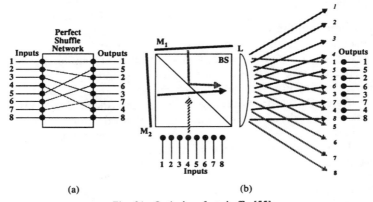

Fig. 21. Optical perfect shuffle [55].

ing a perfect shuffle interconnect using bulk optics [55], [56].

2) Holography: A hologram is a mechanism that can be used to modify and redirect a light wave that is incident upon it [57]. Because of this capability, a hologram or collection of holograms can be thought of as the photonic wires interconnecting optical logic gates [58]. An example of a holographic interconnection network is illustrated in Fig. 22. In this figure, the optical output of a logic device (point a) will be directed by mirror M_1 to the hologram (point b) which will redirect the light via M_2 to one or more other logic devices (points c, d, and e). This redistributed light will be used as the signal beams for the devices they are incident upon. The hologram located at point b could be either space-variant or space-invariant.

There is a price to pay for the flexibility of space-variant connections in terms of the SBWP. Assuming that each of the N^2 subholograms has the capability of addressing any or all of the N^2 pixels of the next stage, this implies a required SBWP $\propto N^4$ [59]. If a computer-generated hologram (CGH) is to be used as the optical interconnect, its SBWP will be determined by dividing the maximum size of the hologram by the minimum feature size and then squaring the result. As an example, for a Fourier hologram, if the minimum linewidth of an elec-

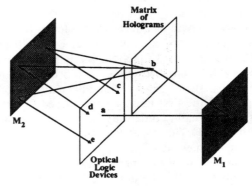

Fig. 22. Holographic interconnection network.

tron-beam system is 0.5 μm and with a maximum hologram size of 10 cm on each side, the maximum SBWP = 4×10^{10}. This implies that the theoretical maximum value of N for a space-variant network must be less than 450. This is the maximum value and for practical systems, N will have to be much smaller than this [59].

For the case of a space-invariant hologram, each point on the hologram should be able to redirect incoming light to all N^2 pixels of the next array and since all the pixels perform the same operation then the SBWP $\propto N^2$. Thus,

473

for space-invariant CGH with the available SBWP of 4×10^{10}, a maximum value of $N \ll 200\,000$ is possible.

A third approach is a hybrid interconnection scheme in which the space-variance requirement is relaxed in an attempt to increase N. If there are N^2 subholograms, each of which can redirect incoming light to $M^2 \ll N^2$ pixels of the next array, then the SBWP $\propto (MN)^2$. Note that M^2 is the effective fan-out of the energy incident on the subhologram. If $M = 3$ (fan-out $= 9$) the CGH will have $N \ll (4 \times 10^{10}/M^2)^{1/2} = 65\,000$. For networks that require a large N either a space-invariant or a hybrid interconnect will be required.

At the current time, most of the work on holographic interconnects has been theoretical and has yet to be engineered to the point of being practical.

C. Potential Applications of Optical Logic Devices

There are many applications that require the capability of sensing and reacting to each and every bit that passes through the system. As an example, packet switching systems require the ability to sense the information present in the headers and then provide the appropriate network routing. One type of control strategy used for packet systems is to sort the input packets by their destination addresses [51], [60].

This is illustrated by the Batcher bitonic sorting network shown in Fig. 23. In this figure, the entering packets, represented by their destination address, encounter the first rank of nodes. When a node has an up arrow, it means that the largest of the two destination addresses will be directed to the upper output. The lower address will then be directed to the lower output. On the other hand, when the arrow is pointing down, the larger (smaller) address will be directed to the lower (upper) output. After each node determines whether it will provide the exchange or bypass function, the incident packet information will be directed to the appropriate output port. The Batcher network begins with adjacent pairs of input lines, entering the first rank, being ordered in either ascending or descending order depending on the configuration of the node. The upper half (lower half) of rank 1 will be interconnected to the upper half (lower half) of rank 2 through a four-element perfect shuffle network. Rank 2 will then sort the upper half (lower half) of its inputs creating an ascending (descending) sequence which is directed through an eight-element perfect shuffle to rank 3. Finally, rank 3 will sort its interleaved ascending and descending four-element sequences into an ascending eight-element sequence which corresponds to sorting the inputs according to their destination address. As an example, the path corresponding to the output destination address 4 has been highlighted in the figure. Thus, for this type of switching network each node will be required to have a modest amount of intelligence. These nodes must be able to read a packet header, and then reconfigure the node so that the entering packet is directed to the proper output channel. The topology of the Batcher bitonic sort-

ing network can be rearranged such that all the interconnects use the same size of perfect shuffle [51]. An example of this is shown in Fig. 24. This adds more space-invariance to the system, thus reducing the overall required SBWP.

An example of a packet switching system based on sorting networks that could eventually be implemented using optical logic and interconnects is the STARLITE wideband digital switch [61]. The basic architecture of this switch is shown in Fig. 25. The STARLITE switch is a self-routing, nonblocking, constant latency packet switch that has the capability of handling gigabit data rates [62]. The concentrator directs the active inputs, which are much less than the total system inputs, to the sort-to-copy subnetwork. The sort-to-copy and copy subnetworks provide the broadcast capability for this switching system. The output of the copy subnetwork is then sorted according to destination. The expander then redirects the data to its final destination. Each of these basic functions (concentrator, sort-to-copy, copy, etc.) can be decomposed into some type of shuffle network that interconnects 2×2 switching nodes.

It is important to understand that optical logic devices and their associated systems are not at the point of development at this time. The devices that have been discussed are research prototypes that are not ready to be manufactured.

D. Applications of Smart Pixel Devices

Smart pixels are devices that attempt to take advantage of the strengths of both the optical and electrical domain. The strength of the electrical domain is that electrons interact easily which allows an electronic signal to control another electronic signal. Conversely, the strength of the optical domain is that photons do not interact with each other which creates an ideal communications environment. Once a collection of photons are encoded with information, they travel directly to their destination without interacting with other photons. The cost of this communications capability is that photons have a difficult time controlling other photons. Combining these two strengths, in that the electronics will be responsible for the processing of information while the photonics will handle the communications between processing elements, two-dimensional processing based on smart pixels should offer some performance advantage.

An example of such a system has been proposed by Midwinter [49]. This system is illustrated in Fig. 26. For this system, the input data enters as a row of information where each element is a single serial channel. This row of information enters the top row of smart pixels which are multiport devices that functionally behave like exchange-bypass modules. They have the capability of comparing packet addresses and directing the incident packets to the appropriate outputs. After processing, the information from the top row is directed via mirror M_1 to the perfect shuffle optics. Following the perfect shuffle operation, the data are directed by mirrors M_2, M_3, and

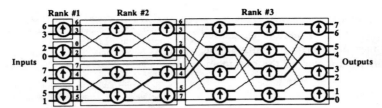

Fig. 23. A Batcher bitonic sorting network [52].

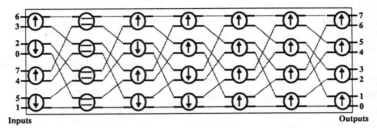

Fig. 24. Sorting network using perfect shuffles.

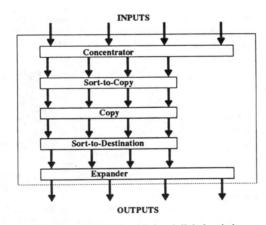

Fig. 25. STARLITE wide-band digital switch.

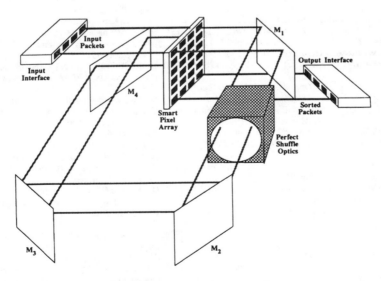

Fig. 26. Photonic sorting network.

M_4 to the second row of the smart pixel array. This looping procedure continues until the information has passed through all rows of the smart pixel array. At this point, the sorted information will be directed to the output interface.

V. Conclusions

This paper has reviewed some of the possible photonic technologies that could become important components of future telecommunications systems. It began by dividing photonic devices and systems into two classes according to the function they perform. The first class, relational, was associated with devices which under external control mapped the input channels to the output channels. The second class, logic, required that the devices perform some type of Boolean logic function. After the classes were defined, the strengths and weaknesses of the photonic domain were shown to be bandwidth and parallelism. Relational devices and their applications were then discussed. It was pointed out that the directional coupler holds the most promise for near-term development. Optical logic devices and systems were then presented. The systems that were outlined were based on SEED, NLFP, or smart pixel devices with either holograms or bulk optics serving as the optical interconnects.

Acknowledgment

I would like to acknowledge the constructive criticism provided by S. J. Hinterlong, A. L. Lentine, and M. E. Prise during the preparation of this paper.

References

[1] P. W. Smith, "On the role of photonic switching in future communications systems," *IEEE Circuits Devices*, pp. 9–14, May 1987.

[2] H. S. Hinton, "Photonic switching technology applications," *AT&T Tech. J.*, vol. 66, pp. 41–53, May/June 1987.

[3] P. W. Smith, "On the physical limits of digital optical switching and logic elements," *Bell Syst. Tech. J.*, vol. 61 pp. 1975–1993, Oct. 1982.

[4] S. L. McCall and H. M. Gibbs, "Conditions and limitations in intrinsic optical bistability," *Optical Bistability*, C. M. Bowden, M. Ciftan, and H. R. Robl, Eds. New York: Plenum, 1981, pp. 1–7.

[5] D. B. Tuckerman and R. F. W. Pease, "High-performance heat sinking for VLSI," *IEEE Electron Device Lett.*, vol. EDL-2, pp. 126–129, May 1981.

[6] J. W. Goodman, *Introduction to Fourier Optics*. New York: McGraw-Hill, 1968.

[7] A. D. Fisher, "A review of spatial light modulators," in *Proc. Top. Meet. Opt. Comput.*, Incline Village, NV, March 18–20, 1985.

[8] W. E. Ross, D. Psaltis, and R. H. Anderson, "2-D magneto optic spatial light modulator for signal processing," in *Proc. SPIE Conf.*, Crystal City-Arlington, VA, May 3–7, 1982.

[9] A. R. Tanguay, "Materials requirements for optical processing and computer devices," *Opt. Eng.*, pp. 2–18, Jan./Feb. 1985.

[10] A. Himeno and M. Kobayashi, "4 × 4 optical-gate matrix switch," *J. Lightwave Technol.*, vol. LT-3, pp. 230–235, Apr. 1985.

[11] D. R. Pape and L. J. Hornbeck, "Characteristics of the deformable mirror device for optical information processing," *Opt. Eng.*, vol. 22, pp. 675–681, 1983.

[12] G. Livescu, D. A. B. Miller, J. E. Henry, A. C. Gossard, and J. H. English, "Spatial light modulator and optical dynamic memory using integrated self electro-optic effect devices," in *Proc. Conf. Lasers Electro-Optics (Postdeadline Paper)*, April 26–May 1, 1987, pp. 283–284.

[13] J. W. Goodman, A. R. Dias, and L. M. Woody, "Fully parallel, high-speed incoherent optical method for performing discrete Fourier transforms," *Opt. Lett.*, vol. 2, pp. 1–3, Jan. 1978.

[14] A. A. Sawchuk, B. K. Jenkins, C. S. Raghavendra, and A. Varma, "Optical crossbar networks," *IEEE Computer*, vol. 20, pp. 50–60, June 1987.

[15] S. Suzuki and K. Nagashima, "Optical broadband communications network architectures utilizing wavelength-division switching technologies," in *Top. Meet. Photonic Switching, Tech. Dig. Series*, vol. 13, Mar. 18–20, 1987, pp. 21–23.

[16] B. S. Glance, K. Pollack, C. A. Burrus, B. L. Kasper, G. Eisenstein, and L. W. Stulz, "WDM coherent optical star network," *J. Lightwave Technol.*, vol. 6, pp. 67–72, Jan. 1988.

[17] T. S. Rzeszewski and A. L. Lentine, "A photonic switch architecture utilizing code-division multiplexing," in *Top. Meet. Photonic Switching, Tech. Dig. Series*, vol. 13, Mar. 18–20, 1987, pp. 144–146.

[18] P. R. Prucnal, D. J. Blumenthal, and P. A. Perrier, "Self-routing switching demonstration with optical control," *Opt. Eng.*, vol. 26, pp. 473–477, May 1987.

[19] H. S. Hinton, "Photonic switching using directional couplers," *IEEE Communications*, vol. 25, pp. 16–26, May 1987.

[20] R. V. Schmidt and R. C. Alferness, "Directional coupler switches, modulators, and filters using alternating $\Delta\beta$ techniques," *IEEE Trans. Circuits Syst.*, vol. CAS-26, pp. 1099–1108, Dec. 1979.

[21] H. S. Hinton, "A non-blocking optical interconnection network using directional couplers," in *Proc. IEEE Global Telecommun. Conf.*, vol. 2, Nov. 1984, pp. 885–889.

[22] P. Granestrand et al., "Strictly nonblocking 8 × 8 integrated optical switch matrix," *Electron. Lett.*, vol. 22, July 17, 1986.

[23] G. A. Bogert, "A low crosstalk 4 × 4 Ti : LiNbO₃ optical switch with permanently attached polarization-maintaining fiber arrays," in *Proc. Top. Meet. Integrated Guided-Wave Opt.*, Atlanta, GA, Feb. 1986, pp. PDP 3.1–3.

[24] R. A Spanke, "Architectures for guided-wave optical space switching networks," *IEEE Communications*, vol. 25, pp. 42–48, May 1987.

[25] S. Kobayashi and T. Kimura, "Semiconductor optical amplifiers," *IEEE Spectrum*, pp. 26–33, May 1984.

[26] H. Goto et al., "An experiment on optical time-division digital switching using bistable laser diodes and optical switches," in *Proc. IEEE Global Telecommun. Conf.*, vol. 2, Nov. 1984, pp. 880–884.

[27] Y. Silberberg, "All-optical repeater," *Opt. Lett.*, vol. 11, pp. 392–394, June 1986.

[28] R. A. Spanke, "Architectures for large nonblocking optical space switches," *IEEE J. Quantum Electron.*, vol. QE-22, pp. 964–967, June 1986.

[29] G. W. Richards and F. K. Hwang, "A two-stage rearrangeable broadcast switching network," *IEEE Trans. Commun.*, vol. COM-33, pp. 1025–1035, Oct. 1985.

[30] S. Suzuki et al., "Thirty-two line optical space-division switching experiment," in *Conf. Opt. Fiber Commun./Int. Conf. Integrated Opt. Opt. Fiber Commun. Tech. Digest Series, 1987*, vol. 3, Reno, NV, Jan. 1987, p. 146.

[31] W. A. Payne and H. S. Hinton, "Design of lithium niobate based photonic switching systems," *IEEE Communications*, vol. 25, pp. 37–41, May 1987.

[32] G. R. Ritchie, "SYNTRAN—A new direction for digital transmission terminals," *IEEE Communications*, vol. 23, pp. 20–25, Nov. 1985.

[33] R. A. Thompson and P. P. Giordano, "Experimental photonic time-slot interchanger using optical fibers as reentrant delay-line memories," in *Proc. Conf. Opt. Fiber Commun.*, Atlanta, GA, Feb. 24–26, 1986, pp. 26–27.

[34] S. D. Smith, "Optical bistability, photonic logic, and optical computation," *Appl. Opt.*, vol. 25, pp. 1550–1564, May 15, 1986.

[35] H. M. Gibbs, *Optical Bistability: Controlling Light with Light*. New York: Academic, 1985.

[36] D. A. B. Miller, J. E. Henry, A. C. Gossard, and J. H. English, "Array of optically bistable integrated self-electrooptic effect devices," in *Proc. Conf. Lasers Electro-optics*, San Francisco, CA, June 1986, pp. 32–33.

[37] J. L. Jewell, A. Scherer, S. L. McCall, A. C. Gossard, and J. H. English, "GaAs-AlAs monolithic microresonator arrays," *Appl. Phys. Lett.*, vol. 51, pp. 94–99, July 13, 1987.

[38] J. L. Jewell et al., "3-picojoule 82 MHz optical logic gates in a room temperature GaAs-AlGaAs multiple-quantum-well etalon," *Appl. Phys. Lett.*, vol. 46, p. 918, 1985.

[39] Y. H. Lee *et al.*, "Speed and effectiveness of windowless GaAs etalons as optical logic gates," *Appl. Phys. Lett.*, vol. 49, p. 486, 1986.

[40] J. L. Jewell, M. C. Rushford, and H. M. Gibbs, "Use of a single nonlinear Fabry-Perot etalon as optical logic gates," *App. Phys. Lett.*, vol. 44, pp. 172–174, Jan. 15, 1984.

[41] Y. H. Lee *et al.*, "Room-temperature optical nonlinearities in GaAs," *Phys. Rev. Lett.*, vol. 57, pp. 2446–2449, Nov. 10, 1986.

[42] N. Streibl and M. E. Prise, "Optical considerations in the design of digital optical computers," *Opt. Quantum Electron.*, to be published.

[43] M. E. Prise, N. Streibl, and M. M. Downs, "Computational properties of nonlinear devices," in *Top. Meet. Photonic Switching, Tech. Dig. Series*, vol. 13, 1987, pp. 110–112.

[44] P. Wheatley and J. E. Midwinter, "Operating curves for optical bistable devices," in *Top. Meet. Photonic Switching, Tech. Dig. Series*, vol. 13, 1987, pp. 113–114.

[45] D. A. B. Miller, U.S. Patent 4 546 244.

[46] P. Wheatley *et al.*, "Novel nonresonant optoelectronic logic device," *Electron. Lett.*, vol. 23, pp. 92–93, Jan. 16, 1987.

[47] P. Wheatley *et al.*, "Three-terminal noninverting optoelectronic logic device," *Opt. Lett.*, vol. 12, pp. 784–786, Oct. 1987.

[48] A. L. Lentine *et al.*, "Symmetric self electro-optic effect device," in *Proc. Conf. Lasers Electro-Opt. (Postdeadline Paper)*, April 26–May 1, 1987, pp. 249–250.

[49] J. E. Midwinter, "A novel approach to the design of optically activated wideband switching matrices," *Proc. IEE*, Part J, Opto-electronics, to be published.

[50] T. Iwama *et al.*, "A 4 × 4 GaAs OEIC switch module," in *Conf. Opt. Fiber Commun./Int. Conf. Integrated Opt. Opt. Fiber Commun. Tech. Dig. Series 1987*, vol. 3, Jan. 19–22, 1987, p. 161.

[51] H. S. Stone, "Parallel processing with the perfect shuffle," *IEEE Trans. Comput.*, vol. C-20, pp. 153–161, Feb. 1971.

[52] C.-L. Wu and T.-Y. Feng, "The universality of the shuffle-exchange network," *IEEE Trans. Comput.*, vol. C-30, pp. 324–332, May 1981.

[53] D. S. Parker, Jr., "Notes on shuffle/exchange-type switching networks," *IEEE Trans. Comput.*, vol. C-29, pp. 213–222, Mar. 1980.

[54] K.-H. Brenner, *Appl. Opt.*, submitted for publication.

[55] A. W. Lohmann, W. Stork, and G. Stucke, "Optical perfect shuffle," *Appl. Opt.*, vol. 25, pp. 1530–1531, May 15, 1986.

[56] G. Eichmann and Y. Li, "Compact optical generalized perfect shuffle," *Appl. Opt.*, vol. 26, pp. 1167–1169, Apr. 1, 1987.

[57] R. J. Collier, C. B. Burkhardt, and L. H. Lin, *Optical Holography*. New York: Academic, 1971.

[58] A. A. Sawchuk and T. C. Strand, "Digital optical computing," *Proc. IEEE*, vol. 72, pp. 758–779, July 1984.

[59] B. K. Jenkins, P. Chavel, R. Forchheimer, A. A. Sawchuk, and T. C. Strand, "Architectural implications of a digital optical processor," *Appl. Opt.*, vol. 23, pp. 3465–3474, Oct. 1, 1984.

[60] C. W. Stirk, R. A. Athale, and C. B. Friedlander, "Optical implementation of the compare-and-exchange operation for applications in symbolic computing," *Proc. SPIE*, submitted for publication.

[61] A. Huang and S. Knauer, "Starlite: A wideband digital switch," in *Proc. IEEE Global Telecommun. Conf.*, Atlanta, GA, vol. 1, pp. 121–125, Nov. 1984.

[62] A. Huang, "The relationship between STARLITE, a wideband digital switch and optics," in *Proc. Int. Conf. Commun.*, Toronto, Ont., Canada, June 22, 1986.

[63] A. de Bosio, C. DeBernardi, and F. Melindo, "Deterministic and statistic circuit assignment architectures for optical switching systems," in *Top. Meet. Photonic Switching, Tech. Dig. Series*, vol. 13, Mar. 18–20, 1987, pp. 35–37.

H. Scott Hinton (S'81–M'82) received the B.S.E.E. degree from Brigham Young University, Provo, UT, in 1981 and the M.S.E.E. degree from Purdue University, Lafayette, IN, in 1982.

He is the Supervisor of the Photonic Switching Technologies Group, AT&T Bell Laboratories, Naperville, IL.

Mr. Hinton is a member of the Optical Society of America.

Time Division Multiplexing Using Optical Switches

ANDERS DJUPSJÖBACKA

Reprinted from *IEEE Journal on Selected Areas in Communications*, Vol. 6, No. 7, August 1988, pages 1227-1231.
Copyright © 1988 by The Institute of Electrical and Electronics Engineers, Inc. All rights reserved.

Abstract—Calculations on an integrated optical time division multiplexing scheme are presented, the bandwidth requirements for the multiplexers and the corresponding drive signals are derived using a time domain approach. It is shown that these requirements are greater than expected from simple considerations. Different types of multiplexing schemes are analyzed.

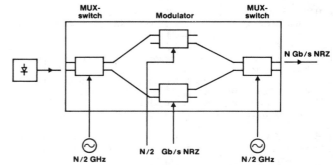

Fig. 1. Schematic of an optical multiplexer system.

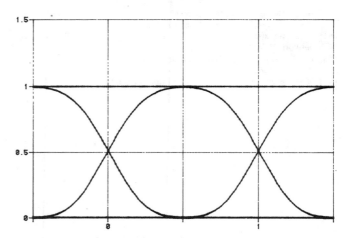

Fig. 2. Eye diagram for an external modulator at $N/2$ Gbits/s. Modulator bandwidth: dc–0.33N GHz.

INTRODUCTION

HIGH-SPEED optical multiplexing has been proposed as an interesting and important application of optical switching. The advantage of optical multiplexing is that it may be possible to generate serial bit streams with higher clock rates than with electronic switching (>5 Gbits/s) [1], [2], or that the requirements on the high-speed electronics are lower. This application has been even suggested as the most imminent one of optical switching. Mostly the modulators/switches have been looked upon as very fast digital devices, although they have an analog response to their drive signals.

This paper presents design considerations for optical time division multiplexing in terms of bandwidth requirements for modulators and switches for a particular multiplexing scheme. We also show that different types of multiplexers behave differently due to varying "chirping" properties; this chirping is also calculated. Comparison to nonmultiplexed systems is performed.

THE OPTICAL MULTIPLEXER

Depicted in Fig. 1 is an integrated optic chip. CW light (left) is divided into two streams of light pulses via a first switch. In the middle of the chip, these two bit streams are modulated by two separated modulators; these two modulators work with no-return-to-zero (NRZ) coded data formats in order to use the minimum possible bandwidth. Finally, a second multiplexer switch to the right combines the two bit streams into one. The electrode type chosen for the multiplexer (MUX) switches as well as for the modulators is the traveling wave (TW) type electrodes [3].

The modulators work with broad-band signals and for practical reasons it is assumed that they have, including their drive stages, a frequency response down to dc. The two types of modulators investigated are Mach–Zehnder modulators and directional couplers. High-frequency response $\Delta\beta$-reversal modulators have not been included. In TW version, $\Delta\beta$-reversal modulators have microwave

Manuscript received November 17, 1987; revised March 15, 1988.
The author is with Ericsson Telecom, S-126 25 Stockholm, Sweden.
IEEE Log Number 8822095.

waveguides that will result in reduced performance at high frequencies, e.g., loaded transmission line—large walkoff [4], coplanar waveguides—large impedance mismatch [5].

The modulators as well as their drive signals are designed to work at the lowest possible bandwidth. The 3 dB frequency response limit (detected power) for the modulators was set to dc–0.33N GHz where N is the output, multiplexed, bit rate expressed in gigabits per second. The drive signal for the modulators is an NRZ-coded bit stream. In order to further reduce the bandwidth for the drive signal, a raised cosine pulse form was chosen. This optimization is essential to facilitate the construction of the drive stages for the modulators.

With a bandwidth of the modulator mentioned above and a drive signal consisting of a NRZ-coded bit stream, a very smooth eye diagram is obtained at $N/2$ Gbits/s, Fig. 2. The jitter for the modulator is almost zero.

The two types of MUX switches investigated in this study are directional couplers and a modified 1 × 2 directional coupler [6], both with uniform TW electrodes. The drive signal for the MUX switches in this study has a lowest frequency of 0.5*N* GHz and in an ideal case it is sinusoidal. In order to yield a more square-wave-like drive signal, the drive signal bandwidth could be extended to 0.5*N*–*N* GHz, or to 0.5*N*–2*N* GHz. The corresponding 3 dB optical bandwidths for the MUX switches, i.e., the bandwidth related to electrode losses and walkoff [7], is dc–0.65*N* GHz, dc–1.3*N* GHz, and dc–2.6*N* GHz.

CALCULATIONS

The calculations of eye diagrams on the multiplexer system are based on coupled wave equations and are performed in the time domain [8]. We thereby take the nonlinear electrooptic behavior of the devices into account as well as coherent addition phenomena which as is shown later significantly affect the properties of the multiplexer. Microwave dispersions effects are, however, not taken into account.

When the modulators are operated at the lowest possible bandwidth (optimized for 0.5*N* Gbits/s) and when the MUX switches work narrow-band, the quality of the output eye is very poor [Fig. 3(a)]. However, when the optical bandwidth for the MUX switches as well as the electrical bandwidth for the drive signal are increased to dc–1.3*N* GHz and 0.5*N*–*N* GHz, respectively, considerable improvements are achieved [Fig. 3(b)].

This result is important from a design point of view, since it is more difficult to increase the total bandwidth of the modulators than that of the MUX switches. The modulators as well as their drive stages have to work from a very low frequency, in most cases from 10 to 100 kHz, depending on code format, and up to several gigahertz in high bit rate systems. The lowest optical frequency limit needed for the MUX switches is, however, 0.5*N* GHz, and they could therefore be more readily increased in bandwidth. In a practical case, this increase could be accomplished in a number of ways:

1) by reducing the length of the MUX switches,
2) by reducing the rise time of the drive signal,
3) by a combination of 1) and 2) above.

Calculations of eye diagrams for different types of optical multiplexers according to Fig. 1 have been performed for a number of bandwidths of the MUX switches as well as their drive signals. In the diagrams, the performance of the multiplexer systems is plotted versus the rise time for the optical output signal from the MUX switches. The *T* in the diagrams stands for the time slot at the bit rate used. The calculations were performed without any absolute level of the optical output power from the multiplexer and no noise levels were included. We therefore define the eye opening by the interval where the output signal is > 0.9 or < 0.1 units of full modulation, this level corresponds to a power penalty of 1 dB. The jitter is defined by the "waist length" in the "crossover" section in the eye diagram (Fig. 4).

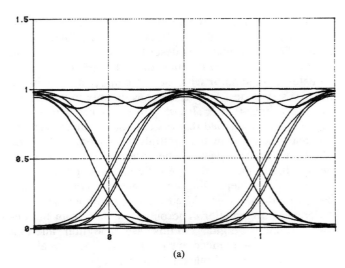

(a)

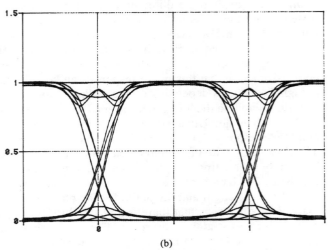

(b)

Fig. 3. Eye diagrams for an optical multiplexer. Modulator performance dc–0.33*N* GHz, where *N* is multiplexed output data rate. (a) Multiplexer bandwidth: dc–0.65*N* GHz. Drive signal bandwidth: narrow-band. (b) Multiplexer bandwidth: dc–1.3*N* GHz. Drive signal bandwidth: 0.5*N*–*N* GHz.

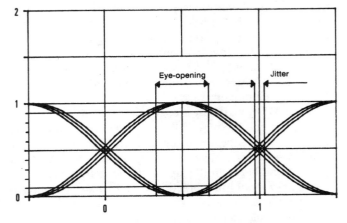

Fig. 4. Definition of eye opening and jitter.

The performance of the multiplexer system is described in Figs. 5 and 6 for different MUX switches and modulators. The performance is described by eye opening and jitter. Fig. 7 shows a comparison to the performance of an external modulator with the same upper bandwidth as the MUX switches. Due to the considerably more opened eye and a jitter that is almost zero (jitter not depicted in Fig. 7), it is concluded that the external modulator is significantly better than the multiplexer system. When the MUX switches as well as the external modulator work narrow-band, there are large differences in performance between them, depending on the analog, nonlinear, behavior of the devices. When they work broad-band, the differences are smaller. The output signals from the devices are more ''digital'' in this case. The jitter, and to a minor extent the reduced eye opening for the optical output signal of the multiplexer systems, depends on an unwanted coherent addition in the second MUX switch.

One might explain the different results for multiplexer systems with the different phase responses of the modulators and the MUX switches involved, which we now analyze. At low frequencies, this phase response has been studied by Koyama and Iga [9]. However, it should be emphasized here that this phase response can be different at high frequencies depending on the walkoff between the electrical signal in the TW electrode and the optical signal in the waveguide. Depicted in Fig. 8 is the phase and amplitude response for a directional coupler with uniform TW electrodes, with the output taken in the cross channel. The phase vector from the output port is plotted on a time abscissa. The length of the vector stands for the output optical field strength and the angle for the phase shift. A vertical vector means that there is a phase shift of $-\pi/2$ rad relative to a straight waveguide. The time scale is adjusted so that 0 is in the middle of the time slot.

The directional coupler is fed with a slow [Fig. 8(a)] and a fast [Fig. 8(b)] drive pulse. The pulse width (FWHM) is 200 μs in Fig. 8(a) and 200 ps in Fig. 8(b). The pulse form is raised cosine and the 3 dB optical bandwidth for the modulator is 3.5 GHz. For the slow pulse, there is no phase modulation (angle shift) seen in the time slot from -100 to 100 μs, but for the fast pulse, there is a phase shift of approximately $\pi/2$ rad. This is equivalent to a $\Delta\lambda \approx 0.07$ Å for $\lambda = 1.3$ μm. For the bar channel of this device, the phase response at high frequencies is similar to that of low frequencies.

The MZ modulator relies on accumulated phase only, and does, therefore, not change its phase behavior in high-frequency modulation.

If the MUX switches are, for practical reasons, fabricated in z cut LiNbO$_3$ (TW electrodes on a directional coupler), the influence of temperature changes is important since pyroelectric effects [10] can induce differential phase shifts between the signals to be multiplexed. This phase shift would in a practical case increase the jitter in the eye diagram.

To avoid the coherent addition problems, one could use short fibers between the first MUX switch and the mod-

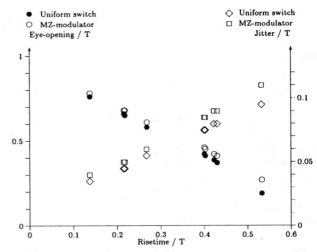

Fig. 5. Eye opening and jitter for an optical multiplexer, with uniform directional couplers as MUX switches, versus the MUX switches optical rise time. ● ◇ uniform switches as bit modulators. ○ □ MZ modulators as bit modulators.

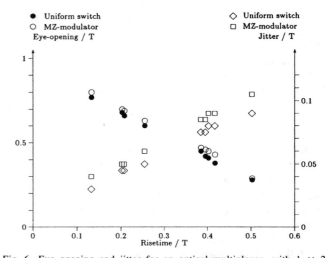

Fig. 6. Eye opening and jitter for an optical multiplexer, with 1 × 2 switches as MUX switches, versus the MUX switches optical rise time. ● ◇ uniform switches as bit modulators. ○ □ MZ modulators as bit modulators.

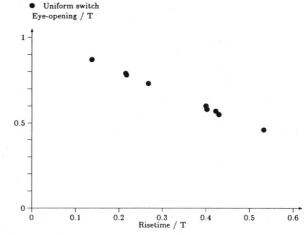

Fig. 7. Eye opening for an external modulator.

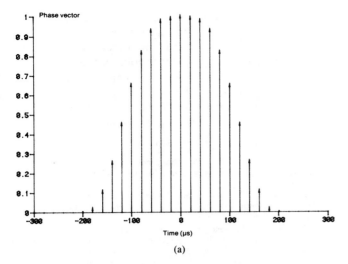

(a)

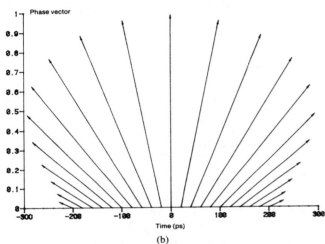

(b)

Fig. 8. Amplitude and phase response for an uniform directional coupler (cross channel). The amplitude is indicated by the length of the vector, the phase is indicated by the angle to the abscissa. (a) Response for a slow drive pulse. FWHM: 200 μs. (b) Response for a fast drive pulse. FWHM: 200 ps.

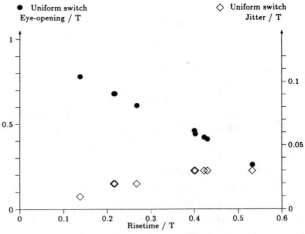

Fig. 9. Eye opening and jitter for an optical multiplexer with suppressed coherent addition.

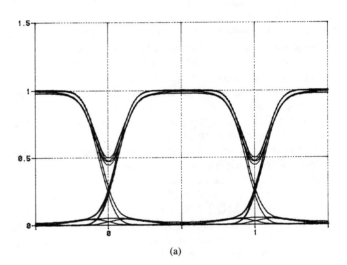

(a)

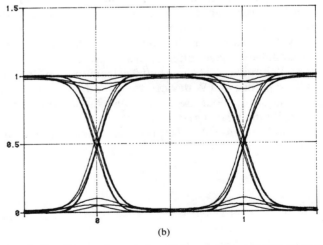

(b)

Fig. 10. Eye diagrams for optical multiplexers with suppressed coherent addition. Multiplexer bandwidth: dc–1.3N GHz. Drive signal bandwidth: 0.5N–N GHz. (a) MUX switches at both sides, different path lengths. (b) The first MUX switch is replaced by a 3 dB power splitter, different path lengths.

ulators. If the relative difference in optical path length is several times larger than the coherence length of the laser, there will be no coherent addition in the second MUX switch. Calculations on such an optical multiplexer with suppressed coherent behavior are presented in Fig. 9. The jitter is below 4 percent for this structure, even for the narrow-band version. However, there is a major drawback: a rather large dip occurs between the sampling points of consecutive ''1'' with a peak amplitude of approximately 50 percent of full modulation [Fig. 10(a)]. This behavior is explained by the 3 dB power splitter mechanism of the first and the second MUX switch when these two switches change state at the edge of each time slot. A 3 dB excess loss from the output of the multiplexer is therefore obtained at these points due to incoherent addition in the second MUX switch.

Another way to suppress the coherent behavior is to replace the first MUX switch by a passive 3 dB power splitter. If the difference in optical path length between the splitter and the modulators is large, several times larger than the coherence length of the laser, there will be no coherent addition in the second MUX switch. One advantage with this concept is that the fiber lengths do not have to be adjusted to the bit rate; one drawback is the 3 dB loss in transmitted power. An eye diagram for this type of multiplexing scheme is shown in Fig. 10(b).

The differences in performance for different types of MUX switches and modulators is rather small in this case since the coherent addition in the second MUX switch is suppressed.

A third way to solve this problem is to use return-to-zero (RZ) coded data format, thereby increasing the bandwidth of the modulator by approximately a factor two. Optical multiplexers employing mode-locked lasers, giving RZ type codes, have also been shown [11], [12].

CONCLUSIONS

The implementation of optical multiplexing requires careful consideration to achieve low system penalties in systems of the general structure of Fig. 1. The power splitting mechanism during the time the MUX switches change states causes coherent addition of the light in the two input channels to the second MUX switch. The coherence is a result of the desirability of using a single laser for dispersion reasons. The coherent addition causes jitter resulting in a reduced eye opening for the output signal. The jitter also makes the clock extraction more difficult, and if the extracted clock is jittering, the eye opening would be further decreased by the amount of jitter in the extracted clock signal. The jitter from the output of the multiplexer should be kept below 5 percent to be practical in system use. From Figs. 5 and 6 it is apparent that a rise time for the MUX switches of $\approx 1/4$ of the time slot will meet the jitter requirements. This is approximately twice the bandwidth required for an external modulator working at the same bit rate. However, the MUX switch does not have to operate from dc, and the bandwidth of the drive signal only has to be one octave. Baseband response is not necessary here, as described above.

It was also shown that different types of MUX switches and modulators have different characteristics due to differing "chirping." The presented calculations on "chirping" properties of TW devices are given for the first time here and should find use in the analysis of "ordinary" externally modulated systems.

Another way to solve the problems concerned with co-

herent addition is to use RZ-coded data format, or to use a difference in optical path length between the first and the second MUX switch, several times the coherence length of the laser used. All these solutions have different drawbacks, as detailed in the text. Thus, a properly designed system as in Fig. 1 could be advantageously employed, the main advantage being the reduction in base bandwidth requirements for the modulators.

REFERENCES

[1] S. D. Personick et al., "Switching technology: Photonic switch applications," Bell Commun. Res., SR-ARH-000015, issue 1, Sept. 1984.

[2] H. F. Taylor, "Technology and design considerations for a very-high-speed fiber-optic data bus," IEEE J. Select. Areas Commun., vol. SAC-1, pp. 500–507, Apr. 1983.

[3] M. Izutsu, Y. Yamane, and T. Sueta, "Broad-band traveling-wave modulator using a LiNbO₃ optical waveguide," IEEE J. Quantum Electron., vol. QE-13, pp. 287–290, Apr. 1977.

[4] F. Auracher, "Design tradeoffs for high-speed directional coupler modulators with $\Delta\beta$-reversal in LiNbO₃," J. Opt. Commun., vol. 5, no. 1, pp. 7–9, 1984.

[5] R. C. Alferness, L. L. Buhl, S. K. Korotky, and R. S. Tucker, "High-speed $\Delta\beta$-reversal directional coupler switch," in Proc. Top. Meet. Photon. Switching, Incline Village, NV, Mar. 1987, paper ThD6.

[6] S. Thaniyavarn, "Modified 1 × 2 directional coupler waveguide modulator," Electron. Lett., vol. 22, pp. 941–942, Aug. 1986.

[7] L. Thylén et al., "High speed LiNbO₃ integrated optics modulators and switches," in Proc. First Int. Conf. Integrated Opt. Circuit Eng., SPIE, vol. 517, Oct. 1984, pp. 249–257.

[8] T. Tamir, Integrated Optics. Berlin, Germany: Springer-Verlag, 1979.

[9] F. Koyama and K. Iga, "Frequency chirping in some types of external modulators," in Proc. OFC/IOOC'87, Reno, NV, Jan. 1987, paper WO4.

[10] C. H. Bulmer and W. K. Burns, "Pyroelectric effects in LiNbO₃ channel-waveguide devices," Appl. Phys. Lett., vol. 48, no. 16, pp. 1036–1038, Apr. 1986.

[11] R. S. Tucker et al., "4 Gb/s optical time-division multiplexed system experiments using Ti:LiNbO₃ switch/modulators," in Proc. Top. Meet. Photon. Switching, Incline Village, NV, Mar. 1987, paper FD3.

[12] G. Eisenstein et al., "Multi-gigabit per second transmission using optical time-division multiplexing aand demultiplexing," in Proc. ECOC '87, Helsinki, Finland, Sept. 1987, Tech. Dig., vol. III, pp. 41–45.

Anders Djupsjöbacka was born in Solna, Sweden, in 1958. He received the M.Sc. degree in electrical engineering from the Royal Institute of Technology, Stockholm, Sweden, in 1982.

In 1982, he joined the Line Transmission and Fibre Optics Department of Ericsson Telecom AB, Stockholm, Sweden, where he is engaged in R&D work on integrated optics, especially high-frequency devices.

Optical Self-Routing Switch Using Integrated Laser Diode Optical Switch

RYOZO KISHIMOTO, MEMBER, IEEE, AND MASAHIRO IKEDA, MEMBER, IEEE

Reprinted from *IEEE Journal on Selected Areas in Communications*, Vol. 6, No. 7, August 1988, pages 1248-1254.

Abstract—This paper proposes an optical self-routing switch using integrated laser diode optical switches. This optical self-routing switch is composed of a Benes network, which can perform large-scale switching functions using less hardware than a crossbar switch. In this self-routing switch, the path each data stream takes through the interconnection network is determined by the binary bits of its destination address. The optical 2 × 2 unit switches are self-routed by monitoring terminal voltage changes in gain guides induced by input optical signals which are injected into a p–n junction. By using the self-routing structure, concentrated control is not necessary; and the large optical multistage switches can be easily constructed, because complicated electrode patterns are not necessary.

I. INTRODUCTION

IN THE developing advanced information society, in addition to voice, data and video are becoming important communication media. Simultaneously, the demands for variety and high quality are also growing. Therefore, many studies on communication networks which can handle multimedia (voice, data, and video) are being conducted [1]–[3]. The most important technology in broadband multimedia communications is a high-performance broad-band switch, which can efficiently support high-speed multimedia on a large scale.

Many types of broad-band switches have been reported, including space-division or time-division switches, and electronic or optical switches. Multistage switching networks have also been studied in the field of communication networks and parallel computers [4], [5]. Multistage switching networks can perform large scale switching functions using less hardware than a crossbar switch. Futhermore, self-routing multistage broad-band switches are very attractive because they have to operate at high speeds and their routing using software-based control is very complicated. In a self-routing switch, the path each data stream takes through the interconnection network is determined by the binary bits of its destination address. Several types of self-routing packet switches have been proposed [2], [6], [7].

This paper proposes an optical self-routing switch constructed from a Benes network using integrated laser diode

Manuscript received October 8, 1987; revised March 21, 1988.

R. Kishimoto is with NTT Transmission Systems Laboratories, Kanagawa-ken, 238-03, Japan.

M. Ikeda is with NTT Opto-Electronics Laboratories, Atsugi-shi 243-01, Japan.

IEEE Log Number 8822102.

(LD) optical switches. The switch topology is a synchronous multistage switching network which is rearrangeable and nonblocking. First, the configuration of the self-routing multistage switching network is discussed. Next, the principle of the optical self-routing switch using integrated LD optical switches is described. Finally, the results of an experiment on terminal voltage changes in LD optical switch modules are presented.

II. CONFIGURATION OF SELF-ROUTING MULTISTAGE SWITCHING NETWORK

In designing a multistage switching network, the following points should be considered.

1) The path setting control should be simple.

2) The number of necessary switch elements should be low.

Self-routing in multistage switching networks is a very attractive switching concept. The following are included in the self-routing switch concept.

1) In a multistage switching network, the path each data stream takes through the interconnection network is determined by the binary bits of its destination address. This destination address may be contained in each bit, or in the data stream header. Each switching node processes this destination information to route the data through the swtich.

2) The path setting calculation is also made distributively and immediately in each 2 × 2 unit switch in a switching network which is an asynchronous packet switching network.

Therefore, optical self-routing switching networks [8] have various merits. They can handle high-speed data streams. Unit switches can be controlled and routed distributively, so switching function stoppages do not occur. In addition, electronic hardware control lines are unnecessary in the integrated optical circuit.

Table I summarizes the characteristics of several proposed self-routing multistage switching networks. A multistage interconnection network can be viewed from the points of operation mode, control strategy, switching methodology, or network topology. From the operation mode viewpoint, there are two types of multistage switching networks: asynchronous blocking networks, and synchronous, rearrangeable, and nonblocking networks.

Asynchronous switching networks are applied only to

TABLE I
PROPOSED SELF-ROUTING MULTISTAGE SWITCHING NETWORK

	Switching network		Routing of switching		Delay time	Proposal (Year)	Application
	Topology	Number of unit switch	Algorithm	Order of process time			
Asynchronous blocking network	Banyan network	$n/2\ \log_2 N$	—	Real time	Large	J. Kulzer (GTE) (1984)	Packet switch
	Benes network	$N(\log_2 N - 1/2)$	Randon algorithm	Real time	Middle	J. S. Turner (ATT) (1986)	
			Dynamic load balance algorithm (Proposal)	Real time	Small	R. Kishimoto (NTT) (1987)	
Synchronous rearrangeable non-blocking network	Benes network	$N(\log_2 N - 1/2)$	Looping algorithm	$N/2 \times \log_2 n$	—	D.C .Opferman (ATT) (1971)	Circuit switch,
			Lee algorithm	$N/2 \times \log_2 N$	—	K. Y. Lee (Ohio State University) (1984)	
			Concatenated network algorithm (Proposal)	$(1\sim N/4) \times \log_2 N$	—	R. Kishimoto (NTT) (1986)	Packet switch
	Batcher-banyan network	$N\ \log_2 N \times (\log_2 N + 1)$	Sorting algorithm	Real time (Hardware)	—	A. Huang (ATT) (1986)	

packet switching, and synchronous switching networks are applied to both circuit and packet switching.

Asynchronous switching networks are suitable for asynchronous high-speed packet switching. Several types of self-routing packet switches for asynchronous networks have been proposed. The first proposed self-routing packet switch was the binary routing network [6], using a basic interconnection network, including the so-called delta, shuffle-exchange, and banyan networks. There is, however, one problem with binary routing networks using the banyan network: the mean time delay becomes very large in the presence of certain prejudiced traffic patterns. If the inputs are heavily loaded, the internal links will be hopelessly overloaded. This problem can be solved by using two networks [2] instead of one. These two networks form a Benes network.

We proposed a self-routing Benes network distributively controlled by dynamic load balance [7]. One of these two networks is called the routing network (RN) and is a basic interconnection network. The other sits in front of the RN and is called the distribution network (DN). It has the same structure as the RN, but instead of routing packets according to their destination address, the DN attempts to distribute packets evenly across all its output ports. The DN is controlled by a method called the dynamic load balance method. The number of loads—the number of input packets—is counted. The number is fed back to the input-side switch elements through the unit switch using the status information lines in the direction opposite to the data stream lines. Thus, the input packets can be distributed very evenly across all the output ports of the distribution network. In each two-input two-output switch element, the load of each output port is compared to the other; these loads correspond to the minimum load value from among the loads of all the output ports in the DN, which are connected to each output port of that switch element. If the loads of the two output ports are equal to each other, the input packets are distributed randomly. If not, the packets are distributed to the output port, whose load is smaller than that of the other port. The RN is self-routed by an output tag. Therefore, the self-routing Benes network can be controlled distributively by dynamic load balance.

In the rearrangeable, nonblocking switching network, the path which the data takes through the network is calculated in a concentrated control circuit from the interconnection between input ports and output ports of the switching network. A Benes network and Batcher-banyan network are proposed as the network topology for the rearrangeable, nonblocking switching network. Only synchronous switching networks are applicable to the circuit switching system. However, when high-speed packets are input asynchronously, it is very difficult to set up the path immediately by using the path setting algorithm. A Batcher-banyan network, whose path can be set up by using sorter hardware, is applied to both circuit and packet switching. The looping [9] and Lee algorithms [10] have been proposed as path setting algorithms employing con-

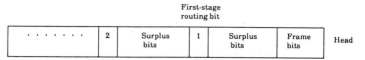

First-stage
routing bit

· · · · · · ·	2	Surplus bits	1	Surplus bits	Frame bits	Head

Fig. 1. Routing frame structure of optical packet signals.

centrated control. We have proposed the concatenated network algorithm [11], which is superior to the previous algorithms from the viewpoint of high-speed path setting.

III. Optical Self-Routing Switch

A. Principle for Optical Self-Routing Switch

The optical switch being investigated here is a synchronous, rearrangeable, and nonblocking multistage switching network. When a new demand for path setting arises, the rearrangement calculation for path setting is made by using a concatenated network algorithm. When all the paths in the switching network are decided, the optical packet signals for routing are first transmitted. Then, in each 2×2 unit switch, the optical pulse signals are converted into electronic pulse signals, and the routing signals are derived from the electronic pulse signals. Fig. 1 shows one example of the routing frame structure of optical packet signals. The routing packet header is discriminated by using frame bits. In a Benes network, there are $2 \log N - 1$ stages of 2×2 optical unit switches. Therefore, $2 \log N - 1$ bits for routing are necessary. Since there is a time delay for header decoding, surplus bits are inserted between each routing bit. The structure of the self-routing Benes network with eight inputs and eight outputs is shown in Fig. 2. Two optical signal data streams are needed: the main signal data stream and information signals for routing. These two types of optical signals are combined with two-input, one-output optical couplers, as shown in Fig. 2(a). Optical couplers are installed in front of each input port of the Benes network, as shown in Fig. 2(b).

B. Integrated Laser-Diode Optical Matrix Switch

1) Laser-Diode Optical Switch: Low-loss and wideband optical switches are expected to be key devices for the evolving fiber-optic communication system of a highly integrated service network system. Many optical switches, such as directional-coupler-type [12] and total-reflection-type [13] integrated optical matrix switches, have recently been reported. These optical switches, however, have some disadvantages for large scale applications. These include large insertion loss and long substrate length. Laser diode optical switches [14]–[16], in which the gain and loss are controlled by an injection current in a p–n junction and the optical signal is directly switched, offer a promising means of solving the above mentioned problems. Laser diode (LD) optical switches operate as optical amplifiers. When the injection current is applied, the optical absorption is reduced and the optical signal exits the LD optical switch through the amplification process.

Therefore, we used LD optical switches as optical self-routing switches.

2) Structure of Integrated Optical Matrix Switch: The structure of an LD integrated optical matrix switch is shown schematically in Fig. 3. Input and output ports are connected to the crossbar network by straight waveguides which are constructed of ridge gain guides. Each gain guide has a DH p–n junction and separate electrodes for current injection. The branching and input waveguides intersect at $45°$ and the branching ratio is controlled by the refractive index of the material filling the intersection gaps between the waveguides. In Fig. 3, a light signal is injected into input port I_1. This signal has a wavelength close to the gain peak wavelength of the active layer. If the forward current is injected into electrode J_1, the cross-shaped guide changes from an absorption to a gain guide. Therefore, the input signal light is amplified and guided through output port O_1. If the injection current is applied to electrode J_2, the branched signal light is amplified and guided through output port O_2. When current is simultaneously injected into both J_1 and J_2, the signal light can be guided through both O_1 and O_2. In particular, by controlling the injection current of the next unit switch, any output port can be selected from any input port without insertion loss. The input and output waveguides can be arranged at right angles, because the branching loss at the intersection part can be compensated for by the gain. A 2×2 unit matrix switch is arranged in a 250 μm square in Fig. 3. Therefore, a 16×16 matrix switch can be constructed on a 4 mm square chip.

C. Principle of Optical 2×2 Unit Switch for Self-Routing

The operation of the optical unit switch is explained in Fig. 4. In an optical unit switch, optical packet signals for routing are assumed to be incident at input port I_1. The optical input signals are branched to optical gain waveguides W_1 and W_2. These branched optical signals induce terminal voltage changes in elecrodes J_1 and J_2 through the optical waveguides when there is no forward injection current. A terminal voltage change is converted into an electronic pulse signal through the pulse reforming circuit. Then, the electronic pulse signal is frame decoded and header decoded using the electronic control circuit. Then, it is determined whether the optical signals passed straight through or curved in an optical unit switch.

If the destination output port is assumed to be O_1, the forward current is injected in electrode J_1, and the cross-shaped branching waveguide W_1 changes from an absorption to a gain guide. The optical signals are amplified and emitted from O_1. If the destination output port is assumed

485

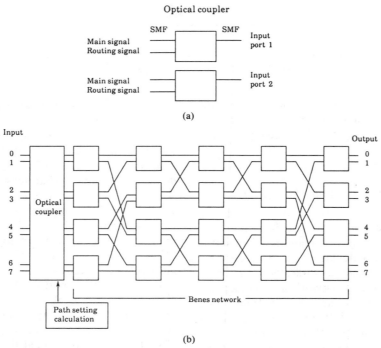

(a)

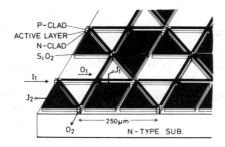

(b)

Fig. 2. Structure of the self-routing Benes network with eight inputs and eight outputs. (a) Optical coupler. (b) Optical self-routing Benes network.

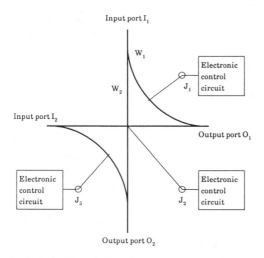

Fig. 3. Structure of LD integrated optical matrix switch.

Fig. 4. Optical self-routing two-input, two-output unit switch.

to be O_2, the forward current is injected in electrode J_2, and the straight waveguide W_2 becomes a gain guide. As mentioned above, the optical signals incident in the optical unit switch can be self-routed. By using the self-routing structure, concentrated control is not necessary; and the large scale optical multistage switches can be easily constructed, because complicated electrode patterns are not necessary.

IV. FABRICATION OF INTEGRATED OPTICAL SWITCH

A. Fabrication

The LD matrix switches were fabricated using a dry process. GaAlAs/GaAs and InGaAsP/InP double-heterojunction wafers, with oscillation wavelengths near 0.87 and 1.53 μm, respectively, were used as the gain guide wafers. An SiO_2 layer was sputtered and patterned to form the electrode pads. After an Au electrode about 0.8 μm thick on the p-side is sputtered, the gain guides and electrode pads were simultaneously patterned by ion beam etching (IBE) using Ar gas. The p-contact was 0.5, p-clad 2.5 and the active layer 0.15 μm thick for both types of wafers. The DH wafers were etched through the active layer about 5 μm. The etching rate was about 0.17 μm/min at a pressure of 3×10^{-4} torr and an acceleration voltage of 200 V. Fig. 5 shows an SEM microphotograph of the 6 μm wide active layer and 5 μm high ridge gain guides. A smooth side wall and sharp-edged intersection gap are achieved.

B. Performance

A 2×2 unit matrix switch was cleaved and wire bonded to measure the fundamental switching character-

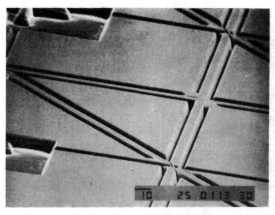

Fig. 5. SEM microphotograph of ridge gain guide in LD matrix switch.

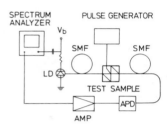

Fig. 6. Measuring setup for isolation characteristics.

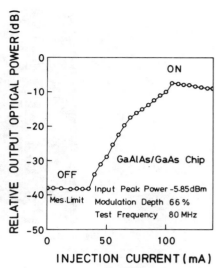

Fig. 7. Relative output power normalized by input power as function of injection current for a GaAlAs/GaAs chip.

istics. An experimental setup for measuring the isolation characteristics of an LD matrix switch is shown in Fig. 6. To separate the spontaneous emission power from the output power, the input light was modulated at 80 MHz with a 66 percent modulation depth. The measured wavelengths of the light source were 1.53 μm for In-GaAsP/InP chips and 0.87 μm for GaAlAs/GaAs chips. Single-mode fibers, having an etched microlens, were coupled to each input and output port. The relative output power normalized by the input power as a function of injected pulse current is shown in Fig. 7 for a Ga-AlAs/GaAs chip in the straight state. The switching pulse width was 800 ns and the repetition period was 200 μs. The oscillation threshold current of this chip was about 100 mA. The insertion loss in the "ON" state, including input and output coupling losses, is about 7.5 dB near the threshold current. Since we first manufactured an integrated LD optical switch, the insertion loss of the 2 × 2 unit optical switch is large. There are following two reasons. 1) Since a gain guide of an integrated LD optical switch is fabricated using a dry etching process, a scattering loss is large. Optimization of the gain guide structure makes the scattering loss small. 2) The coupling loss between unit optical switch and a single-mode fiber is large. Futhermore, the spontaneous emission is compressed by adding an optical filter, whose bandwidth is about 60 Å, to an edge of integrated LD optical switch. The SNR of the LD optical switch would be higher.

In the "OFF" state, the relative output power level is about −39 dB. This is the measurement limit. Therefore,

isolation greater than 31 dB was obtained with this chip. The fundamental 2 × 2 switching functions were observed for both InGaAsP/InP and GaAlAs/GaAs chips.

V. SIGNAL MONITORING CHARACTERISTICS OF LD OPTICAL SWITCH

A. Measurement Setup

A basic experiment was performed to obtain the baseband frequency and pulse respone of the terminal voltage changes in the LD optical switches. The laser diodes used in these measurements were GaAlAs DH lasers. The cavity length of the lasers was about 300 μm. Lasers were used as signal light sources and as LD optical switches. They emitted at almost the same wavelengths at room temperature, that is, they differed by about 5 nm from each other.

Baseband frequency response was measured by a spectrum analyzer with a tracking generator. The modulated signal light was injected into the LD optical switch active layer through an optical isolator and an objective lens. The input signal injection condition was determined to maximize the output optical power through the measured LD optical switch in the "ON" state. The output optical signal was detected through a single-mode fiber to pick up only the single spatial mode in the active layer with an Si-APD.

The terminal voltage change was measured through a capacitor and a 20 dB broad-band preamplifier. The measuring bandwidth of the spectrum analyzer was 1 kHz. The pulse response of the terminal voltage change was observed with a sampling oscilloscope. The input optical pulse waveform had rise time of less than 0.5 ns, and a fall time of less than 1.0 ns.

B. Results

The measured baseband frequency response of the terminal voltage for the buried heterojunction (BH)-type LD

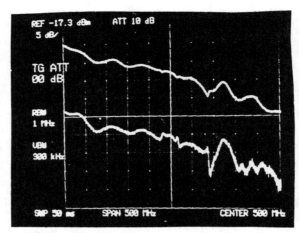

Fig. 8. Baseband frequency response (signal monitoring characteristics).

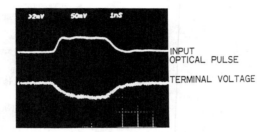

Fig. 9. Pulse response of terminal voltage in the "ON" state.

switch is shown in Fig. 8. The top trace shows the "OFF" state characteristics. The bottom trace shows the "ON" state characteristics when the injection current is just below the oscillation threshold current. The traces are corrected by the modulation response of the light source. In the figure, the horizontal axis represents the baseband frequency (1.0 GHz at full scale), while the vertical axis represents the relative RF power (5 dB/div.). Both frequency responses are wide and similar to each other except for the output level. The 6 dB down frequencies (f_{6dB}) are about 400 MHz in the "ON" state and about 400 MHz in the "OFF" state.

The pulse response of the terminal voltage in the "ON" state is shown in Fig. 9. The top trace shows the input optical pulse waveform and the bottom trace shows the detected pulse waveform at $I = 28$ mA. The figure clearly verifies that the output pulse waveform is counterphase to the input optical pulse. Comparing the input and output optical pulse waveforms, the rise and fall times of the latter are both found to be comparable to 1 ns. The LD optical switches with other structures showed the same phase change tendency.

VI. CONCLUSION

Low-loss and wide-band optical switches are expected to be key devices for the evolving fiber-optic communication system in the highly integrated service network. In this paper, we proposed an optical self-routing switch using integrated laser diode optical switches. This optical self-routing switch is composed of a Benes network, which can perform large scale switching functions using less hardware than a crossbar switch. These optical 2×2 unit switches are self-routed by monitoring terminal voltage changes in gain guides induced by input optical signals which are injected into a p-n junction of gain guides. Optical self-routing switches have various merits. They can handle high-speed data streams. Unit switches can be controlled and routed distributively, so switching function stoppages do not occur. In addition, the large scale optical switches can be easily constructed, because complicated electrodes patterns are not necessary. Since this concept is very attractive for large scale optical system applications, manufacture of the self-routing integrated laser diode optical switches will begin in the near future.

ACKNOWLEDGMENT

The authors wish to thank Dr. S. Shimada, Dr. M. Koyama, T. Aoyama, and Dr. H. Ishio of NTT Laboratories for their encouragement and guidance.

REFERENCES

[1] A. Huang and S. Knauer, "Starlite—A wideband digital switch," in *Proc. GLOBECOM '84.*
[2] J. S. Tuner, "New directions in communications," in *Proc. Zurich Seminar*, 1986.
[3] R. Kishimoto, N. Sakurai, and A. Ishikura, "Bit-rate reduction in the

transmission of high-definition television signals," *SMPTE J.*, Feb. 1987.

[4] C. Clos, "A study of non-blocking switching networks," *Bell Syst. Tech. J.*, vol. 32, pp. 406–424, Mar. 1953.

[5] V. E. Benes, "Optimal rearrangeable multi-stage connecting networks," *Bell Syst. Tech. J.*, vol. 43, pp. 1641–1656, July 1964.

[6] J. Kulzer *et al.*, "Statistical switching architecture for future services," in *Proc. ISS'84*.

[7] R. Kishimoto and N. Sakurai, "Multi-stage switching circuit controlled by dynamic load balance," *IEICE Japan*, July 1987.

[8] P. R. Prucnal, D. J. Blumenthal, and P. A. Perrier, "Self-routing photonic switching demonstration with optical control," *Opt. Eng.*, vol. 26, pp. 473–477, May 1987.

[9] D. C. Opferman and N. T. Tsao-Wu, "On a class of rearrangeable switching networks, Part I: Control algorithm," *Bell Syst. Tech. J.*, vol. 50, pp. 1579–1600, 1971.

[10] K. Y. Lee, "A new Benes network control algorithm," in *Proc. Int. Conf. Parallel Processing*, 1984, pp. 51–58.

[11] R. Kishimoto, "Self-routing control algorithm for concatenated network," *IEICE Japan*, June 1987.

[12] L. McCaughan and G. A. Bogert, "4 × 4 Ti : LiNbO$_3$ integrated-optical crossbar switch array," *Appl. Phys. Lett.*, vol. 47, pp. 348–350, 1985.

[13] S. Sakano, H. Inoue, H. Nakamura, T. Katsuyama, and H. Matsumura, "InGaAsP/InP monolithic integrated circuit with lasers and optical switch," *Electron. Lett.*, vol. 22, pp. 594–596, 1986.

[14] M. Ikeda, "Laser diode switch," *Electron. Lett.*, vol. 17, pp. 899–900, 1981.

[15] ——, "Signal-monitoring characteristics for laser-diode optical switch," *J. Lightwave Technol.*, vol. LT-3, no. 4, pp. 909–913, 1985.

[16] M. Ikeda, O. Ohguchi, and K. Yoshino, "Monolithic LD optical switches," in *Proc. ECOC'87*, 1987, pp. 227–230.

Ryozo Kishimoto (M'79), for a photograph and biography, see this issue, p. 1086.

Masahiro Ikeda (S'68–M'69), for a photograph and biography, see this issue, p. 1086.

HYPASS: An Optoelectronic Hybrid Packet Switching System

EDWARD ARTHURS, MATTHEW S. GOODMAN, HAIM KOBRINSKI, MEMBER, IEEE, AND MARIO P. VECCHI

Reprinted from *IEEE Journal on Selected Areas in Communications*, Vol. 6, No. 9, December 1988, pages 1500-1510.
Copyright © 1988 by The Institute of Electrical and Electronics Engineers, Inc. All rights reserved.

Abstract—A new architecture is presented for an optoelectronic hybrid packet switching system (HYPASS) for the distribution of multiple bit rate broadband services. HYPASS is based on an input buffered/output controlled arbitration protocol. The internal routing and interconnection utilizes a passive optical *transport* network with wavelength tunable laser transmitters and fixed wavelength receivers. This single-stage multiwavelength optical interconnect provides an internally non-blocking network for large-throughput routing of the bit-serial optical signals. An internal optical *control* network, with fixed wavelength transmitters and wavelength tunable receivers, sends output port information to the input nodes for arbitration and control. Packet buffer storage and control processing is performed by word-parallel electronic circuitry. The characteristics and device requirements for this design are presented along with results of a performance analysis of the arbitration and control protocol.

I. INTRODUCTION

FUTURE public switching networks will be based on optical fiber transport and will support a heterogeneous mix of broadband services. In order to enable design flexibility for these networks, statistical packet switching techniques will likely be used to achieve uniform switching and multiplexing of multiple bit rate data streams [1]. A key element of the network is the large capacity, broadband packet switch. A Batcher–banyan switch has recently been proposed [2] to be implemented with CMOS technology. The Batcher–banyan design, like existing electronic switches, requires demultiplexing of an incoming multigigabit optical transmission channel into subrate channels. These lower rate data streams are switched and remultiplexed to the transmission rate streams after the switch. On the other hand, research is in progress within numerous organizations to utilize photonic components and devices to implement switching functions [3]. Wavelength routing has been applied to several optical network designs [4]–[9], including a recent proposal for a packet switch design using multiwavelength interconnections based on the knockout algorithm [10].

In this paper, we present a new architecture for a high performance optoelectronic hybrid packet switching system, referred to as HYPASS, that will incorporate the parallelism and high-throughput characteristics of multi-wavelength optical interconnects together with the flexible processing power of electronic logic. HYPASS is conceived as an *integral system*, with an architecture specifically designed to take advantage of the strengths of both optical and electronic technologies. The underlying assumption is that system architectures should be adapted to the fundamental characteristics of the available technologies (i.e., that it will not necessarily be productive to attempt to repeat electronic designs in the optical domain). Thus, the HYPASS design takes a rather different approach to packet switching than the traditional electronic designs.

HYPASS evolves from previous work on optical network research, in particular, high density multiwavelength network research [4], [6]–[8]. It represents an effort to address complex network functions in a design where both transport and control issues are resolved. The HYPASS architecture should evolve smoothly to benefit from further developments in the performance of both optical and electronic devices.

II. SYSTEM ARCHITECTURE

The overall function performed by HYPASS is to receive packets on optical fiber input trunks, and to switch them to the appropriate optical fiber output trunks (Fig. 1). The input stream of packets is assumed to arrive at a data rate of ≈ 2 Gbits/s, with packet sizes of ≈ 1000 bits. The output stream of packets will be transmitted at the same bit rate, after the routing has been done on the basis of the packet header addressing information.

The HYPASS is formed from two networks: a data transport network that connects the input ports to the output ports, and a control network in which the output ports transmit their status to the input ports (Fig. 2). Packets arriving at an input port are held at the port controller until a ''go'' signal from the desired output port is received indicating that the packet is to be sent to the output port.

Fig. 3 schematically illustrates the HYPASS architecture. In this design, electronic components are used for memory and logic functions, and optical components for routing and transport. The optical transport network in HYPASS is based on a passive optical star coupler. An input–output path through the transport network is established by tuning a single parameter, thereby providing a ''single-stage'' crossconnect. Internally, the passive N × N transmissive star couplers combine the light intensity

Manuscript received August 24, 1987; revised June 15, 1988.

The authors are with Bell Communications Research, Morristown, NJ 07960.

IEEE Log Number 8824392.

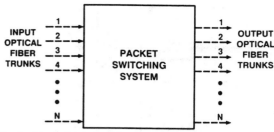

Fig. 1. Block diagram of the packet switch input/output flow.

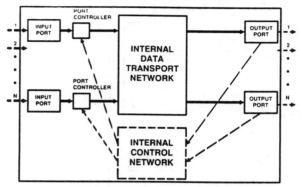

Fig. 2. Schematic diagram of an input buffered/output controled packet switching system.

from every incoming fiber and distribute it uniformly to all output port receivers. The HYPASS design is based on a "broadcast and select" approach in both the data transport and the control networks. A unique optical wavelength is associated with each of the output ports and therefore addressing the output ports is achieved by transmission on this unique wavelength.

In detail, a packet arriving at an input port, after optoelectronic (O/E) conversion, is temporarily stored in a buffer at that input port, and its destination address is decoded. Based on the destination address, a tunable-wavelength laser transmitter is tuned so that its wavelength corresponds to the wavelength of the desired output port. The packet waits at the input port until a poll (i.e., a "request-to-send" signal) is received indicating that this input port should send its packet. The packet is transmitted through the transport network on the wavelength of the destination output port. When it is received at the output port it is stored in an elastic buffer and transmitted over the output optical fibers. The output ports monitor their busy status and when available, they issue polls over the control network to request packets from the input ports. The internal control network is used to transmit both these polls and the "data received" acknowledgments to the input ports. The output port status is broadcast to the input ports using fixed-wavelength transmitters, and tunable-wavelength receivers at the input ports. The packet destination address determines the tuning current of the control network receiver. This control network may operate at lower bit rates than the data transport network.

The HYPASS design is highly parallel. Packets from many inputs are being simultaneously transmitted on different wavelengths to their destination outputs using the same data transport network. The single stage transport network in HYPASS is internally nonblocking due to this parallel use of the multiwavelength optical crossconnect. While no internal collisions can occur in the transport, it is possible to have external collisions, if two transmitters were to simultaneously send packets to the same output node. HYPASS uses an output controlled algorithm for resolving external collisions. A variety of collision arbitration protocols are possible for the HYPASS design. The control network and a specific example of a collision arbitration protocol needed for probing the input ports for packets and resolving possible external collisions in the transport network is described in a later section.

The time-overhead required by the electronic control and optoelectronic tuning operations will limit the throughput. This overhead will depend on the nature of the traffic, the efficiency of the protocol, the conflict arbitration algorithm, and the technology available for the hardware implementation.

In general, the throughput Ω is given by

$$\Omega = NR \frac{\tau_p}{\tau_o} \qquad (1)$$

where N is the number of ports of the switch, R is the data rate per port, τ_p is the packet transmission time exclusive of control overhead, and τ_o is the total packet transmission time including all the overhead for electronic control. Table I indicates a possible time sequence for HYPASS operations using existing (albeit state of the art) control electronics.

A design goal of the HYPASS is to be able to have an overhead time of ≈ 15 percent. The total overhead time listed in Table I is ≈ 100 ns, indicating a peak throughput goal of about 200 Gbits/s for a HYPASS switch with 128 ports and an internal transmission rate of 2 Gbits/s for each port.

III. Hardware Implementation Proposal

The architecture for HYPASS presented in the last section (see Fig. 3) can be subdivided into three subsystems: the input port buffer and control electronics, the transport network in the forward direction, and the control network in the reverse direction. The logical synchronization of packet transmission, polling and collision detection, requires that HYPASS will operate with synchronized slot times. A slot time is defined as the fixed length of time that is sufficient to contain a polling step, the transmission of a packet and the return acknowledgment. The distances between nodes in HYPASS are short (<5 m), therefore direct clock distribution to all the nodes (for instance, using a dedicated wavelength) is possible. HYPASS will be synchronized by a common master clock (not shown in the figures), such that it will be possible to maintain continuous bit-synchronous operation.

A. Input Port Buffer and Control

The temporary storage of input packets, the decoding of the addressing information and the processing of the

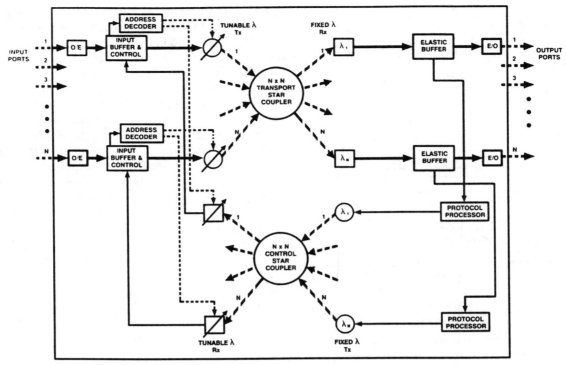

Fig. 3. Global diagram of the HYPASS implementation. Bold dashed lines represent optical paths, bold continuous lines represent serial data paths, dotted lines are the tuning current paths, and thin continuous lines are control signal paths.

TABLE I
EVENT SEQUENCE FOR HYPASS

Tune laser for trans.	Process the Poll	Add preamble (≈ 64 bits)	Transmit (1024 bits)	control and ack
< 10 ns	50 ns	20 ns	560 ns	20 ns
Listen for Poll (Tune Receiver)				

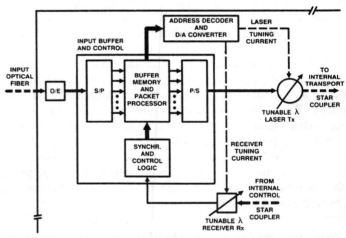

Fig. 4. Schematic representation of the input buffer subsystem of HY-PASS, including the address decoding and wavelength tuning circuits.

acknowledge and poll signals from the output ports may be realized with electronic logic, as shown in Fig. 4. The input packets arriving in bit serial optical streams are converted to electronic signals. An O/E receiver will perform the detection and, if necessary, the reconstruction of the incoming optical signals, and then an S/P shift register will produce word-parallel electronic data of width b.

The packet header address information is decoded, and a D/A converter produces the electrical currents to tune the optical wavelengths of both the laser transmitters of the transport network and the receivers of the control network.

Another fast P/S shift register is used to convert the word-parallel electronic data to a bit serial stream that modulates the laser transmitter tuned to the appropriate output port wavelength. Transmission of the buffered input packet to the corresponding output port is synchronized by the control network poll and acknowledge signals: a poll triggers the packet transmission, and an acknowledgment of either successful packet reception or collision/error is returned to the sending input port.

Electronic processing of the packet information at the input port, such as storage and decoding, is done at speeds that are b times slower than the incoming bit rates. In this way, the critically fast electronic components are limited to the O/E receiver and the $S/P-P/S$ shift registers. Receivers operating at 2 Gbits/s are available commercially. The electronic circuits required by the $S/P-P/S$ conversions could be implemented using commercially available devices based on GaAs nonsaturating logic, as has been demonstrated by similar designs operated at 2.4 Gbits/s [11], [12].

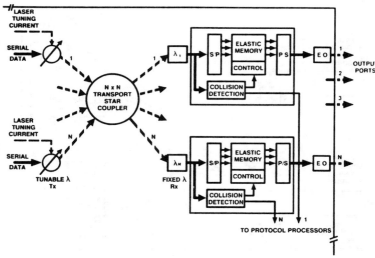

Fig. 5. Schematic representation of the optical multiwavelength transport network.

The memory and address decoding functions are performed on word-parallel data, and hence have significantly reduced speed requirements. It is reasonable to expect that address decoding, based on table lookup, can be done in times much shorter than packet duration, ≈ 500 ns for our example. To achieve fast D/A conversion for the tuning currents of both the tunable laser transmitters and receivers, line selection of preprogrammed current-source values may be used. High-density ($16K \times 1$ or higher) static RAM memories are currently available with access times ≤ 10 ns in ECL logic, and ≤ 25 ns in CMOS technology. Memories of smaller densities are available in GaAs with access times ≤ 2 ns. Assuming a 2 Gbit/s data rate on the fiber trunks, and a parallel word width of $b = 16$, the required internal clock rate for the buffer memory is 125 MHz, and currently available high-density ECL RAM chips would be just at the limit, while GaAs memories already provide some speed margin. Furthermore, a wider bus expansion could bring the speed requirements for the buffer memory chips within the range of CMOS technology.

B. Data Transport Network

The routing and transmission of the packets from the input nodes to the output nodes is done using a large throughput multiwavelength optical interconnection network (Fig. 5). A throughput of 36 Gbits/s has already been demonstrated [13] by broadcasting to 16 nodes 18 wavelength-multiplexed channels at 2 Gbits/s each in a similar optical transport network (the Bellcore LAMB-DANET™).

In the HYPASS transport network, each of the N input nodes has a *tunable* optical transmitter, and each of the N output nodes has a fixed-wavelength optical receiver. Each transmitter is connected through a single mode optical fiber to a *transport* $N \times N$ optical star coupler, and the N outputs of the star coupler are then connected to the output receivers forming a fully connected network. Each output node is associated with a *unique* wavelength, using

an optical bandpass filter (or a fixed heterodyne receiver). The HYPASS design requires a method of wavelength registration such that the center wavelengths associated with given receivers may be successfully (and repeatedly) addressed. Such a registration method, for example, could be based on an electronic comb filter technique to lock the laser frequency.

At the input nodes (see Fig. 5), each transmitter has a tunable laser diode to cover the required wavelength range. The bit serial optical signals are transmitted through the HYPASS at the same rate required by the optical fiber trunks, that is ≈ 2 Gbits/s in our example. The *tuning* speed of the laser diodes depends on the length of the packets. For packet sizes of ≈ 1000 bits, the required time to tune the laser diode between any two arbitrary wavelengths within the tuning range is ≤ 10 ns.

Tunable distributed feedback (DFB) and distributed Bragg reflector (DBR) laser diodes, with separate gain and tuning regions, have been reported by several groups [14]–[16]. Wavelength tuning range up to ≈ 10 nm has been demonstrated for these devices. This tuning range would allow ≈ 100 separate channels with a channel separation of 0.1 nm.

A recent experiment that is particularly relevant to the HYPASS design, has demonstrated high-speed switching among eight distinct wavelengths using a double-section distributed feedback (DS–DFB) laser [15]. In this experiment, the DS–DFB laser was tuned by current-injection, using an eight-level pattern applied to both the forward and rear section electrodes, in such a way that the total current was kept constant. The optical spectrum of the wavelength-switched DS–DFB was observed using a scanning Fabry–Perot interferometer. The total tuning range was 0.32 nm, with ≈ 0.045 nm channel spacing, which is ≈ 2.5 times the filter bandwidth. In the experiment [15], the channel-to-channel switching time was less than 5 ns, with a residency time of 500 ns per channel. In order to demonstrate the ability to switch data, two of the eight selectable wavelengths were chosen for data trans-

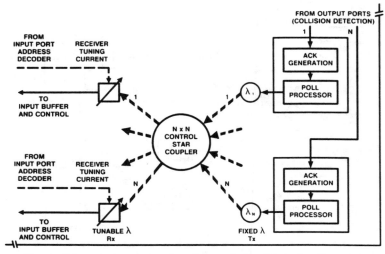

Fig. 6. Schematical representation of the optical multiwavelength control network.

mission. The optical signal was externally modulated at data rates between 0.5 and 1.0 Gbits/s using a commercial LiNbO$_3$ modulator. The Fabry–Perot cavity was used as a fixed filter (no scanning) to choose between the two channels for the wavelength-switched and amplitude-modulated signals. The filtered signals were detected by an InGaAs APD in a custom designed high impedance FET receiver. The Fabry–Perot filter was tuned alternately between each of the two channels, and the amplitude-modulated output of the DS–DFB at each wavelength was observed in the oscilloscope traces measured after the APD receiver. The switching time between the two wavelengths was again measured to be less than 5 ns, and open eye patterns were observed for each of the channels modulated with a pseudorandom sequence at 1.0 Gbits/s.

At the output nodes (see Fig. 5), the fixed-wavelength optical receiver is followed by the required elastic memory and port status detection modules. An output port status signal (i.e., idle, busy, or collision/error) is used to implement the control protocols for the operation of HY-PASS. Collision detection may be performed either by power threshold detection or by multiple bit detection on the packet preamble. The corresponding control signal triggers both the output port external transmission and the protocol processor of the control network (see also Fig. 3). An S/P conversion allows the operation of the elastic memory in word-parallel format, and accurate retiming is performed prior to the final P/S conversion, E/O regeneration, and transmission on the external optical fiber trunk.

C. Control Network

The status of each output port is monitored (see Fig. 5), and the logic circuits generate the required control information to be sent to the input nodes (Fig. 6). This protocol information is converted to an optical signal at a unique wavelength associated with each output port in the control network.[1] A fixed-wavelength laser transmitter broadcasts control information to all input ports using the control $N \times N$ star coupler. At each input node, a tunable receiver selects the control information relevant to the output port, or ports, for which it is preparing for transmission.

The tunable optical receivers for the control network of HYPASS can be realized using Fabry–Perot, DFB or DBR laser structures, *operated below threshold*, as wavelength-selecting devices [17]–[21]. Another type of tunable filter is based on the electrooptic effect in LiNbO$_3$ [22]. This filter has been demonstrated to have a tuning range of at least 8 nm and a filter bandwidth ≈ 1 nm. Although the tuning speed was not reported, it should operate at speeds > 10 MHz since it is based on electronic tuning.

A demonstration of the feasibility of high-speed wavelength selection was recently reported [20], [21]. This experiment used distributed-feedback optical amplifiers (DFB–OA) as the wavelength selective element. Two 1 Gbit/s signal channels with different wavelengths were wavelength–demultiplexed by the optical filter with nanosecond selection times. The DFB–OA device used was a commercially available 1.5 μm BH laser with a first-order grating, one cleaved facet and one AR-coated facet. The DFB–OA was *operated below threshold*, and the resonant wavelength was adjusted by varying the injected current. Two DFB laser transmitters (TX A and TX B), directly modulated and combined in a 2×2 directional coupler, provided the incident light signal through an optical isolator. Fast wavelength-switching and BER measurements were made using three separate pattern generators, all synchronized by a common clock signal. In HYPASS, the short distances make it possible to distribute a central master clock signal throughout the packet

---

[1]Neither the wavelengths, nor their channel spacings in the control network need be the same as in the transport network.

switch for system synchronization. Generators A and B provided data modulation for their respective transmitters, and generator C provided the time-varying square-wave for the electronic tuning of the DFB–OA wavelength. Measurements of high-speed wavelength-selective amplification were made by superimposing a tuning current from pattern generator C on the DFB–OA bias current. This square-wave current had an amplitude $\Delta I_a = 1.2$ mA peak-to-peak, corresponding to a wavelength shift $\Delta\lambda = 0.23$ nm, a fundamental frequency of 50 MHz and a risetime of 0.2 ns. An incident optical signal from TX A, modulated at 1 Gbit/s, was temperature tuned to match one of the two wavelengths to which the DFB–OA is being tuned by the square-wave current modulation. The amplified optical signal was detected by the fast APD receiver (bandwidth > 2 GHz), and a risetime $\tau_{sw} \approx 1$ ns was observed for the wavelength selection by the DFB–OA. The high-speed, dynamic wavelength-tuning coefficient is $\Delta\lambda / \Delta I_a = -0.19$ nm/mA.

In addition, an experiment was performed to simulate the transient conditions of a packet-switching system using the DFB–OA for fast wavelength-selection of high bit rate wavelength-multiplexed signals [20]. The three pattern generators were controlled by the same master clock at 1.2 Gbits/s. The wavelength of the DFB–OA was switched between λ_1 and λ_2, separated by $\Delta\lambda = 0.23$ nm. A BER $< 10^{-9}$ was measured for the total detected pattern of packets arriving alternately at λ_1 and λ_2. These measurements of BER under bursty transmission at 1.2 Gbits/s indicate that pattern integrity is maintained even when the DFB–OA filter is tuning during a single bit interval.

D. Projected Device Parameters

The architectural design of HYPASS, as well as the initial experimental demonstrations of device performance, provide the basis for establishing specifications for the development of a variety of optical and electronic components. These target specifications are summarized in Table II. The column for near-term, refers to values that have been demonstrated, or that could be achieved without the need for technological advances. The second column presents future target parameters that require significant research and development efforts on the respective technologies.

IV. ARBITRATION PROTOCOL AND CONTROL

Future digital networks such as BISDN will be expected to accommodate service demands from a variety of sources such as full-motion video, data, and POTS. The input-buffered/output-controlled arbitration of HYPASS will allow for a variety of specific control algorithms to be implemented. Traffic requirements and detailed device performance will ultimately define the choices. The specific arbitration scheme chosen may be strongly dependent on the type and extent of the traffic demand. We envision two types of traffic, namely, *demand access packets* which arise according to instantaneous traffic requirements, and *preassigned access packets* which arise due to reserved switch bandwidth for high priority, or time-critical traffic.

The arbitration protocol and control algorithm presented in this section permits the preassignment of bandwidth to high bandwidth real-time signals, with strict delay requirements at call setup time. Sources such as typical data and POTS are handled with a queued demand assignment approach. More specifically, the preassigned calls have slots reserved in a periodic frame (in a TDM manner), and the random demand traffic is served by the parallel execution of a multiaccess algorithm with collision resolution. The collision resolution cycles are interrupted by preassigned requests, and then resumed. The protocols are implemented by using group polling with collision detection, which will be described next.

A. Port Polling Protocol

The basic step in the HYPASS arbitration protocol is the slot port polling protocol. Assume that the number of ports is $N = 2^k$, a power of two. We imagine the input ports to be the leaves of a complete binary tree. Starting at the root, the levels of the tree are numbered 0 through k. At the lth level, the nodes are numbered (from the left to right) from 0 to $2^l - 1$ in binary.

A poll from output port i will have three binary fields of the form

$$(\mathfrak{X}, l, m) \text{ where } \begin{cases} \mathfrak{X} \in \{\mathcal{P}, \mathcal{S}, \mathcal{C}\} \\ 0 \leq l \leq k \\ 0 \leq m \leq 2^l - 1. \end{cases}$$

The $\mathfrak{X}$ field identifies the type of protocol (and traffic), and is described below. The (l, m) fields enable all input ports which are located in the subtree of which (l, m) is

TABLE II
TARGET PARAMETER SPECIFICATIONS

	Near-term	Target
HYPASS Switch		
Number of Ports	32	256
Bit Rate/Port	2 Gbit/s	> 2Gbit/s
SINGLE-MODE STAR COUPLER		
Size(N x N)	32 x 32	256 x 256
Uniformity	±6 dB	±3 dB
TUNABLE LASER TRANSMITTER		
Number of Channels	32	256
Output Power	0 dBm	0 dBm
λ Switching Time	5 ns	1 ns
Tuning Range	> 3 nm	> 25 nm
TUNABLE OPTICAL RECEIVER		
Number of Channels	32	256
Filter Bandpass	0.1 nm	< 0.1 nm
λ Selection Time	1 ns	< 1 ns
Tuning Range	> 3 nm	> 25 nm
ELECTRONIC COMPONENTS		
S/P-P/S Clock Rate	2 GHz	> 2 GHz
Memory Chip Capacity (bits)	1K x 16	4K x 16
Memory Access Time	< 8 ns	< 4 ns

a ''root,'' that is, those input ports (leaves) whose binary address has ''m'' as a prefix.

The sequence of events in a typical polling slot are described by considering the logical steps at input port j and output port i. These steps are concurrently occurring in a similar way at each of the N output ports during the operation of the HYPASS.

1) Input Port Protocol:

1) Decode packet header address information to establish the output port wavelength λ_i.

2) Tune the input port laser transmitter to the required output port wavelength λ_i.

3) Tune the input port receiver[2] to the required output port wavelength λ_i.

4) Interpret the control information received at wavelength λ_i from the output port over the control star coupler.

5) When polled by the output, transmit the packet at wavelength λ_i over the transport star coupler.

6) Await Acknowledgment (of no collision or error)

7) Begin a new cycle.

2) Output Port Protocol:

1) Poll input port(s) j by broadcasting output port control information at wavelength λ_i over the control star coupler.

2) Receive a transmission over the transport star coupler from input port(s) j, or detect timeout.

3) Acknowledge correct reception if no collision/error.

4) Begin new slot by polling new input port(s).

This illustrates the basic polling slot. We will next discuss *integrated* packet transport by a protocol combining a periodic individual port polling cycle (for preassigned packets) with a multiaccess type adaptive tree polling cycle (for random traffic).

B. Preassigned Traffic Cycle

Assume that time is divided into fixed length slots. The length of a slot is sufficient to contain a polling step, i.e., an output port can poll an input port and the port can respond by transmitting a packet. Further, the slot sequence is divided into groups of $\mathcal{F}$ consecutive slots called a frame. Bandwidth is assigned to the high-bandwidth real-time traffic during call setup, in units of a number of slots per frame. If the slots in a frame are numbered 1 to $\mathcal{F}$, then during any given slot, a maximum of one preassigned call must be assigned to an output port. The preassigned polls are of the form $(\mathcal{P}, k, l)$ where $\mathcal{X} = \mathcal{P}$ means ''preassigned'' and only input port ''l'' (leaf l) is enabled.

We can represent the state of preassignments by a table with one row for each input port and one column for each slot in a frame. An example shown in Table III illustrates an assignment for four ports and a frame of size four. That is, call A appears in time slot 1 and time slot 3 from input

[2]For simplicity, we will assume that the unique wavelength associated with each output node is the *same* for the *transport* network and the *control* network.

TABLE III
PORT AND TIME SLOTS FOR PREASSIGNMENT PACKET EXAMPLE

Time Slot	1	2	3	4
Input Port 1	A,2	B,3	A,2	
2	C,1		D,4	
3				
4	E,3		E,3	

TABLE IV
PORT AND BANDWIDTH ASSIGNMENTS FOR PREASSIGNED PACKET EXAMPLE

CALL	Input Port	Output Port	Bandwidth
A	1	2	2
B	1	3	1
C	2	1	1
D	2	4	1
E	4	3	2

port 1 to output port 2. This indicates a call setup from input port 1 to output port 2 with a bandwidth of 2. Call C, on the other hand, is setup from input port 2 to output port 1 with a bandwidth of 1. (We note that each output port can only appear once in any column.)

Thus, from this Table III we see that there are 5 preassigned calls, labeled as A, B, C, D, and E with the port and bandwidth assignments shown in Table IV.

The preassigned service is then implemented by high priority polling sequences from each output port. That is, the high priority polling sequence for output port 3 in the above example is: poll input 4 in slot 1, input 1 in slot 2, and input 4 in slot 3. When an output port is *not* executing high priority poll, it executes a tree polling algorithm to satisfy low priority demand access traffic.

C. Tree Protocol with Collision Resolution

The tree protocol [23], [24] consists of a sequence of collision resolution intervals (CRI). Each output port continually executes the tree protocol except when interrupted by a request from the preassigned cycle. When we refer to the length of a CRI, the number of interrupted slots are not included. The first poll of a CRI will consist of $(\mathcal{S}, l, m)$ where the $\mathcal{X} = \mathcal{S}$ signifies ''start of a CRI.'' For example, an input with a demand packet which tunes to output port i, must wait for the start of a CRI before participating in the tree protocol. All other polls during a CRI are of the form $(\mathcal{C}, l, m)$ where $\mathcal{X} = \mathcal{C}$ indicates that a CRI ''continues in progress.''

The basic unit in a tree protocol is the conflict-traversal of a tree. This is the recursive algorithm.

1) Poll (l, m) where (l, m) designates the root of the tree in question. If no packet or a good packet is received (i.e., no collision/error), terminate traversal of the tree; otherwise go to 2).

2) Traverse left subtree then right subtree.

We will now illustrate this algorithm by starting at level $(0, 0)$, the root of the tree. Suppose there are eight ports numbered 0 through 7 (from left to right) on the tree. Output port 5 executes a low priority cycle with input ports 1 and 6 having packets for output port 5. The tree polling time sequence is indicated in Table V.

TABLE V
TREE POLLING TIME SEQUENCE (LEFT TO RIGHT)

poll 0-7	poll 0-3	poll 4-7
Collision	Successful Transmission by port 1	Successful Transmission by port 6

TABLE VI
AVERAGE NUMBER OF PROBES IN TREE POLL

N	$p = \frac{1}{2N}$	$p = \frac{1}{N}$	$p = \frac{2}{N}$
16	1.33	2.10	4.16
32	1.36	2.19	4.41
64	1.38	2.25	4.58
128	1.39	2.29	4.70

Thus, it takes three slots to transmit two packets. If the ports are polled individually, then no collisions ensue, but all low priority polling steps are of length N. That is, it will require eight polls with individual polling to transfer the two packets.

The efficiency of the specific protocol used is a strong function of the load presented. We characterize the probability that a port has a packet as f. Further, we characterize the load balance parameter as γ. When the system is uniformly loaded, $\gamma = 1$, and when $\gamma = 2$, there are twice as many packets at different input ports destined for the same output port and the system is (temporarily) unbalanced. Then the probability that an input port has a packet for a given output port is just

$$p = \frac{f}{2^k} \gamma.$$

Tree polling drastically reduces the average number of polling steps required, even for heavily loaded switches ($f = 1$).

Table VI indicates the average length of a CRI for a given number of ports on the HYPASS switch. Note that the number of simple individual polling steps increases linearly with the number of ports N that need to be polled whereas the tree polling for a given traffic load depends only weakly on the number of ports polled.

The average length of a CRI is given as a function of "p," the probability that a specific input-port "i" has a head-of-line demand access packet for a given output port "j." In normal operation, "p" will be approximately $1/2N$ where N is the number of ports. In this case, the mean number of polls required is ≈ 1.4, slightly larger than 1, the minimum possible number of polls. As seen in Table VI, when the system is uniformly loaded, with $f = 1$ and $\gamma = 1$, the tree polling algorithm still requires only about 2.2 polls to complete a CRI. And even when the heavy ($f = 1$) load is temporarily unbalanced ($\gamma = 2$) the mean length of a CRI is still only ≈ 4.5 slots, and does not strongly depend on the number of nodes being polled.

The tree protocol of HYPASS is the dynamic tree protocol of Capetanakis [23]. Notice that if we start at $(0, 0)$ we have one tree to traverse. If we start at level l, then we can carry out the protocol by traversing the 2^l trees in left to right order. If $l = k$, the algorithm individually polls the input ports (leaves) in a TDM manner. If the traffic for a given output port is light, it will be most efficient to start the CRI at $(0, 0)$. At the other extreme, if almost all input ports have a packet for a given output port i, then port i will be better served by starting its CRI at $(k, 0)$. As described in [23], the dynamic tree protocol interpolates between these extreme cases as the load

changes so as to reduce delay. More specifically, the dynamic tree protocol is defined [23] as follows.

1) Observe L, the length of the previous CRI (excluding interrupt slots).
2) Compute $J = F(L)$, the starting level for the CRI (where $F(\cdots)$ is given in [23]).
3) Start a CRI at $(j, 0)$. Complete CRI and go to 1.

We next look at switch performance for purely random (demand) traffic using the dynamic tree protocol. The integrated control protocol for HYPASS consists of a combination of the high priority protocol with the dynamic version of a tree protocol introduced by Capetanakis [23]. The tree protocol is interrupted to service high priority demands and is then resumed. The dynamic tree protocol chooses the depth of the tree to start polling based on the length of the previous CRI (not including interrupt slots of high priority traffic). If the traffic for a given output port is light, the CRI will be short and polling will start at the root of the tree. If the traffic is very heavy the polling will start at the leaves of the tree and individual input ports will be polled in a TDM manner. The dynamic tree protocol interpolates between these extreme cases as the load changes in order to reduce delay through the switch.

V. SWITCH PERFORMANCE FOR RANDOM TRAFFIC

In this section, we assume a packet arrives at each port with probability "r" per slot, destined with equal probability ($1/N$) to any one of the output ports. Each input port has a buffer for incoming packets, which are served on a first-come first-served basis at the beginning of each slot using dynamic tree polling. Since an exact analysis of the N dimensional Markov chain is out of the question, the approximate engineering approach of Hui and Arthurs [2] will be used. The switch is modeled as N independent single server queues (one for each input port), each queue with an effective service time distribution which accounts for the head-of-line (HOL) contention. More specifically, we assume for each input port that there is a probability q of winning a switch arbitration in a slot where q is a function of r, the traffic carried per port. The analysis is then partitioned into two parts: q is determined as a function of r, and the individual input queues are analyzed. Since the second part of the analysis is identical with that of [2] only the results will be summarized. The first part of the analysis involves bounding the delay of an adaptive tree protocol and requires a quite lengthy and technical analysis. Only the results will be presented in this paper.

For a given carried traffic per port, the mean delay is an increasing function of the number of ports N. We present a conservative analysis by assuming an infinite number of ports ($N \rightarrow \infty$). If, as in [2], we assume a constant probability of a fresh packet arrival to an idle port, the process of fresh packets arriving to HOL, addressed to a given output port, is then a Poisson process. We can use the infinite station dynamic model of [25]. Upper bounding the mean waiting time of this model using the methods of [26], we obtain for the q versus r relation

$$\frac{1}{q} \leq 1 + S(r) \quad 0 \leq r \leq r_{\text{cap}} \quad (2)$$

where

$$S(r) = T + U + e^{-r}(V_1 + V_2 + V_3) \quad (3)$$

where

$$T = \frac{3.497r^2 - 5.582r + 8.990r^4}{(1 - 4.995r^2)(1 - 2.235r + 2r^2)},$$

$$U = \frac{1.827r(e^r - 1)}{(e^r - 2.235r)(1 - 4.995r^2)},$$

$$V_1 = \frac{0.401(e^r - 1)}{(e^r - 2.235r)},$$

$$V_2 = \frac{0.401(1 - 2.235r + 4r^2)}{(1 - 4.995r^2)(1 - 2.235r + 2r^2)},$$

and

$$V_3 = \frac{1.465r^2(e^r - 1)}{(e^r - 2.235r)(1 - 4.995r^2)}.$$

And the capacity per port r_{cap}

$$r_{\text{cap}} = 0.318 \quad (\text{packets/slot}).$$

Note from [2] that with a "no-collision" arbitration the input buffering HOL effect results in a $r_{\text{cap}} = 2 - \sqrt{2} = 0.586$. Here the capacity decreases to 0.318 because of the use of a multiaccess arbitration protocol with collisions.[3]

If we define

$$\omega_r = \frac{r(1 - q)}{(1 - r)q} \quad (4)$$

we have that the mean delay D is given by

$$D = \frac{(1 - r)\omega_r + r}{r(1 - \omega_r)} \quad (\text{slots}). \quad (5)$$

Fig. 7 shows the mean delay D versus r and Fig. 8 plots the mean queue length versus r for an infinite buffer. If the buffer size is finite, of size B packets, then the probability of buffer overflow, $P(\text{loss})$ is bounded by

$$P(\text{loss}) \leq [(1 - r)\omega_r + r](\omega_r)^B. \quad (6)$$

The behavior of $P(\text{loss})$ for several finite buffer sizes B is shown in Fig. 9.

[3]Because of collisions, this capacity of 0.318 packets per slot is smaller than the 0.586 from [2]. However, the time slot in the optical switch is much shorter than the corresponding slot for an electronic switch.

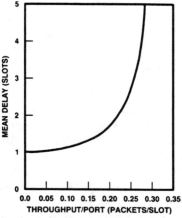

Fig. 7. Calculated mean delay as a function of throughput.

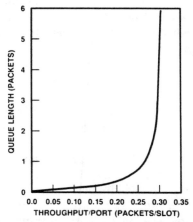

Fig. 8. Calculated queue length as a function of throughput.

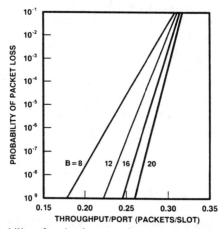

Fig. 9. Probability of packet loss as a function of throughput for several buffers of fixed length B at the input ports.

VI. CONCLUSIONS

A design has been presented for a packet switching system that incorporates both optical and electronic subsystems. Some of the distinct features of HYPASS are the following.

• HYPASS exploits multiwavelength optical interconnections to achieve a functional packet switching system. The proposed architecture is adapted to the properties of optical components, and it is not a duplication of electronic system design with optical devices.

- The optical transport network provides a single-stage, internally nonblocking, large throughput interconnect.
- The optoelectronic control network provides an effective mechanism for the implementation of the required arbitration protocols.
- The input buffer and control processing is realized by electronic components.
- HYPASS is a high-speed packet switching system with flexible parallel control algorithms, which allows implementation of bandwidth reservation, priority assignments, adaptive polling by collision detection, group polling arbitration, etc.

The HYPASS architecture represents a significant departure from conventional electronic packet switches. To explore the potential of this new approach to packet swtiching requires a multidisciplinary research effort. The proposed HYPASS design utilizes a variety of advanced technologies, both for optical and electronic components. In addition, the arbitration and control requires careful understanding of the protocols to be implemented for the different traffic patterns of new broadband services. There are, therefore, a number of research issues to be resolved in order to arrive at a final definition of the detailed design of HYPASS. Several alternative options will have to be examined for the hardware implementation and for the control algorithms, which will depend both on the technology and the broadband service requirements.

REFERENCES

[1] L. T. Wu, E. Arthurs, and W. D. Sincoskie, "A packet network for BISDN applications," B. Platner and P. Gunzburger, Ed., in *Proc. 1988 Zurich Sem. Digital Commun.*, Zurich, Switzerland, 1988, pp. 191–197.

[2] J. Y. Hui and E. Arthurs, "A broadband packet switch for integrated transport," *IEEE J. Select. Areas Commun.*, vol. SAC-5, pp. 1264–1273, 1987.

[3] "Topical Meeting on Photonic Switching," *Opt. Soc. Amer.*, Mar. 1987.

[4] M. S. Goodman, H. Kobrinski, and K. W. Lo, "Application of wavelength division multiplexing to communication network architectures," in *Proc. Int. Conf. Commun.*, Toronto, Canada, 1986, pp. 931–934.

[5] E. Arthurs, M. S. Goodman, H. Kobrinski, and M. P. Vecchi, "HYPASS: An optoelectronic hybrid packet switching system," *Bellcore Tech. Memo.*, TM-ARH-009795 (unpublished), July 1987.

[6] H. Kobrinski, R. M. Bulley, M. S. Goodman, M. P. Vecchi, C. A. Brackett, L. Curtis, and J. L. Gimlett, "Demonstration of high capacity in the LAMBDANET architecture: A multiwavelength optical network," *Electron. Lett.*, vol. 23, p. 824, 1987.

[7] E. Arthurs, R. Bulley, M. S. Goodman, H. Kobrinski, M. P. Vecchi, and J. L. Gimlett, "A fast optical crossconnect for parallel processing computers," presented at 13th Europe. Conf. Opt. Commun., Helsinki, Finland, Sept. 1987.

[8] E. Arthurs, J. M. Cooper, M. S. Goodman, H. Kobrinski, M. Tur, and M. P. Vecchi, "Multiwavelength optical crossconnect for parallel-processing computers," *Electron. Lett.*, vol. 24, pp. 119–120, 1988.

[9] M. S. Goodman, E. Arthurs, J. M. Cooper, H. Kobrinski, and M. P. Vecchi, "Demonstration of fast wavelength tuning for a high performance packet switch," presented at 14th Europe. Conf. Opt. Commun. (to be published), Brighton, England, Sept. 1988.

[10] K. Y. Eng, "A photonic knockout switch for high speed packet networks," in *Proc. IEEE/IEICE Global Telecommun. Conf.*, Tokyo, Japan, Dec. 1987, pp. 47.2.1–47.2.5.

[11] A. Albanese, M. W. Garrett, A. Ippoliti, M. A. Karr, M. Maszczak, and D. Shia, "Overview of Bellcore METROCORE network," presented at IFIP WG6.4 Workshop HSLAN'88, Liege, Belgium, Apr. 1988.

[12] A. Ippoliti, H. Izadpanah, and A. Albanese, "2.4 Gbit/s optical communications network with 16:1 and 1:16 Mux/Demux," to be published.

[13] M. P. Vecchi, R. M. Bulley, M. S. Goodman, H. Kobrinski, and C. A. Brackett, "High-bit-rate measurements in the LAMBDANET multiwavelength optical star network," in *Proc. Opt. Fiber Commun. Conf., WO2*, New Orleans, LA, Jan. 1988, p. 95.

[14] I. Mito, "Recent advances in frequency tunable DFB–DBR lasers," in *Proc. Opt. Fiber Commun. Conf., THK1*, New Orleans, LA, Jan. 1988, p. 176.

[15] J. M. Cooper, J. Dixon, M. S. Goodman, H. Kobrinski, M. P. Vecchi, E. Arthurs, S. G. Menocal, M. Tur, and S. Tsuji, "Nanosecond wavelength switching with a double-section distributed feedback laser," in *Proc. Tech. Digest CLEO'88, WA4*, Anaheim, CA, 1988, pp. 166–169.

[16] T. L. Koch, U. Koren, and B. I. Miller, "Broadly-tunable dynamic-single-mode multiple-quantum-well semi-insulated-blocked distributed-Bragg-reflector lasers at 1.5 μm," presented at Topic. Meet. Integrated Guided Wave Opt., PD8-1, Santa Fe, NM, Mar. 1988.

[17] H. Kawaguchi, K. Magari, K. Oe, Y. Noguchi, Y. Nakano, and G. Motosugi, "Optical frequency-selective amplification in a distributed feedback type semiconductor laser amplifier," *Appl. Phys. Lett.*, vol. 50, pp. 66–67, 1987.

[18] H. Nakajima, "Wavelength selective distributed feedback laser amplifier," in *Proc. 13th Europe. Conf. Opt. Commun.*, Helsinki, Finland, Sept. 1987, pp. 121–124.

[19] K. Magari, H. Kawaguchi, K. Oe, Y. Nakano, and M. Fukuda, "Optical signal selection with a constant gain and a gain bandwidth by a multielectrode distributed feedback laser amplifier," *Appl. Phys. Lett.*, vol. 51–52, pp. 1974–1976, 1987.

[20] M. P. Vecchi, H. Kobrinski, E. L. Goldstein, and R. M. Bulley, "Wavelength selection with nanosecond switching times using distributed-feedback optical amplifiers," presented at 14th Europe. Conf. Opt. Commun., Brighton, England, Sept. 1988, to be published.

[21] E. L. Goldstein, H. Kobrinski, M. P. Vecchi, and R. M. Bulley, "Nanosecond wavelength selection with 10 dB net gain in a subthreshold DFB laser amplifier," presented at 11th IEEE Int. Semiconductor Laser Conf., Boston, MA, Aug. 1988, to be published.

[22] F. Heismann, L. L. Buhl, and R. C. Alferness, "Electro-optically tunable narrowband Ti:LiNbO₃ wavelength filter," *Electron. Lett.*, vol. 23, p. 572, 1987.

[23] J. I. Capetanakis, "Generalized TDMA: The multi-access tree protocol," *IEEE Trans. Commun.*, vol. COM-27, pp. 1476–1484, 1979.

[24] J. F. Hayes, "An adaptive polling technique for local distribution," *IEEE Trans. Commun.*, vol. COM-26, pp. 1178–1186, 1978.

[25] J. I. Capetanakis, "Tree algorithms for packet broadcast channels," *IEEE Trans. Inform. Theory*, vol. IT-25, pp. 505–515, 1979.

[26] N. D. Vvedenskaya and B. S. Tsybakov, "Packet delay for multiaccess stack-algorithm," *Prob. Peredachi Inform.*, vol. 20, pp. 89–101, 1984.

Edward Arthurs received the S.B. and Sc.D. degrees from the Massachusetts Institute of Technology, Cambridge, in 1951 and 1955, respectively.

From 1955 to 1956 he was with RCA, Camden, NJ; from 1956 to 1957 with Hermes, Cambridge, MA; from 1957 to 1962 with M.I.T.; and from 1962 to 1984 with Bell Laboratories, Murray Hill, NJ. Since 1984 he has been with Bell Communications Research, Morristown, NJ, and Columbia University, New York, NY. He has been engaged in theoretical research in computer and communications science. He is coauthor of *A Computer and Communications Network Performance Primer* (Englewood Cliffs, NJ: Prentice-Hall, 1985).

Matthew S. Goodman received his undergraduate education at Indiana University in Bloomington receiving the B.S. degree in physics in 1969. He attended The Johns Hopkins University, Baltimore, MD, for the M.A. (1971) and Ph.D. (1977) degrees in physics.

He did postdoctoral research in elementary particle physics at Harvard University, Cambridge, MA, and was Assistant Professor of Physics there until September 1985 when he joined Bell Communications Research, Morristown, NJ. He is currently engaged in optical network research at Bellcore and fundamental principles of optical communication. His research has centered on applications of multiwavelength optical networks, multiwavelength switching, and quantum optics. He is also the author of more than 45 scientific and technical articles.

Mario P. Vecchi received the B.S. degree in electrical engineering from Cornell University in 1969 and the Ph.D. degree from M.I.T., in 1975.

He was a Researcher in the Semiconductors Laboratory at the Instituto Venezolano de Investigaciones Cientificas, and Professor at the Universidad Metropolitana of Caracas. He also was a Visiting Scientist at IBM Research, Yorktown Heights, NY. He was the founder and Managing Director of XYNERTEK C. A. from 1984 to 1986, the first company to design and manufacture microcomputer systems in Venezuela. He joined Bell Communications Research in 1986. He has published extensively, supervised several graduate and undergraduate thesis students, and has presented numerous contributions in international technical conferences.

Dr. Vecchi is member of Tau Beta Pi, Eta Kappa Nu, Sigma Xi, and Colegio de Ingenieros de Venezuela.

Haim Kobrinski (M'84) received the B.S. degree in physics from Tel-Aviv University, Israel, in 1977, the M.S. degree in applied physics from the Weizmann Institute of Science in Rehovot, Israel, in 1980, and the Ph.D. degree in electrical engineering and physics from Stevens Institute of Technology, NJ, in 1983.

He joined Bell Communications Research in April 1984, working in the Systems Technology Research Division. His particular interests are in design and implementation of optical network.

Bibliography

Chapter 1: Network Architecture

ABRA84   Abrams, M., and Cotton, I., *Computer Networks: A Tutorial*, IEEE Computer Society Press, Los Alamitos, Calif., 1984.

ASCH86   Aschenbrenner, J., "Open Systems Interconnection," *IBM Systems J.*, No. 3-4, 1986.

ATKI80   Atkins, J., "Path Control: The Transport Network of SNA," *IEEE Trans. Comm.*, Apr. 1980.

BURG83   Burg, F., "Design Considerations for Using the X.25 Layer in Data Terminal Equipment," *Proc. Infocom '83*, 1983.

CERF83   Cerf, V., and Cain, E., "The DOD Internet Architecture Model," *Computer Networks*, Oct. 1983.

CHAP82   Chapin, A., "Connectionless Data Transmission," *Computer Comm. Review*, Apr. 1982.

CHAP83   Chapin, A., "Connections and Connectionless Data Transmission," *Proc. IEEE*, Dec. 1983.

CLAR85   Clark, G., and Wong, M., "Verifying Conformance to the X.25 Standard," *Data Comm.*, Apr. 1985.

COLL83   Collie, B., Kayser, L., and Rybczynski, A., "Looking at the ISDN Interfaces: Issues and Answers," *Data Comm.*, June 1983.

CONA83   Conard, J., "Services and Protocols of the Data Link Layer," *Proc. IEEE*, Dec. 1983.

CYPS78   Cypser, R., *Communications Architecture for Distributed Systems*, Addison-Wesley, Reading, Mass., 1978.

DAVI79   Davis, D., Barber, D., Price, W., and Solomonides, C., *Computer Networks and Their Protocols*, Wiley, New York, 1979.

DAY83   Day, J., and Zimmermann, H., "The OSI Reference Model," *Proc. IEEE*, Dec. 1983.

DEAT82   Deaton, G., "OSI and SNA: A Perspective," *J. Telecommunication Networks*, Fall 1982.

DECI82   Decina, M., "Progress Towards User Access Arrangement in Integrated Services Digital Networks," *IEEE Trans. Comm.*, Sept. 1982.

DHAS86   Dhas, C., and Konangi, V., "X.25: An Interface to Public Packet Networks," *IEEE Comm. Magazine*, Sept. 1986.

DORR83   Dorros, I., "Telephone Nets Go Digital," *IEEE Spectrum*, Apr. 1983.

FARO86   Farowich, S., "Communicating in Technical Office," *IEEE Spectrum*, Apr. 1986.

GERL84   Gerla, M., and Pazos-Rangel, R., "Bandwidth Allocation and Routing in ISDN's," *IEEE Comm. Magazine*, Feb. 1984.

GRAY72   Gray, J., "Line Control Procedures," *Proc. IEEE*, Nov. 1972.

GREE80   Green, P., "An Introduction to Network Architectures and Protocols," *IEEE Trans. Comm.*, Apr. 1980.

HEMR84   Hemrick, C., "The Internal Organization of the OSI Network Layer: Concepts, Applications, and Issues," *J. Telecommunication Networks*, Summer 1984.

KAMI86   Kaminski, M., "Protocols for Communicating in the Factory," *IEEE Spectrum*, Apr. 1986.

KONA83   Konangi, V., and Dhas, C., "Introduction to Network Architectures," *IEEE Comm. Magazine*, Oct. 1983.

LAM84   Lam, S., *Principles of Communication and Networking Protocols*, IEEE Computer Society Press, Los Alamitos, Calif., 1984.

LINI83   Linington, P., "Fundamentals of the Layer Service Definitions and Protocol Specifications," *Proc. IEEE*, Dec. 1983.

LIST83   Listanti, M., and Villani, F., "An X.25 Compatible Protocol for Packet Voice Communications," *Computer Comm.*, Feb. 1983.

MART81   Martin, J., *Computer Networks and Distributed Processing*, Prentice-Hall, Englewood Cliffs, N.J., 1981.

MCCL83   McClelland, F., "Services and Protocols of the Physical Layer," *Proc. IEEE*, Dec. 1983.

MEIJ82   Meijer, A., and Peters, P., *Computer Network and Architecture*, Computer Science Press, Rockville, MD, 1982.

POUZ78   Pouzin, L., and Zimmerman, H., "A Tutorial on Protocols," *Proc. IEEE*, Nov. 1978.

ROBE78   Roberts, L., "The Evolution of Packet Switching," *Proc. IEEE*, Nov. 1978.

ROSS83   Ross, M., "Circuit Versus Packet Switching," *Fundamentals of Digital Switching*, Plenum, New York, 1983.

ROTH81   Rothauser, E.H., Janson, P.A., and Mueller, H.R., "Meshed-Star Networks for Local Communication Systems," in *Local Networks for Computer Comm.*, A. West and P. Janson, eds., North-Holland, Amsterdam, 1981, pp. 25-41.

STAL88   Stalling, W., *Data and Computer Communications*, Second Edition, MacMillan, 1988.

TANE81 Tanenbaum, A., "Network Protocols," *Computing Surveys*, Dec. 1981.

TANE89 Tanenbaum, A., *Computer Networks*, Second Edition, Prentice-Hall, Englewood Cliffs, N.J., 1989.

VOEL86 Voelcker, J., "Helping Computers Communicate," *IEEE Spectrum*, Mar. 1986.

WABE82 Waber, K., "Considerations on Customer Access to ISDN," *IEEE Trans. Comm.*, Sept. 1982.

WARE83 Ware, C., "The OSI Network Layer: Standards to Cope with the Real World," *Proc. IEEE*, Dec. 1983.

WECK80 Wecker, S., "DNA: The Digital Network Architecture," *IEEE Trans. Comm.*, Apr. 1980.

YANO81 Yanoschak, V., "Implementing the X.21 Interface," *Data Comm.*, Feb. 1981.

ZIMM80 Zimmermann, H., "OSI Reference Model—The ISO Model of Architecture for Open System Interconnection," *IEEE Trans. Comm.*, Apr. 1980.

Chapter 2: Interconnection Networks

BENE62 Benes, V.E., "On Rearrangeable Three-Stage Connecting Networks," *Bell System Technical J.*, Sept. 1962.

BENE64 Benes, V.E., "Optimal Rearrangeable Multistage Connecting Networks," *Bell System Technical J.*, Vol. 43, July 1964, pp. 1641-1656.

CANT71 Cantor, D.G., "On Nonblocking Switching Networks," *Networks* Vol. 1, No. 4 (Winter), 1971.

CHEN81 Chen, P.Y., Lawrie, D.H., Yew, P.C., and Padua, D.A., "Interconnection Networks Using Shuffles," *Computer*, Vol. 14, Dec. 1981.

CLOS53 Clos, C., "A Study of Non-Blocking Switching Network," *Bell System Technical J.*, Vol. 32, Mar., 1953, pp. 406-424.

DIAS81 Dias, D.M., and Jump, J.R., "Packet Switching Interconnection Networks for Modular Systems," *Computer*, Vol. 14, Dec. 1981, pp. 43-53.

FENG81 Feng, T.Y., "A Survey of Interconnection Networks," *Computer*, Vol. 14, Dec. 1981.

GOKE73 Goke, L.R., and Lipovski, G.J., "Banyan Networks for Partitioning Multiprocessor Systems," *Proc. 1st Annual Int'l Symp. Computer Architecture*, Dec. 1973, pp. 21-28.

LANG76 Lang, T., and Stone H.S., "A Shuffle-Exchange Network with Simplified Control," *IEEE Trans. Computers*, Jan. 1976.

PARK82 Parker, D.S., and Raghavendra, "The Gamma Network: A Multiprocessor Interconnection Network with Redundant Paths," *Proc. 9th Symp. Computer Architecture*, 1982.

PATE79 Patel, J.H., "Processor-Memory Interconnections for Multiprocessors," *Proc. 6th Symp. Computer Architecture*, Apr. 1979, pp. 168-177.

REDD84 Reddy, S.M., and Kumar, V.P., "On Fault-Tolerant Multistage Interconnection Networks," *Proc. 1984 Int'l Conf. on Parallel Processing*, Aug. 1984.

SIEG81 Siegel, H.J., and McMillen, R.J., "The Multistage Cube: A Versatile Interconnection Network," *Computer*, Vol. 14, Dec. 1981.

WAKS68 Waksman, A., "A Permutation Network," *J. ACM*, Jan. 1968.

WU80 Wu, C., and Feng, T., "On a Class of Multistage Interconnection Networks," *IEEE Trans. Computers*, Aug. 1980.

Chapter 3: Experimental Architectures

AMST83 Amstutz, S.R., "Burst Switching—A Method for Dispersed and Integrated Voice and Data Switching," *IEEE Comm. Magazine*, Vol. 21, Nov. 1983, pp. 36-42. Also in *Proc. ICC '83*, June 1983, pp. 288-292.

BARB83 Barberies, G., et al., "Coded Speech in Packet Switched Networks: Models and Experiments," *IEEE J. Selected Areas in Comm.*, Dec. 1983.

BIAL80A Bailly, T., et al., "A Technique for Adaptive Voice Flow Control in Integrated Packet Networks," *IEEE Trans. Comm.*, Mar. 1980.

BIAL80B Bailly T., et al., "Voice Communications in Integrated Voice and Data Networks," *IEEE Trans. Comm.*, Sept. 1980.

BULL59 Bullington, K., and Fraser, J., "Engineering Aspects of TASI," *Bell System Technical J.*, Mar. 1959.

CAMP76 Campanella, S., "Digital Speech Interpolation," *COMSAT Technical Review*, Spring 1976.

COUD87 Coudreuse, J.P., and Servel, M., "Prelude: An Asynchronous Time-Division Switched Network," *Proc. ICC '87*, Seattle, Wash., June 1987, pp. 769-773.

COVI75 Coviello, G., and Vena, P., "Integration of Circuit/Packet Switching in a SENET (Slotted Envelope Network) Concept," *NTC75*.

COVI79 Coviello, G., "Comparative Discussion of Circuit vs. Packet Switched Voice," *IEEE Trans. Comm.*, Aug. 1979.

DIEU87 Dieudonne, M., and Quinquis, M., "Switching Techniques for Asynchronous Time Division Multiplexing for Fast Packet Switching." *Proc. ISS '87*, Phoenix, Ariz., Mar. 1987, pp. 367-371.

FORG77 Forgie, J., and Nemeth, A., "An Efficient Packetized Voice/Data Network Using Statistical Flow Control," *Proc. ICC*, June 1977.

GERK87 Gerke, P., "Fast Packet Switching—A Principle for Future System Generations?" *Proc. ISS '87*, 1987.

GITM81 Gitman, I., et al., "Analysis and Design of Hybrid Switching Networks," *IEEE Trans. Comm.*, Sept. 1981.

GOLD77 Gold, B., "Digital Speech Networks," *Proc. IEEE*, Dec. 1977.

GONE86A Gonet, P., "Fast Packet Approach to Integrated Broadband Networks," *Computer Comm.*, Dec. 1986, pp. 292-298.

GONE86B Gonet, P., Adams, P., and Coudreuse, J.P., "Asynchronous Time-Division Switching: The Way to Flexible Broadband Communication Networks," *Proc. Int'l Zurich Seminar Digital Comm.*, 1986, pp. 141-148.

GOPA87   Gopal, I.S., Cidon, I., and Meleis, H., "Paris: An Approach to Integrated Private Networks," *Proc. ICC '87*, Seattle, Wash., June 1987, pp. 764-773.

GREE84   Green, Jr., P., "Computer Communications: Milestones and Prophecies," *IEEE Comm. Magazine*, May 1984.

HASE83A   Haselton, E.F., "A PCM Frame Switching Concept Leading to Burst Switching Network Architecutre," *IEEE Comm. Magazine*, Vol. 21, Sept. 1983, pp. 13-19. Also in *Proc. ICC '83*, Boston, Mass., June 1983, pp. 1401-1406.

HASES83B   Haselton, E., "A PCM Frame Switching Concept Leading to Burst Switching Network Architecutre," *IEEE Comm. Magazine*, Sept. 1983.

HAUG84   Haugney, J., "Application of Burst-Switching Technology to the Defense Communication System," *IEEE Comm. Magazine*, Oct. 1984.

HEGG84   Heggestad, H., "An Overview of Packet-Switching Communications," *IEEE Comm. Magazine*, Apr. 1984.

HUBES87   Huber, J.F., and Mair, E., "A Flexible Architecture for Small and Large Packet Switching Networks," *Proc. ISS '87*, Phoenix, Ariz., Mar. 1987, pp. B10.4.1-B10.4.6.

JAYA86   Jayant, N.S., "Coding Speech at Low Bit Rates," *IEEE Spectrum*, Aug. 1986.

JOEL83   Joel, Jr., A., "Circuit-Switching Fundamentals," *Fundamentals of Digital Switching*, Plenum, New York, 1983.

KERM79A   Kermani, P., and Kleinrock, L., "A Trade-Off Study of Switching Systems," *Proc. ICC '79*, 1979.

KERM79B   Kermani, P., and Kleinrock, L., "Virtual Cut-through: A New Computer Communication Switching Technique," *Computer Networks*, Vol. 3, 1979, pp. 267-286.

KIRT87   Kirton, P., et al., "Fast Packet Switching for Integrated Network Evolution," *Proc. ISS '87*, 1987.

KULZ84   Kulzer, J.J., and Montgomery, W.A., "Statistical Switching Architecture for Future Services," *Proc. ISS '84*, Florence, Italy, May 1984, pp. 43A.1.1-43A.1.6.

KUMM78   Kummerle, Rudin, H., "Packet and Circuit Switching: Cost Performance Boundaries," *Computer Networks*, 1978.

LUDE87A   Luderer, G., et al., "Wideband Packet Technology for Switching Systems," *Proc. ISS '87*, 1987.

LUDE87B   Luderer, G.W.R., et al., "Wideband Packet Technology for Switching Systems," *Proc. ISS '87*, Phoenix, Ariz., Mar. 1987, pp. 448-454.

MUIS86A   Muise, R., et al., "Experiments in Wideband Packet Technology," *Proc. Int'l Zurich Seminar Digital Comm.*, 1986.

MUIS86B   Muise, R.W., Shonfeld, T.J., and Zimmerman, G.H., "Digital Communications Experiments in Wideband Packet Technology," *Proc. Int'l Zurich Seminar Digital Comm.*, 1986, pp. 135-140.

RETT79   Rettberg, R., et al., "Development of Voice Funnel System: Design Report," Rep. 4098, Bolt, Beranek and Newman, Inc., Aug. 1979.

ROSS80   Ross, M.J., et al., "An Architecture for a Flexible Integrated Voice/Data Network," *Proc. ICC '80*, June 1980, pp. 21.6.1-21.6.5.

RUDI78   Rudin, H., "Studies on the Integration of Circuit and Packet Switching," *Proc. ICC '78*, 1978.

STAN87   Standish, W., and Sistla, S., "Network Switching in the 1990's," *Proc. ISS '87*, 1987.

SUZU86   Suzuki, H., Takeuchi, T., and Yamaguchi, T., "Very High Speed and High Capacity Packet Switching for Broadband ISDN," *Proc. ICC '86*, Toronto, Canada, June 1986, pp. 749-754.

TAKA87   Takami, K., and Takenaka, T., "Architectural and Functional Aspects of a Multi-Media Packet Switched Network," *Proc. ISS '87*, 1987.

TAKE86   Takieuci, T., et al., "An Experimental Synchronous Composite Packet Switching System," *Proc. Int'l Zurich Seminar Digital Comm.*, 1986, pp. 149-153.

THOM84   Thomas, A., Coudreuse, J.P., and Servel, M., "Asynchronous Time Division Techniques: An Experimental Packet Network Integrating Video Communication," *Proc. ISS '84*, Paper 32C2, Florence, Italy, May 1984.

TURN83   Turner, J.S., and Wyatt, L.F., "A Packet Network Architecture for Integrated Services," *Proc. Globecom '83*, San Diego, Calif., Nov. 1983, pp. 2.1.1-2.1.6.

TURN85   Turner, J.S., "Design of an Integrated Services Packet Network, *9th ACM Data Comm. Symp.*, Sept. 1985, pp. 124-133.

TURN86   Turner, J., "New Directions in Communications (Or Which Way to the Information Age?)," *IEEE Comm. Magazine*, Oct. 1986.

VAID88   Vaidya, A.K., and Pashan, M.A., "Technology Advances in Wideband Packet Switching," *Proc. Globecom '88*, Hollywood, Fla., Nov. 1988, pp. 668-671.

WEIN80   Weinstein, C., and Forgie, J., "Experience with Speech Communications in Packet Network," *IEEE J. Selected Areas in Comm.*, Dec. 1983.

Chapter 4: Switch Fabric Design and Analysis

AOYA81   Aoyama, T., et al., "Packetized Service Integration Network for Dedicated Voice/Data Subscribers," *IEEE Trans. Comm.*, Nov. 1981.

BATC68   Batcher, K.E., "Sorting Networks and Their Applications," *Proc. Spring Joint Computer Conf.*, AFIPS, 1968, pp. 307-314.

BROA86   "Broad-band Communications Systems," *IEEE J. Selected Areas in Comm.*, Oct. 1987.

DAY87   Day, C., Giacopelli, J., and Hickey, J., "Application of Self-Routing Switches to LATA Fiber Optic Networks," *Proc. ISS '87*, 1987.

DEPR87   De Prycker, M., and De Somer, M., "Performance of a Service Independent Switching Network with Distributed Control," *IEEE J. Selected Areas in Comm.*, Vol. SAC-5, Oct. 1987, pp. 1293-1301.

DIEU87   Dieudonne, M., and Quinquis, M., "Switching Techniques for Asynchronous Time Division Multiplexing (or Fast Packet Switching)," *Proc. ISS '87*, 1987.

GONE86   Gonet, P., Adam, P., and Coudreuse, J., "Asynchronous Time Division Switching: The Way to Flexible Broadband Communication Networks," *Proc. Int'l Zurich Seminar Digital Comm.*, 1986.

HLUC88   Hluchyj, M.G., and Karol, M., "Queueing in Space-Division Packet Switching," *Proc. Infocom '88*, New Orleans, La., Mar. 1988, pp. 334-343.

HUAN86   Huang, A., "The Relationship Between STARLITE, A Wideband Digital Switch and Optics," *Proc. ICC '86*, Toronto, Canada, June 1986, pp. 1725-1729.

HUI87   Hui, J., "A Broadband Packet Switch for Multi-Rate Services," *Proc. ICC '87*, 1987.

ICHI87   Ichikawa, H., et al., "High-Speed Packet Switching Systems for Multimedia Communications," *IEEE J. Selected Areas in Comm.*, Oct. 1987.

LANG86   Langer, K., Lukanek, G., Vathke, J., and Walf, G., "Broadband Switching Network and TV Switching for 70 Mbps," *Proc. Int'l Zurich Seminar Digital Comm.*, 1986.

LEBR86   LeBris, J., and Servel, M., "Integrated Wideband Networks Using Asynchronous Time Division Techniques," *Proc. ICC '86*, 1986.

LINN86   Linnel, L., "A Wideband Local Access System Using Emerging-Technology-Components," *IEEE J. Selected Areas in Comm.*, July 1986.

REIN86   Reiner, H., "Integrated Circuits for Broadband Communications Systems," *IEEE J. Selected Areas in Comm.*, July 1986.

SCHA86   Schaffer, B., "Switching in the Broad-Band ISDN," *IEEE J. Selected Areas in Comm.*, July 1986.

SUZU86   Suzuki, H., et al., "Very High Speed and High Capacity Packet Switching for Broadband ISDN," *Proc. ICC '86*, 1986.

SWIT87   "Switching Systems for Broad-Band Networks," *IEEE J. Selected Areas in Comm.*, Oct. 1987.

THOM84   Thomas, A., Coudreuse, J.P., and Servel, M., "Asynchronous Time-Division Techniques: An Experimental Packet Network Integrating Video Communication," *Int'l Switching Symp.*, May 1984.

TSUD79   Tsuda, T., et al., "An Approach to Multiservice Subscriber Loop System Using Packetized Voice/Data Terminals," *IEEE Trans. Comm.*, July 1979.

VAN86   van Baardewijk, J., "An Experimental all-in-one Broadband Switch," *Proc. Int'l Zurich Seminar Digital Comm.*, 1986.

WU86   Wu, L.T., and Huang, N.C., "Synchronous Wideband Network—An Interoffice Facility Hubbing Network," *Proc. Int'l Zurich Seminar Digital Comm.*, 1986.

YAMA86   Yamada, H., Kataoka, H., Sampei, T., and Yano, T., "High-Speed Digital Switching Technology Using Space-Division-Switching LSI's," *IEEE J. Selected Areas in Comm.*, July 1986.

YOON88   Yoon, H., Liu, M.T., and Lee, K.Y., "The Knockout Switch Under Nonuniform Traffic," *Proc. Globecom '88*, Hollywood, Fla., Nov. 1988, pp. 1628-1634.

Chapter 5: Switch Architectures

AHMA88   Ahmadi, H., et al., "A High-performance Switch Fabric for Integrated Circuit and Packet Switching," *Proc. Infocom '88*, New Orleans, La., Mar. 1988, pp. 9-18.

DAY87   Day, C.M., Giacopelli, J.N., and Hickey, J., "Applications of Self-routing Switching to LATA Fiber Optic Networks," *Proc. ISS '87*, Phoenix, Ariz., Mar. 1987, pp. 519-523.

DEPR87   De Prycker, M., and Bauwens, J., "A Switching Exchange for an Asynchronous Time Division Based Network," *Proc. ICC '87*, Seattle, Wash., June 1987, pp. 774-781.

ENG87   Eng, K.Y., Hluchyj, M.G., and Yeh, Y.S., "A Knockout Switch for Variable-Length Packets," *IEEE J. Selected Areas in Comm.*, Vol. SAC-5, Dec. 1987, pp. 1426-1435.

HUBE88   Huber, M.N., Rathgeb, E.P., and Theimer, T.H., "Self Routing Banyan Networks in an ATM-Environment," *Proc. ICCC '88*, Tel Aviv, Israel, Oct. 1988, pp. 167-174.

HUI87   Hui, J.Y., and Arthurs, E., "A Broadband Packet Switch for Integrated Transport," *IEEE J. Selected Areas in Comm.*, Vol. SAC-5, Oct. 1987, pp. 1264-1273.

NOJI87   Nojima, S., et al., "Integrated Services Packet Network Using Bus Matrix Switch," *IEEE J. Selected Areas in Comm.*, Vol. SAC-5, Oct. 1987, pp. 1284-1292.

PIER72   J.R. Pierce, "Network for Block Switching of Data," *Bell System Technical J.*, July-Aug. 1972.

RATH88   Rathgeb, E.P., Theimer, T.H., and Huber, M.N., "Buffering Concepts for ATM Switching Networks," *Proc. Globecom '88*, Hollywood, Fla., Nov. 1988, pp. 1277-1281.

SINC87   Sincoskie, W.D., "Frontiers in Switching Technology: Part Two: Broadband Packet Switching," *Bellcore Exchange*, Nov.-Dec. 1987, pp. 22-27.

WU86   Wu, L.T., and Huang, N.C., "Synchronous Wideband Network—An Interoffice Facility Hubbing Network," *Proc. Int'l Zurich Seminar Digital Comm.*, 1986, pp. 33-39.

Chapter 6: Broadcast Switching Networks

BUBE85   Bubenik, R., "Performance Evaluation of a Broadcast Packet Switch," MS Thesis, Computer Science Dept., Washington Univ., Aug. 1985.

CHAN84   Chang, J.M., and Maxemchuk, N.F., "Reliable Broadcast Protocols," *ACM Trans. Computer Systems*, Aug. 1984.

CHER85   Cheriton, D.R., and Deering, S.E., "Host Groups: A Multicast Extension for Datagram

Internetworks," *Proc. 9th Data Comm. Symp.*, Sept. 1985.

DALA78   Dalal, Y.K., and Metcalf, R.M., "Reverse Path Forwarding of Broadcast Packets," *Comm. ACM*, Dec. 1978, pp. 1040-1047.

ENG88   Eng, K.Y., Hluchyj, M.G., and Yeh, Y.S., "Multicast and Broadcast Services in a Knockout Packet Switch," *Proc. Infocom '88*, New Orleans, La., Mar. 1988, pp. 29-34.

HUAN84   Huang, A., and Knauer, S., "Starlite: A Wideband Digital Switch," *Proc. Globecom '84*, Dec. 1984, pp. 121-125.

LEE88   Lee, T.T., et al., "The Architecture of a Multicast Broadband Packet Switch," *Proc. Infocom '88*, New Orleans, La., 1988.

PANT89   Pant, R., "X.25 Broadcast Service," *IEEE Network Magazine*, July 1989.

RICH85   Richards, G.W., and Hwang, F.K., "A Two-stage Rearrangeable Broadcast Switching Network," *IEEE Trans. Comm.* Oct. 1985, pp. 1025-1035.

Chapter 7: Bandwidth Allocation, Flow Control, and Congestion Control

BHAR81   Bharat-Kumar, K., and Jaffe, J.M., "A New Approach to Performance-Oriented Flow Control," *IEEE Trans. Comm.*, Vol. COM-29, No. 4, Apr. 1981, pp. 427-435.

BUX85   Bux, W., and Grillo, D., "Flow Control in Local-Area Networks of Interconnected Token Rings," *IEEE Trans. Comm.*, Vol. COM-33, No. 10, Oct. 1985, pp. 1058-1066.

CHIU89   Chiu, D.M., and Jain, R., "Analysis of Increase and Decrease Algorithms for Congestion Avoidance in Computer Networks," *Comp. Networks and ISDN Syst.*, Vol. 17, 1989, pp. 1-14.

COHE80   Cohen, D., "Flow Control for Real-Time Communication," *Computer Comm. Review*, Vol. 10, No. 1-2, Jan.-Apr. 1980, pp. 41-47.

DAVI72   Davis, D.W., "The Control of Congestion in Packet-Switching Networks," *IEEE Trans. Comm.*, Vol. COM-20, No. 6, June 1972.

DECI90   Decina, M., and Toniatti, T., "On Bandwidth Allocation to Bursty Virtual Connections in ATM Networks," *Proc. ICC '90*, Atlanta, Ga., 1990.

DOSH88   Doshi, B.T., and Nguyen, H.Q., "Congestion Control in ISDN Frame-Relay Networks," *AT&T Technical J.*, Nov.-Dec. 1988, pp. 35-46.

ECKB89   Eckberg, A.E., Luan, D.T., and Lucantoni, D.M., "Bandwidth Management: A Congestion Control Strategy for Broadband Packet Networks—Characterizing the Throughput-Burstiness Filter," *Int'l Teletraffic Congress Special Seminar*, Adelaide, Australia, 1989.

GALL89A   Gallasi, G., Rigolio, G., and Fratta, L., "ATM: Bandwidth Assignment and Bandwidth Enforcement Policies," *Proc. Globecom*, Dallas, Texas, 1989.

GALL89B   Gallasi, G., Rigolio, G., Vacceri, P., and Verri, L., "Resource Allocation in ATM Network,"

3rd RACE 1022 TC Workshop, Paris, France, Oct. 1989.

GERL80   Gerla, M., and Kleinrock, L., "Flow Control" A Comparative Survey," *IEEE Trans. Comm.*, Vol. COM-28, No. 4, Apr. 1980, pp. 553-574.

GERL85   Gerla, M., Chan, H.W., and de Marca, J.R.B., "Fairness in Computer Networks," *Proc. ICC '85*, Chicago, Ill., June 23-26, 1985, pp. 43.5.1-6.

GEOR82   George, F.D., and Young, G.E., "SNA Flow Control: Architecture and Implementation," *IBM Systems J.*, Vol. 21, No. 2, 1982, pp. 179-210.

HAHN86   Hahne, E.L., and Gallager, R.G., "Round Robin Scheduling for Fair Flow Control in Data Commun. Networks," *Proc. ICC '86*, Toronto, Canada, June 22-25, 1986, pp. 4.3.1-5.

HEFF86   Heffes, H., and Lucantoni, M., "A Markov Modulated Characterization of Packetized Voice and Data Traffic and Related Multiplexer Performances," *IEEE J. Selected Areas in Comm.*, 1986.

HIRA88   Hirano, M., and Watanabe, N., "Characteristics of a Cell Multiplexer for Bursty ATM Traffic," *Proc. ICC '86*, Boston, Mass., 1988.

IRLA78   Irland, M., "Buffer Management in a Packet Switch," *IEEE Trans. Comm.*, Vol. COM-26, Mar. 1978, pp. 328-337.

JACO88   Jacobson, V., "Congestion Avoidance and Control," *Proc. ACM SIGCOMM '88*, Stanford, Calif., Aug. 1988., pp. 314-329.

JAFF81   Jaffe, J.M., "Bottleneck Flow Control," *IEEE Trans. Comm.*, Vol. COM-29, No. 7, July 1981, pp. 954-962.

JAIN86   Jain, R., "A Timeout-Based Congestion Control Scheme for Window Flow-Controlled Networks," *IEEE J. Selected Areas in Comm.*, Vol. SAC-4, No. 7, Oct. 1986, pp. 1162-1167.

JAIN88A   Jain, R., and Ramakrishnan, K.K., "Congestion Avoidance in Computer Networks with a Connectionless Network Layer: Concepts, Goals and Methodology," *Proc. IEEE Comp. Networking Symp.*, Washington, D.C., Apr. 1988, pp. 134-143.

JAIN88B   Jain, R., Ramakrishnan, K.K., and Chiu, D.M., "Congestion Avoidance in Computer Networks with a Connectionless Network Layer," in *Innovations in Internetworking*, C. Partridge, ed., Artech House, Norwood, Mass., 1988, pp. 140-156.

JAIN89   Jain, R., "A Delay-Based Approach for Congestion Avoidance in Interconnected Heterogeneous Computer Networks," *Computer Comm. Review*, Vol. 19, No. 5, Oct. 1989, pp. 56-71.

JOSS89   Joss, P., and Verbiest, W., "A Statistical Bandwidth Allocation and Usage Monitoring Algorithm for ATM Networks," *Proc. ICC '89*, Boston, Mass., June 1989, pp. 415-422.

LAM81   Lam, S.S., and Lien, Y.C.L., "Congestion Control of Packet Communication Networks by Input Buffer Limits—A Simulation Study," *IEEE Trans. Computers*, Vol. C-30, No. 10, Oct. 1981, pp. 733-742.

LUTT84 Lutton, J., and Roberts, J., "Traffic Performance on Multi-Slot Call Routing Strategies in an Integrated Services Digital Network," *Proc. ISS '84*, Florence, Italy, 1984.

MAJI79 Majithia, J.C., et al., "Experiments in Congestion Control Techniques," *Proc. Int'l Symp. Flow Control Computer Networks*, Versailles, France, Feb. 1979, pp. 211-234.

MAXE90 Maxemchuk, N.F., and Zarki, M.E., "Routing and Flow Control in High Speed Wide Area Networks," *Proc. IEEE*, Vol. 78, No. 1, Jan. 1990, pp. 204-221.

NAGL87 Nagle, J., "On Packet Switches with Infinite Storage," *IEEE Trans. Comm.*, Vol. COM-35, No. 4, Apr. 1987, pp. 435-438.

NEIS90 Neistegge, G., "The Leaky Bucket Policing Method," *Int'l J. of Digital and Analog Cabled Systems*, June 1990.

RAMA87 Ramakrishnan, K.K., Chiu, D.M., and Jain, R., "Congestion Avoidance in Computer Networks with a Connectionless Network Layer. Part IV: A Selective Binary Feedback Scheme for General Topologies," Tech. Report DEC-TR-510, Digital Equipment Corp., Aug. 1987.

RAMA88 Ramakrishnan, K.K., and Jain, R., "An Explicit Binary Feedback Scheme for Congestion Avoidance in Computer Networks with a Connectionless Network Layer," *Proc. ACM SIGCOMM '88*, Stanford, Calif., Aug. 1988, pp. 303-313.

RATH89 Rathgeb, E., "Comparison of Policing Mechanisms for ATM Networks," *3rd RACE 1022 TC Workshop*, Paris, France, Oct. 1989.

SCHW82 Schwartz, M., "Performance Analysis of the SNA Route Pacing Control," *IEEE Trans. Comm.*, Vol. COM-30, No. 1, Jan. 1982, pp. 172-184.

TURN88 Turner, J., "Buffer Management System," WUCS Internal Report, June 1988.

Chapter 8: Performance Modeling

ARGA83 Agrawal, D.P., "Graph Theoretical Analysis and Design of Multistage Interconnection Network," *IEEE Trans. Computers*, July 1983.

BASK76 Baskett, F., and Smith, A.J., "Interference in Multiprocessor Computer Systems with Interleaved Memory," *Comm. ACM*, June 1976.

BHAN75 Bhandarker, D.P., "Analysis of Memory Interference in Multiprocessors," *IEEE Trans. Computers*, Sept. 1975.

CIDO88 Cidon, I, Gopal, I.S., and Kutten, S., "New Models and Algorithms for Future Networks," *Proc. ACM Principles Distributed Computing*, Toronto, Canada, 1988.

DIAS81 Dias, D.M., and Jump, J.R., "Analysis and Simulation of Buffered Delta Networks," *IEEE Trans. Computers*, Vol. C-30, Apr. 1981, pp. 273-282.

FISH80 Fisher, M.J., "Delay Analysis of TASI with Random Fluctuations in the Number of Voice Calls," *IEEE Trans. Comm.*, Nov. 1980.

GROS74 Gross, D., and Harris, C.M., *Fundamentals of Queueing Theory*, Wiley, New York, 1974.

JAYA80 Jayant, N.S., "Effects of Packet Losses in Waveform-Coded Speech," *Proc. ICCC '80*, 1980.

KARO87 Karol, M.J., Hluchyj, M.G., and Morgan, S.P., "Input Versus Output Queueing on a Space-Division Packet Switch," *IEEE Trans. Comm.*, Vol. COM-35, Dec. 1987, pp. 1347-1356.

KLEI75 Kleinrock, L., *Queueing Systems, Vol. 1: Theory*, Wiley, New York, 1975.

KRUS83 Kruskal, C.P., and Snir, M., "The Performance of Multistage Interconnection Networks for Multiprocessors," *IEEE Trans. Computers*, Dec. 1983.

KUHL83 Kuhl, D.F., "Error Recovery Protocols: Link by Link vs. Edge to Edge," *Proc. Infocom '83*, 1983.

KUMA86 Kumar, M., and Jump, J.R., "Performance of Unbuffered Shuffle-Exchange Networks," *IEEE Trans. Computers*, Vol. C-35, June 1986, pp. 573-577.

LI88 Li, S.Y.R., "Theory of Periodic Contention and its Application to Packet Switching," *Proc. Infocom '88*, New Orleans, La., Mar. 1988, pp. 320-325.

LIM86 Lim, Y., "Performance Analysis of an Integrated Voice and Data TDM Link with Digital Speech Interpolation," *Proc. Globecom '86*, 1986.

MORS83 Morse, J.D., and Kopec, Jr., S.J., "Performance Evaluation of a Distributed Burst Switched Communication System," *Proc. Phoenix Conf. Computer Comm.*, Mar. 1983.

PATE81 Patel, J.H., "Performance of Processor-Memory Interconnections for Multiprocessors," *IEEE Trans. Computers*, Vol. C-30, Oct., 1981, pp. 771-780.

ROSS77 Ross, M.J., et al., "Design Approaches and Performance Criteria for Integrated Voice/Data Switching," *Proc. IEEE*, Vol. 65, Sept. 1977, pp. 1283-1295.

SUDA83 Suda, T., et al., "Performance Evaluation of a Packetized Voice System—Simulation Study," *Proc. ICC '83*, 1983.

WEIN79 Weinstein, C.J., "The Tradeoff Between Delay and TASI Advantage in a Packetized Speech Multiplexer," *IEEE Trans. Comm.*, Nov. 1979.

Chapter 9: Photonic Switching Systems

ACAM87 Acampora, A.S., "Multichannel Multihop Local Lightwave Network," *Proc. Globecom '87*, 1987, pp. 1459-1467.

ACAM88 Acampora, A.S., et al., "Multihop Lightwave Networks: A New Approach to Achieve Terabit Capabilities," *Proc. ICC '88*, 1988, pp. 1478-1484.

ARTH87 Arthurs, E., et al., "A Fast Optical Crossconnect for Parallel Processing Computers," *Proc. 13th European Conf. Opt. Comm.*, Sept. 1987.

BACH86 Bachus, E.J., et al., "Ten-channel Coherent Optical Fiber Transmission," *Electron. Lett.*, Vol. 22, 1986, pp. 1002-1003.

BLAC84   Black, R.W., Stewart, W.J., and Bennion, I., "An Opto-electronic Exchange of the Future," *Proc. ISS '84*, 1984.

BLUM88   Blumenthal, D.J., et al., "Self-routing of 12.5 Data Through an 8x8 LiNbO$_3$ Optical Crossbar Switch Matrix with Gigahertz Switch Speed," *Proc. Opt. Fiber Comm. Conf.* 1988.

BRAC88   Brackett, C.A., "Dense WDM Networks," *Proc. 14th European Conf. Opt. Comm.*, Sept. 1988, pp. 533-538.

BULL87   Bulley, R.M., et al., "Experimental Demonstration of LAMBDANET: A Multiwavelength Optical Network," *Proc. 13th European Conf. Opt. Comm.*, Sept. 1987.

CASS88   Cassidy, S.A., and Yennadhiou, P., "Optimal Switching Architectures Using D-Fiber Optical Space Switches," *IEEE J. Selected Areas in Comm.*, Aug. 1988.

ENG87   Eng, K.Y., "A Photonic Knockout Switch for High Speed Packet Network," *IEEE/IEICE Global Telecom Conf.*, Dec. 1987, pp. 47.2.1-47.2.5.

FUJI88   Fujiwara, M., et al., "A Coherent Photonic Wavelength Division Switching System for Broadband Networks," *Proc. 14th European Conf. Opt. Comm.*, Sept. 1988, pp. 139-141.

GOOD86   Goodman, M.S., et al., "Application of Wavelength Division Multiplexing to Communication Network Architectures," *Proc. ICC '86*, Vol. 2, 1986, pp. 931-934.

GOOD87   Goodman, M.S., et al., "Design and Demonstration of the LAMBDANET System: A Multiwavelength Optical Network," *Proc. Globecom '87*, Nov. 1987.

HILL87   Hill, A.M., "Switching and Distribution Networks for Wideband Optical Signals," *Proc. ISS '87*, 1987.

HLUC88   Hluchyj, M.G., and Karol, M.J., "ShuffleNet: An Application of Generalized Perfected Shuffles to Multihop Lightwave Networks," *Proc. Infocom '88*, 1988.

KAMI87   Kaminow, I.P., et al., "FDM-FSK Star Network with a Tunable Optical Filter Demultiplexer," *Electron. Lett.*, Vol. 23, 1987, pp. 1102-1103.

KOBR87   Kobrinski, H., et al., "Demonstration of High Capacity in the LAMBDANET Architecture: A Wavelength Optical Network," *Electron. Lett.*, Vol. 23, 1987, p. 824.

NISH88   Nishio, M., et al., "Eight Channel Wavelength Division Switching Experiment using Tuning Range DFB LD Filters," *Proc. 14th European Conf. Opt. Comm.*, Sept. 1988, pp. 49-52.

PADM87   Padmanabhan, K., and Netravali, A., "Dilated Networks for Photonic Switching," *1987 Topical Meeting on Photonic Switching*, Reno, Nev., Mar. 1987.

PERS87   Personick, S.D., "Photonic Switch: Technology and Applications," *IEEE Comm.*, May 1987.

SHIM87   Shimazu, Y., Nishi, S., and Yoshikai, N., "Wavelength-division-multiplexing Optical Switch Using Acoustooptic Deflector," *J. Lightwave Technology*, Dec. 1987.

SPAN86   Spanke, R.A., "Architectures for Large Non-blocking Optical Switches," *IEEE Quantum Electron.*, Aug. 1986, pp. 885-889.

SPAN87   Spanke, R.A., "Architectures for Guided-wave Optical Space Switching Systems," *IEEE Comm.*, May 1987.

SPEC84   Special Issue on Architecture of Local Area Networks, *IEEE Comm.*, Aug. 1984.

STER87   Stern, J.R., et al., "Passive Optical Networks for Telephony Applications and Beyond," *Electron. Lett.*, Vol. 23, 1987, 1255-1257.

SU86   Su, S.F., Jou, L., and Lenart, J., "A Review on Classification of Optical Switching Systems," *IEEE Comm.*, May 1986.

TUCK87   Tucker, R.S., et al., "Optical Time-division Multiplexed Fiber Transmission Using Ti:LiNbO$_3$ Waveguide Switch/Modulators," *Proc. Globecom '87*, Nov. 1987.

VECC88   Vecchi, M.P., et al., "High Bit Rate Measurements in the LAMBDANET Multiwavelength Optical Star Network," *Proc. Optical Fiber Comm. Conf.*, Jan. 1988, p. 95.

References

Chapter 1: Network Architecture

LIDI88   Lidinsky, "A Motivation for MANs," Guest Editorial, *IEEE Communications Magazine*, April 1988.

IWAM87   Iwama, W., Timko, J.W., and Day, J.F., "AT&T ISDN Architecture," *International Switching Symposium 1987*, March 1987.

Chapter 2: Interconnection Networks

FENG81   Feng, T.Y., "A Survey of Interconnection Networks," *IEEE Computer Magazine*, December 1981.

DIAS81   Dias, D.M. and Jump, J.R., "Packet Switching Interconnection Networks for Modular Systems," *IEEE Computer Magazine*, December 1981.

Chapter 3: Experimental Architectures

KULZ84   Kulzer, J.J. and Montgomery, M.A., "Statistical Switching Architectures for Future Services," *Proceedings of the International Switching Symposium*, May 1984.

MALE88   Malek, M., "Integrated Voice and Data Communications Overview," *IEEE Communications Magazine*, June 1988.

Chapter 4: Switch Fabric Design and Analysis

TURN86   Turner, J., "New Directions in Communications (or Which Way to the Information Age)," *IEEE Communications Magazine*, October 1986.

FENN87   Fenner, P.D., et al., "Extending the ISDN Horizon," *International Switching Symposium 1987*, March 1987.

Chapter 7: Bandwidth Allocation, Flow Control, and Congestion Control

DOSH88   Doshi, B.T. and Nguyen, H.Q., "Congestion Control in ISDN Frame-Relay Networks," *AT&T Technical Journal*, November/December 1988.

DAVI79   Davis, D., et al., *Computer Networks and Their Protocols*, Wiely 1983.

TANE89   Tanenbaum, A., *Computer Networks*, Second Edition, Prentice-Hall, 1989.

Chapter 9: Photonic Switching Systems

PERS87   Personick, S.D., "Photonic Switching: Technology and Applications," *IEEE Communications Magazine* 1987.

Glossary

A

ADCCP – Advanced Data Communication Control Procedures

ADM – Augmented Data Manipulator

ADPCM – Adaptive Pulse Code Modulation

AI – Access Interface

ANR – Automatic Network Routing

ANSI – American National Standards Institute

Application Layer – Layer 7 of the OSI reference model. Provides an interface to the user.

APS – Automatic Protection Switching

ARPA – Advanced Research Projects Agency

ASR – Address Shift Register

ATD – Asynchronous Time Division

ATM – Asynchronous Transfer Mode

B

BBN – Broadcast Banyan Network

BCN – Broadcast Channel Number

BCSU – Broadband Circuit Switching Unit

BER – Bit-Error Rate

BGT – Broadcast and Group Translator

BH – Buried Heterojunction

BIP – Bit-Interleaved Parity

BISDN – Broadband Integrated Services Digital Network

BISYNC – Binary Synchronous Communication

Blocking – A network is said to be blocking if there exist connection sets that will prevent some additional desired connections from being set up between unused ports, even with rearrangement of existing connections.

BM – Buffer Memory

BMX – Bus Matrix Switch

BOM – Beginning of Message

BSM – Backend Switch Module

Burst – A talk spurt or data message.

Burst Switching – A switching method to switch digitized voice and data characters in an integrated way.

C

CAD – Computer-Aided Design

CAE – Computer-Aided Engineering

Carrier-Division Multiplexing – Concerns the multiplexing of optical carriers.

CCIS – Common Channel Interoffice Signaling

CCITT – Consultative Committee for International Telephony and Telegraphy

CCP – Circuit-Switched Call Auxiliary Processor, Connection Control Processor

Clos Network – A network with three or more stages of relatively small crossbar switches.

CM – Control Memory

CMOS – Complementary Metal-Oxide Semiconductor

CMT – Connection Management

CO – Central Office

COINS – Corporate Information Network System

CP – Connection Processor, Call Processing

CRC – Cyclic Redundancy Check, an error-detection scheme.

CRI – Collision Resolution Interface

Crossbar Switch – Has a crosspoint for each input-output pair, and only one contact pair needs to be closed to establish an input-to-output connection.

CS – Circuit Switching, Consecutive Spreading

CSMA/CD – Carrier Sense Multiple-Access/Collision Detection

CTC – Channel Shift Time Controller

D

DAE – Dummy Address Encoder

DARPA – Department of Defense Advanced Research Projects Agency

Datagram – In the packet switching context, this represents a self-contained packet, independent of other packets, that carries the full destination address used for routing.

DBR – Distributed Bragg Reflector

DCE – Data Circuit Terminating Equipment

DCN – Data Communications Network

DCS – Digital Cross Connect System

DDCMP – Digital Data Communication Message Protocol

DES – Data Encryption Standard

DFB – Distributed Feedback

DFB-OA – DFB Optical Amplifier

Directional Coupler – A device that has two optical inputs, two optical outputs, and one control input. The control input is electrical and can place the device in the bar state or the cross state.
DLCI – Data Link Connection Identifier
DMAC – Direct Memory Access Controller
DNA – Digital Network Architecture
DPC – Distributed Prioritized Concentrator
DQDB – Distributed Queue Dual Bus
DR – Datagram Router
DS – Digital Signal
DSL – Digital Subscriber Line
DTDM – Dynamic Time-Division Multiplexing
DTE – Data Terminal Equipment
DTP – Data Transfer Part of X.25 packet layer procedures.

E

ED – Ending Delimiter
EOM – End of Message
ETSI – European Telecommunications Standards Institute

F

FC – Frame Control
FCS – Frame Check Sequence
FDDI – Fiber Distributed Data Interface
FDM – Frequency-Division Multiplexing
FIFO – First In First Out
FOL – Fiber-Optic Link
FPS – Fast Packet Switching
FR – Frame Relay
FSLC – Fiber Subscriber Line Circuit
FSM – Frond-end Switch Module

G

GHz – Gigahertz
GN – Group Number
GTT – Group Translation Table

H

HDLC – High-Level Data Link Control
HDTV – High-Definition Television
HEP – Heterogeneous Element Processor
HOL – Head of the Line
HQ – High Quality
HRC – Hybrid Ring Control
HSN – High-Speed Network
HST – High-Speed Link

I

IADM – Inverse Augmented Data Manipulator
IBC – Input Buffer Control
IBE – Ion Beam Etching

IC – Input Controller
IDLC – Integrated Digital Loop Carrier
IM – Interface Module
IMP – Interface Message Processor
IMSSI – Inter-Man Switching System Interface
IN – Interconnection Network
IP – Internet Protocol
I-PPU – Input-Packet Processing Unit
ISDN – Integrated Services Digital Network
ISO – International Organization for Standardization
IWF – Interworking Function
IWU – Interworking Unit

J

JK – Two-symbol sequence used in FDDI.

L

LAN – Local Area Network
LAPD – Link Access Protocol on D-Channel
LCLV – Liquid Crystal Light Values
LCN – Logical Channel Identifier
LCXT – Logical Channel Translator Table
LD – Laser Diode
LDN – Local Distribution Network
LED – Light-Emitting Diode
LN – Link Number
LTB – Link Test Buffer

M

MAC – Media Access Control
MAN – Metropolitan Area Network
MCU – Module Control Unit
MF – Message Format
MIMD – Multiple Instruction Multiple Data
MIN – Multistage Interconnection Network
Modulation-Division Multiplexing – Multiplexing methods in which the signals that modulate the optical carriers are multiplexed.
MQ – Medium Quality
MQW – Multiple-Quantum Well
MSS – Man Switching System
MSSR – Multistage Self-Routing
MST – Minimum Spanning Tree
MTDM – Multirate Time-Division Multiplexing
MUX – Multiplexer
MVC – Maintenance Voice Channel

N

NAP-NET – Network Access Processor Network
NCP – Network Control Processor, Network Connection Point
NGS – Next-Generation Switch
NI – Network Interface
NLFP – Nonlinear Fabry-Perot Etalon, an all-optical device where all the energy required to switch is supplied optically.

Nonblocking – Any desired connection between unused ports can be established immediately without interference from any arbitrary existing connections.

NRZ – Nonreturn to zero, a digital signaling technique in which the signal is at a constant level for the duration of time.

NTM – New Transfer Mode

O

OAM – Operation, Administration, and Maintenance

OC – Optical Carrier, Output Controller

O/E – Optical-to-Electrical

OEIC – Optoelectronic Integrated Circuits

OLE – Optical Logic Etalon, a pulsed device that requires separate wavelengths, one for bias and the other for signal.

O-PPU – Output-Packet Processing Unit

Optical Fiber – A transmission medium where signals are transmitted using light technology.

Optical Switching – Accepts and yields optical signals.

Optoelectronic Switching – Switching devices that accept optical signals and yield electrical signals or vice versa.

OS – Operation System

OSI – Open Systems Interconnection

OTC – Operating Telephone Company

P

PA – Preamble

PAD – Packet Assembler/Disassembler

PAL – Programmable Array Logic

PC – Port Circuits, Personal Computer

PCI – Protocol Control Information

PCM – Pulse Code Modulation, a process in which a signal is sampled, and the magnitude of each sample with respect to a fixed reference is quantized and converted by coding to a digital signal.

PDG – Packet Data Group

PDN – Public Data Network

Physical Layer – Layer 1 of the OSI model. Defines the electrical, mechanical, and procedural characteristics of the interface.

PLP – Packet-Level Protocol

PM – Polarization Maintaining

PMD – Physical Layer Medium Dependent

PNI – Packet Network Interface

PP – Packet Processor, Port Processor

PPD – Primary Packet Distributor

PPU – Packet Processing Unit

Presentation Layer – Layer 6 of the OSI model. Concerned with the data format display.

PRNET – Packet Radio Network

PS – Packet Switch, a communication method where messages are segmented into units called

packets and switched in a store-and-forward fashion.

PSPDN – Packet-Switched Public Data Network

PSU – Packet Switch Unit

PTM – Packet Transfer Mode

PVC – Permanent Virtual Circuit.

Q

QOS – Quality of Service

QPSX – Queued Packet and Synchronous Switch

R

RAN – Running Adder Network

RE – Remote Electronics

Rearrangeable – When a desired connection between unused ports may be temporarily blocked, but can be established if one or more existing connections are rerouted or rearranged.

RFNM – Request For Next Message

RL – Ring Latency

RN – Routing Network

RPOA – Recognized Private Operating Agency

RZ – Returning to Zero

S

SAD – Speech Activity Detection

SAP – Service Access Point, a means of identifying a user of the service of a protocol entity.

SAPI – Service Access Point Identifier

SBC – Send Basket Controller

SBWP – Space-Bandwidth Product

SCPS – Synchronous Composite Packet Switching, a switching system that integrates packet and circuit switching in a single switching system.

SD – Starting Delimiter

SDLC – Synchronous Data Link Control, a link-level protocol.

SDU – Service Data Unit

SEED – Self-Electro-optic Effect Device, an electro-optic device where all the energy required to switch is supplied optically.

Self-Routing – Entering from any input, the bits of the desired output address expressed in binary form may be used to control successive stages in the network.

Session Layer – Layer 5 of the OSI model. Manages a logical connection between two communication processes.

SF – Switch Fabric

Shuffle-Exchange Network – Data entering at any input can be permuted to any output by repeated passes through the system.

SI – Switch Interface

SIMD – Single Instruction Multiple Data.

SIP – SMDS Interface Protocol

SLIC – Subscriber Line Interface Circuit

Sliding-Window Protocol – A protocol in which the sender is allowed to have multiple unacknowledged frames outstanding simultaneously.

SLM – Spatial Light Modulators

SM – Switch Module, a high-performance packet switch.

SMDS – Switched Multimegabit Data Service

SMF – Single-Mode Fiber

SMT – Station Management

SNA – Systems Network Architecture

SNI – Subscriber Network Interface

SNR – Signal-to-Noise Ratio

SONET – Synchronous Optical Network

SP – Switching Processor, Supervisor Processor

SPD – Secondary Packet Distributor

SPE – Synchronous Payload Envelope

SPN – Subscriber Premises Network

SRM – Self-Routing Module

Starlite – A switch designed to handle a variety of communications needs in digital form.

STB – Switch Test Buffer

STM – Synchronous Transfer Mode

STM-1 – Synchronous Transport Module–Level 1

Statistical Time-Division Multiplexing – A method of TDM in which time slots on a shared medium are allocated on demand.

STS – Synchronous Transport Signal

Synchronous Time-Division Multiplexing – A method of TDM in which time slots on a shared transmission line are assigned on a fixed predetermined basis.

T

TASI – Time Assigned Speech Interpolation

TC – Trunk Controller

TCP – Transmission Control Protocol

TDM – Time-Division Multiplexing

TDN – Terminal Distribution Network

TH – Transmission Header

TMS – Time-Multiplexed Switch

TNT – Trunk Number Translator

TSI – Time Slot Interchanger, the interchange of time slots within a time-division multiplexed frame.

T-S-T – Time Space Time

TSW – Time Switch

TTR – Timed Token Rotation

TTRT – Target Token Rotation Time

TW – Traveling Wave

U

UNI – User Network Interface

V

VC – Virtual Circuit

VCC – VCI Converter

VCI – Virtual Channel Identifier

VCIT – VCI Conversion Table

VLSI – Very Large Scale Integration

VPI – Virtual Path Identifier

VT – Virtual Tributary

VTP – Virtual Terminal Protocol

W

WAN – Wide Area Network

Wavelength-Division Switching – Input information is used to modulate a light source that has a unique wavelength for each input. All the optical energy is combined and then split, so it can be distributed to all the output channels.

WBC – Wideband Channel

WCN – Wide Area Corporate Network

WDM – Wave-Division Multiplexing

WPN – Wideband Packet Network

X

XMB – Transmit Buffer

XPM – Crosspoint Memory

About the Authors

Chris Dhas

Chris Dhas is a principal in Computer Networks & Software, Inc., a telecommunications consulting and communications software development company located in Fairfax, Va. He has directed activities in telecommunications (network design, planning, management, and performance analysis), product development (telecommunications product planning and design as applied to packet switching, broadband switching, frame relay, and ISDN), and diverse areas of distributed processing, computer architectures, protocol architectures, database management systems, knowledge-based systems, and systems engineering.

Dhas has written a number of technical papers in the fields of distributed systems, computer architectures, modeling, communications networks, and protocol architectures. He has offered seminars on ISDN, X.25, network design and analysis, and broadband switching. He is also active in the IEEE Communications Society, providing technical editorial services to *IEEE Communications* magazine and the *Journal of Selected Areas in Communications*.

Dhas received his PhD from Iowa State University.

Vijaya K. Konangi

Vijaya K. Konangi is an associate professor of electrical engineering at Cleveland State University. His research interests include computer communications networks, local area networks, and distributed systems.

Konangi received his BE and MSc from the University of Madras, India, and his PhD from Iowa State University. He is a member of ACM and IEEE.

M. Sreetharan

M. Sreetharan is a principal in Computer Networks & Software, Inc., where he designs and develops data communications networks. His areas of interest are interconnection networks, high-performance communications systems, and architectures for fault tolerance and parallel processing. He has worked at Dowty Electronics (UK), where he developed full-authority digital controllers for industrial gas turbines, and at Solar Turbines, Inc., San Diego, where he continued his controls and signal processing work.

Sreetharan was born in Jaffna, Sri Lanka, and received his bachelor's degree in electrical engineering with first-class honors from the University of Sri Lanka, Peradeniya campus, in 1974. He attended the University of Manchester Institute of Science and Technology and Brunel University, UK, receiving his MS (in digital electronics) and PhD in 1979 and 1982. Sreetharan is a member of ACM and IEEE, and is a Chartered Engineer (IEE, UK).

IEEE Computer Society

IEEE Computer Society Press Publications

Monographs: A monograph is an authored book consisting of 100-percent original material.

Tutorials: A tutorial is a collection of original materials prepared by the editors, and reprints of the best articles published in a subject area. Tutorials must contain at least five percent of original material (although we recommend 15 to 20 percent of original material).

Reprint collections: A reprint collection contains reprints (divided into sections) with a preface, table of contents, and section introductions discussing the reprints and why they were selected. Collections contain less than five percent of original material.

Technology series: Each technology series is a brief reprint collection — approximately 126-136 pages and containing 12 to 13 papers, each paper focusing on a subset of a specific discipline, such as networks, architecture, software, or robotics.

Submission of proposals: For guidelines on preparing CS Press books, write the Editorial Director, IEEE Computer Society Press, PO Box 3014, 10662 Los Vaqueros Circle, Los Alamitos, CA 90720-1264, or telephone (714) 821-8380.

Purpose

The IEEE Computer Society advances the theory and practice of computer science and engineering, promotes the exchange of technical information among 100,000 members worldwide, and provides a wide range of services to members and nonmembers.

Membership

All members receive the acclaimed monthly magazine *Computer*, discounts, and opportunities to serve (all activities are led by volunteer members). Membership is open to all IEEE members, affiliate society members, and others seriously interested in the computer field.

Publications and Activities

Computer magazine: An authoritative, easy-to-read magazine containing tutorials and in-depth articles on topics across the computer field, plus news, conference reports, book reviews, calendars, calls for papers, interviews, and new products.

Periodicals: The society publishes six magazines and five research transactions. For more details, refer to our membership application or request information as noted above.

Conference proceedings, tutorial texts, and standards documents: The IEEE Computer Society Press publishes more than 100 titles every year.

Standards working groups: Over 100 of these groups produce IEEE standards used throughout the industrial world.

Technical committees: Over 30 TCs publish newsletters, provide interaction with peers in specialty areas, and directly influence standards, conferences, and education.

Conferences/Education: The society holds about 100 conferences each year and sponsors many educational activities, including computing science accreditation.

Chapters: Regular and student chapters worldwide provide the opportunity to interact with colleagues, hear technical experts, and serve the local professional community.

OTHER IEEE COMPUTER SOCIETY PRESS TITLES

MONOGRAPHS

Analyzing Computer Architectures
Written by J.C. Huck and M.J. Flynn
(ISBN 0-8186-8857-2); 206 pages

Desktop Publishing for the Writer:
Designing, Writing, and Developing
Written by Richard Ziegfeld and John Tarp
(ISBN 0-8186-8840-8); 380 pages

Digital Image Warping
Written by George Wolberg
(ISBN 0-8186-8944-7); 340 pages

Integrating Design and Test—CAE Tools for ATE Programming
Written by K.P. Parker
(ISBN 0-8186-8788-6); 160 pages

JSP and JSD—The Jackson Approach to Software Development
(Second Edition)
Written by J.R. Cameron
(ISBN 0-8186-8858-0); 560 pages

National Computer Policies
Written by Ben G. Matley and Thomas A. McDannold
(ISBN 0-8186-8784-3); 192 pages

Physical Level Interfaces and Protocols
Written by Uyless Black
(ISBN 0-8186-8824-2); 240 pages

Protecting Your Proprietary Rights in Computer
and High-Technology Industries
Written by Tobey B. Marzouk, Esq.
(ISBN 0-8186-8754-1); 224 pages

TUTORIALS

Advanced Computer Architecture
Edited by D.P. Agrawal
(ISBN 0-8186-0667-3); 400 pages

Advances in Distributed System Reliability
Edited by Suresh Rai and Dharma P. Agrawal
(ISBN 0-8186-8907-2); 352 pages

Computer and Network Security
Edited by M.D. Abrams and H.J. Podell
(ISBN 0-8186-0756-4); 448 pages

Computer Architecture
Edited by D.D. Gajski, V.M. Milutinovic, H. Siegel, and B.P. Furht
(ISBN 0-8186-0704-1); 602 pages

Computer Arithmetic I
Edited by Earl E. Swartzlander, Jr.
(ISBN 0-8186-8931-5); 398 pages

Computer Arithmetic II
Edited by Earl E. Swartzlander, Jr.
(ISBN 0-8186-8945-5); 412 pages

Computer Communications: Architectures,
Protocols and Standards (Second Edition)
Edited by William Stallings
(ISBN 0-8186-0790-4); 448 pages

Computer Graphics Hardware: Image Generation and Display
Edited by H.K. Reghbati and A.Y.C. Lee
(ISBN 0-8186-0753-X); 384 pages

Computer Graphics: Image Synthesis
Edited by Kenneth Joy, Nelson Max, Charles Grant,
and Lansing Hatfield
(ISBN 0-8186-8854-8); 380 pages

Computer Networks (Fourth Edition)
Edited by M.D. Abrams and I.W. Cotton
(ISBN 0-8186-0568-5); 512 pages

Computers for Artificial Intelligence Applications
Edited by B. Wah and G.J. Li
(ISBN 0-8186-0706-8); 656 pages

Database Management
Edited by J.A. Larson
(ISBN 0-8186-0714-9); 448 pages

Digital Image Processing and Analysis:
Volume 1—Digital Image Processing
Edited by R. Chellappa and A.A. Sawchuk
(ISBN 0-8186-0665-7); 736 pages

Digital Image Processing and Analysis:
Volume 2—Digital Image Analysis
Edited by R. Chellappa and A.A. Sawchuk
(ISBN 0-8186-0666-5); 680 pages

Digital Private Branch Exchanges (PBXs)
by Edwin Coover
(ISBN 0-8186-0829-3); 394 pages

Distributed Computing Network Reliability
Edited by Suresh Rai and Dharma Agrawal
(ISBN 0-8186-8908-0); 357 pages

Distributed Software Engineering
Edited by Sol Shatz and Jia-Ping Wang
(ISBN 0-8186-8856-4); 294 pages

DSP-Based Testing of Analog and
Mixed Signal Circuits
Edited by Matthew Mahoney
(ISBN 0-8186-0785-8); 272 pages

Fault Tolerant Computing
Edited by V. Nelson and B. Carroll
(ISBN 0-8186-8677-4); 428 pages

Formal Verification of Hardware Design
Edited by Michael Yoeli
(ISBN 0-8186-9017-8); 340 pages

Hard Real-Time Systems
Edited by J.A. Stankovic and K. Ramamritham
(ISBN 0-8186-0819-6); 624 pages

Human Factors in Software Development
(Second Edition)
Edited by Bill Curtis
(ISBN 0-8186-0577-4); 736 pages

Integrated Services Digital Networks (ISDN)
(Second Edition)
Edited by William Stallings
(ISBN 0-8186-0823-4); 406 pages

Local Network Technology (Third Edition)
Edited by William Stallings
(ISBN 0-8186-0825-0); 512 pages

Microprogramming and Firmware Engineering
Edited by V. Milutinovic
(ISBN 0-8186-0839-0); 416 pages

Modeling and Control of Automated Manufacturing Systems
Edited by A. A. Desrochers
(ISBN 0-8186-8916-1); 384 pages

Nearest Neighbor Pattern Classification Techniques
Edited by Belur V. Dasarathy
(ISBN 0-8186-8930-7); 464 pages

For Further Information Call 1-800-CS-BOOKS or Write:

IEEE Computer Society Press, 10662 Los Vaqueros Circle, PO Box 3014,
Los Alamitos, California 90720-1264, USA

IEEE Computer Society, 13, avenue de l'Aquilon,
B-1200 Brussels, BELGIUM

IEEE Computer Society, Ooshima Building, 2-19-1 Minami-Aoyama,
Minato-ku, Tokyo 107, JAPAN

New Paradigms for Software Development
Edited by William Agresti
(ISBN 0-8186-0707-6); 304 pages

Object-Oriented Computing, Volume 1: Concepts
Edited by Gerald E. Petersen
(ISBN 0-8186-0821-8); 214 pages

Object-Oriented Computing, Volume 2: Implementations
Edited by Gerald E. Petersen
(ISBN 0-8186-0822-6); 324 pages

Parallel Architectures for Database Systems
Edited by A.R. Hurson, L.L. Miller, and S.H. Pakzad
(ISBN 0-8186-8838-6); 478 pages

Programming Productivity: Issues for '80s
(Second Edition)
Edited by C. Jones
(ISBN 0-8186-0681-9); 472 pages

Reduced Instruction Set Computers (RISC)
(Second Edition)
Edited by William Stallings
(ISBN 0-8186-8943-9); 448 pages

Software Design Techniques (Fourth Edition)
Edited by P. Freeman and A.I. Wasserman
(ISBN 0-8186-0514-6); 736 pages

Software Engineering Project Management
Edited by R. Thayer
(ISBN 0-8186-0751-3); 512 pages

Software Maintenance and Computers
Edited by D.H. Longstreet
(ISBN 0-8186-8898-X); 304 pages

Software Management (Third Edition)
Edited by D.J. Reifer
(ISBN 0-8186-0678-9); 526 pages

Software Quality Assurance: A Practical Approach
Edited by T.S. Chow
(ISBN 0-8186-0569-3); 506 pages

Software Reusability
Edited by Peter Freeman
(ISBN 0-8186-0750-5); 304 pages

Software Reuse—Emerging Technology
Edited by Will Tracz
(ISBN 0-8186-0846-3); 400 pages

Software Risk Management
Edited by B.W. Boehm
(ISBN 0-8186-8906-4); 508 pages

Standards, Guidelines and Examples on System
and Software Requirements Engineering
Edited by Merlin Dorfman and Richard H. Thayer
(ISBN 0-8186-8922-6); 626 pages

System and Software Requirements Engineering
Edited by Richard H. Thayer and Merlin Dorfman
(ISBN 0-8186-8921-8); 740 pages

Test Access Port and Boundary-Scan Architecture
Edited by C. M. Maunder and R. E. Tulloss
(ISBN 0-8186-9070-4); 400 pages

Test Generation for VLSI Chips
Edited by V.D. Agrawal and S.C. Seth
(ISBN 0-8186-8786-X); 416 pages

Visual Programming Environments: Paradigms and Systems
Edited by Ephraim Glinert
(ISBN 0-8186-8973-0); 680 pages

Visual Programming Environments: Applications and Issues
Edited by Ephraim Glinert
(ISBN 0-8186-8974-9); 704 pages

Visualization in Scientific Computing
Edited by G.M. Nielson, B. Shriver, and L. Rosenblum
(ISBN 0-8186-8979-X); 304 pages

VLSI Testing and Validation Techniques
Edited by H. Reghbati
(ISBN 0-8186-0668-1); 616 pages

Volume Visualization
Edited by Arie Kaufman
(ISBN 0-8186-9020-8); 494 pages

REPRINT COLLECTIONS

Dataflow and Reduction Architectures
Edited by S.S. Thakkar
(ISBN 0-8186-0759-9); 460 pages

Distributed Computing Systems: Concepts and Structures
Edited by A. Ananda and B. Srinivasan
(ISBN 0-8186-8975-0); 416 pages

Expert Systems: A Software Methodology
for Modern Applications
Edited by Peter Raeth
(ISBN 0-8186-8904-8); 476 pages

Logic Design for Testability
Edited by C.C. Timoc
(ISBN 0-8186-0573-1); 324 pages

Milestones in Software Evolution
Edited by Paul W. Oman and Ted G. Lewis
(ISBN 0-8186-9033-X); 332 pages

Object-Oriented Databases
Edited by Ez Nahouraii and Fred Petry
(ISBN 0-8186-8929-3); 256 pages

Validating and Verifying Knowledge-Based Systems
Edited by Uma G. Gupta
(ISBN 0-8186-8995-1); 400 pages

ARTIFICIAL NEURAL NETWORKS TECHNOLOGY SERIES

Artificial Neural Networks—Concept Learning
Edited by Joachim Diederich
(ISBN 0-8186-2015-3); 160 pages

Artificial Neural Networks—Electronic Implementation
Edited by Nelson Morgan
(ISBN 0-8186-2029-3); 144 pages

Artificial Neural Networks—Theoretical Concepts
Edited by V. Vemuri
(ISBN 0-8186-0855-2); 160 pages

SOFTWARE TECHNOLOGY SERIES

Computer Algorithms
Edited by Jun-ichi Aoe
(ISBN 0-8186-2123-0); 154 pages

Computer-Aided Software Engineering (CASE)
Edited by E.J. Chikofsky
(ISBN 0-8186-1917-1); 110 pages

Software Reliability Models: Theoretical Development,
Evaluation and Applications
Edited by Y.K. Malaiya and P.K. Srimani
(ISBN 0-8186-2110-9); 136 pages

COMMUNICATIONS TECHNOLOGY SERIES

Multicast Communication in Distributed Systems
Edited by Mustaque Ahamad
(ISBN 0-8186-1970-8); 110 pages

ROBOTICS TECHNOLOGY SERIES

Multirobot Systems
Edited by Rajiv Mehrotra and Murali R. Varanasi
(ISBN 0-8186-1977-5); 122 pages

IEEE COMPUTER SOCIETY PRESS VIDEOTAPES

PUBLISH WITH IEEE COMPUTER SOCIETY PRESS

TODAY'S COMPUTER SCIENCE PROFESSIONALS ARE TURNING TO US TO PUBLISH THEIR TEXTS

BENEFITS OF IEEE COMPUTER SOCIETY PRESS PUBLISHING:

- ☐ **Timely publication schedules**
- ☐ **Society's professional reputation and recognition**
- ☐ **High-quality, reasonably priced books**
- ☐ **Course classroom adoption**
- ☐ **Open options on the type and level of publication to develop**
- ☐ **Built-in mechanisms to reach a strong constituency of professionals**
- ☐ **Peer review and reference**

Enjoy the personal, professional, and financial recognition of having your name in print alongside other respected professionals in the fields of computer and engineering technology. Our rapid turnaround (from approved proposal to the published product) assures you against publication of dated technical material, and gives you the potential for additional royalties and sales from new editions. Your royalty payments are based not only on sales, but also on the amount of time and effort you expend in writing original material for your book.

Your book will be advertised to a vast audience through IEEE Computer Society Press catalogs and brochures that reach over 500,000 carefully chosen professionals in numerous disciplines. Our wide distribution also includes promotional programs through our European and Asian offices, and over-the-counter sales at 50 international computer and engineering conferences annually.

IEEE Computer Society Press books have a proven sales record, and our tutorials enjoy a unique niche in today's fast moving technical fields and help us maintain our goal to publish up-to-date, viable computer science information.

Steps for Book Submittals:

1- Submit your proposal to the Editorial Director of IEEE Computer Society Press. It should include the following data: title; your name, address, and telephone number; a detailed outline; a summary of the subject matter; a statement of the technical level of the book, the intended audience, and the potential market; a table of contents including the titles, authors, and sources for all reprints; and a biography.

2- Upon acceptance of your proposal, prepare and submit copies of the completed manuscript, including xerox copies of any reprinted papers and any other pertinent information, and mail to the Editorial Director of IEEE Computer Society Press.

3- The manuscript will then be reviewed by other respected experts in the field, and the editor-in-charge.

4- Upon publication, you will receive an initial royalty payment with additional royalties based on a percentage of the net sales and on the amount of original material included in the book.

We are searching for authors in the following computer science areas:

ADA	**LOCAL AREA NETWORKS**
ARCHITECTURE	**OPTICAL STORAGE DATABASES**
ARTIFICIAL INTELLIGENCE	**PARALLEL PROCESSING**
AUTOMATED TEST EQUIPMENT	**PATTERN RECOGNITION**
CAD / CAE	**PERSONAL COMPUTING**
COMPUTER GRAPHICS	**RELIABILITY**
COMPUTER LANGUAGES	**ROBOTICS**
COMPUTER MATHEMATICS	**SOFTWARE ENGINEERING**
COMPUTER WORKSTATIONS	**SOFTWARE ENVIRONMENTS**
DATABASE ENGINEERING	**SOFTWARE MAINTENANCE**
DIGITAL IMAGE PROCESSING	**SOFTWARE TESTING AND**
DISTRIBUTED PROCESSING	**VALIDATION**
EXPERT SYSTEMS	**AND OTHER AREAS AT THE**
FAULT-TOLERANT COMPUTING	**FOREFRONT OF**
IMAGING	**COMPUTING TECHNOLOGY !**

Interested ?
For more detailed guidelines please contact:

**Editorial Director, c/o IEEE Computer Society Press,
10662 Los Vaqueros Circle, Los Alamitos, California 90720-1264.**